PREALGEBRA

PREALGEBRA

JAMIE BLAIR

Orange Coast College
Costa Mesa, California

JOHN TOBEY

North Shore Community College
Danvers, Massachusetts

JEFFREY SLATER

North Shore Community College
Danvers, Massachusetts

PRENTICE HALL

Upper Saddle River, NJ 07458

Library of Congress Cataloging-in-Publication Data

Blair, Jamie.
    Prealgebra / Jamie Blair, John Tobey, Jeffrey Slater.
            p.   cm.
    Includes index.
    ISBN 0-13-260936-3 (alk. paper)
    1. Mathematics.   I. Tobey, John (date).   II. Slater, Jeffrey (date).
III. Title.
QA39.2.B587   1999
510—dc21   99-10558
                                                                        CIP

Acquisitions Editor: Karin E. Wagner
Editor-in-Chief: Jerome Grant
Editor-in-Chief, Development: Carol Trueheart
Production Editor: Barbara Mack
Senior Managing Editor: Linda Mihatov Behrens
Executive Managing Editor: Kathleen Schiaparelli
Assistant Vice President of Production and Manufacturing: David W. Riccardi
Marketing Manager: Jolene Howard
Manufacturing Buyer: Alan Fischer
Manufacturing Manager: Trudy Pisciotti
Supplements Editor/Editorial Assistant: Kate Marks
Media Supplements Editor: Audra Walsh
Art Director: Maureen Eide
Associate Creative Director: Amy Rosen
Director of Creative Services: Paula Maylahn
Assistant to Art Director: John Christiana
Art Manager: Gus Vibal
Art Editor: Grace Hazeldine
Interior Designer: Geri Davis, The Davis Group, Inc.
Cover Designer: Maureen Eide
Cover Photo: Kevin Lyman/Amana America, Inc.
Photo Researcher: Melinda Alexander
Photo Research Administrator: Melinda Reo

 © 1999 by Prentice-Hall, Inc.
Upper Saddle River, New Jersey 07458

Photo credits appear on page P-1, which constitutes a continuation of the copyright page.

Printed in the United States of America

Reprinted with corrections July, 2000.

10   9   8   7   6   5   4   3

ISBN 0-13-260936-3

Prentice-Hall International (UK) Limited, *London*
Prentice-Hall of Australia Pty. Limited, *Sydney*
Prentice-Hall Canada Inc., *Toronto*
Prentice-Hall Hispanoamericana, S.A., *Mexico*
Prentice-Hall of India Private Limited, *New Delhi*
Prentice-Hall of Japan, Inc., *Tokyo*
Prentice-Hall Asia Pte. Ltd., *Singapore*
Editora Prentice-Hall do Brasil, Ltda., *Rio de Janeiro*

This book is dedicated to
*my husband, Jerry Blair, and my
two children, Joe and Wendy.
Their support and patience
during the production of the text
are greatly appreciated.*

CONTENTS

Preface   xiii

1   Addition and Subtraction of Whole Number Expressions   1

Pretest

1.1   Understanding Whole Numbers / 4
1.2   Addition and Combining Like Terms / 12
1.3   Subtraction and Combining Like Terms / 24
1.4   Perimeter, Graphs, and Estimation / 41

PUTTING YOUR SKILLS TO WORK:

Redecorating a Room   53

1.5   Understanding Basic Equations / 54
1.6   U.S. and Metric Units of Measurement / 66

CHAPTER 1 REVIEW   79
CHAPTER 1 TEST   84

2   Multiplication and Division of Whole Number Expressions   87

Pretest

2.1   Understanding Multiplication of Expressions / 90
2.2   Multiplication of Whole Numbers and Algebraic Expressions / 98

PUTTING YOUR SKILLS TO WORK:

Cost Analysis: A Vacation Plan   107

2.3   Exponents and Order of Operations / 108
2.4   Performing Operations with Exponents / 116
2.5   Understanding Division of Expressions / 122
2.6   Solving Basic Equations Involving Multiplication and Division / 135
2.7   Geometric Shapes / 145

PUTTING YOUR SKILLS TO WORK:

Reading a Floor Plan   157

2.8   Applied Problems Involving Multiplication and Division / 158

CHAPTER 2 REVIEW   167

CHAPTER 2 TEST   172

CUMULATIVE TEST FOR CHAPTERS 1–2   174

Signed Numbers   177

Pretest

3.1   Understanding Signed Numbers / 179
3.2   Addition of Signed Numbers / 185
3.3   Subtraction of Signed Numbers / 193
3.4   Multiplication and Division of Signed Numbers / 201
3.5   Simplifying Algebraic Expressions / 210
3.6   Order of Operations and Applications Involving Signed Numbers / 214

PUTTING YOUR SKILLS TO WORK:
Universal Time   218

CHAPTER 3 REVIEW   222

CHAPTER 3 TEST   225

CUMULATIVE TEST FOR CHAPTERS 1–3   227

Fractions, Ratio, and Proportion   229

Pretest

4.1   Factoring Whole Numbers / 231
4.2   Understanding Fractions / 239
4.3   Equivalent Fractions / 247
4.4   Simplifying Fractional Expressions / 255
4.5   Ratios and Rates / 261
4.6   Proportions / 270
4.7   Applied Problems Involving Proportions / 276

PUTTING YOUR SKILLS TO WORK:
Planning a Large Event   281

CHAPTER 4 REVIEW   285

CHAPTER 4 TEST   289

CUMULATIVE TEST FOR CHAPTERS 1–4   291

5 Operations on Fractional Expressions   293

Pretest

5.1   Multiplication and Division of Fractional
Expressions / 296
5.2   Multiples and Least Common Multiples
of Expressions / 307
5.3   Addition and Subtraction of Fractional
Expressions / 312
5.4   Operations with Mixed Numbers / 322
5.5   Order of Operations and Complex Fractions / 331
5.6   Multiplication and Division Properties
of Exponents / 337
5.7   Applied Problems Involving Fractions / 345

PUTTING YOUR SKILLS TO WORK:

Investing in the Stock Market   351

CHAPTER 5 REVIEW   355

CHAPTER 5 TEST   358

CUMULATIVE TEST FOR CHAPTERS 1–5   360

6 Equations and Polynomials   363

Pretest

6.1   Solving Equations Using One Principle
of Equality / 365
6.2   Solving Equations Using More Than One
Principle of Equality / 374
6.3   Solving Equations Containing Fractions / 381
6.4   Multiplication of Polynomials / 386
6.5   Addition and Subtraction of Polynomials / 393
6.6   Solving Equations Involving Polynomial
Expressions / 398
6.7   Applied Problems Involving Polynomials / 402

PUTTING YOUR SKILLS TO WORK:

Profit, Cost, and Revenue in Business   410

CHAPTER 6 REVIEW   414

CHAPTER 6 TEST   416

CUMULATIVE TEST FOR CHAPTERS 1–6   418

Decimal Expressions   421

Pretest

7.1 Understanding Decimal Fractions / 424
7.2 Addition and Subtraction of Decimals / 432

> **PUTTING YOUR SKILLS TO WORK:**
> Balancing a Checking Account   435

7.3 Multiplication of Decimals and Applied Problems / 440
7.4 Division of Decimals and Scientific Notation / 447

> **PUTTING YOUR SKILLS TO WORK:**
> Money Exchange   455

7.5 Solving Equations and Applied Problems Involving Decimals / 459
7.6 Percents / 465
7.7 Solving Percent Problems Using Equations / 473
7.8 Applied Problems Involving Percent / 480

> **CHAPTER 7 REVIEW   491**
> **CHAPTER 7 TEST   495**
> **CUMULATIVE TEST FOR CHAPTERS 1–7   497**

 Graphing and Statistics   499

Pretest

8.1 Interpreting and Constructing Graphs / 501

> **PUTTING YOUR SKILLS TO WORK:**
> Analyzing and Constructing Graphs of Production Costs   507

8.2 Mean and Median / 515
8.3 The Rectangular Coordinate System / 522
8.4 Linear Equations with Two Variables / 533

> **PUTTING YOUR SKILLS TO WORK:**
> Using Data and Equations to Make Predictions   540

> **CHAPTER 8 REVIEW   550**
> **CHAPTER 8 TEST   555**
> **CUMULATIVE TEST FOR CHAPTERS 1–8   558**

9 Measurement and Geometric Figures   561

Pretest

9.1 Converting Between U.S. Units; Converting Between Metric Units / 563

9.2 Converting Between the U.S. and Metric Systems (optional) / 573

PUTTING YOUR SKILLS TO WORK:

Traveling and Conversions   578

9.3 Square Roots and Square Root Expressions / 579

9.4 Triangles and the Pythagorean Theorem / 588

9.5 The Circle and Applied Problems / 598

9.6 Volume / 605

9.7 Similar Geometric Figures / 611

CHAPTER 9 REVIEW   622

CHAPTER 9 TEST   625

CUMULATIVE TEST FOR CHAPTERS 1–9   627

APPENDIX A   Square Root Table   A-1

APPENDIX B   Scientific Calculators   A-2

APPENDIX C   Solving Percent Problems Using a Proportion   A-9

APPENDIX D   Congruent Triangles   A-16

SOLUTIONS TO PRACTICE PROBLEMS   SP-1

SELECTED ANSWERS   SA-1

APPLICATIONS INDEX   I-1

SUBJECT INDEX   I-5

PHOTO CREDITS   P-1

PREFACE

TO THE STUDENT

People who enter college have a variety of mathematical backgrounds. Many of you may be looking forward to this course. We hope that our book keeps your enthusiasm high. Others of you may be quite anxious about taking this course. We hope that our book delivers the assistance and support that helps you develop an interest and the skills necessary to do well in this most important field.

Special attention has been given in this text to *teaching you, the student, how to learn* so that you have the best chance of success and develop a good basic foundation of math skills. Once you discover methods of learning that work for your individual learning style, you will be surprised at how much more motivated you will become. Your attitude towards math will become more positive and success in mathematics will follow.

There are many features in this text that will help you *learn mathematics*, one of which is labeled **Developing Your Study Skills**. In these sections different activities that enhance learning are explained, and specific strategies and techniques are systematically developed starting in Chapter 1. These sections include information about the different learning styles each of us have, and will help you discover which ones work best for you so you can learn faster and easier.

SPECIAL FEATURES

- **Developing Your Study Skills**, scattered throughout the text, provides specific strategies and techniques to help you develop good study skills.
- **Pretests** begin each chapter, diagnosing in advance your strengths and weaknesses in the upcoming material.
- **Examples and matched Practice Problems** give you a chance to solve an exercise similar to the preceding example on your own. Answers and solutions appear in a separate section in the back of the text.
- **Chapter Organizers** summarize key concepts and provide additional examples for brief review in a compact grid format.
- **To Think About** material challenges you to think in a logical and organized way, and exposes you to the power of mathematical ideas.
- **Real-life Word Problems and Putting Your Skills to Work Activities** cover topics from everyday situations—maintaining a budget, using a checkbook, figuring mileage on a trip, and so on. These problems along with photos show mathematics at work in real-life situations, answering the question "When will I ever need to use the information in this course?"
- **Calculator Problems and Internet Exploration** sections encourage you to take advantage of this new technology.
- **Practice Chapter Tests** provide tests that can be used as preparation for the actual in-class test.
- **Cumulative Review Problems** review the content of previous chapters within exercise sets.
- **Cumulative Tests**, at the end of each chapter (except the first), cover content from the preceding chapters. Completing these will help you retain the information you have learned.

This book is the second in a series of four developmental math books. In a tightly coordinated sequence of books, *Prealgebra* is followed by *Beginning Algebra* and *Intermediate Algebra*. If you enjoyed this book, ask your instructor whether he or she will be using another book in the series for your next course.

ACKNOWLEDGMENTS

We have been greatly helped by a supportive group of colleagues who not only teach at Orange Coast College but who have provided a number of ideas and class tested this *Prealgebra* book. Thus, a special word of thanks to Carol Chapman, Frank L. Miller, Laura Kaufman, Michele Gulu, and Kathy Dietz.

This book is a product of many years of work and many contributions from faculty across the country. We would like to thank the many reviewers.

Our deep appreciation to each of the following:

Alfredo Alvarez, Palo Alto College

Daphne Bell, Motlow State Community College

Rosanne Benn, Prince George's Community College

Stanley Carter, Central Missouri State University

Patricia Donovan, Delta College—San Joaquin

Irene Doo, Austin Community College

Rebecca Easley, Rose State College

Sharon Edgmon, Bakersfield College

Mark Greenhalgh, Fullerton College

Susan Hahn, Kean University

Will Summers, Saddleback College

Sandra Vrem, College of the Redwoods

Jon Weerts, Triton Coilege

Each textbook is a combination of ideas, writing, and revisions from the authors and wise editorial direction and assistance from editors. We want to thank our Prentice Hall editor, Karin Wagner, for her administrative support and encouragement, and for her helpful insight and perspective on each phase of the production of the textbook. We also express our thanks to the development editor, Tony Palermino, whose expertise, patience, and willingness to listen have been invaluable to the quality of this book. We especially express our thanks to Jerome Grant and his staff at Prentice Hall for their contributions to the production of the *Prealgebra* text.

Book writing is impossible for us without the loyal support of our families. Our deepest thanks and love to Jerry, Joe, Wendy, Nancy, Johnny, Melissa, Marcia, Shelley, Rusty, and Abby. Your understanding, your love and help, and your patience have been a source of great encouragement. Finally, we thank God for the strength and energy to write and the opportunity to help others through this textbook.

We have spent more than 25 years teaching mathematics. Each teaching day we find our greatest joy is helping students learn. We take a personal interest that each student has a good learning experience in taking this course. If you have some personal comments, suggestions, or ideas for the future editions of this textbook, please write us at:

Prof. Blair, Prof. Tobey, and Prof. Slater
Prentice Hall Publishing
Office of the College Mathematics Editor
One Lake Street
Upper Saddle River, N.J. 07458

We wish you success in this course and in the future!

Jamie Blair
John Tobey
Jeffrey Slater

GUIDE FOR STUDENTS

How to use **Prealgebra** to enrich your class experience and prepare for tests.

Each chapter opens with an application. These applications relate to the material you find in the chapter as well as an extended discovery called *Putting Your Skills to Work* which you will encounter later in the chapter.

MULTIPLICATION AND DIVISION OF WHOLE NUMBER EXPRESSIONS

CHAPTER 2

Suppose that you decide to replace the flooring in your residence or office. Do you know how to determine how much floor covering must be purchased? Can you calculate the cost of this flooring? After you have studied the topics in this chapter, you will have an opportunity to consider this situation in Putting Your Skills to Work on page 157.

2.1 Understanding Multiplication of Expressions
2.2 Multiplication of Whole Numbers and Algebraic Expressions
2.3 Exponents and Order of Operations
2.4 Performing Operations with Exponents
2.5 Understanding Division of Expressions
2.6 Solving Basic Equations Involving Multiplication and Division
2.7 Geometric Shapes
2.8 Applied Problems Involving Multiplication and Division

Put your new skills to work. These multi-part projects involve both group and individual work, and are relevant to chapter concepts as well as to problems you encounter from day to day.

READING A FLOOR PLAN

We can solve many real-life problems with the geometric knowledge we now have. This knowledge enables us to find length, area, or volume. How much edging or fencing is required? How much carpet or tile is required? How much can we store? To solve these problems, we often must take measurements, draw or read floor plans, and perform calculations.

PROBLEMS FOR INDIVIDUAL INVESTIGATION

1. How much will it cost to purchase tile for the kitchen floor if the tile sells for $9 a square foot?

2. How much will it cost to purchase vinyl flooring for the bathroom if the flooring sells for $6 a square foot?

PROBLEM FOR GROUP INVESTIGATION AND COOPERATIVE STUDY

1. John wants to lay carpet in the living room and hallway, wood floors in the entry, family room, and dining room, and tile in the kitchen, bathroom, and

closet. John must purchase how many square feet of the following:

(a) Carpet

(b) Tile

(c) Wood floors

(d) How much will the wood floors cost at $8 a square foot?

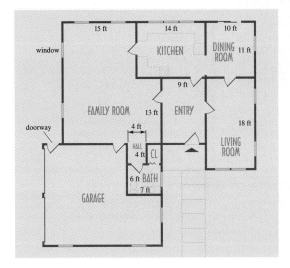

INTERNET: Go to http://www.prenhall.com/blair to explore this application.

157

page 157

These features have been included in this text to help you make connections to mathematics. Use them to explore, connect, and discover.

Explore on your own...

Work cooperatively with fellow students to solve more involved problems. You will find that problems are often solved collaboratively in the workplace.

Explore the Internet for real data that relates to the *Putting Your Skills to Work* explorations. Extend the concepts!

PREPARE YOURSELF

NAME   SECTION   DATE

CHAPTER 2 PRETEST

This test provides a preview of the topics in this chapter. It will help you identify which concepts may require more of your studying time. If you are familiar with the topics in this chapter, take this test now. Check your answers with those in the back of the book. If you are not familiar with the topics in this chapter, begin studying the chapter now.

SECTION 2.1

1. Identify the product and the factors in each equation.

   (a) $8(2) = 16$ **(b)** $3x = 21$ **(c)** $mn = p$

2. Multiply:

   (a) $423(0)$ **(b)** $x(1)$ **(c)** $8 \cdot 4 \cdot 0 \cdot y$

3. Simplify: $6(3x)$.

SECTION 2.2

4. Multiply: $4(3000)$.   5. Multiply: $2(x + 3)$.

6. A cashier earns $8 per hour for the first 40 hours worked and $12 per hour for overtime (hours worked in addition to 40 hours a week). Last week the cashier worked 49 hours. Calculate the cashier's total pay for the 49-hour

1. _____

2. _____

3. _____

4. _____

5. _____

6. _____

7. _____

page 88

MATHEMATICS BLUEPRINT FOR PROBLEM SOLVING

GATHER THE FACTS	WHAT AM I ASKED TO DO?	HOW DO I PROCEED?	KEY POINTS TO REMEMBER
The management position pays $12 per hour for 40 hours. The personnel position pays $2600 per month.	Determine which job pays a higher salary per year.	Multiply $12 × 40 to find the pay for 1 week, then multiply by 52 for yearly pay. Multiply $2600 × 12 to find the yearly pay. Compare yearly pay for both jobs.	The phrases *per week* and *per year* indicate multiplication. I must find *yearly* pay: 12 months = 1 year 52 weeks = 1 year

page 103

■ DEVELOPING YOUR STUDY SKILLS
REVIEWING YOUR TEST

To learn mathematics it is necessary for you to master each skill as you proceed through the course. Learning mathematics is much like building blocks—each block is important. Complete the following activities to assure that each of the mathematics building blocks is mastered.

The Learning Cycle

$$\text{Reading} \longrightarrow \text{Writing}$$
$$\uparrow \qquad\qquad\qquad \downarrow$$
$$\text{Seeing} \leftarrow \text{Verbalizing} \leftarrow \text{Listening}$$

- *Writing.* Try to rework correctly the test problems you missed, without help from others.
- *Verbalizing.* Verify with an instructor, tutor, or classmate that these problems are correct.
- *Listening, seeing, and verbalizing.* Ask for an explanation of why your answers were wrong and how to do the problems you could not correct by yourself.
- .... Review sections of the book and work exercises you did incorrectly on the test.
- .... that you identify how and why you made your ....id making the same types of errors again.

page 93

CHAPTER ORGANIZER

TOPIC	PROCEDURE	EXAMPLES
Key words for multiplication	The key words that represent multiplication are *times, product of, double,* and *triple.*	Translate using symbols. Double a number: $2x$ Triple a number: $3x$ The product of 2 and 3: $2 \cdot 3$ Six times x: $6x$
Properties of multiplication	We can regroup and multiply numbers in any order since multiplication is associative and commutative.	Simplify: $2(x \cdot 3)$. Change the order of multiplication and regroup: $(2 \cdot 3) \cdot x$. Multiply: $(2 \cdot 3) \cdot x = 6x$.
Multiplying numbers with trailing zeros.	We multiply the nonzero numbers and attach the trailing zeros to the right side of the product.	Multiply: $600(5n)$ Regroup: $(600 \cdot 5)n$ Multip.... attach.... produc....

page 164

Have a plan—Develop a problem-solving strategy.

● TO THINK ABOUT

EVALUATE OR SOLVE?

Do you know the difference between evaluating the expression $8x$ for $x = 3$ and solving the equation $8x = 16$?

- *Evaluate an expression.* We **replace the variable** in the expression with the given number, then perform the calculation(s).

   Evaluate $8x$ if $x = 3$. $8 \cdot 3 = 24$.

- *Solve an equation.* We **find the value of the variable** in the equation, that is, the solution to the equation.

   Solve for x: $8x = 16$. $x = 2$.

page 140

CHECK YOUR UNDERSTANDING

Prealgebra includes many different types of exercises—
Become a Problem Solver!

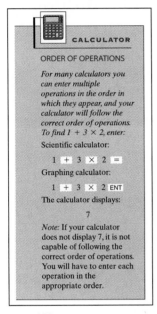

CALCULATOR

ORDER OF OPERATIONS

For many calculators you can enter multiple operations in the order in which they appear, and your calculator will follow the correct order of operations. To find 1 + 3 × 2, enter:

Scientific calculator:

| 1 | + | 3 | × | 2 | = |

Graphing calculator:

| 1 | + | 3 | × | 2 | ENT |

The calculator displays:

7

Note: If your calculator does not display 7, it is not capable of following the correct order of operations. You will have to enter each operation in the appropriate order.

page 112

VERBAL AND WRITING SKILLS

81. Write in words the question being asked by the equation $n^2 = 16$.

82. Write in words the question being asked by the equation $x^3 = 27$.

page 115

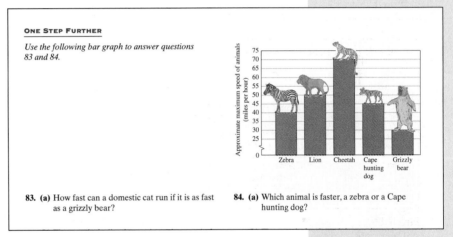

ONE STEP FURTHER

Use the following bar graph to answer questions 83 and 84.

83. (a) How fast can a domestic cat run if it is as fast as a grizzly bear?

84. (a) Which animal is faster, a zebra or a Cape hunting dog?

page 134

NAME   SECTION   DATE

CHAPTER 2 TEST

1. _____

2. _____

3. _____

1. A restaurant sells 4 kinds of sandwiches: turkey, roast beef, veggie, and ham. Customers have a choice of 3 types of bread: wheat, white, or rye. How many different sandwiches are possible?

2. Identify the product and the factors in each equation.
   (a) $7(4) = 28$ **(b)** $8x = 24$ **(c)** $rs = t$

3. Simplify.
   (a) $9(2x)$ **(b)** $7(3 \cdot y)$

page 172

NAME   SECTION   DATE

CUMULATIVE TEST FOR CHAPTERS 1–2

1. _____

2. _____

1. Replace the question mark with the inequality symbol $<$ or $>$: 5 ? 0.

2. The total population of a small town is 5289. Round this population figure to the nearest:
   (a) thousand.   **(b)** hundred.

3. Use the commutative and/or associative property of addition, then simplify.
   (a) $3 + y + 1$ **(b)** $1 + (n + 4)$

page 174

ENHANCE YOUR LEARNING

Prealgebra is more than a textbook; it is an integrated package of instruction. Ask your professor about these supplements, which are part of the **Prealgebra** suite of learning materials.

Items are keyed specifically to this text.

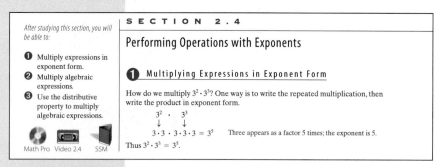

After studying this section, you will be able to:

❶ Multiply expressions in exponent form.
❷ Multiply algebraic expressions.
❸ Use the distributive property to multiply algebraic expressions.

Math Pro   Video 2.4   SSM

SECTION 2.4

Performing Operations with Exponents

❶ **Multiplying Expressions in Exponent Form**

How do we multiply $3^2 \cdot 3^3$? One way is to write the repeated multiplication, then write the product in exponent form.

$$3^2 \cdot 3^3$$
$$\downarrow \qquad \downarrow$$
$$3 \cdot 3 \cdot 3 \cdot 3 \cdot 3 = 3^5 \qquad \text{Three appears as a factor 5 times; the exponent is 5.}$$

Thus $3^2 \cdot 3^3 = 3^5$.

page 116

- Each section of the text begins with a reminder of the additional companion tools that have been designed to enhance your learning experience.

Students Solutions Manual

Prealgebra

Jamie Blair
John Tobey
Jeffrey Slater

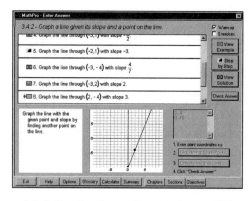

- Contains complete step-by-step solutions for every odd-numbered exercise
- Contains complete step-by-step solutions for all Chapter Review Problems, Chapter Tests, and Cumulative Tests

- MathPro Explorer: Interactive and Tutorial Software
- For Windows and Power Macintosh
- Includes preformatted algebra explorations
- Generates unlimited practice exercises

New York Times *Themes of the Times*
Newspaper-format supplement–
ask your professor about this free supplement

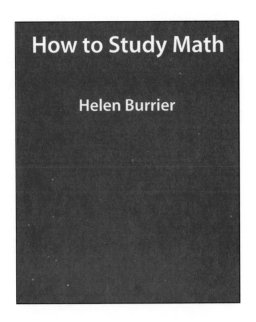

- Tips include how to prepare for class, how to study for and take tests, and how to improve your grades

- Team taught video instruction covers each section of the text

- A guide to navigation strategies through the Internet as well as practice exercises and lists of resources

- Visit the companion website for related and extended applications

www.prenhall.com/blair

PREALGEBRA

ADDITION AND SUBTRACTION OF WHOLE NUMBER EXPRESSIONS

CHAPTER 1

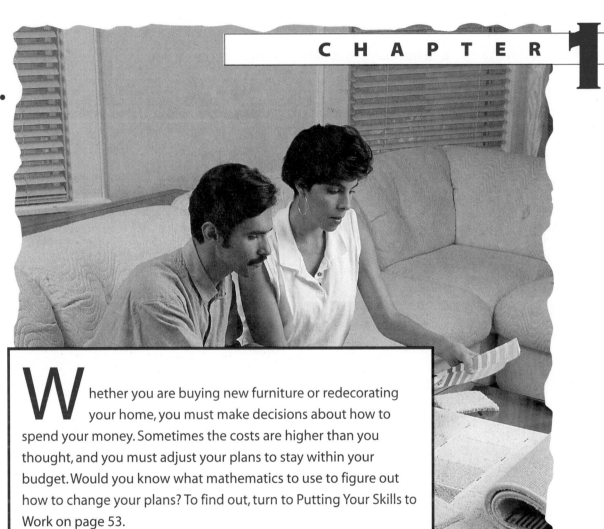

Whether you are buying new furniture or redecorating your home, you must make decisions about how to spend your money. Sometimes the costs are higher than you thought, and you must adjust your plans to stay within your budget. Would you know what mathematics to use to figure out how to change your plans? To find out, turn to Putting Your Skills to Work on page 53.

1.1 Understanding Whole Numbers

1.2 Addition and Combining Like Terms

1.3 Subtraction and Combining Like Terms

1.4 Perimeter, Graphs, and Estimation

1.5 Understanding Basic Equations

1.6 U.S. and Metric Units of Measurement

CHAPTER 1 PRETEST

This test provides a preview of the topics in this chapter. It will help you identify which concepts may require more of your studying time. If you are familiar with the topics in this chapter, take this test now. Check your answers with those in the back of the book. If you are not familiar with the topics in this chapter, begin studying the chapter now.

1. _____

2. _____

SECTION 1.1

1. Write 7032 in expanded notation.

2. Replace the question mark with the inequality symbol $<$ or $>$.
    (a) 10 ? 6   **(b)** 0 ? 2

3. **(a)** Round 439 to the nearest ten.   **(b)** Round 625 to the nearest hundred.

3. _____

4. _____

SECTION 1.2

4. Translate using mathematical symbols: A number is increased by 5.

5. Use the commutative or associative property of addition, then simplify.
    (a) $2 + (6 + y)$ **(b)** $7 + x + 4$

6. Add.
    (a) $301 + 5923$

    (b) $\begin{array}{r} 17 \\ 4 \\ +16 \\ \hline \end{array}$

5. _____

6. _____

7. _____

7. Combine like terms: $y + 8y + 2y + 6xy$.

SECTION 1.3

8. Combine like terms: $4x - x - 4$.

8. _____

9. Translate using mathematical symbols: The difference of 8 and 2.

9. _____

10. Subtract and check: $81 - 19$.

11. Sarah had a balance of \$570 in her checking account. She then wrote checks for \$150 and \$45. What is the new balance in her account?

10. _____

SECTION 1.4

12. Find the perimeter of the rectangle.

11. _____

               3 ft

               ☐   2 ft

12. _____

13. This histogram displays the first test scores of an algebra class. How many students scored 80 or more on the test?

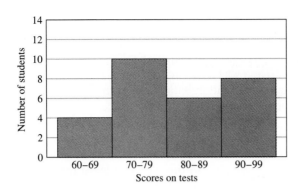

14. John drove his car 15,100 miles the first year he owned it, 14,800 the second year, 12,300 the third, and 13,700 the fourth year. Estimate how many more miles John drove his car the first two years than the second two years.

SECTION 1.5

15. Evaluate $8 - x$:
  (a) If x is 2. **(b)** If x is 0.

16. Solve for the variable and check your answer.
  (a) $2 + x = 15$ **(b)** $(x + 3) + 2 = 12$ **(c)** $17 - x = 13$

17. Translate to an equation and solve: Three added to what number equals 8?

18. Translate to an equation and solve: The sum of 5 and x equals 16.

SECTION 1.6

19. Add: 2 hr 35 min + 3 hr 45 min.

20. Replace the question mark with the inequality symbol $<$ or $>$.
  (a) 2 cm ? 2 m **(b)** 3 mi ? 3 km

13. _____

14. _____

15. _____

16. _____

17. _____

18. _____

19. _____

20. _____

❶ Understand place value of whole numbers.

❷ Write whole numbers in expanded notation.

❸ Understand and use inequality relationships.

❹ Round whole numbers.

Math Pro   Video 1.1   SSM

SECTION 1.1

Understanding Whole Numbers

Often we learn a new concept in stages. First comes learning the new *terms* and basic assumptions. Then we have to master the *reasoning*, or logic, behind the new concept. This often goes hand in hand with learning a *method* for using the idea. Finally, we can move quickly with a *shortcut*.

For example, in the study of stock investments, before tackling the question "What is my profit from this stock transaction?", you must learn the meaning of such terms as *stock*, *profit*, *loss*, and *commission*. Next, you must understand how stocks work (reasoning/logic), so that you can learn the method for calculating your profit. After you master this concept, you can quickly answer many similar questions using shortcuts.

In this book, watch your understanding of mathematics grow through this same process. In the first chapter we review the whole numbers, emphasizing *concepts*, not shortcuts. Do not skip this review even if you feel you have mastered the material, since understanding each stage of the concepts is crucial to learning algebra. With a little patience in looking at the terms, reasoning, and step-by-step methods, you'll find that your understanding of whole numbers has deepened, preparing you to learn algebra.

❶ Understanding Place Value of Whole Numbers

We use a set of numbers called **whole numbers** to count a number of objects.

> The whole numbers are as follows:
> 0, 1, 2, 3, 4, 5, 6, 7, 8, 9, 10, 11, 12, 13, 14, 15, 16, …

There is no largest whole number. The three dots … indicate that the set of whole numbers goes on forever. The numbers 0, 1, 2, 3, 4, 5, 6, 7, 8, 9 are called **digits**. The *position* or *placement* of the digit in a number tells the *value* of the digit and is called **place value**. For example, look at the following three numbers:

632   The "6" has place value 6 hundreds (600).
 61   The "6" has place value 6 tens (60).
  6   The "6" has the place value 6 ones (6).

For this reason, our number system is called a *place-value system*.

To indicate the value of numbers, we can use the following diagram, which shows the names of the digits' place value. Consider the number 47,632, which is entered on the chart.

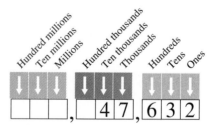

The place value of the digit 4 is ten thousands.
The place value of the digit 7 is thousands.
The place value of the digit 6 is hundreds.
The place value of the digit 3 is tens.
The place value of the digit 2 is ones.

E X A M P L E   1   For the whole number 573,025, state the place value for the digit:

(a) 7   **(b)** 0   **(c)** 2

SOLUTION

(a) 573,025   **(b)** 573,025   **(c)** 573,025
    ↑   ↑   ↑

ten thousands   **hundreds**   **tens**

P R A C T I C E   P R O B L E M   1
For the whole number 3,502,781, state the place value for the digit:

(a) 5   **(b)** 0   **(c)** 7

② Writing Whole Numbers in Expanded Notation

We sometimes write numbers in **expanded notation** to emphasize place value. The number 47,632 can be written in expanded notation as follows:

$$40,000 \; + \; 7000 \; + \; 600 \; + \; 30 \; + \; 2$$

| 4 ten | + | 7 | + | 6 | + | 3 | + | 2 |
| thousands | | thousands | | hundreds | | tens | | ones |

E X A M P L E   2   Write expanded notation for 1,340,765.

SOLUTION

$$1,340,765 = 1,000,000 + 300,000 + 40,000 + 700 + 60 + 5$$

Note: Since there is a zero in the thousands place, we do not write it as part of the sum.

P R A C T I C E   P R O B L E M   2
Write expanded notation for 2,507,235.

E X A M P L E   3   Jon withdraws $493 from his account. He requests the minimum number of bills in one-, ten-, and hundred-dollar bills. Describe the quantity of each denomination of bills the teller must give Jon.

SOLUTION   If we write $493 in expanded notation, we can easily describe the denominations needed.

$$400 \; + \; 90 \; + \; 3$$

4	**9**	**3**
hundred-	ten-	one-
dollar	dollar	dollar
bills	bills	bills

P R A C T I C E   P R O B L E M   3
Christina withdraws $582 from her account. She requests the minimum number of bills in one-, ten-, and hundred-dollar bills. Describe the quantity of each denomination of bills the teller must give Christina.

> **TO THINK ABOUT**
> ....
>
> THE NUMBER ZERO
>
> Not all number systems have had a zero. The Roman numeral number system does not. In our place-value system the zero is necessary so that we can write a number such as 308. This number means three hundreds and eight ones. By putting a zero in the tens place we indicate that there are zero tens. Without a zero symbol we would not be able to indicate this. For example, 38 has a different value than 308. 38 means three *tens* and eight ones, while 308 means three *hundreds* and eight ones. The *zero* is thus a **placeholder**. It holds a position and shows that there is no other digit in that place.

③ Understanding and Using Inequality Relationships

It is often helpful to draw pictures and graphs to help us visualize a mathematical concept. A **number line** is often used for whole numbers. Think of a line in which each point is paired with a number.

The number line has a point matched with zero and with each whole number. Each number is equally spaced, and the "→" arrow at the end indicates that the numbers go on forever. The numbers on the line increase from left to right.

A number *is greater than* a given number if it lies to the *right* of that number on the number line.

4 *is greater than* 2:

4 lies to the *right* of 2 on the number line.

A number *is less than* a given number if it lies to the *left* of that number on the number line.

3 *is less than* 5:

3 lies to the *left* of 5 on the number line.

The symbol $>$ means *is greater than*, and the symbol $<$ means *is less than*. For example,

$$4 > 2 \text{ means} \qquad 3 < 5 \text{ means}$$
$$\downarrow \qquad\qquad\qquad \downarrow$$
$$4 \text{ is greater than 2.} \quad 3 \text{ is less than 5.}$$

The symbols $<$ and $>$ are called **inequality symbols**. $4 > 2$ and $2 < 4$ are both correct statements. Note that the inequality symbol always points to the smaller number.

> **EXAMPLE 4** Replace the question mark with the inequality symbol $<$ or $>$.
>
> **(a)** 1 ? 6 **(b)** 9 ? 7 **(c)** 4 ? 9 **(d)** 9 ? 4

SOLUTION

(a) $1 < 6$ **(b)** $9 > 7$ **(c)** $4 < 9$ **(d)** $9 > 4$

↓   ↓   ↓   ↓

1 is less than 6.   9 is greater than 7.   4 is less than 9.   9 is greater than 4.

PRACTICE PROBLEM 4
Replace the question mark with the inequality symbol $<$ or $>$.

(a) 3 ? 2   **(b)** 6 ? 8   **(c)** 1 ? 7   **(d)** 7 ? 1

EXAMPLE 5   Write using an inequality symbol.

(a) Five is less than eight.   **(b)** Nine is greater than four.

SOLUTION

(a) Five *is less than* eight.   **(b)** Nine *is greater than* four.

    ↓   ↓   ↓   ↓   ↓   ↓

    5   **<**   **8**   **9**   **>**   **4**

PRACTICE PROBLEM 5
Write using an inequality symbol.

(a) Seven is greater than two.   **(b)** Three is less than four.

❹ Rounding Whole Numbers

We often approximate the values of numbers when it is not necessary to know the exact values. These approximations are easier to use and remember. For example, if a hotel bill were $82.00, we might say that we spent about $80. If a car cost $14,792, we would probably say that it cost approximately $15,000.

Why did we approximate the price of the car at $15,000 and not $14,000? To understand why, let's look at the number line.

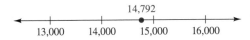

The number 14,792 is closer to 15,000 than to 14,000, so we approximate the cost of the car at $15,000.

It would also be correct to approximate the cost at $14,800 or $14,790, since all these values are close to 14,792 on the number line. How can we assure that everyone gets the same number when approximating? We specify how accurate we would like our approximation. **Rounding** is a process that approximates a number to a specific **round-off place** (ones, tens, hundreds, …). Thus *the value obtained when rounding depends on how accurate we would like our approximation.* To illustrate, we round the price of the car discussed above to the thousands and hundreds place.

14,792 rounded to the nearest thousand is 15,000.

↓

The *round-off place* is thousands.

14,792 rounded to the nearest hundred is 14,800.

↓

The *round-off place* is hundreds.

We can use the following set of rules instead of a number line to round whole numbers.

PROCEDURE TO ROUND A WHOLE NUMBER

1. Identify the round-off place.
2. If the digit to the *right* of the round-off place digit is:
   (a) *Less than 5*, do not change the round-off place digit.
   (b) *5 or more*, increase the round-off place digit by 1.

In either case, replace all digits to the *right* of the round-off place digit with zeros.

EXAMPLE 6   Round 57,441 to the nearest thousand.

SOLUTION

The round-off place digit is the thousands place.

57,④41

The digit to the right of the round-off place is less than 5.

Do not change the round-off place digit.   57,**000**

Replace all digits to the right with zeros.

We have rounded 57,441 to the nearest thousand: **57,000**. This means that 57,441 is closer to 57,000 than to 58,000.

PRACTICE PROBLEM 6
Round 34,627 to the nearest hundred.

EXAMPLE 7   Round 4,254,423 to the nearest hundred thousand.

SOLUTION

The round-off place digit is the hundred thousands place.

4,2⑤4,423

The digit to the right of the round-off place digit is 5 or more.

Increase the round-off place digit by 1.   4,**300,000**

Replace all digits to the right with zeros.

We have rounded 4,254,423 to the nearest hundred thousand: **4,300,000**.

PRACTICE PROBLEM 7
Round 1,335,627 to the nearest ten thousand.

EXAMPLE 8   An airplane flight takes 3 hours and 51 minutes. Approximately how many hours does the flight take?

SOLUTION   We want to round to the nearest *hour*. 3 hours 51 minutes is closer to 4 hours. The flight took approximately **4 hours**.

PRACTICE PROBLEM 8
It takes Sara 1 hour 7 minutes to drive to work. Approximately how many hours does it take her to drive to work?

EXERCISES 1.1

1. For the whole number 9865, state the place value for the digit:

   (a) 8   (b) 5   (c) 9

2. For the whole number 23,981, state the place value for the digit:

   (a) 2   (b) 9   (c) 3

3. For the whole number 754,310, state the place value for the digit:

   (a) 4   (b) 7   (c) 1

4. For the whole number 913,728, state the place value for the digit:

   (a) 9   (b) 1   (c) 7

5. For the whole number 1,284,073, state the place value for the digit:

   (a) 1   (b) 0   (c) 3

6. For the whole number 3,098,269, state the place value for the digit:

   (a) 0   (b) 8   (c) 2

Write in expanded notation.

7. 4967

8. 7632

9. 2493

10. 3562

11. 867,301

12. 913,045

13. Fill in the check with the amount $672.

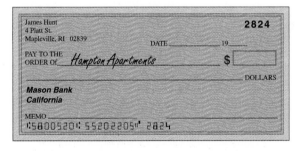

14. Fill in the check with the amount $379.

15. Damian withdraws $562 from his account. He requests the minimum number of bills in one-, ten-, and hundred-dollar bills. Describe the quantity of each denomination of bills the teller must give Damian.

16. Erin withdraws $274 from her account. She requests the minimum number of bills in one-, ten-, and hundred-dollar bills. Describe the quantity of each denomination of bills the teller must give Erin.

17. Describe the denominations of bills for $46:

    (a) Using only ten- and one-dollar bills.

    (b) Using tens, fives, and only 1 one-dollar bill.

18. Describe the denominations of bills for $96:

    (a) Using only ten- and one-dollar bills.

    (b) Using tens, fives, and only 1 one-dollar bill.

Replace the question mark with the inequality symbol < or >.

19. 1 ? 7

20. 2 ? 0

21. 4 ? 8

22. 9 ? 4

23. 11 ? 10

24. 10 ? 11

25. 9 ? 0

26. 0 ? 9

27. 12 ? 6

28. 6 ? 7

29. 8 ? 17

30. 4 ? 1

Write using an inequality symbol.

31. Five is greater than two.

32. Seven is less than ten.

33. Two is less than five.

34. Six is greater than four.

The following table lists the 1998 sticker prices on some popular automobiles. Use this table to answer questions 35 and 36.

TYPE OF AUTOMOBILE	1998 STICKER PRICE
Dodge Neon Highline	$13,765
Ford Escort LX	$13,735
Toyota Corolla LE	$16,092
Honda Civic LX	$16,445

35. Use an inequality symbol to indicate the relationship between the price of the Dodge Neon Highline and the Ford Escort LX.

36. Use an inequality symbol to indicate the relationship between the price of the Toyota Corolla LE and the Honda Civic LX.

Round to the nearest ten.

37. 45

38. 85

39. 661

40. 127

Round to the nearest hundred.

41. 16,462

42. 701,529

43. 823,042

44. 12,799

Round to the nearest thousand.

45. 38,431

46. 671,529

47. 12,577

48. 117,011

Round to the nearest hundred thousand.

49. 5,254,423

50. 3,116,201

51. 9,007,601

52. 1,395,999

Answer each question.

53. A train takes 3 hours and 10 minutes to reach its destination. Approximately how many hours does the trip take?

54. An automobile trip takes 5 hours and 4 minutes. Approximately how many hours does the drive take?

55. About 563,000,000 people in the world speak German. Round this figure to the nearest ten million.

56. There are 3,484,800 inches in 55 miles. Round 3,484,800 to the nearest ten thousand.

57. For fall 1997 the total enrollment of Orange Coast College in California was 25,358. Round this enrollment figure to:

(a) The nearest thousand.

(b) The nearest hundred.

58. Recently, the total enrollment of students in elementary schools in Florida was 356,586. Round this enrollment figure to:

(a) The nearest thousand.

(b) The nearest hundred.

 VERBAL AND WRITING SKILLS

59. Write the word names for:

(a) 8002

(b) 802

(c) 82

(d) What is the place value of the digit "0" in the number "eight hundred twenty"?

60. Write in words.

(a) $2 < 6$

(b) $6 > 2$

(c) What can you say about parts **(a)** and **(b)**?

ONE STEP FURTHER

Very large numbers are used in some disciplines to measure quantities, such as distance in astronomy and the national debt in macroeconomics. We can extend the place-value chart to include these large numbers. The number in the place-value chart below represents the national debt in 1996.

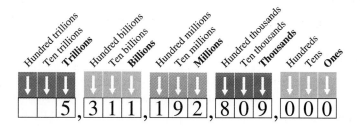

61. Write the national debt in 1996 of $5,311,192,809,000 using the word name.

62. Round the national debt for 1996 to the nearest million.

Round to the nearest hundred.

63. 16,962

64. 44,972

After studying this section, you will be able to:

1 Understand the key words for expressing addition.

2 Use the properties of addition.

3 Add whole numbers when carrying is needed.

4 Combine like terms.

Math Pro   Video 1.2   SSM

S E C T I O N   1 . 2

Addition and Combining Like Terms

1 Understanding the Key Words for Expressing Addition

In mathematics we use symbols such as "+" in place of the words *sum* or *plus*. The English phrase "5 plus 2" written using symbols is "5 + 2." When we do not know the value of a number, we use a letter, such as x, to represent that number. A letter that represents a number is called a **variable.**

Writing English phrases using math symbols is like translating between languages such as Spanish and French. There are several English phrases to describe the operation of addition. The following table gives some English phrases and their translated equivalents written using mathematical symbols.

ENGLISH PHRASE	TRANSLATION INTO SYMBOLS
6 *more than* 9	$6 + 9$
The *sum* of some number and 7	$x + 7$
4 *increased by* 2	$4 + 2$
3 *added* to a number	$3 + n$
1 *plus* a number	$1 + x$

Notice that the variables used in the table are different. We can choose any letter as a variable. Thus we can represent $x + 7$ by $a + 7$, $n + 7$, or $y + 7$, and so on.

E X A M P L E   1   Translate each English phrase using symbols.

(a) The sum of six and eight

(b) A quantity increased by 4

SOLUTION

(a) The sum of six and eight
↓ ↓
6 + 8

6 + 8

(b) A quantity increased by 4
↓ ↓ ↓
x + 4

$x + 4$

Note: Although we used the variable x to represent the unknown quantity in part (b), any letter could have been used to represent the unknown quantity.

P R A C T I C E   P R O B L E M   1
Translate each English phrase using symbols.

(a) Five added to some number

(b) 4 more than 5

2 Using the Properties of Addition

What is addition? We perform addition when we group items together. Consider the manager of a bicycle shop who sells 4 bikes on Saturday and 3 bikes on Sunday. When the manager reviews the inventory on Monday he must group together the bikes sold each day and come up with the total sales for the weekend.

Bikes sold Saturday   + Bikes sold Sunday   is equal to   Total bikes sold

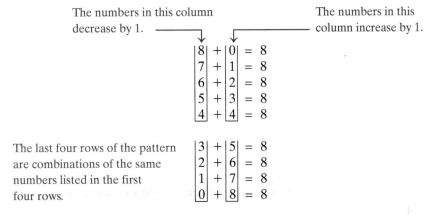

    4   +   3   =   7 bikes

We see that the number 7 is the sum of 4 and 3. That is, $4 + 3 = 7$ is an addition fact. Most of us memorized all the addition facts. Yet if we study these sums, we observe that there are only a few addition facts for each number that we must memorize. For example, when 0 items are added to any number of items, we end up with the same number of items. This will be true no matter what number we add to 0, and leads us to a property of 0 called the **addition property of zero**.

ADDITION PROPERTY OF ZERO

$a + 0 = a$ and $0 + a = a$ $5 + 0 = 5$ and $0 + 5 = 5$

When zero is added to any number, the sum is that number.

EXAMPLE 2   Express 8 as the sum of two whole numbers. Write all possibilities. How many arithmetic facts must we memorize? Why?

SOLUTION   From the addition property of zero we know one sum equal to 8: $8 + 0 = 8$. Now starting with 1, we write all the sums equal to 8 and see what we observe.

The numbers in this column decrease by 1. ─────┐   ┌───── The numbers in this column increase by 1.

$$8 + 0 = 8$$
$$7 + 1 = 8$$
$$6 + 2 = 8$$
$$5 + 3 = 8$$
$$4 + 4 = 8$$

The last four rows of the pattern are combinations of the same numbers listed in the first four rows.

$$3 + 5 = 8$$
$$2 + 6 = 8$$
$$1 + 7 = 8$$
$$0 + 8 = 8$$

We need to learn only **four addition facts** for the number 8: $7 + 1$, $6 + 2$, $5 + 3$, and $4 + 4$. The remaining facts are either a repeat of these, or use the addition property of zero.

PRACTICE PROBLEM 2
Express 6 as the sum of two whole numbers. Write all possibilities. How many arithmetic facts must we memorize? Why?

In Example 2 we saw that the order in which we add numbers doesn't affect the sum. This is true for all numbers and leads us to a property called the **commutative property of addition**.

COMMUTATIVE PROPERTY OF ADDITION

$$a + b = b + a \qquad 4 + 9 = 9 + 4$$
$$13 = 13$$

Two numbers can be added in either order with the same result.

EXAMPLE 3   Use the commutative property of addition to rewrite each sum.

(a) $8 + 2$ **(b)** $7 + n$ **(c)** $x + 3$

SOLUTION

(a) $8 + 2 = \mathbf{2 + 8}$ **(b)** $7 + n = \mathbf{n + 7}$ **(c)** $x + 3 = \mathbf{3 + x}$

Notice that we applied the commutative property of addition to the expressions with variables n and x. That is because variables represent numbers, even though they are unknown numbers.

PRACTICE PROBLEM 3
Use the commutative property of addition to rewrite each sum.

(a) $x + 3$ **(b)** $9 + w$ **(c)** $4 + 0$

EXAMPLE 4
If $\ \ 2566 + \ \ 159 = 2725,\ \$ then
$\ \ \ \ \ 159 + 2566 = ?$

SOLUTION

$\ \ \ \ 159 + 2566 = \mathbf{2725}$ Why? The commutative property states that the order in which we add numbers doesn't affect the sum.

PRACTICE PROBLEM 4
If $\ \ 5663 + \ \ 412 = 6075,\ \$ then
$\ \ \ \ \ 412 + 5663 = ?$

To **simplify** the expression $8 + 1 + x$, we find the sum of 8 and 1.

$$8 + 1 + x = 9 + x \text{ or } x + 9$$

Simplifying $8 + 1 + x$ is similar to rewriting the English phrase "8 plus 1 plus some number" as a more simplified phrase: "9 plus some number." We can write this simplification as either $9 + x$ or $x + 9$ since addition is commutative. We usually write this sum as $x + 9$, since it is standard to *write the variable first in the expression.*

EXAMPLE 5   Simplify: $3 + 2 + n$.

SOLUTION   To simplify, we find the sum:

$$3 + 2 + n = 5 + n \text{ or } \mathbf{n + 5}.$$

Note: We cannot add the variable n and the number 5 because n represents an unknown quantity; thus we have no way of knowing what quantity to add to the number 5.

PRACTICE PROBLEM 5
Simplify: $6 + 3 + x$.

Addition of more than two numbers may be performed in more than one manner. To add $5 + 2 + 1$ we can first add the 5 and 2, or we can add the 2 and 1 first. We indicate which sum we add first by using parentheses. *We perform the operation inside the parentheses first.*

$$5 + 2 + 1 = (5 + 2) + 1 = 7 + 1 = 8$$
$$5 + 2 + 1 = 5 + (2 + 1) = 5 + 3 = 8$$

In both cases the sum is the same. This illustrates the **associative property of addition**.

ASSOCIATIVE PROPERTY OF ADDITION

$$(a + b) + c = a + (b + c) \qquad (4 + 9) + 1 = 4 + (9 + 1)$$
$$13 + 1 = 4 + 10$$
$$14 = 14$$

When we add two or more numbers, the addition may be grouped in any order.

EXAMPLE 6   Use the associative property of addition to rewrite the sum, then simplify: $(x + 3) + 6$.

SOLUTION

$$(x + 3) + 6 = x + (3 + 6) \qquad \text{The associative property allows us to regroup.}$$
$$= x + 9 \qquad \text{Simplify: } 3 + 6 = 9.$$

PRACTICE PROBLEM 6
Use the associative property of addition to rewrite the sum, then simplify: $(w + 1) + 4$.

Sometimes we must use both the associative and commutative properties of addition to rewrite a sum and simplify. In other words, we can *change the order in which we add* (commutative property) and *regroup the addition* (associative property) to simplify an expression.

EXAMPLE 7   Use the associative and/or commutative property as necessary to simplify the expression: $5 + (n + 7)$.

SOLUTION

$$5 + (n + 7) = 5 + (7 + n) \qquad \text{The commutative property allows us to change the order of addition.}$$
$$= (5 + 7) + n \qquad \text{Regroup the sum using the associative property.}$$
$$= 12 + n \qquad \text{Simplify.}$$
$$5 + (n + 7) = n + 12$$

PRACTICE PROBLEM 7
Use the associative and/or commutative property as necessary to simplify each expression:

(a) $(2 + x) + 8$ **(b)** $(4 + x + 3) + 1$

TO THINK ABOUT

ADDITION FACTS MADE SIMPLE

There are many methods that can be used to add one-digit numbers. For example, if you can't remember that $7 + 8 = 15$ but can remember that $7 + 7 = 14$, just add 1 to 14 to get 15.

$$7 + \quad 8 = ?$$
$$\downarrow \qquad \downarrow$$
$$7 + (7 + 1) \qquad \text{Since 8 is the sum } 7 + 1.$$
$$= (7 + 7) + 1$$
$$= \qquad 14 + 1 = 15$$

Using this idea, you can add $6 + 7$ as follows:

$$6 + \quad 7 \quad = ?$$
$$\downarrow$$
$$6 + (6 + 1) = 12 + 1 = 13$$

Another quick way to add is to use the sum $5 + 5 = 10$, since it is easy to remember. Let's use this to add $7 + 5$.

$$7 \; + 5 = ? \qquad \text{Write 7 as the sum } 2 + 5.$$
$$\downarrow$$
$$= 2 + \underline{5 + 5} \qquad \text{Add } 5 + 5.$$
$$= 2 + \underline{10} \quad = 12 \qquad \text{Add } 10 + 2.$$

Thus $7 + 5 = 12$.

EXERCISE

1. Use the fact that $5 + 5 = 10$ to add $8 + 5$.

---

❸ Adding Whole Numbers When Carrying Is Needed

Of course, many numbers that require addition are more than single-digit numbers. In such cases we must:

1. Arrange the numbers vertically, lining up the digits according to place value.
2. Add first the digits in the ones column, then the digits in the tens column, then those in the hundreds column, and so on, moving from *right to left*.

Sometimes the sum of a column is a two-digit number—that is, a number larger than 9. When this happens we evaluate the place values of the digits to find the sum.

EXAMPLE 8   Add: $68 + 25$.

SOLUTION   We arrange numbers vertically and begin adding in the ones column:

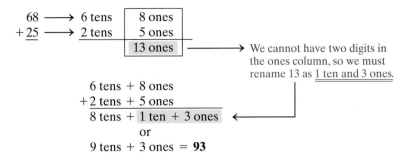

$$68 \longrightarrow 6 \text{ tens} \quad \boxed{8 \text{ ones}}$$
$$+\underline{25} \longrightarrow 2 \text{ tens} \quad 5 \text{ ones}$$
$$\boxed{13 \text{ ones}} \longrightarrow$$

We cannot have two digits in the ones column, so we must rename 13 as <u>1 ten and 3 ones</u>.

$$6 \text{ tens} + 8 \text{ ones}$$
$$+\underline{2 \text{ tens} + 5 \text{ ones}}$$
$$8 \text{ tens} + \boxed{1 \text{ ten} + 3 \text{ ones}} \longleftarrow$$
or
$$9 \text{ tens} + 3 \text{ ones} = \mathbf{93}$$

A shorter way to do this problem involves a process called "carry." Instead of rewriting 13 ones as *1 ten and 3 ones* we would carry the *1 ten* to the tens column by placing a 1 above the 6 and writing the 3 in the ones column of the sum.

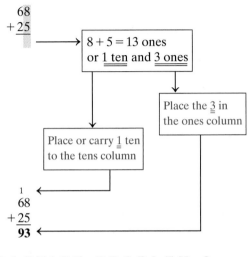

$$68$$
$$+\underline{25}$$

$8 + 5 = 13$ ones or <u>1 ten</u> and <u>3 ones</u>

Place the <u>3</u> in the ones column

Place or carry <u>1</u> ten to the tens column

$$\overset{1}{6}8$$
$$+\underline{25}$$
$$93$$

PRACTICE PROBLEM 8
Add: $247 + 38$.

It may be necessary to carry several times, and the number being carried can be any digit from 1 to 9.

EXAMPLE 9
A market research company surveyed 1870 people to determine the type of beverage they order most often at a restaurant. The results of the survey are shown in the table. Find the total number of people whose responses were iced tea, soda, and coffee.

SOLUTION   We add the responses for iced tea, 357; soda, 577; and coffee, 84.

Type of Beverage		Number of Responses
Soda		577
Orange juice		475
Coffee		84
Iced tea		357
Milk		286
Other		91

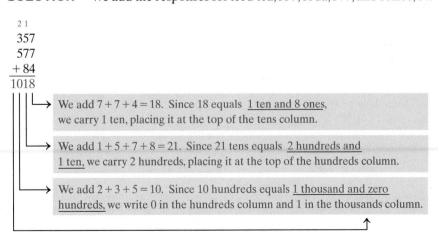

$$\overset{2\ 1}{357}$$
$$577$$
$$+\underline{84}$$
$$1018$$

We add $7 + 7 + 4 = 18$. Since 18 equals <u>1 ten and 8 ones</u>, we carry 1 ten, placing it at the top of the tens column.

We add $1 + 5 + 7 + 8 = 21$. Since 21 tens equals <u>2 hundreds and 1 ten</u>, we carry 2 hundreds, placing it at the top of the hundreds column.

We add $2 + 3 + 5 = 10$. Since 10 hundreds equals <u>1 thousand and zero hundreds</u>, we write 0 in the hundreds column and 1 in the thousands column.

$$357 + 577 + 84 = \mathbf{1018}$$

PRACTICE PROBLEM 9

Use the survey results from Example 9 to answer the following: Find the total number of people whose reponses were milk, orange juice, and other.

④ Combining Like Terms

In algebra we often deal with terms such as $4y$ or $7x$. What do we mean by *terms*?

> A **term** is a number, variable, or a product of a number and one or more variables.

Usually a term has a *numerical part* and a *variable part*.

Term: $7x$

numerical part   variable part
of term   of term

In an expression, terms are separated by addition and subtraction signs. A term that has no variable is called a **constant term**, and a term that has a variable is called a **variable term**.

$$9 + 3n + 4x$$

constant term   variable terms

What do the expressions $3n$ and $4x$ mean? $3n$ is the term that represents the sum $n + n + n$, and $4x$ is the term that represents the sum $x + x + x + x$. As we will see later, $3n$ and $4x$ also indicate multiplication; 3 times n and 4 times x.

$n + n + n$ three n's added $x + x + x + x$ four x's added
$\downarrow$ $\downarrow$
$3n$ $4x$

EXAMPLE 10   Write a term that represents each of the following.

(a) Two y's   **(b)** $a + a + a + a$ **(c)** Seven   **(d)** One x

SOLUTION

(a) Two y's = **$2y$**   **(b)** $a + a + a + a$ = **$4a$**
(c) Seven = **7**   **(d)** One x = $1x$ or x

PRACTICE PROBLEM 10

Write a term that represents each of the following.

(a) Four n's   **(b)** $y + y + y$ **(c)** Eight   **(d)** One y

We see many examples of adding quantities that are like quantities. This is called **combining like quantities**.

3 feet + 7 feet = 10 feet   7 trucks + 2 trucks = 9 trucks

However, we cannot combine things that are not the same:

7 trucks + 4 feet   (cannot be done!)

Similarly, in algebra we cannot combine terms that are not like terms. **Like terms** are terms that have identical variable parts. For example, in the expression $3x + 7y + 2x$, the $3x$ and $2x$ are like terms since they have the same variable parts.

Like Terms

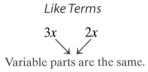

$3x \qquad 2x$

Variable parts are the same.

There are no like terms for $7y$, since none of the terms have exactly the same variable part as $7y$.

E X A M P L E   1 1   Identify like terms: $7ab + 4a + 2ab + 3y$.

SOLUTION

There are **no** terms like $4a$;
none have the **exact** variable part, a.

There are **no** terms like $3y$;
none have the **exact** variable part, y.

$7\boldsymbol{ab} + 4a + 2\boldsymbol{ab} + 3y$

$7ab$ and $2ab$ are like terms; the variable parts, ab, are the same.

The like terms are **7ab** and **2ab**.

P R A C T I C E   P R O B L E M   1 1
Identify like terms: $2mn + 5y + 4mn + 6n$.

The numerical part of a term is called the **coefficient** of a term. To combine like terms we add the coefficients of like terms. Let's combine like terms in the expression $3x + 7y + 2x$.

$$3x + 7y + 2x = (3x + 2x) + 7y \qquad \text{We identify and group like terms.}$$
$$= (3 + 2)x + 7y \qquad 3x + 2x; \text{"three } x\text{'s plus two } x\text{'s" can be}$$
restated as $(3 + 2)x$'s; "three plus two x's."
$$= 5x + 7y$$

Thus $3x + 7y + 2x = \mathbf{5x + 7y}$.

COMBINING LIKE TERMS
To combine like terms, add the numerical coefficients of like terms. The variable part stays the same.

$$6x + 4x = 10x$$

E X A M P L E   1 2   Identify like terms, then combine like terms.

(a) $4xy + 8y + 2xy$ **(b)** $5a + 6 + a$ **(c)** $8y + 2a + 3x$

SOLUTION

(a) $4xy + 8y + 2xy$

There are **no** terms like $8y$;
none have the **exact** variable part.

$4xy$ $+ 8y +$ **$2xy$**

$4xy$ and $2xy$ are like terms;
the variable parts are the same.

$$4xy + 8y + 2xy = 4xy + 2xy + 8y \qquad \text{Rearrange the terms.}$$
$$= (4 + 2)xy + 8y \qquad \text{Add the numerical coefficients of}$$
$$\text{like terms.}$$
$$= \mathbf{6xy + 8y}$$

Note: We write $8y$ as a separate term. We cannot combine it with $6xy$ since the variable parts are not the same.

(b) $5a + 6 + a = 5a + 6 + 1a$ Write the numerical coefficient 1.

There are **no** terms like 6 since it is the only constant term.

$5a + 6 + 1a$

The variable parts are the same; $5a$ and $1a$ are like terms.

$$\mathbf{5a + 6 + 1a} = \mathbf{5a + 1a} + 6$$
$$= (5 + 1)a + 6$$
$$= \mathbf{6a + 6}$$

Note: The term a does not have a visible numerical coefficient. We can write "1" as the numerical coefficient since a represents the quantity "one a."

(c) $8y + 2a + 3x = \mathbf{8y + 2a + 3x}$ Terms cannot be combined; there are no like terms.

P R A C T I C E   P R O B L E M   1 2
Combine like terms.

(a) $2ab + 4a + 3ab$ **(b)** $4y + 5x + y + x$ **(c)** $7x + 3y + 3z$

EXERCISES 1.2

Translate using symbols.

1. Ten more than x

2. A quantity is increased by 4

3. A number plus 2

4. Three added to a number

5. The sum of 5 and y

6. The sum of 8 and 7

7. A number added to 3

8. Twelve more than y

9. A number increased by 7

10. Six plus 4

Write each number as a sum of two whole numbers. Write all possibilities. How many addition facts must be memorized?

11. 6

12. 5

Use the commutative property of addition to rewrite each sum.

13. $2 + x$

14. $y + 6$

15. $8 + n$

16. $5 + x$

17. If $3542 + 216 = 3758$, then $216 + 3542 = ?$

18. If $8791 + 156 = 8947$, then $156 + 8791 = ?$

19. If $5 + n = 12$, then $n + 5 = ?$

20. If $8 + x = 31$, then $x + 8 = ?$

Simplify.

21. $n + 7 + 3$

22. $a + 5 + 3$

23. $6 + 2 + x$

24. $7 + 1 + y$

25. $x + 0 + 2$

26. $x + 3 + 0$

Use the associative property of addition to rewrite each sum, and simplify.

27. $(x + 2) + 1$

28. $(x + 4) + 2$

29. $5 + (3 + n)$

30. $6 + (4 + x)$

31. $(n + 3) + 8$

32. $(a + 4) + 6$

Use the associative and/or commutative property as necessary to simplify each expression.

33. $(x + 2) + 8$

34. $5 + (3 + a)$

35. $(6 + n) + 3$

36. $(4 + x) + 5$

37. $8 + (1 + x)$

38. $(y + 1) + 4$

39. $(3 + n) + 6$

40. $4 + (n + 2)$

41. $(7 + a + 1) + 3$

42. $(6 + x + 4) + 4$

43. $(1 + x + 7) + 2$

44. $(2 + n + 8) + 5$

Add.

45. $\begin{array}{r} 36 \\ +23 \\ \hline \end{array}$

46. $\begin{array}{r} 71 \\ +12 \\ \hline \end{array}$

47. $\begin{array}{r} 142 \\ +\ 34 \\ \hline \end{array}$

48. $\begin{array}{r} 331 \\ +\ 57 \\ \hline \end{array}$

49.
```
  21
  14
   8
+  7
```

50.
```
  33
  11
   6
+  4
```

51.
```
 105
   8
 133
+ 98
```

52.
```
 308
   7
 245
+ 75
```

53. $236 + 467 + 26$

54. $431 + 217 + 18$

55. $397 + 29 + 467$

56. $562 + 65 + 133$

57. $7287 + 273 + 522$

58. $3366 + 152 + 485$

59. $200 + 54 + 1287 + 5000$

60. $300 + 99 + 2413 + 4000$

61. $3121 + 8050 + 16 + 1667$

62. $2902 + 9050 + 12 + 3337$

Answer each question.

63. Angelica's check register indicates the deposits and debits (checks written or ATM withdrawals) for a 1-month period.

Date	Deposits	Debits
12/3/97	$152	
12/9/97		$ 63
12/13/97	$241	
12/15/97		$121
12/22/97		$ 44

  (a) What is the total of the deposits made to Angelica's checking account?

  (b) What is the total of the debits made to Angelica's checking account?

64. The bookkeeper for the Spaulding Appliance Company examined the following record from the company account for the month of March.

Date	Deposits	Debits
3/6/97	$3400	
3/9/97		$ 120
3/13/97		$3500
3/15/97	$4000	
3/22/97		$1300

  (a) What is the total of the deposits in this time period?

  (b) What is the total of the debits in this time period?

65. The rent on an apartment was $875. To move in, Charles and Vincent were required to pay the first and last month's rent, a security deposit of $500, a connection fee with the utility company for $24, and a telephone installation fee of $35. How much money did they need to move into the apartment?

66. Shawnee found that for a six-month period, in addition to gasoline, she had the following car expenses: insurance, $562; repair to brakes, $276; new tires, $142. If gasoline for her car cost $495 for this time period, what was the total amount she spent on her car?

Write a term that represents each expression.

67. Two x's

68. Four y's

69. $a + a + a$

70. $x + x + x + x + x$

Identify like terms.

71. $5x + 3y + 2x + 8m + 6y$

72. $6m + 4b + 7m + 3x + 8b$

73. $2mn + 3y + 4mn + 2$

74. $6x + 3xy + 8 + 2xy$

Combine like terms.

75. $3x + 2x + 6x$

76. $5a + 3a + 7a$

77. $8x + 4a + 3x + 2a$

78. $9y + 2b + 2y + 4b$

79. $6xy + 4b + 3xy$

80. $3ab + 5x + 9ab$

81. $2x + 3xy + 4n + 6$

82. $3a + 6ab + 9y + 7$

83. $5mn + 6m + 1 + 2mn$

84. $6xy + 3x + 9 + 9xy$

85. $4ax + 8x + 6ax + 9x$

86. $8xy + 4y + 6xy + 6y$

 CALCULATOR EXERCISES

87. Approximately 30 percent of the total surface area of the earth is land and about 70 percent is water. This means that the earth has approximately 57,259,000 square miles of land and 139,692,000 square miles of water. Find the total square miles of land and water.

88. A Boeing 747 jet flew 2988 miles on Monday, 3455 miles on Tuesday, 1997 miles on Wednesday, and 2440 miles on Thursday. How many total miles did it fly on those 4 days?

 VERBAL AND WRITING SKILLS

89. Can you think of an easy way to remember the addition of two numbers if one of the numbers is 9? For example, $9 + 4, 9 + 7, 9 + 26, \ldots$. Explain.

90. Write in your own words how the associative and commutative properties of addition allow us to simplify expressions. You may use an example in your explanation.

91. Give a written explanation as to why we cannot add terms that are not alike. You may use an example in your explanation.

CUMULATIVE REVIEW

1. State the place value of the digit 8 in the number 20,891.

2. State the place value of the digit 0 in the number 7,903,122.

3. Write 408 using expanded notation.

4. Describe the denomination of bills for $67 using only ten- and one-dollar bills.

❶ Understand subtraction.

❷ Use subtraction to combine like terms.

❸ Understand the key words for expressing subtraction.

❹ Use the properties of subtraction.

❺ Subtract numbers with two or more digits.

❻ Solve application problems involving addition and subtraction.

Math Pro   Video 1.3   SSM

S E C T I O N   1 . 3

Subtraction and Combining Like Terms

❶ Understanding Subtraction

What is subtraction? This is the basic idea of subtraction: *From a whole amount a part is "taken away" or subtracted.* When we subtract we find "how many are left" or the **difference**. The symbol used to indicate subtraction is called a *minus sign* "−." We illustrate below.

Six   take away   two   =   four left
↓   ↓   ↓   ↓
6   −   2   =   4

It is helpful if you can subtract quickly. See if you can do Example 1 in 10 seconds or less. Repeat again with Practice Problem 1. Strive to obtain all answers correctly in 10 seconds or less.

E X A M P L E   1   Subtract.

(a) $9 - 5$ **(b)** $7 - 2$ **(c)** $5 - 4$ **(d)** $7 - 3$

SOLUTION

(a) $9 - 5 = \mathbf{4}$ **(b)** $7 - 2 = \mathbf{5}$ **(c)** $5 - 4 = \mathbf{1}$ **(d)** $7 - 3 = \mathbf{4}$

P R A C T I C E   P R O B L E M   1
Subtract.

(a) $5 - 2$ **(b)** $6 - 3$ **(c)** $8 - 7$ **(d)** $3 - 1$

Do we need to learn subtraction facts in order to perform subtraction? No, it is not necessary since we can use addition facts to subtract. To illustrate this idea, let's consider the following subtraction situation. A car dealer has 6 trucks available for sale when the dealership opens. If 2 of these trucks are sold during that day, how many are left? We represent this situation in two different ways. One way uses subtraction facts, and the other uses addition facts.

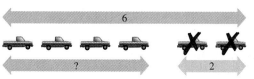

Using subtraction facts we see that $6 - 2 = 4$.

Trucks at the start of day	Minus	Trucks sold	Equals	Trucks left
6	−	2	=	?

Using addition facts we see that $6 = 4 + 2$.

Trucks at the start of day	Equals	Trucks left	Plus	Trucks sold
6	=	?	+	2

Therefore, to subtract $6 - 2 = ?$, we can use addition facts by writing the equivalent addition problem $6 = ? + 2$ and thinking, "What number plus 2 equals 6?"

$6 - 2 = ?$ Subtraction problem.

$6 = ? + 2$ Equivalent addition problem: What number plus 2 equals 6? 4.

Thus, $6 - 2 = 4$.

To write an equivalent addition problem we write the subtraction as addition on the opposite side of the equal sign.

Writing equivalent addition problems not only helps us with subtraction facts, but as we will see later, is also used when we are solving equations.

EXAMPLE 2 $12 - 4 = ?$

(a) Write an equivalent addition problem.
(b) Use addition facts to complete the subtraction problem.

SOLUTION

(a) To write the equivalent addition problem, we write **subtraction** on the **left side** of the equal sign as **addition** on the **right side** of the equal sign.

$$12\ \boxed{-\ 4}\ = ?$$ Subtraction problem.

$$12 = ?\ \boxed{+\ 4}$$ Equivalent addition problem.

The equivalent addition problem is **$12 = ? + 4$**.

(b) $12 = \boxed{?} + 4$ 12 equals what number plus 4? 8.
$$12 = \boxed{8} + 4$$

Thus $12 - 4 = \mathbf{8}$

PRACTICE PROBLEM 2
$18 - 11 = ?$

(a) Write an equivalent addition problem.
(b) Use addition facts to complete the subtraction problem.

② Using Subtraction to Combine Like Terms

As we saw earlier, to combine like terms we add coefficients of like terms. How do we combine like terms when we are subtracting terms? We subtract coefficients of like terms as illustrated below.

If Rapid Florist had 5 baskets of flowers and sold 3 baskets, they would have 2 baskets of flowers left:

5 baskets minus 3 baskets   =   2 baskets left

Similarly, in algebra we subtract like terms:

Five x's minus three x's   =   two x's
$\cancel{x} + \cancel{x} + \cancel{x} + x + x$ = $x + x$
$5x - 3x$ $= (5 - 3)x$ or $2x$

EXAMPLE 3   Combine like terms.

(a) $7y - 4y$ **(b)** $9m - 5m - 8$ **(c)** $6x - x - 4a - 1$

SOLUTION

(a) $7y - 4y = (7 - 4)y = 3y$ Subtract numerical coefficients of like terms.
$7y - 4y = \mathbf{3y}$

(b) $9m - 5m - 8 = (9 - 5)m - 8$ Subtract numerical coefficients of like terms.
$9m - 5m - 8 = \mathbf{4m - 8}$

Note: We cannot combine the terms $4m$ and 8, since they are not like terms. $4m$ is a variable term and 8 is a constant term.

(c) $6x - x - 4a - 1 = 6x - 1x - 4a - 1$ Write the numerical coefficient 1.

$= 6x - 1x - 4a - 1$ Identify like terms.

$= (6 - 1)x - 4a - 1$ Subtract numerical coefficients of like terms.

$6x - x - 4a - 1 = \mathbf{5x - 4a - 1}$ Terms cannot be combined; $5x$, $4a$, and 1 are not like terms.

PRACTICE PROBLEM 3
Combine like terms.

(a) $8m - 3m$ **(b)** $5x - 4x - 2$ **(c)** $7y - y - 3b - 2$

Understanding the Key
3 Words for Expressing Subtraction

In an earlier section we translated English phrases into math symbols using key words for addition. In this section and some of the following sections we will add to this list of key words. Translating and solving equations are skills needed to solve word problems. As you develop these skills you will find that solving real-life word problems will become easier.

There are three parts to a subtraction problem: minuend, subtrahend, and difference.

$$\begin{array}{rl} 6 & \text{minuend} \\ -\,4 & \text{subtrahend} \\ \hline 2 & \text{difference} \end{array}$$

There are several English phrases to describe the operation of subtraction. The following table presents some English phrases and their translated equivalents written using mathematical symbols.

ENGLISH PHRASE	TRANSLATION INTO SYMBOLS
The *difference* of 3 and x	$3 - x$
Eight *minus* a number equals two.	$8 - n = 2$
Two *subtracted from* seven	$7 - 2$
A number *decreased by* 4 is the same as 3.	$n - 4 = 3$
Five *less than* nine	$9 - 5$

E X A M P L E   4   Translate using symbols.

(a) The difference between 5 and x **(b)** Six minus two equals four.
(c) Four less than seven

SOLUTION

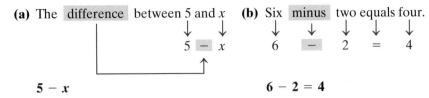

(a) The difference between 5 and x **(b)** Six minus two equals four.

$$5 - x$$ $$6 - 2 = 4$$

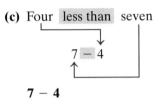

(c) Four less than seven

$$7 - 4$$

Note: We do not always write the math symbols in the same order as they are read in the statement, as illustrated in part (c).

P R A C T I C E   P R O B L E M   4
Translate using symbols.

(a) The difference of 9 and n is 1.   **(b)** x minus 3
(c) x subtracted from eight

❹ Using the Properties of Subtraction

As we saw in the previous section, the order in which we write the numbers in the subtraction is *reversed* when we use the phrase *less than* and *subtracted from*. It is important to write these numbers in the correct order because, in general, *subtraction is not commutative*. In other words, $30 - 20$ is not the same as $20 - 30$. To show this, let's see what happens when we change the order of the numbers in subtraction.

$30 - $20   You have \$30 in your checking account and write a check for \$20; your balance will be \$10.

$20 - $30   You have \$20 in your checking account and write a check for \$30; you will be overdrawn!

Obviously, the results are not the same.

The way we group numbers when subtracting is also important because subtraction is not associative. For example, if you had \$25 in the checking account and wrote a check for \$10, you would have \$15 left. Then if you wrote another check for \$5, you would end up with a balance of \$10 in the checking account.

$(25 - 10) - 5 = 15 - 5$ Parentheses show the first operation to be performed.

$= \mathbf{10}$ \$10 left, *which is correct.*

Now, let's see what happens if we change the grouping symbols and thus the order of subtraction.

$25 - (10 - 5) = 25 - 5$

$= \mathbf{20}$ \$20 left, *which is NOT correct.*

We summarize as follows.

SUBTRACTION IS NOT COMMUTATIVE
If a and b are not the same number, then

$\qquad a - b$ does not equal $b - a$. $30 - 20$ does not equal $20 - 30$.

SUBTRACTION IS NOT ASSOCIATIVE
If c does not equal 0, then

$\qquad (a - b) - c$ is not the same as $a - (b - c)$.

$\qquad (25 - 10) - 5$ is not the same as $25 - (10 - 5)$.

EXAMPLE 5   Subtract.

(a) $(13 - 5) - 2$ **(b)** $13 - (5 - 2)$

SOLUTION

(a) $(13 - 5) - 2$ Parentheses show the first operation to be performed.

$\qquad = 8 - 2$ Subtract: $13 - 5 = 8$.

$\qquad = 6$ Subtract: $8 - 2 = 6$.

$\qquad \mathbf{(13 - 5) - 2 = 6}$

(b) $13 - (5 - 2)$ Parentheses show the first operation to be performed.

$\qquad = 13 - 3$ Subtract: $5 - 2 = 3$.

$\qquad = 10$ Subtract: $13 - 3 = 10$.

$\qquad \mathbf{13 - (5 - 2) = 10}$

PRACTICE PROBLEM 5
Subtract.

(a) $(11 - 4) - 1$ **(b)** $11 - (4 - 1)$

TO THINK ABOUT

EXCEPTIONS TO THE RULE

We found that in general subtraction is neither commutative nor associative. We say "in general" since there are some cases when subtraction is in fact commutative and associative. For example, the associative property applies to subtraction as follows: If a and b are whole numbers, $(a - b) - 0 = a - (b - 0)$. That is, $(5 - 2) - 0 = 5 - (2 - 0)$.

EXERCISE

1. Can you think of other cases when subtraction is associative and commutative, that is, when $(a - b) - c = a - (b - c)$ and $a - b = b - a$?

What about the number zero and subtraction? If we start with 4 items and subtract or "take away" 0 items, we are left with the same number of items we started with. Now, if we start with 4 items and "take away" all 4 items, we are left with 0 items. These facts lead us to the following subtraction properties of zero.

SUBTRACTION PROPERTIES OF ZERO

$a - 0 = a$ $4 - 0 = 4$

When 0 is subtracted from a number, the result is the number.

$a - a = 0$ $4 - 4 = 0$

When a number is subtracted from itself, the result is zero.

From the subtraction properties of zero we know that $6 - 0 = 6$. Now, starting with 0, let's write all subtraction facts and see what we observe.

The numbers in this column increase by 1. ———→   ←——— The numbers in this column decrease by 1.

$$6 - 0 = 6$$
$$6 - 1 = 5$$
$$6 - 2 = 4$$
$$6 - 3 = 3$$
$$6 - 4 = 2$$
$$6 - 5 = 1$$
$$6 - 6 = 0$$

Thus each time you subtract the next larger whole number, the result decreases by 1. We can use this subtraction pattern to subtract mentally.

EXAMPLE 6 $800 - 50 = 750$. Use this fact to find to find $800 - 53$.

SOLUTION   Since we know $800 - 50 = 750$, we can use subtraction patterns to find $800 - 53$.

Increase numbers in this column by 1.   Decrease numbers in this column by 1.
↓   ↓

$$800 - 50 = 750$$
$$800 - 51 = 749$$
$$800 - 52 = 748$$
$$800 - 53 = \mathbf{747}$$

PRACTICE PROBLEM 6
$600 - 50 = 550$. Use this fact to find $600 - 54$.

⑤ Subtracting Numbers with Two or More Digits

Often, we cannot subtract mentally, especially if the numbers being subtracted involve more than two digits. In this case we follow the same procedure as we did in addition, except we subtract digits instead of add. Therefore, we must:

1. Arrange the numbers vertically.
2. Subtract the digits in the ones column first, then the digits in the tens column, then those in the hundreds column, and so on, moving from *right to left*.

Many times, however, a digit in the lower number (subtrahend) is greater than the digit in the upper number (minuend) for that place value.

$$5\boxed{1}\!\!\!\nwarrow\!\!\!7 \quad 7 > 1 \quad \text{We cannot subtract } 1 - 7.$$
$$-2\boxed{7}$$

When this happens, we must *rename* place values so that we can subtract.

E X A M P L E   7   Subtract: $72 - 38$.

SOLUTION

$$\begin{array}{r} 72 \\ -\,38 \end{array}$$ We cannot subtract $2 - 8$ so rewrite 7 tens as "tens and ones."

$$\begin{array}{r} 72 \\ -\,38 \end{array} \qquad \begin{array}{r} \text{7 tens} + \text{2 ones} \\ -\,\text{3 tens} + \text{8 ones} \end{array} \longrightarrow \quad \begin{array}{c} \underline{\text{7 tens}} \; + \; \text{2 ones} \\ \swarrow \qquad \searrow \qquad \searrow \\ \underline{\text{6 tens}} \; \underline{\text{10 ones}} \; + \; \text{2 ones} \end{array}$$

$$\begin{array}{r} 72 \\ -\,38 \end{array} \qquad \begin{array}{r} \text{6 tens} + \text{12 ones} \\ -\,\text{3 tens} + \;\;\text{8 ones} \end{array} \longrightarrow \text{We add:} \qquad \underline{\text{10 ones} + 2 \text{ ones}} = \text{12 ones.}$$

$$\begin{array}{r} 34 \end{array} \qquad \text{3 tens} + \;\;\text{4 ones} \longrightarrow \text{12 ones} - \text{8 ones} = \textbf{4 ones}; \quad \text{6 tens} - \text{3 tens} = \textbf{3 tens}$$

A shorter way to do this is called **borrowing**. Instead of rewriting *7 tens + 2 ones* as *6 tens + 12 ones*, we would borrow 1 ten from the 7 tens by crossing out the 7 and placing the 6 above the 7. Then we would cross out the 2 and place the 12 above the 2.

$$\begin{array}{r} {\scriptstyle 6\;12} \\ \cancel{7\,2} \\ -\,3\,8 \\ \hline \mathbf{3\,4} \end{array} \quad \begin{array}{l} \text{6 tens} + \text{12 ones} \\ \longrightarrow \text{7 tens} + \text{2 ones} \\ \\ \longrightarrow \text{We subtract: 12 ones} - \text{8 ones} = \text{4 ones.} \\ \\ \longrightarrow \text{We subtract: 6 tens} - \text{3 tens} = \text{3 tens.} \end{array}$$

P R A C T I C E   P R O B L E M   7
Subtract: $93 - 46$.

Sometimes we cannot borrow from the digit directly to the left, because this digit is 0. In this case we borrow from the next nonzero digit to the left of 0, as illustrated in the next example.

E X A M P L E   8   Subtract: $304 - 146$.

SOLUTION

$$\begin{array}{r} 304 \\ -\,146 \end{array}$$ We must borrow since we cannot subtract 4 ones $-$ 6 ones.

We cannot borrow a "ten" since there are **0 tens**, so we must borrow from 3 hundreds.

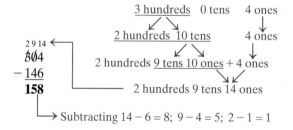

$$\begin{array}{c} \underline{\text{3 hundreds}} \quad \text{0 tens} \quad \text{4 ones} \\ \swarrow \quad \searrow \qquad \downarrow \\ \underline{\text{2 hundreds}}\;\underline{\text{10 tens}} \qquad \text{4 ones} \\ \swarrow \quad \searrow \qquad \downarrow \\ \text{2 hundreds }\underline{\text{9 tens 10 ones}} + \text{4 ones} \\ \searrow \swarrow \\ \text{2 hundreds 9 tens 14 ones} \end{array}$$

$$\begin{array}{r} {\scriptstyle 2\;9\;14} \\ \cancel{3\,0\,4} \\ -\,1\,4\,6 \\ \hline \mathbf{1\,5\,8} \end{array}$$

Subtracting $14 - 6 = 8$; $9 - 4 = 5$; $2 - 1 = 1$

P R A C T I C E   P R O B L E M   8
Subtract: $603 - 278$.

TO THINK ABOUT
··

MONEY AND BORROWING

Converting money (changing $100 bills to $10 and $1 bills) illustrates the process "borrow," similar to the one used in Example 8. To see this, let's look at the following:

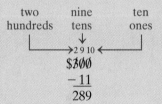

Mathematical Process "Borrow"

Real-Life Situation

A cashier in a gift shop is out of small bills and has only 3 hundred-dollar bills left in the cash register. The cashier must give a customer $11 change for a purchase. Since the cashier only has 3 hundred-dollar bills in the register, it is necessary to ask another cashier to convert a hundred-dollar bill to tens and ones.

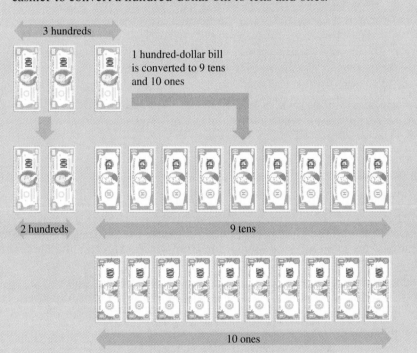

$$\begin{array}{r} \$\,300 \\ -\ \ 11 \\ \end{array}$$

two hundreds nine tens ten ones

$$\begin{array}{r} 2\ 9\ 10 \\ \$\cancel{300} \\ -\ \ 11 \\ \hline 289 \end{array}$$

The cashier now has 2 hundreds, 9 tens, and 10 ones and can give the customer $11 change.

EXERCISES

1. Can you see what happens when we must borrow from 0? That is, when subtracting $400 - 68$, why must we change 0's to 9, then borrow 1 from the first whole number to the left of the 0('s)? Explain.

2. Explain why changing 13 ones to 1 ten-dollar and 3 one-dollar bills is similar to "carrying" in addition?

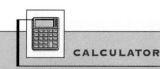

CALCULATOR

SUBTRACTING WHOLE NUMBERS

The calculator can be used to check your answer. You can use your calculator to subtract whole numbers. To find $401 - 245$, enter:

Scientific calculator:

$$401 \; \boxed{-} \; 245 \; \boxed{=}$$

Graphing calculator:

$$401 \; \boxed{-} \; 245 \; \boxed{\text{ENT}}$$

The calculator displays

$$156$$

We can check our subtraction problems using equivalent addition problems. For example, to check that $7 - 2 = 5$, we verify that $7 = 5 + 2$.

EXAMPLE 9 Subtract: $7004 - 3675$.

SOLUTION

→ We cannot subtract $4 - 5$, so we must change 700 to 699 to borrow 10 ones.

$$\begin{array}{r} {}^{6}\cancel{7}{}^{9}\cancel{0}{}^{9}\cancel{0}{}^{14}\cancel{4} \\ -3\;6\;7\;5 \\ \hline 3\;3\;2\;9 \end{array}$$

→ Then we add: 10 ones + 4 ones = 14 ones.

→ Subtract: $14 - 5 = \underline{9}$; $9 - 7 = \underline{2}$; $9 - 6 = \underline{3}$; $6 - 3 = \underline{3}$.

Check your answers:

Subtraction	Check by Addition
$7\;0\;0\;4$ ✓	
$-3\;6\;7\;5$ →	$3\;6\;7\;5$
$3\;3\;2\;9$ →	$+3\;3\;2\;9$
	$7\;0\;0\;4$ ✓

It checks.

PRACTICE PROBLEM 9
Subtract and check: $8006 - 4237$.

Solving Application Problems
6 Involving Addition and Subtraction

Key words and phrases found in applied problems often help determine which operations should be used for computations. Subtraction is often used in real-life problems when we are comparing more than one amount. Often we want to know *how much more* or *how much less* one amount is than another. Subtraction is also necessary when we want to know *how much is left* or when the problem uses the key words for subtraction, such as difference, minus, subtracted from, decreased by, or less than.

EXAMPLE 10 The California Department of Fish and Game reported that a record number of yellowtail were caught in local waters in 1997. The estimated number of yellowtail caught was more than 458,000. The department also reported that the largest number of yellowtail caught in a single day were by charter boats from the following California landings:

California Landing	Record Number of Yellowtail Caught
Dana Point	1810
Balboa Peninsula	1367
Newport	917

How many more yellowtail were caught by the charter boat from the Dana Point landing than by the charter boat from the Balboa Peninsula landing?

SOLUTION   The key phrase "*how many more*" indicates that the operation used is subtraction.

Number of yellowtail caught at Dana Point	minus	Number of yellowtail caught at Balboa Peninsula

$$1810 \quad - \quad 1367$$

$$\begin{array}{r} {\scriptstyle 0\ 10} \\ 1\ 8\ \overset{\scriptstyle}{1}\ \cancel{0} \\ -\ 1\ 3\ 6\ 7 \\ \hline 3 \end{array}$$ → We must borrow.

→ We cannot subtract $0 - 6$, so we must borrow again.

$$\begin{array}{r} {\scriptstyle 7\ 10} \\ {\scriptstyle \cancel{8}\ 10} \\ 1\ 8\ \overset{\scriptstyle}{1}\ \cancel{0} \\ -\ 1\ 3\ 6\ 7 \\ \hline \mathbf{4\ 4\ 3} \end{array}$$ → We borrow: write 8 hundreds as 7 hundreds 10 tens.

Check:
$$\begin{array}{r} 1810 \\ -1367 \\ \hline 443 \end{array} \quad \longrightarrow \quad \begin{array}{r} 1367 \\ +\ 443 \\ \hline 1810 \end{array} \quad \checkmark$$

The charter from Dana Point caught 443 more yellowtail.

PRACTICE PROBLEM 10
Use the information in Example 10 to answer the following. How many fewer fish were caught by the charter boat from the Newport landing than the charter boat from the Dana Point landing?

Bar graphs are helpful in seeing changes over a period of time, especially when the same type of data are studied repeatedly. Since graphs are encountered so frequently, one of the main tasks for a student of statistics is to interpret and draw proper conclusions from various kinds of graphs.

The profits for the Excelsior Engineering Co. are displayed in the following bar graph.

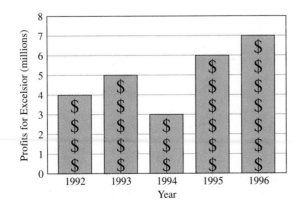

E X A M P L E 1 1   Use the bar graph to answer the following questions.

(a) What was the profit in 1992?
(b) How much more profit was made in 1995 than in 1994?
(c) In which year was the profit the lowest?
(d) Were the total profits for the two-year period 1992–1993 more, less, or the same as those for the two year period 1994–1995?

SOLUTION

(a) For the year 1992, the bar rises to 4. This represents a profit of $4 million. **The profit in 1992 was $4,000,000.**
(b) For the year 1994, the bar graph rises to $3 million in profits, and it rises to $6 million in 1995. We subtract to find how much more profit there was in 1995.

$$\begin{array}{rl} 6,000,000 & 1995 \\ -3,000,000 & 1994 \\ \hline 3,000,000 & \end{array}$$

There was $3,000,000 more in profits in 1995 than in 1994.
(c) The bar for the year 1994 has the *lowest* height, so it represents the year that the profits were the lowest. **The profits were lowest in 1994.**
(d) To find the profits for each two-year period, we add.

$$\begin{array}{rl} 4\text{ million} & 1992 \\ +5\text{ million} & 1993 \\ \hline 9\text{ million} & \end{array} \qquad \leftarrow \text{Total for two-year period} \rightarrow \qquad \begin{array}{rl} 3\text{ million} & 1994 \\ +6\text{ million} & 1995 \\ \hline 9\text{ million} & \end{array}$$

The profits were the same for each two-year period.

P R A C T I C E P R O B L E M 1 1
Use the bar graph to answer the following questions.

(a) What was the profit in 1996?
(b) How much more profit was made in 1992 than in 1994?
(c) In which year was the profit highest?
(d) Were the total profits for the two-year period 1993–1994 more, less, or the same as those for the two-year period 1995–1996?

To solve a word problem it is helpful to *organize* the information, then *plan* the process that you will use. This is very similar to what we do when we use a daily planner or date book to *organize* and *plan* our days. A **Mathematics Blueprint for Problem Solving** will be used to organize the information and plan the method to be used. From the blueprint we will be able to see clearly the following three steps for solving real-life situations.

- *Step 1: Understand the problem.* Read the problem and organize the information.

- *Step 2: Solve and state the answer.* Use arithmetic and algebra to find the answer.

- *Step 3: Check the answer.* Use estimation and other techniques to test your answer.

E X A M P L E   1 2   Alisha had a balance of $350 in her checking account when she went to the bank to deposit the following checks: $156, $130, $83, and $105. Then she wrote checks for $278, $59, and $357. What is her new balance?

SOLUTION

1. *Understand the problem.* We read the problem carefully and fill in the information on the following Mathematics Blueprint.

MATHEMATICS BLUEPRINT FOR PROBLEM SOLVING			
GATHER THE FACTS	WHAT AM I ASKED TO DO?	HOW DO I PROCEED?	KEY POINTS TO REMEMBER
Old balance is $350. The *deposits* are $156, $130, $83, $105. The *checks* written are $278, $59, $357	Find the *new balance* in the checking account.	1. Find the *total deposits* and the *total checks* written. 2. *Add* the total *deposits* to the old balance. 3. From this figure *subtract* the total of the *checks*.	*Add* deposits, then *subtract* checks.

2. *Solve and state the answer.* From the information organized in the blueprint, we can write out the process to find the answer.

Old balance	+	total deposits	−	total checks	=	new balance
		$156		$278		
		130		59		
		83		357		
		+ 105		+		
$350	+	$474	−	$694	=	$130

The new balance in the checking account is $130.

3. *Check the answer.* We first make sure that we copied all the information correctly and answered the question. Then we rework the problem a different way to see if we get the same answer. We add the total deposits to the old balance:

Old balance + Total deposits
$$\$350 + 156 + 130 + 83 + 105 = \$824$$

Now we subtract each check from this total:

$$\$824 - \$278 = \$546; \quad \$546 - \$59 = \$487; \quad \$487 - \$357 = \$130$$

We get a balance of $130, which is the same as our original answer.

P R A C T I C E   P R O B L E M   1 2
Last month George had $569 in a savings account. He made two deposits: one for $706 and one for $234. The bank credited him with $22. Since last month he made four withdrawals: $42, $132, $341, and $202. What is his balance this month?

EXERCISES 1.3

Subtract.

1. $6 - 2$ **2.** $5 - 3$ **3.** $9 - 5$ **4.** $8 - 4$

5. $12 - 3$ **6.** $13 - 8$ **7.** $17 - 8$ **8.** $11 - 9$

9. $16 - 7$ **10.** $19 - 6$ **11.** $15 - 7$ **12.** $24 - 18$

13. $18 - 0$ **14.** $29 - 0$ **15.** $20 - 20$ **16.** $15 - 15$

For the following equations:

    (a) *Write an equivalent addition problem.*

    (b) *Use addition facts to complete the subtraction problem.*

17. $19 - 7 = ?$ **18.** $17 - 9 = ?$ **19.** $18 - 12 = ?$

20. $21 - 11 = ?$ **21.** $15 - 8 = ?$ **22.** $18 - ? = 3$

23. $16 - ? = 4$ **24.** $25 - ? = 19$ **25.** $33 - ? = 27$

Combine like terms.

26. $8x - 2x$ **27.** $9y - 3y$ **28.** $10x - 4x$

29. $11a - 9a$ **30.** $7x - x$ **31.** $8m - m$

32. $5a - 2a - 2$ **33.** $6y - 3y - 1$ **34.** $12a - 3a - 4$

35. $14x - 2x - 3$ **36.** $8x - 2x - 5a - 1$ **37.** $9a - 3a - 2x - 2$

Translate using symbols.

38. Seven decreased by three   **39.** Eight minus two equals 6

40. The difference of 8 and y **41.** Nine less than 12

42. Ten subtracted from 17   **43.** Three minus a number

44. Fifteen minus a number equals 8   **45.** Seven subtracted from a number

46. Two less than some number   **47.** The difference of three and one

Evaluate.

48. **(a)** $(16 - 11) - 1$ **49.** **(a)** $(18 - 9) - 5$

    (b) $16 - (11 - 1)$ **(b)** $18 - (9 - 5)$

50. (a) $(11 - 3) - 2$

(b) $11 - (3 - 2)$

51. (a) $15 - (6 - 3)$

(b) $(15 - 6) - 3$

We can use subtraction patterns to subtract large numbers mentally; for example, if we know $500 - 400 = 100$, we can find $500 - 403$ as follows: $500 - 400 = 100$

$$500 - 401 = 99$$
$$500 - 402 = 98$$
$$500 - 403 = 97$$

52. If $600 - 500 = 100$, find $600 - 504$ using subtraction patterns.

53. If $900 - 800 = 100$, find $900 - 806$ using subtraction patterns.

54. If $300 - 200 = 100$, find $300 - 205$ using subtraction patterns.

55. If $800 - 700 = 100$, find $800 - 705$ using subtraction patterns.

Subtract and check.

56. $87 - 46$

57. $99 - 26$

58. $69 - 34$

59. $76 - 41$

60. $83 - 67$

61. $56 - 37$

62. $759 - 613$

63. $873 - 551$

64. $966 - 177$

65. $761 - 542$

66. $598 - 343$

67. $753 - 329$

68. $9922 - 2667$

69. $8721 - 6654$

70. $5301 - 2185$

71. $8801 - 4583$

72. $7008 - 4839$

73. $9002 - 3667$

74. $4300 - 256$

75. $6400 - 457$

Use the Mathematics Blueprint for Problem Solving to help you solve the word problems.

76. Jim and Anna each held various fundraisers for their city's Cancer Foundation. Anna raised $2445 and Jim raised $1981. How much more money did Anna raise than Jim?

77. The Thomas Trucking Company made a bank deposit of $7231 and wrote checks equaling $2987. How much money was left of their deposit?

78. The down payment on a boat costing $9130 is $2899. Find the amount that is left to be paid.

79. The Dow Jones average opened at 7837 one day and closed that day at 7894. How much did it rise that day?

80. A market research company reported that in 1997 apartment rents in Orange County, California, rose faster than in any metropolitan area in the United States. The rent for a one-bedroom apartment averaged $811, a two-bedroom averaged $1111, and a three-bedroom $1352. Studios averaged $788.

(a) How much more is the average rent for a one-bedroom than for a studio apartment?

(b) How much cheaper is the average rent on a two-bedroom than on a three-bedroom apartment?

81. Fill in the balance in Pedro's check register.

Check Number	Amount	Balance $1364
Check # 123	$238	
Check # 124	$137	
Check # 125	$ 69	
Check # 126	$ 98	
Check # 127	$369	

82. The moon is about 400 times smaller than the sun.

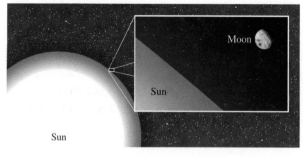

The diameter of the sun is approximately 865,000 miles, and the diameter of the moon is approximately 2160 miles. How many more miles is the diameter of the sun than the moon?

83. If the moon were next to the earth it would like a tennis ball next to a basketball.

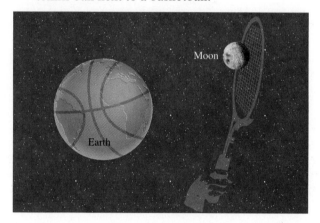

The approximate polar diameter of the earth is 7900 miles (distance through the earth from North Pole to South Pole). The diameter of the moon is approximately 2160 miles. Find the difference in the diameters of the earth and the moon.

84. A family's monthly budget for household expenses is $3033. After $1295 is spent for the house payment, $469 is spent for food, $387 is spent for clothing, and $287 is spent for entertainment, how much is left in the budget?

85. Joan has a $500 monthly expense allowance from her parents while she is living in the dorm at college. Her expenses this month were: food $190, gas $43, telephone $42, school supplies $96, and entertainment $55. How much expense money did she have left after expenses?

86. Jordan must drive 1342 miles back to school in 3 days. If he travels 562 miles the first day and 487 miles the second day, how far must he travel on the third day?

87. A teacher's assistant receives a total salary per month of $2320. Deducted from her paycheck are taxes of $399, Social Security of $118, and retirement of $87. What was the total of her check after the deductions?

For exercises 88 and 89, refer to the following bar graph, which displays the total sales for a large corporation.

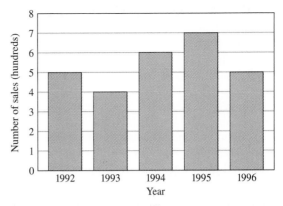

88. (a) What were the number of sales in 1994?

(b) How many more sales were there in 1996 than in 1993?

(c) In which year were the sales the highest?

(d) Are the total sales for the three-year period 1993–1995 more, less, or the same as those for the three-year period 1994–1996?

89. (a) What were the number of sales in 1995?

(b) How many fewer sales were there in 1992 than in 1994?

(c) In which year were the sales lowest?

(d) Are the total sales for 1992 and 1994 more, less, or the same as the total sales for 1993 and 1995?

For exercises 90 and 91, refer to the following bar graph, which displays the number of new building permits granted in a large county.

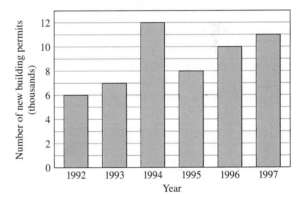

90. (a) How many fewer permits were issued in 1995 than in 1997?

(b) How many more permits were issued for the two-year period 1996–1997 than for the two-year period 1992–1993?

91. (a) How many more permits were issued in 1996 than in 1993?

(b) How many fewer permits were issued in the two-year period 1995–1996 than in 1993–1994?

 CALCULATOR EXERCISES

92. The distance between the sun and the earth varies because the earth travels around the sun in an orbit that has an elliptical (oval) shape.

The shortest distance from the earth to the sun is approximately 91,400,000 miles, and the greatest distance is about 94,500,000. What is the difference between the greatest and shortest distance?

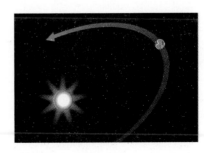

93. The national debt in 1996 was $5,311,192,809,000. In 1991 the national debt was $3,502,432,800,000. How much more was the national debt in 1996 than in 1991?

94. The Census Bureau reported that Las Vegas was the largest-growing metropolitan area in the country from 1990 to 1996. The Las Vegas population grew during that period from 852,646 to 1,201,073. How much did the population grow during that period?

95. The U.S. population at the beginning of 1998 was 268,922,000. In April 1990 the population was 248,765,000. How much did the population grow during this period?

ONE STEP FURTHER

96. Can 2 hundred-dollar bills and 1 fifty-dollar bill be converted into all twenty-dollar bills? Why or why not?

97. For what value(s) of x and y will $x - y = y - x$?

CUMULATIVE REVIEW

Replace the question mark with an inequality symbol.

1. 5,117,206 ? 13,842

2. 2,386,702 ? 117,401

Subtract.

3. 12,066 − 237

4. 31,007 − 579

SECTION 1.4

Perimeter, Graphs, and Estimation

1 Using Addition to Calculate a Perimeter

Geometry has a visual aspect that many students find helpful to their learning. Numbers and abstract quantities may be hard to visualize, but we can take pen in hand and actually draw a picture of a rectangle that represents a room with certain dimensions. We can easily visualize problems such as "How many feet are around the outside edges of the room (perimeter)?" In this section we study rectangles, squares, triangles, and complex shapes that are made up of these figures.

A **rectangle** is a four-sided figure like the ones shown here.

A rectangle has the following two properties:

1. Any two adjoining sides are perpendicular.
2. Opposite sides are equal.

By "any two adjoining sides are perpendicular" we mean that any two sides that are next to each other form an angle (called a *right angle*) that measures 90 degrees and forms one of these shapes:

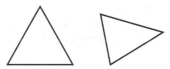

When we say that "opposite sides are equal" we mean that the measure of a side is equal to the measure of the side across from it. When all sides of a rectangle are the same length, we call the rectangle a **square**.

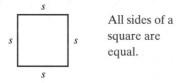

All sides of a square are equal.

A **triangle** is a three-sided figure with three angles.

> The distance around an object (such as a rectangle or triangle) is called the **perimeter**. To find the perimeter of an object, add the lengths of all its sides.

EXAMPLE 1   Find the perimeter of the rectangle. (The abbreviation ft means feet.)

5 ft

2 ft

SOLUTION   Since opposite sides of a rectangle are equal, we have

We add the length of the sides: 5 ft + 5 ft + 2 ft + 2 ft = 14 ft

The perimeter is 14 ft.

P R A C T I C E   P R O B L E M   1
Find the perimeter of the square.

15 ft

E X A M P L E   2   Find the perimeter of the shape consisting of a rectangle and a square.

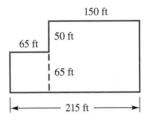

SOLUTION   We want to find the distance around the object. Therefore, we look only at the outside edges. Dashed lines indicate inside lengths.

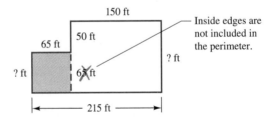

Inside edges are not included in the perimeter.

First, we must find the values of the missing sides. The shaded figure is a square since the length and width have the same measure. Thus all sides of the shaded figure have a measure of 65 ft.

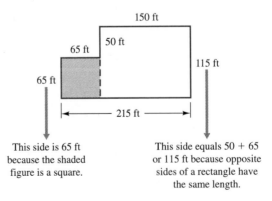

This side is 65 ft because the shaded figure is a square.

This side equals 50 + 65 or 115 ft because opposite sides of a rectangle have the same length.

Next, we add the six sides to find the perimeter:

$$150 \text{ ft} + 115 \text{ ft} + 215 \text{ ft} + 65 \text{ ft} + 65 \text{ ft} + 50 \text{ ft} = 660 \text{ ft}$$

The perimeter is 660 ft.

PRACTICE PROBLEM 2

Find the perimeter of the shape consisting of a rectangle and a square.

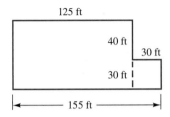

DEVELOPING YOUR STUDY SKILLS

CLASS ATTENDANCE: THE LEARNING CYCLE

Did you know that an important part of the learning process happens in the classroom? Statistics show that class attendance and good grades go together. People learn by *reading, writing, listening, verbalizing,* and *seeing* activities in the **learning cycle**. These activities always occur in class:

- *Listening and seeing:* hearing and watching the instructor's lecture
- *Reading:* reading the information on the board and in handouts
- *Verbalizing:* asking questions and participating in class discussions
- *Writing:* taking notes and working problems assigned in class

The Learning Cycle

Reading   ⟶   Writing
↑   ↓
Seeing ← Verbalizing ← Listening

    Attendance in class completes the entire learning cycle once. Completing assignments activates the entire learning cycle one more time:

- Reading class notes and the text
- Writing your homework
- Seeing, listening, and talking about your strategies with other students

    Attending class and completing assignments are both critical learning opportunities. The more times you complete the entire learning cycle, the more you learn.

② Interpreting Graphs

Suppose that an English professor posts the scores for the final exam. If 50 students took the final and scored between 60 and 99, the results can be displayed in four groups on the following table:

SCORES ON THE FINAL EXAM	NUMBER OF FINAL EXAMS
60–69	6
70–79	14
80–89	20
90–99	10

The table indicates the number of final exams scores that lie within four different intervals: 60–69, 70–79, 80–89, 90–99. For example, 14 students had a final exam score between 70 and 79.

These results can also be displayed on a special type of graph called a *histogram*. On a histogram, the height of bar indicates the number of exam scores that lie within the intervals written below the bar.

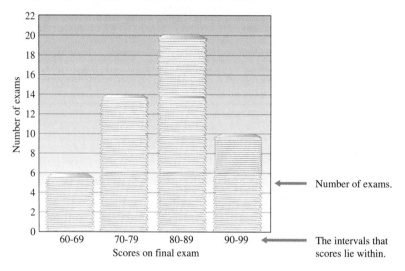

Number of exams.

The intervals that scores lie within.

EXAMPLE 3

(a) How many students scored a B on the final if the interval for a B is 80–89?
(b) How many students scored less than 80 on the final?
(c) How many more students scored between 90 and 99 than between 60 and 69?

SOLUTION

(a) The 80–89 bar rises to a height of 20. **Thus there are 20 students who scored a B on the final**.
(b) From the histogram we see that there are two different bar heights to be considered. Six tests were 60–69 and 14 tests were 70–79. When we add, $6 + 14 = 20$, we see that **20 students scored less than 80 on the final**.
(c) The histogram indicates that 10 students scored between 90 and 99 and 6 students scored between 60 and 69. We subtract: $10 - 6 = 4$. **Therefore, 4 more students scored in the 90–99 interval than in the 60–69 interval**.

PRACTICE PROBLEM 3

(a) How many students scored a C on the final if the interval for a C is 70–79?
(b) How many students scored less than 90 on the final?
(c) How many more students scored between 80 and 89 than between 70 and 79?

3 Using Estimation to Solve Application Problems

Often, it is not necessary to know the exact sum or difference; in this case we can estimate. Estimating is also helpful when it is necessary to do mental calculations. There are many ways to estimate, but in this book we use the following rule to make estimations:

To estimate, round each number to the same round-off place, then find the sum or difference.

EXAMPLE 4   Some sample sale prices for 1997 Ford motor vehicles are listed below.

(a) Estimate the difference in the cost if you purchase a Ford Probe instead of a Mustang GT.

(b) Calculate the exact difference in cost. Is your estimate reasonable?

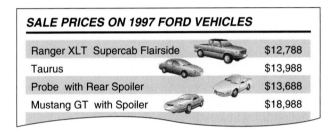

SALE PRICES ON 1997 FORD VEHICLES	
Ranger XLT Supercab Flairside	$12,788
Taurus	$13,988
Probe with Rear Spoiler	$13,688
Mustang GT with Spoiler	$18,988

SOLUTION

(a) To estimate, we round each number to the thousands place.

Price of the Mustang GT: $18,988 $\longrightarrow$ $19,000
The price of the Probe: $13,688 $\longrightarrow$ $14,000

We subtract the *rounded figures* to *estimate* the difference in the cost of the two vehicles.

$19,000 - 14,000 = 5000$

The estimated difference in price is $5000.

(b) We subtract the *original figures* to find the *exact* difference in the cost of the two vehicles.

$18988 - 13688 = 5300$

The exact difference in cost, $5300, is close to the estimated difference, $5000, so our estimate is reasonable.

PRACTICE PROBLEM 4

Use the sale prices listed in Example 4 to answer the following.

(a) Estimate the difference in cost if you purchase the Ranger XLT instead of the Taurus.

(b) Calculate the exact difference in cost. Is your estimate reasonable?

E X A M P L E 5 The Van Tassel family kept the following record of their basic living expenses for the month of October 1998 in order to determine a monthly budget.

Rent for townhouse	$890
Cable and utilities	106
Food	494
Gas and insurance	186
Car payment	191

(a) Estimate the Van Tassels' monthly basic expenses.

(b) If the take-home income for the family is $2745, estimate how much money is left after all the basic expenses are paid.

SOLUTION

(a) To estimate, round each number to the hundreds place.

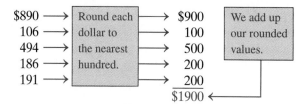

The estimated monthly basic expenses are $1900.

(b) To estimate how much is left after expenses, we subtract the estimated expenses from the estimated income.

$$\begin{array}{rl} \$2745 \longrightarrow \$2700 & \text{We round the income to the nearest hundred.} \\ -\quad 1900 & \text{We subtract: income minus expenses.} \\ \hline \$\ 800 & \end{array}$$

The estimated amount of money left after basic expenses is $800.

Note: If you rounded each number to the nearest ten instead of hundred, your estimation is not wrong, just a little closer to the exact amount. When we estimate we want to make calculations with numbers that are easy to work with. In this case it is easier to add numbers rounded to the hundreds place than the tens place; adding $900 + 100 + 500 + \cdots$ is easier than adding $890 + 110 + 490 + \cdots$.

P R A C T I C E P R O B L E M 5

Refer to Example 5 to answer the following. The Van Tassels bought a cheaper car, which reduced their car payments to $105 per month, and moved to a more expensive townhome, increasing their rent to $1020.

(a) Estimate the monthly basic expenses for the Van Tassels.

(b) If the take-home income for the family remained the same, estimate how much money is left after all the basic expenses are paid.

TO THINK ABOUT

ESTIMATING AND ROUNDING

Is there a difference in the method used and the answer you get when you round compared to when you make an estimate? There is definitely a difference in the method, and usually in the answer, too. To illustrate, consider the following: Suppose that A&J Used Cars had 23 station wagons, 81 trucks, 14 full-size cars, and 47 compact cars on the lot. *Estimate the total vehicles on the lot.*

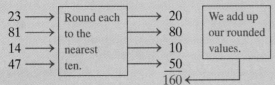

Our estimate is 160.

To round, find the total vehicles and round the sum to the nearest ten.

Total: 23 + 81 + 14 + 47 = 165.

Round to the nearest ten: 165 rounds to 170. **Our rounded sum is 170.**

We see that there is a difference in the method used when we estimate and round, and in this case, in the answer, too.

To *estimate*, we *round first*, then calculate the answer.

To *round*, we *calculate* the answer *first*, then round.

We won't confuse the different methods if we just read the directions when we solve the problem.

Note: There is one more difference between estimating and rounding: *Estimations can differ* depending on which round-off place is chosen. *Rounded answers don't differ* since the directions, or the standard used in many fields, state what round-off place must be used.

EXERCISE

1. In the example above, is it possible to change one or two of the vehicle numbers so that the estimated and rounded amounts are the same? Why or why not?

EXERCISES 1.4

Find the perimeter of each rectangle.

1.
11 in.
6 in.

2.
7 in.
1 in.

Find the perimeter of each square.

3. 4 ft

4. 8 ft

Find the perimeter of each triangle.

5.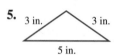
3 in.   3 in.
5 in.

6.
8 ft
3 ft
8 ft

Find the perimeters of the shapes made of rectangles, squares, and triangles.

7.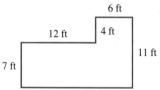
6 ft
12 ft   4 ft
11 ft
7 ft

8.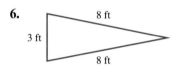
7 ft
8 ft   24 ft
25 ft
17 ft

9.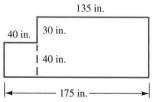
135 in.
40 in.   30 in.
40 in.
175 in.

10.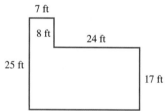
190 in.
100 in.
120 in.
150 in.
310 in.

11. John Tulson wants to put a fence around the back and sides of his property (see the diagram). How many feet of fence must he purchase?

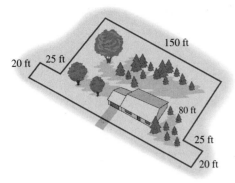

12. Rosa would like to put molding along the edge of the ceiling in her kitchen (see the diagram). How many feet of molding will she need?

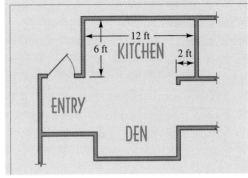

The eighth-grade social studies teacher at Fulton Middle School made the following histogram of the scores on the first test of the school year. Use the histogram to answer questions 13 and 14.

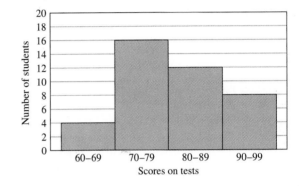

13. **(a)** How many students scored a C on the test if the interval for a C is 70–79?

   (b) How many students scored less than 90 on the test?

   (c) How many more students scored between 70 and 79 than between 80 and 89?

14. **(a)** How many students scored less than 80 on the test?

   (b) How many students scored an A on the test if the interval for an A is 90–100?

   (c) How many more students scored between 80 and 89 than between 60 and 69?

The Pine County Electric Company reported the use of electricity for July on the following histogram. The intervals represent the number of kilowatthours of electricity used in a single-family home during the month of July. Use the histogram to answer questions 15 and 16.

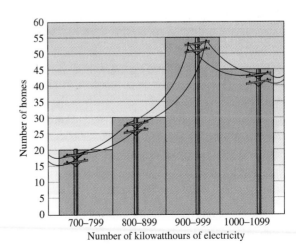

15. Determine the number of homes that used:

(a) Fewer than 900 kilowatthours of electricity.

(b) 900 or more kilowatthours of electricity.

(c) Between 700 and 799 kilowatthours of electricity.

The head coach of South County High School made the following histogram indicating the height of all the players on the varsity and junior varsity basketball teams. Use the histogram to answer questions 17 and 18.

16. Determine the number of homes that used:

(a) Fewer than 1000 kilowatthours of electricity.

(b) 800 or more kilowatthours of electricity.

(c) Between 900 and 999 kilowatthours of electricity.

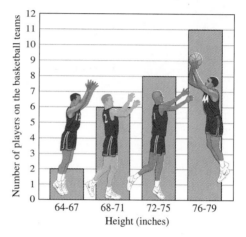

17. Find the total number of players on the varsity and junior varsity teams.

18. Find the total number of players whose heights are less than 68 inches or more than 75 inches.

19. Mr. Hensen purchased supplies to paint his kitchen, family room, and dining room. The following store receipt indicates the supplies he purchased.

```
        7/17/98
     Sale Transaction
Satin enamel
paint, 5 gal.        $81

Paint brushes         36
Drop cloths, 2        22
Paint roller and tray 14

Total
Tax
Total Sale

      THANK YOU
```

(a) Excluding tax, estimate how much money Mr. Hensen spent.

(b) Find the total money spent.

(c) Is your estimate reasonable?

20. Julio Arias bought school books and school supplies. The following store receipt indicates his purchases.

```
        8/27/98
     Sale Transaction

Math book          $41
History book        37
Notebooks, 2        13
Graphing calculator 89

   Total
   Tax
   Total Sale

     THANK YOU
  FOR SHOPPING AT THE
  CORNER BOOK STORE
```

(a) Excluding tax, estimate how much money Julio spent.

(b) Find the total money spent.

(c) Is your estimate reasonable?

21. Natalie was comparing prices of the 1997 end-of-year sales for a multimedia computer with a 266-MHz Intel Pentium processor with MMX technology. She found a Supercon on sale for $1599 and a Hewlett-Packard for $1788. Estimate the difference in price between the two computers.

22. The first of the 1998 new year sales advertised a Fujitsu 133-MHz Pentium processor with MMX technology notebook for $1499. A similar Hitachi notebook is not on sale and is priced at $2199. Estimate the difference in price of the two multimedia notebooks.

23. Mason drove his Toyota SR5 truck 14,200 miles the first year he owned it, 15,980 the second year, 10,100 the third year, and 13,950 the fourth year. Estimate how many more miles Mason drove his truck the first two years than the second two years.

24. The Tsutsumida family is taking a scenic drive across the country. From Phoenix, Arizona, they drove east 597 miles the first day, 412 miles the second day, 389 miles the third day, and 410 miles the fourth day. Estimate how many more miles the Tsutsumida family drove the first two days than the second two days.

CALCULATOR EXERCISES

25. A mutual funds investment specialist received the following funds to invest for his clients this quarter.

Client	Amount Invested	Client	Amount Invested
Jason Santor	$435,750	Silva Auto Repair	$125,500
Mason Construction	144,800	Dr. R. L. Marks	213,125

(a) Estimate how much money the specialist invested for his clients.

(b) Use your calculator to find the total funds invested.

(c) Is your estimate accurate?

26. The costs for attending a university for an entire school year are listed below.

Tuition	$6550	Books and supplies	950
Room and board	3930	Personal expenses	1330

(a) Estimate the total cost for attending the university for an entire school year.

(b) Use your calculator to find the total cost for attending the university.

(c) Is your estimate accurate?

VERBAL AND WRITING SKILLS

27. Write in words the difference between the methods used to estimate and round.

28. Write in words the definition of perimeter and give one example of when you would calculate the perimeter.

29. Earlier we saw that there can be a difference between an estimation and a rounded sum. Name three numbers whose rounded sum <u>does not</u> differ from the estimated total.

30. Construct a histogram that displays the following information. The data in the table shows the results of a telephone survey asking women in different age groups, "Do you listen to the radio on an average of 2 hours a day?"

Age Interval (years)	Number of "Yes" Responses
15–19	45
20–24	20
25–29	30
30–34	5

CUMULATIVE REVIEW

1. Round to the nearest thousand: 28,465.

2. Simplify: $(2 + x) + 9$.

3. Translate to an algebraic expression: Nine more than x.

4. Replace the ? with the inequality symbol $<$ or $>$.

    (a) 53,476 ? 53,467   **(b)** 32,768 ? 32,868

REDECORATING A ROOM

Often we must plan projects, both at home and at work. There are many decisions that must be made and details to consider when developing a plan. Listed below is a six-step process to develop a well-defined plan.

1. Determine a budget for the project.
2. Clearly define the project (what you plan to do).
3. List items and/or labor needed to complete the project.
4. Determine the cost of each item and any labor expenses.
5. Determine if the total cost is within the budget for the project.
6. If the cost of the project exceeds the budget, adjust the extent of the project and reevaluate the budget.

PROBLEM FOR INDIVIDUAL INVESTIGATION

Sara and Jesse have $550 saved for home improvements. They decide to use this money to redecorate their bathroom. Sara and Jesse limit the redecorating project to replacing the mirror, vanity, toilet, and sink faucet. They take all the appropriate measurements, then visit a local decorating store to look at the different types of bathroom fixtures available and inquire about prices. Sara and Jesse then decide to purchase the following items for the bathroom:

> 1 mirror (4 ft by 3 ft)
>
> stained oak wood to make a frame for the mirror (buy 1 extra foot to fit corners)
>
> 1 oak vanity (48 in.)
>
> 1 toilet (almond)
>
> 1 brass sink faucet

If Jesse performs all the labor required to redecorate the bathroom, what is the total cost? Is the total cost within the budget for the redecorating project? (See the price list below and do not include sales tax.)

INLAND DECORATING CENTER PRICE LIST

Faucets			Mirrors			Toilets	
Chrome (plain)	$ 5		4 ft by 3 ft	$24		White	$ 98
Chrome (decorative)	34		5 ft by 3 ft	44		Almond	122
Brass (plain)	65						
Light Fixtures (ceiling)			**Wood (price is per foot)**			**Vanities (48 in.)**	
Two-light (plain)	$10		Oak (stained)	$2		Oak	$298
Two-light (decorative)	25		Pine (stained)	1		Pine	220
Two-light (elite)	40					Hardboard	150
Medicine chests with mirror			**Sinks and counter tops (48-in. vanity)**			**Bath tubs**	
Metal with stainless steel frame	$15		White with plain sink	$106		Steel (white)	$ 74
Plastic with chrome frame	26		Onyx with plain sink	184		Steel (almond)	84
Metal with oak frame	64		Onyx with decorative sink	206		Cast iron (white)	198
						Cast iron (almond)	288

PROBLEM FOR GROUP INVESTIGATION AND COOPERATIVE STUDY

Suppose that Sara and Jesse are able to increase their decorating budget from $550 to $1100. They decide to use the extra money to replace the light fixture, medicine chest, sink and counter, and bathtub in addition to the items listed previously in their project plan. Is it possible for Sara and Jesse to purchase all the items (previous list of fixtures plus those added to the list) and stay within the $1100 budget? If so, which types of bath fixtures can be purchased so that Sara and Jesse use the entire amount of money in the budget? (List each type of fixture along with its price; do not include sales tax in the total cost of the project.)

Internet: Go to   http://www.prenhall.com/blair   to explore this application.

After studying this section, you will be able to:

❶ Evaluate a variable expression.

❷ Solve equations by inspection.

❸ Solve equations involving subtraction.

❹ Solve equations involving addition.

❺ Use translation to solve problems.

Math Pro   Video 1.5   SSM

S E C T I O N   1 . 5

Understanding Basic Equations

❶ ### Evaluating a Variable Expression

We have already learned that when we do not know the value of a number, we designate the number by a letter. We call this letter a *variable*. We use a variable to represent an unknown number until such time as its value can be determined. For example, if 5 is added to a number but we do not know the number, we could write,

$n + 5$ where *n* is the unknown number.

Combinations of variables and numbers such as $n + 5$, $4 - x$, and $a + b$ are called **algebraic expressions**. Consider the algebraic expression

$n + 6$

If we were told that *n* has the value 9, we can *replace n* with 9.

$n + 6$

9 $+ 6$ Replace *n* with 9.

15   Simplify by adding.

Thus $n + 6$ has the value 15 when *n* is replaced by 9.

> To evaluate an algebraic expression, we replace the variable in the expression with the given value and simplify.

Algebraic expressions have different values depending on the value we use to replace the variable.

E X A M P L E   1   Evaluate $7 - x$:

(a) If *x* is 2.   **(b)** If *x* is 4.

SOLUTION

(a) $7 - x$
    ↓
$7 - \boxed{2}$ Replace *x* with 2.
  5   Simplify.

(b) $7 - x$
    ↓
$7 - \boxed{4}$ Replace *x* with 4.
  3   Simplify.

$7 - x$ evaluated when *x* is 2 is 5.   **$7 - x$ evaluated when *x* is 4 is 3.**

P R A C T I C E   P R O B L E M   1
Evaluate $8 - n$:

(a) If *n* is 3.   **(b)** If *n* is 6.

We can evaluate a formula if we know the value of the variables. For example, the perimeter of a rectangle is found by adding the lengths of all its sides. We can also find the perimeter by using the formula

$$P = L + L + W + W$$

where L is the length, W is the width, and P is the perimeter (see the figure). If we know the values of L and W, we can replace the variables with these values to obtain P.

$$P = L + L + W + W$$

EXAMPLE 2 Use the formula $P = L + L + W + W$ to find the perimeter of a rectangle with a length of 5 feet and a width of 3 feet.

SOLUTION

$$
\begin{aligned}
P = L \; + \; L \; + \; W \; + \; W & \\
\;\;\downarrow \quad \downarrow \quad \downarrow \quad \downarrow & \\
= 5\,\text{ft} + 5\,\text{ft} + 3\,\text{ft} + 3\,\text{ft} & \qquad \text{Replace } L \text{ with 5 ft and } W \text{ with 3 ft.} \\
= 16\,\text{ft} & \qquad \text{Simplify.}
\end{aligned}
$$

The perimeter is 16 ft.

PRACTICE PROBLEM 2

Use the formula $P = L + L + W + W$ to find the perimeter of a rectangle with a length of 6 yards and a width of 5 yards.

DEVELOPING YOUR STUDY SKILLS

GETTING THE MOST FROM YOUR TIME STUDYING

Did you know that there are many things you can do to increase your learning when you study? If you use the following strategies, you can improve the way you study and learn more while studying less.

1. Read the material and review your class notes on the same day as your class meets.
2. Do homework in more than one sitting so that you are fresh for the later problems, which are usually the hardest.
3. Check your answer **after** you complete a problem. Put a * beside any problem that you get wrong or don't know how to start.
4. Follow up wrong answers. Check your work for errors, or look in the book for a similar problem. Compare your solution with the book's and, if necessary, rework the problem using the book's solution as a guide. Use this process to solve the problems you didn't know how to start. If you still can't solve the problem, reread the section or ask for help.
5. Revisit * problems. After finishing the assignment, work another problem that is like each * problem. In the text, an even-numbered problem is similar to the preceding odd-numbered problem.
6. Review or rewrite your notes at the end of each week. Work a few problems in the sections covered since the last test. Review past tests periodically, especially if you are having difficulty or can't remember earlier material.

② Solving Equations by Inspection

Two expressions separated by an equal sign are called an **equation**. When we use an equal sign (=) we are indicating that two expressions are equal in value. To illustrate, $2 + 6 = 8$ indicates that the sum of 2 and 6 is *equal in value* to 8.

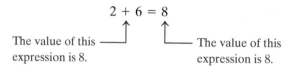

$$2 + 6 = 8$$

The value of this expression is 8.   The value of this expression is 8.

Some English phrases for the symbol "=" are:

is	is the same as	equals
is equal to	the result is	

EXAMPLE 3   Translate the English sentences into equations.

(a) Three plus what number is equal to nine?
(b) Eight minus what number equals two?
(c) Kari's savings increased by $100 equals $500.

SOLUTION

(a) Three plus what number *is equal to* nine?

| ↓ | ↓ | | ↓ | ↓ | ↓ |
| 3 | + | | n | = | 9 |

$$3 + n = 9$$

(b) Eight minus what number *equals* two?

| ↓ | ↓ | | ↓ | ↓ | ↓ |
| 8 | − | | x | = | 2 |

$$8 - x = 2$$

(c) We let x represent Kari's savings, the unknown value.

Kari's savings increased by $100 equals $500.

| ↓ | | ↓ | ↓ | ↓ | ↓ |
| x | | + | $100 | = | $500 |

$$x + \$100 = \$500$$

PRACTICE PROBLEM 3

Translate each English sentence into an equation.

(a) What number plus four is the same as seven?
(b) Three subtracted from six is equal to what number?
(c) The number of baseball cards in a collection plus 20 new cards equals 75 cards.

In Example 3a we ask the question, "Three plus what number is equal to nine?" The answer to this question is 6, since *three plus six is equal to nine*. The number 6 is called the **solution** to the equation $x + 3 = 9$ and is written $x = 6$: "the value of x is 6." In other words, an *equation* is like a *question* and the *solution* is the *answer* to this question.

Question	Equation
Three plus what number is equal to nine?	$3 + x = 9$
Answer to the Question	*Solution*
Three plus **six** is equal to nine.	$x = 6$

The solution to the equation must make the equation a true statement. For example, if 6 is a solution to $3 + x = 9$, we must get a true statement when we evaluate the equation for $x = 6$.

$$3 + x = 9$$
$$\downarrow$$
$$3 + \mathbf{6} = 9 \qquad \text{We evaluate } 3 + x = 9 \text{ for } x = 6.$$
$$9 = 9 \qquad \text{We get a true statement.}$$

> To *solve an equation* we must find a value for the variable in the equation that makes the equation a true statement.

EXAMPLE 4   Is 2 a solution to $6 - x = 9$?

SOLUTION   If 2 is a solution to $6 - x = 9$, when we evaluate $6 - x = 9$ for the value $x = 2$ we will get a true statement.

$$6 - x = 9 \qquad \text{"Six minus what number equals nine?"}$$
$$6 - \mathbf{2} \stackrel{?}{=} 9 \qquad \text{Replace the variable with 2 and simplify.}$$
$$4 \stackrel{?}{=} 9 \qquad \text{This is a false statement.}$$

Since $4 = 9$ is *not* a true statement, **2 is *not* a solution to $6 - x = 9$.**

PRACTICE PROBLEM 4
Is 5 a solution to $x + 8 = 11$?

EXAMPLE 5   Solve the equation $3 + n = 10$ and check your answer.

SOLUTION   To solve the equation $3 + n = 10$, we answer this question:

"Three plus what number is equal to ten?"

Using addition facts we see that the *answer*, or *solution*, is 7. To check the solution we evaluate the equation for $x = 7$ and verify that we get a true statement.

$$3 + n = 10 \qquad \text{Write the equation.}$$
$$\downarrow$$
$$3 + \mathbf{7} \stackrel{?}{=} 10 \qquad \text{Replace the variable with 7 and simplify.}$$
$$10 = 10 \qquad \text{Verify that we get a true statement.}$$

The solution to the equation is 7 and is written $n = 7$.

PRACTICE PROBLEM 5
Solve the equation $4 + n = 9$ and check your answer.

TO THINK ABOUT

EDUCATED GUESSES

When we solve basic equations mentally using addition or subtraction facts, we are actually "guessing" the answer. Then we check our guess to verify that this guess is correct. We are not just guessing any random number for the solution. We make what is called an **educated guess**. By this we mean that we are basing our guess on facts that we know—in this case, addition or subtraction facts. As we will see later in the section, when problems are more complex and we cannot accurately make educated guesses, we use more advanced methods to find the solutions to equations.

EXERCISE

1. We often use the *guess and check* approach to solve everyday problems. Can you think of situations in which you make educated guesses?

Sometimes, we must first use the associative and commutative properties to simplify an equation, then find the solution.

EXAMPLE 6   Simplify using the associative and commutative properties, then find the solution to the equation $(5 + n) + 1 = 8$.

SOLUTION   First, we simplify:

$$(5 + n) + 1 = 8$$
$$(\boldsymbol{n + 5}) + 1 = 8 \qquad \text{Commutative property.}$$
$$n + (\boldsymbol{5 + 1}) = 8 \qquad \text{Associative property.}$$
$$n + 6 = 8 \qquad \text{Simplify.}$$

Then we solve $n + 6 = 8$:

$$n + 6 = 8 \qquad \text{What number plus 6 is equal to 8?}$$
$$n = 2$$

Note: We leave the check to the student.

The solution to the equation is 2 and is written $n = 2$.

PRACTICE PROBLEM 6

Simplify using the associative and commutative properties, then find the solution to the equation $(3 + x) + 1 = 7$.

❸ Solving Equations Involving Subtraction

When an equation is too complicated to solve mentally, we can use various methods to find the solutions. For example, for equations in the form $x - a = b$, we can write equivalent addition problems, then use addition facts to find the solution. That is, we write the subtraction on one side of the equal sign as addition on the opposite side of the equal sign. This is the same idea that we used to subtract whole numbers.

> To solve an equation in the form $x - a = b$, we write the subtraction on one side of the equal sign as addition on the opposite side. Then we use subtraction facts to find the solution.
> We write $x - a = b$ as $x = b + a$.

E X A M P L E   7

(a) Change to an equivalent addition problem: $n - 922 = 486$.
(b) Solve for the variable.

SOLUTION

(a) $n - 922 = 486$ We must answer the question "What number minus 922 equals 486?"

$n = 486 + 922$ Write subtraction on the left side of "=" as addition on the right side.

(b) $n = 486 + 922$

$n = 1408$ Use addition facts to find the value of n.

$n = \textbf{1408}$ **is the solution to the equation** $n - \textbf{922} = \textbf{486.}$

P R A C T I C E   P R O B L E M   7

(a) Change to an equivalent addition problem: $x - 567 = 349$.
(b) Solve for the variable.

E X A M P L E   8   Solve $18 - x = 2$ for x.

SOLUTION

$18 - \boldsymbol{x} = 2$ "Eighteen minus what number is equal to 2?"

$18 = 2 + \boldsymbol{x}$ Write subtraction on the left side of "=" as addition on the right side.

$18 = 2 + 16$

Thus $\boldsymbol{x} = \textbf{16 is the solution to the equation } \textbf{18} - \boldsymbol{x} = \textbf{2.}$

P R A C T I C E   P R O B L E M   8
Solve $13 - x = 4$ for x.

4   Solving Equations Involving Addition

We have seen how to write a subtraction problem as an equivalent addition problem. In Example 8 we wrote $18 - x = 2$ as $18 = 2 + x$. Now let's reverse the process.

1. We start with a subtraction problem.

$18 - x = 2$ Subtraction problem

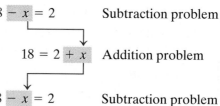

2. We write an equivalent addition problem.

$18 = 2 + x$ Addition problem

3. Now we rewrite the equivalent subtraction. We *undo* what we just did.

$18 - x = 2$ Subtraction problem

As we can see, to write an addition problem as an equivalent subtraction problem, we write the addition on one side of the equal sign as subtraction on the opposite side of the equal sign. We use this idea to solve equations involving addition, as stated in the following rule.

> To solve an equation in the form $x + a = b$, we write the addition on one side of the equal sign as subtraction on the opposite side. Then we use subtraction facts to find the solution.
> We write $x + a = b$ as $x = b - a$.

EXAMPLE 9   Solve: $x + 31 = 112$.

SOLUTION

$$x + 31 = 112$$
$$x = 112 - 31 \qquad \text{Write addition as subtraction.}$$
$$x = 81$$

PRACTICE PROBLEM 9

Solve: $x + 46 = 220$.

We can summarize the two rules for solving equations as follows:

We write $x + a = b$ as $x = b - a$, then use subtraction facts to solve.

We write $x - a = b$ as $x = b + a$, then use addition facts to solve.

⑤ Using Translation to Solve Problems

In most real-life situations we must translate an English statement into an equation, then solve the equation.

EXAMPLE 10 Translate into an equation, then solve.

(a) What number minus twelve equals fifty?
(b) What number plus twenty equals fifty-two?

SOLUTION

(a) What number minus twelve equals fifty?

$$
\begin{array}{ccccc}
\downarrow & \downarrow & \downarrow & \downarrow & \downarrow \\
n & - & 12 & = & 50
\end{array}
$$ Translate into an equation.

$n - 12 = 50$

$\quad n = 50 + 12$ Write subtraction on the left side of "=" as addition on the right side.

$\quad \boldsymbol{n = 62}$ Use addition facts to find the value of x.

(b) What number plus twenty equals fifty-two?

$$
\begin{array}{ccccc}
\downarrow & \downarrow & \downarrow & \downarrow & \downarrow \\
x & + & 20 & = & 52
\end{array}
$$ Translate into an equation.

$x + 20 = 52$

$\quad x = 52 - 20$ Write addition on the left side of "=" as subtraction on the right side.

$\quad \boldsymbol{x = 32}$ Use subtraction facts to find the value of x.

PRACTICE PROBLEM 10

Translate into an equation, then solve.

(a) What number decreased by thirty-four is the same as sixty?
(b) What number plus fifteen equals forty?

EXERCISES 1.5

Evaluate 9 − n:

1. If *n* is 0.   **2.** If *n* is 6.   **3.** If *n* is 8.   **4.** If *n* is 1.

Evaluate x + 3:

5. If *x* is 3.   **6.** If *x* is 5.   **7.** If *x* is 7.   **8.** If *x* is 0.

Translate using symbols, then evaluate.

9. Five more than *n*, if *n* is 6.

10. Six minus *x*, if *x* is 2.

11. Eight minus *y*, if *y* is 3.

12. Seven increased by *x*, if *x* is 1.

Use the formula P = L + L + W + W to find the perimeter of a rectangle with:

13. A length of 4 feet and a width of 2 feet.

14. A length of 8 inches and a width of 5 inches.

15. A length of 51 miles and a width of 13 miles.

16. A length of 21 miles and a width of 16 miles.

For exercises 17 and 18 use the formula x + y + 250 to calculate the yearly bonus for MJ Industry employees.

Bonus = x + y + 250

x represents the number of productivity units earned.

y represents the number of years of employment.

Employee name	Employee number	Years of employment	Productivity units earned
Julio Sanchez	00315	15	150
Mary McCab	00316	12	180
Jamal March	00317	18	125
Leo J. Cornell	00318	10	175

17. Calculate the yearly bonus for:
   (a) Mary McCab.
   (b) Leo J. Cornell.

18. Calculate the yearly bonus for:
   (a) Julio Sanchez.
   (b) Jamal March.

Translate into an equation.

19. When 24 is added to a number, the result is 50.

20. Four plus some number equals 16.

21. What number subtracted from 88 is equal to 40?

22. If a number is subtracted from 45 the result is 6.

23. Let *J* represent James' age. James' age plus 10 years equals 25.

24. Let *S* represent Sherie's checking account balance. Sherie's checking account balance plus $14 equals $56.

25. Let C represent Chuong's monthly salary. Chuong's monthly salary decreased by $50 equals $1480.

26. Let P represent the price of the ticket. The price of the ticket decreased by $5 equals $16.

Answer yes or no.

27. Is 4 a solution to the equation $8 - x = 3$?

28. Is 3 a solution to the equation $5 - x = 3$?

29. Is 15 a solution to the equation $x + 4 = 19$?

30. Is 20 a solution to the equation $x + 6 = 26$?

Solve and check your answer.

31. $x + 3 = 9$

32. $x + 4 = 10$

33. $9 - n = 7$

34. $13 - n = 10$

35. $n + 7 = 8$

36. $a + 5 = 6$

37. $4 + n = 8$

38. $5 + x = 9$

39. $x - 6 = 0$

40. $x - 2 = 0$

41. $1 + x = 13$

42. $1 + x = 16$

Simplify using the associative and/or commutative property, then find the solution. Check your answer.

43. $(x + 1) + 3 = 7$

44. $(x + 6) + 5 = 13$

45. $2 + (5 + a) = 14$

46. $7 + (2 + y) = 10$

47. $(3 + x) + 2 = 7$

48. $(6 + x) + 1 = 10$

49. $(6 + y) + 0 = 11$

50. $(7 + y) + 0 = 12$

51. $1 + (a + 3) = 7$

52. $2 + (a + 5) = 8$

53. $3 + (n + 5) = 10$

54. $2 + (8 + x) = 12$

55. $1 + (x + 3) + 3 = 11$

56. $2 + (x + 1) + 7 = 15$

57. $4 + (a + 1) + 5 = 17$

58. $5 + (a + 2) + 6 = 19$

For the following equations:
  (a) *Write the subtraction problem as an equivalent addition problem.*
  (b) *Solve for the variable and check your answer.*

59. $x - 133 = 200$

60. $n - 193 = 346$

61. $n - 988 = 122$

62. $x - 455 = 299$

63. $21 - x = 13$

64. $25 - x = 14$

65. $44 - x = 28$

66. $38 - x = 16$

For the following equations:
   (a) *Write the addition problem as an equivalent subtraction problem.*
   (b) *Solve for the variable and check your answer.*

67. $x + 29 = 66$ **68.** $x + 31 = 56$ **69.** $x + 121 = 365$ **70.** $x + 144 = 436$

Solve.

71. (a) $x - 45 = 98$ **72. (a)** $x - 15 = 57$ **73. (a)** $n - 13 = 56$ **74. (a)** $n - 21 = 98$

   (b) $x + 45 = 98$ **(b)** $x + 15 = 57$ **(b)** $n + 13 = 56$ **(b)** $n + 21 = 98$

75. (a) $a - 141 = 220$ **76. (a)** $x - 331 = 541$

   (b) $a + 141 = 220$ **(b)** $x + 331 = 541$

For the following English sentences:
   (a) *Translate into an equation.*   **(b)** *Solve the equation*

77. Four plus what number equals eight?   **78.** Two added to what number equals nine?

79. Seventeen minus what number is equal to seven?   **80.** Fourteen minus what number is equal to five?

81. What number plus 152 equals 944?   **82.** What number increased by 131 equals 544?

 CALCULATOR EXERCISES

83. $x + 14,000 = 23,456$ **84.** $x - 13,221 = 14,768$

ONE STEP FURTHER

85. Find the missing side of the following triangle if the perimeter is 170 feet.

86. Find the missing side of the following triangle if the perimeter is 110 yards.

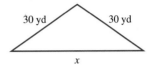

No matter what size the triangle, the sum of all the interior angles always equals 180°. Find the missing angle for each triangle.

87.

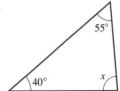

88.

CUMULATIVE REVIEW

Combine like terms.

1. $7x - 2x$

2. $7x + 2x$

The following bar graph illustrates the number of customers at Carol's Cafe for each quarter in 1998. Use this graph to answer questions 3 and 4.

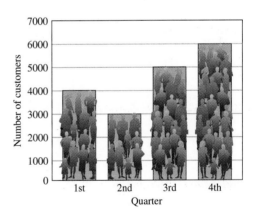

3. (a) State the total number of customers at the cafe in the third and fourth quarters.

(b) How many more customers were there in the fourth quarter than in the third quarter?

4. (a) State the total number of customers at the cafe in the first and second quarters.

(b) How many fewer customers were there in the second quarter than in the first quarter?

After studying this section, you will be able to:

1 Understand U.S. units.

2 Understand metric units.

Math Pro   Video 1.6   SSM

SECTION 1.6

U.S. and Metric Units of Measurement

The two systems of measurement most common in the world are the metric system and the U.S. system. Nearly all countries in the world use the metric system, but in the United States we use the U.S. system, except in fields such as science and medicine. It is important that you learn both systems, since you will deal with both to some extent.

Most of us have seen or used units of measurement in our everyday life. We know that our height is measured in feet and inches and our weight in pounds. Many prescription doses are measured in milligrams (mg), and track and field races are often measured in kilometers. Yet do we understand what each measurement means, and the relationships between the units and the systems? In this section we introduce the U.S. and metric systems and give a general presentation of each system. That is, we answer questions such as "How many inches are in a yard?" and "How long is a kilometer?" We also introduce the idea of "thinking metric." By this we mean the ability to estimate how big or little a metric unit is and to compare metric units to U.S. units.

1 Understanding U.S. Units

The table below indicates relationships between units that are used often in our daily lives. Your instructor may require that you memorize these relationships.

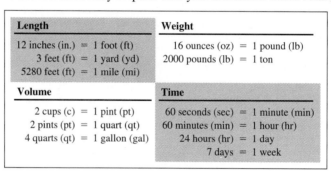

Length	Weight
12 inches (in.) = 1 foot (ft)	16 ounces (oz) = 1 pound (lb)
3 feet (ft) = 1 yard (yd)	2000 pounds (lb) = 1 ton
5280 feet (ft) = 1 mile (mi)	
Volume	**Time**
2 cups (c) = 1 pint (pt)	60 seconds (sec) = 1 minute (min)
2 pints (pt) = 1 quart (qt)	60 minutes (min) = 1 hour (hr)
4 quarts (qt) = 1 gallon (gal)	24 hours (hr) = 1 day
	7 days = 1 week

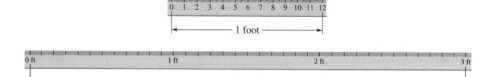

It is important that you are familiar with the equivalent units, since we use these units in many aspects of our daily lives.

EXAMPLE 1

(a) 3 ft = __?__ yd   **(b)** 1 ton = __?__ lb

SOLUTION

(a) 3 feet has the same measure of length as 1 yard: **3 ft = 1 yd**.
(b) 1 ton has the same weight as 2000 pounds: **1 ton = 2000 lb**.

PRACTICE PROBLEM 1

(a) 60 sec = __?__ min   **(b)** 1 gal = __?__ qt

Although we will learn a method to convert from one unit to another in Chapter 9, for simple conversions we can use addition facts.

E X A M P L E   2

(a) 24 in. = _?_ ft

(b) 21 days = _?_ weeks

SOLUTION

(a) We know that 12 in. = 1 ft.

24 in. = 12 in. + 12 in.

$\downarrow$ $\downarrow$

= 1 ft + 1 ft

= **2 ft**

(b) There are 7 days in 1 week.

21 days = 7 days + 7 days + 7 days

$\downarrow$ $\downarrow$ $\downarrow$

= 1 week + 1 week + 1 week

= **3 weeks**

P R A C T I C E   P R O B L E M   2

(a) 4 pt = _?_ qt

(b) 2 tons = _?_ lb

Often, we must write measurements using more than one unit. For example, 6 hours and 10 minutes can be written 6 hr + 10 min, or 6 hr 10 min (we usually do not write the + sign).

E X A M P L E   3   Perform the operations indicated.

(a) 3 hr 40 min + 5 hr 30 min

(b) 5 yd 1 ft − 3 yd 2 ft

SOLUTION   Only measurements of the same units can be added or subtracted. We line up the common units and either add or subtract, starting in the column on the right.

(a)   3 hr 40 min

+5 hr 30 min

8 hr 70 min   Since 70 min = 60 min + 10 min,

$\downarrow$ $\downarrow$ $\downarrow$

8 hr (1 hr 10 min)   we can write 70 min as 1 hr 10 min.

9 hr 10 min   We add: 8 hr + 1 hr = 9 hr.

Note: This is an example of carrying units. We carried an hour to the hours column in order to change the 70 minutes to hours and minutes. This form gives us a better idea of the size of the measurement.

(b)   5 yd 1 ft   We cannot subtract 1 ft − 2 ft.

−3 yd 2 ft

Since 1 yd = 3 ft we can write 5 yd as follows:

5 yd = 4 yd + 1 yd   or   4 yd + 3 ft

5 yd 1 ft	(4 yd + 3 ft) 1 ft	4 yd **4 ft**	3 ft + 1 ft = 4 ft.
−3 yd 2 ft	−3 yd 2 ft	−3 yd 2 ft	
		1 yd 2 ft	

Note: This is an example of borrowing units. We borrowed a yard from the yards column and added it to the feet column so that we could subtract.

PRACTICE PROBLEM 3

Perform the operation indicated.

(a) 12 min 43 sec + 4 min 33 sec   **(b)** 4 gal 1 qt − 2 gal 3 qt

DEVELOPING YOUR STUDY SKILLS

REVIEWING FOR THE EXAM

Reviewing for an exam enables you to connect concepts you learned over several classes. Your review activities should cover all the components of the learning cycle.

The Learning Cycle

Reading $\longrightarrow$ Writing

Seeing ← Verbalizing ← Listening

1. Reread your textbook. List and study important terms, rules, and formulas.
2. Reread your notes. Study returned homework and quizzes, and redo problems you got wrong.
3. Practice a few problems of each type that you will be tested on. Pay particular attention to the problems that gave you trouble (the * problems).
4. Read the Chapter Organizer and solve some of the review problems at the end of the chapter. Check your answers and redo problems you got wrong.
5. Start reviewing several days before the test so that you have time to review completely and get help if you need it.

It is not a good idea to complete all five steps at one time. For best results, complete each step at a separate sitting, starting the process early so that you are done at least one day before the test.

EXERCISE

1. Can you think of other ways of preparing for a test that include activities in the learning cycle?

❷ Understanding Metric Units

Meters, grams, and liters are the *basic units* in the metric system. All other units in the metric system are based on the number 10 and these basic units. The *meter* measures length, and the *liter* measures volume. The *gram* is a unit of measure that is related to the weight of an object. We can use the basic unit gram when we are referring to weight.

Measurement	Basic Unit	Abbreviation
length	meter	m
weight	gram	g
volume	liter	L

Just as in the U.S. system, where 12 inches equals 1 foot, various size units have standard names. The prefixes used for these standard names are *kilo, hecto, deka, deci, centi,* and *milli.* For example, 10 *centi*meters = 1 *deci*meter. Units that are larger than the basic unit use the prefixes kilo, meaning 1000; hecto, meaning 100; and deka, meaning 10. For units smaller than the basic unit we use the prefixes deci, meaning 1/10; centi, meaning 1/100; and milli, meaning 1/1000.

Prefixes	Basic unit	Prefixes
kilo hecto deka	gram liter meter	deci centi milli
These prefixes identify units that are larger than the basic unit.		These prefixes identify units that are smaller than the basic unit.

We say

1 *kilo*gram for weight   1 *kilo*liter for volume   1 *kilo*meter for length

Length.   The advantage of the metric system is that it is easier to convert from one unit to another since it is based on the number 10. For example, every 10 units of a smaller unit of measure equals 1 unit of a larger unit of measure. The table below states the relationship between units of measure for the meter.

METRIC MEASURES OF EQUIVALENCE

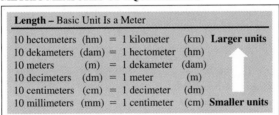

Length – Basic Unit Is a Meter

10 hectometers	(hm)	= 1 kilometer	(km) **Larger units**
10 dekameters	(dam)	= 1 hectometer	(hm)
10 meters	(m)	= 1 dekameter	(dam)
10 decimeters	(dm)	= 1 meter	(m)
10 centimeters	(cm)	= 1 decimeter	(dm)
10 millimeters	(mm)	= 1 centimeter	(cm) **Smaller units**

The following illustrations can help you see the relationships between the units.

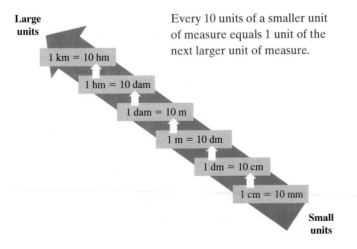

Every 10 units of a smaller unit of measure equals 1 unit of the next larger unit of measure.

Large units

1 km = 10 hm

1 hm = 10 dam

1 dam = 10 m

1 m = 10 dm

1 dm = 10 cm

1 cm = 10 mm

Small units

There are 10 millimeters (mm)
in 1 centimeter (cm).

1 cm

There are 10 centimeters (cm)
in 1 decimeter (dm).

1 Meter

1 dm   2 dm   3 dm   4 dm   5 dm   6 dm   7 dm   8 dm   9 dm   10 dm

There are 10 decimeters (dm) in 1 meter (m).
There are 1000 meters in a kilometer.

As you can see, a kilometer is a fairly large measure.

E X A M P L E   4

(a) 10 meters = 1 ____   **(b)** 10 millimeters = 1 ____

SOLUTION

(a) 10 meters = 1 **dekameter**   **(b)** 10 millimeters = 1 **centimeter**

P R A C T I C E   P R O B L E M   4

(a) 10 cm = 1 ____   **(b)** 1 dm = 10 ____

E X A M P L E   5   Place the appropriate symbol, < or >, in the blank.

(a) 1 cm ____ 1 m   **(b)** 2 dm ____ 2 cm

SOLUTION

(a) 1 cm < 1 m   **(b)** 2 dm > 2 cm

P R A C T I C E   P R O B L E M   5

Place the appropriate symbol, < or >, in the blank.

(a) 1 mm ____ 1 km   **(b)** 5 m ____ 5 hm

E X A M P L E   6   Which of the following represents a unit that is larger
than the basic unit, meter: 1 millimeter or 1 dekameter?

SOLUTION   If you memorize the prefixes in the order listed below, it is easier to determine which ones represent the larger or smaller units. The size of
the units decrease as you move from left to right on the chart.

Prefixes

kilo   hecto   deka   basic unit   deci   centi   milli

▲   (meter)   ▲

**The prefix deka is located to the left of the meter, so it represents a unit larg-
er than a meter.**

P R A C T I C E   P R O B L E M   6

Which of the following represents a unit that is smaller than the basic unit,
meter: 1 centimeter or 1 hectometer?

Thinking Metric

A *millimeter* is very small. The thickness of the edge of a paper clip is about 1 millimeter.

1 mm

A *centimeter* is smaller than 1 inch. Approximately $2\frac{1}{2}$ centimeters equal 1 inch.

Centimeters

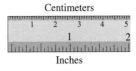

Inches

A *meter* is a little longer than a yard.

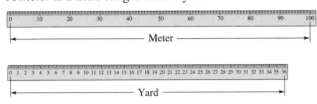

Meter

Yard

A *kilometer* is slightly more than $\frac{1}{2}$ mile.

1 mi

1 km

E X A M P L E   7   Place the appropriate symbol, $<$ or $>$, in the blank.

(a) 7 in. ____ 7 cm

(b) 1 mi ____ 1 km

SOLUTION

(a) 7 in. $>$ 7 cm

(b) 1 mi $>$ 1 km

P R A C T I C E   P R O B L E M   7

Place the appropriate symbol, $<$ or $>$, in the blank.

(a) 1 mm ____ 1 in.

(b) 4 m ____ 4 yd

Volume and Weight.   Another advantage of the metric system is that the prefixes and the conversions from one unit to another are the same for each measurement; length, volume, and weight. After each prefix we write *gram* if we are dealing with weight, and *liter* for volume. The table below states the relationships between units of measure for the liter and the gram.

METRIC MEASURES OF EQUIVALENCE

Volume – Basic Unit Is a Liter	**Weight – Basic Unit Is a Gram**
10 hectoliters (hL) = 1 kiloliter (kL)	10 hectograms (hg) = 1 kilogram (kg)
10 dekaliters (daL) = 1 hectoliter (hL)	10 dekagrams (dag) = 1 hectogram (hg)
10 liters (L) = 1 dekaliter (daL)	10 grams (g) = 1 dekagram (dag)
10 deciliters (dL) = 1 liter (L)	10 decigrams (dg) = 1 gram (g)
10 centiliters (cL) = 1 deciliter (dL)	10 centigrams (cg) = 1 decigram (dg)
10 milliliters (mL) = 1 centiliter (cL)	10 milligrams (mg) = 1 centigram (cg)

A *milliliter* is a very small measurement and often used in laboratories.

1 L   1000 mL

1 *liter* contains a little more fluid than a quart.

1 L Milk   1 qt Milk

A *kiloliter* is used to measure large volumes.

1 kiloliter (1000 liters)

A *milligram* is a small weight. Medicine is measured in milligrams.

1 *gram* weighs about the same as two paper clips.

1 *kilogram* weighs slightly more than a 2-pound steak.

E X A M P L E   8   Place the appropriate symbol, $<$ or $>$, in the blank.

(a) 1 mg ____ 1 g   **(b)** 5 kL ____ 5 qt

SOLUTION

(a) 1 mg **<** 1 g   **(b)** 5 kL **>** 5 qt

P R A C T I C E   P R O B L E M   8

Place the appropriate symbol, $<$ or $>$, in the blank.

(a) 1 L ____ 1 dL   **(b)** 1 kg ____ 1 lb

E X A M P L E   9   Select the most reasonable measurement for the weight of a backpack filled with books: 3 mg, 3 L, 3 kg, or 3 m.

SOLUTION   We do not choose 3 L or 3 m as a reasonable weight since these units do not represent weight; 3 L is the measure of a volume, and 3 m is the measure of a length. Since 1 kg weighs a little more than 2 lb, the most appropriate choice is **3 kg** (between 6 and 7 lb).

P R A C T I C E   P R O B L E M   9

Select the most reasonable measurement for the volume of a bottle of soda: 1 L, 1 m, 1 g, or 1 mL.

DEVELOPING YOUR STUDY SKILLS

THE DAY BEFORE THE EXAM

If you have been following the advice in the other Developing Your Study Skills sections, you should be almost ready for the exam. If you have not read these sections, reading them now will provide valuable advice.

The day before the exam is not a good time to start reviewing. Use this day to skim your chapter review, homework, quizzes, and other review material. On this day you can "fine tune" what you already know and review what you are unsure of. Starting your review early reduces anxiety so that you can think clearly during the test. Often, low test scores are related to high anxiety. Plan ahead so that you can relax on the day of the exam.

A few days before the exam, complete the chapter test at the end of the chapter. Take the test as if it were the real exam. Do not refer to notes or to the text while completing the chapter test. Grade the test. The problems you missed on this test are the type of problems that you should get help with and review the day before the test.

EXERCISES 1.6

Fill in the blank.

1. 1 mi = _____ ft
2. 1 ft = ____ in.
3. 2 pt = ____ qt
4. 4 qt = ____ gal

5. 1 ton = _____ lb
6. 1 lb = ____ oz
7. 60 sec = ____ min
8. 24 hr = ____ day

9. 14 days = ____ weeks
10. 48 hr = ____ days
11. 2 gal = ____ qt
12. 2 pt = ____ c

13. 36 in. = ____ ft
14. 9 ft = ____ yd
15. 2 mi = _____ ft
16. 3 mi = _____ ft

17. 120 sec = ____ min
18. 120 min = ____ hr
19. 8 qt = ____ gal
20. 4 pt = ____ qt

21. 32 oz = ____ lb
22. 4000 lb = ____ tons
23. 28 days = ____ weeks
24. 72 hr = ____ days

Perform the operation indicated.

25. 5 ft 4 in. + 4 ft 3 in.
26. 16 ft 6 in. + 5 ft 3 in.
27. 12 ft 7 in. + 3 ft 8 in.
28. 8 ft 9 in. + 2 ft 4 in.

29. 26 yd 1 ft − 9 yd 2 ft
30. 14 yd 1 ft − 6 yd 2 ft

31. While golfing, Todd made birdie putts of 4 ft 3 in., 9 ft 10 in., 11 ft 5 in., and 12 ft 9 in. during a round. What was the total length of all the birdie putts?

32. Jessica worked out at the health spa for 2 hr 20 min on Monday, 1 hr 15 min on Wednesday, and 2 hr 45 min on Friday. What was Jessica's total workout time for the week?

33. Natalie bought a 1-pint can of tomato sauce. She used 1 cup to make a recipe. How much tomato sauce did she have left?

34. Joel has 1 quart of milk. If he uses 1 pint to prepare mashed potatoes for a family holiday gathering, how much milk does he have left?

35. It takes Isaac 2 hours to detail a car. Sean can detail the same car in 130 minutes. Who takes the least amount of time to detail the car?

36. A 3-pound can of Joe's Chili costs the same price as a 45-ounce can of the store brand of chili. If both brands of chili taste good, which one is a better buy?

Fill in the blank.

37. 10 centimeters = ____ decimeters
38. 10 hectometers = ____ kilometers

39. 1 meter = ____ decimeters
40. 1 dekameter = ____ meters

41. 10 milligrams = ____ centigram
42. 10 liters = ____ dekaliter

Write the abbreviation for the metric unit.

43. decimeter ____

44. kiloliter ____

45. milligram ____

46. milliliter ____

47. centigram ____

48. centimeter ____

Place the appropriate symbol, < or >, in the blank.

49. 1 mm ____ 1 m

50. 1 dg ____ 1 kg

51. 1 hL ____ 1 cL

52. 1 kg ____ 1 dag

53. 2 mL ____ 2 L

54. 3 km ____ 3 m

Which of the following represents a unit that is smaller than the basic unit liter?

55. mL or kL

56. hL or cL

Which of the following represents a unit that is larger than the basic unit gram?

57. dg or kg

58. hg or mg

Place the appropriate symbol, < or >, in the blank.

59. 1 km ____ 1 mi

60. 3 yd ____ 3 m

61. 2 L ____ 2 qt

62. 3 mg ____ 3 lb

63. 2 in. ____ 2 mm

64. 1 cm ____ 1 yd

Fill in the blank.

65. The _____ is the metric basic unit for weight.

66. The meter is the metric basic unit for _____ .

67. The _____ is the metric basic unit for volume.

68. The gram is the metric basic unit for _____ .

69. Jose bought a container of fresh-squeezed juice at the store. Choose the most appropriate unit to describe its contents.

(a) kilometer (b) gram
(c) liter (d) milliliter

70. Lisa measured the width of a doorway. Choose the most appropriate unit for this measurement.

(a) millimeter (b) centiliter
(c) meter (d) liter

The basic metric units are missing from the following measurements. What should they be?

71. A large London broil steak weighs 2 _____ .

(a) meters (b) kilograms
(c) kiloliter (d) grams

72. The thickness of the tip of a felt-tip pen is approximately 2 _____ .

(a) centigrams (b) millimeters
(c) meters (d) liters

Use the following figure for exercises 73 and 74. This figure shows a comparison between common speed limits in miles per hour (mph) and kilometers per hour (km/h). These comparisons are approximated to the nearest kilometer per hour.

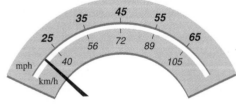

73. The speed limit posted on a road in France is 89 km/h. Sam Johnson drives an American car with a speedometer that only displays speed in miles per hour. If Sam's speedometer reads 65 mph, is he driving above or below the posted speed limit?

74. In most parts of the United States the speed limit within a school zone is 25 mph. If the speedometer reading on Edward's foreign car is 30 km/h as he drives through a school zone, is he driving above or below the speed limit?

The temperature scale used in the metric system is Celsius (°C). The following figure shows a comparison between degrees Fahrenheit (°F) and degrees Celsius (°C). Use the figure to answer questions 75 and 76.

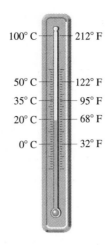

100° C —— 212° F

50° C —— 122° F

35° C —— 95° F

20° C —— 68° F

0° C —— 32° F

75. Michele is planning a trip to Italy. The weather service predicts that the temperature will be between 30 and 32°C during her visit. Should Michele plan to bring clothes for warm or cold weather?

76. Water freezes at 32°F and boils at 212°F. If the temperature of water is 20°Celsius, how many degrees Fahrenheit does it need to drop in order to freeze?

ONE STEP FURTHER

Use the following information to answer questions 77 and 78. Water for the Spensers' camping trip was stored in 1-quart, 1-pint, and 1-cup containers. The total quantity of water brought on the trip was 8 gallons. The Spensers recorded the amount of water used each day as follows:

Water used each day

Day 1	3 qt 3 pt		Day 4	3 qt 3 c
Day 2	4 qt 1 c		Day 5	4 qt 1 c
Day 3	3 qt 1 pt		Day 6	3 qt 3 c

77. How much water was left after 4 days?

78. How much water was left after 6 days?

CUMULATIVE REVIEW

1. Evaluate $12 + x$ if $x = 3$; if $x = 5$.

2. Combine like terms: $9x - 4x + 1$.

3. Add: $288 + 3671 + 900$.

4. Subtract: $4811 - 766$.

CHAPTER ORGANIZER

TOPIC	PROCEDURE	EXAMPLES
Inequality symbols	< and > are called inequality symbols. The symbol ">" means *greater than*, and "<" means *less than*. The inequality symbol always points to the smaller number.	Replace ? with the inequality symbol < or >. 9 ? 3 15 ? 20 9 > 3 15 < 20

Rounding whole numbers	1. Identify the round-off place.
	2. If the digit to the right of the round-off place is:
	  (a) Less than 5, do not change the round-off place digit.
	  (b) 5 or more, increase the round-off place digit by 1.
	3. In either case, replace all digits to the right of the round-off place digit with zeros.
		Round 27,468 to the nearest hundred.
		
		The round-off place digit is 4.
		2 7,4 6 8
		The digit to the right is 5 or more.
		
		Increase the round-off place digit.
		2 7,5 0 0
		Replace digits to the right with zero.

Using properties of addition to simplify	The associative property states that we can regroup numbers when adding. The commutative property states that we can change the order of numbers when adding. We use both of these properties to simplify expressions.	Use the associative and/or commutative property as necessary to simplify: $3 + (n + 2)$. Change the order of addition: $= 3 + (2 + n)$ Regroup: $= (3 + 2) + n$ Add: $= 5 + n$
Adding whole numbers	Starting with the right column, add each column separately. If a two-digit sum occurs, carry the first digit over to the next column to the left.	$\begin{array}{r} {}^{2\;1} \\ 3\,8\,2 \\ 1\,5\,6 \\ 7\,3 \\ +\quad 5 \\ \hline 6\,1\,6 \end{array}$
Subtracting whole numbers	Starting with the right column, subtract each column separately. If necessary, borrow a unit from the column to the left and bring it to the right as a "10."	$\begin{array}{r} {}^{5\;15\;7\;11} \\ 2\,6\,5\,8\,1 \\ -\quad 4\,8\,3\,2 \\ \hline 2\,1,7\,4\,9 \end{array}$
Combining like terms	We either add or subtract numerical parts of like terms. The variable part stays the same.	Combine like terms: $2xy + 3x + 5xy$. $7xy + 3x$

TOPIC	PROCEDURE	EXAMPLES

Finding the perimeter	The perimeter is the distance around an object. We add the lengths of all sides to find the perimeter.	Find the perimeter of the shape consisting of rectangles.
		
		(figure: shape made of rectangles with labels 8 m, 9 m, 6 m, 15 m, and two sides marked ?)
		
		1. We find the missing sides:
		   9 m − 6 m = 3 m; missing vertical side.
		   15 m − 8 m = 7 m; missing horizontal side.
		2. We add all sides:
		   $6 + 7 + 3 + 8 + 9 + 15 = 48$ m

Estimating	We round each number to the same round-off place, then find the sum or difference.	A Ford dealership reduced the price of a Ford Taurus from \$13,988 to \$12,950 for the Labor Day weekend sale. Estimate the savings on the Taurus. We round \$13,988 to \$14,000 and the sale price of \$12,950 to \$13,000. To estimate the savings, we subtract: \$13,000 − \$12,000 = \$1000 savings

Evaluating an expression	To evaluate an expression, we replace the variable in the expression with the given value, then simplify.	Evaluate $9 - x$ if $x = 3$.
		  We replace x with 3: $9 - 3 = 6$.
		  The value of $9 - x$ if $x = 3$ is 6.

Solving equations in the form $x - a = b$; $x + a = b$	To solve more complicated equations, we proceed as follows: If the equation is in the form $x - a = b$, we write the subtraction on one side of the equal sign as addition on the opposite side. If the equation is in the form $x + a = b$, we write the *addition* on one side as *subtraction* on the *opposite side*.	Solve: $x - 156 = 256$. $x - \mathbf{156} = 258$ $x = 258 + \mathbf{156}$ $x = 414$ Solve: $x + \mathbf{102} = 347$ $x = 347 - \mathbf{102}$ $x = 245$

Adding or subtracting measurements	Only measurements with the same units can be added. So we line up common units and either add or subtract.	Add: 4 hr 45 min + 2 hr 30 min.
		
		  4 hr 45 min
		+2 hr 30 min
		  6 hr 75 min = 6 hr + (1 hr 15 min)
		          = 7 hr 15 min

TOPIC	PROCEDURE	EXAMPLES
U.S. and metric units	The meter is a little longer than a yard, and a kilometer is slightly less than $\frac{1}{2}$ mile. A liter is a little more than a quart, and a gram weighs as much as two paper clips.	Place the appropriate symbol, < or >, in the blank. 1 m ___ 1 mi; 1 L ___ 1 gal < <

USING THE MATHEMATICS BLUEPRINT FOR PROBLEM SOLVING

When solving problems, students often find it helpful to complete the following steps. You will not use all the steps all the time. Choose the steps that best fit the condition of the problem.

1. *Understand the problem.*
   (a) Read the problem carefully, then draw a picture or chart.
   (b) Think about the facts you are given and what you are asked for.
   (c) Use the Mathematics Blueprint for Problem Solving to organize your work.
2. *Solve and state the answer.* Perform the necessary calculations and state the answer, including the units of measure.
3. *Check.*
   (a) Estimate your answer and check it with the value calculated to see if your answer is reasonable, or
   (b) Repeat the calculations working the problem a different way.

The Austin Department of Housing purchased several Pentium computer stations for a total cost of $7850, and software to update their system for $1055. The department had $9450 in their supply budget for the quarter. After this purchase, how much money is left for supplies?

1. *Understand the problem.*

MATHEMATICS BLUEPRINT FOR PROBLEM SOLVING			
GATHER THE FACTS	WHAT AM I ASKED TO DO?	HOW DO I PROCEED?	KEY POINTS TO REMEMBER

Cost of purchases:
   Computers: $7850
   Software: $1055
Money available:
   $9450
	Find the amount of money left after the purchases.	Find the total cost of the computers and software. Subtract this amount from the money available in the supply budget.	Be sure that you copied all the facts correctly.

2. *Solve and state the answer.* Calculate the cost of the computer and the software.

$$\begin{array}{r} \$7850 \\ +\ 1055 \\ \hline \end{array} \qquad \begin{array}{r} \$9450 \\ -\ 8905 \\ \hline \end{array}$$

Cost of supplies: $8905   Money left: $ 545

3. *Check.* We estimate the amount of money left in the budget.

$7850 rounds to: $7900 $9450 rounds to: $9500
$1055 rounds to: + 1100   − 9000

Estimated cost of supplies: $9000   Estimate of money left: $ 500

The estimated balance of $500 is close to the answer, $545. We determine that the answer is reasonable.

CHAPTER 1 REVIEW

Do all odd 1-99

SECTION 1.1

1. For the whole number 175,493, state the place value for the digit:

    (a) 7   **(b)** 5   **(c)** 9

2. For the whole number 458,013, state the place value for the digit:

    (a) 8   **(b)** 0   **(c)** 5

Write each number in expanded notation.

3. 7694

4. 4325

Fill in the check for the amount indicated.

5. $341

6. $187

Replace the question mark with the inequality symbol < or >.

7. 2 ? 8

8. 12 ? 11

Write using an inequality symbol.

9. Six is greater than one.

10. Three is less than five.

Round to the nearest hundred.

11. 61,269

12. 382,240

Round to the nearest hundred thousand.

13. 6,365,534

14. 8,118,701

Answer each question.

15. A car takes 4 hours and 15 minutes to reach its destination. Approximately how many hours does the trip take?

16. About 793,000,000 people in the world speak Italian. Round this figure to the nearest ten million.

SECTION 1.2

Translate using symbols.

17. Seven more than x.

18. A number increased by 4.

Use the commutative property of addition to rewrite each sum.

19. $x + 3$ **20.** $x + 6$ **21.** $b + c$ **22.** $y + z$

Use whichever properties (associative and/or commutative) are necessary to simplify each expression.

23. $7 + (9 + x)$ **24.** $(n + 4) + 1$ **25.** $(2 + n) + 9$

26. $5 + (n + 2)$ **27.** $(5 + x + 3) + 2$ **28.** $(3 + x + 4) + 1$

Add.

29. $\begin{array}{r} 63 \\ +32 \\ \hline \end{array}$ **30.** $\begin{array}{r} 52 \\ +43 \\ \hline \end{array}$ **31.** $\begin{array}{r} 159 \\ + \ 46 \\ \hline \end{array}$ **32.** $\begin{array}{r} 267 \\ + \ 56 \\ \hline \end{array}$

33. $326 + 647 + 62$ **34.** $793 + 92 + 764$ **35.** $8398 + 372 + 255$ **36.** $7456 + 213 + 982$

Answer each question.

37. Mr. Thomas purchased supplies to paint his family room. He spent $159 on paint, $22 on brushes, and $14 for drop cloths. How much money did Mr. Thomas spend?

38. A private college has 1434 freshmen, 1596 sophomores, 1423 juniors, and 1565 seniors. How many students are attending the college?

Identify like terms.

39. $3xy + 5y + 2xy + 8y$

40. $4ab + 3b + 2ab + 8b$

Combine like terms.

41. $4x + 3x + 6x$ **42.** $8a + 6a + 2a$ **43.** $3x + 2y + 6x$ **44.** $5b + 2a + 9b$

SECTION 1.3

Combine like terms.

45. $6x - 2x - 5$ **46.** $9m - 3m - 2$ **47.** $14y - 3y - x$ **48.** $15a - 3a - b$

Translate using symbols.

49. Eight decreased by 3. **50.** The difference of 3 and y. **51.** Ten subtracted from a number.

Evaluate.

52. (a) $(14 - 4) - 2$
   (b) $14 - (4 - 2)$

53. (a) $(17 - 5) - 3$
   (b) $17 - (5 - 3)$

Subtract and check.

54.
$$\begin{array}{r} 85 \\ -23 \\ \hline \end{array}$$

55.
$$\begin{array}{r} 96 \\ -33 \\ \hline \end{array}$$

56.
$$\begin{array}{r} 763 \\ -219 \\ \hline \end{array}$$

57.
$$\begin{array}{r} 841 \\ -316 \\ \hline \end{array}$$

58. 8502 − 2957

59. 9021 − 5862

60. 29,104 − 4988

61. 37,405 − 6877

Use the Mathematics Blueprint for Problem Solving for exercises 62 and 63.

62. A teacher's assistant receives a total salary per month of $3560. Deducted from her paycheck are taxes of $499, social security of $218, and retirement of $97. What was the total of her check after the deductions?

63. Jean's savings account had a balance of $5021. Over the course of a year, she made deposits of $759, $2534, and $532. She made withdrawals of $799, $533, and $87. What was her ending balance?

California is the nation's leading state for high-tech jobs, employing 664,325 in 1995. The bar graph lists the states that are also in the top 5 for high-tech employment. Use the graph to answer questions 64 and 65.

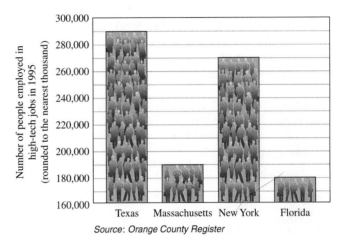

Source: Orange County Register

64. How many more people are employed in high-tech jobs in Texas and Massachusetts than in New York and Florida?

65. How many more people are employed in high-tech jobs in Texas and New York than in Massachusetts and Florida?

SECTION 1.4

Find the perimeter.

66.

19 m
8 m

67.
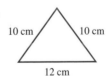
10 cm   10 cm
12 cm

68.

7 in.
7 in.

69. Find the perimeter of the shape made up of two rectangles.
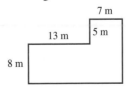
7 m
13 m   5 m
8 m

Use the following histogram to answer questions 70 and 71.

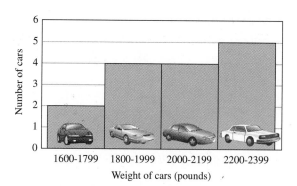

70. (a) How many cars weighed 2000–2199 pounds?

(b) How many cars weighed less than 2000 pounds?

71. (a) How many cars weighed 2200–2399 pounds?

(b) How many cars weighed less than 2200 pounds?

72. Joseph must purchase supplies for his home office. The catalog from the supply warehouse lists the following prices for the items he plans to purchase. Estimate the amount of money he will pay for these supplies.

Price List	
Statistical calculator	$25
Color inkjet print cartridge	26
Four drawer metal file cabinet	87
Computer chair	156

SECTION 1.5

Evaluate.

73. $7 + n$, if n is 6.

74. Eight minus x, if x is 3.

Translate using symbols.

75. Two plus what number is 9?

76. When 3 is subtracted from a number, the result is 8.

Solve and check your answer.

77. $x + 2 = 9$ **78.** $a + 5 = 11$ **79.** $8 - n = 2$ **80.** $9 - n = 6$

Simplify using the associative and/or commutative property, then find the solution.

81. $(3 + x) + 1 = 8$ **82.** $2 + (n + 7) = 10$

For the following equations:
    (a) *Change the subtraction problem to an equivalent addition problem.* **(b)** *Solve.*

83. $x - 34 = 53$ **84.** $y - 78 = 93$ **85.** $15 - n = 8$ **86.** $16 - y = 12$

For the following equations:
    (a) *Change the addition problem to an equivalent subtraction problem.* **(b)** *Solve.*

87. $n + 44 = 98$ **88.** $x + 48 = 89$ **89.** $x + 123 = 455$ **90.** $y + 321 = 412$

For the following equations:
    (a) *Translate the English statement into an equation.* **(b)** *Solve.*

91. What number subtracted from 18 equals 3?   **92.** What number minus 12 equals 5?

93. Seven plus what number equals 12?   **94.** What number increased by 5 equals 11?

SECTION 1.6

Fill in the blank.

95. 16 oz = ____ lb   **96.** 1 mi = _____ ft   **97.** 2 ft = ____ in.

98. 2 pt = ____ qt   **99.** 1 day = ____ hr   **100.** 1 min = ____ sec

Perform the operation indicated.

101. 6 yd 1 ft − 3 yd 2 ft   **102.** 4 yd 1 ft − 2 yd 2 ft

103. Brandon jogged for 1 hr 10 min on Monday, 1 hr 12 min on Tuesday, 45 min on Wednesday, and 1 hr 5 min on Friday. What was Brandon's total jogging time for the week?

104. Justin has 1 quart of milk. A recipe he is making requires 3 cups of milk. Does Justin have more or less milk than the recipe requires? How much more or less does he have?

Fill in the blank.

105. 10 millimeters = ____ centimeters   **106.** 10 grams = ____ dekagrams

107. 1 dekaliter = ____ liters   **108.** 1 decimeter = ____ centimeters

Place the appropriate symbol, < or >, in the blank.

109. 1 km ____ 1 mm   **110.** 1 g ____ 1 kg

111. 2 yd ____ 2 cm   **112.** 3 in. ____ 3 m

Answer each question.

113. Juan bought Tylenol Geltabs. The label reads as follows: 100 solid geltabs—500 ____ each. Fill in the blank with the most appropriate unit.
    (a) kilometers   **(b)** grams
    (c) liters   **(d)** milligrams

114. Anita measured the length and width of her bedroom. Choose the most appropriate unit for this measurement.
    (a) millimeters   **(b)** centigrams
    (c) meters   **(d)** liters

The basic metric units are missing from the following measurements. What should they be?

115. A large package of hamburger meat weighs 3 _____ .
    (a) meters   **(b)** kilograms
    (c) kiloliters   **(d)** grams

116. A large bottle of Coke is sold as a _____ bottle.
    (a) 2-centigram   **(b)** 2-millimeter
    (c) 2-meter   **(d)** 2-liter

CHAPTER 1 TEST

1. _____

2. _____

3. _____

4. _____

5. _____

6. _____

7. _____

8. _____

9. _____

10. _____

11. _____

12. _____

13. _____

1. Write 1525 in expanded notation.

2. Replace the question mark with the appropriate symbol, $<$ or $>$.
   (a) 7 ? 2   **(b)** 5 ? 0

3. The total population of a small town is 2925. Round this population figure to:
   (a) The nearest thousand.   **(b)** The nearest hundred.

4. Use the commutative and/or associative property of addition, then simplify.
   (a) $3 + (8 + x)$ **(b)** $5 + y + 2$ **(c)** $1 + (n + 2)$

5. Add: $311 + 4 + 2302$.

6. Combine like terms.
   (a) $3xy + 2y + 4xy$ **(b)** $2m + 5 + m + 6mn$

7. Combine like terms: $5x - 2x - x - 3$.

8. Translate using mathematical symbols: 7 subtracted from a number.

9. The rent on an apartment was $525. To move in, Fred was required to pay the first and last month's rent, a security deposit of $200, and a telephone installation fee of $40. How much money did he need to move into the apartment?

10. Subtract and check.
   (a) $(10 - 2) - 4$ **(b)** $613 - 75$

11. A store clerk receives a total salary per month of $1540. Deducted from her paycheck are taxes of $265, social security of $78, and retirement of $57. What was the total of her check after the deductions?

12. Find the perimeter.

(a)

5 in.
Square

(b)
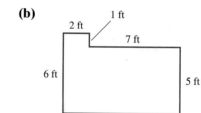
2 ft 1 ft
7 ft
6 ft
5 ft

13. This bar graph shows the surfing forecast for 2/2/98 in southern California. What is the forecasted wave height at Huntington Beach?

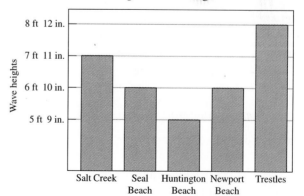

Wave heights

8 ft 12 in.
7 ft 11 in.
6 ft 10 in.
5 ft 9 in.

Salt Creek   Seal Beach   Huntington Beach   Newport Beach   Trestles

14. Sylvia kept the following record of her living expenses for the month of February 1998.

Rent	$790	Car payment	210
Phone and utilities	114	Gas and insurance	187
Food	318		

(a) Estimate Sylvia's monthly expenses for February.

(b) If her take-home (net) income for February is $1921, estimate how much money is left after all the expenses are paid.

15. Evaluate $4 - x$:

(a) If x is 1.   (b) If x is 0.

16. Solve for the variable and check your answer.

(a) $7 + x = 22$ (b) $(y + 1) + 4 = 8$

(c) $34 - x = 27$ (d) $5 + (b + 2) = 18$

Translate into an equation.

17. Let B represent Fred's checking account balance. Fred's checking account balance decreased by $155 equals $275.

18. Seven plus what number equals nine?

19. Three subtracted from thirty equals what number?

Translate into an equation and solve.

20. The sum of two and what number equals 36?

21. What number minus six equals nine?

Add or subtract.

22. (a) 3 yd 2 ft + 1 yd 2 ft   (b) 5 ft 2 in. − 2 ft 4 in.

23. Five quarts of Joe's motor oil cost the same as 1 gallon of W5Z motor oil. If both products are of the same quality, which is the better buy?

24. Replace the question mark with the appropriate symbol, $<$ or $>$.

(a) 3 mg ? 3 g   (b) 6 in. ? 6 cm

14. _____

15. _____

16. _____

17. _____

18. _____

19. _____

20. _____

21. _____

22. _____

23. _____

24. _____

MULTIPLICATION AND DIVISION OF WHOLE NUMBER EXPRESSIONS

S uppose that you decide to replace the flooring in your residence or office. Do you know how to determine how much floor covering must be purchased? Can you calculate the cost of this flooring? After you have studied the topics in this chapter, you will have an opportunity to consider this situation in Putting Your Skills to Work on page 157.

2.1 Understanding Multiplication of Expressions

2.2 Multiplication of Whole Numbers and Algebraic Expressions

2.3 Exponents and Order of Operations

2.4 Performing Operations with Exponents

2.5 Understanding Division of Expressions

2.6 Solving Basic Equations Involving Multiplication and Division

2.7 Geometric Shapes

2.8 Applied Problems Involving Multiplication and Division

CHAPTER 2 PRETEST

This test provides a preview of the topics in this chapter. It will help you identify which concepts may require more of your studying time. If you are familiar with the topics in this chapter, take this test now. Check your answers with those in the back of the book. If you are not familiar with the topics in this chapter, begin studying the chapter now.

1. _____

2. _____

3. _____

4. _____

5. _____

6. _____

7. _____

8. _____

9. _____

10. _____

11. _____

12. _____

13. _____

14. _____

SECTION 2.1

1. Identify the product and the factors in each equation.

   (a) $8(2) = 16$ **(b)** $3x = 21$ **(c)** $mn = p$

2. Multiply:

   (a) $423(0)$ **(b)** $x(1)$ **(c)** $8 \cdot 4 \cdot 0 \cdot y$

3. Simplify: $6(3x)$.

SECTION 2.2

4. Multiply: $4(3000)$.   **5.** Multiply: $2(x + 3)$.

6. A cashier earns $8 per hour for the first 40 hours worked and $12 per hour for overtime (hours worked in addition to 40 hours a week). Last week the cashier worked 49 hours. Calculate the cashier's total pay for the 49-hour week.

SECTION 2.3

7. Write in exponent form: $3 \cdot 3 \cdot 3 \cdot 3$.   **8.** Find the value: 2^3.

9. Translate using symbols.

   (a) x to the fifth power   **(b)** 4 squared

10. Find: 10^6.   **11.** Find: $5^2 - 4 + 1$.

SECTION 2.4

12. Multiply:

   (a) $x^2 \cdot x$ **(b)** $a^3 \cdot a^4$

13. Multiply: $(2x)(x^4)(x^5)$.   **14.** Multiply: $z(z^2 + 2)$.

S E C T I O N 2 . 5

15. Evaluate: $18 \div 2 - 3 \cdot 2$.

16. Divide and check your answer: $279 \div 13$.

17. Marco drove 312 miles and used 12 gallons of gas. How many miles per gallon did he get?

S E C T I O N 2 . 6

18. Evaluate $6x - 1$ if $x = 2$.

19. Solve each equation and check your answer.

    (a) $6y = 42$ **(b)** $\dfrac{x}{3} = 17$ **(c)** $3(5x) = 45$

20. Double a number plus seven equals nineteen. What is the number?

S E C T I O N 2 . 7

21. Use the formula $P = 4s$ to find the perimeter of a square with a side 6 feet in length.

22. Find the area of a flower garden that has a square shape and a side that measures 4 feet.

23. An aquarium (fish tank) 3 feet long and 1 foot wide is filled to a depth of 2 feet. How many cubic feet of water are in the aquarium?

S E C T I O N 2 . 8

24. A small grove of lemon trees had 20 rows of trees, with 15 trees in each row. If each tree produces approximately 220 pounds of lemons, how many pounds of lemons would the grower produce?

15. _____

16. _____

17. _____

18. _____

19. _____

20. _____

21. _____

22. _____

23. _____

24. _____

After studying this section, you will be able to:

❶ Understand the meaning of multiplication.

❷ Use the symbols and key words for multiplication.

❸ Use the properties of multiplication.

❹ Use multiplication to simplify variable expressions.

Math Pro   Video 2.1   SSM

S E C T I O N   2 . 1

Understanding Multiplication of Expressions

❶ Understanding the Meaning of Multiplication

Multiplication of whole numbers can be thought of as repeated addition. For example, suppose that a parking lot has 6 rows of parking spaces with 8 spaces in each row. How many parking spaces are in the lot?

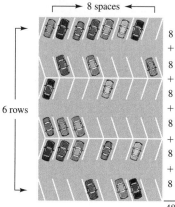

48 spaces or
6 times 8 = 48

To get the total we add 8 six times, $8 + 8 + 8 + 8 + 8 + 8 = 48$, or we can use a shortcut: 6 rows of 8 is the same as 6 times 8, which equals 48. This is multiplication, a shortcut for repeated addition. When numbers are large, multiplication is easier than addition, but for smaller numbers, you can—if you are stuck—do a multiplication problem by working the equivalent addition problem.

The illustration of the parking lot is an example of an *array*, a rectangular figure that consists of rows and columns. Since the parking lot has 6 rows and 8 columns, it is a 6 by 8 array (always write the rows first). We can use dots, squares, or any figure to display an array.

E X A M P L E   1   Draw an array that represents the multiplication 4 times 5.

SOLUTION   We draw an array with 4 rows and 5 columns. Is there another array that represents 4 times 5? Explain.

P R A C T I C E   P R O B L E M   1

Draw an array that represents the multiplication 5 times 3.

It is often helpful to use arrays for many real-life multiplication problems.

E X A M P L E   2   L&M's Print Shop makes business cards in 3 colors: white, beige, and light blue. The shop has 4 types of print to choose from: bold-face, italic, fine line, and Roman.

(a) Set up an array that describes all possible business cards that can be made.
(b) Determine how many different types of cards can be made.

SOLUTION

(a) We set up a 4 by 3 array where *each row corresponds to a type of print* and *each column corresponds to a color*. Each item in the array represents one possible business card.

	White	Beige	Light blue
Bold face	**Jesse Willettes** **Sales Manager** **(312) 123-5462**	**Jesse Willettes** **Sales Manager** **(312) 123-5462**	**Jesse Willettes** **Sales Manager** **(312) 123-5462**
Italic	*Jesse Willettes* *Sales Manager* *(312) 123-5462*	*Jesse Willettes* *Sales Manager* *(312) 123-5462*	*Jesse Willettes* *Sales Manager* *(312) 123-5462*
Fine line	Jesse Willettes Sales Manager (312) 123-5462	Jesse Willettes Sales Manager (312) 123-5462	Jesse Willettes Sales Manager (312) 123-5462
Roman	Jesse Willettes Sales Manager (312) 123-5462	Jesse Willettes Sales Manager (312) 123-5462	Jesse Willettes Sales Manager (312) 123-5462

(b) We have a 4 by 3 array that corresponds to the multiplication 4 times 3, or **12 business cards**.

PRACTICE PROBLEM 2

A manufacturer makes 3 different types of bikes: dirt, racer, and road. Each type comes in 5 different colors: red, blue, green, pink, and black.

(a) Set up an array that describes all possible bikes that can be made.
(b) Determine how many different bikes can be made.

② Using the Symbols and Key Words for Multiplication

In mathematics there are several ways of indicating multiplication. We write the multiplication *4 times 5* as follows:

$$4 \times 5 \quad (4)(5) \quad 4(5) \quad (4)5 \quad 4 \cdot 5$$

Since $\times$ can be confused with the varible x, the symbols most often used are a *single set of parentheses* or a *dot*.

If two numbers are denoted by the variables a and b, their multiplication can be indicated by ab, with *no symbols between the a and b*. If a number is multiplied by a variable, we write the number first with no symbols between the number and the variable. Thus $6a$ indicates "six times a number."

The numbers we multiply are called **factors**. The *result* of the multiplication is called the **product**.

$$
\begin{array}{ccc}
6 & \cdot \quad 8 & = \quad 48 \\
\downarrow & \downarrow & \downarrow \\
\text{factor} & \text{factor} & \text{product}
\end{array}
$$

EXAMPLE 3   Identify the product and the factors in each equation.

(a) $5(4) = 20$ **(b)** $3x = 12$

SOLUTION

(a) $5(4) = 20$
   5 and 4 are the factors and 20 is the product.

(b) $3x = 12$
   3 and x are factors and 12 is the product.

PRACTICE PROBLEM 3
Identify the product and the factors in each equation.

(a) $9 \cdot 7 = 63$ **(b)** $xy = z$

The word *product* is also used to indicate the operation multiplication. There are several other English phrases used to describe multiplication. The following table gives some English phrases and their translated equivalents written using mathematical symbols.

ENGLISH PHRASE	TRANSLATION INTO SYMBOLS
The *product* of 2 and 3	2(3) or 2 · 3
The *product* of 7 and 5	7(5) or 7 · 5
The *product* of *x* and *y*	*xy*
Six *times* a number	6*x*
Double a number	2*x*
Triple a number	3*x*
Five *times* a number is 20	5*n* = 20

EXAMPLE 4   Translate using symbols.

(a) The product of four and a number
(b) Triple a number
(c) What number times 5 equals 30?

SOLUTION

(a) The product of four and a number   **(b)** Triple a number

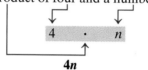

4*n* $\qquad\qquad$ **3*n***

(c) *What number* times 5 equals 30?

$$x \quad \cdot \quad 5 \quad = \quad 30$$

$$x \cdot 5 = 30 \quad \text{or} \quad \mathbf{5x = 30}$$

It is common practice to write the number before the variable.

PRACTICE PROBLEM 4

Translate using symbols.

(a) The product of *a* and *b*
(b) Double a number
(c) Two times what number equals ten?

EXAMPLE 5   Translate the mathematical symbols into words.

(a) 3*n*   **(b)** 4*x* = 12

SOLUTION

(a) 3*n*   **(b)** 4*x* = 12
  Three times a number   **Four times what number equals twelve?**

Note: There are many correct answers. For example, we can translate 3*n* as "*the product of three and n*," and 4*x* = 12 as "*the product of four and x is equal to twelve*."

PRACTICE PROBLEM 5

Translate the mathematical symbols into words.

(a) 5*y*   **(b)** 3*a* = 12

DEVELOPING YOUR STUDY SKILLS

REVIEWING YOUR TEST

To learn mathematics it is necessary for you to master each skill as you proceed through the course. Learning mathematics is much like building blocks—each block is important. Complete the following activities to assure that each of the mathematics building blocks is mastered.

The Learning Cycle

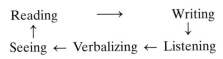

- *Writing.* Try to rework correctly the test problems you missed, without help from others.
- *Verbalizing.* Verify with an instructor, tutor, or classmate that these problems are correct.
- *Listening, seeing, and verbalizing.* Ask for an explanation of why your answers were wrong and how to do the problems you could not correct by yourself.
- *Reading and writing.* Review sections of the book and work exercises similar to all problems you did incorrectly on the test.

Note: It is important that you identify how and why you made your errors so that you can avoid making the same types of errors again.

❸ Using the Properties of Multiplication

Like addition, multiplication is **commutative**. By this we mean that the order in which we multiply factors does not change the product. We use an array to illustrate this fact.

Both arrays represent multiplication of 3 and 4; 3(4) = 12 and 4(3) = 12, illustrating that multiplication is commutative.

The multiplication facts for 0 and 1 are simple because when multiplying a number by 1, the product is that number, and when multiplying by 0, the product is zero. This illustrates the **multiplication properties of zero and 1**.

$$0(7) = 0 \qquad \text{Adding 7 zero times gives us 0.}$$

$$1(7) = 7 \qquad \text{Adding 7 one time gives us 7.}$$

Multiplication is also **associative**, meaning that we can regroup the factors when multiplying and the product does not change.

$$2 \cdot (4 \cdot 1) = 2(4) = 8 \qquad \text{Group the factors 4 and 1; the result is } \mathbf{8}.$$

$$(2 \cdot 4) \cdot 1 = 8(1) = 8 \qquad \text{Group the factors 2 and 4; the result is } \mathbf{8}.$$

Associative Property of Multiplication	$(ab)c = a(bc)$
Changing the grouping of factors does not change the product.	$(7 \cdot 3) \cdot 2 = 7 \cdot (3 \cdot 2)$
	$21(2) = 7(6)$
	$42 = 42$
Commutative Property of Multiplication	$ab = ba$
Changing the order of factors does not change the product.	$5(6) = 6(5)$
	$30 = 30$
Multiplication Property of Zero	$a \cdot 0 = 0$
When 0 is a factor, the product is zero.	$8(0) = 0$
Multiplication Property of 1	$1 \cdot a = a$
When 1 is multiplied by a number, the result is the number.	$1(9) = 9$

We list a few other facts that can help us with multiplication.

1. Multiplying by 2 is the same as doubling a number or counting by two: 2, 4, 6, 8,....
2. Multiplying by 5 is the same as counting by 5, which is easy since all the numbers end with 0 or 5: 5, 10, 15, 20, 25,....
3. Multiplying any number by 10 can be done simply by attaching a 0 to the end of that number. We explain why later.

$$3(10) = 3\mathbf{0} \qquad 4(10) = 4\mathbf{0} \qquad 5(10) = 5\mathbf{0}$$

E X A M P L E 6   Multiply: $4 \cdot 2 \cdot 4 \cdot 5$.

SOLUTION

$4 \cdot 2 \cdot 4 \cdot 5 =$

$4 \cdot 4 \cdot 2 \cdot 5 =$ Use the commutative property to change the order of factors so that we can have 10 as factor.

$16 \cdot 10 \quad =$ Use the associative property and regroup, then multiply.

$\quad\;\; \mathbf{160}$ Multiply by 10.

P R A C T I C E   P R O B L E M   6

Multiply.

(a) $2 \cdot 6 \cdot 0 \cdot 3$ **(b)** $2 \cdot 3 \cdot 1 \cdot 5$

④ Using Multiplication to Simplify Variable Expressions

We also use the multiplication facts and the properties of multiplication to simplify expressions.

E X A M P L E 7   Simplify: $5(7n)$.

SOLUTION

$5(7n) = (5 \cdot 7)n$ Use the associative property to regroup.

$5(7n) = \mathbf{35n}$ Simplify by multiplying.

PRACTICE PROBLEM 7

Simplify: $8(6x)$.

EXAMPLE 8   Simplify.

(a) $6(x \cdot 9)$ **(b)** $2(3)(n \cdot 7)$

SOLUTION

(a) $6(x \cdot 9) = 6(9 \cdot x)$ Use the commutative property.

    $= (6 \cdot 9)x$ Use the associative property to regroup.

$6(x \cdot 9) = 54x$ Simplify by multiplying.

(b) $2(3)(n \cdot 7) = 6(n \cdot 7)$ Multiply: $2(3) = 6$.

        $= 6(7n)$ Change the order of factors.

        $= (6 \cdot 7)n$ Regroup.

$2(3)(n \cdot 7) = 42n$ Multiply.

PRACTICE PROBLEM 8

Simplify.

(a) $4(x \cdot 3)$ **(b)** $2(4)(n \cdot 5)$

TO THINK ABOUT
....

MEMORIZATION OF MULTIPLICATION FACTS

If we think of multiplication as repeated addition, very little memorization is needed to learn the multiplication facts. Once we know the 2, 5, and 10 times tables, which are fairly easy, we can get the other multiplication facts as follows.

From the 2 times table we can get the 3's by multiplying a number by 2 and adding that number.

$\boxed{2(8) \quad + 8 \text{ is the same as } 3(8)}$

$\quad\downarrow \qquad \downarrow \qquad\qquad\qquad \downarrow$

$(8 + 8) + 8 \qquad\qquad 8 + 8 + 8 = 24$

$\quad 16 \quad + 8 = 24$

From the 5 times table we can get the 4 and 6 times table using a similar idea.

$\boxed{5(7) - 7 \text{ is the same as } 4(7)}$

$\qquad\qquad \downarrow \quad\, \downarrow \qquad\qquad \downarrow$

$(7 + 7 + 7 + 7 + 7) - 7 \qquad 7 + 7 + 7 + 7 = 28$

$\qquad\qquad 35 - 7 = 28$

$\boxed{5(7) + 7 \text{ is the same as } 6(7)}$

$\qquad\qquad\quad \downarrow \qquad\qquad\qquad \downarrow$

$(7 + 7 + 7 + 7 + 7) + 7 \qquad 7 + 7 + 7 + 7 + 7 + 7 = 42$

$\qquad\qquad 35 + 7 = 42$

Similarly, from the 10 times table we can get the 9 times table.

EXERCISE

1. Use the techniques discussed to find the product

    (a) $3(7)$ **(b)** $4(8)$ **(c)** $6(8)$ **(d)** $9(8)$

EXERCISES 2.1

Draw an array that represents each multiplication.

1. 2 times 3

★ ★ ★
★ ★ ★

2. 2 times 2

★ ★
★ ★

3. Anthony has 4 ties: brown, black, gray, and dark blue, and 3 shirts: white, pink, and blue.

(a) Set up an array which shows all the possible outfits that Anthony can make.

(b) How many different outfits are possible?

4. Gerry has a choice of 4 carpet colors: beige, gray, blue, and light brown; and 3 colors of blinds: white, pale blue, and rose.

(a) Set up an array which shows all the possible color combinations of carpet and blinds that Gerry can choose from.

(b) How many different combinations are possible?

5. The Ice Cream Palace has 8 flavors of ice cream: vanilla, French vanilla, chocolate, strawberry, coffee, pecan, chocolate chip, and mint chip. There are 5 toppings for the ice cream: fudge, cherry, candy sprinkle, caramel, and nut. How many different ice cream dishes can you order?

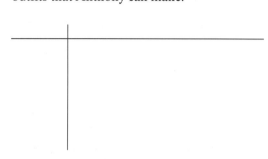

Identify the product and the factors in each equation.

6. $6(3) = 18$ **7.** $5(7) = 35$ **8.** $22x = 88$ **9.** $7a = 28$

Translate each phrase using symbols.

10. Seven times a number

11. What number times 7 equals 49?

12. The product of x and y

13. The product of 9 and a number

14. Six times a number is 42

15. The product of a and b

Translate the symbols into words.

16. **(a)** $4x$

   (b) $2n = 8$

17. **(a)** $7y$

   (b) $5x = 25$

Use the properties of multiplication to answer each question.

18. If $x \cdot y = 0$ and $x = 6$, then $y = ?$

19. If $a \cdot b = 0$ and $a = 2$, then $b = ?$

20. If $x(y \cdot z) = 40$, then $(x \cdot y)z = ?$ **21.** If $b(a \cdot c) = 30$, then $(a \cdot b) \cdot c = ?$

Multiply.

22. $3(6)$	**23.** $4(5)$	**24.** $6(3)$	**25.** $5(4)$
26. $9(0)$	**27.** $8(0)$	**28.** $1(6)$	**29.** $1(7)$
30. $2 \cdot 4 \cdot 5 \cdot 0$	**31.** $9 \cdot 0 \cdot 3 \cdot 8$	**32.** $2 \cdot 2 \cdot 3 \cdot 5$	**33.** $3 \cdot 2 \cdot 4 \cdot 5$

Simplify.

34. $9(6c)$	**35.** $7(5b)$	**36.** $5(z \cdot 8)$	**37.** $4(x \cdot 6)$
38. $8(a \cdot 7)$	**39.** $3(9 \cdot c)$	**40.** $3(a \cdot 6)$	**41.** $5(8 \cdot x)$
42. $3(1)(x \cdot 8)$	**43.** $5(3)(2 \cdot z)$	**44.** $9(2)(0 \cdot y)$	**45.** $0(5)(z \cdot 9)$
46. $6(3)(1 \cdot b)$	**47.** $7(4)(x \cdot 1)$	**48.** $2 \cdot 3(4x)(5y)$	**49.** $6 \cdot 3(5y)(2b)$
50. $(2y)(6x)3 \cdot 7$	**51.** $(9m)(2a)5 \cdot 4$	**52.** $3 \cdot 6(8y)(2x)$	**53.** $4 \cdot 5(3a)(4b)$

 CALCULATOR EXERCISES

Multiply.

54. $3 \cdot 5 \cdot 6 \cdot 6 \cdot 7$	**55.** $3 \cdot 9 \cdot 9 \cdot 8 \cdot 5$	**56.** $5 \cdot 5 \cdot 5 \cdot 5 \cdot 5$	**57.** $8 \cdot 8 \cdot 8 \cdot 8 \cdot 8$

ONE STEP FURTHER

Multiplication facts can be listed on the following table. The product of 9 and 3 is placed where row 9 and column 3 meet.

58. Fill in the multiplication table using the following step-by-step directions:

	0	1	2	3	4	5	6	7	8	9
0										
1										
2										
3										
4										
5										
6										
7										
8										
9										

(a) Use the multiplication property of zero and fill in the second row. Now, use the commutative property and fill in the second column.

(b) Use the multiplication property of 1 and fill in the third row. Now, use the commutative property and fill in the third column.

(c) Complete the 2 times tables: $2 \cdot 1, 2 \cdot 2, 2 \cdot 3$, and so on. Place the products in the fourth row. Now, use the commutative property and place the products in the fourth column.

(d) Complete the 5 times tables: $5 \cdot 1, 5 \cdot 2, 5 \cdot 3$, and so on. Place the products in the seventh row. Now, use the commutative property and place the products in the seventh column.

(e) How many multiplication facts are blank in the table?

(f) Since the 0, 1, 2, and 5 times tables are fairly simple to learn, what does this process tell you about the amount of memorization necessary to learn all the multiplication facts?

CUMULATIVE REVIEW

1. Evaluate $x - 9$ if $x = 16$.   **2.** 12 inches = ____ foot(feet).   **3.** 1 hour = ____ minute(s)

After studying this section, you will be able to:

1 Multiply by numbers with trailing zeros.

2 Use the distributive property to multiply.

3 Multiply a single-digit number by a number with several digits.

4 Multiply a several-digit number by a several-digit number.

5 Solve applied problems involving multiplication.

Math Pro   Video 2.2   SSM

Multiplication of Whole Numbers and Algebraic Expressions

1 Multiplying by Numbers with Trailing Zeros

The numbers 10, 100, 200, and 2000 have **trailing zeros** (zeros at the end). We can multiply these numbers fairly easily. For example, to find 4 times 200 we use repeated addition.

$$4(200) = 200 + 200 + 200 + 200 \longrightarrow$$
$$\begin{array}{r} 200 \\ 200 \\ 200 \\ +200 \\ \hline \mathbf{800} \end{array}$$

We add 2 four times, or $2(4) = 8.$ $\longleftarrow$ $\longrightarrow$ No matter how many times we add 0, the result will be 0.

We see that to get the product we need only *multiply the nonzero numbers* (numbers that are not equal to zero) and attach the number of trailing zeros to the right side of the product. We follow the same process with variable expressions.

E X A M P L E   1   Multiply: $70(9n)$.

SOLUTION

$$\begin{aligned} 70(9n) &= (70 \cdot 9)n && \text{Regroup so that the we can multiply } 70(9). \\ &= \mathbf{630n} && \text{Multiply: } 7(9) = 63, \text{ and attach one zero to the} \\ & && \text{product.} \end{aligned}$$

P R A C T I C E   P R O B L E M   1

Multiply: $8y(40)$.

2 Using the Distributive Property to Multiply

A property that is often used to simplify and multiply is the **distributive property**. This property states that we can distribute multiplication over addition or subtraction. The following example will help you understand what we mean by *distribute* multiplication over addition.

"4 *times* $(n + 7)$" is written $4(n + 7)$. We can find this product using repeated addition.

$$\begin{aligned} 4(n+7) &= (n+7) + (n+7) + (n+7) + (n+7) && \text{We write } 4(n+7) \text{ as} \\ & && \text{repeated addition.} \\[6pt] &= (n+n+n+n) + (7+7+7+7) && \text{We change the order of} \\ & && \text{addition and group the} \\ & && n\text{'s and 7's together.} \\[6pt] &= \quad 4n \quad + \quad 4 \cdot 7 && \text{We have 4 } n\text{'s plus 4 7's.} \end{aligned}$$
$$4(n+7) = 4n + 4 \cdot 7 \quad \text{or} \quad 4n + 28$$

A shorter way to do this is to **distribute** the 4 by multiplying each number or variable inside the parentheses by 4.

$$4(n + 7) = 4(n + 7) = 4 \cdot n + 4 \cdot 7 \quad \text{or} \quad 4n + 28$$

We can state the distributive property as follows:

DISTRIBUTIVE PROPERTY

If a, b, and c are numbers or variables, then

$$a(b + c) = ab + ac \qquad a(b - c) = ab - ac$$

We distribute a by first multiplying every number or variable inside the parentheses by a, then adding or subtracting the result.

EXAMPLE 2   Use the distributive property to simplify: $3(x + 2)$.

SOLUTION

$$3(x + 2) = 3(x + 2) = 3 \cdot x + 3 \cdot 2$$

Multiply 3 times x.

Multiply 3 times 2.

$$3(x + 2) = \mathbf{3x + 6}$$

Write as addition.

PRACTICE PROBLEM 2

Use the distributive property to simplify.

(a) $2(x - 5)$ **(b)** $4(y + 3)$

Using the distributive property makes multiplying numbers mentally easier. For example, to multiply $5 \cdot 13$, we can think of 13 as the sum $(10 + 3)$ and use the distributive property.

$$5(13) = 5(10 + 3) = 5 \cdot 10 + 5 \cdot 3 = 50 + 15 = 65$$

EXAMPLE 3   Use the distributive property to multiply: $6(12)$.

SOLUTION

$$6(12) = 6(10 + 2) = 6(10) + 6(2) = 60 + 12 = \mathbf{72}$$

PRACTICE PROBLEM 3

Use the distributive property to multiply: $7(11)$.

DEVELOPING YOUR STUDY SKILLS

TIME MANAGEMENT

Planning and organizing your schedule is an efficient, low-stress way to juggle school, work, family, and social activities. It allows you to set realistic goals and priorities. You will also be less likely to forget assignments or appointments. A time management schedule provides you with a road map to achieving your goals.

1. Make a list of your daily activities.
2. Make a list of exam and assignment due dates.
3. Place this information in a weekly planner.
4. Plan your study time for each day. Since exam and assignment schedules vary, these activities may vary from day to day.
5. Leave space to insert last-minute things that come up, or specific questions that you want to ask during class.
6. Review the planner periodically during the day so that you don't forget a task and can adjust for unexpected changes.

Think of your time management plan as a contract with yourself. You will find that by adhering to the contract, your grades will improve and you will have more free time.

	MON.	TUES.
7–9ᵃᵐ	Jogging and breakfast	→
9ᵃᵐ	Preview math lecture material	Prepare test review questions
10ᵃᵐ	Math class	→ *Homework due
11ᵃᵐ	English class, questions for final draft of paper	→ *Term paper due
12ᵖᵐ	Lunch	→
1ᵖᵐ	Review math class notes and begin homework	Do practice test—math
2:30–5:30ᵖᵐ	Work	→
5:30–8ᵖᵐ	Dinner and social time	→
8–10ᵖᵐ	Finish math homework, review term paper	Review math and read history

Multiplying a Single-Digit Number by a Number with Several Digits

When we multiply large numbers, we can write the multiplication using the distributive property in a condensed form.

Distributive process: $3(23) = 3(3 + 20) = 3(3) + 3(20)$
$= 9 + 60 = 69$

Simplified version:
$$\begin{array}{r} 23 \\ \times\ 3 \\ \hline 69 \end{array}$$

$3(3) = 9$

$3(20) = 60$

Note: The 6 in the tens column represents 60.

E X A M P L E   4   Multiply: $(547)(6)$.

SOLUTION

$$\begin{array}{r} \overset{2\,4}{547} \\ \times\ \ 6 \\ \hline 3282 \end{array}$$

Multiply: $6(7) = 42$; place the 2 here and carry the 4.

Multiply: $6(4) = 24$. Then add the carry: $24 + 4 = 28$. Place the 8 here, and carry the 2.

Multiply: $6(5) = 30$. Then add the carry: $30 + 2 = 32$. Place the 32 here.

P R A C T I C E   P R O B L E M   4
Multiply: $(436)(7)$.

Multiplying a Several-Digit Number
4 by a Several-Digit Number

E X A M P L E   5   Multiply: $593(400)$.

SOLUTION   Since the number 400 has trailing zeros, we use the same method stated earlier: Multiply the nonzero digits and attach the trailing zeros to the right side of the product.

$$\begin{array}{r} 593 \\ \times\ \ 400 \\ \hline \mathbf{237{,}200} \end{array}$$ Multiply: $4(593)$ and attach the trailing zeros.

P R A C T I C E   P R O B L E M   5
Multiply: $600(872)$.

We multiply $72(34)$ using the distributive property as follows:

$$72(4 + 30) = 72(4) + 72(30)$$

$$\begin{array}{cc} 72 & + & 72 \\ \times\ 4 & & \times\ 30 \\ \hline 288 & +\ \ 2160 & = 2448 \end{array}$$

We shorten the process by combining the two steps:

$$\begin{array}{r} 72 \\ \times\ 34 \\ \hline 288 \\ +\ 2160 \\ \hline 2448 \end{array}$$

$72(4) = 288$

$72(30) = 2160$

The products 288 and 2160 are called **partial products**.

E X A M P L E   6   Multiply: 857(43).

SOLUTION

$$
\begin{array}{r}
857 \\
\times\ \ 43 \\
\hline
2571 \\
34280 \\
\hline
36851
\end{array}
$$

Step 1: Multiply: 3(857) = 2571.

Step 2: Multiply: 40(857) = 34280.
       We multiply 4(857) and add one trailing zero.

Step 3: Add.

Note: It is a good idea to place our carries above each partial product since there is more than one set of carries. It could be confusing to place them all above 857.

P R A C T I C E   P R O B L E M   6

Multiply: 936(38).

E X A M P L E   7   Multiply: 3679(132).

SOLUTION

$$
\begin{array}{r}
3679 \\
\times\ \ 132 \\
\hline
7358 \\
110370 \\
367900 \\
\hline
485628
\end{array}
$$

Multiply: 2(3679).
Multiply: 30(3679), or 3(3679) and attach 1 trailing zero.
Multiply: 100(3679), or 1(3679) and attach 2 trailing zeros.
Add.

$$3679(132) = 485{,}628$$

Note: We can eliminate the trailing zeros in the partial products if we line up the partial products correctly. Place the last nonzero digit of each partial product in the same column as the number by which we are multiplying.

$$
\begin{array}{r}
3679 \\
\times\ \ 132 \\
\hline
7358 \\
11037 \\
3679 \\
\hline
485628
\end{array}
$$

Place the 8 under the 2.
Place the 7 under the 3.
Place the 9 under the 1.

P R A C T I C E   P R O B L E M   7

Multiply: 203(4651).

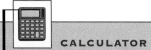

CALCULATOR

You can use your calculator to multiply. To find 43(378), enter:

Scientific calculator:

43 $\times$ 378 $=$

Graphing calculator:

43 $\times$ 378 ENT

The calculator displays

16,254

⑤   Solving Applied Problems Involving Multiplication

Applied problems that require the operation multiplication often state the word *per* followed by a noun, such as: cost *per item*, money earned *per hour*, miles driven *per hour*, or number of items produced *per unit* of time. Other problems that require multiplication deal with arrays (rows and columns).

E X A M P L E   8   Koursh was offered two different jobs: a 40-hour-a-week store management position that pays $12 per hour, and a personnel assistant position paying a monthly salary of $2600. Which job pays more per year?

SOLUTION   When problems have a lot of information, it is a good idea to organize your work in the Mathematics Blueprint for Problem Solving.

1. *Understand the problem.*

MATHEMATICS BLUEPRINT FOR PROBLEM SOLVING			
GATHER THE FACTS	WHAT AM I ASKED TO DO?	HOW DO I PROCEED?	KEY POINTS TO REMEMBER
The management position pays $12 per hour for 40 hours. The personnel position pays $2600 per month.	Determine which job pays a higher salary per year.	Multiply $12 × 40 to find the pay for 1 week, then multiply by 52 for yearly pay. Multiply $2600 × 12 to find the yearly pay. Compare yearly pay for both jobs.	The phrases *per week* and *per year* indicate multiplication. I must find *yearly* pay: 12 months = 1 year 52 weeks = 1 year

2. *Solve and state the answer.* From the information organized in the blueprint, we can write out a process to find the answer.

$12 × 40 = $480   Pay for 1 week (management)

$480 × 52 = $24,960   Pay for 1 year (management)

$2600 × 12 = $31,200   Pay for 1 year (personnel)

The yearly pay is $24,960 for the management position and $31,200 for the personnel position. The **personnel assistant position** pays more per year.

3. *Check the answer.* Since there is a large difference in salaries, we can estimate to check our answer. Estimate the manager's pay per year by rounding $12 per hour to $10 and 52 weeks to 50 weeks.

$10 × 40 hr = $400 per week; $400 × 50 weeks = $20,000 per year

Estimate the personnel assistant's pay per year by rounding 12 months per year to 10. $10 × $2600 = $26,000 per year

Since $26,000 > $20,000, the personnel assistant position pays more.   ✓

PRACTICE PROBLEM 8

Emily is a salesperson for A&E Appliance. For the last two years she has averaged about 7 sales per week, and she is paid solely on commission—$55 per sale. The store manager has decided to offer all salespersons the option of accepting a salary of $1770 per month, or remaining on commission. If Emily continues to maintain her past sales record, which option would earn her more money per year?

EXERCISES 2.2

Multiply.

1. $3(5000)$

2. $7(3000)$

3. $8000 \cdot 4$

4. $4000 \cdot 8$

5. $700 \cdot 9n$

6. $800 \cdot 8b$

7. $900(4z)$

8. $500(4z)$

Use the distributive property to simplify.

9. $4(x + 2)$

10. $2(x + 1)$

11. $3(n - 5)$

12. $6(n - 4)$

13. $3(x - 6)$

14. $4(x - 3)$

15. $4(x + 4)$

16. $5(x + 9)$

The distributive property is used to perform the multiplication in the following problems. Fill in the blanks.

17. $8(13) = 8(\underline{\quad} + 10) = 8(\underline{\quad}) + 8(10) = \underline{\quad} + 80 = \underline{\quad}$

18. $7(14) = 7(\underline{\quad} + 10) = 7(\underline{\quad}) + 7(10) = \underline{\quad} + 70 = \underline{\quad}$

19. $2(104) = 2(\underline{\quad} + 100) = 2(\underline{\quad}) + 2(100) = \underline{\quad} + 200 = \underline{\quad}$

20. $4(103) = 4(\underline{\quad} + 100) = 4(\underline{\quad}) + 4(100) = \underline{\quad} + 400 = \underline{\quad}$

Multiply.

21. $9(637)$

22. $8(926)$

23. $7(602)$

24. $6(405)$

25. $\begin{array}{r} 398 \\ \times\ \ 300 \\ \hline \end{array}$

26. $\begin{array}{r} 578 \\ \times\ \ 500 \\ \hline \end{array}$

27. $\begin{array}{r} 793 \\ \times\ \ 600 \\ \hline \end{array}$

28. $\begin{array}{r} 871 \\ \times\ \ 300 \\ \hline \end{array}$

29. $\begin{array}{r} 76 \\ \times 68 \\ \hline \end{array}$

30. $\begin{array}{r} 81 \\ \times 34 \\ \hline \end{array}$

31. $\begin{array}{r} 99 \\ \times 94 \\ \hline \end{array}$

32. $\begin{array}{r} 44 \\ \times 68 \\ \hline \end{array}$

33. $56(847)$

34. $95(668)$

35. $455(86)$

36. $322(74)$

37. $762(309)$

38. $632(201)$

39. $409(432)$

40. $(201)631$

41. $8324(922)$

42. $4456(578)$

43. $2009(651)$

44. $9002(563)$

Solve each applied problem.

45. A restaurant cook earns $8 per hour and works 40 hours per week. Calculate the cook's total pay for the week.

46. An airplane with an average speed of 450 miles per hour travels for 6 hours. How far does it travel?

47. An orange grove has 25 trees in each row, and there are 15 rows of trees in the grove. How many orange trees are in the grove?

48. John places 6 rows of impatiens flower plants with 12 small plants in each row in his garden. How many plants does he have?

49. The Earth is approximately 12,800 kilometers in diameter. The diameter of Jupiter, the largest planet, is approximately 11 times the diameter of the Earth. What is the approximate diameter of Jupiter?

50. Robert is laying tile on each floor of a two-story department store. He determined that on each floor of the store he would need to lay 50 rows of tile with 35 tiles in each row. How many tiles should he purchase?

51. A 15-story Victorian-style hotel has 60 rooms on each floor. The owners are purchasing 50 boxes of curtains at a discount. If there are 20 curtains in each box, can the owners replace one curtain in every room of the hotel?

52. Round-trip bus fare is $2. Justin rides the bus 5 days a week to work and 2 nights a week to school. He can buy a pass at school that allows 6 months of unlimited bus rides for $400. If Justin only rides the bus round trip to work and school, is it cheaper for Justin to buy the pass or to pay each time he rides the bus?

53. Ricardo's current job as a computer technician at Com-Tec pays a salary of $2200 per month. BLM Accountants offered him a programmer's position that pays $14 per hour for a 40-hour week. Which job pays more per year?

54. Myra sells new memberships for a Total Flex Fitness Center chain. She is paid only on commission—$35 for each new membership. For the last three years she has signed up an average of 11 new members a week. She has been offered an alternative pay option—a salary of $1800 per month. Which pay option pays more per year?

Use the bar graph to answer questions 55 and 56.

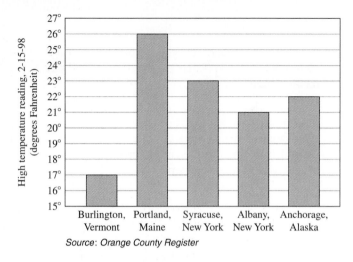

High temperature reading, 2-15-98 (degrees Fahrenheit)

Burlington, Vermont | Portland, Maine | Syracuse, New York | Albany, New York | Anchorage, Alaska

Source: Orange County Register

55. (a) What was the high temperature in Portland on February 15, 1998?

(b) On February 15, 1998 the high temperature reading in Honolulu, Hawaii was four times the high temperature in Albany, New York. What was the high in Honolulu that day?

56. (a) What was the high temperature in Syracuse on February 15, 1998?

(b) On February 15, 1998 the high temperature reading in Buffalo, New York was two times the high temperature in Burlington. What was the high in Buffalo that day?

 CALCULATOR EXERCISES

57. The Earth is approximately 150 million kilometers from the sun. Pluto is approximately 39 times as far as the Earth from the sun. About how far is Pluto from the sun?

58. An average house payment for a home in Los Angeles, California is $1300 per month. How much money would be paid at the end of a 30-year loan for an average house payment?

 VERBAL AND WRITING SKILLS

59. Explain why the distributive property helps you find products such as 6(16) mentally.

60. If you do not write the trailing zero when you multiply numbers with several digits, why is it important that you line up the partial products correctly?

CUMULATIVE REVIEW

1. Simplify: $8(2)(x \cdot 4)$

2. Evaluate $4 + x$ if x is 2.

3. Subtract: $2001 - 463$.

COST ANALYSIS: A VACATION PLAN

When planning a vacation you must consider many things: the places you want to visit, the length of time you can be gone, and the cost of the entire vacation. Many options for vacations are usually available, but not all will fit into the budget available. Therefore, careful planning is necessary.

The Chapman family is planning a vacation across the country. There are 2 adults and 2 children in their family. They researched prices for various travel options and expenses. These figures are as follows:

Plane fare: round-trip Los Angeles to New York: $450 per person.

Train fare: a round-trip vacation package: $350 per adult, $175 per child.

Car rental: $40 per day (including gas).

Hotel: $75 per day.

Meals: $80 per day.

Family car expenses: driving across country costs $20 per day (including gas).

Based on these figures, the Chapman family made the following three plans:

Plan 1. Fly from Los Angeles to New York City, rent a car and make a round-trip to Boston, then a round-trip to Washington, D.C. Finally, fly back to Los Angeles from New York City. This trip will take 14 days and will require renting a car for 13 days. Lodging and meal expenses will be needed for 13 days.

Plan 2. Take the train vacation package. This package includes the fare from Los Angeles to Boston, then on to New York City, Washington, D.C. and back to Los Angeles. There will be a layover of several days in Boston, New York, and Washington. The Chapmans plan to rent a car for 2 days of their stay in each of the three cities to do sightseeing. This trip will take 17 days. Lodging will be needed for 12 days and meals for all 17 days.

Plan 3. Drive the family car on the entire trip. The Chapmans will begin the trip in Los Angeles and drive to Boston. They will then visit New York and Washington, D.C. before returning to Los Angeles. This trip will take 22 days, and lodging and meals will be needed for 21 days.

PROBLEMS FOR INDIVIDUAL INVESTIGATION

1. Find the cost of each plan.

2. Which plan is cheapest?

3. A sleeper car on the train costs an additional $65 per night. If the Chapmans include the sleeper car in plan 2 for 4 nights, what would the new cost of this plan be? Which plan is cheapest now?

PROBLEMS FOR GROUP INVESTIGATION AND COOPERATIVE STUDY

1. Mr. Chapman has 10 days of vacation time. For every day off in addition to the 10 days, he loses $240 of pay. Assuming that the Chapman family leaves on a Saturday, what is the cost of plans 1, 2, and 3?

2. Which plan is cheapest?

INTERNET: Go to http://www.prenhall.com/blair to explore this application.

After studying this section, you will be able to:

❶ Write a whole number in exponent form.

❷ Understand key words involving exponents.

❸ Evaluate powers of 10.

❹ Follow the order of operations.

Math Pro   Video 2.3   SSM

Exponents and Order of Operations

❶ Writing a Whole Number in Exponent Form

We can write the repeated multiplication $3 \cdot 3 \cdot 3 \cdot 3 \cdot 3$ a shorter way: 3^5, because there are five factors of 3 in the multiplication. We say that 3^5 is written in **exponent form**.

3^5 is read "three to the fifth power."

EXPONENT FORM
The small number 5 is called the **exponent**. Whole-number exponents, except zero, tell us how many factors are in the multiplication. The number 3 is called the **base**. The base is the number that is multiplied.

$$3 \cdot 3 \cdot 3 \cdot 3 \cdot 3 = 3^5 \longrightarrow \text{The exponent is 5.}$$

3 appears as a factor 5 times.   The base is 3.

Note that *5 is not part of the multiplication*, it just tells us how many 3's are in the multiplication.

If a whole number does not have an exponent visible, the exponent is understood to be 1.

$$9 = 9^1 \quad \text{and} \quad 10 = 10^1$$

EXAMPLE 1   Write in exponent form.

(a) $2 \cdot 2 \cdot 2 \cdot 2 \cdot 2 \cdot 2$ **(b)** $4 \cdot 4 \cdot 4 \cdot x \cdot x$

(c) 7

SOLUTION

(a) $2 \cdot 2 \cdot 2 \cdot 2 \cdot 2 \cdot 2 = \mathbf{2^6}$ **(b)** $4 \cdot 4 \cdot 4 \cdot x \cdot x = 4^3 \cdot x^2$ or $\mathbf{4^3 x^2}$

(c) $7 = \mathbf{7^1}$

PRACTICE PROBLEM 1
Write in exponent form.

(a) n **(b)** $6 \cdot 6 \cdot y \cdot y \cdot y \cdot y$

(c) $5 \cdot 5 \cdot 5 \cdot 5 \cdot 5 \cdot 5 \cdot 5 \cdot 5$

EXAMPLE 2   Write as a repeated multiplication.

(a) n^3 **(b)** 6^5

SOLUTION

(a) $n^3 = \mathbf{n \cdot n \cdot n}$ **(b)** $6^5 = \mathbf{6 \cdot 6 \cdot 6 \cdot 6 \cdot 6}$

PRACTICE PROBLEM 2
Write as a repeated multiplication.

(a) x^6 **(b)** 1^7

To find the *value* of an expression in exponent form, we first write the expression as repeated multiplication, then multiply the factors.

E X A M P L E   3   Find the value of each expression.

(a) 3^3 **(b)** 1^9 **(c)** 2^4

SOLUTION

(a) $3^3 = 3 \cdot 3 \cdot 3 = \mathbf{27}$

(b) $1^9 = \mathbf{1}$. We do not need to write out this multiplication because repeated multiplication of 1 will always equal 1.

(c) $2^4 = 2 \cdot 2 \cdot 2 \cdot 2 = \mathbf{16}$

P R A C T I C E   P R O B L E M   3

Find the value of each expression.

(a) 4^3 **(b)** 8^1 **(c)** 10^2

Note: Sometimes we are asked to express the answer in *exponent form* and other times to *find the value* of the expression. Therefore, it is important that you read the question carefully and express the answer in the correct form.

Write $5 \cdot 5 \cdot 5$ in *exponent form*: $5 \cdot 5 \cdot 5 = 5^3$

Find the value of $5 \cdot 5 \cdot 5$: $5 \cdot 5 \cdot 5 = 125$

Any whole number (except 0) can be raised to the zero power. The result is 1. Thus $10^0 = 1$, $4^0 = 1$, $x^0 = 1$, and $23^0 = 1$, and so on.

> For any whole number a other than zero, $a^0 = 1$.

② Understanding Key Words Involving Exponents

How do you say 10^2 or 5^3? We can say: 10 raised to the power 2, or 5 raised to the power 3, but the following phrases are more commonly used.

> If the value of the exponent is 2, we say the base is **squared**.
>
> 6^2 is read "six squared."
>
> If the value of the exponent is 3, we say the base is **cubed**.
>
> 6^3 is read "six cubed."
>
> If the value of the exponent is *greater than* 3, we say "to the **(exponent)th power**."
>
> 6^5 is read "six to the fifth power."

E X A M P L E   4   Translate using symbols.

(a) Five cubed   **(b)** Seven squared   **(c)** y to the eighth power

SOLUTION

(a) Five cubed $= \mathbf{5^3}$.   **(b)** Seven squared $= \mathbf{7^2}$.

(c) y to the eighth power $= \mathbf{y^8}$.

PRACTICE PROBLEM 4

Translate using symbols.

(a) Four to the sixth power   **(b)** x cubed   **(c)** Ten squared

EXAMPLE 5   Translate each statement into an equation and solve.

(a) What whole number squared is equal to 25?
(b) Three cubed is equal to what number?

SOLUTION

(a) *What whole number squared* is equal to *25?*

$$n^2 \qquad\qquad = \qquad 25$$
$$n^2 = 25$$

$n = 5$ because $5^2 = 25$

(b) *Three cubed* is equal to *what number?*

$$3^3 \qquad = \qquad x$$
$$3^3 = x$$
$$3 \cdot 3 \cdot 3 = 27$$
$$x = 27 \text{ because } 3^3 = 27.$$

PRACTICE PROBLEM 5

Translate each statement into an equation and solve.

(a) What whole number cubed is equal to 8?
(b) Four squared is equal to what number?

❸ Evaluating Powers of 10

In today's world, large numbers are often expressed as a power of 10. $10^1, 10^2, 10^3$, and 10^4 are powers of 10. Let's look for a pattern to find an easy way to calculate powers of 10.

$$10^1 = 10 \qquad\qquad 10^3 = (10)(10)(10) = 1000$$
$$10^2 = (10)(10) = 100 \qquad 10^4 = (10)(10)(10)(10) = 10,000$$

Notice that when the exponent is 1 there is 1 trailing zero, when the exponent is 2 there are 2 trailing zeros, when it is 3 there are 3 trailing zeros, and so on. Thus to calculate a power of 10, we write 1 and attach the number of trailing zeros named by the exponent.

EXAMPLE 6   Find the value of 10^7.

SOLUTION

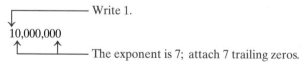

$$10^7 = \mathbf{10,000,000}$$

PRACTICE PROBLEM 6

Find the value of 10^5.

④ Following the Order of Operations

It is often necessary to do more than one operation to solve a problem. For example, if you bought 2 pairs of socks at $3 a pair and 4 undershirts for $6 a shirt, you would multiply then add to find the total cost. In other words, the order in which we performed the operations (order of operations) was multiply first, then add. In most calculations, however, the order is not as clear. The problem $4 + 3(5)$ can be tricky. Do we add then multiply, or multiply before adding? Let's work this calculation both ways.

Add First		Multiply First

$$4 + \mathbf{3(5)} = 7(5) = 35 \quad \text{Wrong!} \qquad 4 + \mathbf{3(5)} = 4 + 15 = 19 \quad \text{Correct}$$

Since $4 + 3(5)$ can be written $4 + (5 + 5 + 5) = 4 + 15,\ 19$ is correct. Thus we see that the order of operations makes a difference. The following rule tells which operations to do first: the correct **order of operations**. We call this a *list of priorities*.

ORDER OF OPERATIONS
Follow this order of operations:

Do first. 1. Perform operations inside *parentheses*.
  ↑   2. Simplify any expressions with *exponents*.
  ↓   3. *Multiply*.
Do last. 4. *Add or subtract* from left to right.

    Parentheses → exponent → multiply → add or subtract

Now, following the order of operations, we can clearly see that to find $4 + 3(5)$, we multiply then add. You will find it easier to follow the order of operations if you keep your work neat and organized, perform one operation at a time, and follow the sequence *identify, calculate, replace*.

1. *Identify* by underlining the operation that has the highest priority.
2. *Calculate* the operation underlined.
3. *Replace* the operation with your result.

EXAMPLE 7 Evaluate: $2^3 - 6 + 4$.

SOLUTION

$$2^3 - 6 + 4 = 8 - 6 + 4 \qquad \textit{Identify: } \text{The highest priority is } \textbf{exponents.}$$
$$\textit{Calculate: } 2 \cdot 2 \cdot 2 = 8.\ \textit{Replace: } 2^3 \text{ with 8.}$$

$$= \underline{8 - 6} + 4 \qquad \textit{Identify: } \textbf{Subtraction} \text{ has the highest priority.}$$
$$\textit{Calculate: } 8 - 6 = 2.\ \textit{Replace: } 8 - 6 \text{ with 2.}$$

$$= 2 + 4 \qquad \textit{Identify: } \textbf{Addition} \text{ is last. } \textit{Calculate: } 2 + 4 = 6.$$

$$2^3 - 6 + 4 = \mathbf{6} \qquad \textit{Replace: } 2 + 4 \text{ with 6.}$$

Note: Addition and subtraction have equal priority. We do the operations as they appear, reading from left to right. In this problem the subtraction appears first, so we subtract before we add.

PRACTICE PROBLEM 7

Evaluate: $3^2 + 2 - 5$.

CALCULATOR

ORDER OF OPERATIONS

For many calculators you can enter multiple operations in the order in which they appear, and your calculator will follow the correct order of operations. To find 1 + 3 × 2, enter:

Scientific calculator:

1 + 3 × 2 =

Graphing calculator:

1 + 3 × 2 ENT

The calculator displays:

7

Note: If your calculator does not display 7, it is not capable of following the correct order of operations. You will have to enter each operation in the appropriate order.

E X A M P L E   8   Evaluate: $2 \cdot 3^2$.

SOLUTION

$2 \cdot 3^2 = 2 \cdot 9$ *Identify:* The highest priority is **exponents**. *Calculate:* $3 \cdot 3 = 9$.
                        Replace: 3^2 with 9.

$\quad\quad = 2 \cdot 9$ *Identify:* **Multiplication** is last. *Calculate:* $2 \cdot 9 = 18$.
                        Replace: $2 \cdot 9$ with 18.

$2 \cdot 3^2 = \mathbf{18}$

Note: We must follow the rule for order of operations and simplify the exponent 3^2 before we multiply; otherwise, we will get the wrong answer. $2 \cdot 3^2$ *does not equal* 6^2.

P R A C T I C E   P R O B L E M   8

Evaluate: $4 \cdot 2^3$.

E X A M P L E   9   Evaluate: $4 + 3(6 - 2^2) - 7$.

SOLUTION   We always do the calculations inside the parentheses first. Once inside the parentheses, we proceed using the order of operations rule.

$4 + 3(6 - 2^2) - 7$

$= 4 + 3(6 - \mathbf{4}) - 7$ Within the parentheses, **exponents** have the highest priority: $2^2 = 4$.

$= 4 + 3(\mathbf{2}) - 7$ We must finish all operations inside the parentheses, so we **subtract**: $6 - 4 = 2$.

$= 4 + \mathbf{6} - 7$ The highest priority is multiplication: $3 \cdot 2 = 6$.

$= \mathbf{10} - 7$ Add first: $4 + 6 = 10$.

$= 3$ Subtract last: $10 - 7 = 3$.

$4 + 3(6 - 2^2) - 7 = \mathbf{3}$

P R A C T I C E   P R O B L E M   9

Evaluate: $2 + 7(10 - 3 \cdot 2) - 4$.

EXERCISES 2.3

Write each product in exponent form.

1. $3 \cdot 3 \cdot 3 \cdot 3$ **2.** $9 \cdot 9 \cdot 9$ **3.** $a \cdot a \cdot a \cdot a \cdot a$ **4.** $z \cdot z$

5. 8 **6.** y **7.** $2 \cdot 2$ **8.** $5 \cdot 5$

Write each product in exponent form.

9. $5 \cdot 5 \cdot a \cdot a \cdot a$ **10.** $3 \cdot 3 \cdot x \cdot x \cdot x$ **11.** $3 \cdot 3 \cdot z \cdot z \cdot z \cdot z \cdot z$

12. $2 \cdot 2 \cdot y \cdot y \cdot y \cdot y$ **13.** $7 \cdot 7 \cdot 7 \cdot y \cdot y$ **14.** $6 \cdot 6 \cdot x$

15. $n \cdot n \cdot n \cdot n \cdot n \cdot 9 \cdot 9$ **16.** $x \cdot x \cdot x \cdot x \cdot x \cdot 7 \cdot 7$

Write as a repeated multiplication.

17. (a) y^3 **(b)** 7^5 **18. (a)** 8^6 **(b)** x^2

Find the value.

19. 2^3 **20.** 3^3 **21.** 5^2 **22.** 6^2

23. 1^6 **24.** 1^9 **25.** 7^0 **26.** a^0

27. 4^4 **28.** 9^3 **29.** 10^1 **30.** 10^0

31. 5^3 **32.** 2^4 **33.** 10^5 **34.** 10^4

35. x^2 if $x = 5$ **36.** y^3 if $y = 2$

Translate using symbols.

37. Seven to the third power.   **38.** Three cubed.

39. Nine squared.   **40.** Four to the seventh power.

Translate each statement into an equation and solve.

41. Five squared is equal to what number?

42. Three cubed is equal to what number?

43. Two to the fourth power is equal to what number?

44. Five to the third power is equal to what number?

45. What number cubed is equal to 27?

46. What number squared is equal to 49?

Evaluate.

47. $2 \cdot 4 - 1$

48. $3 \cdot 5 - 2$

49. $7^2 + 5 - 3$

50. $6^3 + 4 - 8$

51. $3^4 - 7 + 9$

52. $9^3 - 7 + 4$

53. $5 \cdot 3^2$

54. $4 \cdot 2^2$

55. $2 \cdot 2^2$

56. $4 \cdot 4^2$

57. $5^2 - 7 + 3$

58. $4^3 - 8 + 7$

59. $9 + 2 \cdot 2$

60. $5 + 3 \cdot 9$

61. $2 \cdot 5^2 + 1$

62. $4 \cdot 3^2 + 5$

63. $7 \cdot 5^2 - 9$

64. $5 \cdot 4^2 - 7$

65. $8 + (7 + 4^3)$

66. $9 + (6 + 2^2)$

67. $7 + 2(7^2 + 4)$

68. $3 + 4(5^2 + 1)$

69. $5 + 2(3^2 + 8)$

70. $1 + 4(2^2 + 6)$

71. $7 + 5(3 \cdot 4 + 7) - 2$

72. $3 + 4(5 \cdot 2 + 8) - 3$

73. $59 - 4(1 + 5 \cdot 2) + 4$

74. $88 - 3(2 + 6 \cdot 4) + 6$

75. $32 \cdot 6 - 4(4^3 - 5 \cdot 2^2) + 3$

76. $63 \cdot 4 - 5(3^2 + 4 \cdot 2^3) + 5$

CALCULATOR EXERCISES

77. $49 + 36 \cdot 37 + 9^6$

78. $17^4 + 68 - 54$

79. $15 + 23(75 + 6^9)$

80. $72 \cdot 83 - 2(8^3 + 3 \cdot 9^2) - 125$

 VERBAL AND WRITING SKILLS

81. Write in words the question being asked by the equation $n^2 = 16$.

82. Write in words the question being asked by the equation $x^3 = 27$.

ONE STEP FURTHER

83. Fred wanted to evaluate $3 \cdot 2 + 4$. He multiplied 3 times 6 to get 18. What is wrong with his reasoning? What is the correct answer?

84. Sara wanted to evaluate $2 \cdot 3^2$. She squared 6 to get 36. What is wrong with her reasoning? What is the correct answer?

85. Multiply: $21 \cdot 10^1$; $21 \cdot 10^2$; $21 \cdot 10^3$; $21 \cdot 10^4$. Do you see a pattern that might suggest a quick way to multiply a number by the powers of 10? Explain.

86. Multiply: $10^1 \cdot 10^2$; $10^1 \cdot 10^3$; $10^1 \cdot 10^4$. Do you see a pattern that might suggest a quick way to multiply 10 by powers of 10? Explain.

CUMULATIVE REVIEW

1. Add: $4079 + 2762$.

2. Subtract: $8900 - 477$.

3. Multiply: $(387)(196)$.

4. Translate using symbols: The product of two times some number is 72.

After studying this section, you will be able to:

❶ Multiply expressions in exponent form.

❷ Multiply algebraic expressions.

❸ Use the distributive property to multiply algebraic expressions.

Math Pro   Video 2.4   SSM

SECTION 2.4

Performing Operations with Exponents

❶ Multiplying Expressions in Exponent Form

How do we multiply $3^2 \cdot 3^3$? One way is to write the repeated multiplication, then write the product in exponent form.

$$3^2 \quad \cdot \quad 3^3$$
$$\downarrow \qquad \downarrow$$
$$3 \cdot 3 \cdot 3 \cdot 3 \cdot 3 = 3^5 \qquad \text{Three appears as a factor 5 times; the exponent is 5.}$$

Thus $3^2 \cdot 3^3 = 3^5$.

EXAMPLE 1   Write $5^4 \cdot 5^3$ as repeated multiplication, then write the product in exponent form.

SOLUTION

$$5^4 \qquad \cdot \qquad 5^3$$
$$\downarrow \qquad\qquad \downarrow$$
$$5 \cdot 5 \cdot 5 \cdot 5 \cdot 5 \cdot 5 \cdot 5 = 5^7 \qquad \text{Since five appears as a factor 7 times, the exponent is 7.}$$

$$\mathbf{5^4 \cdot 5^3 = 5^7}$$

PRACTICE PROBLEM 1

Write $4^2 \cdot 4^4$ as repeated multiplication, then write the product in exponent form.

We can shorten this process by using rules that tell us how to multiply numbers in exponent form with the same base. Try to notice a pattern in the following multiplication.

$$\overset{\text{3 factors} \quad \text{4 factors} \quad \text{7 factors}}{\underset{\downarrow \qquad\quad \downarrow \qquad\quad \downarrow}{}}$$
$$4^3 \cdot 4^4 = (4 \cdot 4 \cdot 4)(4 \cdot 4 \cdot 4 \cdot 4) = 4^7 \qquad \text{The exponent is 7 since there are 7 factors.}$$

$$\overset{\text{2 factors} \quad \text{3 factors} \quad \text{5 factors}}{\underset{\downarrow \qquad\quad \downarrow \qquad\quad \downarrow}{}}$$
$$x^2 \cdot x^3 = (x \cdot x)(x \cdot x \cdot x) = x^5 \qquad \text{The exponent is 5 since there are 5 factors.}$$

Let's summarize these products: $(4^3)(4^4) = 4^7$; $(x^2)(x^3) = x^5$. What is the pattern? We state the rule as follows.

> **MULTIPLYING EXPRESSIONS IN EXPONENT FORM**
> To multiply constants or variables in exponent form that have the *same base*, add the exponents but keep the base unchanged.
> $$x^a \cdot x^b = x^{a+b}$$

There are three things we must remember when we perform multiplication with exponents:

1. We can only use the rule for multiplication with exponents if the bases are the same.
2. We add exponents.
3. We do not change the base.

E X A M P L E   2   Multiply and write the product in exponent form.

(a) $x^3 \cdot x^6$ **(b)** $x^7 \cdot x$ **(c)** $4^5 \cdot 3^4$ **(d)** $2^2 \cdot 2^4$

SOLUTION

(a) $x^3 \cdot x^6 = x^{3+6} = x^9$ $\boldsymbol{x^3 \cdot x^6 = x^9}$

(b) $x^7 \cdot x = x^7 \cdot x^1 = x^{7+1} = x^8$ $\boldsymbol{x^7 \cdot x = x^8}$

Note: The exponent is 1. Every variable that does not have a written exponent is understood to have an exponent of 1.

(c) $4^5 \cdot 3^4 = \boldsymbol{4^5 \cdot 3^4}$ The rule for multiplying numbers with the *same base* does not apply since the bases are *different*.

(d) $2^2 \cdot 2^4 = 2^{2+4} = 2^6$ $\boldsymbol{2^2 \cdot 2^4 = 2^6}$

Notice that in part (d), the base does not change; only the exponents change.

This is correct: $2^2 \cdot 2^4 = 2 \cdot 2 \cdot 2 \cdot 2 \cdot 2 \cdot 2 = 2^6$ There are 6 factors of 2 in the product.

This is wrong!: $2^2 \cdot 2^4 = 2 \cdot 2 \cdot 2 \cdot 2 \cdot 2 \cdot 2 \neq 4^6$ There are *not* 6 factors of 4 in the product.

P R A C T I C E   P R O B L E M   2

Multiply.

(a) $y^5 \cdot y$ **(b)** $a^4 \cdot a^5$ **(d)** $5^3 \cdot 3^4$ **(b)** $6^5 \cdot 6^6$

❷ Multiplying Algebraic Expressions

For the algebraic expression $3x^4$, the number 3 is called the **numerical coefficient**. A numerical coefficient is a number that is multiplied by a variable.

$$3x^4$$
$$\downarrow$$

3 is a numerical coefficient.

When we multiply two expressions such as $3x^4$ and $4x^7$, we first multiply the numerical coefficients, then we multiply the variable expressions separately.

$$(3x^4)(4x^7) = 3 \cdot x^4 \cdot 4 \cdot x^7$$
$$= (3 \cdot 4) \cdot (x^4 \cdot x^7) \quad \text{Change the order of multiplication and regroup factors.}$$
$$= \quad 12 \quad \cdot \quad x^{11} \quad \text{Multiply the numerical coefficients. Multiply variables separately by adding their exponents: } x^4 \cdot x^7 = x^{4+7} = x^{11}.$$

$$(3x^4)(4x^7) = \boldsymbol{12x^{11}}$$

MULTIPLICATION OF ALGEBRAIC EXPRESSIONS WITH EXPONENTS
First multiply the numerical coefficients. Then use the rule for multiplying variables with exponents.

$$(5a^2)(3a^4) = 15a^6$$

Not only should you know how to work with exponent notation, but it is critical that you know clearly which part is the *numerical coefficient*, which parts are the *bases*, and which parts are *exponents*. These words are used extensively in algebra.

As stated earlier, any variable that does not have a *visible exponent* is understood to have an *exponent of 1*: $x = x^1$ $y = y^1$

Every variable that does not have a visible *numerical coefficient* is understood to have a numerical *coefficient of 1*: $y^4 = 1y^4$ $a^6 = 1a^6$

E X A M P L E   3   Multiply: $(7a)(a^3)(4a^6)$.

SOLUTION

$$(7a)(a^3)(4a^6) = (7a^1)(1a^3)(4a^6)$$

$$= (7 \cdot 1 \cdot 4)(a^1 \cdot a^3 \cdot a^6) \quad \text{Change the order of multiplication and regroup factors.}$$

$$= \quad 28 \quad \cdot \quad a^{10} \qquad \text{Multiply the numerical coefficients and variables separately: } 7 \cdot 1 \cdot 4 = 28$$

$$(7a)(a^3)(4a^6) = \mathbf{28a^{10}} \qquad \text{and } a^1 \cdot a^3 \cdot a^6 = a^{1+3+6} = a^{10}.$$

P R A C T I C E   P R O B L E M   3
Multiply: $(4y)(5y^2)(y^5)$.

We must sometimes multiply algebraic expressions that involve more than one variable.

E X A M P L E   4   Multiply: $(5x^6)(7y^4)(2x^4)$.

SOLUTION

$$(5x^6)(7y^4)(2x^4) = (5 \cdot 7 \cdot 2)(x^6 \cdot y^4 \cdot x^4) = \mathbf{70\,x^{10}y^4}$$

Note: The rule for multiplying expressions in exponent form does not apply to $x^{10}y^4$ because the bases, x and y, are not the same.

P R A C T I C E   P R O B L E M   4
Multiply: $(4y^3)(3x^2)(5y^2)$.

TO THINK ABOUT
· ·

DO I ADD OR MULTIPLY COEFFICIENTS?

When we combine the like terms $4n + 3n$, do we get the same answer as when we multiply $(4n)(3n)$? Let's simplify both and see what happens.

$4n + 3n$	$(4n)(3n)$
$(n+n+n+n) + (n+n+n) = 7n$	$4 \cdot n \cdot 3 \cdot n = 4 \cdot 3 \cdot n \cdot n = 12n^2$
$4n + 3n = 7n$	$(4n)(3n) = 12n^2$

We *add* the numerical coefficients. The variable stays the same.

We *multiply* the numerical coefficients. We must add the exponents of the variable.

We can see that the answers are not the same. When we simplify we must remember the difference between adding and multiplying algebraic expressions.

EXERCISE
Simplify.

(a) $3x + 5x$ **(b)** $(3x)(5x)$ **(c)** $7xy^2 - 5xy^2$ **(d)** $(7xy^2)(5xy^2)$

Using the Distributive Property
3 to Multiply Algebraic Expressions

Recall that to multiply $5 \cdot (x + 1)$ we use the distributive property: multiplying $5 \cdot x$, then $5 \cdot 1$, to obtain $5x + 5$. We follow the same process when we have exponent expressions such as $3(x^2 - 4)$, as illustrated in the next example.

EXAMPLE 5   Use the distributive property to simplify: $3(x^2 - 4)$.

SOLUTION

$$3(x^2 - 4) = 3 \cdot x^2 - 3 \cdot 4 = 3x^2 - 12$$

$$3(x^2 - 4) = \mathbf{3x^2 - 12}$$

PRACTICE PROBLEM 5

Use the distributive property to simplify: $2(x^3 - 6)$.

EXAMPLE 6   Use the distributive property to simplify: $x^2(x^5 + 8)$.

SOLUTION

$$x^2(x^5 + 8) = x^2 \cdot x^5 + x^2 \cdot 8$$

$$= x^{2+5} + 8x^2 = \mathbf{x^7 + 8x^2}$$

$$x^2(x^5 + 8) = \mathbf{x^7 + 8x^2}$$

PRACTICE PROBLEM 6

Use the distributive property to simplify: $x^4(x^3 + 6)$.

EXERCISES 2.4

Multiply and write the product in exponent form.

1. (a) $(z \cdot z \cdot z) \cdot (z \cdot z)$ **(b)** $z^3 \cdot z^2$ **2. (a)** $(y \cdot y \cdot y \cdot y) \cdot (y \cdot y \cdot y)$ **(b)** $y^4 \cdot y^3$

3. (a) $(x \cdot x) \cdot (x \cdot x \cdot x \cdot x)$ **(b)** $x^2 \cdot x^4$ **4. (a)** $(b \cdot b \cdot b) \cdot (b \cdot b \cdot b)$ **(b)** $b^3 \cdot b^3$

5. (a) $(z \cdot z) \cdot z$ **(b)** $z^2 \cdot z$ **6. (a)** $(x \cdot x \cdot x) \cdot x$ **(b)** $x^3 \cdot x$

Multiply and write the product in exponent form.

7. $x^7 \cdot x^2$ **8.** $y^6 \cdot y^6$ **9.** $x^4 \cdot x^5$ **10.** $y^4 \cdot y^2$

11. $x^0 \cdot x^2$ **12.** $y^4 \cdot y^0$ **13.** $x^4 \cdot x$ **14.** $x^7 \cdot x$

15. $3^2 \cdot 3^3$ **16.** $2^3 \cdot 2^4$ **17.** $4 \cdot 4^5$ **18.** $6 \cdot 6^3$

19. $8^2 \cdot 7^5$ **20.** $3^9 \cdot 4^6$ **21.** $x^5 \cdot y^3$ **22.** $x^2 \cdot y^4$

23. $7^4 \cdot 7^3$ **24.** $9^5 \cdot 9^6$ **25.** $3^5 \cdot 3^3$ **26.** $2^6 \cdot 2^7$

27. $x^5 \cdot x^2 \cdot x^7$ **28.** $y^7 \cdot y^3 \cdot y^5$ **29.** $y^2 \cdot y^4 \cdot y^5$ **30.** $x^4 \cdot x^3 \cdot x^5$

31. $3^3 \cdot 3^2 \cdot 3^5$ **32.** $4^7 \cdot 4^4 \cdot 4^2$ **33.** $2^5 \cdot 3^2 \cdot 4^7$ **34.** $4^7 \cdot 5^3 \cdot 2^4$

Simplify.

35. $(4y^5)(6y^7)$ **36.** $(7y^3)(3y^3)$ **37.** $(6a^6)(9a^8)$ **38.** $(2a^4)(4a^4)$

39. $(8x^8)(6x^5)$ **40.** $(4x^9)(6x^6)$ **41.** $(x^4)(2x^3)$ **42.** $(x^6)(3x^7)$

43. $(9a)(8a^3)(2a^6)$ **44.** $(3x)(2x^4)(4x^5)$ **45.** $(x)(4x^6)(7x^5)$ **46.** $(4x)(x^7)(9x^6)$

47. $(9x)(4x^9)(2x^5)$ **48.** $(2x)(5x^7)(3x^7)$ **49.** $(4x^0)(2x^4)(2x^5)$ **50.** $(3x^0)(2x^7)(4x^3)$

51. $(4y^3)(2x^4)(2x^5)$ **52.** $(3y^4)(2x^7)(4x^3)$ **53.** $(2a^3)(2b^4)(3a^5)$ **54.** $(4a^3)(2b^2)(4a^3)$

55. $(y^3)(5y^4)(4x^2)$ **56.** $(3y^4)(2x^7)(4x^3)$ **57.** $(2y^2)(3y^5)(2x^2)(4x^4)$ **58.** $(4a^3)(2a^4)(2b^6)(2b^2)$

Use the distributive property to simplify.

59. $4(x + 1)$ **60.** $2(x + 1)$ **61.** $3(x^2 + 1)$ **62.** $5(x^3 + 1)$

63. $4(x^3 + 2)$ **64.** $3(x^6 + 3)$ **65.** $7(x^5 + 4)$ **66.** $6(x^6 + 8)$

67. $8(x^2 - 5)$ **68.** $5(x^3 - 6)$ **69.** $x^2(x^2 + 1)$ **70.** $x^3(x^3 + 1)$

71. $x^3(x^4 + 2)$ **72.** $x^4(x^3 + 2)$ **73.** $x^4(x^6 + 5)$ **74.** $x^6(x^2 + 3)$

75. $x^2(x + 1)$ **76.** $x^3(x + 1)$ **77.** $x(x^2 - 2)$ **78.** $x(x^3 - 2)$

 CALCULATOR EXERCISES

To simplify expressions with large coefficients, the calculator can be a useful tool to add, subtract, or multiply the numerical parts of the expression.

79. $(43x^2)(85x^4)(23x^5)$ **80.** $(33x^4)(12x^7)(44x^3)$ **81.** $125x + 245x$ **82.** $321y - 188y$

 VERBAL AND WRITING SKILLS

Do each exercise first using the calculator, then without using the calculator.

83. $(3x^3)(2x^4)(2x^3)$ Which way was easier? Why?   **84.** $(4x^4)(2x^7)(3x^3)$ Which way was easier? Why?

ONE STEP FURTHER

Perform the operation indicated.

85. (a) $4cd + 9cd$ **(b)** $(9cd)(4cd)$ **86. (a)** $7xz + 2xz$ **(b)** $(7xz)(2xz)$

87. (a) $9ab^5 - 7ab^5$ **(b)** $(9ab^5)(7ab^5)$ **88. (a)** $6y^7z - 3y^7z$ **(b)** $(6y^7z)(3y^7z)$

CUMULATIVE REVIEW

1. Simplify.
    (a) $2 + (x + 7)$ **(b)** $2(x \cdot 7)$

2. Subtract: $700 - 18$.

3. Evaluate: $6 + 2(7)$.

S E C T I O N 2 . 5

Understanding Division of Expressions

After studying this section, you will be able to:

❶ Understand the meaning of division.

❷ Identify the key words indicating division.

❸ Form equivalent multiplication problems.

❹ Use the order of operations with division.

❺ Perform long division.

❻ Solve applied problems involving division.

Math Pro   Video 2.5   SSM

When is division necessary to solve real-life problems? How do I divide whole numbers? Both these questions are answered in this section. It is just as important to know when a situation requires division as it is to know how to divide. Even if we use a calculator, we must know when the situation requires us to divide.

❶ Understanding the Meaning of Division

Contestants in the annual Rose Festival are allowed to enter 12 roses in the competition for the "Prize Rose." Martha would like to display her 12 roses in bouquets of 3. To determine the number of bouquets Martha can make, we can count out 12 roses and repeatedly take out sets of 3.

$$12 - 3 = 9 \qquad 9 - 3 = 6$$
$$\downarrow \qquad\qquad \downarrow$$
$$\text{1 bouquet} \qquad \text{1 bouquet}$$

$$6 - 3 = 3 \qquad 3 - 3 = 0$$
$$\downarrow \qquad\qquad \downarrow$$
$$\text{1 bouquet} \qquad \text{1 bouquet}$$

4 bouquets can be made.

By repeatedly subtracting 3, we found how many groups of 3 are in 12. In mathematics we express this as division:

"12 divided by 3 equals 4."

The symbols used for division are ⌐ or ÷ or / or ——. To write "12 divided by 3 equals 4" using symbols, we have

$$3\overline{\smash{\big)}12}^{\;4} \; , \qquad 12 \div 3 = 4, \qquad 12/3 = 4, \qquad \frac{12}{3} = 4$$

E X A M P L E 1 Write the division that corresponds to the following situation. You need not carry out the division. 180 chairs in an auditorium are arranged so that there are 12 chairs in each row. How many rows of chairs are there?

SOLUTION It is helpful to draw a picture:

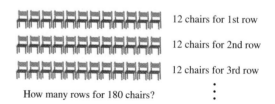

12 chairs for 1st row

12 chairs for 2nd row

12 chairs for 3rd row

How many rows for 180 chairs?

We want to know *how many groups of 12* are in 180. The division that corresponds to this situation is **180 ÷ 12**.

PRACTICE PROBLEM 1

Write the division that corresponds to the following situation. You need not carry out the division. John has $150 to spend on paint that costs $15 per gallon. How many gallons of paint can John purchase?

We also divide when we want to split an amount equally into a certain number of parts . For example, if we split the 12 roses into 3 equal groups, how many roses would be in each group? There would be 4 roses in each group.

The division that represents this situation is

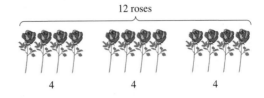

    12 divided by 3 equals 4 or 12 ÷ 3 = 4

EXAMPLE 2   Write the division that corresponds to the following situation. You need not carry out the division. 120 students in a band are marching in 8 rows. How many students are in each row?

SOLUTION   We draw a picture.

We want to *split 120 into 8 equal groups*. The division that corresponds to this situation is **120 ÷ 8**.

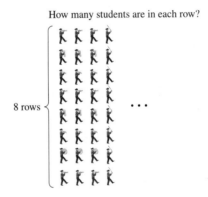

PRACTICE PROBLEM 2

Write the division that corresponds to the following situation. You need not carry out the division. Rita would like to donate $170 to 5 charities, giving each charity an equal amount of money. How much money will each charity receive?

❷ Identifying the Key Words Indicating Division

When referring to division we sometimes use the words **quotient**, **divisor**, and **dividend** to identify the three parts.

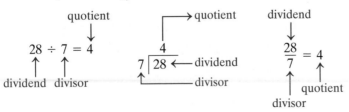

There are also several phrases to describe division. The following table gives some English phrases and their mathematical equivalents.

ENGLISH PHRASE	TRANSLATION INTO SYMBOLS
n divided by 6	$n \div 6$
The *quotient* of 7 and 35	$7 \div 35$
The *quotient* of 35 and 7	$35 \div 7$
15 items *divided equally* among 5 groups	$15 \div 5$
15 items *shared equally* among 5 groups	$15 \div 5$

E X A M P L E 3   Translate each phrase using symbols.

(a) 7 divided by *n*   **(b)** 55 items shared equally among *n* people
(c) The quotient of 46 and 2   **(d)** The quotient of 2 and 46

SOLUTION

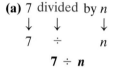

(a) 7 divided by *n*
$$7 \div n$$

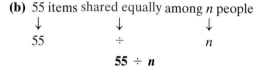

(b) 55 items shared equally among *n* people
$$55 \div n$$

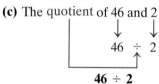

(c) The quotient of 46 and 2
$$46 \div 2$$

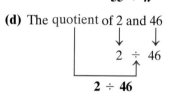

(d) The quotient of 2 and 46
$$2 \div 46$$

P R A C T I C E P R O B L E M 3

Translate each phrase using symbols.

(a) A number divided by 15   **(b)** 42 items divided equally among *n* groups
(c) The quotient of 72 and 8   **(d)** The quotient of 8 and 72

TO THINK ABOUT
••••••••••••••••••••••••••

THE COMMUTATIVE PROPERTY AND DIVISION

Example 3 illustrates that the order in which we write the numbers in the division is different when we use the phrases

"The quotient of **46** and **2**"   and   "The quotient of **2** and **46**"
$$46 \div 2 \qquad\qquad 2 \div 46$$

It is important to write these numbers in the correct order because *division is not commutative*, as illustrated below.

The division: $2 \div 46$

The situation: \$2 divided equally among 46 people

The division: $46 \div 2$

The situation: \$46 divided equally between 2 people

We can see that these are **not** the same situations, thus in general, *division is not commutative*: $2 \div 46 \neq 46 \div 2$.

EXERCISE

1. Can you think of one case where division is commutative?

❸ Forming Equivalent Multiplication Problems

To divide, we use multiplication facts. We illustrate this relationship between multiplication and division using the following array. Earlier we saw that the number of items in an array is equal to the number of rows $\times$ the number of columns. We can also use the same array to find how many groups of 3 are in 6.

3 columns

? rows   6 items

$6 = $ number of rows $\times$ number of columns

$$6 = ? \times 3 \qquad ? = 2$$

How many groups of 3 are in 6?

$$6 \div 3 = ? \qquad ? = 2$$

Therefore, to divide $6 \div 3 = ?$, we can use multiplication facts by writing the equivalent multiplication problem $6 = ? \times 3$ and thinking "what number times 3 equals 6?"

> To write an equivalent multiplication problem, we write the division as multiplication on the opposite side of the equal sign.

EXAMPLE 4   Write the equivalent multiplication problem, then use multiplication facts to complete the division: $\dfrac{48}{6} = ?$

SOLUTION   Recall that $\dfrac{48}{6}$ represents the division $48 \div 6$.

$\dfrac{48}{6} = ?$ ⌐   Division problem: $48 \div 6$.

$48 = ? \cdot 6$ Multiplication problem.

$48 = ? \cdot 6$ What number times 6 equals 48? 8
$48 = 8 \cdot 6$

Thus $\dfrac{48}{6} = 8$.

PRACTICE PROBLEM 4

Write the equivalent multiplication problem, then use multiplication facts to complete the division: $\dfrac{56}{7} = ?$

❹ Using the Order of Operations with Division

Earlier we saw that division is not commutative—that is, in general, $a \div b$ is not the same as $b \div a$. Does the associative property apply to division? Can we regroup numbers when we divide? Let's see if $(16 \div 4) \div 2 = 16 \div (4 \div 2)$.

We must do the operation inside the parentheses first.

$$(16 \div 4) \div 2 = \boxed{4} \div 2 \qquad 16 \div (4 \div 2) = 16 \div \boxed{2}$$
$$= 2 \qquad\qquad\qquad\qquad = 8$$

In general, *division is not associative.* Can you think of one case where division is associative?

In Section 2.3 we learned about the correct *order of operations.* We can now add division to this list of priorities. Division has the same priority as multiplication; we multiply or divide working from left to right.

ORDER OF OPERATIONS
Follow this order of operations:

Do first. 1. Perform operations inside *parentheses.*
   ↑   2. Simplify any expressions with *exponents.*
   ↓   3. *Multiply or divide* from left to right.
Do last.   4. *Add or subtract* from left to right.

As we stated earlier, it is easier to follow the order of operations if we keep our work neat and organized, perform one operation at a time, and follow the sequence *identify, calculate, replace.*

EXAMPLE 5   Evaluate: $\dfrac{6 + 6 \div 3}{5 - 1}$.

SOLUTION   Although we usually do not write parentheses in the numerator and denominator, it is understood they exist. We *must include parentheses* before we evaluate.

$$(6 + \boxed{6 \div 3}) \div (\boxed{5 - 1})$$

We perform operations inside each parentheses first.

$$(6 + 2) \quad \div \quad 4 \qquad 6 \div 3 = 2; \; 5 - 1 = 4.$$

$$8 \quad \div \quad 4 = 2 \qquad \text{Divide.}$$

PRACTICE PROBLEM 5

Evaluate: $\dfrac{4 + 8 \div 2}{7 - 3}$.

EXAMPLE 6   Evaluate: $5(x + 2) + 10 \div 2$.

SOLUTION   Since we cannot add x and 2, the expression inside the parentheses is simplified as much as possible. Thus we must move on to the next priority, which is multiplication: $5 \cdot (x + 2)$.

5(x + 2) + 10 ÷ 2   We use the distributive property to multiply.

= 5(x + 2) + 10 ÷ 2

= 5 · x + 5 · 2 + 10 ÷ 2   We multiply: 5 times x plus 5 times 2.
= 5x + 10 + 10 ÷ 2   We identify division as the next priority.
= 5x + 10 + 5   Calculate: 10 ÷ 2 = 5
= **5x + 15**

PRACTICE PROBLEM 6

Evaluate: $2(x + 7) + 15 ÷ 3$.

⑤ Performing Long Division

Suppose that we want to split 17 items equally between 2 people.

| 8 items | 8 items | 1 item |

Each person would get 8 items with 1 left over. We call this 1 the **remainder** (R) and write

$17 ÷ 2 = 8\,R1$

We use multiplication facts and the division symbol ⌐ when division involves large numbers, or remainders. For example,

Division problem: $17 ÷ 2 = ?$

$$2\overline{)17}^{\,?}$$

Multiplication problem: $17 = ? · 2$

$$2\overline{)17}^{\,?} \qquad 2 · ? = 17$$

$17 = \mathbf{8} · 2 + R$

$$2\overline{)17}^{\,\mathbf{8R}} \qquad \begin{array}{l} 2 · 8 = 16, \text{ which is close to} \\ 17, \text{ so we have a remainder} \end{array}$$

$17 = \mathbf{8} · 2 + 1$

$$\begin{array}{r} \mathbf{8R1} \\ 2\overline{)17} \\ \underline{-16} \\ 1 \end{array}$$

Thus to divide, we *guess* the quotient and *check* by multiplying the quotient × the divisor. If the guess is too large or too small, we *adjust* it and continue the process until we get a remainder that is less than the divisor.

E X A M P L E   7   Divide and check your answer: $38 \div 6$.

SOLUTION   We *guess* that 6×6 is close to 38.

$$
\begin{array}{r}
6 \\
6\overline{)38} \\
-36
\end{array}
$$

Our guess, 6, is placed here.

Check: $6 \times 6 = 36$;   36 *must be less than* 38. ✓

Since $36 < 38$, we do not need to *adjust* our guess to a smaller number.

$$
\begin{array}{r}
6\text{R}2 \\
6\overline{)38} \\
-36 \\
2
\end{array}
$$

We subtract: $38 - 36 = 2$.

Check: 2 *must be less than* 6. ✓ We write R2 in the quotient.

Since $2 < 6$, we do not need to *adjust* our guess to a larger number.

To verify that this is correct, we multiply the divisor times the quotient, then add the remainder:

$$
\begin{array}{r}
6\ R2 \\
6\overline{)38}
\end{array}
\qquad
\begin{array}{l}
\text{Multiply } 6 \cdot 6 = 36 \\
 + \underline{\ 2\ } \quad \text{Then add remainder.} \\
 38
\end{array}
$$

$$38 = 38 \ ✓$$

$$\mathbf{38 \div 6 = 6\ R2}$$

P R A C T I C E   P R O B L E M   7

Divide and check your answer: $43 \div 6$.

Let's see what we do if our guess is either too large or too small.

E X A M P L E   8   Divide and check your answer: $293 \div 41$.

SOLUTION

First guess (too large):

$$
\begin{array}{r}
8 \\
41\overline{)293} \\
-328
\end{array}
$$

Guess: 41 times what number is close to 293? **8**. Write 8 in the quotient.

Check: $41(8) = 328$; 328 *is not less than* 293, so we must *adjust*; our guess is too large.

Second guess (too small):

$$
\begin{array}{r}
6 \\
41\overline{)293} \\
-246 \\
47
\end{array}
$$

Guess: Try 6

Check: $41(6) = 246$; 246 *is less than* 293. ✓

47 *is not less than* 41, so we must *adjust*; our guess is too small.

Third guess:

$$
\begin{array}{r}
7\ \text{R}6 \\
41\overline{)293} \\
-287 \\
6
\end{array}
$$

Guess: Try 7.

Check: $41(7) = 287$; 287 *is less than* 293. ✓

6 *is less than* 41. ✓ We *do not* need to *adjust* our guess, and 6 is the remainder. Write R6 in the quotient.

We verify that the answer is correct: divisor $\cdot$ quotient $+$ remainder

$$41 \quad \cdot \quad 7 \quad + \quad 6 \quad = 293$$

$$\mathbf{293 \div 41 = 7\ R6}$$

P R A C T I C E   P R O B L E M   8

Divide and check your answer: $354 \div 36$.

EXAMPLE 9   Divide and check your answer: $70\overline{)3672}$.

SOLUTION   Accurate guesses can shorten the division process. If we consider only the *first digit of the divisor* and the *first two digits of the dividend*, it is easier to get accurate guesses.

First set of steps:

$$
\begin{array}{r}
5 \\
70\overline{)3672} \\
-350 \\
\hline
17
\end{array}
$$

Guess: We look at 7 and 36 to make our guess.
7 times what number is close to 36? **5.**
Check: 5(70) = 350; 350 *is less than* 367. ✓

17 *is less than* 70. ✓   We *do not adjust* our guess.

Second set of steps: We bring down the next number in the dividend: 2. Then we continue the guess, check, and adjust process until there are no more numbers in the dividend to bring down.

$$
\begin{array}{r}
52 \text{ R32} \\
70\overline{)3672} \\
-350\downarrow \\
\hline
172 \\
-140 \\
\hline
32
\end{array}
$$

Guess: We look at 7 and 17 to make our guess. Try **2.**

Check: 2(70) = 140; 140 *is less than* 172. ✓

32 *is less than* 70. ✓

Remainder: 32 is the remainder because there are no more numbers to bring down.

3672 ÷ 70 = 52 R32

$$\text{divisor} \cdot \text{quotient} + \text{remainder}$$

Check: 70   ·   52   +   32   = 3672. ✓

PRACTICE PROBLEM 9

Divide and check your answer: $80\overline{)2611}$.

EXAMPLE 10   Divide and check your answer: 33,897 ÷ 56.

SOLUTION

First set of steps:

$$
\begin{array}{r}
60 \\
56\overline{)33897} \\
-336\downarrow \\
\hline
29
\end{array}
$$

Guess: We look at 5 and 33 to make our guess. Try 6.
Check: 6(56) = 336; 336 *is less than* 338. ✓

2 *is less than* 56. ✓
Bring down the 9. Since 56 cannot be divided into 29, we write 0 in the quotient.

Second set of steps: We bring down the 7.

$$
\begin{array}{r}
605 \text{ R17} \\
56\overline{)33897} \\
-336\downarrow \\
\hline
297 \\
-280 \\
\hline
17
\end{array}
$$

Guess: We look at 5 and 29 to make our guess. Try 5.
Check: 5(56) = 280; 280 *is less than* 297. ✓
17 *is less than* 56. ✓

Remainder: 17 is the remainder because there are no more numbers to bring down.

CALCULATOR

You can use your calculator to divide. To find 33,897 ÷ 56, enter:

Scientific calculator:

    33897 ÷ 56 =

Graphing calculator:

    33897 ÷ 56 ENT

The calculator displays

    605.30357

The last few digits may differ.

Note: The answer is in decimal form, which we will discuss later in the book. For now, it is best to use the calculator to check your answer:

    56 × 605 + 17 = *or* ENT

The calculator displays

    33897

$$33{,}897 \div 56 = 605 \text{ R}17$$

$$\text{Check:} \quad \underset{\text{divisor}}{56} \cdot \underset{\text{quotient}}{605} + \underset{\text{remainder}}{17} = 33{,}897. \checkmark$$

PRACTICE PROBLEM 10

Divide and check your answer: $14{,}911 \div 37$.

⑥ Solving Applied Problems Involving Division

As we have seen, there are various key words, phrases, and situations that indicate when we perform the operation division. Knowing these can help us solve real-life applications.

EXAMPLE 11   Twenty-six students in Ellis High School entered their class project in a contest sponsored by the Falls City Baseball Association. The class won first place and received 250 tickets to the baseball play-offs. The teacher gave each student in the class an equal number of tickets, then donated the extra tickets to a local boys and girls club. How many tickets were donated to the boys and girls club?

SOLUTION   Since we must split 250 equally among 26 students, we divide

$$\begin{array}{r} 9 \text{ R}16 \\ 26 \overline{\smash{\big)}\ 250} \\ \underline{234} \\ 16 \end{array}$$

Since there are 16 tickets left over, **16 tickets** are donated to the boys and girls club.

PRACTICE PROBLEM 11

Twenty-two players on a recreational basketball team won second place in a tournament sponsored by Meris and Mann 3DMax Movie Theater. The team won 100 movie passes and divided these passes equally among players on the team. The extra tickets were donated to a local children's home. How many tickets were donated to the children's home?

EXERCISES 2.5

Write the division that corresponds to the situation. You need not carry out the division.

1. 220 paintings are arranged for display so that 20 paintings are in each row. How many paintings are in each row?

2. In the school gym, 320 chairs must be arranged with 16 chairs in each row. How many rows of chairs are there?

3. Ellen would like to split 66 pieces of candy equally among 22 students. How much candy should she give to each student?

4. John would like to split 32 carrots among 4 horses. How many carrots will each horse receive?

5. A dinner bill totaling n was split among 5 people?

6. 225 tickets to the Dodgers' first game of the year will be distributed equally among n people.

Translate each phrase using symbols.

7. 27 divided by x

8. 36 divided by a

9. Forty-two dollars divided equally among 6 people

10. Sixty-three jelly beans divided equally among 3 children

11. The quotient of 36 and 3

12. The quotient of 44 and 11

13. The quotient of 3 and 36

14. The quotient of 11 and 44

For each division problem:
    (a) *Write the equivalent multiplication problem.*
    (b) *Use multiplication facts to complete the division.*

15. $45 \div 5 = ?$

16. $32 \div 4 = ?$

17. $\dfrac{27}{3} = ?$

18. $\dfrac{30}{6} = ?$

Evaluate.

19. (a) $(32 \div 4) \div 2$ **(b)** $32 \div (4 \div 2)$

20. (a) $(48 \div 6) \div 2$ **(b)** $48 \div (6 \div 2)$

21. (a) $(16 \div 4) \div 2$ **(b)** $16 \div (4 \div 2)$

22. (a) $(64 \div 8) \div 4$ **(b)** $64 \div (8 \div 4)$

Evaluate.

23. $3 \times 12 \div 4 + 2$

24. $2 \times 15 \div 5 + 10$

25. $40 \div 5 \times 2 + 3^2$

26. $6^2 \div 6 \times 2 + 1$

27. $2^2 + 8 \div 4$

28. $3^3 + 6 \div 3$

29. $\dfrac{4 + 4 \div 2}{5 - 3}$

30. $\dfrac{5 + 15 \div 5}{9 - 5}$

31. $\dfrac{12 - 2}{25 \div 5 \times 2}$

32. $\dfrac{16 - 4}{36 \div 6 \times 2}$

33. $3(x + 6) + 8 \div 2$

34. $4(x + 5) + 6 \div 3$

35. $2(x + 8) + 4 \div 4$

36. $5(x + 7) + 10 \div 2$

37. $2(x + 1) + 12 \div 2$

38. $3(x + 1) + 15 \div 5$

Divide and check your answer.

39. $50 \div 6$

40. $60 \div 9$

41. $7\overline{)2597}$

42. $5\overline{)3105}$

43. $4\overline{)2097}$

44. $6\overline{)4046}$

45. $1268 \div 3$

46. $863 \div 2$

47. $30\overline{)632}$

48. $20\overline{)783}$

49. $19\overline{)5817}$

50. $32\overline{)6436}$

51. $2093 \div 41$

52. $1301 \div 24$

53. $1369 \div 19$

54. $1350 \div 16$

55. $18,985 \div 27$

56. $10,923 \div 42$

57. $11,571 \div 34$

58. $21,945 \div 29$

59. $\dfrac{13,317}{23}$

60. $\dfrac{24,624}{36}$

61. $\dfrac{70,141}{136}$

62. $\dfrac{43,317}{117}$

Solve each applied problem.

63. The fourteen members of the Carver High School Chess Club team won first place in a tournament sponsored by the Carver Convention Center. The chess team won 60 tickets to the Worldwide Computer Conference. The team decided to divide the tickets equally among all 14 team members and to donate the extra tickets to the PTA. How many tickets were donated to the PTA?

64. The twenty-one members of the Laurel High School track team won first place in a tournament sponsored by the Laurel Recreation Center. The team won 75 tickets to the county fair. The team decided to divide the tickets equally among all 21 team members and to donate the extra tickets to the homeless shelter. How many tickets were donated to the shelter?

At the beginning of 1998, the U.S. population was close to 269 million. It was estimated that at the time there was 1 birth every 9 seconds, 1 death every 13 seconds, and a net gain of 1 person every 17 seconds. Use this information to answer questions 65 and 66. Round your answers to the nearest person.

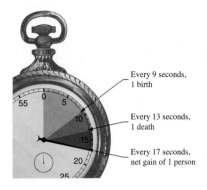

Every 9 seconds, 1 birth

Every 13 seconds, 1 death

Every 17 seconds, net gain of 1 person

65. **(a)** How many babies were born in 36 seconds?

   (b) How many babies were born in 1 hour? (Hint: 1 hour = 3600 sec)

   (c) What was the net gain of people in 1 hour?

66. **(a)** How many deaths occurred in 39 seconds?

   (b) How many deaths occurred in 1 hour?

   (c) What was the net gain of people in 2 hours?

67. The bill for dinner at Lido's Restaurant, including tip, was $85. If five people split the bill evenly, how much will each person have to pay?

68. The members of the Elks Club are planning a banquet. The cost of the entire banquet is $1071. If 63 members plan to attend, how much should be charged for each ticket to cover the cost of the banquet?

69. JoAnn received a travel allowance of $1050 from her employer for food and lodging. If her business trip is 6 days long, how much money should she budget herself to spend each day so that she will not go over her travel allowance?

70. A rancher plans to have 250 square feet of pasture for each cow on his field. If the field has 156,250 square feet, how many cows should the rancher allow on the field?

CALCULATOR EXERCISES

71. Due to a prolonged dry spell, Orange County, California homeowners watered their lawns more often than normal and used approximately 210 billion gallons of water in November 1997. This is enough water to fill the Rose Bowl in Pasadena more than 1400 times. Approximately how many gallons of water would fill the Rose Bowl?

Source: Orange County Register.

72. The approximate size of the earth's diameter is the sun's diameter divided by 109. If the sun's diameter is about 1,391,193 kilometers, what is the diameter of the earth? Round your answer to the nearest ten.

73. Harry reasoned that division is associative because $12 \div (12 \div 1) = (12 \div 12) \div 1$ What is wrong with his thinking?

74. The product of 2 times any whole number is always an even number. If $\dfrac{x}{2} = b$, where b is a whole number, what can be concluded about the value of the variable x? Why?

ONE STEP FURTHER

75. A photographer sets a telephoto lens so that he can be twice as far away as with a regular lens. He is taking pictures of deer that are 124 feet from his camera. How far would the photographer have to be to get the same shot with a regular lens?

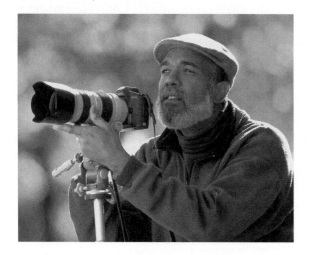

76. Janice is making a cross stitch pattern on 14-count material. This means that there are 14 squares to the inch. If Janice's pattern is 98 squares across, how many inches wide will it be?

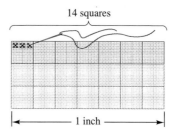

CUMULATIVE REVIEW

1. Translate: Seven plus x equals eleven.

2. Subtract: $1060 - 114$.

3. Use the formula $P = 2L + 2W$ to find the perimeter of a rectangle with length 8 feet and width 3 feet.

4. Round 556,432 to the nearest thousand.

Solving Basic Equations Involving Multiplication and Division

After studying this section, you will be able to:

1 Evaluate expressions involving multiplication and division.
2 Solve equations by inspection.
3 Solve equations involving division.
4 Solve equations involving multiplication.
5 Translate and solve equations involving multiplication and division.

1 Evaluating Expressions Involving Multiplication and Division

We can evaluate algebraic expressions such as $5y + 1$ by replacing the variable with the given value and simplifying. The value of the algebraic expression will depend on the value we use to replace the variable.

Math Pro   Video 2.6   SSM

E X A M P L E 1   Evaluate.

(a) $7n + 1$ if $n = 3$.

(b) $\dfrac{x}{4}$ if $x = 16$.

SOLUTION

(a) $7n + 1$
             $\downarrow$
             $7(3) + 1$ Replace n with 3.
         $= 21 + 1$
         $= 22$ Simplify.   **$7n + 1$ evaluated when $n = 3$ is 22.**

(b) $\dfrac{x}{4} = \dfrac{16}{4}$ Replace x with 16.

         $= 4$ Divide: $16 \div 4 = 4$. $\dfrac{x}{4}$ **evaluated when $x = 16$ is 4.**

P R A C T I C E   P R O B L E M   1
Evaluate.

(a) $9y - 2$ if $y = 4$.

(b) $\dfrac{x}{3}$ if $x = 12$.

2 Solving Equations by Inspection

When we solve the equation $5n = 35$, we answer the question

   "Five times what number is equal to 35?"

The answer to the question, and thus the solution to the equation, is 7, because 5 times 7 equals 35. For simple equations we can use multiplication or division facts to find the solution.

E X A M P L E 2   Solve the equation $9n = 45$ and check your answer.

SOLUTION   To solve the equation $9n = 45$, we answer this question:

   "Nine times what number equals forty-five?"

The answer or solution is $n = 5$.

To check the answer, we evaluate $9n = 45$ for $n = 5$ and verify that we get a true statement.

$$9n = 45$$
$$\downarrow$$
$$9(5) \overset{?}{=} 45 \qquad \text{Replace the variable with 5 and simplify.}$$
$$45 = 45 \checkmark \qquad \text{Verify that this is a true statement.}$$

PRACTICE PROBLEM 2

Solve the equation $6x = 48$ and check your answer.

EXAMPLE 3   Solve the equation $\dfrac{21}{x} = 3$ and check your answer.

SOLUTION

$$\frac{21}{x} = 3 \quad \text{``Twenty-one divided by what number equals 3?''}$$

The answer or solution is $n = 7$.
   Check:

$$\frac{21}{x} = 3 \rightarrow \frac{21}{7} \overset{?}{=} 3 \qquad \text{Replace the variable with 7 and simplify.}$$
$$3 = 3 \checkmark \qquad \text{Verify that this is a true statement.}$$

PRACTICE PROBLEM 3

Solve the equation $\dfrac{x}{4} = 2$ and check your answer.

❸   Solving Equations Involving Division

When equations contain large numbers or are too complicated to solve mentally, we can use various methods to help us find the solutions. For example, to solve an equation in the form $\dfrac{x}{a} = b$, we can write an equivalent multiplication problem, then use multiplication facts to find the solution.

> **Procedure to Solve an Equation in the Form $\dfrac{x}{a} = b$.**
>
> 1. Write the division as multiplication on the opposite side of the equal sign.
> 2. Use multiplication facts to find the solution.
>
>   We write $\dfrac{x}{a} = b$ as $x = b \cdot a$.

EXAMPLE 4   For the following equation, write an equivalent multiplication problem and solve.

$$\frac{a}{12} = 9$$

SOLUTION

$$\frac{a}{12} = 9 \qquad \text{"What number divided by 12 equals 9?"}$$

$$a = 9(12) \qquad \text{Write an equivalent multiplication problem.}$$

$$a = 108 \qquad \text{Use multiplication facts to find } a.$$

$$\text{Check:} \quad \frac{a}{12} = 9 \rightarrow \frac{108}{12} \overset{?}{=} 9$$

$$9 = 9 \checkmark$$

PRACTICE PROBLEM 4

For the following equation, write an equivalent multiplication problem and solve.

$$\frac{n}{7} = 52$$

We often have to simplify an equation before we can find the solution.

EXAMPLE 5 Solve: $\frac{x}{2^3} = 12 \cdot 5 + 1$.

SOLUTION   We simplify each side of the equation first, then we find the solution.

$$\frac{x}{2^3} = 12 \cdot 5 + 1$$

$$\frac{x}{8} = 60 + 1 \qquad \text{Simplify: } 2^3 = 8; 12 \cdot 5 = 60.$$

$$\frac{x}{8} = 61 \qquad 60 + 1 = 61.$$

$$x = 61 \cdot 8 \qquad \text{Write an equivalent multiplication problem.}$$

$$x = 488 \qquad \text{Multiply to find the solution.}$$

Note: We leave the check for the student.

PRACTICE PROBLEM 5

Solve: $\frac{x}{3^2} = 11 \cdot 4 + 9$.

4 Solving Equations Involving Multiplication

So far we have written division as multiplication to solve division equations: $\frac{10}{2} = 5 \rightarrow 10 = 5 \cdot 2$. Now, if we reverse the process, writing multiplication as division (undo what we just did), we can solve multiplication equations.

$$10 = 5 \cdot 2 \rightarrow \frac{10}{2} = 5$$

We *reverse* the process, writing multiplication as division on the opposite side of the equal sign.

We use this idea to solve equations involving multiplication as stated in the following rule.

> **Procedure to Solve an Equation of the Form** $ax = b$.
>
> 1. Write the multiplication on one side of the equal sign as division on the opposite side.
> 2. Divide to find the solution.
>
> We write $a \cdot x = b$ as $x = \dfrac{b}{a}$.

E X A M P L E 6 For the following equation, write an equivalent division problem and solve: $15x = 75$.

SOLUTION

$$15 \cdot x = 75 \qquad \text{``15 times what number equals 75?''}$$

$$x = \frac{75}{15} \qquad \text{Write an equivalent division problem.}$$

$$x = 5 \qquad \text{Divide to find the solution.}$$

We leave the check for the student.

P R A C T I C E   P R O B L E M   6

For the following equation, write an equivalent division problem and solve: $9x = 108$.

Remember, we must simplify each side of the equation first, then solve for the variable.

E X A M P L E 7 Simplify, then solve: $4(2n) = 96$. Check your solution.

SOLUTION

$$4(2n) = 96$$

$$(4 \cdot 2)n = 96 \qquad \text{Regroup the multiplication.}$$

$$8n = 96 \qquad \text{Simplify.}$$

$$n = \frac{96}{8} \qquad \text{Write the equivalent division.}$$

$$n = 12 \qquad \text{Divide to find the solution.}$$

$$\textit{Check:} \qquad 4(2n) = 96$$
$$\downarrow$$
$$4(2 \cdot 12) \overset{?}{=} 96$$
$$4(24) \overset{?}{=} 96$$
$$96 = 96 \; \checkmark \qquad \text{We get a true statement.}$$

We always check the solution in the original equation, $4(2n) = 96$, not the simplified form, $8n = 96$. Otherwise, we are not assured that our solution is correct, since we could have made an error in the simplification.

PRACTICE PROBLEM 7

Simplify, then solve: $2(3n) = 30$. Check your solution.

EXAMPLE 8   Simplify, then solve: $3(x \cdot 5) = \dfrac{450}{5}$.

SOLUTION

$$3(x \cdot 5) = \frac{450}{5}$$

$3(x \cdot 5) = 90$	Divide: $450 \div 5 = 90$.
$3(\mathbf{5x}) = 90$	Change the order of multiplication.
$(3 \cdot 5)x = 90$	Regroup the multiplication.
$15x = 90$	Simplify.
$x = \dfrac{90}{15}$	Write the equivalent division.
$x = \mathbf{6}$	Divide to find the solution.

We leave the check to the student.

PRACTICE PROBLEM 8

Simplify, then solve: $8(n \cdot 5) = \dfrac{320}{2}$.

⑤ Translating and Solving Equations Involving Multiplication and Division

Sometimes an equation is written using English statements and must be translated into an algebraic expression before solving.

EXAMPLE 9   Translate into an equation and solve: Double a number times three equals seventy-eight.

SOLUTION

Double a number times three equals seventy-eight.

$\downarrow$	$\downarrow$	$\downarrow$	$\downarrow$	$\downarrow$	
$(2x)$	$\cdot$	3	$=$	78	Translate.

$(2x) \cdot 3 = 78$	
$(x \cdot 2) \cdot 3 = 78$	Change the order of multiplication.
$x(2 \cdot 3) = 78$	Regroup the multiplication.
$x \cdot 6 = 78$	Simplify.
$x = \dfrac{78}{6}$	Write as division.
$x = \mathbf{13}$	Divide to find the solution.

PRACTICE PROBLEM 9

Translate into an equation and solve: The quotient of x and 35 is equal to 16.

..... Jurassic Park 2

TO THINK ABOUT

EVALUATE OR SOLVE?

Do you know the difference between evaluating the expression $8x$ for $x = 3$ and solving the equation $8x = 16$?

- *Evaluate an expression.* We **replace the variable** in the expression with the given number, then perform the calculation(s).

  Evaluate $8x$ if $x = 3$. $8 \cdot 3 = 24$.

- *Solve an equation.* We **find the value of the variable** in the equation, that is, the solution to the equation.

  Solve for x: $8x = 16$. $x = 2$.

  We can illustrate this idea with the following situations.

1. Evaluating
   a. *Fact.* You are given directions to the movie theater where *Jurassic Park 2* is playing.
   b. *Evaluate.* You follow these directions to the movie theater.
2. Solving
   a. *Fact.* You know the name of the theater where *Jurassic Park 2* is playing.
   b. *Solve.* You must find the directions yourself.

   In summary, an equation has an equal sign, and an expression does not. We find the solutions to equations, and we evaluate expressions as directed.

EXERCISE

1. Can you think of other real-life situations that illustrate the difference between evaluating and solving?

EXERCISES 2.6

1. Evaluate $5x$: **(a)** If x is 8. **(b)** If x is 3. **2.** Evaluate $9y$: **(a)** If y is 7. **(b)** If y is 9.

3. Evaluate $\dfrac{n}{6}$: **(a)** If n is 30. **(b)** If n is 18. **4.** Evaluate $\dfrac{x}{8}$: **(a)** If x is 40. **(b)** If x is 24.

5. Evaluate $6x + 1$: **(a)** If x is 3. **(b)** If x is 7. **6.** Evaluate $7n + 2$: **(a)** If n is 2. **(b)** If n is 7.

Solve and check your answer.

7. $8x = 16$ **8.** $7y = 14$ **9.** $4y = 12$ **10.** $9x = 63$

11. $5x = 20$ **12.** $8x = 56$ **13.** $10y = 30$ **14.** $7y = 28$

15. $\dfrac{30}{x} = 15$ **16.** $\dfrac{18}{x} = 9$ **17.** $\dfrac{14}{x} = 2$ **18.** $\dfrac{20}{x} = 2$

19. $\dfrac{28}{n} = 7$ **20.** $\dfrac{32}{n} = 8$

For the following equations:
   (a) *Write the equivalent multiplication problem.*
   (b) *Solve.*

21. $\dfrac{x}{10} = 6$ **22.** $\dfrac{x}{12} = 3$ **23.** $\dfrac{x}{8} = 9$ **24.** $\dfrac{x}{9} = 11$

25. $\dfrac{x}{6} = 7$ **26.** $\dfrac{x}{7} = 9$ **27.** $\dfrac{x}{33} = 15$ **28.** $\dfrac{x}{42} = 9$

29. $\dfrac{x}{3^2} = 4 \cdot 2$ **30.** $\dfrac{x}{2^2} = 5 \cdot 3$ **31.** $\dfrac{x}{4^2} = 7 + 2 \cdot 5$ **32.** $\dfrac{x}{2^3} = 3 + 5 \cdot 2$

141

For the following equations:
  (a) *Write the equivalent division problem.*
  (b) *Solve.*

33. $12x = 48$

34. $9x = 99$

35. $17n = 51$

36. $22n = 66$

37. $19x = 38$

38. $7x = 84$

39. $16x = 192$

40. $15x = 105$

41. $12n = 84$

42. $11n = 121$

Simplify, then solve.

43. $2(3x) = 54$

44. $4(2x) = 64$

45. $5(4x) = 40$

46. $2(2x) = 16$

47. $5(x \cdot 1) = 15$

48. $3(x \cdot 5) = 30$

49. $4(x \cdot 2) = 96$

50. $6(x \cdot 5) = 120$

51. $7(4 \cdot x) = 364$

52. $5(8 \cdot x) = 440$

53. $4(5 \cdot x) = \dfrac{260}{13}$

54. $7(2 \cdot x) = \dfrac{196}{14}$

55. **(a)** $12x = 60$ **(b)** $\dfrac{x}{11} = 5$

56. **(a)** $13x = 52$ **(b)** $\dfrac{x}{11} = 6$

57. **(a)** $8x = 96$ **(b)** $\dfrac{x}{13} = 3$

58. **(a)** $9x = 108$ **(b)** $\dfrac{x}{14} = 4$

59. **(a)** $7x = 84$ **(b)** $\dfrac{x}{13} = 6$

60. **(a)** $5x = 60$ **(b)** $\dfrac{x}{14} = 3$

142

Translate each statement into an equation and solve.

61. Nine times what number is equal to eighty-one?

62. The product of seven and what number is thirty-five?

63. The quotient of what number and eleven is equal to fifteen?

64. What number divided by sixteen is equal to seven?

65. The product of 9 and a number is 72.

66. What number times 7 equals 49?

67. Double what number times 5 equals thirty?

68. Triple a number times 4 equals thirty-six.

69. Triple a number times 2 equals twelve.

70. Double what number times 3 equals twenty-four?

71. (a) Evaluate $6x$ for $x = 2$.

(b) Solve $6x = 18$ for x.

72. (a) Evaluate $9x$ for $x = 3$.

(b) Solve $9x = 45$ for x.

73. (a) Evaluate $\dfrac{n}{2}$ for $n = 50$.

(b) Solve $\dfrac{n}{2} = 24$ for n.

74. (a) Evaluate $\dfrac{n}{8}$ for $n = 40$.

(b) Solve $\dfrac{n}{8} = 10$ for n.

 CALCULATOR EXERCISES

Solve and check your answer.

75. $42x = 1890$

76. $38x = 1938$

77. $\dfrac{x}{1200} = 16$

78. $\dfrac{x}{2440} = 7$

 VERBAL AND WRITING SKILLS

Translate the mathematical symbols using words.

79. $7x$

80. $3x$

81. $8x = 40$

82. $5x = 30$

Use the following bar graph to answer questions 83 and 84.

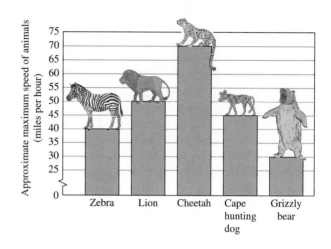

83. (a) How fast can a domestic cat run if it is as fast as a grizzly bear?

(b) The speed of a lion is two times the speed of an elephant. How fast is the elephant?

84. (a) Which animal is faster, a zebra or a Cape hunting dog?

(b) The speed of a cheetah is two times the speed of a rabbit. How fast is the rabbit?

A right triangle is a special kind of triangle in which one angle is 90°. This angle is identified using the symbol "⌐". The sum of the interior angles of a triangle is 180°.

85. Find the missing angle for the following right triangle.

86. Find the missing angle for the following right triangle.

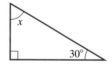

1. Add: $7016 + 923$.

2. Subtract: $2006 - 874$.

3. Add: 2 hr   35 min
       $+$3 hr   45 min
       6 hr   20 min

Geometric Shapes

After studying this section, you will be able to:

❶ Find the perimeter of a geometric shape.

❷ Find the area of a geometric shape.

❸ Find the volume and surface area of a geometric solid.

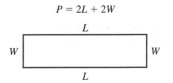

Math Pro   Video 2.7   SSM

❶ Finding the Perimeter of a Geometric Shape

Earlier we learned that to find the perimeter of a rectangle we can use the formula $P = L + L + W + W$ or $P = 2L + 2W$. Since a square is a special case of a rectangle in which all sides are equal, we find the perimeter using the formula $P = s + s + s + s$ or $P = 4s$.

> **PERIMETER**
> The perimeter (P) of a rectangle is twice the length plus twice the width.
>
> $$P = 2L + 2W$$
>
> The perimeter (P) of a square is four times the length of a side.
>
> $$P = 4s$$

E X A M P L E   1   Use the formula $P = 2L + 2W$ to find the perimeter of a rectangle with $L = 8$ feet and $W = 6$ feet.

SOLUTION

$$P = \ 2L \ \ + \ 2W$$
$$= 2(8 \text{ ft}) + 2(6 \text{ ft}) \qquad \text{Replace } L \text{ with 8 ft and } W \text{ with 6 ft.}$$
$$= \ 16 \text{ ft} \ + \ 12 \text{ ft}$$
$$= \ 28 \text{ ft} \qquad \text{The perimeter of the rectangle is } \textbf{28 feet}.$$

P R A C T I C E   P R O B L E M   1
Use the formula $P = 4s$ to find the perimeter of a square with sides 11 yards in length.

❷ Finding the Area of a Geometric Shape

Arrays can be used to illustrate the area of a rectangular region. Earlier in the chapter we learned that to find the number of items in an array we multiply the number of rows times the number of columns. Thus, the number of items in a 3 by 5 array is 3×5, or 15.

If each object in the array is a *square unit*, the area is equal to the number of square units needed to fill the rectangular region.

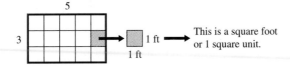

Each object in this 3×5 array is a square with sides 1 foot. The number of squares needed to fill the region is 3 times 5, or 15 square feet. Thus the **area** of the rectangular array is 15 square feet. We can check this by counting the number of squares in the array.

EXAMPLE 2 How many squares with 1-centimeter sides can be placed on a surface that has an area of 6 square centimeters? Why?

SOLUTION An area of 6 square centimeters describes a space that consists of **6 squares** with sides equal to 1 centimeter.

PRACTICE PROBLEM 2

How many squares with 1-foot sides can be placed on a surface that has an area of 25 square feet?

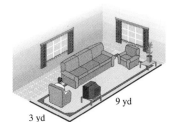

EXAMPLE 3 What is the area of the rug pictured in the margin?

SOLUTION Think of an array with 3 rows and 9 columns.

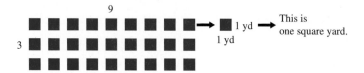

Just as we multiplied the number of rows times the number of columns to find the number of items in an array, we multiply the length times the width to find the area of the rug. The area is $3(9) = $ **27 square yards**.

PRACTICE PROBLEM 3

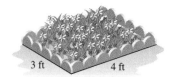

What is the area of the flower garden pictured in the margin?

The area of a rectangle is the product of the length times the width: $A = L \cdot W$, where A represents the value of the area, L the value of the length, and W the value of the width.

Since the length and width of a square are equal, we can use the formula $A = s \cdot s$ or $A = s^2$ to find the area of a square.

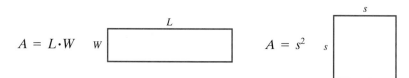

Parallelograms are figures that are related to rectangles. Actually, they are in the same "family," the **quadrilaterals** (four-sided figures).

> A **parallelogram** is a four-sided figure with both pairs of opposite sides parallel.

Parallel lines are two straight lines that are always the same distance apart. The opposite sides of a parallelogram are equal in length. The figures in the margin are parallelograms.

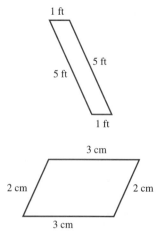

To find the area of a parallelogram, we multiply the base times the height. Any side can be considered the base. The height is the shortest distance between the base and the side opposite the base. The height is a line segment perpendicular to the base. When we write the formula for area we use the length of the base b and the height h.

$A = bh$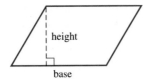

The formulas for area are summarized as follows:

> The area of a rectangle is the length times the width.
>
> $A = L \cdot W$
>
> The area of a square is the length of one side squared.
>
> $A = s^2$
>
> The area of a parallelogram is the base b times the height h.
>
> $A = bh$

If the length of a rectangle is 3 feet and the width 2 feet, how do we write the *units* when we multiply to find area? We perform calculations with units in the same way as variables.

$$A = L \cdot W$$
$$= 3 \text{ ft} \cdot 2 \text{ ft}$$
$$= (3 \cdot 2)(\text{ft} \cdot \text{ft}) \qquad \text{We multiply the units "ft": } ft \text{ times } ft = ft^2.$$
$$= 6 \text{ ft}^2 \qquad \text{This is read "six square feet," abbreviated as 6 ft}^2.$$

We see that the units for area can be expressed two ways:

1. Using exponents: ft^2 , yd^2 , in^2 , and so on.

2. Using abbreviations: sq ft ,   sq yd ,   sq in.

(square feet)   (square yards)   (square inches)

EXAMPLE 4   Find the area of a parallelogram with base $= 8$ meters and height $= 3$ meters.

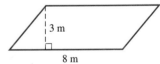

SOLUTION   The formula for the area of a parallelogram is $A = bh$.

$$A = b \times h$$
$$= (8 \text{ m})(3 \text{ m}) \qquad \text{Evaluate the formula for the values given.}$$
$$= (8 \cdot 3)(\text{m} \cdot \text{m}) \qquad \text{Write the multiplication for units and numbers separately.}$$
$$= 24 \text{ m}^2 \qquad \text{Multiply the units "m": m times m } = m^2.$$

The area of the parallelogram is **24 m²**.

PRACTICE PROBLEM 4

Find the area of a parallelogram with base $= 13$ feet and height $= 9$ feet.

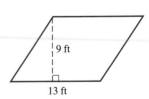

When we calculate area, perimeter, or solve any problem that deals with geometric figures and formulas, it is important that we know the definition of the figure and the correct formula associated with the problem.

EXAMPLE 5   Find the area of the following shape.

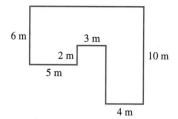

SOLUTION   Divide the figure into three rectangles, then find the area of each rectangle separately. Next, add these three areas together to find the area of the figure.

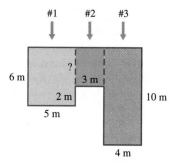

Area of rectangle 1: $A = L \cdot W \rightarrow A = 5\,\text{m} \cdot 6\,\text{m} = 30\,\text{m}^2$

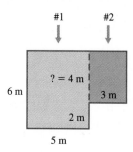

If the width of the left side of rectangle 1 is 6 m, then the width of the right side is 6 m. Thus the width of rectangle 2 is 4 m, since $2 + 4 = 6$.

Area of rectangle 2: $A = L \cdot W \rightarrow A = 3\,\text{m} \cdot 4\,\text{m} = 12\,\text{m}^2$

Area of rectangle 3: $A = L \cdot W \rightarrow A = 4\,\text{m} \cdot 10\,\text{m} = 40\,\text{m}^2$

$$30\,\text{m}^2 + 12\,\text{m}^2 + 40\,\text{m}^2 = 82\,\text{m}^2$$

The area of the shape is **82 m²**.

PRACTICE PROBLEM 5

Find the area of the following shape.

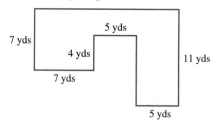

EXAMPLE 6   The blueprints for an office building show that the length of the entryway is 9 feet and the width is 6 feet. The tile for the entryway costs $16 per square yard.

(a) How many square yards of tile must be purchased for the entryway?
(b) How much will the tile for the entryway cost?

SOLUTION

(a) First we draw a 6-foot by 9-foot rectangle. Then we convert feet to yards, since the tile is sold in square yards: 3 feet = 1 yard.

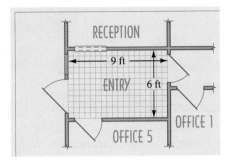

$9 \text{ ft} = 3 \text{ ft} + 3 \text{ ft} + 3 \text{ ft}$ $6 \text{ ft} = 3 \text{ ft} + 3 \text{ ft}$

$\phantom{9 \text{ ft}} = 3 \times \boxed{3 \text{ ft}}$ $\phantom{6 \text{ ft}} = 2 \times \boxed{3 \text{ ft}}$

$\phantom{9 \text{ ft} = 3 \times } \downarrow$ $\phantom{6 \text{ ft} = 2 \times } \downarrow$

$\phantom{9 \text{ ft}} = 3 \times \boxed{1 \text{ yd}}$ $\phantom{6 \text{ ft}} = 2 \times \boxed{1 \text{ yd}}$

$\phantom{9 \text{ ft}} = 3 \text{ yd}$ $\phantom{6 \text{ ft}} = 2 \text{ yd}$

We relabel the figure and find the area in square yards.

6 ft = 2 yd

9 ft = 3 yd

$A = L \times W$
$ \downarrow \downarrow$
$ = 3 \text{ yd} \times 2 \text{ yd}$
$ = 3 \times 2 \times \text{yd} \times \text{yd}$
$ = 6 \text{ yd}^2$

6 yd^2 of tile must be purchased.

(b) The tile sells for $16 per square yard and 6 square yards must be purchased: $16 \times 6 = \$96$. The tile will cost **$96**.

PRACTICE PROBLEM 6

Jesse is purchasing carpet for his living room. The measurements he brought to the store are in feet, $L = 12$ feet and $W = 15$ feet. The carpet he plans to purchase is priced at $11 per square yard.

(a) How many square yards of carpet must be purchased for the living room?
(b) How much will the carpet for the living room cost?

③ Finding the Volume and Surface Area of a Geometric Solid

How much water can a pool hold? How much air is inside a box? These are questions of *volume*. We use volume to measure the space enclosed by solid geometric figures that have three dimensions. We call the three dimensions length (L), width (W), and height (H).

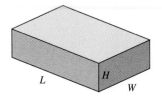

Recall that the area of a rectangular region is the number of square units needed to fill the rectangular region. The volume of a rectangular solid is the number of unit cubes needed to fill the figure completely. A unit cube is a rectangular solid with all the edges 1 unit in length.

Unit cube:

 1 cubic centimeter (cm³)

Volume of a rectangular solid:

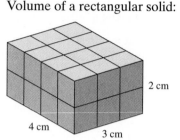

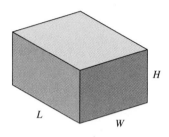

The first layer of this rectangular solid can be viewed as a 3 by 4 array (3 rows of cubes with 4 cubes in each row). There are 3(4) or 12 cubes in the first layer. Since we have two layers, there are 3(4)(2) cubes or 24 cubes in the solid figure. Therefore, the volume of the figure is 24 cubic centimeters (cm³).

As you can see, to find the number of cubes in the solid figure, we can either count the cubes needed to fill the figure, or multiply the length times the width times the height.

> **VOLUME**
> The **volume of a rectangular solid** is the product of the length times the width times the height.
>
> $V = LWH$

EXAMPLE 7   A swimming pool with uniform depth is 25 feet long and 11 feet wide. If the pool is filled to a depth of 6 feet, how many cubic feet of water are in the pool?

SOLUTION   We must find the volume of water in the pool. The formula is

$$V = L \times W \times H$$
$$= 25 \text{ ft} \times 11 \text{ ft} \times 6 \text{ ft}$$
$$= (25 \times 11 \times 6)(\text{ft} \times \text{ft} \times \text{ft})$$
$$= 1650 \text{ ft}^3$$

There are **1650 ft³** of water in the pool.

PRACTICE PROBLEM 7

A contractor's crew must dig a hole and haul away dirt in a space 25 feet wide, 35 feet long, and 8 feet deep. How much dirt will they need to haul away?

Surface Area.   To find the surface area of a rectangular solid, we add the areas of each of the six surfaces. For example, to find the surface area of a box, we add the areas of each of the 6 sides.

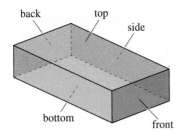

The *back and front faces* have the same dimensions.

Both *sides* have the same dimensions.

The *top and bottom bases* have the same dimensions.

We use the surface area when we want to calculate how much material we need to make the rectangular solid (box).

EXAMPLE 8   How much cardboard do we need to make the box illustrated? Assume the box has a top and bottom.

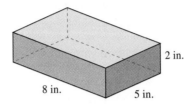

SOLUTION   We must find the surface area to determine how much cardboard we need. Since opposite surfaces of the box have the same size, we only need to find one of the areas—then double it.

$$\begin{aligned}
\text{Surface area} &= \text{side} + \text{side} \; + \; \text{front} + \text{back} \; + \; \text{top} + \text{bottom} \\
&\quad\;\; \boxed{2 \cdot \text{side area}} \; + \; \boxed{2 \cdot \text{face area}} \; + \; \boxed{2 \cdot \text{base area}} \\
&\quad\;\; 2(8 \text{ in.} \cdot 2 \text{ in.}) \; + \; 2(5 \text{ in.} \cdot 2 \text{ in.}) \; + \; 2(8 \text{ in.} \cdot 5 \text{ in.}) \\
&\quad\;\; 2(16 \text{ in}^2) \qquad + \quad 2(10 \text{ in}^2) \qquad + \quad 2(40 \text{ in}^2) \\
&= 32 \text{ in}^2 \qquad + \quad 20 \text{ in}^2 \qquad + \quad 80 \text{ in}^2 \qquad = 132 \text{ in}^2
\end{aligned}$$

We need **132 in^2** of cardboard.

PRACTICE PROBLEM 8

How much metal material do we need to make the box illustrated? Assume the box has a top and bottom.

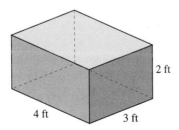

EXERCISES 2.7

Use the formula P = 2L + 2W to find the perimeter of a rectangle with:

1. $L = 2$ feet and $W = 7$ feet

2. $L = 18$ feet and $W = 25$ feet

Use the formula P = 4s to find the perimeter of a square with sides of length:

3. 8 feet

4. 38 inches

5. 54 yards

6. 28 centimeters

Use the appropriate formula to find the perimeter.

7. A square with sides 9 centimeters in length.

8. A rectangle of length 4 feet and width 2 feet.

9. A rectangle of length 12 yards and width 8 yards.

10. A square with sides 78 feet in length.

11. The perimeter of a preschool play yard is 70 yards. State the perimeter in feet (3 ft = 1 yd).

12. A table has a perimeter of 20 feet. State the perimeter in inches (12 in. = 1 ft).

13. A square patio has a perimeter of 16 meters. State the perimeter in decimeters. (Recall from Chapter 1 that 1 m = 10 dm.)

14. A square tile has a perimeter of 8 decimeters. State the perimeter in centimeters. (Recall from Chapter 1 that 1 dm = 10 cm.)

Find the area of:

15. A driveway.

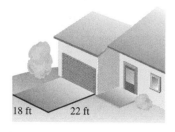

18 ft 22 ft

16. An Oriental rug.

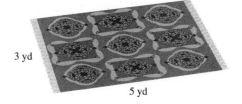

3 yd

5 yd

17. How many squares with 1-inch sides can be placed in a space that is 50 square inches?

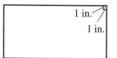

1 in.
1 in.

18. How many squares with 1-foot sides can be placed in a space that is 22 square feet.

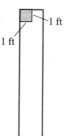

1 ft

1 ft

19. Squares with sides 1 yard in length are placed in a space that is 40 square yards. How many of the 1-yard squares are needed to fill the space?

20. How many squares of linoleum floor tiles with sides 1 foot in length would be needed to cover a floor space that is 60 square feet.

Use the appropriate formula to find the area.

21. A square with sides of 8 inches.

22. A rectangle of length 27 yards and width 21 yards.

23. A parallelogram with a base of 12 feet and a height of 9 feet.

24. A square with sides of 10 meters.

25. A parallelogram with a base of 87 meters and a height of 57 meters.

26. A parallelogram with a base of 92 inches and a height of 74 inches.

Find the area of each shape.

27.

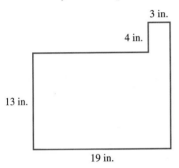

28.

29.

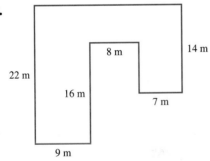

30.

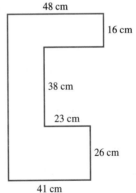

Answer each applied problem.

31. Damien has a 2-foot-square piece of marble.
   (a) State the dimensions of the marble in inches.

   (b) State the area of the marble in square inches.

32. Marcy has a 3-foot-square card table.
   (a) State the dimensions of the table in inches.

   (b) State the area of the table in square inches.

33. A rectangular vegetable garden has a length of 4 yards and a width of 3 yards.

(a) State the dimensions of the garden in feet.

(b) State the area of the garden in square feet.

34. A rectangular room has a length of 6 yards and a width of 4 yards.

(a) State the dimensions of the room in feet.

(b) State the area of the room in square feet.

35. Carmen is purchasing outdoor carpet for her patio enclosure. The measurements of the patio that she brought to the store are in feet, $L = 12$ feet and $W = 9$ feet. The carpet she plans to purchase is priced at $8 per square yard.

(a) How many square yards of carpet must she purchase?

(b) How much will the outdoor carpet for Carmen's patio cost?

36. The blueprints for a warehouse show that the length of the main office is 15 feet and the width is 12 feet. The carpet for the office costs $11 per square yard.

(a) How many square yards of carpet must be purchased for the office?

(b) How much will the office carpet cost?

37. A 1-pound container of rose fertilizer sells for $3. Each 1-pound container will fertilize 100 square feet. If the length of the rose garden is 5 yards and the width is 4 yards, how much will it cost to fertilize the entire rose garden?

38. One roll of felt material at the fabric store sells for $12 and can cover 10 square feet of area. Daisy wishes to place felt on the base of the interior of a large wooden hope chest that measures 2 yards by 1 yard. How much will it cost Daisy to purchase the felt for the hope chest?

Find the area of the shaded region.

39.

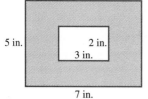

40.

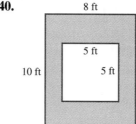

41. Find the area of the region that is *not* shaded.

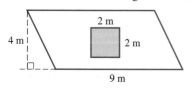

42. Find the area of the region that is *not* shaded.

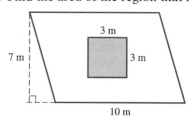

Answer each question.

43. **(a)** How many cubes with 1-inch sides can be placed in a rectangular solid of length = 4 inches, width = 5 inches, and height = 2 inches?

(b) What is the volume of the rectangular solid?

44. **(a)** How many cubes with 1-centimeter sides can be placed in a rectangular solid of length = 5 centimeters, width = 2 centimeters, and height = 3 centimeters?

(b) What is the volume of the rectangular solid?

Find the volume of each rectangular solid.

45. Length = 6 inches, width = 5 inches, height = 9 inches.

46. Length = 13 feet, width = 7 feet, height = 10 feet.

47. Length = 27 yards, width = 10 yards, height = 16 yards.

48. Length = 120 meters, width = 32 meters, height = 37 meters.

Solve each applied problem.

49. A fish tank is 2 feet wide, 4 feet long, and 3 feet high. How much water can be placed in the tank?

50. The ceiling in a room measures 8 feet high. The width of the floor is 10 feet and the length is 15 feet. How much airspace is in the room?

Find the surface area of each box. Assume each box has a top and bottom.

51.

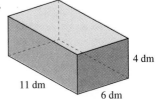

11 dm   4 dm   6 dm

52.

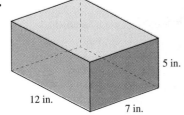

12 in.   5 in.   7 in.

53. A 269-gram box of Triscuit crackers is approximately 8 inches high, 6 inches long, and 2 inches wide. How much material will it take to make the box?

54. You want to stain a cedar chest that is 4 feet long, 2 feet wide, and 3 feet high. How many square feet of the exterior would need to be stained (including the bottom)?

 VERBAL AND WRITING SKILLS

55. To find out how much water is needed to fill a rectangular pool, do you need to find the perimeter, the area, or the volume? Why?

56. To lay one layer of brick along the edge of the pool, do you need to find the perimeter, the area, or the volume? Why?

The walls of a family room are illustrated below. All walls are 8 feet high, and the rear and front walls are each 22 feet long. Each of the side walls is 16 feet long. The French door is 3 feet wide and 7 feet high, while the sliding door is 6 feet wide and 7 feet high. The window on the first side wall is 4 feet by 3 feet; the window on the second side wall is 2 feet by 4 feet.

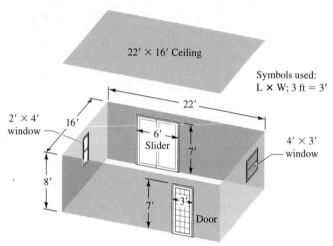

22' × 16' Ceiling

Symbols used:
L × W; 3 ft = 3'

22'

2' × 4' window

16'

6' Slider

7'

4' × 3' window

8'

7'

3'

Door

Ben's Paint Store Price List

TYPE	PRICE	COVERAGE
One-coat	$18 per gal	1 gal per 400 ft²
	$5 per qt	1 qt per 100 ft²
Primer	$12 per gal	1 gal per 300 ft²
	$3 per qt	1 qt per 75 ft²

Use the information described above to answer problems 57 and 58.

57. Anita's landlord is painting the family room of the home she is renting. He has agreed to let Anita purchase the paint so that she can choose the brand and color of paint she wants. The landlord will reimburse Anita only for the cost of the paint used. Anita must buy only as much paint as she needs since the extra paint cannot be returned. Describe the quantity of paint that Anita should purchase if the landlord will put 1 coat of paint on all walls and the ceiling of the family room.

58. Refer to problem 57 to answer the following. If the landlord will put 1 coat of primer on the walls and ceiling of the family room, describe the quantity of primer that Anita must purchase.

1. Find the sum, then round to the nearest thousand: $3426 + 2{,}510{,}777$.

2. Combine like terms: $7x - 3x$.

3. Simplify: $(2)(3x)(5)$.

4. Multiply: $(5x^2)(4x^3)$.

READING A FLOOR PLAN

We can solve many real-life problems with the geometric knowledge we now have. This knowledge enables us to find length, area, or volume. How much edging or fencing is required? How much carpet or tile is required? How much can we store? To solve these problems, we often must take measurements, draw or read floor plans, and perform calculations.

PROBLEMS FOR INDIVIDUAL INVESTIGATION

1. How much will it cost to purchase tile for the kitchen floor if the tile sells for $9 a square foot?

2. How much will it cost to purchase vinyl flooring for the bathroom if the flooring sells for $6 a square foot?

PROBLEM FOR GROUP INVESTIGATION AND COOPERATIVE STUDY

1. John wants to lay carpet in the living room and hallway, wood floors in the entry, family room, and dining room, and tile in the kitchen, bathroom, and closet. John must purchase how many square feet of the following:

(a) Carpet

(b) Tile

(c) Wood floors

(d) How much will the wood floors cost at $8 a square foot?

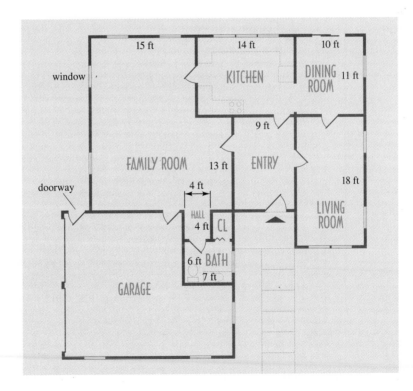

INTERNET: Go to http://www.prenhall.com/blair to explore this application.

After studying this section, you will be able to:

1 Solve problems involving purchases.

2 Solve problems involving incentives.

Math Pro   Video 2.8   SSM

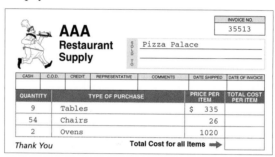

Applied Problems Involving Multiplication and Division

How much material do I need to fence my yard? How much gasoline will I need for my trip? How much profit did my business make? One important use of mathematics is to answer these types of questions. In this section we combine problem-solving skills with the mathematical operations add, subtract, multiply, and divide to solve everyday problems.

1 Solving Problems Involving Purchases

EXAMPLE 1   The 3 owners of the Pizza Palace redecorated their business. The items purchased are listed on the following invoice. If the cost of these purchases is divided equally among the owners, excluding tax, how much did each owner pay?

QUANTITY	TYPE OF PURCHASE	PRICE PER ITEM	TOTAL COST PER ITEM
9	Tables	$ 335	
54	Chairs	26	
2	Ovens	1020	

AAA Restaurant Supply

SOLD TO: Pizza Palace

INVOICE NO. 35513

Thank You   **Total Cost for all Items** ➡

SOLUTION   Read the problem carefully and study the invoice, then fill in the Mathematics Blueprint.

1. *Understand the problem.*

MATHEMATICS BLUEPRINT FOR PROBLEM SOLVING			
GATHER THE FACTS	WHAT AM I ASKED TO DO?	HOW DO I PROCEED?	KEY POINTS TO REMEMBER
See invoice for items and prices. There are 3 owners who must split expenses.	Determine each owner's share of expenses.	1. Find the total *cost per item* by *multiplying*: number of items × cost per item. 2. Find the *total cost* of all items by *adding* all the numbers in step 1. 3. Find each owner's *share of expenses* by *dividing* the total cost by 3.	Fill in the invoice as I complete each step to help keep the facts organized.

2. *Solve and state the answer.* Multiply to get the total cost per item.

9 tables	54 chairs	2 ovens
9(335)	54(26)	2(1020)
↓	↓	↓
$3015	$1404	$2040

Place these amounts on the invoice as the total cost per item. Add to find the total cost for all items.

$3015 + $1404 + $2040 = $6459

Divide to find the amount that each owner paid. 6459 ÷ 3 = 2153
Each owner paid **$2153**.

3. *Check your answer.* We estimate and compare the estimate with the answer above.

Round: number of items ×   price per item   =   total cost

9 → 10	$335 → 300	10 · 300 = $3000
54 → 50	$26 → 30	50 · 30 = 1500
2 → 2	$1020 → 1000	2 · 1000 = 2000
			$6500

Divide the estimated total by 3:

$$\underset{3\overline{)6500}}{2166\,R2 \approx \$2167}$$

Our estimate of $2167 per owner is close to our calculation of $2153. *Our answer is reasonable.*

   Note: Although 2 rounded to the nearest ten is 0, why in this situation would it be wrong to round 2 to 0?

PRACTICE PROBLEM 1

The 2 owners of a Chinese restaurant redecorated their place of business. The items purchased are listed on the following invoice. If the cost of these purchases was divided equally between the owners, excluding tax, how much did each owner pay?

INVOICE NO. 25104

SOLD TO The China Palace   WHOLESALE SUPPLY STORE

CASH	C.O.D.	CREDIT	REPRESENTATIVE	COMMENTS	DATE SHIPPED	DATE OF INVOICE

QUANTITY	TYPE OF PURCHASE	PRICE PER ITEM	TOTAL COST PER ITEM
8	Tables	$ 230	
50	Chairs	25	
3	Ovens	910	

Thank You   Total Cost for all Items ➡

② Solving Problems Involving Incentives

EXAMPLE 2   A frequent-flyer program offered by many major airlines to first-class passengers awards 3 frequent-flyer mileage points for every 2 miles flown. When customers accumulate a certain number of frequent-flyer points, they can cash in these points for free air travel, ticket upgrade, or other awards. How many frequent-flyer points would Louie accumulate if he flew 3500 miles?

SOLUTION

1. *Understand the problem.* Sometimes, drawing charts or pictures can help us understand the problem, as well as plan our approach to solving the problem.

2 miles	+	2 miles	....	3500 miles
↓		↓		↓
3 points	+	3 points	....	? points

How many groups of 2's are in 3500?

We organize our plan in the Mathematics Blueprint.

MATHEMATICS BLUEPRINT FOR PROBLEM SOLVING			
GATHER THE FACTS	WHAT AM I ASKED TO DO?	HOW DO I PROCEED?	KEY POINTS TO REMEMBER
A customer is awarded 3 frequent-flyer points for every 2 miles flown.	Determine how many frequent-flyer points Louie earned.	1. *Divide* 3500 by 2. 2. *Multiply* 3 times the number obtained in step 1.	Frequent-flyer points are determined by the number of miles flown.

2. *Solve and state the answer.*

   Step 1. We divide to find how many groups of 2 are in 3500.

   $3500 \div 2 = 1750$

   Step 2. We multiply 1750 times 3 to find the total points earned.

   $1750 \cdot 3 = 5250$ points   Louie would earn **5250 points**.

3. *Check.* If we earned 4 points (instead of 3) for every 2 miles traveled, we could just double our mileage to find the points earned.

   2 miles $\rightarrow$ 4 points   or $2 \cdot 3500 = 7000$ points

   Since we earned a little less than 4 points, our total should be less than 7000. It is, $5250 < 7000$. We also earned more points than miles traveled (3 points $\rightarrow$ 2 miles), so our total points should be more than the total miles traveled. It is, $5250 > 3500$. *Our answer is reasonable.*

PRACTICE PROBLEM 2

Referring to Example 2, how many frequent-flyer points would Louie accumulate if he flew 4500 miles?

DEVELOPING YOUR STUDY SKILLS

WHY IS HOMEWORK NECESSARY?

You learn mathematics by practicing, not by watching. Your instructor may make solving a mathematics problem look easy, but to learn the necessary skills you must practice them over and over again, just as your instructor once had to do. There is no other way. Learning mathematics is like learning how to play a musical instrument or to play a sport. *You must practice, not just observe, to do well.* Homework provides this practice. The amount of practice varies for each person. The more problems you do, the better you get.

Many students underestimate the amount of time each week that is required to learn math. In general, 2 to 3 hours per week per unit is a good rule of thumb. This means that for a 3-unit class you should spend 6 to 9 hours a week studying math. Spread this time throughout the week, not just in a few sittings. Your brain gets overworked just as your muscles do!

EXERCISE

1. Start keeping a log of the time that you spend studying math. If your performance is not up to your expectations, increase your study time.

EXERCISES 2.8

Solve the following problems, which require only one operation.

1. Martin wants to pay off a $5592 car loan in 12 months. How much will his monthly payments be?

2. Mike wants to pay off a guitar that he has put on layaway in 15 payments. He owes $1140. How much will his payments be?

3. East Gate Academy purchased 327 spelling workbooks at $12 per book. What was the total cost of the workbooks?

4. A football player averages 116 yards a game rushing. At this average, how many rushing yards will be gained in a 9-game season?

5. The Interlogic Electronics assembly plant assembles 24 electronic units per minute. They must assemble 696 units. How many minutes will it take to assemble the units?

6. Superior TV factory produces 12 television sets per hour, and must produce 1152 sets. How many hours will it take to produce the sets?

Solve the following problems, which require more than one operation.

7. A restaurant cook earns $8 per hour for the first 40 hours worked, and $12 per hour for overtime (hours worked in addition to the 40 hours a week). Last week the cook worked 52 hours. Calculate the cook's total pay for that week.

8. Four roommates share expenses for their apartment. How much is each roommate's share of the following monthly expenses?

Apartment expenses

Rent: $920
Utilities: 96
Telephone: 56

9. A dairy cow produces an average of 7 gallons of milk a day. If a farmer has a herd of 35 cows, how much milk will they produce in 1 day? In 1 week?

10. If a ranch has 225 hens that produce about 1 egg per day, how many eggs can the owner expect to produce in 30 days? In 60 days?

11. T.B. Etron's Company made $782,535 in 1997. The expenses for that year were $600,333.

    (a) How much profit did the company make?

    (b) If the 2 owners divided the profits equally, how much money did each owner receive?

12. R.L. Saunders High School PTA sold $2568 in raffle tickets. The expenses for the prizes were $1062.

    (a) How much profit did the PTA make?

    (b) If the profits were divided equally among 3 clubs, how much money did each club receive?

13. Happytime Theater received $3798 from ticket sales for a holiday musical. The expenses for the musical were $2124.

(a) How much profit did the theater make?

(b) If the 3 owners divided the profits equally, how much money did each owner receive?

14. Ramon's Drugstore made $823,222 in 1998. The expenses for that year were $603,000.

(a) How much profit did the store make?

(b) If the 2 owners divided the profits equally, how much money did each owner receive?

15. Janice and her family went to the Middletown Amusement Park. They purchased 2 adult, 4 child, and 1 senior citizen ticket. How much did they spend on the tickets?

Middletown Ticket Prices	
Adult –	$ 13
Child –	$ 5 (under 12 years)
Senior citizen –	$ 7 (over 55 years)

16. Dave and his friends went to an outdoor jazz concert. They purchased 4 adult, 6 student, and 2 child tickets. How much money did they spend on concert tickets?

Outdoor Jazz Concert Ticket Prices	
Adult –	$ 17
Child –	$ 8 (under 12 years)
Student discount –	$ 9 (college ID required)

17. The 5 owners of Mei's Restaurant remodeled their business. They bought 7 tables, 20 chairs, and 2 crystal light fixtures. The cost of these purchases was divided equally among the owners. Excluding tax, how much did each owner pay?

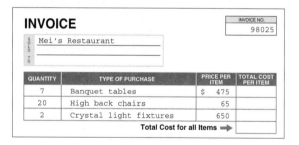

INVOICE

INVOICE NO. 98025

SOLD TO: Mei's Restaurant

QUANTITY	TYPE OF PURCHASE	PRICE PER ITEM	TOTAL COST PER ITEM
7	Banquet tables	$ 475	
20	High back chairs	65	
2	Crystal light fixtures	650	
	Total Cost for all Items ➡		

18. Last weekend, May's Appliance Store sold 6 washing machines, 7 dryers, and 12 dishwashers to the corporation that owns the Whitten apartment complex. The owner will divide the expense for upgrades equally among the 240 tenants by charging each tenant a one-time assessment fee. How much will the assessment fee be for each tenant?

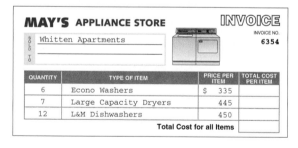

MAY'S APPLIANCE STORE   **INVOICE**

INVOICE NO. 6354

SOLD TO: Whitten Apartments

QUANTITY	TYPE OF ITEM	PRICE PER ITEM	TOTAL COST PER ITEM
6	Econo Washers	$ 335	
7	Large Capacity Dryers	445	
12	L&M Dishwashers	450	
	Total Cost for all Items		

19. A small grove of orange trees had 15 rows of trees with 12 trees in a row. If each tree produces approximately 250 pounds of oranges, how many pounds of oranges will the grower have?

20. Jessica is making a cross stitch that is 70 squares wide and 84 squares high. There are 14 squares per inch on her material. If she needs 2 inches extra on each side of the stitchery to frame it, what are the dimensions of the material she needs?

21. On Monday, Katie had the following servings of food: 3 fruit, 2 vegetables, 3 milk, 6 grains, 4 meat, and 2 fat. How many calories did she consume that day?

1 Serving	No. of Calories
Fruit	60
Vegetables	25
Milk	90
Grains	80
Meat	55
Fat	45

162

22. The following chart lists the meter reading from the gas meters of apartments on Adams Street.
   (a) Which apartment used the most gas this month?
   (b) Which apartment used the least gas this month?

Apartment number	100	101	102	103
Reading previous month	7425	8671	6539	7126
Reading current month	7439	8702	6551	7179

23. To make the dresses for a wedding, Pamela needs the following amounts of material:

   4 yards for 1 dress for the bride   2 yards for each of the 3 flower girls' dresses
   3 yards for each of the 2 bridesmaids' dresses

   Bows, buttons, zippers, and lace for all the dresses will cost $122.
   (a) If the material for the dresses is $8 per yard, what is the total cost to make the dresses?

   (b) If the bride's dress could be purchased for $120, the bridesmaids' dresses for $55 each, and the flower girls dresses for $35 each, which is cheaper, making or purchasing the dresses? How much cheaper?

24. Refer to the chart of custom window blind prices for the following problem. A window blind of size 30 × 36 means 30 inches wide and 36 inches high. We always state width first. Determine the total cost to purchase the following window blinds: two 30 × 36, one 36 × 48, and two 48 × 42.

WIDTH TO:	24″	30″	36″	42″	48″	54″
HEIGHT TO:						
36″	282	316	353	387	423	460
42″	297	335	373	413	452	492
48″	313	354	397	438	481	523
54″	327	372	419	464	509	557
60″	342	392	441	490	537	591

 CALCULATOR EXERCISES

25. In 1995 the state of Washington had a population of 5,430,940 and an average per capita (per person) income of $23,709 per year. What is the approximate total yearly income in the state? Round your answer to the nearest 10 million.

26. In 1995 the state of Tennessee had a population of 5,256,051 and an average per capita (per person) income of $21,060 per year. What is the approximate total yearly income in the state? Round your answer to the nearest 10 million.

ONE STEP FURTHER

27. A $5000 debt on a credit card will take 32 years to repay if the minimum bank monthly payment is made, and will cost the borrower about $7800 in interest. The borrower could be out of debt in 3 years by paying $175 per month. Find the amount of interest paid at the end of 3 years.

28. The amount of force applied by a torque wrench is equal to the amount of force applied to the wrench multiplied by the length of the handle of the wrench. How much more force can be delivered by exerting 5 pounds of pressure on a wrench with a handle of 24 inches than one with a handle of 12 inches?

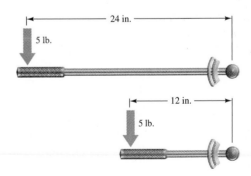

CUMULATIVE REVIEW

1. Multiply: $4 \cdot 3 \cdot 2 \cdot 5$.   **2.** Divide: $215 \div 5$.   **3.** Solve: $\dfrac{x}{26} = 15$.

CHAPTER ORGANIZER

TOPIC	PROCEDURE	EXAMPLES
Key words for multiplication	The key words that represent multiplication are *times*, *product of*, *double*, and *triple*.	Translate using symbols. Double a number: $2x$ Triple a number: $3x$ The product of 2 and 3: $2 \cdot 3$ Six times x: $6x$
Properties of multiplication	We can regroup and multiply numbers in any order since multiplication is associative and commutative.	Simplify: $2(x \cdot 3)$. Change the order of multiplication and regroup: $(2 \cdot 3) \cdot x$. Multiply: $(2 \cdot 3) \cdot x = 6x$.
Multiplying numbers with trailing zeros.	We multiply the nonzero numbers and attach the trailing zeros to the right side of the product.	Multiply: $600(5n)$ Regroup: $(600 \cdot 5)n$ Multiply $6 \cdot 5 = 30$, and attach 2 zeros to the product: $3000n$
Distributive property	To multiply $a(b + c)$ and $a(b - c)$ we distribute the a by multiplying every number or variable inside the parentheses by a, then simplifying.	Simplify: $4(x + 2)$ $4 \cdot x + 4 \cdot 2 = 4x + 8$
Multiplying whole numbers	Multiply top factor by ones digit, then tens digit, then by hundreds digit. Add the partial products.	Multiply: $567 \cdot 238$ $\begin{array}{r} 567 \\ \times\ 238 \\ \hline 4536 \\ 1701 \\ 1134 \\ \hline 134{,}946 \end{array}$
Exponents	2^3 is written in exponent form. The exponent is 3 and the base is 2. 2^3 is read "two to the third power" and means that there are 3 factors of 2.	Write in exponent form: $4 \cdot 4 \cdot 4 \cdot 4 \cdot x \cdot x$ $4^4 x^2$ Find the value: 3^3; 7^0. $3^3 = 3 \cdot 3 \cdot 3 = 27$; $7^0 = 1$
Order of operations	1. Perform operations inside parentheses. 2. Then raise to a power. 3. Then do multiplication and division in order from left to right. 4. Then do addition and subtraction in order from left to right.	Evaluate: $2^3 + 16 \div 4^2 \cdot 5 - 3$. Raise to a power first. $8 + 16 \div 16 \cdot 5 - 3$ Then do multiplication and division from left to right. $8 + 1 \cdot 5 - 3$ $8 + 5 - 3$ Then do addition and subtraction. $13 - 3 = 10$

TOPIC	PROCEDURE	EXAMPLES
Multiplying in exponent form	If bases are the same, we add exponents but keep the base unchanged.	(a) $x^3 \cdot x^6 = x^{3+6} = x^9$ (b) $3 \cdot 3^4 = 3^1 \cdot 3^4 = 3^{1+4} = 3^5$ (c) $4^8 \cdot 2^2$ The rule for multiplying in exponent form does not apply—the bases are not the same.
Multiplying algebraic expressions	We multiply the numerical coefficients. Then we multiply the variable expressions by adding exponents.	Multiply: $(4x^4)(3x^2)$. $(4x^4)(3x^2) = (4 \cdot 3) \cdot (x^4 \cdot x^2) = 12x^6$
The distributive property and exponent form	The process used to multiply using the distributive property is the same when numbers and variables are in exponent form.	Multiply: $x^4(x^2 + 3)$. $x^4 \cdot x^2 + x^4 \cdot 3 = x^6 + 3x^4$
Key words for division	The key words that represent division are *divided, shared equally, divided equally,* and *quotient.*	Translate using symbols. 9 divided by n: $9 \div n$ The quotient of 15 and 5: $15 \div 5$ The quotient of 5 and 15: $5 \div 15$
Long division	We *guess* the quotient and *check* by multiplying the quotient by the divisor. We *adjust* our guess if it is too large or too small and continue the process until we get a remainder less than the divisor.	Divide: $1278 \div 25$. $$\begin{array}{r} 51 \text{ R}3 \\ 25\overline{\smash{)}1278} \\ \underline{-125} \\ 28 \\ \underline{-25} \\ 3 \end{array}$$
Evaluating expressions	We replace the variable with the given value and simplify.	Evaluate $\dfrac{x}{2}$ if $x = 18$. $\dfrac{18}{2} = 9$
Solving equations	We solve simple equations by inspection. Otherwise, we can write equivalent multiplication and division problems to help us solve the equation.	Solve. (a) $\dfrac{n}{3} = 11 \ \rightarrow\ n = 11 \cdot 3 = 33$ (b) $8x = 96 \rightarrow x = \dfrac{96}{8} = 12$
Simplifying and solving equations	We simplify using the commutative and associative properties before we solve equations.	Solve: $(2x) \cdot 4 = 72$. $(2x) \cdot 4 = (2 \cdot 4)x = 72$ $8x = 72$ $x = \dfrac{72}{8} = 9$
Area of a rectangle	$A = LW$	Find the area of a rectangle with length $= 5$ m and width $= 3$ m. $A = (5 \text{ m})(3 \text{ m}) = 15 \text{ m}^2$
Area of a square	$A = s^2$	Find the area of a square with a side of 3 in. $A = (3 \text{ in.})^2 = 9 \text{ in}^2$

T O P I C	P R O C E D U R E	E X A M P L E S
Area of a parallelogram	$A = bh$ b = length of base h = height	Find the area of a parallelogram with a base of 12 m and a height of 9 m. $\begin{aligned} A &= (12\,\text{m})(9\,\text{m}) \\ &= 108\,\text{m}^2 \end{aligned}$
Volume of a rectangular solid (box)	$V = LWH$	Find the volume of a box with dimensions $L = 8$ in., $W = 3$ in., and $H = 4$ in. $V = (8\,\text{in.})(3\,\text{in.})(4\,\text{in.}) = 96\,\text{in}^3$
Surface area of a rectangular solid.	Add the areas of each of the six surfaces.	Find the surface area of a box with dimensions $L = 4$ ft, $W = 2$ ft, and $H = 3$ ft. $2 \cdot$ area of side $+ 2 \cdot$ area of face $+ 2 \cdot$ area of base $2(4\,\text{ft} \cdot 3\,\text{ft}) + 2(2\,\text{ft} \cdot 3\,\text{ft}) + 2(4\,\text{ft} \cdot 2\,\text{ft})$ $24\,\text{ft}^2 + 12\,\text{ft}^2 + 16\,\text{ft}^2 = 52\,\text{ft}^2$

CHAPTER 2 REVIEW

SECTION 2.1

What multiplication does each array suggest?

1.

2.

Identify the product and the factors in each equation.

3. $4x = 32$

4. $xy = z$

Translate the phrase using symbols.

5. Six times a number.

Translate the mathematical symbols to words.

6. $7y = 63$

Multiply.

7. $7 \cdot 2 \cdot 3$

8. $5 \cdot 3 \cdot 2 \cdot 2$

Simplify.

9. $5(10z)$

10. $2(y \cdot 7)$

11. $3(7)(x \cdot 2)$

12. $2(5)(y \cdot 7)$

SECTION 2.2

Multiply.

13. $4(7000)$

14. $(800)6n$

Use the distributive property to simplify.

15. $2(x + 1)$

16. $4(x + 1)$

Multiply.

17. $572(71)$

18. $406(32)$

19. $(4251)352$

20. $6424(903)$

Solve each applied problem.

21. Ken has a truck that averages 17 miles per gallon on the highway. Approximately how far can he travel if he has 18 gallons of gas in his tank?

22. J&R Doors & Windows is replacing all the interior doors in a 6-unit apartment complex that has 21 apartments in each unit. If each apartment has 4 interior doors, how many doors will need to be replaced?

Use the bar graph to answer questions 23 and 24.

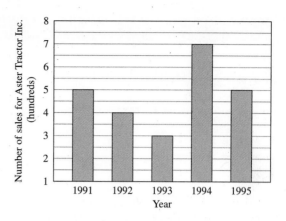

23. The number of sales in 1997 was double the number of sales in 1992. Find the number of sales in 1997.

24. The number of sales in 1998 was triple the number of sales in 1993. Find the number of sales in 1998.

SECTION 2.3

Write each product in exponent form.

25. $x \cdot x \cdot x \cdot x$

26. $6 \cdot 6 \cdot 6$

27. $2 \cdot 2 \cdot 2 \cdot n \cdot n$

28. $z \cdot z \cdot z \cdot z \cdot 5 \cdot 5 \cdot 5$

Write as a repeated multiplication.

29. a^4

30. 6^5

Find the value.

31. 10^3

32. a^0

33. 2^5

34. 9^2

Translate each statement to an equation and solve.

35. Four squared is equal to what number?

36. What number cubed is equal to 8?

Evaluate.

37. $3^3 - 2 + 4$

38. $2 \cdot 4^3$

39. $2 + 5(2^3 + 5 \cdot 6)$

SECTION 2.4

Multiply.

40. $a^4 \cdot a^5$

41. $3^4 \cdot 3^6$

42. $2^4 \cdot 4^3$

43. $x^4 \cdot x$

44. $(3y^4)(5y^4)$

45. $(4x^2)(3x^6)$

46. $(3a)(a^3)(7a^6)$

47. $(4z)(y^8)(3z^3)(2y^3)$

Use the distributive property to simplify.

48. $x(x^2 + 2)$

49. $x(x^3 - 4)$

SECTION 2.5

Write the division that corresponds to each situation.

50. 300 desks are arranged so that 20 desks are in each row. How many rows are there?

51. A $500 prize is divided equally between n people. How much will each person receive?

Translate each phrase into symbols.

52. 35 divided by y.

53. The quotient of 26 and 13.

Divide.

54. (a) $(24 \div 4) \div 2$

(b) $24 \div (4 \div 2)$

55. $6 + 14 \div 2 - 2^2$

56. $\dfrac{15 + 25 \div 5}{8 - 4}$

57. $4(x + 1) + 6 \div 3$

Divide.

58. $4\overline{)1804}$

59. $7\overline{)1701}$

60. $2485 \div 31$

61. $1456 \div 29$

62. $369,757 \div 922$

63. $\dfrac{510,144}{846}$

Solve each applied problem.

64. The Dalton City Music Club fundraising committee raised $447. The club divided the funds equally between four youth groups and deposited the rest of the funds in their club account. How much money did the club deposit in their club account?

65. Lisa wishes to pay off a loan of $3528 in 24 months. How large will her monthly payments be?

SECTION 2.6

66. Evaluate $5n - 6$:   **(a)** If n is 2.

(b) If n is 4.

Solve each equation and check your answer.

67. $9x = 27$

68. $\dfrac{20}{x} = 5$

Write the equivalent multiplication problem and solve.

69. $\dfrac{x}{44} = 21$

70. $\dfrac{y}{16} = 19$

Simplify, then solve the equation.

71. $5(3x) = 45$

72. $3(y \cdot 4) = 24$

73. $6(x \cdot 3) = \dfrac{72}{2}$

Translate each statement into an equation and solve.

74. Triple a number times 5 equals thirty.

75. What number divided by twelve is equal to forty?

76. (a) Evaluate $\dfrac{x}{6}$ if $x = 54$.   **(b)** Solve: $\dfrac{x}{6} = 30$

SECTION 2.7

Find the perimeter.

77. A rectangle with $L = 8$ feet and $W = 5$ feet.

78. A square with a side equal to 98 feet.

79. Find the area of the square garden.

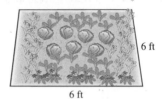

6 ft

6 ft

80. Find the area of the rectangular table cloth.

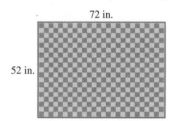

72 in.

52 in.

81. A parallelogram with base $= 9$ inches and height $= 11$ inches.

82. Find the area of the following shape made up of rectangles.

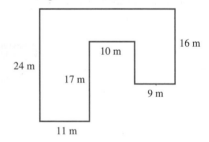

24 m

10 m

16 m

17 m

9 m

11 m

83. Find the area of the shaded region.

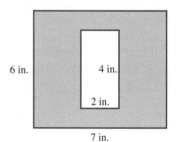

6 in.

4 in.

2 in.

7 in.

Solve each application problem.

84. A surface has a perimeter that measures 15 feet. Find the perimeter in inches.

85. Douglas is purchasing carpet for his office. The measurements he brought to the store are in feet, $L = 18$ feet and $W = 12$ feet. The carpet he plans to purchase is priced at \$15 per square yard.

(a) How many square yards of carpet must he purchase for the office?

(b) How much will the carpet cost?

170

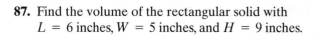

86. The McBeths' front yard measures 16 yards in length and 12 yards in width. A 1-pound container of weed killer sells for $4 and covers an area of 1000 square feet. How much will it cost the McBeths to purchase enough weed killer for their entire front yard?

87. Find the volume of the rectangular solid with $L = 6$ inches, $W = 5$ inches, and $H = 9$ inches.

88. A rectangular children's wading pool has the following measurements; $L = 25$ feet, $W = 15$ feet, and $H = 2$ feet. How much water is needed to fill the pool to capacity?

89. What is the maximum amount of fluid that a 5-inch cube can hold?

90. Find the surface area of a rectangular box with $L = 9$ in., $W = 6$ in., and $H = 3$ in.

SECTION 2.8

91. The Cafe Royale made $190,200 in 1997. The cafe's expenses were $80,400.

  (a) How much money profit did the company make?

  (b) If 4 owners divide the profits equally, how much money does each owner receive?

92. Justin earns $12 per hour for the first 40 hours worked and $18 per hour for overtime (hours worked in addition to the 40 hours a week). If Justin works 51 hours in one week, how much does he earn?

In solving problem 93, consider the following price list.

Adaden Office Supplies			
Office chair	$ 122	4-drawer file cabinet	$ 185
Deluxe swivel chair	200	2-drawer file cabinet	94
Computer desk	375	Desk lamp	45

93. Miriam Investments Advisors made the following purchases from Adaden Office Supplies: 4 office chairs, 3 deluxe swivel chairs, 7 computer desks, 2 4-drawer file cabinets, 3 2-drawer file cabinets, and 5 desk lamps. The cost of these purchases was divided equally among the 5 owners of the Investment Company. Excluding tax, how much is each owner's share of the cost?

CHAPTER 2 TEST

1. _____

2. _____

3. _____

4. _____

5. _____

6. _____

7. _____

8. _____

9. _____

10. _____

11. _____

12. _____

13. _____

14. _____

15. _____

16. _____

17. _____

18. _____

1. A restaurant sells 4 kinds of sandwiches: turkey, roast beef, veggie, and ham. Customers have a choice of 3 types of bread: wheat, white, or rye. How many different sandwiches are possible?

2. Identify the product and the factors in each equation.
 (a) $7(4) = 28$ **(b)** $8x = 24$ **(c)** $rs = t$

3. Simplify.
 (a) $9(2x)$ **(b)** $7(3 \cdot y)$

Multiply.

4. (a) $816(0)$ **(b)** $y(1)$ **(c)** $x \cdot 0 \cdot 5 \cdot 2$

5. $8(4000)$ **6.** $5(y + 6)$

7. Tickets to a play were $25 for adults and $18 for children. 412 adult tickets were sold and 280 children's tickets were sold.
 (a) Find the total income from the sale of tickets.
 (b) If the expenses for the play were $7350, how much profit was made?

8. Write in exponent form: **9.** Find the value: 5^3.
 $6 \cdot 6 \cdot 6 \cdot 6 \cdot 6$.

10. Translate using symbols.
 (a) y to the fourth power   **(b)** 7 cubed

Find the value.

11. 10^5 **12.** $6^2 - 7 + 3$ **13.** $3 \cdot 2 + 4$

Multiply. Leave your answer in exponent form.

14. (a) $y^3 \cdot y^2$ **(b)** $z \cdot z^3$ **(c)** $a^3 b^2$

15. (a) $(5x)(x^3)(x^4)$ **(b)** $(y^4)(y^0)$

16. Multiply: $6(x + 3)$.   **17.** Evaluate: $24 \div 4 - 2 \cdot 3$.

18. Divide and check your answer: $492 \div 12$.

19. The bill for dinner at the Palm Tree Restaurant, including tip, was $64. If 4 people split the bill evenly, how much will each person have to pay?

20. Evaluate $5x - 3$, if $x = 4$.

21. Solve each equation and check your answer.

    (a) $5x = 35$ **(b)** $\dfrac{x}{4} = 12$ **(c)** $2(4x) = 72$

22. Triple a number plus four equals twenty-two. What is the number?

23. Use the formula $P = 2L + 2W$ to find the perimeter of a rectangle with a length 8 inches and a width 3 inches.

24. The blueprints for an office building show that the length of the entryway is 7 feet and the width is 5 feet. How many square feet of tile must be purchased for the entryway?

25. A contractor's crew must dig a hole and haul away dirt in a space 30 feet wide, 40 feet long, and 10 feet deep. How much dirt will they need to haul away?

26. A frequent-flyer program offered by many major airlines to first-class passengers awards 3 frequent-flyer mileage points for every 2 miles flown. When customers accumulate a certain number of frequent-flyer points, they can cash in these points for free air travel, ticket upgrade, or other awards. How many frequent-flyer points would Elizabeth accumulate if she flew 5000 miles?

19. _____

20. _____

21. _____

22. _____

23. _____

24. _____

25. _____

26. _____

CUMULATIVE TEST FOR CHAPTERS 1–2

1. _____

2. _____

3. _____

4. _____

5. _____

6. _____

7. _____

8. _____

9. _____

10. _____

11. _____

12. _____

13. _____

1. Replace the question mark with the inequality symbol $<$ or $>$: 5 ? 0.

2. The total population of a small town is 5289. Round this population figure to the nearest:

   (a) thousand.   **(b)** hundred.

3. Use the commutative and/or associative property of addition, then simplify.

   (a) $3 + y + 1$ **(b)** $1 + (n + 4)$

4. Combine like terms.

   (a) $5x + 2 + 3x + 5xy$ **(b)** $7x - 4x - x - 4$

5. The rent on an apartment was $425. To move in, John was required to pay first and last months' rent, a security deposit of $150, and a telephone installation fee of $35. How much money did he need to move into the apartment?

6. A store clerk receives a total salary per month of $1230. Deducted from her paycheck are taxes of $212, social security of $63, and retirement of $45. What was the total of her check after the deductions?

7. Use the formula $P = 2L + 2W$ to find the perimeter of a rectangle with a length of 20 miles and a width of 12 miles.

Use the following diagram of a patio for problems 8 and 9.

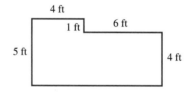

4 ft
1 ft   6 ft
5 ft   4 ft

8. Find the perimeter of the patio.   **9.** Find the area of the patio.

10. Hannah kept the following record of her living expenses for the month of April 1998.

Rent	$550	Car payment	190
Phone/utilities	89	Gas and Insurance	145
Food	273		

   (a) Estimate Hannah's monthly expenses for April.

   (b) If her take-home (net) income for April is $1518, estimate how much money is left after all the expenses are paid.

11. Translate into an equation: Let x represent Jeff's checking account balance. Jeff's checking account balance increased by $385 equals $795.

12. Solve for the variable and check your answer.

   (a) $20 - x = 13$ **(b)** $9 + (y + 8) = 21$

13. Sam bought a gallon of milk. If he used 1 quart, how many quarts did he have left?

Multiply.

14. $r(1)$

15. $2 \cdot 0 \cdot x \cdot 12$

16. $5(2000)$

17. $4(x + 5)$

18. $x^4 \cdot x^2$

19. Write in exponent form: $4 \cdot 4 \cdot 4$.

Evaluate.

20. $2^3 \div 2 - 1$

21. $2y - 8$, if $y = 14$.

22. Carol drove 352 miles using 16 gallons of gas. How many miles per gallon of gas did her car get?

23. Use the formula $P = 4s$, to find the perimeter of a square with a side of 7 inches.

24. Neil's car payment is $250 per month for 30 months. What is the total amount of his payments?

25. Ana plans to recarpet her living room. Her living room has a length of 14 feet and a width of 10 feet. The carpet she selected costs $3 per square foot. Find the cost to recarpet her living room.

14. _____

15. _____

16. _____

17. _____

18. _____

19. _____

20. _____

21. _____

22. _____

23. _____

24. _____

25. _____

SIGNED NUMBERS

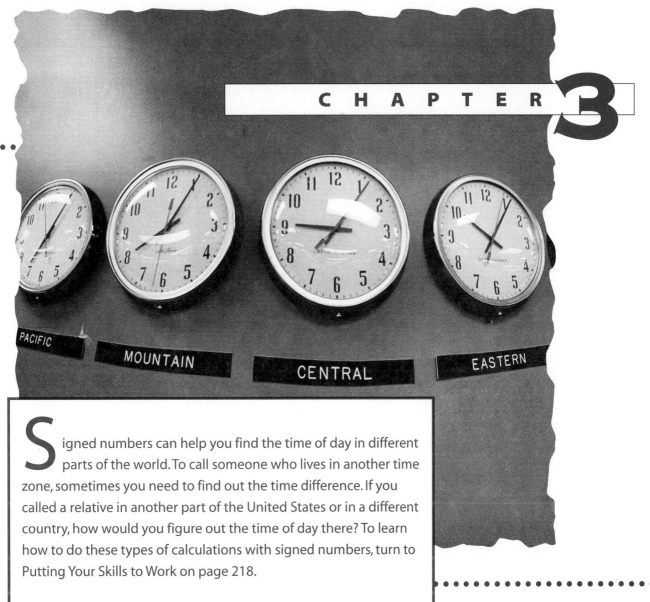

CHAPTER **3**

PACIFIC   MOUNTAIN   CENTRAL   EASTERN

S igned numbers can help you find the time of day in different parts of the world. To call someone who lives in another time zone, sometimes you need to find out the time difference. If you called a relative in another part of the United States or in a different country, how would you figure out the time of day there? To learn how to do these types of calculations with signed numbers, turn to Putting Your Skills to Work on page 218.

3.1   Understanding Signed Numbers

3.2   Addition of Signed Numbers

3.3   Subtraction of Signed Numbers

3.4   Multiplication and Division of Signed Numbers

3.5   Simplifying Algebraic Expressions

3.6   Order of Operations and Applications Involving Signed Numbers

CHAPTER 3 PRETEST

This test provides a preview of the topics in this chapter. It will help you identify which concepts may require more of your studying time. If you are familiar with the topics in this chapter, take this test now. Check your answers with those in the back of the book. If you are not familiar with the topics in this chapter, begin studying the chapter now.

SECTION 3.1

1. Replace the ? with the inequality symbol $<$ or $>$:

    (a) -13 ? 13 **(b)** 6 ? -85

2. Evaluate the absolute value.

    (a) $|4|$ **(b)** $|-7|$

3. The opposite of -18 is ____.

SECTION 3.2   *Add.*

4. $-5 + (-2)$ **5.** $(-1) + 8 + (-6)$

SECTION 3.3   *Subtract.*

6. $3 - 5$ **7.** $-8 - 2$ **8.** $5 - (-14)$

9. Find the difference in altitude between a mountain 1200 feet high and a desert valley that is 250 feet below sea level.

10. Evaluate $x + 8$ for:

    (a) $x = -1$ **(b)** $x = -9$

SECTION 3.4   *Multiply.*

11. $(-5)(4)$ **12.** $(-6)(-3)$

13. Evaluate:

    (a) $(-4)^2$ **(b)** -4^2

14. Divide:

    (a) $12 \div (-3)$ **(b)** $(-12) \div (-3)$

15. Divide: $\dfrac{-42}{6}$.

SECTION 3.5

16. Simplify by combining like terms: $5x + (-3x)$.

17. Simplify by multiplying: $(-6x^2)(-3x^3)$.

SECTION 3.6   *Simplify.*

18. $5 - 12(9 - 10)$ **19.** $7 - 24 \div 6(-2)^2 - 3$

1. _____

2. _____

3. _____

4. _____

5. _____

6. _____

7. _____

8. _____

9. _____

10. _____

11. _____

12. _____

13. _____

14. _____

15. _____

16. _____

17. _____

18. _____

19. _____

SECTION 3.1

Understanding Signed Numbers

After studying this section, you will be able to:

1 Use the inequality relationship with signed numbers.

2 Find the absolute value of a number.

3 Find the opposite of a number.

4 Read a line graph.

Math Pro   Video 3.1   SSM

Using the Inequality
1 Relationship with Signed Numbers

We often encounter real-life applications that require us to consider numbers that are less than zero. These numbers are called *negative numbers*. For example, a weather report states that the temperature is 20 degrees below zero. How will we write this temperature? We can use negative numbers. Thus the temperature reading on the thermometer, 20 degrees below zero, is −20°F (−20 is a negative number).

We can also represent negative numbers using a *number line*. **Positive numbers** are to the *right of zero* on the number line. **Negative numbers** are to the *left of zero* on the number line. The **origin** is at *zero*, and the number 0 is neither positive nor negative. All these numbers are called **signed numbers**.

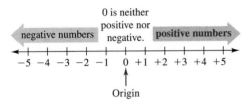

The symbols "−" and "+" indicate the *sign* of the number. For positive numbers we usually do not write the plus sign.

Numbers decrease in value as we move from right to left on the number line. Therefore, 1 is less than 3 (1 < 3) since 1 lies to the left of 3 on the number line, and −5 is less than −2 (−5 < −2) since −5 lies to the left of −2 on the number line.

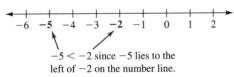

We see that we must consider the *sign* of the number as well as the *numerical part* when determining less than (<), or greater than (>).

EXAMPLE 1   Replace the ? with the inequality symbol < or >.

(a) −3 ? −1   **(b)** 4 ? −5   **(c)** −10 ? −345

SOLUTION

(a) −3 **<** −1   −3 lies to the left of −1 on the number line.

(b) 4 **>** −5   Positive numbers are always greater than negative numbers. This reads: "4 is greater than −5" or "−5 is less than 4."

(c) −10 **>** −345   The inequality symbol always points to the smaller number.

Temperature (degrees Fahrenheit)

30
20
10
0
−10
−20
−30

PRACTICE PROBLEM 1

Replace the ? with the inequality symbol $<$ or $>$.

(a) $-5\ ?\ 2$ **(b)** $-3\ ?\ -6$ **(c)** $-53\ ?\ -218$

In everyday situations we use the concept of signed numbers to represent many different things. We can use "+" to represent an increase, a rise, or whenever something goes up, and a "−" to represent a decrease, a decline, or whenever something goes down. For example, when you use a checking account we can associate a deposit with a "+" since your balance goes *up*, and a check written with a "−" since your balance goes *down*.

EXAMPLE 2   Fill in the blank with the appropriate symbol, + or −, to describe either an increase or decrease.

(a) A discount of $5: ____ $5
(b) The temperature rises 10°F: ____ 10°F

SOLUTION

(a) A discount of $5 results in the price decreasing: **−$5**
(b) The temperature rises 10°F: **+10°F**

PRACTICE PROBLEM 2

Fill in the blank with the appropriate symbol, + or −, to describe either an increase or decrease.

(a) A property tax increase of $130: ____ $130
(b) A dive of 7 ft below the surface of the sea: ____ 7 ft

❷ Finding the Absolute Value of a Number

Suppose that we want to find the distance from 0 to −4 and from 0 to +4. We can use the number line to measure this distance just as we use a ruler to measure feet or inches.

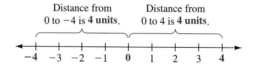

Distance from 0 to −4 is **4 units**.   Distance from 0 to 4 is **4 units**.

In either case, the distance is the same. One distance is just on the left (negative) side of 0 and the other on the right (positive) side of 0. Notice that the number of units, 4, is the same as the numerical parts of both 4 and −4. Since this numerical part is important in mathematics, we call it the **absolute value**. We place the symbols "| |" around the number to indicate that we want the absolute value of that number. We write $|-4| = 4$ and $|4| = 4$.

> The *absolute value* of a number a is the number of units between 0 and a on the number line.
>
> To find the absolute value of a number, we select only the numerical part of the number: $|-5| = 5$.

E X A M P L E 3   State the absolute value.

(a) $|-9|$ **(b)** $|3|$

SOLUTION

(a) $|-9| = 9$ Since the number of units between -9 and 0 is 9, we select the
numerical part of the number.

(b) $|3| = 3$ The number of units between 3 and 0 is 3.

P R A C T I C E   P R O B L E M 3

State the absolute value.

(a) $|-67|$ **(b)** $|8|$

E X A M P L E 4   Replace the ? with the inequality symbol $<$ or $>$.

$|-15| \; ? \; |6|$

SOLUTION

$$
\begin{array}{ccc}
|-15| & ? & |6| \\
\downarrow & & \downarrow \\
15 & ? & 6 \\
15 & > & 6 \\
|{-}\mathbf{15}| & > & |\mathbf{6}|
\end{array}
$$

We find the absolute values.

-15 has the larger absolute value, because the numerical
part, 15, is larger than 6.

P R A C T I C E   P R O B L E M 4

Replace the ? with the inequality symbol $<$ or $>$.

$|-12| \; ? \; |2|$

❸ Finding the Opposite of a Number

Numbers that are the same distance from zero but lie on the opposite side of zero
on the number line are called **opposites**. For example, 2 and -2 are opposites. By
this we mean that the opposite of 2 is -2, and the opposite of -2 is 2.

opposite sides of zero but the same distance from zero

E X A M P L E 5   State the opposite.

(a) 6 **(b)** -9

SOLUTION

(a) The opposite of 6 is $-\mathbf{6}$.   **(b)** The opposite of -9 is $\mathbf{9}$.

P R A C T I C E   P R O B L E M 5

State the opposite.

(a) -6 **(b)** -1 **(c)** 12 **(d)** 1

④ Reading a Line Graph

We can use a line graph to display information similar to the way that we use a bar graph. On a line graph we use a dot, instead of a bar, to display information. Then we connect the dots with straight lines. The vertical number line on a graph is sometimes extended to include negative numbers.

EXAMPLE 6   The line graph indicates the low temperatures for March 9, 1998.

(a) In which city was the temperature colder, Bismarck or Fargo?
(b) Which cities recorded a positive temperature for the day, and which cities recorded a negative temperature?

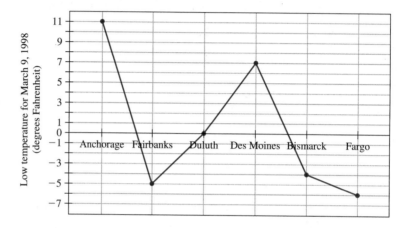

SOLUTION

(a) The temperature was −4°F in Bismarck and −6°F in Fargo. **It was colder in Fargo**.

　　　Note: The dot representing Fargo's temperature is lower on the line graph than Bismarck, indicating a lower temperature.

(b) The positive temperatures are listed above 0 and negative temperatures below 0; therefore, Duluth's temperature, 0°F, was neither positive nor negative.

　　　Positive temperature: **Anchorage, Des Moines**
　　　Negative temperature: **Fairbanks, Bismarck, Fargo**

PRACTICE PROBLEM 6

The line graph indicates the low temperatures for March 9, 1998.

(a) In which city was the temperature colder, Fairbanks or Bismarck?
(b) Name the city that recorded the highest temperature and the city that recorded the lowest temperature.

EXERCISES 3.1

Replace the ? with the inequality symbol < or >.

1. −5 ? −2   **2.** −7 ? −4   **3.** 9 ? −3   **4.** 6 ? −4

5. −5 ? 5   **6.** −6 ? 6   **7.** −291 ? −5   **8.** −312 ? −2

9. −1250 ? 5   **10.** −4122 ? 6   **11.** 298 ? −3   **12.** 765 ? −7

13. Which dot represents a larger number on the following number line, A or B?

14. Which dot represents a larger number on the following number line, X or Y?

Fill in the blank with the appropriate symbol, + or −, to describe either an increase or decrease.

15. ____ Tax increase   **16.** ____ Plane descending   **17.** ____ Loss

18. ____ Temperature rising   **19.** ____ Discount   **20.** ____ Profit

21. ____ Plane ascending   **22.** ____ Tax decrease

State the absolute value.

23. |8| **24.** |6| **25.** |−5| **26.** |−7|

27. |−16| **28.** |−19| **29.** |44| **30.** |56|

Replace the ? with the inequality symbol < or >.

31. |−3| ? |1| **32.** |−9| ? |5| **33.** |5| ? |−8| **34.** |2| ? |−6|

35. |16| ? |−9| **36.** |19| ? |−13| **37.** |−35| ? |−8| **38.** |−71| ? |−6|

39. Which of the two numbers has the larger absolute value: −33 or 12?

40. Which of the two numbers has the larger absolute value: −43 or 11?

41. Which of the two numbers has the larger absolute value: 129 or −112?

42. Which of the two numbers has the larger absolute value: 231 or −98?

Fill in the blank.

43. The opposite of −5 is ____.

44. The opposite of −8 is ____.

45. The opposite of 16 is ____.

46. The opposite of 19 is ____.

Solve each applied problem.

47. The line graph indicates the low temperature for March 10, 1998.

   (a) In which city was the temperature colder, Duluth or Rapid City?

   (b) Which cities recorded a positive temperature for the day, and which cities recorded a negative temperature?

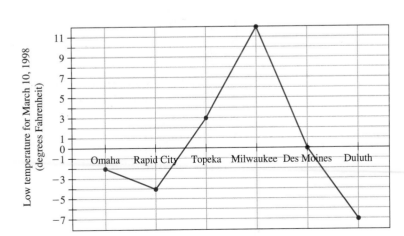

48. Use the line graph in exercise 47 to answer the following.

(a) In which city was the temperature colder, Omaha or Des Moines?

(b) Name the cities that recorded the highest and lowest temperatures.

  VERBAL AND WRITING SKILLS

49. Explain why $|6| = |-6|$.

50. The opposite of a negative number is a _____ number.

CUMULATIVE REVIEW

Perform the operation indicated.

1. $5009 - 258$

2. $5699 + 351$

3. $(256)(91)$

4. $456 \div 3$

DEVELOPING YOUR STUDY SKILLS

PREVIEWING NEW MATERIAL

Does the pace of the lecture seem too fast for you? Do you miss parts of the instructor's explanation? Previewing the new material can help you with these problems as well as enhance the amount of learning that happens while you are *listening* to the lecture.

The Learning Cycle

Reading $\longrightarrow$ Writing

$\uparrow$ $\downarrow$

Seeing $\leftarrow$ Verbalizing $\leftarrow$ Listening

Part of your study time each day should consist of looking ahead to those sections in your text that are to be covered the following day. You do not necessarily have to learn the material on your own. Survey the concepts, terminology, diagrams, and examples, so that you are familiar with the new ideas when the instructor presents them.

To help yourself in class:

1. Take note of concepts that appear confusing or difficult as you read.
2. Listen carefully for your instructor's explanation of material that gave you difficulty.
3. Be prepared to ask questions.

Previewing new material enables you to see what is coming and prepares you to learn.

EXERCISE

1. Review your time management schedule and insert time to preview the new material.

S E C T I O N   3 . 2

Addition of Signed Numbers

❶ Add signed numbers with the same sign.

❷ Add signed numbers with different signs.

❸ Evaluate expressions involving addition of signed numbers.

❹ Solve applied problems involving addition of signed numbers.

Math Pro Video 3.2   SSM

❶ Adding Signed Numbers with the Same Sign

When we associate the $+$ and $-$ symbols with positive and negative situations, we can find the sum of signed numbers by considering the outcome of these situations. For example, a salary *increase* of \$10 ($+10$) followed by a salary *increase* of \$20 ($+20$), results in an *increase* of \$30 ($+30$).

Since signed numbers are often used to indicate *direction* and *distance*, we can also use the number line to find the sum of numbers, such as $-1 + (-3)$. We say that we move in the *negative direction* on the number line when we move in the direction to the left of 0 and the *positive direction* when we move to the right of 0. The direction we move on the number line is indicated by the sign of the number, and the distance is indicated by the numerical part of the number.

We see that
$$-1 + (-3) = -4$$

E X A M P L E   1

(a) Place a marker at 0 on the number line, then move the marker *left* 3 units followed by another move *left* 2 units.

(b) Is the marker in the positive or the negative region?

(c) Write the math symbols that represent the situation.

(d) Use the number line to find the sum.

SOLUTION

(a)

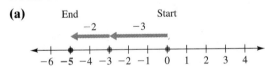

(b) From the illustration we see that the marker is in the **negative region**, since we move 3 units in the negative direction (left), followed by another move in the negative direction 2 units.

left 3	followed by	left 2
↓	↓	↓

(c) The math symbols are: -3 $+$ (-2)

(d) We end at -5, which is the sum.

$$-3 + (-2) = -5$$

PRACTICE PROBLEM 1

(a) Place a marker at 0 on the number line, then move the marker left 4 units followed by another move left 1 unit.

(b) Is the marker in the positive or the negative region?

(c) Write the math symbols that represent the situation.

(d) Use the number line to find the sum.

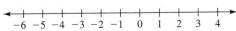

Example 1 shows that when we add two negative numbers the answer is a negative number. Of course, we know that when we add two positive numbers the answer is a positive number.

Now, without using a number line, how do we find the sum when we add numbers with the same sign? Let's look at the results from Example 1.

We add two negative numbers, so the sum is negative.

$$-3 + (-2) = -5$$

$2 + 3 = 5$; we add absolute values (numerical parts) to get 5.

We state the formal rule.

ADDITION RULE FOR TWO NUMBERS WITH THE SAME SIGN

To add two numbers with the *same* sign:

1. Use the common sign in the answer.
2. Add the absolute values of the numbers.

In other words, keep the common sign and add the absolute values.

EXAMPLE 2   Add: $-1 + (-3)$.

SOLUTION   We are adding two numbers with the same sign, so we keep the common sign and add the absolute values.

$-1 + (-3) = -$ The answer is *negative*, since the common sign is negative.
$-1 + (-3) = \mathbf{-4}$ Add: $1 + 3 = 4$.

PRACTICE PROBLEM 2
Add: $-2 + (-4)$.

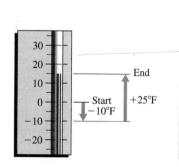

② Adding Signed Numbers with Different Signs

So far we have seen how to add numbers with the *same sign*. We use a similar approach to see how we add numbers that have *different signs*. Addition of numbers with *different signs* often involves situations such as a decrease followed by an increase, or something rising followed by going down.

We can also use a vertical number line to illustrate these types of situations. A move *up* is considered the positive direction, and a move *down* is considered the negative direction.

EXAMPLE 3   One night the temperature on Long Island, New York was $-10°F$. At dawn the temperature had risen $25°F$.

(a) Write the math symbols that represent the situation.
(b) At dawn, was the temperature a positive or a negative reading?
(c) Use the thermometer at the left to find the sum.

SOLUTION

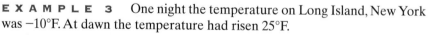

down 10   followed by   up 25
  ↓   ↓   ↓
(a) $-10°F$ + $(+25°F)$

(b) From the chart we see that **the temperature reading at dawn was positive** since it went up (+) more degrees than it went down (−).

(c) The temperature ends up at +15°F, which is the sum.

$$-10°F + 25°F = 15°F$$

PRACTICE PROBLEM 3

Last night the temperature in Boston, Massachusetts dropped to −15°F. At dawn it had risen 30°F.

(a) Write the math symbols that represent the situation.
(b) At dawn, was the temperature a positive or a negative reading?
(c) Use the thermometer to find the sum.

Example 3 involves addition of signed numbers with *different signs*. How do we perform this addition without using a chart? We first determine the *sign* of the sum, then the *numerical part* of the sum. Let's look at the results from Example 3.

$-10 + (+25) = $ '**+**'   The sign of the sum is positive, since we move a larger distance in the positive direction.

$-10 + (+25) = $ **+15**

$25 - 10 = $ **15**;   We subtract absolute values (numerical parts) to get 15.

We state the formal rule.

ADDITION RULE FOR TWO NUMBERS WITH DIFFERENT SIGNS

To add two numbers with *different* signs:

1. Use the *sign* of the number with the larger absolute value in the answer.
2. Subtract the absolute value of the numbers.

In other words, we keep the sign of the larger absolute value, and subtract.

EXAMPLE 4   Add.

(a) $2 + (-3)$ **(b)** $-2 + 3$

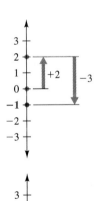

SOLUTION   We are adding numbers with different signs, so we keep the sign of the larger absolute value and subtract.

(a) $2 + (-3) = -$ The answer is *negative*, since −3 is negative and it has a larger absolute value.

$2 + (-3) = $ **−1**   Subtract: $3 - 2 = 1$.

(b) $-2 + 3 = +$ The answer is *positive*, since 3 is positive and it has a larger absolute value.

$-2 + 3 = $ **+1**   Subtract: $3 - 2 = 1$.

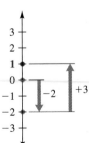

Note: The *numerical part* of the sum, 1, is the same for both parts (a) and (b); only the *sign* is different. Why?

PRACTICE PROBLEM 4

Add.

(a) $-4 + 7$ **(b)** $4 + (-7)$.

We summarize the rules for adding signed numbers as follows:

Adding numbers with the same sign: Keep the common sign and add the absolute values.

Adding numbers with different signs: Keep the sign of the larger absolute value and subtract the absolute values.

EXAMPLE 5   Add.

(a) $8 + (-5)$ **(b)** $-8 + (-5)$

SOLUTION

(a) $8 + (-5)$ We have *different* signs.

$8 + (-5) = +$ 8 is larger than 5, so the answer is positive.

$8 + (-5) = +3$ or **3**   Subtract: $8 - 5 = 3$.

(b) $(-8) + (-5)$ We have the *same* signs.

$(-8) + (-5) = -$ Keep the common sign: negative

$(-8) + (-5) = -\mathbf{13}$ Add: $8 + 5 = 13$.

PRACTICE PROBLEM 5

Add.

(a) $-4 + 7$ **(b)** $-4 + (-7)$

If there are three or more numbers to add, it may be easier to add positive numbers and negative numbers separately, and then combine the results. Why? We can do this because, just like with whole numbers, addition of signed numbers is commutative and associative.

EXAMPLE 6   Add: $-3 + 9 + (-4) + 12$.

SOLUTION

$$-3 + 9 + (-4) + 12 = -\mathbf{3} + 9 + (\mathbf{-4}) + 12$$

$$= -\mathbf{7} + 9 + 12 \qquad \text{Add the negative numbers;}$$
$$-3 + (-4) = -7.$$

$$= -7 + 21 \qquad \text{Add the positive numbers;}$$
$$9 + 12 = 21.$$

$$= \mathbf{14} \qquad \text{Add the result: } -7 + 21 = 14.$$

PRACTICE PROBLEM 6

Add: $-8 + 6 + (-2) + 5$.

CALCULATOR

ADD NEGATIVE NUMBERS

There are a few different ways to enter negative numbers in the calculator. Usually, either a $+/-$ *or the* $(-)$ *key is used. You should read the manual for directions. To find* $(-119) + 85$*, enter:*

Scientific calculator:

119 $+/-$ $+$ 85 $=$

Graphing calculator:

$(-)$ 119 $+$ 85 ENT

The calculator displays

-34

Evaluating Expressions
3 Involving Addition of Signed Numbers

We evaluate expressions involving signed numbers just as we did in earlier chapters: We replace the variable with the given number and perform the operation indicated.

EXAMPLE 7 Evaluate.

(a) $x + 13$ for $x = -2$ **(b)** $-1 + x$ for $x = -3$

SOLUTION

(a) $x + 13$
  ↓
    $-2 + 13 = \mathbf{11}$

(b) $-1 + x$
  ↓
    $-1 + (-3) = \mathbf{-4}$

PRACTICE PROBLEM 7

Evaluate.

(a) $-7 + x$ for $x = -5$.   **(b)** $x + 8$ for $x = -2$.

Solving Applied Problems
4 Involving Addition of Signed Numbers

EXAMPLE 8   The results of Micro Firm Computer Sales' profit and loss situation are listed on the graph. What was the company's overall profit or loss at the end of the third quarter?

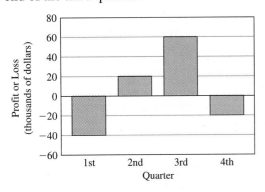

SOLUTION

| 1st quarter loss | + | 2nd quarter gain | + | 3rd quarter gain | = | net gain |

  $-\$40,000$ + $\$20,000$ + $\$60,000$ = $\$40,000$

  $-40,000 + (20,000 + 60,000) = -40,000 + 80,000 = 40,000$

At the end of the third quarter the company had a net gain of $40,000.

PRACTICE PROBLEM 8

What was Micro Firm Computer Sales' overall profit or loss at the end of the second quarter?

EXERCISES 3.2

Express the outcome of each situation as a signed number.

1. A decrease of 10°F followed by a decrease of 5°F.

2. A discount of $4 followed by a discount of $2.

3. A profit of $100 followed by a profit of $50.

4. An increase of 120 units of inventory followed by an increase of 50 units of inventory.

5. (a) Place a marker at 0 on the number line, then move the marker left 2 units, followed by another move left 2 units.

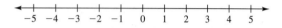

(b) Is the marker in the positive or negative region?

(c) Write the math symbols that represent the situation.

(d) Use the number line to find the sum.

6. (a) Place a marker at 0 on the number line, then move the marker left 1 unit, followed by another move left 2 units.

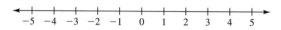

(b) Is the marker in the positive or negative region?

(c) Write the math symbols that represent the situation.

(d) Use the number line to find the sum.

Add.

7. (a) $(-11) + (-13)$ (b) $11 + 13$

8. (a) $(-14) + (-19)$ (b) $14 + 19$

9. (a) $(-29) + (-39)$ (b) $29 + 39$

10. (a) $(-24) + (-44)$ (b) $24 + 44$

11. (a) $(-43) + (-18)$ (b) $43 + 18$

12. (a) $(-32) + (-12)$ (b) $32 + 12$

Express the outcome of each situation as a signed number.

13. A 300-foot increase in altitude followed by a 400-foot decrease in altitude.

14. Diving 10 feet downward followed by rising 2 feet.

15. A loss of $400 followed by a profit of $500.

16. An increase of 10 pounds in weight followed by a decrease of 5 pounds.

SECTION 3.3

Subtraction of Signed Numbers

After studying this section, you will be able to:

1 Subtract signed numbers.

2 Perform several signed-number operations.

3 Solve applied problems involving the subtraction of signed numbers.

Math Pro   Video 3.3   SSM

1 Subtracting Signed Numbers

How do we subtract signed numbers? What is the value of $-4 - 2$? Before we define a rule for subtracting signed numbers, let's look at a few subtraction problems that we can do mentally.

Suppose that you have $20 in the bank and you write a check for $30. The bank will not be able to pay the $30 because you are *short $10*. If the bank cashes the check, your balance would represent a "debt of $10," or $-$10. Thus we can see that **$20 − $30 = −$10**.

Now, what do you think the value of $6 - 7$ is? We can think of this as a situation in which we have 6 items and want to take away 7 items. We are *short 1 item*, or -1. Thus **6 − 7 = −1**.

EXAMPLE 1   Subtract.

(a) $15 − $20

(b) 3 − 4

SOLUTION

(a) $15 − $20 = **−$5**   If we have $15 and want to spend $20, we are short $5, or −$5.

(b) 3 − 4 = **−1**   If we have 3 items and try to take away 4 items, we are short 1 item, or −1.

PRACTICE PROBLEM 1
Subtract.

(a) $10 − $20

(b) 5 − 6

It is not always possible to subtract mentally, so we must find an efficient way to do more complicated subtraction problems. As we saw in Chapter 1, to subtract whole numbers we do not need to learn subtraction facts. Instead, we write equivalent addition problems, then use addition facts. We use a similar approach to subtract signed numbers. *We can rewrite our subtraction as an addition problem, then use the rules for adding signed numbers we learned in Section 3.2.* To illustrate, we write an addition problem that gives the same result as the subtraction problem in Example 1a. Look for a pattern.

Same result

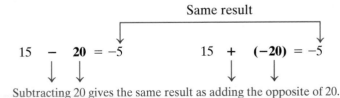

$$15 - 20 = -5 \qquad 15 + (-20) = -5$$

Subtracting 20 gives the same result as adding the opposite of 20.

We see that $15 - 20$ is equivalent to $15 + (-20)$. Both give the same result, -5. Subtracting 20 seems to give the same result as adding the opposite of 20. We see it is reasonable to generalize that subtracting is equivalent to adding the opposite .

E X A M P L E   2   Rewrite each subtraction as addition of the opposite.

(a) $40 - 10 = 30$ **(b)** $6 - 2 = 4$ **(c)** $25 - 5 = 20$

SOLUTION

Subtraction	*Addition of the Opposite*
(a) $40 - 10 = 30$	$40 + (-10) = 30$
(b) $6 - 2 = 4$	$6 + (-2) = 4$
(c) $25 - 5 = 20$	$25 + (-5) = 20$

P R A C T I C E   P R O B L E M   2

Rewrite each subtraction as addition of the opposite.

(a) $20 - 10 = 10$ **(b)** $5 - 2 = 3$ **(c)** $20 - 5 = 15$

We state the rule for subtraction:

SUBTRACTION RULE FOR SIGNED NUMBERS

$$a - b = a + (-b)$$

To subtract signed numbers, add the opposite of the second number to the first.

The rule tells us to do three things when we subtract signed numbers:

1. Change the subtraction to addition.
2. Replace the second number by its opposite.
3. Add using the rules for addition of signed numbers.

E X A M P L E   3   Subtract.

(a) $-8 - 3$ **(b)** $-6 - (-4)$

SOLUTION

(a) -8 $-$ 3
        -8 $+$ (-3) $= -11$

Change subtraction to addition.	Write the opposite of the second number.	Add using the rule for adding numbers with the *same sign*.

$$-8 - 3 = -8 + (-3) = -11$$

(b) -6 $-$ (-4)
        -6 $+$ (4) $= -2$

Change subtraction to addition.	Write the opposite of the second number.	Add using the rule for adding numbers with *different signs*.

$$-6 - (-4) = -6 + 4 = -2$$

PRACTICE   PROBLEM   3

Subtract.

(a) $-5 - 4$ **(b)** $-9 - (-5)$.

At this point you should be able to do several subtraction problems quickly.

> Remember that in performing subtraction of two numbers:
>
> 1. The first number does not change.
> 2. The subtraction sign is changed to addition.
> 3. We write the opposite of the second number.
> 4. We find the result of this addition problem.

When you see $7 - 10$, the sign "$-$" means subtraction. Then you should think: "$7 + (-10)$." Try to think of each subtraction problem as a problem of **adding the opposite**.

> If you see $-3 - 19$, think $-3 + (-19)$.
> If you see $8 - (-2)$, think $8 + 2$.

EXAMPLE 4   Subtract.

(a) $8 - 9$ **(b)** $-3 - 16$ **(c)** $5 - (-4)$ **(d)** $-4 - (-2)$

SOLUTION

(a) $8 - 9 = 8 + (-9) = \mathbf{-1}$ **(b)** $-3 - 16 = -3 + (-16) = \mathbf{-19}$

(c) $5 - (-4) = 5 + 4 = \mathbf{9}$ **(d)** $-4 - (-2) = -4 + 2 = \mathbf{-2}$

PRACTICE   PROBLEM   4

Subtract.

(a) $7 - 10$ **(b)** $(-4) - 15$ **(c)** $8 - (-3)$ **(d)** $(-5) - (-1)$

② Performing Several Signed-Number Operations

Subtraction of signed numbers is not commutative or associative, but addition is. Thus if we first rewrite *all* subtraction as addition of the opposite, we can perform the addition in any order.

EXAMPLE 5   Perform the necessary operations: $4 - 7 - 5 - 3$.

SOLUTION

$$4 - 7 - 5 - 3 = 4 \underbrace{+ (-7) + (-5) + (-3)}$$ First, write all subtraction as addition of the opposite.

$$= 4 + \quad (-15)$$ Then add all *like signs*: $(-7) + (-5) + (-3) = (-15)$.

$$= \mathbf{-11}$$ Next, add *unlike signs*: $4 + (-15) = (-11)$.

Note: Do you see that you would obtain the same answer if you first added $4 + (-7) = -3$ and then added the remaining numbers from left to right?

CALCULATOR

NEGATIVE NUMBER OPERATIONS

The key used to enter a negative number on a calculator is marked $+/-$ *or* $(-)$ *. This key changes the sign of a number from + to − or − to +. To enter the number −3, press the key 3 and then the key* $+/-$ *. The display should read −3. To find* $-32 + (-46)$, *enter*

Scientific calculator:

32 $+/-$ $+$ 46 $+/-$ $=$

Graphing calculator:

$(-)$ 32 $+$ $(-)$ 46 ENT

The calculator displays

$$-78$$

Try (a) $-756 + 129$;
(b) $-256 - (-302)$.

That is, $(-3) + (-5) + (-3) = -11$. Although it is easier to add like signs first, the commutative property of addition states that we may add the numbers in any order.

PRACTICE PROBLEM 5

Perform the necessary operations: $6 - 9 - 2 - 8$.

EXAMPLE 6   Perform the necessary operations: $-9 - (-3) + (-4)$.

SOLUTION

$$-9 - (-3) + (-4) = -9 + 3 + (-4)$$ Write subtraction as addition of the opposite.

$$= -13 + 3$$ Add like signs: $-9 + (-4) = -13$.

$$= -10$$ Add unlike signs: $-13 + 3 = -10$.

Note: You obtain the same answer if you first add $-9 + 3 = -6$, then add the result to -4. Which way do you find easier?

PRACTICE PROBLEM 6

Perform the necessary operations: $-3 - (-5) + (-11)$.

Solving Applied Problems

3   Involving the Subtraction of Signed Numbers

When we subtract $3000 - (-50) = 3050$, we obtain a result that is larger than 3000. Why is the result larger than 3000 if we are subtracting a number from 3000? Because we are subtracting a negative number. We illustrate this idea next.

Suppose that we want to find the difference in altitude between the two mountains illustrated below. We subtract the lower altitude from the higher altitude. The difference in altitude between the two mountains is 3000 feet − 1000 feet = 2000 feet.

$3000 \text{ ft} - 1000 \text{ ft} = 2000 \text{ ft}$
Subtract a positive number and the result is **less than** 3000.

Land that is below sea level is considered to have a negative altitude. A valley that is 50 feet below sea level is said to have an altitude of −50 feet. The difference in altitude between the mountain and the valley is found by subtracting, 3000 feet − (−50) feet.

$3000 \text{ ft} - (-50 \text{ ft}) = 3050 \text{ ft}$
Subtract a negative number and the result is **more than** 3000.

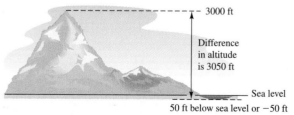

EXAMPLE 7   A portion of the Dead Sea is 1286 feet below sea level. What is the difference in altitude between Mount Carmel in Israel, which has an altitude of 1791 feet, and the Dead Sea?

SOLUTION

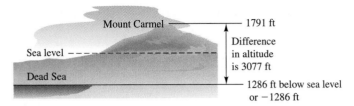

We want to find the difference, so we must subtract:

higher altitude	minus	lower altitude	
↓	↓	↓	
1791 ft	−	(−1286 ft)	
= 1791 ft	+	1286 ft	= 3077 ft

The difference in altitude is **3077 ft**.

PRACTICE PROBLEM 7

Find the difference in altitude between a mountain 3800 feet high and a desert valley 895 feet below sea level.

EXAMPLE 8   The following graph represents the operating profit for Hanover Glass Company. What is the difference between the net loss in the third quarter and the net loss in the second quarter?

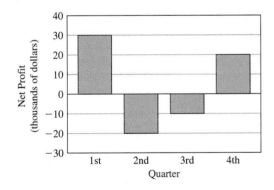

SOLUTION   We want to find the difference, so we subtract.

net loss 3rd quarter	minus	net loss 2nd quarter	equals	difference in net loss
↓	↓	↓	↓	↓
(−10,000)	−	(−20,000)	=	10,000

The difference in net loss is **$10,000**.

PRACTICE PROBLEM 8

Use the graph in Example 8 to answer the following. What is the difference between the net gain in the first quarter and the net loss in the second quarter?

EXERCISES 3.3

Subtract.

1. (a) $20 − $30   **(b)** 5 − 8   **(c)** 4 − 6   **2. (a)** $5 − $10   **(b)** 2 − 7   **(c)** 8 − 10

3. (a) $7 − $8   **(b)** 10 − 11   **(c)** 2 − 3   **4. (a)** $5 − $6   **(b)** 9 − 10   **(c)** 15 − 16

Rewrite each subtraction as addition of the opposite.

Subtraction	Addition of the Opposite
5. (a) 7 − 4 = 3	
(b) 15 − 7 = 8	
(c) 10 − 8 = 2	

Subtraction	Addition of the Opposite
6. (a) 5 − 3 = 2	
(b) 12 − 6 = 6	
(c) 7 − 1 = 6	

Subtract.

7. −6 − 4   **8.** −8 − 3   **9.** −8 − (−6)   **10.** −6 − (−3)

11. −5 − 4   **12.** −4 − 3   **13.** −8 − (−3)   **14.** −7 − (−5)

15. 2 − 7   **16.** 8 − 11   **17.** 4 − (−2)   **18.** 8 − (−4)

19. 3 − 7   **20.** 7 − 9   **21.** 5 − (−9)   **22.** 6 − (−7)

23. 5 − 19   **24.** 9 − 13   **25.** −7 − (−8)   **26.** −6 − (−9)

27. −4 − 18   **28.** −8 − 56   **29.** 5 − (−1)   **30.** 8 − (−1)

31. 8 − 9   **32.** 7 − 8   **33.** −8 − (−2)   **34.** −7 − (−11)

Perform the necessary operations.

35. 7 − 9 − 3 − 8   **36.** 5 − 2 − 6 − 10   **37.** 8 − 1 − 9 − 5   **38.** 3 − 7 − 5 − 16

39. 9 − 10 − 2 − 3   **40.** 8 − 11 − 4 − 1   **41.** 5 − 8 − 6 − 4   **42.** 6 − 4 − 8 − 22

43. 2 − 1 − 9 − 7   **44.** 9 − 3 − 7 − 25   **45.** −6 − (−3) + (−7)   **46.** −5 − (−2) + (−7)

47. −7 − (−2) + (−5)   **48.** −5 − (−9) + (−4)   **49.** −3 − (−8) + (−6)   **50.** −7 − (−2) + (−9)

Evaluate.

51. $x - 12$ for $x = -8$ **52.** $x - 15$ for $x = -9$ **53.** $x - 11$ for $x = -3$ **54.** $x - 10$ for $x = -1$

55. $14 - y$ for $y = -5$ **56.** $19 - y$ for $y = -6$ **57.** $21 - y$ for $y = -1$ **58.** $14 - y$ for $y = -2$

59. $-8 - x$ for $x = -4$ **60.** $-7 - x$ for $x = -3$ **61.** $-1 - x$ for $x = -6$ **62.** $-2 - x$ for $x = -5$

Solve each applied problem.

63. Find the difference in altitude between a mountain that has an altitude of 3556 feet and a desert valley that is 150 feet below sea level.

64. Find the difference in altitude between a mountain that has an altitude of 5889 feet and a desert valley that is 175 feet below sea level.

65. How far above the floor of the basement is the roof of the office building?

326 ft → ── Roof

0 ft → ── Ground floor
−18 ft → ── Basement

66. On the same day in January 1998, the hottest spot in the nation was Gila Bend, Arizona, 78°F, while the coldest spot was Grand Forks, North Dakota, −14°F. What was the difference in temperature between the two cities?

Use the following graph to answer questions 67 and 68.

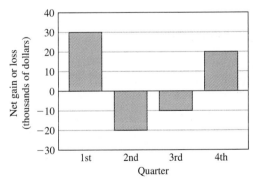

67. What is the difference between the net gain in the first quarter and the net loss in the third quarter?

68. What is the difference between the net gain in the fourth quarter and the net loss in the second quarter?

The following chart displays the hottest and coldest spots for selected days in the winter of 1998. Use this chart to answer questions 69 and 70.

	Record High			Record Low	
Day 1	Gila Bend, Arizona	72°F		Presque Isle, Maine	−18°F
2	Lajitas, Texas	79°F		Ely, Minnesota	−8°F
3	Indio, California	77°F		Devil's Lake, North Dakota	−9°F
4	Brownsville, Texas	88°F		Bodie State Park, California	−13°F
5	Del Rio, Texas	84°F		Presque Isle, Maine	−9°F

69. (a) Where was the temperature the hottest during the five days listed on the chart?

(b) What was the difference in temperature between the record high and the record low on day 3?

70. (a) Where was the temperature the coldest during the five days listed on the chart?

(b) What was the difference in temperature between the record high and the record low on day 5?

CALCULATOR EXERCISES

71. $-345 - 768$ **72.** $-3009 - 893$ **73.** $632 - (-1346)$ **74.** $-2001 - (-987)$

VERBAL AND WRITING SKILLS

75. To subtract two signed numbers, we change the subtraction sign to _____ and take the _____ of the second number, then we _____ .

76. When we subtract $25 - (-10)$, we get a number that is larger then 25. Explain why. You can use an illustration in your explanation.

CUMULATIVE REVIEW

Simplify.

1. $2 + 3(5)$ **2.** $12 - 3(4 - 1)$ **3.** $3^2 + 4(2) - 5$ **4.** $3 + [3 + 2(8 - 6)]$

SECTION 3.4

Multiplication and Division of Signed Numbers

After studying this section, you will be able to:

1 Multiply two signed numbers.
2 Multiply more than two signed numbers.
3 Use exponents with signed numbers.
4 Divide with signed numbers.

Math Pro   Video 3.4   SSM

We are familiar with multiplying and dividing positive numbers. In this section we learn how to multiply and divide signed numbers.

1 Multiplying Two Signed Numbers

Recall the different ways we can indicate multiplication. You should be able to identify and use all of them.

$$-3 \times 5 \qquad -3 \cdot 5 \qquad 3(-5) \qquad (-3)(-5)$$

How do we determine whether a product is positive or negative? We follow a set of rules for multiplying signed numbers. Before we state these rules, let's look at some situations involving multiplication of signed numbers.

Since multiplication represents repeated addition, we can express situations involving addition as a multiplication problem.

Situation 1: Your business has a profit of $1000 a month for 3 months.

Math symbols that represent situation 1:

$$\$1000 + \$1000 + \$1000 = \$3000 \quad \text{or} \quad (3)(\$1000) = \$3000$$

Situation 2: Your business suffers a loss of $1000 a month for 3 months.

Math symbols that represent situation 2:

$$-\$1000 + (-\$1000) + (-\$1000) = -\$3000 \quad \text{or} \quad (3)(-\$1000) = -\$3000$$

EXAMPLE 1   To find the product, write as repeated addition: $4(-5)$.

SOLUTION

$$4(-5) = -5 + (-5) + (-5) + (-5) = \mathbf{-20}$$

PRACTICE PROBLEM 1

To find the product, write as repeated addition: $3(-1)$.

Let's summarize the results of Example 1. $4(-5) = -20$. The product of a *positive* number and a *negative* number is *negative*. Since multiplication is commutative, we know that $(-5)4 = -20$, and thus we see that a *negative* number times a *positive* number is also *negative*.

Since we know that the product of two *positive* numbers is a positive number, we might see the first part of a general rule.

PROCEDURE TO FIND THE PRODUCT OF SIGNED NUMBERS

1. Determine the *sign* of the product as follows.

   Positive number $\times$ positive number is positive: $[+] \cdot [+] = [+]$

   Negative number $\times$ positive number is negative: $[-] \cdot [+] = [-]$

   Positive number $\times$ negative number is negative: $[+] \cdot [-] = [-]$

2. Multiply the absolute values (numerical parts).

EXAMPLE 2 Multiply.

(a) $6(7)$ **(b)** $6(-7)$ **(c)** $(-6)(7)$

SOLUTION

(a) $6(7) = \mathbf{42}$ 1. Determine the sign of the product: $[+] \cdot [+] = [+]$.
     2. Multiply the absolute values: $6 \cdot 7 = 42$.

(b) $6(-7) = \mathbf{-42}$ 1. Determine the sign of the product: $[+] \cdot [-] = [-]$.
     2. Multiply the absolute values: $6 \cdot 7 = 42$.

(c) $(-6)(7) = \mathbf{-42}$ 1. Determine the sign of the product: $[-] \cdot [+] = [-]$.
     2. Multiply the absolute values: $6 \cdot 7 = 42$.

Observe that the numerical part of the products in parts (a), (b), and (c) is 42. The only difference in each answer is the sign.

PRACTICE PROBLEM 2
Multiply.

(a) $3(8)$ **(b)** $3(-8)$ **(c)** $(-3)(8)$

How do we multiply two negative numbers? Consider the following. Multiplying a number by -1 gives the same result as taking the opposite of the number.

(-1) times 5 $\rightarrow (-1)(5) = -5$ Since $[-] \cdot [+] = [-]$.
The opposite of 5 $\rightarrow$ -5

Thus, to multiply a number by (-1), we can take the opposite of the number. That is, to multiply $(-1)(-2)$, we take the opposite of -2.

$(-1)(-2) = $ the opposite of $(-2) = 2$

This seems to suggest that a *negative* number times a *negative* number gives a *positive* result, and this is the case.

> Negative number $\times$ negative number is positive: $[-] \cdot [-] = [+]$
> Positive number $\times$ positive number is positive: $[+] \cdot [+] = [+]$
> Negative number $\times$ positive number is negative: $[-] \cdot [+] = [-]$
> Positive number $\times$ negative number is negative: $[+] \cdot [-] = [-]$

Another way to state the rule for multiplication is as follows: *If two numbers have the same sign, the product is positive; if they have different signs, the product is negative.*

EXAMPLE 3 Multiply: $(-9)(-8)$.

SOLUTION

$(-9)(-8) = \mathbf{72}$ $[-] \cdot [-] = [+]$ and $9 \cdot 8 = 72$.

PRACTICE PROBLEM 3
Multiply: $(-2)(-4)$.

DEVELOPING YOUR STUDY SKILLS

WHEN TO USE A CALCULATOR

A calculator is an important tool and therefore we benefit by learning how to use it. You may be thinking, "Why learn math if I can use a calculator?" Well, often it is not practical or convenient to use a calculator. Many times we must perform calculations unexpectedly and do not have a calculator available, as in the following situations.

- You receive change for a purchase made. Is the change correct?
- You are shopping and notice that an item you'd like to buy is marked down 30 percent. You must calculate the reduced price to determine if you can afford to buy it.
- You have lunch with four friends and the bill is on one ticket. How much do you owe for your lunch?

These are just a few of the many situations that require you to use your knowledge of mathematics. Besides, even if you had a calculator in the situations above, you must still have the ability to know how to go about solving the problem. Do I add, subtract, multiply, or divide? The calculator does only what you tell it to do; it does not plan the approach! Learning how to do mathematics develops problem-solving skills, and the calculator assists us in solving the problem.

EXERCISES

Do each of the following problems without, then with a calculator. Which way was faster, with or without a calculator?

1. Add $-2 + 3 + 1$.

2. Combine like terms: $4xy + 2x + 3xy$.

3. If you didn't know the rules, could you do these problems with a calculator?

---

②   Multiplying More Than Two Signed Numbers

When multiplying more than two signed numbers, multiply any pair of numbers first, then multiply the result by another number. Continue until each factor has been used once.

EXAMPLE 4   Multiply.

(a) $(-3)(-1)(-2)$ **(b)** $(-2)(-4)(-3)(-1)$

SOLUTION

(a) $(-3)(-1)\,(-2) = 3\,(-2)$ First, we multiply: $(-3)(-1) = 3$.

$\qquad\qquad\qquad = -6$ Then, we multiply: $3(-2) = -6$.

(b) $(-2)(-4)\,(-3)(-1) = 8\,(-3)(-1)$ First, we multiply: $(-2)(-4) = 8$.

$\qquad\qquad\qquad\quad = 8(3)$ Next, we multiply: $(-3)(-1) = 3$.

$\qquad\qquad\qquad\quad = 24$ Then, we multiply: $8(3) = 24$.

Observe that the factors could be multiplied in a different order. The result would be the same.

PRACTICE PROBLEM 4

Multiply.

(a) $(-2)(-1)(-4)$ **(b)** $(-3)(-2)(-1)(-3)$

Let's summarize the answers to Example 4 and look for a pattern. Recall that when we multiply *two negative numbers*, the answer is *positive*.

$(-3)(-1)(-2) = -6$ When we multiply *three negative numbers*, the answer is *negative*.

$(-2)(-4)(-3)(-1) = 24$ When we multiply *four negative numbers*, the answer is *positive*.

What do you think would happen if we multiplied *five negative numbers*? If you guessed *negative*, you probably see the pattern. The pattern can be summarized briefly as follows:

> When you multiply two or more signed numbers:
>
> 1. The result is always *positive* if there are an *even* number of negative signs.
> 2. The result is always *negative* if there are an *odd* number of negative signs.

Why is this true? Because every pair of two negative signs yields a positive result. Finding the product is simplified if we multiply as stated below.

> To multiply signed numbers:
>
> 1. Determine the sign of the product.
> 2. Multiply the absolute values (numerical parts).

EXAMPLE 5   Multiply: $(-3)(2)(-1)(4)(-3)$.

SOLUTION

$(-3)(2)(-1)(4)(-3)$ The *answer is negative* since there are 3 *negative* signs and 3 is an *odd* number.

$= -\left[3(2)(1)(4)(3)\right]$ Now we can multiply the absolute values.

$= -72$

PRACTICE PROBLEM 5

Multiply: $(-3)(2)(-1)(6)(-3)$.

③ Using Exponents with Signed Numbers

In Chapter 2 we saw that we use exponents as a way to abbreviate repeated multiplication.

$$
\begin{array}{cc}
\text{repeated} & \text{exponent} \\
\text{multiplication} & \text{form} \\
(-3) \cdot (-3) \cdot (-3) & = (-3)^3
\end{array}
$$

The numerical value of the exponent determines the number of factors, and thus the number of negative signs in the multiplication. Therefore, we can evaluate a negative number written in exponent form in the same way that we do when we multiply.

The number of negative signs is **even**.	The exponent is **even**.	The product is **positive**.

$$(-3)(-3) \quad = \quad (-3)^2 \quad = \quad 9$$

The number of negative signs is **odd**.	The exponent is **odd**.	The product is **negative**.

$$(-3)(-3)(-3) \quad = \quad (-3)^3 \quad = \quad -27$$

We see that the sign of the product depends on whether the exponent is odd or even. This can be generalized as follows.

SIGN RULE FOR INTEGERS IN EXPONENT FORM

Suppose that a number is written in exponent notation and the *base is negative*. The expression is *positive* if the exponent is *even*. The expression is *negative* if the exponent is *odd*.

EXAMPLE 6   Evaluate: $(-4)^3$.

SOLUTION

$$(-4)^3 = (-4)(-4)(-4) = \mathbf{-64}$$ The answer is negative since the exponent 3 is odd.

PRACTICE PROBLEM 6

Evaluate: $(-4)^4$.

EXAMPLE 7   Evaluate.

(a) $(-1)^2$ **(b)** $(-1)^3$ **(c)** $(-1)^{31}$

SOLUTION

(a) $(-1)^2 = \mathbf{1}$ The answer is positive since the exponent 2 is even.
(b) $(-1)^3 = \mathbf{-1}$ The answer is negative since the exponent 3 is odd.
(c) $(-1)^{31} = \mathbf{-1}$ The answer is negative since the exponent 31 is odd.

PRACTICE PROBLEM 7

Evaluate.

(a) $(-2)^3$ **(b)** $(-2)^6$ **(c)** $(-2)^7$

We must be sure to use parentheses when the base is a negative number. For example, $(-2)^4$ is not the same as -2^4. *The base of a number in exponent form does not include the negative sign unless we use parentheses.*

$(-2)^4$ means "-2 raised to the fourth power," since -2 is the base.

-2^4 means "the opposite of 2 raised to the fourth power," since 2 is the base.

$$(-2)^4 = (-2)(-2)(-2)(-2) = 16$$

↑   ↑

The base is -2 We use the -2 as the factor for repeated multiplication.

$$-2^4 = -(2 \cdot 2 \cdot 2 \cdot 2) = -(16) = -16$$

↑   ↑

The base is 2   We use 2 as the factor for repeated multiplication and take the opposite of the product.

E X A M P L E   8   Evaluate.

(a) -3^2 **(b)** $(-3)^2$

SOLUTION

(a) $-3^2 = -(3 \cdot 3)$ The base is 3; we use 3 as the factor for repeated multiplication.

$\qquad = -9$ We take the opposite of the product.

(b) $(-3)^2 = (-3)(-3)$ The base is -3; we use -3 as the factor for repeated multiplication.

$\qquad = 9$

P R A C T I C E   P R O B L E M   8

Evaluate.

(a) -5^2 **(b)** $(-5)^2$

④ Dividing with Signed Numbers

What about division? Any division problem can be rewritten as a multiplication problem. Therefore, the rules for division are very much like those for multiplication.

Division problem: $(-20) \div (-4) = n$
Equivalent multiplication problem: $(-20) = n(-4)$

└──→ n must be positive 5, since $\mathbf{5}(-4) = -20$.

Therefore, $(-20) \div (-4) = 5$

Division problem: $(-20) \div 4 = n$
Equivalent multiplication problem: $(-20) = n(4)$

└──→ n must be negative 5, since $(-\mathbf{5})(4) = -20$.

Therefore, $(-20) \div 4 = -5$. Similarly, $20 \div (-4) = -5$ because $20 = (-5)(-4)$.

As we can see, the rules for division are the same as those for multiplication. We will state them together. When you multiply or divide two numbers and the *signs* are the *same*, the answer is *positive*. When you multiply or divide two numbers and the *signs* are *different*, the answer is *negative*.

MULTIPLICATION AND DIVISION OF SIGNED NUMBERS

1. Determine the sign of the answer as follows:

$$[-] \cdot [-] = [+] \qquad [-] \div [-] = [+]$$
$$[+] \cdot [+] = [+] \qquad [+] \div [+] = [+]$$
$$[+] \cdot [-] = [-] \qquad [+] \div [-] = [-]$$
$$[-] \cdot [+] = [-] \qquad [-] \div [+] = [-]$$

2. Multiply or divide the absolute values.

EXAMPLE 9 Divide.

(a) $36 \div 6$ **(b)** $36 \div (-6)$ **(c)** $(-36) \div 6$ **(d)** $(-36) \div (-6)$

SOLUTION

(a) $36 \div 6 = \mathbf{6}$ $[+] \div [+] = [+]$

(b) $36 \div (-6) = \mathbf{-6}$ $[+] \div [-] = [-]$

(c) $(-36) \div 6 = \mathbf{-6}$ $[-] \div [+] = [-]$

(d) $(-36) \div (-6) = \mathbf{6}$ $[-] \div [-] = [+]$

PRACTICE PROBLEM 9

Divide.

(a) $42 \div 7$ **(b)** $42 \div (-7)$ **(c)** $(-42) \div 7$ **(d)** $(-42) \div (-7)$

EXAMPLE 10 Perform each indicated operation.

(a) $56 \div (-8)$ **(b)** $9(-5)$ **(c)** $(-20)(-3)$ **(d)** $\dfrac{-72}{-8}$

SOLUTION

(a) $56 \div (-8) = \mathbf{-7}$ **(b)** $9(-5) = \mathbf{-45}$

(c) $(-20)(-3) = \mathbf{60}$ **(d)** $\dfrac{-72}{-8} = (-72) \div (-8) = \mathbf{9}$

PRACTICE PROBLEM 10

Perform each indicated operation.

(a) $49 \div (-7)$ **(b)** $4(-9)$ **(c)** $(-30)(-4)$ **(d)** $\dfrac{-54}{-9}$

EXERCISES 3.4

Find the product by writing as repeated addition.

1. $3(-4)$ **2.** $4(-1)$ **3.** $4(-6)$

4. $2(-5)$ **5.** $2(-3)$ **6.** $3(-2)$

Multiply.

7. $4(-1)$ **8.** $5(-8)$ **9.** $(-5)(9)$ **10.** $(-7)(1)$

11. $(-2)(-9)$ **12.** $(-5)(-4)$ **13.** $(-3)(-6)$ **14.** $(-4)(-3)$

15. $(5)(-6)$ **16.** $(2)(-11)$ **17.** $(-8)(3)$ **18.** $(-7)(3)$

19. (a) $4(2)$ **(b)** $4(-2)$ **(c)** $(-4)(2)$ **(d)** $(-4)(-2)$

20. (a) $11(7)$ **(b)** $11(-7)$ **(c)** $(-11)(7)$ **(d)** $(-11)(-7)$

21. (a) $5(2)$ **(b)** $(-5)(-2)$ **(c)** $(-5)(2)$ **(d)** $5(-2)$

22. (a) $1(8)$ **(b)** $(-1)(-8)$ **(c)** $(-1)(8)$ **(d)** $1(-8)$

23. $(-3)(-2)(-3)(-4)$ **24.** $(-5)(-3)(-2)(-2)$ **25.** $3(-7)(-2)$ **26.** $2(-4)(-6)$

27. $3(-1)(5)(-6)$ **28.** $9(-1)(2)(-3)$ **29.** $(-2)(-1)(4)(-5)$ **30.** $(-1)(-3)(2)(-4)$

31. $(-5)(4)(-3)(2)(-1)$ **32.** $(-4)(5)(-2)(1)(-4)$

Evaluate.

33. $(-5)^2$ **34.** $(-7)^2$ **35.** $(-5)^3$ **36.** $(-7)^3$

37. (a) $(-3)^2$ **(b)** $(-3)^3$ **38. (a)** $(-2)^2$ **(b)** $(-2)^3$

39. (a) $(-1)^{11}$ **(b)** $(-1)^{24}$ **40. (a)** $(-1)^{21}$ **(b)** $(-1)^{16}$

41. (a) -4^2 **(b)** $(-4)^2$ **42. (a)** -6^2 **(b)** $(-6)^2$

Divide.

43. $20 \div (-5)$ **44.** $10 \div (-2)$ **45.** $\dfrac{-36}{6}$ **46.** $\dfrac{-24}{4}$

47. $-16 \div (-8)$ **48.** $-12 \div (-6)$ **49.** $\dfrac{-49}{-7}$ **50.** $\dfrac{-50}{-10}$

51. (a) $35 \div 7$ **(b)** $35 \div (-7)$ **(c)** $(-35) \div 7$ **(d)** $(-35) \div (-7)$

52. (a) $50 \div 5$ **(b)** $50 \div (-5)$ **(c)** $(-50) \div 5$ **(d)** $(-50) \div (-5)$

53. (a) $40 \div 8$ **(b)** $40 \div (-8)$ **(c)** $(-40) \div 8$ **(d)** $(-40) \div (-8)$

54. (a) $20 \div 4$ **(b)** $20 \div (-4)$ **(c)** $(-20) \div 4$ **(d)** $(-20) \div (-4)$

Perform each indicated operation.

55. (a) $22 \div (-2)$ **(b)** $22(-2)$ **56. (a)** $18 \div (-3)$ **(b)** $18(-3)$

57. (a) $-4 \div (-2)$ **(b)** $-4(-2)$ **58. (a)** $-8 \div (-4)$ **(b)** $-8(-4)$

59. Baker Sporting Goods marked \$2 off the price of all baseball gloves in stock. If there are 350 gloves in stock, write the total reduction of all gloves as a signed number.

60. During a cold front in Minnesota the temperature dropped 3°F each hour for 4 hours. Express the total drop in temperature as a signed number.

The following formula is used to calculate the distance an object has traveled at a given rate and time. Use this formula to answer questions 61 and 62.

$$\text{Distance} = \text{rate} \times \text{time}$$

61. The velocity (rate) of a projectile is -30 meters per second, indicating that it is moving to the left on a number line. Currently, it is at time $t = 0$ and at the zero mark on the number line. Find where it will be on the number line in 3 seconds.

62. Find where the projectile in problem 61 will be on the number line after 4 seconds.

 CALCULATOR EXERCISES

63. $(-578)(-698)$ **64.** $(986)(-421)$ **65.** $\dfrac{-1357}{23}$ **66.** $\dfrac{-1235}{-65}$

 VERBAL AND WRITING SKILLS

67. If two numbers have the same sign, the product is a _____ number.

68. If two numbers have different signs, the product is a _____ number.

69. The quotient of a positive and a _____ number is negative.

70. The quotient of a negative and a _____ number is positive.

ONE STEP FURTHER

Determine the value of x.

71. $\dfrac{x}{-3} = 8$ **72.** $\dfrac{x}{2} = -10$

Answer true or false.

73. If you multiply fifteen negative numbers, the product will be a positive number.

74. If you multiply twelve negative numbers, the product will be a positive number.

CUMULATIVE REVIEW

1. Evaluate $x + 2$ if x is 7.

2. Evaluate $3y$ if y is 6.

3. Simplify: $(3x^4)(5x^2)$.

4. Simplify: $3(x + 6)$.

After studying this section, you will be able to:

❶ Combine like terms.

❷ Multiply algebraic expressions.

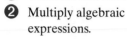

Math Pro   Video 3.5   SSM

SECTION 3.5

Simplifying Algebraic Expressions

❶ **Combining Like Terms**

Simplifying algebraic expressions with signed numbers differs from whole numbers only in that we must consider the sign of the number when simplifying.

$$5x + 3x = \;(5 + 3)x = 8x$$

↓   ↓

Add the   The variable part
numerical coefficients.   stays the same.

$$-5x + 3x = (-5 + 3)x = -2x$$

↓   ↓

Add the   The variable part
numerical coefficients.   stays the same.

EXAMPLE 1   Simplify by combining like terms.

(a) $-9x + 4x$ **(b)** $-4x + 7y + 2x$

SOLUTION

(a) $-9x + 4x = (-9 + 4)x = \mathbf{-5x}$

(b) $-4x + 7y + 2x = \mathbf{-4x + 2x + 7y}$ Rearrange terms.

$$= (-4 + 2)x + 7y \qquad \text{Combine like terms.}$$

$$= \mathbf{-2x + 7y} \qquad\qquad 7y \text{ is not a like term.}$$

PRACTICE PROBLEM 1

Simplify by combining like terms.

(a) $-3b + 5b$ **(b)** $-6y + 8x + 4y$

In Example 1 we were able to rearrange terms because addition is commutative. Since subtraction is *not commutative*, we must first change all subtractions to additions of the opposite and then rearrange the terms.

EXAMPLE 2   Simplify: $3b + 4a - 9b$.

SOLUTION

$$3b + 4a - 9b = 3b + 4a + \mathbf{(-9b)} \quad \begin{array}{l}\text{First write the subtraction as addi-}\\ \text{tion of the opposite.}\end{array}$$

$$= 3b + (-9b) + 4a \qquad \text{We can now rearrange the addition.}$$

$$= [3 + (-9)]b + 4a$$

$$= \mathbf{-6b + 4a}$$

PRACTICE PROBLEM 2

Simplify: $7x + 5y - 8x$.

There are various ways to write the same expression. Let's consider the answer to Example 2, $-6b + 4a$. This expression can be written as

$$-6b + 4a \quad = \quad 4a + (-6b) \quad = \quad 4a - 6b$$
$$\downarrow \qquad\qquad\qquad\qquad\qquad \downarrow$$

Use the commutative   Write as an equivalent
property to rearrange terms.   subtraction.

The expression $4a + (-6b)$ is **not** considered to be in simplified form.

E X A M P L E   3   Perform each indicated operation.

(a) $2 - 3 + 6$ **(b)** $2x - 3x + 6x$

SOLUTION

(a) $2 - 3 + 6 = 2 + (-3) + 6$ Write subtraction as addition of
                                                                the opposite.

$\qquad\qquad = 8 + (-3)$ Add: $2 + 6 = 8$.

$\qquad\qquad = \mathbf{5}$ Add: $8 + (-3) = 5$.

(b) $2x - 3x + 6x = 2x + (-3x) + 6x$ Write subtraction as addition of
                                                                the opposite.

$\qquad\qquad = 8x + (-3x)$ Add: $2x + 6x = 8x$.

$\qquad\qquad = \mathbf{5x}$ Add: $8x + (-3x) = 5x$.

What can you say about the answers to parts (a) and (b)?

PRACTICE PROBLEM 3
Perform each indicated operation.

(a) $4 - 6 + 8$ **(b)** $4x - 6x + 8x$

---

2 Multiplying Algebraic Expressions

Just as with whole numbers, we use the associative and commutative properties of multiplication to simplify algebraic expressions with signed numbers. We state the rule presented in Section 2.4. To multiply algebraic expressions:

1. Multiply the numerical coefficients.
2. If variables in exponent form have the same base, add exponents and keep the base unchanged.

$$4(2x) = (4 \cdot 2)x = 8x \qquad\qquad -4(2x) = (-4 \cdot 2)x = -8x$$

E X A M P L E   4   Multiply.

(a) $(5x)(-8x)$ **(b)** $(-7y^2)(-4y^4)$

SOLUTION

(a) $(5x)(-8x) = (5)(-8)(x \cdot x)$ We have $x \cdot x = x^1 \cdot x^1 = x^2$.

$\qquad\qquad = \mathbf{-40x^2}$

(b) $(-7y^2)(-4y^4) = (-7)(-4)(y^2 \cdot y^4)$ We have $y^{2+4} = y^6$.

$\qquad\qquad = \mathbf{28y^6}$

PRACTICE PROBLEM 4
Multiply.

(a) $(-6a)(-8a)$ **(b)** $(5x^3)(-2y^5)$

EXERCISES 3.5

Simplify by combining like terms.

1. $-8x + 3x$

2. $-6x + 2x$

3. $2x + (-3x)$

4. $6y + (-5y)$

5. $-5x + 7x$

6. $-5a + 8a$

7. $-7a + (-2a)$

8. $-9b + (-3b)$

9. $9x + (-7x)$

10. $5x + (-2x)$

11. $-7x + (-6x)$

12. $-6y + (-8y)$

13. $-8y + 5x + 2y$

14. $-7x + 3y + 5x$

15. $6x + 2y + (-8x)$

16. $9a + 4b + (-3a)$

17. $9x + 3y + (-5x)$

18. $10y + 2x + (-4y)$

19. $-8x - 4x$

20. $-7a - 2a$

21. $-6y - 8y$

22. $-5y - 9y$

23. $4x - (-6x)$

24. $5x - (-8x)$

25. $3a + 2x - 5a$

26. $5y + 4x - 8y$

27. $6x + 5y - 10x$

28. $6x + 3y - 9x$

29. $4x + 2y - 6x$

30. $8x + 4y - 11x$

31. $4 + 3a - 2$

32. $7 - 5y + 9$

33. $5 + 5x - 9$

34. $5 + 8x - 11$

Perform the operations indicated.

35. **(a)** $2 - 7 + 3$
  (b) $2x - 7x + 3x$

36. **(a)** $4 - 9 + 2$
  (b) $4x - 9x + 2x$

37. **(a)** $3 - 8 + 4$
  (b) $3x - 8x + 4x$

38. **(a)** $6 - 10 + 3$
  (b) $6x - 10x + 3x$

39. **(a)** $2 - 6 + 1$
  (b) $2x - 6x + 1x$

40. **(a)** $3 - 8 + 1$
  (b) $3x - 8x + 1x$

Simplify by multiplying.

41. $(3y)(-4y)$

42. $(5x)(-3x)$

43. $(-9a)(-3a)$

44. $(-2a)(-7a)$

45. $(-3x)(8x)$ **46.** $(-7x)(5x)$ **47.** $(-4y)(3y)$ **48.** $(6y)(-4y)$

49. $(3x^3)(-4x^2)$ **50.** $(6y^6)(-2y^2)$ **51.** $(-7m^5)(-8m)$ **52.** $(-5x^3)(-2x)$

53. $(-5n^2)(-2n^5)$ **54.** $(-6n^3)(-2n^5)$ **55.** $(-4x)(x)$ **56.** $(-3x)(x)$

57. $(5x)(3x)(-2x)$ **58.** $(-4x)(2x)(7x)$ **59.** $(4x)(-3x)(8x)$ **60.** $(5x)(-2x)(6x)$

 CALCULATOR EXERCISES

61. $(-251)(192x)$ **62.** $(-356)(134x)$

63. $(34x)(-442x^3)(-32x^5)$ **64.** $(89x)(-198x^6)(-312x^4)$

VERBAL AND WRITING SKILLS

65. Combine like terms.

    (a) $-2x + 3x$ **(b)** $2x + 3x$

    Explain the difference between the processes you used for parts (a) and (b).

66. Multiply.

    (a) $(-4x)(5x)$ **(b)** $(4x)(5x)$

    Explain the difference between the processes you used for parts (a) and (b).

ONE STEP FURTHER

67. Evaluate $2x + 3y + 6$ if $x = -2$ and $y = -5$.   **68.** Evaluate $5a + 6b + 1$ if $a = -4$ and $b = -3$.

CUMULATIVE REVIEW

  1. $2^2 + 3(5) - 1$ **2.** $8 + 2(9 \div 3)$ **3.** $2^3 + (4 \div 2 + 6)$ **4.** $3^2 + (6 \div 2 + 8)$

After studying this section, you will be able to:

❶ Follow the order of operations with signed numbers.

❷ Solve applied problems involving more than one operation.

Math Pro   Video 3.6   SSM

S E C T I O N   3 . 6

Order of Operations and Applications Involving Signed Numbers

When there is more than one operation in a problem, we must follow the order of operations presented in Chapter 2.

Do first	1. Perform operations inside parentheses.
	2. Simplify any expressions with exponents.
	3. Multiply and divide from left to right.
Do last	4. Add and subtract from left to right.

We perform one operation at a time and follow the sequence *identify, calculate, replace*. We identify and underline the highest priority, do the calculation underlined, and then replace the operation underlined with the calculated amount.

We must be careful when working with signed numbers, paying special attention to the sign of the number.

❶ Following the Order of Operations with Signed Numbers

E X A M P L E   1   Simplify: $12 - 30 \div 5(-3)^2 - 2$.

SOLUTION

$$12 - 30 \div 5\underline{(-3)^2} - 2$$

Identify: The highest priority is exponents. *Calculate:* $(-3)^2 = 9$. *Replace:* $(-3)^2$ with 9.

$$= 12 - \underline{30 \div 5}(9) - 2$$

Identify: The highest priority is division. *Calculate:* $30 \div 5 = 6$. *Replace:* $30 \div 5$ with 6.

$$= 12 - \underline{6(9)} - 2$$

Identify: The highest priority is multiplication. *Calculate:* $6 \cdot 9 = 54$. *Replace:* $6 \cdot 9$ with 54.

$$= 12 - 54 - 2$$

We subtract last, changing all subtraction to addition of the opposite.

$$= 12 + (-54) + (-2)$$

We add: $12 + (-54) + (-2) = -44$.

$$= \mathbf{-44}$$

P R A C T I C E   P R O B L E M   1
Simplify: $-6 + 20 \div 2(-2)^2 - 5$.

E X A M P L E   2   Simplify: $\dfrac{-15 + 5(-3)}{13 - 18}$.

SOLUTION   We simplify the numerator and the denominator separately. Then we divide the numerator by the denominator.

$$\frac{-15 + 5(-3)}{13 - 18} = \frac{-15 + (-15)}{-5}$$

We multiply: $5(-3) = -15$.
We subtract: $13 - 18 = -5$.

$$= \frac{-30}{-5}$$

We add: $-15 + (-15) = -30$.

$$= \mathbf{6}$$

We divide last: $-30 \div (-5) = 6$.

PRACTICE PROBLEM 2

Simplify: $\dfrac{-10 + 4(-2)}{11 - 20}$.

② Solving Applied Problems Involving More Than One Operation

Since real-life applications often require that we perform more than one operation, we must take care to follow the order of operations when solving these problems.

EXAMPLE 3   To find the speed of a free-falling skydiver, we use the formula given below. Find the speed of the skydiver if the initial downward velocity(v) is 9 feet per second and $t = 5$ seconds.

speed of skydiver		initial velocity		time it takes to reach speed
↓		↓		↓
s	$=$	v	$+$	$32t$

SOLUTION   We evaluate the formula for the values given: $v = 9$ and $t = 5$.

$$s = v + 32t$$
$$= 9 + 32(5)$$
$$= 9 + 160$$
$$= 169$$

The speed of the skydiver is **169 feet per second**.

PRACTICE PROBLEM 3

Use the formula in Example 3 to find the speed of the skydiver if $v = -8$ and $t = 4$.

EXAMPLE 4   An oxide ion has an electrical charge of -2, while the magnesium ion has $+2$. Find the total charge of 8 oxide and 3 magnesium ions.

SOLUTION   We summarize the information: oxide, -2; magnesium, $+2$. For the total charge:

8 oxide + 3 magnesium

$8\,(-2)\ +\ 3\,(+2) = -16 + 6 = -10$

The total charge is **−10**.

PRACTICE PROBLEM 4

Find the total charge in Example 6 of 9 oxide and 4 magnesium ions.

EXERCISES 3.6

Simplify.

1. $7 - 2(3 - 5)$ **2.** $8 - 4(6 - 9)$ **3.** $-10 + 5(2 - 8)$ **4.** $-9 + 3(4 - 6)$

5. $6 - 9(5 - 8)$ **6.** $1 - 7(5 - 9)$ **7.** $5 + 7(2 - 6)$ **8.** $5 + 3(2 - 5)$

9. $5(-3)(4 - 7) + 9$ **10.** $6(-2)(3 - 9) + 4$ **11.** $-3(4)(3 - 6) + 7$ **12.** $-7(3)(2 - 8) + 6$

13. $3(-2)(9 - 5) + 10$ **14.** $5(-3)(5 - 2) + 3$ **15.** $-2(2)(1 - 8) + 5$ **16.** $-4(2)(3 - 7) + 9$

17. $(-3)^2 + 6(-9)$ **18.** $(-2)^2 + 5(-7)$ **19.** $(-2)^3 + 7(8)$ **20.** $(-3)^3 + 6(2)$

21. $(-6)^2 + 8(-7)$ **22.** $(-5)^2 + 5(-9)$ **23.** $(-2)^3 + 3(-8)$ **24.** $(-3)^3 + 6(-4)$

25. $12 - 20 \div 4(-4)^2 + 9$ **26.** $-15 - 50 \div 10(-3)^2 + 2$

27. $-3 - 8 \div (-2)(-2)^2 + 1$ **28.** $35 - 15 \div (-5)(-3)^2 + 4$

29. $9 + 22 \div (-2)(-3)^2 + 2$ **30.** $5 + 24 \div (-3)(-3)^2 + 2$

31. $\dfrac{32 - 16 \div 4}{7 - 9}$ **32.** $\dfrac{30 - 15 \div 3}{5 - 10}$ **33.** $\dfrac{-45 \div 5 + 1}{2 - (-2)}$ **34.** $\dfrac{-50 \div 2 + 3}{20 - 9}$

35. $\dfrac{2^2 + 6(-3)}{-2 + (-5)}$ **36.** $\dfrac{3^2 + 4(-6)}{-3 + (-2)}$

Solve each applied problem.

37. A projectile is fired straight up with an initial velocity of 72 feet per second. It is known that the subsequent velocity of the projectile is given by the formula $v = 72 - 32t$, where t represents time in seconds. If $v > 0$, the object is rising, and if $v < 0$, it is descending. At which of the following times is the object descending: $t = 1, 2,$ or 3 seconds?

38. During a storm in Anchorage, Alaska the temperature was 8°F at noon. Then it dropped 2°F each hour for the next 4 hours, followed by an additional 5°F the fifth hour. What was the temperature at 5 P.M.?

Ions are atoms or groups of atoms with positive or negative electrical charges. The charges of some ions are given below.

Aluminum +3   Chloride −1
Phosphate −3   Silver   +1

Use these values to find the total charge in problems 39–44.

39. 14 phosphate and 9 silver

40. 11 chloride and 2 aluminum

41. 7 aluminum, 5 chloride, and 4 silver

42. 15 silver, 9 phosphate, and 8 chloride

Removing 5 ions from a substance can be represented by the number −5. If 5 chloride ions are removed from a substance, we multiply $(−5)(−1) = +5$ to find the change in the charge of the remaining substance.

43. Four phosphate ions are removed from a substance. What is the change in the charge of the remaining substance?

44. Seven chloride ions are removed from a substance. What is the change in the charge of the remaining substance?

 CALCULATOR EXERCISES

45. $(−9)^2 − 17(−85)$

46. $(−12)^2 − 15(−90)$

47. $−65 − 144 ÷ 9(−14) + 64$

48. $−25 − 53 + 13(11) − 21$

 VERBAL AND WRITING SKILLS

49. If you multiply −4 times 0, your answer is neither positive nor negative. Why?

50. Is $2 + 3(−1) = 5(−1) = −5$? Why or why not?

ONE STEP FURTHER

Find the value of x.

51. $3 + x − 2(−4) = 7 − (−13)$

52. $−2 + x + 3(−4) = −6 + (−4)$

CUMULATIVE REVIEW

Simplify.

1. $2(x + 3)$ **2.** $4(a + 2)$ **3.** $4x + 3y + 2x + 8$ **4.** $5a + 2b + 6a + 4$

UNIVERSAL TIME

The basis of time for almost all locations in the world is considered the mean solar time for the meridian that passes through Greenwich, England. This time is called *Universal Time* or *Greenwich Time*. To determine the time in another location in the world, a time line similar to the one illustrated below can be used.

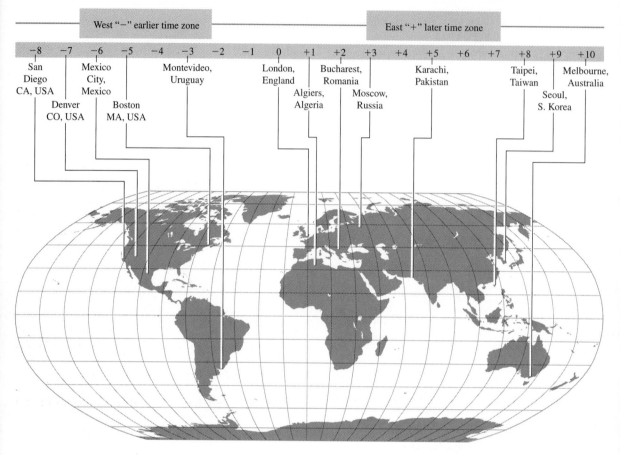

As we travel toward the east the time is later, and toward the west it is earlier. The time zone changes with a "+" indicate that the time is later than London, England, and "−" indicates that the time is earlier than London. For example, if it is 4 P.M. in London, it is 5 hours earlier, or 11 A.M., in Boston, Massachusetts.

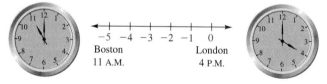

To find the time of day between any two time zones, we can proceed as follows:

1. Find the difference in hours between time zones.
2. Depending on whether you are changing to a later or earlier time zone, add or subtract this amount to the given time.
3. If necessary, change between A.M. and P.M. and days of the week.

For example, if it is 1 P.M. in Mexico City, Mexico, how do we find the time in Moscow, Russia? First, we find the difference in hours between time zones by subtracting time zone values:

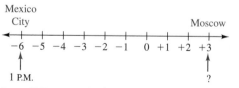

The time difference is: $|-6 - 3| = 9$ hours later in Moscow. Note that we find the absolute value of the difference since it does not make sense to consider negative hours. Since it is later in Moscow, we add: 1 P.M. + 9 hours is 10 P.M.

PROBLEMS FOR INDIVIDUAL INVESTIGATION

1. It is 10 A.M. in London. What time is it in Montevideo, Uruguay?

2. It is 2 P.M. in London. What time is it in Seoul, S. Korea?

3. It is 3 P.M. in Moscow, Russia. What time is it in Denver, Colorado?

4. It is 6 A.M. in Taipei, Taiwan. What time is it in Karachi, Pakistan?

PROBLEMS FOR GROUP INVESTIGATION AND COOPERATIVE STUDY

1. Victor Arias of Mexico City, Mexico watched an event televised live from the Summer Olympics in Sydney, Australia. The event took place at 9 A.M. on Monday morning in Sydney, Australia. If Sydney is in the same time zone as Melbourne, what day and time did Victor view the event in Mexico City?

2. Refer to the situation described in problem 1 to answer the following. If Victor must work from 9 A.M. to 6 P.M. Monday through Friday, will he be home to watch an event that takes place at the Olympics at 2 P.M. on Tuesday? Why or why not?

3. Justin Pommier must set up a 30-minute conference call with business associates in Bucharest, Romania and Montevideo, Uruguay. He wants to have the conference call between the hours of 8 A.M. and 5:30 P.M. so that each associate will be at the office. At what time should he make the call from his office in Boston?

4. Three senior engineers for the Misor engineering firm are on assignment overseas. One engineer is in Karachi, Pakistan, another is in Montevideo, Uruguay, and the third is in Algiers, Algeria. The three must set up a 30-minute conference call between the hours of 8 A.M. and 4:30 P.M. At what time should the engineer in Algeria make the call to his associates?

INTERNET: Go to http://www.prenhall.com/blair to explore this application.

CHAPTER ORGANIZER

TOPIC	PROCEDURE	EXAMPLES
Absolute value	The *absolute value* of a number *a* is the number of units between 0 and *a* on the number line.	Find: (a) $\lvert -6 \rvert$ (b) $\lvert 8 \rvert$ (a) $\lvert -6 \rvert = 6$ (b) $\lvert 8 \rvert = 8$
Opposites of numbers	Numbers that are the same distance from zero but lie on the opposite side of zero on the number line are called *opposites*. We use "−" to indicate the opposite of a number.	Give the opposites of each of the following numbers: 1, −3: The opposite of 1 is −1. The opposite of −3 is 3.
Adding two numbers with the same sign	To add two numbers with the same sign: 1. Use the common sign, 2. Add the absolute value of the numbers.	Add: $-4 + (-6)$. $-4 + (-6) = -10$
Adding two numbers with different signs	To add two numbers with different signs: 1. Use the sign of the larger absolute value in the answer. 2. Subtract the absolute values of the numbers.	Add: (a) $9 + (-5)$ (b) $-8 + 6$ (a) $9 + (-5) = 4$ (b) $-8 + 6 = -2$

Reading graphs involving signed numbers	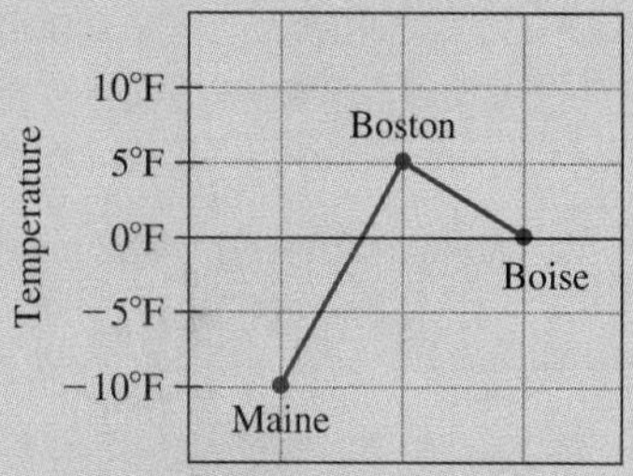	1. Where was the temperature the coldest?
		
		   Maine
		
		2. Where was the temperature a positive number?
		
		   Boston

Subtracting signed numbers	To subtract two signed numbers, add the opposite of the second number to the first.	Subtract: (a) $-7 - 3$ (b) $4 - (-8)$ (a) $-7 - 3 = -7 + (-3) = -10$ (b) $4 - (-8) = 4 + 8 = 12$
Evaluating expressions with signed numbers	To evaluate an expression, we replace the variable with the given number and simplify.	Evaluate: (a) $-3 + x$ for $x = -4$. (b) $-6 - x$ for $x = -1$. (a) $-3 + x = -3 + (-4) = -7$ (b) $-6 - x = -6 - (-1) = -5$
Multiplying signed numbers	To find the product of signed numbers we determine the *sign* of the product from the chart below, then multiply absolute values. $[+] \cdot [+] = [+]$ $[-] \cdot [+] = [-]$ $[-] \cdot [-] = [+]$ $[+] \cdot [-] = [-]$ If two numbers have the same sign, the product is positive; if they have different signs, the product is negative.	Multiply: (a) $(-2)(-5)$ (b) $8(-3)$ (a) $(-2)(-5) = 10$ (b) $8(-3) = -24$

TOPIC	PROCEDURE	EXAMPLES
Using exponents with signed numbers	If a number is written in exponent notation and the base is negative: 1. The expression is positive if the exponent is even. 2. The expression is negative if the exponent is odd.	Evaluate: (a) $(-3)^3$ (b) $(-5)^2$ (c) -2^2 (a) $(-3)^3 = -27$ (b) $(-5)^2 = 25$ (c) $-2^2 = -(2 \cdot 2) = -4$ *Note:* The base in -2^2 is 2, so we square 2, then take the opposite.
Dividing signed numbers	To find the quotient of signed numbers we determine the *sign* of the quotient from the chart below, then divide absolute values. $[-] \div [-] = [+]$ $[+] \div [+] = [+]$ $[+] \div [-] = [-]$ $[-] \div [+] = [-]$ If two numbers have the same sign, the quotient is positive; if they have different signs, the quotient is negative.	Divide: (a) $42 \div (-6)$ (b) $-54 \div (-9)$ (a) $42 \div (-6) = -7$ (b) $-54 \div (-9) = 6$
Combining like terms	To combine like terms, we add or subtract the numerical coefficients of like terms. The variable part stays the same.	Simplify: $4x + 5y - 7x$. We change subtraction to addition of the opposite, then simplify. $$4x + 5y - 7x = 4x + 5y + (-7x)$$ $$= 4x + (-7x) + 5y$$ $$= -3x + 5y$$
Multiplying algebraic expressions	To multiply expressions: 1. First we multiply the numerical coefficients. 2. If variables in exponent form have the same base, add exponents and keep the base unchanged.	Multiply: $(-8x^4)(5x^3)$. $$(-8x^4)(5x^3) = (-8 \cdot 5)(x^4 \cdot x^3)$$ $$= -40x^7$$
Order of operations with signed numbers	We follow the same order of operations presented in Chapter 2: 1. Perform operations inside parentheses. 2. Simplify any expressions with exponents. 3. Multiply and divide from left to right. 4. Add and subtract from left to right.	Simplify: $\dfrac{-12 + 3(-2)}{3 - 9}$. Simplify the numerator and denominator separately. Then divide. $$\frac{-12 + 3(-2)}{3 - 9} = \frac{-12 + (-6)}{3 - 9}$$ $$= \frac{-18}{-6} = 3$$

CHAPTER 3 REVIEW

SECTION 3.1

Replace the ? with the inequality symbol < or >.

1. −3 ? −1

2. −7 ? −2

3. 4 ? −4

4. 5 ? −5

Replace the ? with a + or − symbol.

5. ? Profit

6. ? Loss

Evaluate the absolute value.

7. |5|

8. |−7|

9. Which of the two numbers has the larger absolute value: −23 or 15?

10. Which of the two numbers has the larger absolute value: 12 or −10?

11. The opposite of −16 is ____?

12. The opposite of 8 is ____?

Justin invested in various stocks from January through May. At the end of each month he calculated how much net gain or loss he made on all his stock transactions, then recorded this information on the following line graph. Use the graph to answer questions 13 and 14.

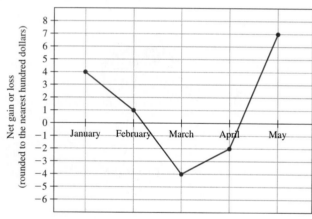

13. (a) In which month did Justin make the most money on his investments?

(b) In which month did Justin lose the most money on his investments?

14. (a) List the months in which Justin had a net gain.

(b) List the months in which Justin had a net loss.

SECTION 3.2

Express the outcome of the situation as a signed number.

15. A loss of $500 followed by a loss of $200.

16. A tax increase of $100 followed by a tax increase of $200.

Evaluate.

17. (a) (−43) + (−16)   **(b)** 43 + 16

18. (a) (−27) + (−39)   **(b)** 27 + 39

19. A company lost $25,000 in May and had a net gain (profit) of $15,000 in June. At the end of these 2 months, was there a net profit or a loss?

20. Terry lost $14 in the slot machines on Thursday. Later that evening he won $25 playing roulette. At the end of the evening, was there a net profit or a loss?

21. Yesterday the temperature in Yosemite dropped to $-32°F$. Today it had risen $20°F$.

    (a) Write the math symbols that represent the situation.

    (b) Today, was the temperature a positive or a negative reading?

    (c) Use the thermometer to find the sum.

Add using the rules for addition of signed numbers.

22. (a) $2 + (-8)$ **(b)** $(-2) + 8$ **(c)** $(-2) + (-8)$

23. (a) $32 + (-18)$ **(b)** $(-32) + 18$ **(c)** $(-32) + (-18)$

24. $3 + (-5) + 8 + (-2)$ **25.** $24 + (-52) + (-12) + (-56)$

26. Evaluate $x + 6$ for $x = -1$.   **27.** Evaluate $-2 + x$ for $x = -3$.

28. At noon a U-boat dives 900 feet below the surface. Then at 2 P.M. it rises 220 feet. Express the depth of the U-boat at 2 P.M. as a signed number.

29. To avoid turbulence due to a storm, the pilot is changing altitudes until he finds an altitude that is not quite so turbulent. How many feet above or below the initial elevation of 35,000 feet is the plane at 5 P.M.? Express your answer as a signed number.

Flight Recordings of Altitude

4:15 P.M.	240 descent
4:30 P.M.	350 ascent
4:45 P.M.	400 ascent
5:00 P.M.	800 descent

SECTION 3.3

Perform the indicated operations.

30. $-7 - 5$ **31.** $-9 - (-4)$ **32.** $-5 - 3$

33. $-6 - (-2)$ **34.** $-3 - 8 + 6$ **35.** $6 - (-4) + (-5)$

36. $-4 - (-2)$ **37.** $6 - 9 - 2 - 8$ **38.** $-6 - (-9) + (-1)$

Use the following graph to answer questions 39 and 40.

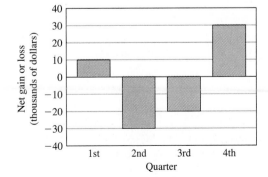

39. What is the difference between the net gain in the fourth quarter and the net loss in the third quarter? Use the graph on page 223.

40. What is the difference between the net gain in the first quarter and the net loss in the second quarter? Use the graph on page 223.

SECTION 3.4

Multiply.

41. (a) $6(3)$ **(b)** $6(-3)$ **(c)** $(-6)(3)$ **(d)** $(-6)(-3)$

42. (a) $5(2)$ **(b)** $5(-2)$ **(c)** $(-5)(2)$ **(d)** $(-5)(-2)$

43. $(-7)(-5)$ **44.** $(-2)(5)$ **45.** $3(-4)$ **46.** $(-4)(-1)$

47. $(-2)(-5)(-9)$ **48.** $(-2)(-8)(-1)(-4)$ **49.** $(-5)(1)(-2)(4)(-6)$

Evaluate.

50. $(-5)^3$ **51.** -2^2 **52.** $(-2)^2$

Divide.

53. (a) $49 \div 7$ **(b)** $49 \div (-7)$ **54. (a)** $(-30) \div 5$ **(b)** $(-30) \div (-5)$

Perform the operations indicated.

55. (a) $-44 \div (-4)$ **(b)** $9(-5)$ **(c)** $(-11)(-3)$ **(d)** $\dfrac{25}{-5}$

56. (a) $12 \div (-4)$ **(b)** $5(-8)$ **(c)** $(-12)(-2)$ **(d)** $\dfrac{45}{-9}$

SECTION 3.5

Simplify by combining like terms.

57. $-5y + 3y$ **58.** $-4y + 3x + 9y$ **59.** $-3a - 6a$

60. $7x + 9y - 6x$ **61.** $3 + 5z - 7$ **62.** $-8 + 7y + 7 - 2y$

Simplify by multiplying.

63. $(3x)(-7x)$ **64.** $(-9y)(2y)$ **65.** $(-3x)(-9x)$ **66.** $(-5x)(2x)$

67. $(4x^5)(-3x^2)$ **68.** $(-7z^7)(5z^3)$ **69.** $(-3a^4)(-4a^{10})$ **70.** $(3y^4)(-6y^{11})$

SECTION 3.6

Simplify.

71. $4 - 1(6 - 9)$ **72.** $3(-5)(2 - 6) + 8$ **73.** $-2^2 + 3(-4)$ **74.** $\dfrac{-32 \div 8 + 4}{7 - 9}$

75. The temperature of a small lake in Michigan was 12°F at 8 P.M. If the temperature of the lake dropped 5°F every hour for the next 3 hours, then dropped another 2°F the fourth hour, what was the temperature of the lake at midnight?

CHAPTER 3 TEST

Replace the ? with the inequality symbol $<$ or $>$.

1. -234 ? -5

2. $|4|$? $|-18|$

3. Replace the ? with the appropriate symbol, $+$ or $-$: The Dow Jones Industrial Average falls 14 points. __?__ 14 points.

4. Evaluate the absolute value.

    (a) $|12|$

    (b) $|-3|$

5. The opposite of -8 is ____.

6. The opposite of 10 is ____.

Simplify.

7. $-6 + 8$

8. $-6 + (-4)$

9. $-20 + 5$

10. Last night the temperature dropped to $-10°F$. At dawn the temperature had risen $15°F$.

    (a) Write the math symbols that represent the situation.

    (b) At dawn, what was the temperature?

11. For the first quarter of 1998, Earth Systems had a $20,000 profit. For the second quarter, the company had a $5000 loss. What was the company's overall profit or loss at the end of the second quarter?

Perform the operations indicated.

12. $12 - 18$

13. $-1 - 11$

14. $3 - (-10)$

15. $-14 - 3 + (-6)$

16. Evaluate $-7 + x$ for:

    (a) $x = 16$

    (b) $x = -7$

Multiply.

17. $(7)(-3)$

18. $(-8)(-4)$

19. $(-5)(-2)(-1)(3)$

Evaluate.

20. (a) $(-5)^2$

    (b) $(-5)^3$

    (c) -5^2

Divide.

21. (a) $(-8) \div 2$

    (b) $(-8) \div (-2)$

22. $\dfrac{-22}{11}$

1. _____

2. _____

3. _____

4. _____

5. _____

6. _____

7. _____

8. _____

9. _____

10. _____

11. _____

12. _____

13. _____

14. _____

15. _____

16. _____

17. _____

18. _____

19. _____

20. _____

21. _____

22. _____

23. Perform the indicated operations: $5 - 7 + 3 - 8$.

23. _____

24. Simplify: $-5x - 2x + 8y$.

24. _____

25. Multiply: $(-8x^2)(-9x^4)$

25. _____

Simplify.

26. $-4(11 - 13) + 5$ **27.** $2 - 35 \div 5(-3)^2 - 6$

26. _____

28. $\dfrac{-8 + 2(-3)}{14 - 21}$

27. _____

29. The formula to determine a person's intelligence quotient (IQ) is

28. _____

$\text{IQ} = \dfrac{100M}{C}$, where M stands for mental age as measured on a particular test. C is a person's chronological (actual) age. Find the IQ for a person with $C = 20$ and $M = 23$.

29. _____

CUMULATIVE TEST FOR CHAPTERS 1-3

1. There are 5280 feet in a mile. Round 5280 to the nearest hundred.

2. Combine like terms: $3 + 6m + 2 + 4m + m$.

3. According to *Webster's New Collegiate Dictionary*, Early Modern English was used from 1475 to 1650. For how many years was Early Modern English used?

4. Solve for x: $25 + x = 85$.

5. Evaluate 3 more than x, if x is 10.

6. Solve for x and check your answer: $\dfrac{28}{x} = 4$.

7. Translate into an equation: The sum of x and 9 equals 16.

8. Which one of the following metric units would be the most reasonable measurement for the height of an adult male: grams, kilometers, centimeters, or liters?

Simplify.

9. $6(7x)$ 10. $3(y \cdot 8)$ 11. $400(7n)$

Translate each English statement into an equation and solve.

12. Double a number plus 14 equals twenty-eight.

13. Eight squared equals a number.

Multiply.

14. $(219)(67)$ 15. $(5y^2)(3y^5)$ 16. Solve: $9y = 63$.

17. Find the area of a square with a side that measures 12 feet.

18. VideoTime made $167,350 in 1998. The expenses for that year were $86,000.
    (a) How much profit did VideoTime make?
    (b) If the 2 owners divided the profits equally, how much money did each owner receive?

Replace the ? with the inequality symbol $<$ or $>$.

19. -8 ? -617 20. $|-17|$? $|-2|$

Perform the operations indicated.

21. $5 + (-6)$ 22. $-10 - 8$ 23. $(16)(-2)$

24. $(-18) \div (-9)$ 25. Evaluate: $(-2)^5$.

26. Combine like terms: $3mn - 7mn + 4m$.

27. Multiply: $(-3x^2)(4x^6)$

28. Simplify: $-4 + 15 \div 5(-3)^2 - 1$.

1. _____

2. _____

3. _____

4. _____

5. _____

6. _____

7. _____

8. _____

9. _____

10. _____

11. _____

12. _____

13. _____

14. _____

15. _____

16. _____

17. _____

18. _____

19. _____

20. _____

21. _____

22. _____

23. _____

24. _____

25. _____

26. _____

27. _____

28. _____

FRACTIONS, RATIO, AND PROPORTION

We use proportions when we plan the menu for a large event such as a wedding, birthday celebration, or family reunion. We must adjust recipes and order materials based on the number of people attending the event. To find out how to do this, turn to Putting Your Skills to Work on page 281.

4.1   Factoring Whole Numbers

4.2   Understanding Fractions

4.3   Equivalent Fractions

4.4   Simplifying Fractional Expressions

4.5   Ratios and Rates

4.6   Proportions

4.7   Applied Problems Involving Proportions

CHAPTER 4 PRETEST

1. _____

2. _____

3. _____

4. _____

5. _____

6. _____

7. _____

8. _____

9. _____

10. _____

11. _____

12. _____

13. _____

14. _____

15. _____

16. _____

This test provides a preview of the topics in this chapter. It will help you identify which concepts may require more of your studying time. If you are familiar with the topics in this chapter, take this test now. Check your answers with those in the back of the book. If you are not familiar with the topics in this chapter, begin studying the chapter now.

SECTION 4.1

1. Determine if 216 is divisible by 2, 3, or 5.

2. Identify as prime, composite, or neither.

   (a) 0   **(b)** 17   **(c)** 142

3. Express 120 as a product of prime factors.

SECTION 4.2

4. Carol correctly answered 7 out of 10 questions on a quiz. Write the fraction that describes the number of quiz questions she answered correctly.

5. Identify as a proper fraction, improper fraction, or mixed number.

   (a) $5\frac{2}{3}$ **(b)** $\frac{6}{7}$ **(c)** $\frac{9}{3}$

6. Write $\frac{12}{5}$ as a mixed number.   **7.** Change $2\frac{3}{4}$ to an improper fraction.

SECTION 4.3

8. Multiply the numerator and the denominator of $\frac{3}{5}$ by 2 to find an equivalent fraction.

9. Replace the ? with the appropriate symbol: =, <, or >: $\frac{5}{8}$? $\frac{35}{56}$.

SECTION 4.4

10. Simplify: $\frac{24}{44}$.

SECTION 4.5

11. A class has 16 males and 28 females. Write the ratio of males to females as a fraction in simplest form.

12. Write as a rate in simplest form: $5 for a 25-pound bag of fertilizer.

13. A car travels 364 miles on 14 gallons of gas. Find the miles per gallon unit rate.

SECTION 4.6

14. Determine if $\frac{4}{9} = \frac{18}{36}$ is a proportion.

15. Find the value of x: $\frac{x}{5} = \frac{8}{20}$.

SECTION 4.7

16. A bag of lawn fertilizer states that you need to use 1 pound of fertilizer per 250 square feet. How many pounds of fertilizer will you need to use to cover 1750 square feet of lawn?

230

Factoring Whole Numbers

The set of whole numbers includes many special types of numbers, such as *prime numbers* and *composite numbers*. In this section you will learn the difference between prime and composite numbers and see how to write a whole number as a product of prime factors.

After studying this section, you will be able to:

① Use the divisibility tests.
② Identify prime and composite numbers.
③ Find the prime factors of whole numbers.

Math Pro   Video 4.1   SSM

① Using the Divisibility Tests

A number *x* is said to be *divisible* by a number *y* if *y* divides *x* exactly, without a remainder (a remainder equal to 0). We say "25 is divisible by 5" because there is no remainder when we divide 25 by 5: $25 \div 5 = 5$. We say that "26 is not divisible by 5" because there is a remainder when we divide 26 by 5: $26 \div 5 = 5 \, \text{R}1$.

Some students find the following rules helpful when deciding if a number is divisible by 2, 3, or 5.

DIVISIBILITY TESTS

1. A number is divisible by 2 if it is even. This means that the last digit is 0, 2, 4, 6, or 8.
2. A number is divisible by 3 if the sum of the digits is divisible by 3.
3. A number is divisible by 5 if the last digit is 0 or 5.

E X A M P L E   1   Determine if the number is divisible by 2, 3, or 5.

(a) 234   **(b)** 910   **(c)** 711   **(d)** 38,910

SOLUTION

(a) 234	**2** and **3**	2 because 234 is even, and 3 since the sum of the digits is divisible by 3.
(b) 910	**5** and **2**	5 because the last digit is 0, and 2 since 910 is even.
(c) 711	**3**	The sum of the digits is divisible by 3.
(d) 38,910	**2, 3, 5**	2 because 38,910 is even, 3 since the sum of the digits is divisible by 3, and 5 since the last digit is 0.

P R A C T I C E   P R O B L E M   1

Determine if the number is divisible by 2, 3, or 5.

(a) 975   **(b)** 122   **(c)** 420   **(d)** 11,121

② Identifying Prime and Composite Numbers

A **prime number** is a whole number greater than 1 that is divisible only by itself and 1.

The number 5 is *prime* since it is divisible only by the numbers 5 and 1.

The number 6 is *not prime* since it is divisible by 2 and 3, in addition to 6 and 1.

> A **composite number** is a whole number greater than 1 that is divisible by whole numbers other than itself and 1.

The number 10 is **composite** since it is divisible by 2 and 5 as well as 10 and 1.

A whole number (except 0 and 1) that is not prime is composite. The numbers 0 and 1 are neither prime nor composite numbers.

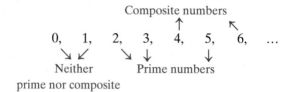

E X A M P L E   2   Which numbers are prime, composite, or neither? 1, 4, 7, 11, 14, 15, 17, 22, 27, 31, 120.

SOLUTION

1 is neither prime nor composite.

4, 14, 15, 22, 27 and 120 are composite.

7, 11, 17, and 31 are prime.

P R A C T I C E   P R O B L E M   2

Which numbers are prime, composite, or neither? 0, 3, 9, 13, 16, 19, 23, 32, 37, 41, 50.

> The first few prime numbers are 2, 3, 5, 7, 11, 13, 17, 19, 23, 29, ...

③ Finding the Prime Factors of Whole Numbers

Factors are numbers that are multiplied together. In the multiplication $6 \cdot 4 = 24$, 6 and 4 are called factors of 24.

$$24 = 6 \cdot 4 \qquad \text{These factors are } not \text{ prime numbers.}$$
$$\downarrow \quad \downarrow$$
$$\text{factor} \quad \text{factor}$$

Prime factors are factors that are prime. To write a number as a product of prime factors, we must break the multiplication down until each factor is prime. Thus, 24 written as a product of prime factors is

$$24 = 2 \cdot 3 \cdot 2 \cdot 2 \qquad \text{These factors are prime numbers.}$$

E X A M P L E   3   Express as a product of prime factors.

(a) 9   **(b)** 20

SOLUTION

(a) $9 = 3 \cdot 3$ or $\mathbf{3^2}$

(b) $20 = 2 \cdot 2 \cdot 5$ or $\mathbf{2^2 \cdot 5}$ $20 = 4 \cdot 5$ is *not correct* because 4 is not a prime number.

P R A C T I C E   P R O B L E M   3

Express as a product of prime factors.

(a) 14   **(b)** 27

For large numbers we can use division to find prime factors. When you divide two whole numbers and get a remainder of 0, both the divisor and the quotient are factors. Thus *we can divide to find prime factors*.

We use the division $15 \div 3 = 5$ to find the prime factors of 15.

Division Problem   *Related Multiplication*

$$
\begin{array}{r}
5 \longrightarrow \text{quotient} \\
3\,\overline{)\,15} \\
\downarrow \\
\text{divisor}
\end{array}
\qquad
\begin{array}{c}
3 \cdot 5 = 15 \\
\downarrow \; \downarrow \\
\text{Divisor and quotient are factors.}
\end{array}
$$

The divisor (3) and quotient (5) are factors of 15. We often refer to this method as using a **division ladder** to find prime factors.

PROCEDURE TO FIND PRIME FACTORS USING A DIVISION LADDER

1. Determine if the original number is divisible by a prime number. If so, perform the division and find the quotient.
2. Divide the quotient by prime numbers until the final quotient is a prime number.
3. Write all the divisors and the final quotient as a product of prime numbers.

E X A M P L E   4   Express 28 as a product of prime factors.

SOLUTION   Since 28 is even, it is divisible by the prime number 2. We start the division ladder by dividing 28 by 2.

Step 1: $2\,\overline{)\,28}\;\;^{14}$ The quotient 14 is not a prime number.

We must continue to divide until the quotient is a prime number.

Step 2: $2\,\overline{)\,14}\;\;^{7}$ The quotient 7 is a prime number.

The quotient is a prime number. Thus all the factors are prime. We are finished dividing. This process is simplified if we write the divisions as follows:

Step 2: $2\,\overline{)\,14}\;\;^{7}$
$\uparrow$
Step 1: $2\,\overline{|\,28}$

Now we write all the divisors and the quotient as follows: $\mathbf{28 = 2 \cdot 2 \cdot 7}$ or $\mathbf{2^2 \cdot 7}$.

PRACTICE PROBLEM 4

Express 50 as a product of prime factors.

It is important to note that all the divisors and the final quotient must be prime numbers to assure that all factors are prime.

We can check our answer by multiplying the prime factors.

EXAMPLE 5   Express 60 as a product of prime factors and check your answer.

SOLUTION   We must divide by prime numbers to assure that all factors are prime. We start by dividing 60 by the prime number 5.

Step 3: $2\overline{\smash{)}4}$ → 2 2 is prime, so we are finished dividing.

Step 2: $3\overline{\smash{)}12}$

Step 1: $5\overline{\smash{)}60}$

$$60 = 5 \cdot 3 \cdot 2^2$$

Check: $60 \stackrel{?}{=} 5 \cdot 3 \cdot 2^2$

$60 \stackrel{?}{=} 15 \cdot 4$

$60 = 60$ ✓   The answer checks.

PRACTICE PROBLEM 5

Express 96 as a product of prime factors.

TO THINK ABOUT

In Example 5, if we started the division process with 2 instead of 5, the result would have been the same. In fact, it does not matter what prime number is used to start the division—the results will be equivalent. Why? To illustrate, let's compare the following results.

Step 3: $3\overline{\smash{)}15}$ → 5 $5\overline{\smash{)}10}$ → 2 $2\overline{\smash{)}4}$ → 2

Step 2: $2\overline{\smash{)}30}$ $2\overline{\smash{)}20}$ $3\overline{\smash{)}12}$

Step 1: $2\overline{\smash{)}60}$ $3\overline{\smash{)}60}$ $5\overline{\smash{)}60}$

$60 = 2 \cdot 2 \cdot 3 \cdot 5$ $60 = 3 \cdot 2 \cdot 5 \cdot 2$ $60 = 5 \cdot 3 \cdot 2 \cdot 2$

$$60 = 5 \cdot 3 \cdot 2^2$$

Since multiplication is commutative, the order in which we write the factors does not matter.

E X A M P L E   6   Express 210 as a product of prime factors.

SOLUTION

$$\begin{array}{r} 7 \\ 2\overline{)14} \end{array}$$ 7 is prime, so we are finished dividing.

Step 3: $2\overline{)14}$

Step 2: $3\overline{)42}$

Step 1: $5\overline{)210}$

$210 = 5 \cdot 3 \cdot 2 \cdot 7$

P R A C T I C E   P R O B L E M   6

Express 315 as a product of prime factors.

We can also use a **factor tree** to find prime factors. This method uses the related multiplication instead of division to find the factors. Let's see how we would have factored the number 210 from Example 6 using a factor tree.

Division Ladder   *Factor Tree*

Step 3: $2\overline{)14}$ → $14 = 2 \cdot 7$

Step 2: $3\overline{)42}$ → $42 = 3 \cdot 14$

Step 1: $5\overline{)210}$ → $210 = 5 \cdot 42$

210

$5 \cdot 42$

$③ \cdot 14$

$② \cdot ⑦$

$210 = 5 \cdot 3 \cdot 2 \cdot 7$

PROCEDURE TO BUILD A FACTOR TREE TO FIND PRIME FACTORS

210
$\wedge$
$5 \cdot 42$
$⑤ \cdot 42$
$\wedge$
$3 \cdot 14$

$③ \cdot 14$
$\wedge$
$② \cdot ⑦$

$210 = 5 \cdot 3 \cdot 2 \cdot 7$

1. Write the number to be factored as a product of any two numbers other than 1 and itself.
2. In this product, circle any prime factor(s).
3. Write all factors that are *not prime* as products.
4. Circle any prime factor(s).
5. Repeat steps 3 and 4 until *all factors* are prime.
6. Write the numbers that are circled as a product of prime numbers.

E X A M P L E   7   Use a factor tree to express 48 as a product of prime factors.

SOLUTION

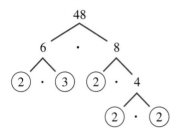

$$48 = 2 \cdot 3 \cdot 2 \cdot 2 \cdot 2 \text{ or } 2^4 \cdot 3$$

1. Write 48 as the product of any two factors other than 1 and itself.
2. We *do not* circle 6 or 8 since neither are prime. We must write both 6 and 8 as products: $6 = 2 \cdot 3$ and $8 = 2 \cdot 4$.
3. Circle the prime numbers 2 and 3. Since 4 is *not prime* we must write it as a product: $4 = 2 \cdot 2$.
4. Circle both factors of 4 since they are both prime.
5. Write 48 as a product of the prime numbers that are circled: $48 = 2 \cdot 3 \cdot 2 \cdot 2 \cdot 2$.

P R A C T I C E   P R O B L E M   7

Use a factor tree to express 36 as a product of prime factors.

DEVELOPING YOUR STUDY SKILLS

TAKING NOTES IN CLASS

During a lecture you are *listening, seeing, reading,* and *writing* all at the same time. Although an important part of mathematics studying is taking notes, you must also focus on what the instructor is saying so you can follow the logic.

The Learning Cycle

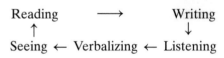

GETTING THE MOST OUT OF TAKING NOTES

1. Preview the lesson before class so that the important concepts will be more familiar and you will have a better idea of what to write down.
2. Write down only the important ideas and examples as the instructor lectures, making sure that you're also listening so you can follow the logic.
3. Include helpful hints and references to the text that your instructor gives. You will be amazed how easily you forget these if you don't write them down.
4. Review your notes, clarifying whatever appears vague, on the *same day* sometime after class, so that you will be able to recall material from the lecture that you did not include in your notes.

   You may find that you learn more by *seeing* and *listening* than by the others modes of learning. You may prefer to take fewer notes, focusing more on what the instructor is saying. This is fine as long as you write a brief outline during class of the instructor's lecture. Then immediately after class, you should add the details.

EXERCISES 4.1

1. Is 155 divisible by 2? Why or why not?

2. Is 185 divisible by 2? Why or why not?

3. Is 232 divisible by 3? Why or why not?

4. Is 156 divisible by 3? Why or why not?

5. Is 485 divisible by 5? Why or why not?

6. Is 680 divisible by 5? Why or why not?

Determine if the number is divisible by 2, 3, or 5.

7. 324   **8.** 732   **9.** 805   **10.** 640

11. 330   **12.** 955   **13.** 22,971   **14.** 11,391

15. 991,500   **16.** 700,550

Which numbers are prime, composite, or neither?

17. 0, 9, 1, 23, 40, 8, 15, 33

18. 1, 32, 7, 12, 50, 6, 13, 41

19. Write 56 as a product of:
   (a) Any two factors   **(b)** Prime factors

20. Write 72 as a product of:
   (a) Any two factors   **(b)** Prime factors

Express as a product of prime factors.

21. 9   **22.** 4   **23.** 12   **24.** 18   **25.** 24

26. 32   **27.** 36   **28.** 49   **29.** 21   **30.** 64

31. 70   **32.** 80   **33.** 75   **34.** 81   **35.** 45

36. 55   **37.** 99   **38.** 63   **39.** 105   **40.** 200

41. 110   **42.** 155   **43.** 136   **44.** 126

45. 220   **46.** 135   **47.** 810   **48.** 630

CALCULATOR EXERCISES

The calculator can be a useful tool for finding prime factors when 2, 3, or 5 are not factors. You can quickly check for factors by dividing by the prime numbers 7, 11, 13, 17, 19, or higher if needed. Use your calculator to help you express the following numbers as a product of prime factors.

49. 91   **50.** 1309   **51.** 561   **52.** 2737

53. We can also use a calculator to check that the prime factors in an answer are correct. Use your calculator to multiply the prime factors in exercises 49–52, and verify that your answers are correct.

VERBAL AND WRITING SKILLS

54. Explain why you must divide by prime numbers when you use a division ladder to find prime factors.

55. Write in words the difference between a composite and a prime number.

ONE STEP FURTHER

56. List the multiplication facts for the number 9: $1 \cdot 9 = 9$, $2 \cdot 9 = 18$, $3 \cdot 9 = 27$, and so on. Can you see a pattern that might help determine if a number is divisible by 9? What is it? Try your idea on a large number that is divisible by 9, such as 2,148,543.

57. Find the following: $430 \div 10$; $560 \div 10$; $33,440 \div 10$; $5550 \div 10$. Can you see a pattern that might help determine if a number is divisible by 10? What is it?

58. Find a five-digit number that is divisible by the number 3 and the number 5.

59. Find a six-digit number that is divisible by the number 2 and the number 3.

CUMULATIVE REVIEW

Simplify.

1. $(2x^2)(5x^3 y)$ **2.** $6y^2 + 3y^2$ **3.** $5x + 3x + 2$ **4.** $(5x)(3x)(2)$

SECTION 4.2

Understanding Fractions

After studying this section, you will be able to:

① Understand the meaning of fractions.
② Identify proper fractions, improper fractions, and mixed numbers.
③ Change improper fractions to mixed numbers.
④ Change mixed numbers to improper fractions.

① Understanding the Meaning of Fractions

Whole numbers are used to describe whole objects, or entire quantities. However, often we have to represent parts of whole quantities. In mathematics, **fractions** are a set of numbers used to describe parts of whole quantities. The *whole* can be an object (a pizza), or, just as often, the whole can be a set of things that we choose to consider as a whole unit. Here are some examples.

A whole pizza   Part of a pizza

Whole   Part of a whole

Math Pro Video 4.2   SSM

In the object shown in the margin there are eight equal parts. The five shaded parts represent *part of the whole* and are represented by the fraction 5/8 or $\frac{5}{8}$.

In the fraction $\frac{5}{8}$ the number 5 is called the **numerator** and the number 8 is called the **denominator**.

$\frac{5}{8}$ → The *numerator* specifies how many of these parts.

→ The *denominator* specifies the total number of parts.

> The *denominator* of a fraction shows the number of equal parts in the whole. The *numerator* shows the number of parts being talked about or being used.

When you say "$\frac{3}{4}$ of a pizza has been eaten," what you are indicating is that three of four equal parts of a pizza have been eaten.

Remember that the *numerator* is always the *top number* and the *denominator* is always the *bottom number*.

3 pieces eaten

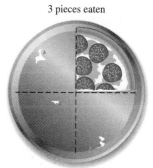

4 pieces in one pizza

EXAMPLE 1   Use a fraction to represent the shaded part of the object.

(a)   (b)    (c)

SOLUTION

(a) One out of four parts are shaded, or $\frac{1}{4}$.

(b) Seven out of nine parts are shaded, or $\frac{7}{9}$.

(c) Three out of three parts are shaded, or $\frac{3}{3} = \mathbf{1}$.

PRACTICE PROBLEM 1

Use a fraction to represent the shaded part of the object.

(a)  **(b)**  **(c)**

What does it mean when a fraction has the value zero, or when there is a zero in the denominator? Let's look at the following situations to see how to answer these questions.

If $20 is divided among 5 people, each person receives $4.

$$20 \div 5 = 4 \quad \text{or} \quad \frac{20}{5} = 4$$

If $0 is divided among 5 people, each person receives $0.

$$0 \div 5 = 0 \quad \text{or} \quad \frac{0}{5} = 0$$

If $20 is divided among 0 people?—This cannot be done.

$$20 \div 0 \quad \text{and} \quad \frac{20}{0} \quad \text{cannot be done.}$$

We summarize as follows.

DIVISION INVOLVING THE NUMBER ZERO

1. Zero can never be the divisor in a division problem:

   $5 \div 0 = \dfrac{5}{0}$ cannot be done. *We say division by 0 is undefined.*

2. Zero may be divided by any number except zero; the result is always zero:

   $0 \div 4 = \dfrac{0}{4} = 0$

   In other words, *any fraction with 0 in the numerator and a nonzero denominator equals 0.*

EXAMPLE 2 Divide, if possible.

(a) $\dfrac{23}{0}$ **(b)** $\dfrac{0}{23}$

SOLUTION

(a) $\dfrac{23}{0}$ Division by 0 is undefined.

(b) $\dfrac{0}{23} = \mathbf{0}$ Any fraction with 0 in the numerator and a nonzero denominator equals 0.

PRACTICE PROBLEM 2

Divide, if possible.

(a) $4 \div 0$ **(b)** $0 \div 18$ **(c)** $\dfrac{0}{65}$ **(d)** $\dfrac{65}{0}$

Circle graphs are especially helpful for showing the relationship of parts to a whole. The circle represents the whole, and the pie-shaped pieces represent parts of the whole.

EXAMPLE 3

The approximate number of inches of rain that occurred during each quarter of one year in Seattle, Washington is shown by the circle graph.

(a) What fractional part of the total yearly rainfall occurs from October to December?

(b) What fractional part of the total yearly rainfall occurs from July to September?

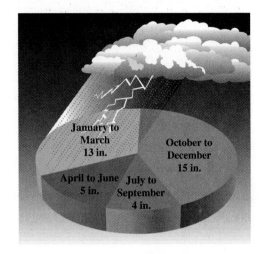

SOLUTION   First we must find the total rainfall for 1 year.

13 in. + 15 in. + 4 in. + 5 in. = 37 in.

(a) From October to December there were 15 inches of rain out of a total of 37.

$$\frac{15}{37}$$

(b) From July to September there were 4 inches of rain out of a total of 37.

$$\frac{4}{37}$$

PRACTICE PROBLEM 3

Use the circle graph in Example 3 to answer the following.

(a) What fractional part of the total yearly rainfall occurs from January to March?

(b) What fractional part of the total yearly rainfall occurs from April to June?

❷ Identifying Proper Fractions, Improper Fractions, and Mixed Numbers

We use three names to identify fractions: *proper fraction*, *improper fraction*, and *mixed number*. The word we use to name the fraction depends on the value of the fraction.

Value	Illustration	Fraction	Name of Fraction
Less than 1		$\frac{3}{4}$	proper fraction
Equal to 1		$\frac{4}{4}$ or 1	improper fraction
Greater than 1		$\frac{7}{4}$ or $1\frac{3}{4}$	improper fraction or mixed number

> A **proper fraction** is used to describe a quantity less than 1. If the numerator is less than the denominator, the fraction is a proper fraction.

The fraction $\frac{3}{4}$ is a proper fraction.

> An **improper fraction** is used to describe a quantity greater than or equal to 1.

The fraction $\frac{7}{4}$ is an improper fraction because the numerator is larger than the denominator. Since $\frac{4}{4}$ describes a quantity equal to 1, it is also an improper fraction.

> A **mixed number** consists of a whole number and a proper fraction, and is used to describe a quantity greater than 1.

The last figure on page 241 can also be represented by 1 whole added to $\frac{3}{4}$ of a whole, or $1 + \frac{3}{4}$. This is written $1\frac{3}{4}$ (we do not write the addition symbol). The fraction $1\frac{3}{4}$ is a mixed number. Thus the improper fraction $\frac{7}{4}$ is equivalent to the mixed number $1\frac{3}{4}$. This suggests that we can change from one form to the other without changing the value of the fraction.

EXAMPLE 4   Identify as a proper fraction, an improper fraction, or a mixed number.

(a) $\frac{9}{8}$ **(b)** $\frac{8}{9}$ **(c)** $7\frac{3}{4}$ **(d)** $\frac{3}{3}$

SOLUTION

(a) $\frac{9}{8}$ **improper fraction**   The numerator is larger than the denominator.

(b) $\frac{8}{9}$ **proper fraction**   The numerator is less than the denominator.

(c) $7\frac{3}{4}$ **mixed number**   A whole number is added to a proper fraction.

(d) $\frac{3}{3}$ **improper fraction**   The numerator is equal to the denominator.

PRACTICE PROBLEM 4
Identify as a proper fraction, an improper fraction, or a mixed number.

(a) $\frac{6}{5}$ **(b)** $\frac{x}{x}$ **(c)** $6\frac{2}{9}$ **(d)** $\frac{1}{2}$

③ Changing Improper Fractions to Mixed Numbers

By drawing a picture it is easy to see how to change improper fractions to mixed numbers. For example, if we start with the fraction $\frac{13}{5}$ and represent it by the shaded boxes (where 13 of the pieces that are $\frac{1}{5}$ of a box are shaded), we see that $\frac{13}{5} = 2\frac{3}{5}$ since 2 whole boxes and $\frac{3}{5}$ of a box are shaded.

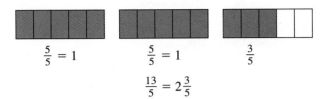

$$\frac{5}{5} = 1 \qquad \frac{5}{5} = 1 \qquad \frac{3}{5}$$

$$\frac{13}{5} = 2\frac{3}{5}$$

Now since $\frac{13}{5} = 13 \div 5$, we can divide to change an improper fraction to a mixed number.

$$\frac{13}{5} = 13 \div 5 = 5\overline{)13}^{\,2R3} = 2\frac{3}{5}$$

We state that process in the following procedure.

PROCEDURE TO CHANGE AN IMPROPER FRACTION TO A MIXED NUMBER

1. Divide the numerator by the denominator.
2. The quotient is the whole-number part of the mixed number.
3. The remainder from the division will be the numerator of the fraction. The denominator of the fraction remains unchanged.

A mixed number is in the form: quotient $\dfrac{\text{remainder}}{\text{denominator}}$

EXAMPLE 5   Write $\frac{19}{7}$ as a mixed number.

SOLUTION   The answer is in the form: quotient $\dfrac{\text{remainder}}{\text{denominator}}$.

$$\frac{19}{7} \rightarrow 7\overline{)\begin{array}{l}2 \leftarrow \text{quotient} \\ 19 \\ \underline{14} \\ 5 \leftarrow \text{remainder}\end{array}} \qquad 2\frac{5}{7} \leftarrow$$

Use the same denominator as the original fraction.

$$\frac{19}{7} = 2\frac{5}{7}$$

PRACTICE PROBLEM 5

Write $\frac{23}{6}$ as a mixed number.

④ Changing Mixed Numbers to Improper Fractions

It is not difficult to see how to change mixed numbers to improper fractions using a picture. For example, suppose that you wanted to write $2\frac{1}{4}$ as an improper fraction. We can illustrate the quantity $2\frac{1}{4}$ as two whole quantities plus $\frac{1}{4}$ of a third quantity. Now if we count the shaded squares, we see that we have $\frac{9}{4}$.

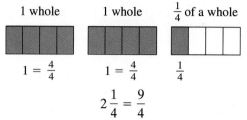

$$2\frac{1}{4} = \frac{9}{4}$$

You can do the same procedure without a picture using the following method.

> **PROCEDURE TO CHANGE A MIXED NUMBER TO AN IMPROPER FRACTION**
>
> 1. Multiply the whole number by the denominator.
> 2. Add this product to the numerator. The result is the numerator of the improper fraction. The denominator does not change.
>
> Improper fraction: $\dfrac{(\text{denominator} \cdot \text{whole number}) + \text{numerator}}{\text{denominator}}$

EXAMPLE 6   Change $6\frac{1}{2}$ to an improper fraction.

SOLUTION

Improper fraction: $\dfrac{(\text{denominator} \cdot \text{whole number}) + \text{numerator}}{\text{denominator}}$

| Multiply the whole number by the denominator. | Add the numerator to the product. | | Write the sum over the denominator. |

$$6\frac{1}{2} \longrightarrow \frac{(6 \cdot 2) + 1}{2} = \frac{12 + 1}{2} = \frac{13}{2}$$

The denominator does not change.

We can also write the process as follows:

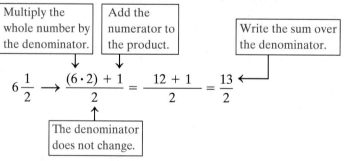

$$\frac{2 \text{ times } 6 \text{ plus } 1}{2} = \frac{13}{2}$$

$$6\frac{1}{2} = \frac{\mathbf{13}}{\mathbf{2}}$$

PRACTICE PROBLEM 6

Change $8\frac{2}{3}$ to an improper fraction.

EXERCISES 4.2

Use a fraction to represent the shaded area.

1. **2.** **3.** **4.**

Divide if possible.

5. $\dfrac{7}{0}$ **6.** $\dfrac{9}{0}$ **7.** $\dfrac{0}{z}$ **8.** $\dfrac{0}{a}$

9. $\dfrac{44}{44}$ **10.** $\dfrac{a}{a}$ **11.** $\dfrac{y}{y}$ **12.** $\dfrac{87}{87}$

13. $\dfrac{8}{0}$ **14.** $\dfrac{6}{0}$ **15.** $\dfrac{0}{8}$ **16.** $\dfrac{0}{6}$

Answer each question.

17. A baseball player had 7 base hits in 15 times at bat. Write the fraction that describes the number of times the player had a base hit.

18. An archer hit the target 3 times out of 11 shots. Write the fraction that describes the number of times the archer hit the target.

19. There are 13 marbles. Six are blue. Write the fraction that describes the number of marbles that are blue.

20. There are 7 cupcakes. Two are vanilla. Write the fraction that describes the number of cupcakes that are vanilla.

21. There are 26 dancers in the dance production class at a high school. Nine of the dancers are juniors. Write the fraction that describes the dancers who are juniors.

22. At a salad bar, there are 29 different items to choose from. Eleven of the choices contain pasta. Write the fraction that describes the choices that contain pasta?

23. There are 57 first graders and 37 second graders in the schoolyard. What fractional part of the children on the playground consists of second graders?

24. There are 87 men and 63 women working for a small corporation. What fractional part of the employees consists of men only?

25. Ralph has 57 baseball trading cards and 29 soccer trading cards. What fractional part of the trading cards consists of soccer cards?

26. Tina has 31 silver coins and 87 bronze coins. What fractional part of the coins consists of bronze coins?

The deductions from Arnold's pay check are shown on the following circle graph. Use this circle graph to answer questions 27–30.

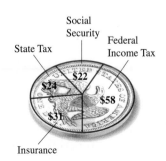

Social Security
State Tax
Federal Income Tax
$22
$24
$58
$31
Insurance

Refer to the circle graph on page 245 to answer the following questions.

27. What fractional part of the deductions is for federal income tax?

28. What fractional part of the deductions is for social security?

29. What fractional part of the deductions is for insurance?

30. What fractional part of the deductions is for state tax?

Identify as a proper fraction, improper fraction, or mixed number.

31. $\dfrac{7}{9}$

32. $1\dfrac{1}{2}$

33. $\dfrac{9}{9}$

34. $\dfrac{9}{7}$

35. $\dfrac{6}{6}$

36. $\dfrac{z}{z}$

37. $7\dfrac{1}{9}$

38. $\dfrac{3}{5}$

Change each improper fraction to a mixed number or a whole number.

39. $\dfrac{13}{7}$

40. $\dfrac{19}{4}$

41. $\dfrac{73}{4}$

42. $\dfrac{92}{7}$

43. $\dfrac{41}{2}$

44. $\dfrac{25}{3}$

45. $\dfrac{32}{5}$

46. $\dfrac{79}{7}$

47. $\dfrac{47}{5}$

48. $\dfrac{54}{7}$

49. $\dfrac{44}{44}$

50. $\dfrac{89}{89}$

Change each mixed number to an improper fraction.

51. $8\dfrac{3}{7}$

52. $6\dfrac{2}{3}$

53. $24\dfrac{1}{4}$

54. $10\dfrac{1}{9}$

55. $15\dfrac{2}{3}$

56. $13\dfrac{3}{5}$

57. $33\dfrac{1}{3}$

58. $41\dfrac{1}{2}$

59. $8\dfrac{9}{10}$

60. $3\dfrac{1}{50}$

61. $8\dfrac{21}{30}$

62. $5\dfrac{19}{20}$

CALCULATOR EXERCISES

Change each mixed number to an improper fraction.

63. $1344\dfrac{29}{32}$

64. $2188\dfrac{114}{120}$

65. $39\dfrac{533}{612}$

66. $42\dfrac{345}{486}$

VERBAL AND WRITING SKILLS

Fill in the blanks.

67. To change a mixed number to an improper fraction, we multiply the _____ times the whole number, then add the _____ to this product. The denominator of the improper fraction is _____ .

68. To change an improper fraction to a mixed number, we divide the numerator by the _____ . The quotient becomes the _____ of the mixed number. The remainder from this division becomes the _____ . The denominator of the mixed number _____ .

CUMULATIVE REVIEW

Simplify.

1. $(-5)(-8)$

2. $(-7)(9)$

3. $\dfrac{63}{-7}$

4. $\dfrac{-54}{-6}$

246

SECTION 4.3

Equivalent Fractions

When we compare whole numbers we can easily see which number is larger and which number is smaller. For example, 2 is smaller than 20 and 13 is larger than 4. Comparing fractions is not as easy to see. Is $\frac{3}{4}$ smaller or larger than $\frac{2}{3}$? It is larger. In this section you will learn how to compare fractions as well as how to recognize and write equivalent fractions.

After studying this section, you will be able to:

❶ Multiply prime fractions.

❷ Identify and use equivalent fractions.

❸ Use the test for equality of fractions.

Math Pro   Video 4.3   SSM

❶ Multiplying Prime Fractions

Before we begin our discussion of equivalent fractions we must introduce multiplication with fractions, since we use this skill to write equivalent fractions.

> To multiply fractions, we multiply numerator times numerator and denominator times denominator.

Multiply numerator
times numerator.
↓
$$\frac{2}{7} \cdot \frac{3}{5} = \frac{2 \cdot 3}{7 \cdot 5} = \frac{6}{35}$$
↑
Multiply denominator
times denominator.

EXAMPLE 1   Multiply: $\frac{1}{9} \cdot \frac{4}{5}$.

SOLUTION

Multiply numerator
times numerator.
↓
$$\frac{1}{9} \cdot \frac{4}{5} = \frac{1 \cdot 4}{9 \cdot 5} = \frac{4}{45}$$
↑
Multiply denominator
times denominator.

PRACTICE PROBLEM 1

Multiply: $\frac{1}{3} \cdot \frac{2}{7}$.

❷ Identifying and Using Equivalent Fractions

There are many fractions that name the same quantity. For example, in the following illustration two pieces of wood are the same length. One piece of wood is

cut into two equal pieces and the other into four. As you can see, 1 of the 2 pieces of wood represents the same quantity as 2 of the 4 pieces of wood.

We say that the fraction $\frac{1}{2}$ is **equivalent** to $\frac{2}{4}$ or $\frac{1}{2} = \frac{2}{4}$. There are many fractions that represent the same value as $\frac{1}{2}$ and thus are equivalent to $\frac{1}{2}$. In fact, $\frac{1}{2}, \frac{2}{4}, \frac{4}{8}$, and $\frac{8}{16}$ are all equivalent fractions.

$$\frac{1}{2} = \frac{2}{4} = \frac{4}{8} = \frac{8}{16}$$

Equivalent fractions *look different* but have the *same value* because they represent the *same quantity*.

How can we find equivalent fractions without using a diagram or picture? We multiply the fraction by a fraction equivalent to 1. We can do this since the *multiplication property of 1* states that if we multiply a number by 1, the value of that number does not change. Now since $1 = \frac{1}{1} = \frac{2}{2} = \frac{3}{3}$, and so on, *we can multiply a fraction by any fraction equivalent to 1 and the value of that fraction does not change*. That is, the fractions are equivalent. For example, to rewrite $\frac{1}{3}$ as $\frac{2}{6}$, we proceed as follows:

Step 1:
We multiply the fraction by 1.

Step 2:
We rewrite $\frac{1}{1}$ as $\frac{2}{2}$.

Step 3:
We multiply fractions.

We have equivalent fractions.

$$\frac{1}{3} = \frac{1}{3} \cdot \boxed{\frac{1}{1}} = \frac{1}{3} \cdot \boxed{\frac{2}{2}} = \frac{1 \cdot \boxed{2}}{3 \cdot \boxed{2}} = \frac{2}{6} \qquad \frac{1}{3} = \frac{2}{6}$$

We often skip the first two steps in the process above (we use a shortcut). That is, we simply multiply the numerator and denominator by the *same number* to obtain an equivalent fraction. It is important that we understand the entire process when we use a shortcut. This concept, multiplying by a fraction equivalent to 1, is important, especially later in the chapter when we learn how to reduce fractions.

> To find an equivalent fraction we must multiply *both* the numerator and denominator by the *same* nonzero number.
>
> $$\frac{a}{b} = \frac{a \cdot c}{b \cdot c} \qquad \text{where } b \text{ and } c \text{ are not 0}$$

EXAMPLE 2

(a) Multiply the numerator and the denominator of $\frac{3}{4}$ by 2 to find an equivalent fraction.

(b) Multiply the numerator and the denominator of $\frac{3}{4}$ by 3 to find an equivalent fraction.

SOLUTION

(a) $\frac{3}{4} = \frac{3 \cdot 2}{4 \cdot 2} = \frac{6}{8}$

(b) $\frac{3}{4} = \frac{3 \cdot 3}{4 \cdot 3} = \frac{9}{12}$

PRACTICE PROBLEM 2

(a) Multiply the numerator and the denominator of $\frac{4}{7}$ by 3 to find an equivalent fraction.

(b) Multiply the numerator and the denominator of $\frac{4}{7}$ by 5 to find an equivalent fraction.

When there is a variable in the denominator, we will assume that the variable does not equal zero, since division by zero is not defined.

EXAMPLE 3   Write $\frac{3}{4}$ as an equivalent fraction with a denominator of $16x$.

SOLUTION

$$\frac{3}{4} = \frac{}{16x}$$

$$\frac{3 \cdot \boxed{?}}{4 \cdot \boxed{?}} = \frac{}{16x}$$

4 times what number equals $16x$? $4x$

Since we must multiply the denominator by $4x$ to obtain $16x$, we must also multiply the numerator by $4x$.

$$\frac{3 \cdot \boxed{4x}}{4 \cdot \boxed{4x}} = \frac{\mathbf{12x}}{\mathbf{16x}}$$

PRACTICE PROBLEM 3

Write $\frac{2}{9}$ as an equivalent fraction with a denominator of $36x$.

---

3 **Using the Test for Equality of Fractions**

When we build an equivalent fraction, how can we *check* to see if our answer is an equivalent fraction? We use the **equality test for fractions**, which states: If two fractions are equal, their diagonal products are equal.

If $\frac{3}{7} = \frac{9}{21}$, then $21 \cdot 3$ must be equal to $7 \cdot 9$.

Products are equal.

$$\boxed{21 \cdot 3 = 63} \qquad \boxed{7 \cdot 9 = 63}$$

$$\frac{3}{7} \times \frac{9}{21}$$

Since $63 = 63$, we know that $\frac{3}{7} = \frac{9}{21}$.

We use the equality test for fractions to determine if fractions are equal. If two fractions are unequal (we use the symbol $\neq$), their diagonal products are unequal. We often refer to diagonal products as *forming a* **cross product**.

The test can be described in this way:

> **EQUALITY TEST FOR FRACTIONS**
> For any two fractions where $a, b, c,$ and d are whole numbers and $b \neq 0$, $d \neq 0$, if and only if
>
> $$\frac{a}{b} = \frac{c}{d}, \text{ then } d \cdot a = b \cdot c.$$
>
> The cross products are equal.

EXAMPLE 4   Use the equality test for fractions to see if the fractions are equal.

(a) $\dfrac{2}{11} \overset{?}{=} \dfrac{18}{99}$

(b) $\dfrac{3}{16} \overset{?}{=} \dfrac{12}{62}$

SOLUTION   We form the cross products to determine if the fractions are equal.

(a) $\dfrac{2}{11} \overset{?}{=} \dfrac{18}{99}$

(b) $\dfrac{3}{16} \overset{?}{=} \dfrac{12}{62}$

Products are equal.

$\boxed{99 \cdot 2 = 198}$ $\boxed{11 \cdot 18 = 198}$

$$\frac{2}{11} \times \frac{18}{99}$$

Since 198 = 198, we know that $\dfrac{2}{11} = \dfrac{18}{99}$.

Products are *not* equal.

$\boxed{62 \cdot 3 = 186}$ $\boxed{16 \cdot 12 = 192}$

$$\frac{3}{16} \times \frac{12}{62}$$

Since 186 $\neq$ 192, we know that $\dfrac{3}{16} \neq \dfrac{12}{62}$.

PRACTICE PROBLEM 4

Use the equality test for fractions to see if the fractions are equal.

(a) $\dfrac{4}{22} \overset{?}{=} \dfrac{12}{87}$

(b) $\dfrac{84}{108} \overset{?}{=} \dfrac{7}{9}$

We can also use the equality test for fractions to determine which fraction is smaller and which fraction is larger. To illustrate, let's look at the fractions in Example 4b. Using the equality test for fractions, we determined that the fractions $\dfrac{3}{16}$ and $\dfrac{12}{62}$ are not equal. If these fractions are not equal, which is smaller?

This product is *less than* this product.

$\downarrow$ $\downarrow$

$\boxed{62 \cdot 3 = 186}$ $\boxed{16 \cdot 12 = 192}$

$$\frac{3}{16} \times \frac{12}{62}$$

This fraction is *less than* this fraction.

$\downarrow \quad \downarrow \quad \downarrow$

$$\frac{3}{16} < \frac{12}{62}$$

$\frac{3}{16}$ is smaller because the cross product $62 \cdot 3 = 186$ is smaller than the cross product $16 \cdot 12 = 192$.

PROCEDURE TO DETERMINE THE RELATIONSHIP BETWEEN TWO FRACTIONS

When $a, b, c,$ and d are whole numbers and $b \neq 0$ and $d \neq 0$, then

1. For each fraction cross-multiply denominator $\times$ numerator, and write this product above the numerator.
2. The smaller product will be written above the smaller fraction and the larger product above the larger fraction.

Note, we must multiply the cross products from bottom to top for this method to give the correct result.

E X A M P L E   5   Replace the ? with the appropriate symbol: $=, <,$ or $>$.

(a) $\frac{7}{11} ? \frac{4}{9}$ 　　　　　　　　　　**(b)** $\frac{2}{7} ? \frac{4}{13}$

SOLUTION

(a) $\frac{7}{11} ? \frac{4}{9}$ 　　　We multiply the cross products from bottom to top.

This product is greater than this product.

$\boxed{9 \cdot 7 = 63}$ 　　$\boxed{11 \cdot 4 = 44}$

$$\frac{7}{11} \times \frac{4}{9}$$

This fraction is greater than this fraction.

$$\frac{7}{11} > \frac{4}{9}$$

(b) $\frac{2}{7} ? \frac{4}{13}$

This product is less than this product.

$\boxed{13 \cdot 2 = 26}$ 　　$\boxed{7 \cdot 4 = 28}$

$$\frac{2}{7} \times \frac{4}{13}$$

This fraction is less than this fraction.

$$\frac{2}{7} < \frac{4}{13}$$

P R A C T I C E   P R O B L E M   5

Replace the ? with the appropriate symbol: $=, <,$ or $>$.

(a) $\frac{6}{13} ? \frac{2}{7}$ 　　　　　　　　　　**(b)** $\frac{11}{30} ? \frac{9}{20}$

E X A M P L E 6   Johanna owns two rectangular plots of land, with each plot having the same dimensions. She subdivides one into 5 equal parcels and the other into 7 equal parcels. For the same price you can buy either 3 of the 5 parcels or 4 of the 7 parcels. Which purchase yields more land?

SOLUTION   We write the fractions that represent the situation, then determine which fraction is larger in value.

$$7 \cdot 3 = 21 \qquad\qquad 5 \cdot 4 = 20$$

$$\frac{3}{5} \times \frac{4}{7}$$

Since $21 > 20$, then $\dfrac{3}{5} > \dfrac{4}{7}$.

3 out of 5 parcels yields more land.

P R A C T I C E P R O B L E M 6

Brett owns two square plots of land, with each plot having the same dimensions. He subdivides one into 8 equal parcels and the other into 11 equal parcels. For the same price you can buy either 7 of the 8 parcels or 10 of the 11 parcels. Which purchase yields more land?

DEVELOPING YOUR STUDY SKILLS

PROBLEMS WITH ACCURACY

Strive for accuracy. Mistakes are often made because of human error rather than lack of understanding. Such mistakes are frustrating. A simple arithmetic or sign error can lead to an incorrect answer. These six steps will help you cut down on errors.

1. Work carefully, and take your time. Do not rush through a problem just to get it done.
2. Concentrate on the problem. Sometimes problems become mechanical, and your mind begins to wander. You become careless and make a mistake. Concentrating on the problem will help you avoid this.
3. Check your problem. Be sure that you copied it correctly from the book.
4. Check each step of the problem for sign errors as well as computation errors. Does your answer make sense?
5. Make a mental note of the types of errors you make most often. Becoming aware of where you make errors will help you avoid making the same mistake.
6. Keep practicing new skills. Remember the old saying that "practice makes perfect." Many errors are due simply to a lack of practice.

EXERCISES 4.3

Multiply.

1. $\dfrac{2}{3} \cdot \dfrac{4}{7}$

2. $\dfrac{1}{8} \cdot \dfrac{2}{3}$

3. $\dfrac{1}{4} \cdot \dfrac{1}{3}$

4. $\dfrac{2}{5} \cdot \dfrac{3}{7}$

5. $\dfrac{4}{5} \cdot \dfrac{1}{7}$

6. $\dfrac{3}{4} \cdot \dfrac{1}{3}$

7. $\dfrac{3}{4} \cdot \dfrac{1}{2}$

8. $\dfrac{6}{7} \cdot \dfrac{1}{7}$

9. $\dfrac{1}{2} \times \dfrac{1}{x}$

10. $\dfrac{1}{3} \times \dfrac{1}{y}$

11. $\dfrac{5}{4} \times \dfrac{y}{6}$

12. $\dfrac{6}{7} \times \dfrac{x}{5}$

Multiply the numerator and the denominator of the given fractions by the following numbers to find two different equivalent fractions: **(a)** 4   **(b)** 5

13. $\dfrac{7}{9}$

14. $\dfrac{6}{7}$

15. $\dfrac{4}{11}$

16. $\dfrac{9}{13}$

Multiply the numerator and the denominator of the given fractions by the following numbers to find two different equivalent fractions: **(a)** 2   **(b)** 3

17. $\dfrac{11}{17}$

18. $\dfrac{5}{8}$

19. $\dfrac{3}{14}$

20. $\dfrac{7}{20}$

Find an equivalent fraction with the given denominator.

21. $\dfrac{2}{3} = \dfrac{?}{24}$

22. $\dfrac{9}{11} = \dfrac{?}{33}$

23. $\dfrac{7}{12} = \dfrac{?}{60}$

24. $\dfrac{3}{5} = \dfrac{?}{45}$

25. $\dfrac{3}{7} = \dfrac{?}{49}$

26. $\dfrac{10}{15} = \dfrac{?}{60}$

27. $\dfrac{3}{4} = \dfrac{?}{20}$

28. $\dfrac{7}{8} = \dfrac{?}{40}$

29. $\dfrac{9}{13} = \dfrac{?}{39}$

30. $\dfrac{8}{11} = \dfrac{?}{44}$

31. $\dfrac{35}{40} = \dfrac{?}{80}$

32. $\dfrac{45}{50} = \dfrac{?}{100}$

33. $\dfrac{8}{9} = \dfrac{?}{9y}$

34. $\dfrac{7}{13} = \dfrac{?}{13n}$

35. $\dfrac{3}{7} = \dfrac{?}{28y}$

36. $\dfrac{3}{12} = \dfrac{?}{60y}$

37. $\dfrac{3}{6} = \dfrac{?}{18a}$

38. $\dfrac{4}{9} = \dfrac{?}{81x}$

39. $\dfrac{5}{7} = \dfrac{?}{21x}$

40. $\dfrac{7}{8} = \dfrac{?}{16x}$

Replace the ? with the appropriate symbol: $=, <,$ *or* $>$.

41. $\dfrac{7}{8} \, ? \, \dfrac{28}{32}$

42. $\dfrac{4}{7} \, ? \, \dfrac{28}{49}$

43. $\dfrac{9}{11} \, ? \, \dfrac{42}{66}$

44. $\dfrac{11}{16} \, ? \, \dfrac{33}{36}$

45. $\dfrac{5}{11} \, ? \, \dfrac{7}{12}$

46. $\dfrac{8}{13} \, ? \, \dfrac{7}{15}$

47. $\dfrac{1}{8} \, ? \, \dfrac{1}{15}$

48. $\dfrac{1}{11} \, ? \, \dfrac{1}{7}$

49. $\dfrac{4}{21} \, ? \, \dfrac{20}{105}$

50. $\dfrac{3}{16} \, ? \, \dfrac{9}{48}$

51. $\dfrac{22}{22} \, ? \, \dfrac{55}{55}$

52. $\dfrac{56}{56} \, ? \, \dfrac{65}{65}$

Which distance is farther?

53. A 7-Eleven store $\dfrac{4}{7}$ mile from your house, or a 7-Eleven store $\dfrac{3}{4}$ mile from your house.

54. A bus stop that is $\dfrac{9}{10}$ mile from where you are located, or a bus stop that is $\dfrac{6}{7}$ mile from your location.

Which amount is smaller?

55. A $\frac{3}{4}$-pound package of cheese, or a $\frac{5}{6}$-pound package of cheese.

56. Sliced ham that weighs $\frac{7}{8}$ pound, or sliced ham that weighs $\frac{2}{3}$ pound.

Solve each applied problem.

57. Kathy can buy $\frac{2}{3}$ pound of vanilla nut coffee at the store for $7. For the same price she can buy $\frac{3}{4}$ pound of French roast coffee. Which choice will give Kathy more coffee for her money?

58. Jason owns two square plots of land, with each plot having the same dimensions. He subdivides one into 4 equal parcels and the other into 5 equal parcels. For the same price you can buy either 3 of the 4 parcels or 4 of the 5 parcels. Which purchase yields more land?

 CALCULATOR EXERCISES

Replace the ? with the appropriate symbol: =, <, or >.

59. $\frac{244}{356} \; ? \; \frac{209}{322}$

60. $\frac{421}{559} \; ? \; \frac{540}{653}$

61. $\frac{355}{431} \; ? \; \frac{1420}{1724}$

62. $\frac{242}{323} \; ? \; \frac{1210}{1615}$

 VERBAL AND WRITING SKILLS

Fill in the blanks.

63. When we use the equality test for fractions, we multiply the _____ products from _____ to top.

64. To multiply two fractions, we multiply the _____ times the numerator and the _____ times denominator.

ONE STEP FURTHER

65. Why should we be able to see that $\frac{55}{55} = \frac{621}{621}$ without doing any calculations?

66. Why should we be able to see that $\frac{1}{8} < \frac{7}{9}$ without doing any calculations?

CUMULATIVE REVIEW

Express as a product of prime numbers.

1. 66

2. 72

3. 210

4. 112

SECTION 4.4

Simplifying Fractional Expressions

After studying this section, you will be able to:

❶ Reduce fractions.
❷ Simplify algebraic fractions.

Math Pro   Video 4.4   SSM

❶ Reducing Fractions

A fraction is considered **reduced** if the numerator and denominator have no common factors other than 1. The fraction $\frac{2}{3}$ is a *reduced fraction* since 2 and 3 only have 1 as a common factor. The fraction $\frac{4}{6}$ is *not a reduced fraction*; the 4 and the 6 both have a common factor of 2.

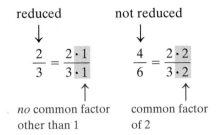

When we multiply a fraction by $\frac{2}{2}$, or $\frac{3}{3}$, or $\frac{4}{4}$, and so on …, we obtain an equivalent fraction that is not reduced. We reverse this process to reduce a fraction.

Building an equivalent fraction	**Reducing a fraction**
reduced ⟶ not	not ⟶ reduced
form reduced	reduced form
$\frac{2}{3} = \frac{2 \cdot 2}{3 \cdot 2} = \frac{4}{6}$	$\frac{4}{6} = \frac{2 \cdot 2}{3 \cdot 2} = \frac{2 \cdot 1}{3 \cdot 1} = \frac{2}{3}$

We can use slashes ╱ to indicate that we are rewriting common factors as the equivalent fraction $\frac{1}{1}$. For example, we can shorten the process to reduce $\frac{4}{6}$ as follows.

$$\frac{4}{6} = \frac{2 \cdot \cancel{2}^{\,1}}{3 \cdot \cancel{2}_{\,1}} = \frac{2}{3} \qquad \text{Slashes indicate that we are rewriting } \frac{2}{2} \text{ as } \frac{1}{1}.$$

PROCEDURE TO REDUCE A FRACTION TO LOWEST TERMS

1. Write the numerator and denominator of the fraction as products of prime numbers. For example, $\dfrac{4}{6} = \dfrac{2 \cdot 2}{3 \cdot 2}$

2. Any factor that appears in both the numerator and denominator is a common factor. Rewrite the fraction so that common factors are a fraction equivalent to $\dfrac{1}{1}$, and multiply. $\dfrac{2 \cdot 2}{3 \cdot 2} = \dfrac{2 \cdot \cancel{2}^{\,1}}{3 \cdot \cancel{2}_{\,1}} = \dfrac{2}{3}$

The direction *simplify* or *reduce* means to reduce to lowest terms.

EXAMPLE 1   Simplify: $\dfrac{15}{35}$.

SOLUTION

$\dfrac{15}{35} = \dfrac{3 \cdot 5}{7 \cdot 5}$ First, write the numerator and denominator as products of prime numbers.

$= \dfrac{3 \cdot \overset{1}{\cancel{5}}}{7 \cdot \underset{1}{\cancel{5}}}$ Then, rewrite $\dfrac{5}{5}$ as the equivalent fraction $\dfrac{1}{1}$.

$= \dfrac{3 \cdot 1}{7 \cdot 1} = \dfrac{\mathbf{3}}{\mathbf{7}}$

PRACTICE PROBLEM 1

Simplify: $\dfrac{18}{54}$.

Be careful! Students sometimes apply slashes incorrectly as follows:

$\dfrac{\cancel{3} + 4}{\cancel{3}} = 4$ THIS IS WRONG!

$\dfrac{3 + 4}{3} = \dfrac{7}{3}$ THIS IS RIGHT! $\dfrac{\cancel{3} \cdot 4}{\cancel{3}} = \dfrac{4}{1} = 4$ THIS IS RIGHT!

We may not use slashes between addition or subtraction signs. We may use slashes only if we are multiplying factors. Why?

EXAMPLE 2   Write the numerator and denominator of $\dfrac{42}{60}$ as a product of prime numbers, then simplify.

SOLUTION

$\dfrac{42}{60} = \dfrac{\mathbf{7 \cdot 2 \cdot 3}}{\mathbf{2 \cdot 2 \cdot 3 \cdot 5}} = \dfrac{7 \cdot \overset{1}{\cancel{2}} \cdot \overset{1}{\cancel{3}}}{2 \cdot \underset{1}{\cancel{2}} \cdot \underset{1}{\cancel{3}} \cdot 5} = \dfrac{7 \cdot 1 \cdot 1}{2 \cdot 1 \cdot 1 \cdot 5} = \dfrac{\mathbf{7}}{\mathbf{10}}$

You may be thinking: "I can reduce the fraction $\dfrac{42}{60}$ by dividing by the common factor: 6 goes into 42 seven times, and 6 goes into 60 ten times. Thus $\dfrac{42}{60} = \dfrac{7}{10}$. So why should I use prime factors as we did in Example 2?" There are two reasons:

1. Dividing by common factors can be difficult with larger fractions and may not reduce the fraction completely.
2. We must use methods and develop skills that can be used easily with algebraic expressions. When reducing an algebraic expression such as $\dfrac{120xy}{150yz}$, it is more difficult to find the common factor and divide: $30y$ goes into $120xy$ how many times?

PRACTICE PROBLEM 2

Write the numerator and denominator of $\dfrac{28}{60}$ as a product of prime numbers, then simplify.

❷ Simplifying Algebraic Fractions

E X A M P L E 3 Write the numerator and denominator of $\dfrac{21xy}{66x}$ as a product of prime numbers and variables, then simplify.

SOLUTION

$$\frac{21xy}{66x} = \frac{3 \cdot 7 \cdot x \cdot y}{3 \cdot 2 \cdot 11 \cdot x} = \frac{\overset{1}{\cancel{3}} \cdot 7 \cdot \overset{1}{\cancel{x}} \cdot y}{\underset{1}{\cancel{3}} \cdot 2 \cdot \underset{1}{\cancel{x}} \cdot 11} = \frac{1 \cdot 7 \cdot 1 \cdot y}{1 \cdot 2 \cdot 1 \cdot 11} = \frac{7y}{22}$$

P R A C T I C E P R O B L E M 3

Write the numerator and denominator of $\dfrac{90y}{150xy}$ as a product of prime numbers and variables, then simplify.

E X A M P L E 4 Simplify: $\dfrac{150n^2}{200n}$.

SOLUTION   If we recognize that both the numerator and denominator have *common factors that are not prime*, we can use these factors. Since 25 is a common factor of both the numerator and denominator, we can write each as a product of "25 times prime numbers."

$$\frac{150n^2}{200n} = \frac{25 \cdot 3 \cdot 2 \cdot n \cdot n}{25 \cdot 2 \cdot 2 \cdot 2 \cdot n}$$ Write all other factors as a product of prime numbers to assure that the fraction is reduced to lowest terms.

$$= \frac{\overset{1}{\cancel{25}} \cdot 3 \cdot \overset{1}{\cancel{2}} \cdot n \cdot \overset{1}{\cancel{n}}}{\underset{1}{\cancel{25}} \cdot 2 \cdot \underset{1}{\cancel{2}} \cdot 2 \cdot \underset{1}{\cancel{n}}}$$

$$= \frac{3 \cdot n}{2 \cdot 2}$$

$$\frac{150n^2}{200n} = \frac{3n}{4}$$

Note: We could also use 50 as a common factor in both the numerator and denominator of $\dfrac{150n^2}{200n}$.

P R A C T I C E P R O B L E M 4

Simplify: $\dfrac{80x^2}{140x}$.

Fractions are sometimes negative quantities. For example, $-\dfrac{1}{2}$ could mean that the price of stock dropped $\dfrac{1}{2}$, or that the amount of sugar in a recipe is reduced by $\dfrac{1}{2}$ cup. Now $-\dfrac{1}{2}$ can be written $\dfrac{-1}{2}$ or $\dfrac{1}{-2}$. These fractions are equivalent.

We can reduce a fraction with a negative number in either the numerator or the denominator by writing the negative sign in front of the fraction. The value of the fraction will not change, as illustrated below.

$$\frac{-15}{5} = -15 \div 5 = -3 \qquad \frac{15}{-5} = 15 \div (-5) = -3$$

$$-\frac{15}{5} = -(15 \div 5) = -(3) = -3$$

As you can see, we can write the negative sign in the numerator, in the denominator, or in front of the fraction. The value of the fraction does not change.

When we reduce a fraction, we write the negative sign in front of the fraction and do not include it as part of the prime factors.

EXAMPLE 5 Simplify: $\dfrac{-135}{25}$.

SOLUTION We write the negative sign in front of the fraction, then we factor.

$$\frac{-135}{25} = -\frac{135}{25} = -\frac{3 \cdot 3 \cdot 3 \cdot 5}{5 \cdot 5}$$

$$= -\frac{3 \cdot 3 \cdot 3 \cdot \cancel{5}}{5 \cdot \cancel{5}} = -\frac{27}{5} \quad \text{or} \quad \frac{-27}{5}$$

PRACTICE PROBLEM 5

Simplify: $\dfrac{-84}{14}$.

EXERCISES 4.4

Write each numerator and denominator as a product of prime numbers, then simplify.

1. $\dfrac{15}{25}$ **2.** $\dfrac{14}{21}$ **3.** $\dfrac{12}{16}$ **4.** $\dfrac{24}{30}$

5. $\dfrac{30}{36}$ **6.** $\dfrac{12}{32}$ **7.** $\dfrac{24}{28}$ **8.** $\dfrac{18}{27}$

9. $\dfrac{24}{36}$ **10.** $\dfrac{32}{64}$ **11.** $\dfrac{30}{85}$ **12.** $\dfrac{33}{55}$

13. $\dfrac{42}{54}$ **14.** $\dfrac{63}{81}$ **15.** $\dfrac{36}{72}$ **16.** $\dfrac{62}{54}$

17. $\dfrac{49}{35}$ **18.** $\dfrac{81}{72}$ **19.** $\dfrac{75}{60}$ **20.** $\dfrac{46}{23}$

21. $\dfrac{48}{16}$ **22.** $\dfrac{15}{75}$ **23.** $\dfrac{12}{72}$ **24.** $\dfrac{125}{25}$

25. $\dfrac{130}{13}$ **26.** $\dfrac{12}{120}$ **27.** $\dfrac{10}{140}$ **28.** $\dfrac{95}{35}$

29. $\dfrac{86}{42}$ **30.** $\dfrac{120}{165}$ **31.** $\dfrac{108}{148}$ **32.** $\dfrac{150}{250}$

33. $\dfrac{200}{300}$ **34.** $\dfrac{44}{220}$ **35.** $\dfrac{36}{180}$

Write each numerator and denominator as a product of prime numbers and variables, then simplify. Assume that any variable in a denominator is nonzero.

36. $\dfrac{24xy}{42x}$ **37.** $\dfrac{20nx}{45n}$ **38.** $\dfrac{21x}{28xy}$ **39.** $\dfrac{12x}{18nx}$

40. $\dfrac{27x^2}{45x}$ **41.** $\dfrac{28x^2}{49x}$ **42.** $\dfrac{20y}{24y^2}$ **43.** $\dfrac{21y}{24y^2}$

44. $\dfrac{16x}{18x}$ **45.** $\dfrac{25n}{55n}$ **46.** $\dfrac{20y}{35y}$ **47.** $\dfrac{14x}{21x}$

48. $\dfrac{36n^2}{42n}$ **49.** $\dfrac{64y^2}{72y}$ **50.** $\dfrac{35x}{45x^2}$ **51.** $\dfrac{20y}{30y^2}$

Simplify.

52. $\dfrac{-24}{36}$

53. $\dfrac{-35}{40}$

54. $\dfrac{-42}{48}$

55. $\dfrac{-40}{50}$

56. $\dfrac{30}{-42}$

57. $\dfrac{25}{-40}$

58. $-\dfrac{16}{18}$

59. $-\dfrac{14}{18}$

Simplify each fraction.

60. (a) $\dfrac{-12}{18}$ **(b)** $\dfrac{12}{-18}$ **(c)** $-\dfrac{12}{18}$

61. (a) $\dfrac{-15}{25}$ **(b)** $\dfrac{15}{-25}$ **(c)** $-\dfrac{15}{25}$

CALCULATOR EXERCISES

Simplify. The calculator can be used to help factor larger numbers so that the fraction may be reduced.

62. $\dfrac{91}{133}$

63. $\dfrac{1105}{1955}$

64. $\dfrac{-627}{3553}$

65. $\dfrac{-253}{319}$

ONE STEP FURTHER

Simplify. Assume that any variable in a denominator is nonzero.

66. $\dfrac{25x^2\,y^2\,z^4}{135x^3\,y}$

67. $\dfrac{40a^2\,b^2\,c^4}{88ab}$

68. $\dfrac{156ab^3}{144bc^4}$

69. $\dfrac{256xy^3}{300yz^5}$

CUMULATIVE REVIEW

1. Add: $12 + 13 + 25 + 7$.

2. Subtract: $9 - (6 - 1)$.

3. Combine like terms: $-7a + 3a + 5$.

SECTION 4.5

Ratios and Rates

After studying this section, you will be able to:

❶ Write two quantities with the same units as a ratio.
❷ Write two quantities with different units as a rate.
❸ Find unit rates.
❹ Solve applied problems involving ratios and rates.

Math Pro   Video 4.5   SSM

❶ Writing Two Quantities with the Same Units as a Ratio

Suppose that we wanted to compare two quantities with the same units. For example, comparing a 5-foot width of a garden to a 22-foot width of the back yard. The ratio of the lengths would be 5 to 22.

$w = 5$ ft

$w = 22$ ft

We can express the ratio three ways.

Writing the words: the ratio of 5 to 22

Using a colon: $5 : 22$

Using a fraction: $\dfrac{5}{22}$

Each of the ways of expressing a ratio is read "5 to 22."

Although a ratio can be written in different forms, it is a fraction and therefore should always be simplified (reduced to lowest terms).

EXAMPLE 1   Write each ratio in simplest form. Express your answer as a fraction.

(a) The ratio of 20 dollars to 35 dollars
(b) $14 : 21$

SOLUTION

(a) 20 dollars to 35 dollars $= \dfrac{20 \ \cancel{\text{dollars}}}{35 \ \cancel{\text{dollars}}} = \dfrac{5 \cdot 4}{5 \cdot 7} = \dfrac{\mathbf{4}}{\mathbf{7}}$

(b) $14 : 21 = \dfrac{14}{21} = \dfrac{7 \cdot 2}{7 \cdot 3} = \dfrac{\mathbf{2}}{\mathbf{3}}$

PRACTICE PROBLEM 1
Write each ratio in simplest form. Express your answer as a fraction.

(a) The ratio of 28 feet to 49 feet   **(b)** $27 : 81$

It is important that you read the problem carefully since the order of quantities is important, as shown in the next example.

EXAMPLE 2   A mixture contains of 20 milliliters of water and 8 milliliters of alcohol. Write each ratio as a fraction and reduce to lowest terms.

(a) The ratio of alcohol to water   **(b)** The ratio of water to alcohol

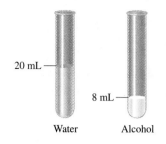

20 mL —

8 mL —

Water   Alcohol

SOLUTION

(a) $\dfrac{\text{alcohol}}{\text{water}} = \dfrac{8 \text{ mL}}{20 \text{ mL}} = \dfrac{4 \cdot 2}{4 \cdot 5} = \dfrac{2}{5}$ **(b)** $\dfrac{\text{water}}{\text{alcohol}} = \dfrac{20 \text{ mL}}{8 \text{ mL}} = \dfrac{4 \cdot 5}{4 \cdot 2} = \dfrac{5}{2}$

PRACTICE PROBLEM 2

15 women and 21 men are enrolled in a physical science class. Write each ratio as a fraction and reduce to lowest terms.

(a) The ratio of men to women   **(b)** The ratio of women to men

EXAMPLE 3   Not all regions of Alaska receive the largest annual snowfall. Barrow, Alaska, located near the Arctic Ocean, receives only 29 inches annually on average, while Chicago's Midway Airport receives an annual average of 46 inches. What is the ratio of the annual average snowfall in Barrow, Alaska to the annual average snowfall at Midway Airport in Chicago?

SOLUTION

$$\frac{\text{Barrow}}{\text{Midway Airport}} = \frac{29 \text{ inches}}{46 \text{ inches}} = \frac{29}{46}$$

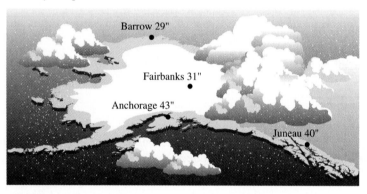

Barrow 29"

Fairbanks 31"

Anchorage 43"

Juneau 40"

PRACTICE PROBLEM 3

Refer to Example 3 to answer the following: Write the ratio of the annual average snowfall in Midway Airport, Chicago to the annual average snowfall in Barrow, Alaska.

②   Writing Two Quantities with Different Units as a Rate

A *rate* is a special type of ratio that compares two quantities with *different units*. We write the units when we express a rate as a fraction.

Same units—*ratio:* $\dfrac{4 \text{ liters}}{10 \text{ liters}} = \dfrac{2}{5}$

Different units—*rate:* $\dfrac{90 \text{ students}}{4 \text{ teachers}} = \dfrac{45 \text{ students}}{2 \text{ teachers}}$ We write the units.

We always reduce to lowest terms.

EXAMPLE 4   The calories and fat content for 100 grams of french fries (about 1 medium bag) is given on the chart below.

(a) Write the rate of calories to fat in a medium bag of Burger King french fries.

(b) Write the rate of calories to fat in a medium bag of McDonald's french fries.

SOLUTION

(a) Burger King: $\dfrac{\textbf{318 calories}}{\textbf{17 grams fat}}$

(b) McDonald's: $\dfrac{308 \text{ calories}}{14 \text{ grams fat}} = \dfrac{\textbf{22 calories}}{\textbf{1 gram fat}}$

	CALORIES	FAT
Burger King	318	17 g
McDonald's	308	14 g

Source: Orange County Register.

PRACTICE PROBLEM 4

C&R Construction must pay $26 in fees for every 400 pounds they dispose at the local dump. Write the rate of fees to pounds.

③ Finding Unit Rates

We are familiar with the phrase *55 miles per hour*. What does this mean? Since the word *per* means *for every*, we have 55 miles traveled for every 1 hour traveled.

$$\frac{55 \text{ miles}}{1 \text{ hour}} \quad \text{or} \quad 55 \text{ miles per hour (mph)}$$

This type of rate, with a denominator of 1, is called a **unit rate**. The rate in Example 4b is also a unit rate. It can also be written as 22 calories per gram.

EXAMPLE 5   Bertha drove her car 417 miles in 8 hours. Find the unit rate.

SOLUTION

$$\frac{417 \text{ miles}}{8 \text{ hours}} \quad \text{We divide} \quad 8\overline{)417} \quad \frac{52\text{R}1 = 52\tfrac{1}{8}}{}$$

$$\frac{417 \text{ miles}}{8 \text{ hours}} = \frac{52\tfrac{1}{8} \text{ miles}}{1 \text{ hour}} \quad \text{or} \quad \textbf{52}\tfrac{1}{8} \textbf{ miles per hour}$$

PRACTICE PROBLEM 5

Iris travels 92 miles on 5 gallons of gas. Find the unit rate.

④ Solving Applied Problems Involving Ratios and Rates

We work with **unit rates** in many areas, such as sports, business, budgeting, and science. Sometimes we want to find the number of boards needed for each wall, or we may need to find the number of people needed to complete a task. Often we want to find the best buy for the dollar. These are all applications of unit rates.

E X A M P L E   6   University of Chicago tornado researcher Tetsuya Theodore Fujita cataloged 31,054 tornados in the United States during the 70 years 1916–1985, and found that $\frac{7}{10}$ of the tornados occur in the spring and early summer.

(a) Write as a ratio: the number of tornado occurrences in December to the number of occurrences in April.

(b) Write as a unit rate: the average number of tornados per year that occur in the month of May. Round your answer to the nearest whole number.

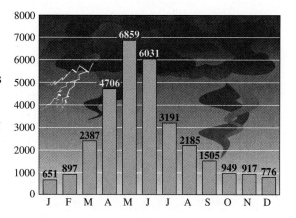

SOLUTION

(a) $\dfrac{\text{December}}{\text{April}} = \dfrac{776}{4706} = \dfrac{\textbf{388}}{\textbf{2353}}$

(b) $\dfrac{6859 \text{ tornados in May}}{70 \text{ years}}$

We divide to find the unit rate.

$$70\,\overline{)6859}\quad 97\tfrac{69}{70} \quad \text{or approximately } \textbf{98 tornados per year in May}$$

Note: We rounded $97\frac{69}{70}$ to 98 because the fraction $\frac{69}{70}$ is close to 1.

P R A C T I C E   P R O B L E M   6

Refer to Example 6.

(a) Write as a ratio: the number of tornado occurrences in September to the number of occurrences in August.

(b) Write as a unit rate: the average number of tornados per year that occur in the month of June. Round your answer to the nearest whole number.

E X A M P L E   7   Sunshine Preschool has a staffing policy requiring that for every 60 children there are 3 preschool teachers, and for every 24 children there are 2 aides.

(a) How many children per teacher does the preschool have?

(b) How many children per aide does the preschool have?

(c) If there are 60 students at the preschool, how many aides must there be to satisfy the staffing policy?

SOLUTION

(a) Children per teacher:

$$\frac{\text{children}}{\text{teacher}} = \frac{60 \text{ children}}{3 \text{ teachers}} = \frac{20 \text{ children}}{1 \text{ teacher}} \text{ or } \textbf{20 children per teacher}$$

(b) Children per aide:

$$\frac{\text{children}}{\text{aide}} = \frac{24 \text{ children}}{2 \text{ aides}} = \frac{12 \text{ children}}{1 \text{ aide}} \text{ or } \textbf{12 children per aide}$$

(c) Since there are 12 children for 1 aide, we divide $60 \div 12$ to find how many aides are needed for 60 children.

$$60 \div 12 = \textbf{5 aides for 60 children}$$

PRACTICE PROBLEM 7

Autumn Home, a private nursing home, has a medical staffing policy requiring that for every 40 patients there are 2 registered nurses (RNs), and for every 30 patients there are 2 nurse's aides.

(a) How many patients per RN does Autumn Home have?
(b) How many patients per aide does Autumn Home have?
(c) If there are 60 patients at Autumn Home, how many aides must there be to satisfy the staffing policy?

We often ask ourselves, "Which package is the better buy, the pack of 3 or the pack of 7?" "Is it cheaper to buy the 12-ounce box, or the 16-ounce box?" We find the *unit price* (price per item) to answer these types of questions.

EXAMPLE 8 The Computer Warehouse is having a sale on black print cartridges, a package of 6 for $96, and the same brand in a package of 8 for $136.

(a) Find each unit price.   **(b)** Which is the better buy?

page 12

ON SALE
at
Computer Warehouse

PRINT CARTRIDGES (black)

8 for $136
6 for $96

SOLUTION

(a) $\dfrac{\$96}{6} = \textbf{\$16 per cartridge;}$ $\dfrac{\$136}{8} = \textbf{\$17 per cartridge.}$

(b) The **package of 6 cartridges** is the better buy.

PRACTICE PROBLEM 8

The Linen Factory is having a sale on their designer hand towels. The Hazelette Collection is on sale at 6 for $78, and the Springview Collection is on sale at 9 for $108.

(a) Find each unit price.   **(b)** Which is the better buy?

EXERCISES 4.5

Write each ratio as a fraction in simplest form.

1. 15 to 65

2. 12 to 32

3. 19 : 47

4. 21 : 59

5. 35 : 10

6. 46 : 14

7. 25 : 70

8. 30 : 45

9. 34 minutes to 12 minutes

10. 24 dollars to 16 dollars

11. 14 gallons to 35 gallons

12. 20 feet to 45 feet

13. 17 hours to 41 hours

14. 33 inches to 11 inches

15. $121 to $423

16. $85 to $151

Write each ratio as a fraction and simplify.

17. A mixture contains 35 milliliters of water and 15 milliliters of chlorine.

    (a) State the ratio of chlorine to water.

    (b) State the ratio of water to chlorine.

18. The marine science field trip consisted of 30 juniors and 22 seniors.

    (a) State the ratio of seniors to juniors.

    (b) State the ratio of juniors to seniors.

19. The Willow Brook recreational basketball team had a season record of 23 wins and 14 losses.

    (a) State the ratio of wins to losses.

    (b) State the ratio of losses to wins.

Candy sales are generally extremely high at Easter. In fact, if the 15 billion jelly beans eaten at Easter were lined up end to end, they would circle the Earth $4\frac{1}{2}$ times. Use the chart displaying the amount of money spent on candy at Easter to answer questions 20 and 21.

Easter Candy Sales

1995	$800 million
1996	$837 million
1997	$875 million

Source: Orange County Register, 3/12/98.

Write as a ratio and simplify.

20. Candy sales in 1996 to candy sales in 1997

21. Candy sales in 1995 to candy sales in 1996

Use the following information to answer questions 22 and 23. Recently several major newspapers reported the following information. The word's longest suspension bridge is Japan's Akashi Kaikyo Bridge. The bridge is 6066 feet long and links Kobe and Awaji Island.

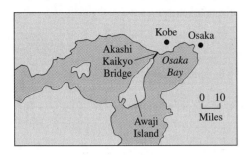

10 Longest Suspension Bridges

Bridge	Length
Akashi Kaikyo, Japan	6066 ft
Great Belt Link, Denmark	5328 ft
Humber River, England	4626 ft
Verrazano Narrows, New York City	4260 ft
Golden Gate, San Francisco	4200 ft
Mackinac Straits, Michigan	3800 ft
Minami Bian-Seto, Japan	3668 ft
Second Bosporus, Turkey	3576 ft
First Bosporus, Turkey	3524 ft
George Washington, NYC	3500 ft

Write as a ratio and simplify.

22. The length of the George Washington Bridge to the length of the Golden Gate Bridge

23. The length of the Great Belt Link Bridge to the length of the Akashi Kaikyo Bridge

Write as a rate in simplest form.

24. $50 for 13 plants

25. 425 feet in 7 seconds

26. 155 students for 6 teachers

27. 47 pies for 340 people

28. 171 words in 2 minutes

29. 253 guests for 122 parking spaces

30. 410 calories for 19 grams of fat

31. 205 calories for 7 grams of fat

Use the following information to answer questions 32 to 35. In 1996 there were 50,047 people in the United States on a waiting list for a transplant, yet only 9410 transplants could be performed.

Transplants Performed in the United States in 1996

Kidney	1099	Heart	2342
Kidney–pancreas	850	Heart–Lung	39
Pancreas	172	Lung	805
Liver	4058	Intestine	45

Source: Dallas Morning News.

Write as a rate in simplest form.

32. The number of heart transplants to the number of lung transplants

33. The number of kidney–pancreas transplants to the number of kidney transplants

34. The number of intestine transplants to the number of heart–lung transplants

35. The number of pancreas transplants to the number of liver transplants.

Find the unit rate.

36. Traveling 320 miles in 6 hours

37. Pumping 2760 gallons in 16 hours

38. Earning $304 in 38 hours

39. Traveling 315 miles on 14 gallons of gas

40. Traveling 405 miles on 18 gallons of gas

41. Earning $455 in 35 hours

Solve each applied problem.

42. A car travels 616 miles on 28 gallons of gas. Find how many miles the car can be driven on one gallon of gas.

43. A pulley makes 63 complete rotations in 18 seconds. How many rotations per second does the pulley make?

44. Marci paid $3380 for 65 shares of stock. What was the cost per share?

45. Delroy joined a CD club that charges $189 for 21 CDs. How much per CD did Delroy pay?

46. The J. D. Robertson Academy of Arts school has a staffing policy requiring that for every 90 students there are 5 instructors, and for every 30 students there are 2 tutors.

   (a) How many students per instructor does the academy have?

   (b) How many students per tutor does the academy have?

   (c) How many tutors are needed to satisfy the staffing policy if there are 90 students in the academy?

47. The All Point Insurance Group requires each office to have 4 insurance agents for every 620 clients and 5 clerical staff members for every 310 clients.

   (a) How many clients per agent does the group have?

   (b) How many clients per clerical staff member does the group have?

   (c) How many clerical staff members would be required for an office that has 930 clients?

48. The Crystal Shop has their Gold Lace crystal wine glasses on sale. A box of 8 glasses is $96 and a box of 6 is $78.

   (a) Find each unit price.

   (b) Which is the better buy?

49. The tanning salon has a special on their tanning sessions: 12 sessions for $96 or 15 sessions for $135.

   (a) Find each unit price.

   (b) Which is the better deal?

50. The music shop sells used CDs 4 for $32 or 6 for $48.

   (a) Find each unit price.

   (b) Which is the better deal?

51. Computer World is having a sale on printer paper: 4 reams of paper for $12 or 7 reams of paper for $21.

   (a) Find each unit price.

   (b) Which is the better deal?

Use the following information to answers questions 52 and 53. Johnson and Brothers Suits recorded the following sales for the third quarter of the year.

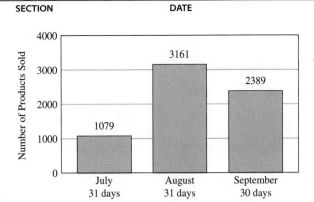

52. (a) Write as a ratio: the number of sales in July to the number of sales in September.

(b) Write as a unit rate: the average number of sales per day in the month of August. Round your answer to the nearest whole number.

53. (a) Write as a ratio: the number of sales in August to the number of sales in September.

(b) Write as a unit rate: the average number of sales per day in the month of July. Round your answer to the nearest whole number.

 CALCULATOR EXERCISES

54. 4500 shares of stock cost $562,500. What is the price of stock per share?

55. Jason drove 3800 miles across the country and used 190 gallons of gas. How many miles per gallon did Jason's car average?

VERBAL AND WRITING SKILLS

56. Explain the difference between a ratio and a rate.

57. Explain the difference between a rate and a unit rate.

CUMULATIVE REVIEW

Solve and check your answer.

1. $12x = 84$

2. $15n = 180$

3. $64 = 4n$

4. $48 = 3x$

After studying this section, you will be able to:

❶ Write a proportion.

❷ Determine if a state-ment is a proportion.

❸ Find the missing number in a proportion.

Math Pro   Video 4.6   SSM

Proportions

❶ Writing a Proportion

Recall from Section 4.3 that the fractions $\frac{4}{8}$ and $\frac{1}{2}$ are equivalent fractions: $\frac{4}{8} = \frac{1}{2}$. Now if these fractions represent ratios, we say that the ratios are **proportional**. In other words, if two ratios or two rates are equal, they are called a *proportion*. For example, $\frac{4}{8} = \frac{1}{2}$ is a proportion and $\frac{2 \text{ trees}}{7 \text{ feet}} = \frac{4 \text{ trees}}{14 \text{ feet}}$ is also a proportion. The proportion $\frac{4}{8} = \frac{1}{2}$ is read "four *is to* eight as one *is to* two."

> A *proportion* states that two ratios or two rates are equal.
>
> If $\frac{a}{b}$ and $\frac{c}{d}$ are two equal ratios, then $\frac{a}{b} = \frac{c}{d}$ is called a proportion.

When we write a proportion we must be sure that the units of each ratio are in the appropriate position. One way to write a proportion is for the numerators to have the same units and the denominators to have the same units. In this book we write proportions in this manner.

E X A M P L E   1   Write the proportion to express the following: If 6 pounds of flour cost \$2, then 18 pounds will cost \$6.

SOLUTION

$$\frac{6 \text{ pounds}}{2 \text{ dollars}} = \frac{18 \text{ pounds}}{6 \text{ dollars}}$$

We write pounds in the numerator.
We write dollars in the denominator.

P R A C T I C E   P R O B L E M   1
Write the proportion to express the following: If it takes 4 hours to drive 144 miles, it will take 6 hours to drive 216 miles.

❷ Determining if a Statement Is a Proportion

The statement $\frac{3}{8} = \frac{6}{16}$ is a proportion because the fractions are equivalent. There-fore, to determine whether or not a statement is a proportion, we can use the *equality test for fractions* that we learned in Section 4.3. This test states that *two fractions are equal if and only if their cross products are equal.*

E X A M P L E   2   Determine whether the statement is a proportion.

(a) $\frac{16}{35} \overset{?}{=} \frac{48}{125}$ (b) $\frac{22}{29} \overset{?}{=} \frac{88}{116}$

SOLUTION   We form the two cross products.

(a) $125 \times 16 = \mathbf{2000}$ $35 \times 48 = \mathbf{1680}$

$$\frac{16}{35} \diagdown \frac{48}{125}$$

The two cross products are *not* equal.

Thus $\dfrac{16}{35} \neq \dfrac{48}{125}$. **This is not a proportion.**

(b) $116 \times 22 = \mathbf{2552}$ $29 \times 88 = \mathbf{2552}$

$\dfrac{22}{29}$ $\dfrac{88}{116}$

The two cross products are equal. Thus $\dfrac{22}{29} = \dfrac{88}{116}$. **This is a proportion.**

PRACTICE PROBLEM 2

Determine whether the statement is a proportion.

(a) $\dfrac{14}{45} \overset{?}{=} \dfrac{42}{135}$

(b) $\dfrac{32}{72} \overset{?}{=} \dfrac{128}{144}$

EXAMPLE 3   Is the rate $\dfrac{48 \text{ points}}{56 \text{ games}}$ equal to the rate $\dfrac{40 \text{ points}}{45 \text{ games}}$?

SOLUTION   We must check to see if $\dfrac{48}{56} \overset{?}{=} \dfrac{40}{45}$, so we form the cross products.

$45 \times 48 = \mathbf{2160}$ $56 \times 40 = \mathbf{2240}$

$\dfrac{48}{56}$ $\dfrac{40}{45}$

The cross products, 2160 and 2240, are not equal. Thus, *the two rates are not equal.* **This is not a proportion.**

PRACTICE PROBLEM 3

Is the rate $\dfrac{240 \text{ words typed}}{3 \text{ minutes}}$ equal to the rate $\dfrac{1240 \text{ words typed}}{18 \text{ minutes}}$?

❸   Finding the Missing Number in a Proportion

Sometimes one of the quantities in a proportion is unknown. We can find this unknown quantity by finding the cross products and solving the resulting equation.

PROCEDURE TO SOLVE FOR A MISSING NUMBER IN A PROPORTION

1. Find the cross products and form an equation.

$$\dfrac{x}{21} = \dfrac{2}{7}$$
$$\mathbf{7} \cdot x = 21 \cdot 2$$

2. Write the multiplication as division on the opposite side of the equal sign.

$$x = \dfrac{21 \cdot 2}{\mathbf{7}}$$

3. Simplify the result.

$$x = 6$$

4. Check your answer.

$$\dfrac{6}{21} \overset{?}{=} \dfrac{2}{7}$$
$$42 = 42 \quad \checkmark$$

E X A M P L E   4   Find the value of n in $\dfrac{n}{24} = \dfrac{15}{6}$.

SOLUTION

$$\frac{n}{24} = \frac{15}{6}$$

$6 \cdot n = 24 \cdot 15$ Find the cross products and form an equation.

$6 \cdot n = 360$

$$n = \frac{360}{6}$$

$n = 60$ Divide.

Check whether the proportion is true.

$$\frac{60}{24} \stackrel{?}{=} \frac{15}{6}$$

$6 \cdot 60 \stackrel{?}{=} 24 \cdot 15$

$360 = 360$ ✓

P R A C T I C E   P R O B L E M   4

Find the value of n in $\dfrac{n}{18} = \dfrac{28}{72}$.

We use the same process if the numbers in the proportions are fractions.

E X A M P L E   5   Find the value of x in $\dfrac{\frac{1}{6}}{x} = \dfrac{1}{\frac{1}{7}}$.

SOLUTION

$$\frac{\frac{1}{6}}{x} = \frac{1}{\frac{1}{7}}$$

$\dfrac{1}{7} \cdot \dfrac{1}{6} = x \cdot 1$ Find the cross products and form an equation.

$\dfrac{\mathbf{1}}{\mathbf{42}} = x$ Since $1 \cdot x = x$, we do not need to divide.

Check whether the proportion is true.

$$\frac{\frac{1}{6}}{\frac{1}{42}} \stackrel{?}{=} \frac{1}{\frac{1}{7}} \qquad \frac{1}{7} \cdot \frac{1}{6} \stackrel{?}{=} \frac{1}{42} \cdot 1$$

$$\frac{1}{42} = \frac{1}{42} \quad ✓$$

P R A C T I C E   P R O B L E M   5

Find the value of x in $\dfrac{\frac{1}{3}}{x} = \dfrac{1}{\frac{1}{5}}$.

EXERCISES 4.6

Write the proportion.

1. 4 is to 9 as 28 is to 63.

2. 6 is to 11 as 30 is to 55.

3. 12 is 7 as 48 is to 28.

4. 16 is to 5 as 48 is to 15.

5. 3 is to 8 as 18 is to 48.

6. 2 is to 9 as 8 is to 36.

7. $\frac{1}{3}$ is to $\frac{1}{8}$ as $\frac{1}{4}$ is to $\frac{3}{32}$.

8. $\frac{1}{7}$ is to $\frac{1}{9}$ as $\frac{1}{6}$ is to $\frac{7}{54}$.

Write the proportion.

9. If 2 printers are needed for 4 secretaries, 12 printers are needed for 24 secretaries.

10. If 6 benches are needed to seat 24 people, 48 benches are needed to seat 192 people.

11. If 2 cups of cereal contain 50 grams of carbohydrates, then 6 cups of cereal contain 150 grams of carbohydrates.

12. If 3 inches on a map represents 270 miles, 6 inches represents 540 miles.

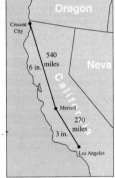

13. If 6 American dollars has a value of 4 British pounds, then 264 American dollars has a value of 176 British pounds.

14. If Jaime can drive 176 miles in his Toyota truck on 8 gallons of gas, then he should be able to drive 528 miles on 24 gallons of gas.

15. If a pulley can complete $3\frac{1}{2}$ rotations in 2 minutes, it should complete 14 rotations in 8 minutes.

16. If it takes $2\frac{1}{4}$ yards of material to make 1 skirt, it will take 9 yards to make 4 skirts.

17. If Matt averages 4 baskets out of 7 free throws attempted in a basketball game, he should make 12 out of 21 free throws.

18. If Sara averages 2 goals for every 5 shots on goal, she should make 12 out of 30 shots on goal.

Determine whether the statement is a proportion.

19. $\dfrac{2}{7} \overset{?}{=} \dfrac{8}{28}$

20. $\dfrac{5}{8} \overset{?}{=} \dfrac{25}{40}$

21. $\dfrac{13}{19} \overset{?}{=} \dfrac{26}{29}$

22. $\dfrac{15}{37} \overset{?}{=} \dfrac{18}{39}$

23. $\dfrac{2 \text{ American dollars}}{11 \text{ French francs}} \overset{?}{=} \dfrac{65 \text{ American dollars}}{135 \text{ French francs}}$

24. $\dfrac{6 \text{ defective parts}}{109 \text{ parts produced}} \overset{?}{=} \dfrac{20 \text{ defective parts}}{401 \text{ parts produced}}$

25. $\dfrac{180 \text{ miles}}{3 \text{ hours}} \overset{?}{=} \dfrac{360 \text{ miles}}{6 \text{ hours}}$

26. $\dfrac{54 \text{ seniors}}{33 \text{ juniors}} \overset{?}{=} \dfrac{108 \text{ seniors}}{66 \text{ juniors}}$

27. Marc can type 400 words in 5 minutes, and Jessica can type 675 words in 9 minutes. Do they type at the same rate?

28. Les scored 4 goals in 7 soccer games, and Joe scored 6 goals in 9 soccer games. Who is scoring at a higher rate?

Find the value of x in each proportion. Check your answer.

29. $\dfrac{x}{8} = \dfrac{5}{2}$

30. $\dfrac{x}{10} = \dfrac{6}{5}$

31. $\dfrac{x}{12} = \dfrac{8}{3}$

32. $\dfrac{x}{15} = \dfrac{6}{5}$

33. $\dfrac{12}{x} = \dfrac{3}{5}$

34. $\dfrac{4}{x} = \dfrac{2}{7}$

35. $\dfrac{18}{x} = \dfrac{3}{4}$

36. $\dfrac{8}{x} = \dfrac{20}{15}$

37. $\dfrac{9}{75} = \dfrac{6}{x}$

38. $\dfrac{12}{18} = \dfrac{x}{21}$

39. $\dfrac{7}{5} = \dfrac{x}{25}$

40. $\dfrac{5}{4} = \dfrac{35}{x}$

41. $\dfrac{15}{x} = \dfrac{30}{14}$

42. $\dfrac{18}{x} = \dfrac{36}{20}$

43. $\dfrac{\frac{1}{2}}{x} = \dfrac{1}{\frac{1}{9}}$

44. $\dfrac{\frac{1}{3}}{x} = \dfrac{1}{\frac{1}{7}}$

45. $\dfrac{x}{\frac{1}{3}} = \dfrac{\frac{2}{7}}{1}$

46. $\dfrac{x}{\frac{3}{4}} = \dfrac{\frac{1}{5}}{1}$

Find the value of n.

47. $\dfrac{80 \text{ gallons}}{24 \text{ acres}} = \dfrac{20 \text{ gallons}}{n \text{ acres}}$

48. $\dfrac{70 \text{ women}}{25 \text{ men}} = \dfrac{14 \text{ women}}{n \text{ men}}$

49. $\dfrac{10 \text{ miles}}{16 \text{ kilometers}} = \dfrac{n \text{ miles}}{8 \text{ kilometers}}$

50. $\dfrac{n \text{ grams}}{15 \text{ liters}} = \dfrac{12 \text{ grams}}{45 \text{ liters}}$

51. $\dfrac{n \text{ miles}}{15 \text{ gallons}} = \dfrac{16 \text{ miles}}{3 \text{ gallons}}$

52. $\dfrac{6 \text{ hours}}{300 \text{ miles}} = \dfrac{2 \text{ hours}}{n \text{ miles}}$

ONE STEP FURTHER

Determine whether the statement is a proportion.

53. $\dfrac{\frac{1}{5}}{\frac{1}{10}} \stackrel{?}{=} \dfrac{\frac{1}{4}}{\frac{1}{8}}$

54. $\dfrac{\frac{2}{3}}{\frac{1}{5}} \stackrel{?}{=} \dfrac{\frac{3}{7}}{\frac{1}{9}}$

CUMULATIVE REVIEW

Translate each sentence using mathematical symbols.

1. Two times a number is added to 6.

2. Twenty divided by a number is equal to five.

3. Eight subtracted from some number is equal to nine.

4. Four plus three times some number is equal to nineteen.

After studying this section, you will be able to:

1 Solve applied problems involving proportions.

Math Pro   Video 4.7   SSM

S E C T I O N   4 . 7

Applied Problems Involving Proportions

1 Solving Applied Problems Involving Proportions

When a situation involves a ratio or rate, we can use proportions to find the solution. Let us examine a variety of applied problems that can be solved with a proportion.

EXAMPLE 1   A large automobile dealership has found that for every 14 cars sold, 2 are brought back for major repairs. If the dealership sells 112 cars this month, approximately how many cars will be brought back for major repairs?

SOLUTION   We set up the proportion, then solve for the missing number in the proportion. We let n represent the total number of major repairs.

$$\frac{14 \text{ cars sold}}{2 \text{ need major repairs}} = \frac{112 \text{ total cars sold}}{n \text{ total major repairs}}$$

$$\frac{14}{2} = \frac{112}{n}$$

$$14n = 2 \times 112 \qquad \text{Form a cross product.}$$

$$14n = 224 \qquad \text{Simplify.}$$

$$n = \frac{224}{14} \qquad \text{Divide by 14.}$$

$$n = 16$$

Approximately 16 cars will be brought back for major repairs.

PRACTICE PROBLEM 1

Mary Lou's Catering has a policy that when planning a buffet there should be 18 desserts for every 15 people who will be attending the buffet. How many desserts should the catering company plan to serve at a buffet if 180 people are expected to attend?

EXAMPLE 2   Estelle is planning to place a fence around her vegetable garden, which is 12 feet wide and 16 feet long. She wants the fence's dimensions to be proportional to the garden's. What must the length of the fence be if the width is 24 feet?

SOLUTION   First, we set up the proportion, letting the letter x represent the length of the fence.

$$\frac{12 \text{ ft width of garden}}{16 \text{ ft length of garden}} = \frac{24 \text{ ft width of fence}}{x \text{ ft length of fence}}$$

Now we solve for x.

$$\frac{12}{16} = \frac{24}{x}$$

$$12x = 16 \times 24$$

$$12x = 384$$

$$x = \frac{384}{12} = 32 \qquad \text{The length of the fence is } \textbf{32 feet}.$$

PRACTICE PROBLEM 2

Refer to Example 2 to answer the following. If the width of the fence is 18 feet, what must the length be for the dimensions of the fence to be proportional to the garden?

TO THINK ABOUT
· ·

Can we reduce a proportion before we solve for the missing number in the proportion? Yes. If you can see that the ratio without the variable can be reduced, you may reduce it, and the answer will still be correct. Let's look at the proportion in Example 2 and observe what happens when we reduce the ratio $\frac{12}{16}$.

$$\frac{12}{16} = \frac{24}{x}$$

$$\frac{3}{4} = \frac{24}{x} \qquad \text{Reduce } \frac{12}{16} \text{ to } \frac{3}{4}.$$

$$3x = 4 \times 24 \qquad \text{Cross-multiply.}$$

$$3x = 96$$

$$x = \frac{96}{3} = 32 \qquad \text{We see that the answer is the same.}$$

This step is simplified when we reduce. Do you see why?

EXERCISE

1. Why can we reduce a ratio and still get the correct answer when we solve a proportion?

EXAMPLE 3   Two partners, Cleo and Julie, invest money in their small business at the ratio 3 to 5, with Cleo investing the smaller amount. If Cleo invested $6000, how much did Julie invest?

SOLUTION   The ratio *3 to 5* represents Cleo's investment *to* Julie's investment.

$$\frac{3 \text{ Cleo's investment}}{5 \text{ Julie's investment}} = \frac{\$6000 \text{ is Cleo's share of investment}}{\$x \text{ is Julie's share of investment}}$$

$$\frac{3}{5} = \frac{6000}{x}$$

$$3x = 30{,}000$$

$$x = \frac{30{,}000}{3} = 10{,}000$$

Julie invested **$10,000** in their business.

PRACTICE PROBLEM 3

Refer to Example 3 to answer the following. Cleo and Julie also split the profits from the partnership in the same ratio, 3 to 5. If Cleo receives $2400 for her share of the profit, how much does Julie receive in profits?

1. Jack can drive 350 miles in 5 hours. How long will it take him to drive 840 miles?

2. A baseball player gets 12 hits in the first 18 games. At this rate, how many hits will he get in 150 games?

3. If 2 cups of cereal contains 50 grams of carbohydrates, how many grams of carbohydrates does 5 cups of cereal contain?

4. If a 200-pound man can have 1000 milligrams of a medicine a day, how much can a 120-pound woman have?

5. If shirts are on sale at 3 for $25, how much will it cost to buy 12 shirts?

6. A school has a student to teacher ratio of 35 to 2. If the school has 875 students, how many teachers will they need?

7. If it takes Mason 25 minutes to water 2 rows of plants in the field, how long will it take him to water a field of 12 rows?

8. A baseball player gets 20 hits out of 50 times at bat. How many hits must he get in his next 150 times at bat to keep his batting average the same?

9. The amount of water in punch is 5 cups for every 2 cups of punch concentrate. How much water is needed with 8 cups of concentrate?

10. If 100 grams of ice cream contains 15 grams of fat, how much fat is in 260 grams of ice cream?

11. If 5 shares of a certain stock cost $160, how much will 12 shares cost?

12. Emily is traveling in London. She can exchange $5 for 3 British pounds. How many pounds will she receive for $65?

13. In a scale drawing a 210-foot-tall building is drawn 3 inches high. If another building is drawn 5 inches high, how tall is that building?

14. On a tour guide map of Canada, 2 inches on the map represents 260 miles. How many miles does 3 inches represent?

15. In a stock split, each person received 8 shares for each 5 shares that he or she held. If a person had 850 shares of stock in the company, how many shares will she receive in the stock split?

16. If Wendy pedals her bicycle at 84 revolutions per minute, she travels at 14 miles per hour. How fast does she go if she pedals at 96 revolutions per minute?

17. A 100-watt stereo system needs copper speaker wire that is 30 millimeters thick to handle the output of sound clearly. How thick would the speaker wire need to be if you had a 140-watt stereo and you wanted the same ratio of watts to millimeters?

18. A bottle of spurge and oxalis killer for your lawn states that you need to use 2 tablespoons to treat 300 square feet of lawn. How many tablespoons will you need to use to treat 1500 square feet of lawn?

19. Julio wants to put a fence around his rectangular pool, which is 12 feet wide and 18 feet long. If the size of the yard will only allow for a fence that is 30 feet long and Julio wants the fence to be proportional to the size of the pool, how wide should the fence be?

20. Devon has a small cement patio 5 feet wide and 7 feet long in his yard. He wants to enlarge the patio, keeping the dimensions of the new patio proportional to the old patio. If he has room to increase the length to 21 feet, how wide should the patio be?

Justin received a promotion from office clerk to office manager at Elen Insurance Group. He earns a monthly salary of $1950 at his new position as an office manager instead of $325 weekly salary as an office clerk. Assume that Justin's deductions as office manager remain proportional to his deductions as an office clerk to answer questions 21–26.

Weekly paycheck

Employee	Position	*ELEN* ELEN INSURANCE GROUP
Justin Dow	Office Clerk	

Total Gross Pay	Federal Withholding	State Withholding	Retirement	Insurance	Net Pay
$ 325	$ 40	$ 22	$ 32	$ 16	$ 215

Monthly paycheck

Employee	Position	*ELEN* ELEN INSURANCE GROUP
Justin Dow	Office Manager	

Total Gross Pay	Federal Withholding	State Withholding	Retirement	Insurance	Net Pay
$ 1950					

Determine the following information about his new position as an office manager.

21. Find the federal withholding.

22. Find the state withholding.

23. Find the retirement deduction.

24. Find the insurance deduction.

25. Find Justin's take-home pay.

26. When Justin worked as an office clerk, he placed $20 a week in his savings account. How much should Justin place in his savings account each month so that his monthly savings contribution is proportional to the amount he saved as a clerk?

Use the following information to answer questions 27 and 28. Two partners, John Ling and Kelvey Marks, each invest money in their business at a ratio of 6 to 7, with Kelvey investing the larger amount.

27. If John invested $2400, how much did Kelvey invest?

28. If the profits from the partnership are distributed to John and Kelvey based on the ratio of their investment, how much profit would John receive if Kelvey receives $798 for profits?

 CALCULATOR EXERCISES

29. Koursh conducted a science experiment and found that sound travels 34,720 feet in air in 31 seconds. How many feet would sound travel in 50 seconds?

30. Natalie became a millionaire by making very profitable investment choices. She told other investors that for every $5 she invested, she earned $800. How much did she invest to earn $1 million?

31. In a small city located in the midwest, 62 out of every 100 registered voters cast a vote in the last election. If there were 22,550 registered voters, how many people voted?

32. If the property tax is $1600 on a home valued at $256,000, how much will be the property tax on a home valued at $192,000?

ONE STEP FURTHER

33. A box has the dimensions, $L = 2$ inches, $W = 3$ inches, and $H = 5$ inches. If you increase the length of this box to 6 inches, what do the width and height have to be so that each dimension of the new box is proportional to each dimension of the original box?

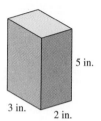

5 in.

3 in.   2 in.

34. Helena is making three frames for her living room wall. She wants three different-sized frames with dimensions that are proportional. If the smallest frame is 5 inches wide by 7 inches high, and the largest frame is 21 inches high, what must the remaining dimensions be? Assume that measurements for both frames are whole numbers.

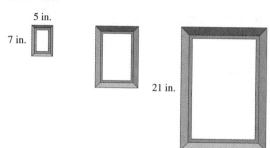

5 in.

7 in.

21 in.

CUMULATIVE REVIEW PROBLEMS

Multiply.

1. $\dfrac{1}{3} \cdot \dfrac{1}{4}$

2. $\dfrac{2}{7} \cdot \dfrac{1}{5}$

3. $x^4 \cdot x^5$

4. $y^6 \cdot y^5$

PLANNING A LARGE EVENT

Frequently when planning the menu for large events, the ingredients for recipes must be increased to make larger amounts. This type of adjustment also applies to other things, such as materials needed for the event. Ratios and proportions can be used to help us make these adjustments.

Janine is making the food and flower arrangements for her daughter's wedding reception. She is planning to serve rolls with a plate of ham, turkey, and cheese to make sandwiches. Janine will also have plates of dipped strawberries, cake, and punch. She plans to have bowls with a mixture of candy and nuts on the tables. For the flower arrangements she is using yellow and white daisies and yellow and white ribbon.

Listed below are the recipes that Janine must adjust to serve food to 192 people at the reception.

Ingredients for Menu

48 Rolls	4 cups flour, 2 cups milk, 1 egg, 1 package yeast, 2 tbsp shortening, 4 tbsp sugar, 1 tsp salt
Ham	2 pounds for 24 people
Turkey	2 pounds for 16 people
Cheese	2 pounds for 16 people
Cake	1 cake serves 24
Strawberries	1 basket for 6 people
Punch (24 servings)	2 cups concentrate, 5 cups water, 4 cups strawberry ice, 3 cups ginger ale
Candy/nuts	3 parts nuts to 2 parts candy

Materials for Flower Arrangements

Flowers	6 yellow daisies for every 9 white daisies
Ribbon for bows	2 yards of white ribbon for every 3 yards yellow ribbon

PROBLEMS FOR INDIVIDUAL INVESTIGATION

Determine how much of the following Janine must purchase to serve 192 people.

1. Rolls (3 rolls per person)

2. (a) Ham;　　**(b)** Turkey;　　**(c)** Cheese

3. Baskets of strawberries

4. Cake

5. If Janine buys 15 bags of nuts, how many bags of candy should she buy?

6. Janine needs 72 yellow daisies to make her arrangements. How many white daisies should she buy?

7. If the flower arrangement requires 30 yards of yellow ribbon, how much white ribbon will be needed?

8. Flour to make the rolls

9. Milk to make the rolls

1. If one recipe of rolls costs $3 to make, how much will it cost to make 3 rolls per person?

2. Ham is $6 a pound, turkey is $5 a pound, and cheese is $4 a pound. What is the total cost of the ham, turkey, and cheese?

3. If Janine can buy strawberries at $5 for eight baskets, how much will the strawberries cost?

4. If the cake is on sale 2 for $28, how much will the cakes cost?

5. If 4 servings of punch are needed for each person, find the amounts needed of (a) concentrate for the punch, (b) strawberry ice, and (c) ginger ale.

6. If one recipe of punch costs $3, how much will the punch ingredients cost?

7. Nuts are $3 a bag and candy is $2 a bag. What is the total cost of nuts and candy?

INTERNET: Go to http://www.prenhall.com/blair to explore this application.

CHAPTER ORGANIZER

TOPIC	PROCEDURE	EXAMPLES
Divisibility tests	1. A number is divisible by 2 if it is even. 2. A number is divisible by 3 if the sum of the digits is divisible by 3. 3. A number is divisible by 5 if the last digit is 0 or 5.	Determine if 6740 is divisible by 2, 3, or 5. 6740 is even, so it is divisible by 2. The sum of the digits is 17, which is not divisible by 3. 6740 is not divisible by 3. Since the last digit of 6740 is 0, it is divisible by 5.
Prime factors	We write a number as a product of prime factors using a division ladder or a factor tree.	Express 75 as a product of prime numbers. Division Ladder: $\begin{array}{r} 3 \\ 5\overline{\smash{)}15} \\ 5\overline{\smash{)}75} \end{array}$ $75 = 5 \cdot 5 \cdot 3$ or $3 \cdot 5^2$ Factor tree: 75 $75 = 3 \cdot 5^2$
Changing an improper fraction to a mixed number	1. Divide the numerator by the denominator. 2. The quotient is the whole number. 3. The fraction is the remainder over the divisor.	Change to a mixed number: $\frac{37}{7}$. $\begin{array}{r} 5 \\ 7\overline{\smash{)}37} \\ \underline{35} \\ 2 \end{array} = 5\frac{2}{7}$

282

TOPIC	PROCEDURE	EXAMPLES
Changing a mixed number to an improper fraction	1. Multiply the whole number by denominator. 2. Add the product to the numerator. 3. Place the sum over the denominator.	Write as an improper fraction: $4\frac{2}{3}$. $$4\frac{2}{3} = \frac{(3 \times 4) + 2}{3} = \frac{12 + 2}{3} = \frac{14}{3}$$
Multiplying prime fractions	1. Multiply numerators. 2. Multiply denominators.	Multiply: $\frac{2}{3} \times \frac{1}{5}$. $$\frac{2}{3} \times \frac{1}{5} = \frac{2 \times 1}{3 \times 5} = \frac{2}{15}$$
Building fractions	1. Find the number or expression by which you must multiply the denominator to get the new denominator. 2. Multiply both the numerator and denominator of the fraction by this number or expression.	Write $\frac{4}{7}$ as an equivalent fraction with a denominator of $21x$. $$\frac{4}{7} = \frac{?}{21x}$$ 7 times what expression equals $21x$? $$7 \times 3x = 21x$$ $$\frac{4 \times 3x}{7 \times 3x} = \frac{12x}{21x}$$
Using the test for equality of fractions	To determine the relationship between two fractions: 1. For each fraction cross-multiply the *denominator* $\times$ *numerator* and write this product above the numerator (form cross products). 2. The smaller product will be written above the smaller fraction and the larger product above the larger fraction. 3. If the products are equal, the fractions are equal.	Replace the ? with the appropriate symbol $=, <,$ or $>$: $\frac{6}{7} \overset{?}{=} \frac{3}{4}$. $$4 \times 6 = 24 \qquad 7 \times 3 = 21$$ $$\frac{6}{7} \quad\times\quad \frac{3}{4}$$ $24 > 21$, so $$\frac{6}{7} > \frac{3}{4}.$$
Reducing fractions	To reduce fractions to lowest terms: 1. Write the numerator and denominator of the fraction as products of prime numbers. 2. Rewrite factors that appear in both the numerator and denominator as $\frac{1}{1}$. 3. Multiply.	Simplify: $\frac{30xy}{42x}$. $$\frac{30xy}{42x} = \frac{\overset{1}{2} \cdot \overset{1}{3} \cdot 5 \cdot \overset{1}{x} \cdot y}{\underset{1}{2} \cdot \underset{1}{3} \cdot 7 \cdot \underset{1}{x}} = \frac{5y}{7}$$
Forming a rate	A *rate* is a comparison of two quantities that have different units. A rate is usually expressed as a fraction in reduced form.	A college dormitory has 10 washing machines for every 85 students living in the dorm. What is the rate of washing machines to students? $$\frac{10 \text{ machines}}{85 \text{ students}} = \frac{2 \text{ machines}}{17 \text{ students}}$$

TOPIC	**PROCEDURE**	**EXAMPLES**
Forming a unit rate	A *unit rate* is a rate with a denominator of 1. To find a unit rate, divide the denominator into the numerator.	Leslie drove her car 360 miles in 7 hours. Find the unit rate. $$\frac{360 \text{ miles}}{7 \text{ hours}} = 51\frac{3}{7} \text{ miles per hour}$$
Writing proportions	A *proportion* is a statement that two rates or ratios are equal. The proportion statement "*a* is to *b* as *c* is to *d*" can be written $$\frac{a}{b} = \frac{c}{d}$$	Write the proportion 33 is to 44 as 15 is to 20. $$\frac{33}{44} = \frac{15}{20}$$
Determining if a statement is a proportion	For any proportion where $b \neq 0$, $d \neq 0$, $\frac{a}{b} = \frac{c}{d}$ if and only if $d \times a = b \times c$. In words, two proportions are equal if and only if their cross products are equal.	Is $\frac{9}{31} = \frac{7}{28}$ a proportion? $$28 \times 9 \stackrel{?}{=} 31 \times 7$$ $$252 \neq 217$$ This is not a proportion.
Finding the missing number in a proportion	To solve a proportion for the value of the variable: 1. Form the cross product. 2. Write the multiplication as division on the opposite side of the equal sign. 4. Simplify the result.	Solve for n: $\frac{13}{n} = \frac{52}{8}$. $8 \times 13 = n \times 52$ Form the cross product. $104 = n \times 52$ $\frac{104}{52} = n$ Divide by 52. $2 = n$
Solving applied problems	1. Write a proportion with n representing the unknown value. 2. Solve the proportion.	A hockey player can score 3 goals in every 7 games. At this rate, how many goals should this hockey player score in 35 games? $\frac{3 \text{ goals}}{7 \text{ games}} = \frac{n \text{ goals}}{35 \text{ games}}$ Write the proportion. $35 \times 3 = 7 \times n$ $105 = 7 \times n$ $\frac{105}{7} = n$ $15 = n$ He should score 15 goals.

CHAPTER 4 REVIEW

SECTION 4.1

Determine if the number is divisible by 2, 3, or 5.

1. 322,970

2. 41,592

Which numbers are prime, composite, or neither?

3. 0, 7, 21, 50, 11, 25, 51

4. 1, 32, 7, 12, 50, 6, 13, 41

Express as a product of prime factors.

5. 36

6. 56

7. 425

8. 312

9. 900

10. 880

SECTION 4.2

Evaluate if possible.

11. $\dfrac{1}{0}$

12. $\dfrac{0}{6}$

13. $\dfrac{z}{z}$

14. $\dfrac{a}{a}$

Solve each applied problem.

15. At a potluck dinner, 7 of the 20 dishes are desserts. Write the fraction that describes the part of the dishes that are desserts.

16. At Steven L. Smith Elementary School, 27 of the 69 third graders are boys. Write the fraction that describes the part of the third grade that are boys.

Write the improper fraction as a mixed number or a whole number.

17. $\dfrac{42}{5}$

18. $\dfrac{55}{6}$

19. $\dfrac{56}{7}$

20. $\dfrac{42}{7}$

Change the mixed number to an improper fraction.

21. $2\dfrac{1}{3}$

22. $4\dfrac{3}{5}$

23. $10\dfrac{2}{5}$

24. $11\dfrac{1}{4}$

SECTION 4.3

Multiply.

25. $\dfrac{1}{3} \times \dfrac{2}{7}$

26. $\dfrac{4}{5} \cdot \dfrac{1}{3}$

Find an equivalent fraction with the given denominator.

27. $\dfrac{2}{3} = \dfrac{?}{27}$

28. $\dfrac{3}{4} = \dfrac{?}{36}$

29. $\dfrac{4}{5} = \dfrac{?}{35x}$

30. $\dfrac{6}{11} = \dfrac{?}{33y}$

Replace the ? with the appropriate symbol: =, <, or >.

31. $\dfrac{3}{4} ? \dfrac{39}{70}$

32. $\dfrac{15}{22} ? \dfrac{12}{19}$

33. $\dfrac{12}{17} ? \dfrac{9}{11}$

34. $\dfrac{16}{31} ? \dfrac{64}{124}$

Which is a shorter distance?

35. A bus stop $\dfrac{3}{8}$ mile from your home, or a bus stop $\dfrac{2}{3}$ mile from your home

36. A nature trail that is $\dfrac{11}{12}$ mile long, or a nature trail that is $\dfrac{9}{10}$ mile long

SECTION 4.4

Simplify. Assume that all variables in the denominator are nonzero.

37. $\dfrac{55}{75}$

38. $\dfrac{48}{54}$

39. $\dfrac{108}{36}$

40. $\dfrac{175}{75}$

41. $\dfrac{25x}{60x}$

42. $\dfrac{84x}{105xy}$

43. $\dfrac{-16}{18}$

44. $\dfrac{24}{-36}$

SECTION 4.5

Write each ratio as a fraction in simplest form.

45. 20 to 46

46. 15 : 25

47. 35 yards to 55 yards

Use the following graph to answer questions 48 and 49.

Number of Marinas in California

Sources: City of Dana Point, Orange County, California Department of Boating and Waterways

48. Write the ratio of the number of marinas in California in 1986 to the number of marinas in California in 1995. Simplify your answer.

49. Write the ratio of the number of marinas in California in 1995 to the number of marinas in California in 1986. Simplify your answer.

Write as a rate in simplest form.

50. $31 for 7 washcloths

51. 12 inches in 28 hours

52. 210 miles in 8 hours

The nutrition facts for a single serving of Frosted Cheerios is given in the chart. Use this information to answer questions 53–54.

Amount/serving

Calories	110	Carbohydrate	24 g
Fat calories	10	Sugars	13 g
Sodium	200 mg	Protein	2 g

53. Write the rate of protein to fat calories. Simplify your answer.

54. Write the rate of fat calories to sugars. Simplify your answer.

Find the unit rate.

55. Traveling 500 miles in 10 hours

56. Traveling 451 miles on 22 gallons of gas

57. The Johnson and Associates law firm has 32 legal secretaries for every 16 lawyers and 12 paralegals for every 4 lawyers.

   (a) How many legal secretaries per lawyer does the law firm have?

   (b) How many paralegals per lawyer does the law firm have?

   (c) How many paralegals would be required if the law firm has 60 lawyers?

58. The Whitbread Round the World Yacht Race has taken place every 4 years since 1973. All boats must be Whitbread 60s, which are 64 feet in length and weigh 29,700 pounds. Find the pound per foot rate of the Whitbread 60s.

59. The KB Music Store is having a special on their 60s CDs: 6 for $72 or 8 for $96.
   (a) Find each unit price.

   (b) Which is the better buy?

60. Jenny's Clothing Store recorded the following profits in the second quarter of the year.
   (a) Write as a ratio: the profit in April to the profit in May.

   (b) Write as a unit rate: the average profit per day in the month of June.

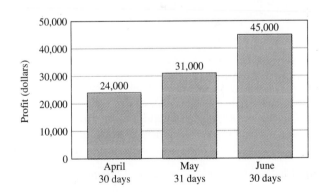

SECTION 4.6

Write the proportion.

61. 2 is to 7 as 14 is to 49.

62. $\frac{1}{5}$ is to $\frac{1}{9}$ as $\frac{1}{2}$ is to $\frac{5}{18}$.

Write the proportion to express the statement.

63. If 2 inches on a map represents 190 miles, 6 inches represents 570 miles.

64. If Tuan can drive 234 miles in his Honda Accord on 9 gallons of gas, he should be able to drive 468 miles on 18 gallons of gas.

Determine whether the statement is a proportion.

65. $\dfrac{3}{5} \overset{?}{=} \dfrac{12}{20}$

66. $\dfrac{4}{7} \overset{?}{=} \dfrac{20}{46}$

67. $\dfrac{14}{23} \overset{?}{=} \dfrac{36}{47}$

68. Darin can type 300 words in 4 minutes, and Jessica can type 720 words in 9 minutes. Do they type at the same rate?

Find the value of x in each proportion. Check your answer.

69. $\dfrac{x}{8} = \dfrac{3}{2}$

70. $\dfrac{x}{11} = \dfrac{4}{22}$

71. $\dfrac{14}{x} = \dfrac{7}{3}$

72. $\dfrac{8}{x} = \dfrac{12}{15}$

73. $\dfrac{\frac{1}{5}}{x} = \dfrac{1}{\frac{1}{2}}$

74. $\dfrac{\frac{1}{4}}{x} = \dfrac{1}{\frac{1}{8}}$

Find the value of n.

75. $\dfrac{17 \text{ quarts}}{47 \text{ square feet}} = \dfrac{n \text{ quarts}}{94 \text{ square feet}}$

76. $\dfrac{10 \text{ miles}}{2 \text{ hours}} = \dfrac{25 \text{ miles}}{n \text{ hours}}$

SECTION 4.7

77. If 3 cups of cereal contains 75 grams of carbohydrates, how many grams of carbohydrates does 7 cups of cereal contain?

78. If 4 desk lamps cost $80, how much will it cost to buy 6 desk lamps?

79. On a map of England, 1 inch on the map represents 120 miles. How many miles does 3 inches represent?

80. If Dale pedals his bicycle at 75 revolutions per minute, he travels 12 miles per hour. How fast does he go if he pedals at 100 revolutions per minute?

81. Lacey has a small rectangular patio cover 4 feet wide and 7 feet long in his yard. He wants to enlarge the cover, keeping the dimensions of the new patio cover proportional to the old one. If he has room to increase the length to 14 feet, how wide should the patio cover be?

82. Leslie and Gloria work at the same hospital but earn different salaries. Assume that Gloria's deductions for taxes are proportional to Leslie's deductions, and answer the following.

(a) Find Gloria's federal withholding.

(b) Find Gloria's state withholding.

Weekly paycheck

MIRA HOSPITAL

Employee: Leslie Brook

Total Gross Pay	Federal Withholding	State Withholding
$ 400	$ 60	$ 20

Weekly paycheck

MIRA HOSPITAL

Employee: Gloria Smart

Total Gross Pay	Federal Withholding	State Withholding
$ 340		

CHAPTER 4 TEST

1. Determine if 230 is divisible by 2, 3, or 5.

2. Identify as prime, composite, or neither:

    (a) 27   **(b)** 1   **(c)** 19

Express as a product of prime factors.

3. 84   **4.** 49

Divide, if possible.

5. $\dfrac{0}{4}$ **6.** $\dfrac{t}{t}$ **7.** $\dfrac{12}{0}$

8. There are 36 students in a prealgebra class. Seventeen of these students are male. What fraction of the students are male?

9. The local parks and recreation department offers 12 classes for children and 16 classes for adults. What fraction of the classes are for adults?

Identify as a proper fraction, improper fraction, or mixed number.

10. $\dfrac{15}{8}$ **11.** $\dfrac{7}{9}$ **12.** $3\dfrac{5}{6}$

Change to a mixed number or whole number.

13. $\dfrac{12}{3}$ **14.** $\dfrac{8}{5}$

15. Change $7\dfrac{1}{6}$ to an improper fraction.   **16.** Write $\dfrac{4}{9}$ as an equivalent fraction with a denominator of 36.

17. Find an equivalent fraction with the given denominator:

$$\dfrac{4}{9} = \dfrac{?}{27y}.$$

18. Replace the ? with the appropriate symbol: $=, <,$ or $>$:

$$\dfrac{2}{15} \; ? \; \dfrac{9}{45}.$$

19. _____

20. _____

21. _____

22. _____

23. _____

24. _____

25. _____

26. _____

19. Neil can choose between a socket wrench with a $\frac{3}{4}$-inch head and one with a $\frac{2}{3}$-inch head. Which socket wrench has the larger head?

20. Simplify.

(a) $\frac{18}{56}$

(b) $\frac{16x}{32x^2}$

21. A class is attended by 16 males and 28 females. Write the ratio of males to students in the class as a fraction in simplest form.

22. Write as a rate in simplest form: 250 calories for 5 grams of fat.

23. Joe played tennis for 20 minutes and burned 150 calories. How many calories did he burn per minute?

24. Determine if $\frac{20}{52} = \frac{5}{13}$ is a proportion.

25. Find the value of x: $\frac{4}{6} = \frac{20}{x}$.

26. A pound of fertilizer covers 1200 square feet of lawn. How many pounds are needed to cover a lawn measuring 6000 square feet?

1. _____

This illustration shows the salaries of the highest-paid baseball players in 1997. Use the illustration to answer Exercises 1 and 2.

2. _____

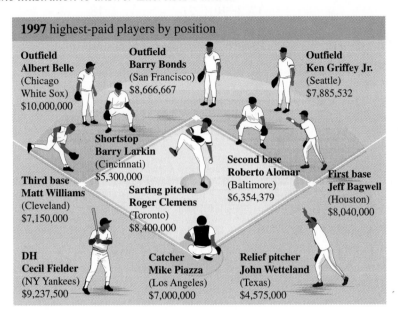

1997 highest-paid players by position

**Outfield
Albert Belle**
(Chicago
White Sox)
$10,000,000

**Outfield
Barry Bonds**
(San Francisco)
$8,666,667

**Outfield
Ken Griffey Jr.**
(Seattle)
$7,885,532

**Shortstop
Barry Larkin**
(Cincinnati)
$5,300,000

**Second base
Roberto Alomar**
(Baltimore)
$6,354,379

**First base
Jeff Bagwell**
(Houston)
$8,040,000

**Third base
Matt Williams**
(Cleveland)
$7,150,000

**Sarting pitcher
Roger Clemens**
(Toronto)
$8,400,000

**DH
Cecil Fielder**
(NY Yankees)
$9,237,500

**Catcher
Mike Piazza**
(Los Angeles)
$7,000,000

**Relief pitcher
John Wetteland**
(Texas)
$4,575,000

Copyright 1997, *USA Today*. Reprinted with permission.

3. _____

4. _____

1. How much more did starting pitcher Roger Clemens earn than relief pitcher John Wetteland?

2. If a team's 1997 outfield included Albert Belle, Barry Bonds, and Ken Griffey, Jr., what is the total amount that they would be paid in salaries?

5. _____

6. _____

3. (a) Translate this English sentence into an equation and **(b)** solve: Twenty-five minus a number equals 18.

Find the value.

7. _____

8. _____

4. 10^3 **5.** 8^0 **6.** 12^1

7. Multiply and write the product in exponent form: $(3x)(5x^2)$.

Use the distributive property to simplify.

9. _____

8. $2(x + 3)$ **9.** $4(y - 2)$

10. _____

10. $a(b + c)$ **11.** Evaluate $6n$ if $n = 3$.

11. _____

12. Evaluate $6n$ if $n = -3$.   **13.** Evaluate $-6n$ if $n = -3$.

12. _____

13. _____

14. _____

15. _____

16. _____

17. _____

18. _____

19. _____

20. _____

21. _____

22. _____

23. _____

24. _____

25. _____

26. _____

27. _____

14. Evaluate $\dfrac{8}{x}$ if $x = 2$.

15. Evaluate $\dfrac{8}{x}$ if $x = -2$.

Perform each indicated operation.

16. $10(-2)$

17. $(-10)(-2)$

18. $-12 \div 4$

19. $-12 \div (-4)$

20. $-8 + 7$

21. $-8 + (-7)$

22. $-8 - 7$

23. $-8 - (-7)$

24. A pollster surveyed 350 registered voters in a city. Two hundred of those surveyed said they had voted in the last election. What fraction of the people surveyed said they had voted in the last election? Simplify your answer.

25. Determine which is the larger quantity: $\dfrac{2}{3}$ cup of flour or $\dfrac{3}{4}$ cup of flour.

26. Write as a rate in simplest form: 225 words in 3 minutes.

27. If a 4-ounce serving of yogurt contains 9 grams of carbohydrates, how many grams of carbohydrates does 12 ounces of yogurt contain?

OPERATIONS ON FRACTIONAL EXPRESSIONS

CHAPTER 5

I f you have the opportunity to invest in the stock market, would you know how to determine the value of a particular stock on any given day? Can you do the necessary calculations involving fractions to determine the profit or loss on your investment? If you would like to find out, turn to Putting Your Skills to Work on page 351.

5.1 Multiplication and Division of Fractional Expressions

5.2 Multiples and Least Common Multiples of Expressions

5.3 Addition and Subtraction of Fractional Expressions

5.4 Operations with Mixed Numbers

5.5 Order of Operations and Complex Fractions

5.6 Multiplication and Division Properties of Exponents

5.7 Applied Problems Involving Fractions

This test provides a preview of the topics in this chapter. It will help you identify which concepts may require more of your studying time. If you are familiar with the topics in this chapter, take this test now. Check your answers with those in the back of the book. If you are not familiar with the topics in this chapter, begin studying the chapter now.

1. _____

2. _____

SECTION 5.1

3. _____

1. Find: $\dfrac{2}{3}$ of $\dfrac{5}{7}$.

2. Multiply and simplify: $\dfrac{3x^4}{20} \cdot \dfrac{16}{9x^3}$.

4. _____

3. Find the reciprocal: -9.

5. _____

4. Divide and simplify: $-\dfrac{5}{8} \div \dfrac{3}{4}$.

5. A shipment of candy is delivered to a market in an 18-pound bag. The store

6. _____

owner wants to package the candy in $\dfrac{2}{3}$-pound bags. How many $\dfrac{2}{3}$-pound

7. _____

bags can the owner package?

8. _____

SECTION 5.2

6. List the first four multiples of 7.

9. _____

Find the least common multiple (LCM).

7. 20, 15

10. _____

8. $3x, 6x^2, 9x$

9. 7, 35, 91

11. _____

SECTION 5.3

12. _____

10. Find the least common denominator (LCD): $\dfrac{6}{7}, \dfrac{1}{3}$.

13. _____

11. Add: $\dfrac{5}{12} + \dfrac{3}{12}$ **12.** Add: $\dfrac{3}{7x} + \dfrac{5}{x}$ **13.** Subtract: $\dfrac{3x}{4} - \dfrac{7x}{12}$.

SECTION 5.4

14. Add: $1\dfrac{7}{10} + 8\dfrac{9}{10}$

15. Add: $25\dfrac{3}{4} + 13\dfrac{5}{8}$

16. Subtract: $6\dfrac{9}{14} - 2\dfrac{3}{7}$.

17. Multiply: $3\dfrac{1}{4} \cdot (-2)$.

18. Divide: $7\dfrac{1}{2} \div 3\dfrac{3}{4}$.

SECTION 5.5

Simplify.

19. $\dfrac{1}{18} + \dfrac{2}{3} \cdot \dfrac{1}{3}$

20. $\dfrac{3}{4} + \dfrac{1}{2} \div \dfrac{2}{3}$

21. $\dfrac{\dfrac{1}{2}}{\dfrac{3}{8}}$

SECTION 5.6

Simplify.

22. $\left(y^2\right)^4$

23. $(3x)^2$

24. $\dfrac{9x^4}{18x^3}$

SECTION 5.7

25. Adam planted a vegetable garden in the center of a $35\dfrac{1}{2}$-foot by 24-foot area. There is a sidewalk around the garden that is $2\dfrac{1}{2}$ feet wide. The garden and sidewalk take up the entire $35\dfrac{1}{2}$-foot by 24-foot area.

(a) What are the dimensions of the vegetable garden?

(b) How much will it cost to put a fence around the garden if the fencing costs $\$2\dfrac{1}{4}$ per linear foot?

14. _____

15. _____

16. _____

17. _____

18. _____

19. _____

20. _____

21. _____

22. _____

23. _____

24. _____

25. _____

After studying this section, you will be able to:

1 Multiply fractions.

2 Divide fractions.

3 Solve applied problems involving fractions.

Math Pro   Video 5.1   SSM

SECTION 5.1

Multiplication and Division of Fractional Expressions

1 Multiplying Fractions

Let's look at multiplication with fractions. We begin with an illustration representing the fraction $\frac{1}{3}$.

One whole divided into 3 parts with each part equal to the fraction $\frac{1}{3}$ (1 out of 3 parts)

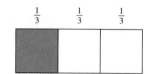

Let's see what it means to have one-half of $\frac{1}{3}$. Imagine cutting each of the 3 pieces in the figure above in half.

One whole divided into 6 parts with each part equal to the fraction $\frac{1}{6}$ (1 out of 6 parts)

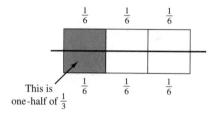

This is one-half of $\frac{1}{3}$

From the illustration above we see that one-half of $\frac{1}{3}$ is $\frac{1}{6}$.

We often use multiplication of fractions to describe *taking a fraction of something*. In Section 4.3 we saw that to multiply fractions, we multiply numerator times numerator and denominator times denominator. Therefore, to find *one-half of* $\frac{1}{3}$, *we multiply* $\frac{1}{2} \cdot \frac{1}{3} = \frac{1}{6}$.

> **RULE FOR MULTIPLICATION OF FRACTIONS**
> In general, for all numbers $a, b, c,$ and d, with $b \neq 0, d \neq 0,$
> $$\frac{a}{b} \cdot \frac{c}{d} = \frac{a \cdot c}{b \cdot d}$$

EXAMPLE 1   Find: $\frac{3}{7}$ of $\frac{2}{9}$.

SOLUTION   We multiply:
$$\frac{3}{7} \cdot \frac{2}{9} = \frac{3 \cdot 2}{7 \cdot 9} = \frac{3 \cdot 2}{7 \cdot 3 \cdot 3} = \frac{\cancel{3} \cdot 2}{7 \cdot \cancel{3} \cdot 3} = \frac{2}{21}$$

PRACTICE PROBLEM 1

Find: $\frac{5}{8}$ of $\frac{2}{3}$.

We should always *remove common factors before we multiply* the numerators and denominators; otherwise, we must reduce the product to lowest terms. This is a lot of extra work! Let's see what would happen if in Example 1 we multiplied the numbers in the numerator and denominator before we removed common factors.

The fraction is partially factored.	We multiply numerator and denominator.	To simplify we must **refactor**.
↓	↓	↓

$$\frac{3}{7} \cdot \frac{2}{9} = \frac{3 \cdot 2}{7 \cdot 9} \qquad = \qquad \frac{6}{63} \qquad = \qquad \frac{3 \cdot 2}{7 \cdot 3 \cdot 3} = \frac{2}{21}$$

↑
We should skip this step.

> When multiplying fractions, always simplify before performing the multiplication in the numerator and the denominator of the fraction. This will assure that the product is reduced to lowest terms.

E X A M P L E   2   Find: $\dfrac{9}{20} \cdot \dfrac{5}{21}$.

SOLUTION

$$\frac{9}{20} \cdot \frac{5}{21} = \frac{9 \cdot 5}{20 \cdot 21} = \frac{3 \cdot 3 \cdot 5}{2 \cdot 2 \cdot 5 \cdot 3 \cdot 7} = \frac{3 \cdot \cancel{3} \cdot \cancel{5}}{2 \cdot 2 \cdot \cancel{3} \cdot \cancel{5} \cdot 7} = \frac{\mathbf{3}}{\mathbf{28}}$$

P R A C T I C E   P R O B L E M   2

Find: $\dfrac{8}{15} \cdot \dfrac{12}{14}$.

E X A M P L E   3   Find: $\dfrac{-10}{18} \cdot \dfrac{9}{11}$.

SOLUTION   When multiplying positive and negative fractions, we determine the sign of the product, then multiply and simplify.

$$\frac{-10}{18} \cdot \frac{9}{11} = (-)$$

> The product of a negative and a positive fraction is negative.

$$= -\frac{10 \cdot 9}{18 \cdot 11}$$

$$= -\frac{2 \cdot 5 \cdot 3 \cdot 3}{2 \cdot 3 \cdot 3 \cdot 11}$$ Write as a product of prime factors.

$$= -\frac{\cancel{2} \cdot \cancel{3} \cdot \cancel{3} \cdot 5}{\cancel{2} \cdot \cancel{3} \cdot \cancel{3} \cdot 11}$$ Simplify.

$$= -\frac{\mathbf{5}}{\mathbf{11}}$$

P R A C T I C E   P R O B L E M   3

Find: $\dfrac{-12}{24} \cdot \dfrac{-8}{13}$.

When multiplying a fraction by a whole number, it is more convenient to express the whole number with a denominator of 1. We can do this since $\frac{6}{1} = 6$, and $\frac{8}{1} = 8$, and so on.

E X A M P L E 4   Multiply: $12x^3 \cdot \dfrac{5x^2}{4}$.

SOLUTION

$$\frac{12x^3}{1} \cdot \frac{5x^2}{4} = \frac{4 \cdot 3 \cdot x^3 \cdot 5 \cdot x^2}{1 \cdot 4} = \frac{\cancel{4} \cdot 3 \cdot 5 \cdot x^3 \cdot x^2}{\cancel{4} \cdot 1} = \frac{3 \cdot 5 \cdot x^5}{1} = \mathbf{15x^5}$$

Why did we factor $12 = 4 \cdot 3$ instead of $12 = 2 \cdot 2 \cdot 3$?

P R A C T I C E P R O B L E M 4

Multiply: $\dfrac{3x^6}{7} \cdot \left(14x^2\right)$.

DEVELOPING YOUR STUDY SKILLS

A POSITIVE ATTITUDE TOWARD FRACTIONS

Often, students panic when they begin to study fractions in their math class. They may think, "I never understood fractions, and I never will!" Not true. You experience situations involving fractions every day. You may not actually perform mathematical calculations with fractions, but you find the results. Consider these two situations.

Situation: You know that if you slice an apple into quarters, you can take 1 of the 4 pieces or $\frac{1}{4}$ of the apple.

Calculation: $1 \div 4 = \dfrac{1}{4}$

Situation: If you take $\frac{1}{2}$ of $6, you know you have $3.

Calculation: $\dfrac{1}{2} \times 6 = 3$

As you can see, you already know how to work with fractions. Now you are ready to see how to perform the calculations so that you can find the outcome of more complex situations. How can you do this? Follow these suggestions.

1. You must have a *positive attitude* toward fractions.
2. Try to *understand the process* used to perform the calculations. You can do this by:
   (a) Drawing pictures and diagrams.
   (b) Asking questions when you do not understand.
   (c) Learning all the rules.
   (d) Practicing by completing your assignments.

➋ Dividing Fractions

Before we discuss division with fractions, we introduce **reciprocal fractions**. The fractions $\frac{2}{3}$ and $\frac{3}{2}$ are called *reciprocals*, and $\frac{x}{a}$ and $\frac{a}{x}$ are also reciprocals. Let's see what happens when we multiply reciprocal fractions.

$$\frac{2}{3} \cdot \frac{3}{2} = \frac{2 \cdot 3}{3 \cdot 2} = \frac{\cancel{2} \cdot \cancel{3}}{\cancel{2} \cdot \cancel{3}} = 1 \qquad \frac{x}{a} \cdot \frac{a}{x} = \frac{x \cdot a}{a \cdot x} = \frac{\cancel{x} \cdot \cancel{a}}{\cancel{x} \cdot \cancel{a}} = 1$$

Notice that both products are equal to 1. If the product of two numbers is 1, we say that these two numbers are **reciprocals** of each other.

> To find the reciprocal of a fraction, we *interchange* the numerator and denominator. This is often referred to as **inverting the fraction.**

EXAMPLE 5   Find the reciprocal.

(a) $\dfrac{-7}{8}$ 　　　　　　　　　　　　　　**(b)** 6

SOLUTION   To find the reciprocal, we invert each fraction.

(a) $\dfrac{-7}{8} \rightarrow \dfrac{\mathbf{8}}{\mathbf{-7}} = -\dfrac{8}{7}$ 　　　　　　**(b)** $6 = \dfrac{6}{1} \rightarrow \dfrac{\mathbf{1}}{\mathbf{6}}$

PRACTICE PROBLEM 5
Find the reciprocal.

(a) $\dfrac{a}{-y}$ 　　　　　　　　　　　　　　**(b)** 4

Why would you divide fractions? Consider this problem.

A total of $\frac{3}{4}$ pound of peanuts is to be placed in $\frac{1}{4}$-pound bags.

How many $\frac{1}{4}$-pound bags will there be?

We must find how many $\frac{1}{4}$'s are in $\frac{3}{4}$. This is the division situation $\frac{3}{4} \div \frac{1}{4}$. To illustrate, we draw a picture.

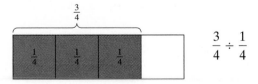

$$\frac{3}{4} \div \frac{1}{4}$$

Notice that there are three $\frac{1}{4}$'s in $\frac{3}{4}$. Thus, $\frac{3}{4} \div \frac{1}{4} = 3$.

Now, how do we perform this division? We *invert the second fraction* and *multiply*.

Invert second fraction
and multiply.
↓

$$\frac{3}{4} \div \frac{1}{4} = \frac{3}{4} \cdot \frac{4}{1} = \frac{3 \cdot \cancel{4}}{\cancel{4} \cdot 1} = \frac{3}{1} = 3$$

Thus dividing $\frac{3}{4}$ by $\frac{1}{4}$ is the same as multiplying $\frac{3}{4}$ by the reciprocal of $\frac{1}{4}$.

RULE FOR DIVISION OF FRACTIONS
To divide two fractions, we *invert* the divisor (second fraction) and *multiply*.

$$\frac{a}{b} \div \frac{c}{d} = \frac{a}{b} \cdot \frac{d}{c} \qquad \text{(when } b, c, \text{ and } d \text{ are not 0)}$$

EXAMPLE 6   Divide: $\frac{-4}{11} \div \left(\frac{-3}{5}\right)$.

SOLUTION

$$\frac{-4}{11} \div \left(\frac{-3}{5}\right) = \frac{-4}{11} \cdot \left(\frac{5}{-3}\right) \qquad \text{Invert the second fraction and multiply.}$$

The product of a negative times a negative is positive, so the product will be positive.

$$= \frac{4 \cdot 5}{11 \cdot 3} = \frac{\mathbf{20}}{\mathbf{33}}$$

PRACTICE PROBLEM 6
Divide: $\frac{-7}{8} \div \frac{5}{13}$.

TO THINK ABOUT
Why do we invert the second fraction and multiply when we divide? To see why this makes sense, let's write the fraction $\frac{8}{2}$ as both a division and a multiplication problem.

$$\frac{8}{2} = 8 \div 2 = 4$$

$$\frac{8}{2} = 8 \cdot \frac{1}{2} = 4$$

Now, we know that if we invert 2, we obtain its reciprocal $\frac{1}{2}$. Thus we can see that whether we divide 8 by 2 or multiply 8 by $\frac{1}{2}$, we get the same result.

To obtain the right answer, it is important that you *invert the second fraction*, not the first. Since we cannot divide by zero, we will assume that all variables in the denominator are nonzero.

E X A M P L E 7   Divide: $\dfrac{7x^4}{20} \div \left(\dfrac{-14}{45x^2}\right)$.

SOLUTION

$$\dfrac{7x^4}{20} \div \left(\dfrac{-14}{45x^2}\right) = \dfrac{7x^4}{20} \cdot \left(\dfrac{45x^2}{-14}\right)$$ Invert the second fraction and multiply.

The product of a negative and positive is negative, so the answer is negative.

$$= -\dfrac{7 \cdot 5 \cdot 3 \cdot 3 \cdot x^4 \cdot x^2}{2 \cdot 2 \cdot 5 \cdot 2 \cdot 7}$$ Factor and simplify.

$$= -\dfrac{9x^6}{8}$$ Multiply terms in exponent form and simplify.

Note: We can also determine the sign of the answer by observing the signs of the fractions in the division problem. Why?

P R A C T I C E P R O B L E M 7

Divide: $\dfrac{9x^6}{-21} \div \left(\dfrac{-42}{18x^4}\right)$.

E X A M P L E 8   Divide: $16x^2 \div \dfrac{8x}{11}$.

SOLUTION

$$16x^2 \div \dfrac{8x}{11} = \dfrac{16x^2}{1} \cdot \dfrac{11}{8x} = \dfrac{2 \cdot 8 \cdot 11 \cdot x^2}{1 \cdot 8 \cdot x} = \dfrac{22 \cdot x \cdot x}{x} = \mathbf{22x}$$

P R A C T I C E P R O B L E M 8

Divide: $28x^5 \div \dfrac{4x}{19}$.

❸ Solving Applied Problems Involving Fractions

One of the most important steps in solving a real-life application is determining what operation to use. We often ask ourselves, "Should I multiply or divide?" The multiplication and division situations for fractions are similar to those for whole numbers and are stated below for your review. Drawing pictures and making charts can also help determine whether to divide or multiply.

> **MULTIPLICATION SITUATIONS**
> We multiply in situations that require *repeated addition*, or *taking a fraction of something*.

1. *Repeated addition.* A recipe requires $\dfrac{1}{4}$ cup of flour for each serving. How many cups are needed to make 3 servings?

1 serving   1 serving   1 serving = 3 servings
↓   ↓   ↓

$$\frac{1}{4} \quad + \quad \frac{1}{4} \quad + \quad \frac{1}{4} \quad \text{or} \quad \mathbf{3 \cdot \frac{1}{4}} = \frac{3}{4} \text{ cup of flour}$$

2. *Taking a fraction of something.* A recipe requires $\frac{3}{4}$ cup of flour. How much flour is needed to make $\frac{1}{2}$ of the recipe?

We want to find $\frac{1}{2}$ of $\frac{3}{4}$ or $\mathbf{\frac{1}{2} \cdot \frac{3}{4}} = \frac{3}{8}$ cup of flour

---

DIVISION SITUATIONS

We divide when we want to *split an amount into a certain number of equal parts*, or to *find how many groups of a number are in another number*.

---

1. *Split an amount into equal parts.* A pipe $\frac{3}{5}$ foot long must be cut into 2 equal parts. How long is each part?

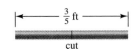

$$\mathbf{\frac{3}{5} \div 2} = \frac{3}{10} \text{ foot}$$

2. *Find how many groups of a number are in another number.* A scarf requires $\frac{4}{5}$ yards of material. How many scarves can be made from 8 yards of material?
We must find how many $\frac{4}{5}$-yard segments are in 8 yards. This is

$$\mathbf{8 \div \frac{4}{5}} = 10 \text{ scarves.}$$

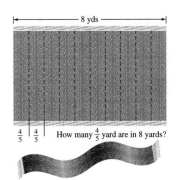

How many $\frac{4}{5}$ yard are in 8 yards?

EXAMPLE 9   Samuel Jensen has $\frac{9}{40}$ of his income withheld for taxes and retirement. What amount is withheld each month if he earns \$1440 per month?

SOLUTION   We must find a fraction of his income, so we multiply.

$\frac{9}{40}$ of *monthly pay* is withheld for taxes and and retirement
↓

$$\frac{9}{40} \text{ of } \$1440 = \frac{9}{40} \cdot \frac{\$1440}{1} = \$324$$

\$324 is withheld for federal taxes each month.

PRACTICE PROBLEM 9

Nancy Levine places $\frac{2}{13}$ of her income in a savings account each month. How much money does she place in a savings account if her income is $1703 per month?

EXAMPLE 10   Harry must install 44 feet of baseboard along the edge of the floor of a library. If he places a nail in the baseboard every $\frac{2}{3}$ foot, how many nails will he need?

SOLUTION   We draw a picture.

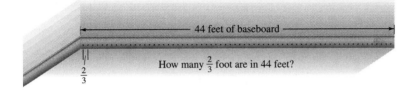

We divide: $44 \div \frac{2}{3} = 44 \cdot \frac{3}{2} = \frac{44 \cdot 3}{2} = \frac{\cancel{2} \cdot 22 \cdot 3}{\cancel{2}} =$ **66 nails**

PRACTICE PROBLEM 10

Alice must place $\frac{3}{4}$ pound of sugar in 2 equal-sized containers. How much sugar should Alice place in each container?

EXERCISES 5.1

In this exercise set assume that all variables in any denominator are nonzero.

1. Find $\dfrac{1}{3}$ of $\dfrac{1}{5}$.

2. Find $\dfrac{1}{5}$ of $\dfrac{1}{7}$.

3. Find $\dfrac{5}{21}$ of $\dfrac{7}{8}$.

4. Find $\dfrac{2}{16}$ of $\dfrac{8}{9}$.

Multiply. Be sure your answer is simplified.

5. $\dfrac{7}{12} \cdot \dfrac{8}{28}$

6. $\dfrac{6}{21} \cdot \dfrac{9}{18}$

7. $\dfrac{3}{42} \cdot \dfrac{6}{15}$

8. $\dfrac{4}{35} \cdot \dfrac{5}{24}$

9. $\dfrac{-2}{48} \cdot \left(\dfrac{32}{-6}\right)$

10. $\dfrac{-5}{28} \cdot \left(\dfrac{22}{-30}\right)$

11. $\dfrac{16}{56} \cdot \left(\dfrac{-18}{36}\right)$

12. $\dfrac{4}{27} \cdot \left(\dfrac{-45}{18}\right)$

13. $\dfrac{6}{35} \cdot 5$

14. $\dfrac{2}{21} \cdot 15$

15. $-14 \cdot \dfrac{1}{28}$

16. $-13 \cdot \dfrac{2}{26}$

17. $\dfrac{-2}{63} \cdot \left(\dfrac{-14}{18}\right)$

18. $\dfrac{-8}{20} \cdot \left(\dfrac{-25}{32}\right)$

19. $\dfrac{1}{5} \cdot 25$

20. $\dfrac{1}{5} \cdot 15$

21. $\dfrac{2x^2}{9} \cdot \dfrac{3x^2}{8}$

22. $\dfrac{3x^3}{21} \cdot \dfrac{7x^2}{9}$

23. $\dfrac{6x}{25} \cdot \dfrac{15}{12x^2}$

24. $\dfrac{4x}{35} \cdot \dfrac{7}{6x^2}$

25. $\dfrac{-3y^3}{20} \cdot \dfrac{12}{21y^2}$

26. $\dfrac{15y^3}{26} \cdot \left(\dfrac{-13}{10y}\right)$

27. $\dfrac{3x^2}{15} \cdot \dfrac{18x^3}{20}$

28. $\dfrac{5x^4}{12} \cdot \dfrac{32x^2}{25}$

29. $\dfrac{6x^4}{7} \cdot 14x^3$

30. $\dfrac{6x^5}{15} \cdot 21x^7$

31. $\dfrac{-6y^2}{14} \cdot \dfrac{21}{24y^3}$

32. $\dfrac{-5y^3}{6} \cdot \dfrac{16}{15y}$

Find the reciprocal.

33. $\dfrac{3}{7}$

34. $\dfrac{2}{3}$

35. $\dfrac{1}{8}$

36. $\dfrac{1}{9}$

37. 8

38. 9

39. $\dfrac{2}{-5}$

40. $\dfrac{7}{-8}$

41. $\dfrac{-x}{y}$

42. $\dfrac{-a}{b}$

43. -6

44. -3

Divide. Be sure your answer is simplified.

45. $\dfrac{6}{14} \div \dfrac{3}{8}$

46. $\dfrac{8}{12} \div \dfrac{5}{6}$

47. $\dfrac{7}{24} \div \dfrac{9}{8}$

48. $\dfrac{9}{28} \div \dfrac{4}{7}$

49. $\dfrac{-1}{12} \div \dfrac{3}{4}$

50. $\dfrac{-1}{15} \div \dfrac{2}{3}$

51. $\dfrac{-7}{24} \div \left(\dfrac{9}{-8}\right)$

52. $\dfrac{-9}{28} \div \left(\dfrac{4}{-7}\right)$

53. $15 \div \dfrac{3}{7}$

54. $18 \div \dfrac{2}{3}$

55. $\dfrac{7}{22} \div 14$

56. $\dfrac{8}{26} \div 16$

57. $\dfrac{-7}{15} \div \left(\dfrac{-5}{25}\right)$

58. $\dfrac{-5}{24} \div \left(\dfrac{-8}{32}\right)$

59. $\dfrac{5}{14} \div \dfrac{2}{21} \div \left(\dfrac{15}{-3}\right)$

60. $\dfrac{8}{21} \div \left(\dfrac{4}{-7}\right) \div \dfrac{4}{3}$

61. $\dfrac{8x^6}{15} \div \dfrac{16x^2}{20}$

62. $\dfrac{6y^4}{35} \div \dfrac{36y^2}{25}$

63. $\dfrac{7x^4}{12} \div \dfrac{28}{36x^2}$

64. $\dfrac{3x^4}{45} \div \dfrac{27}{45x^5}$

65. $-9x \div \dfrac{6x^2}{15}$

66. $-6x \div \dfrac{12x^2}{11}$

Solve each applied problem.

67. Ross and Lilly Smith have $\dfrac{3}{12}$ of their income withheld for taxes, union dues, and medical coverage. What amount is withheld each month if they earn \$3300 per month?

68. Elliott has $\dfrac{2}{16}$ of his monthly income placed in a savings account. What amount is placed in his savings account if he earns \$1600 per month?

69. Harold must cut pipes into lengths of $\dfrac{3}{4}$ foot. How many pipes can he make from a pipe that is 12 feet long?

70. The distance around a track is $\dfrac{1}{4}$ mile. Julie runs 8 miles for her daily workout. How many laps around the track must Julie run to complete her 8-mile workout?

71. James is planning a party at which he intends to serve pizza. If James estimates that each guest will eat $\dfrac{3}{8}$ of a pizza, how many pizzas should he order if 16 will attend the party?

72. The propeller on the Ipswich River Cruise Boat turns 320 revolutions per minute. How fast would it turn at $\frac{3}{4}$ of that speed?

73. In the Westerfield Factory, products are made in vats that have the capacity to hold 120 quarts. If each bottle of a product contains $\frac{3}{4}$ quart, how many bottles can be made from each vat?

74. Dunday Building Company purchased 56 acres of land. The company subdivides the land into $\frac{2}{5}$-acre parcels. How many $\frac{2}{5}$-acre parcels does the company have?

CALCULATOR EXERCISES

75. The records at a private college with 9600 students indicate that $\frac{1}{6}$ of these students are from California and $\frac{1}{5}$ are from Texas.
 (a) How many students are from California?
 (b) How many students are from Texas?

76. It is recommended that your house payment be no more than $\frac{1}{3}$ of your monthly income. If the Hensens' monthly income is $4590, how much should they budget for a house payment?

ONE STEP FURTHER

77. Multiply: $\frac{12}{21} \cdot \frac{15}{16} \cdot \frac{4}{5} \cdot \frac{9}{10}$.

78. Divide: $\frac{1}{2} \div \frac{3}{16} \div \frac{6}{7} \div \frac{2}{3} \div 14$.

CUMULATIVE REVIEW

Find the equivalent fraction.

1. $\frac{2}{3} = \frac{?}{15}$

2 $\frac{3}{4} = \frac{?}{20}$

3. $\frac{9}{10} = \frac{?}{50}$

4. $\frac{12}{17} = \frac{?}{51}$

Multiples and Least Common Multiples of Expressions

Recall that to combine like terms, the variable parts must be the same. Well, the rules for fractions are similar: to add or subtract fractions, the denominators must be the same. In this section we see how to find this common denominator so that we can add or subtract fractions.

After studying this section, you will be able to:

1 Find multiples of expressions.

2 Find the least common multiple of expressions.

Math Pro   Video 5.2   SSM

1 Finding Multiples of Expressions

If we multiply a number by 1, and then by 2, and then by 3, we generate a list of multiples. For example, we can list some multiples of 4 by multiplying 4 by 1, 2, 3, 4, and so on:

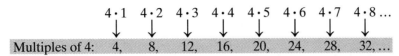

$$
\begin{array}{cccccccc}
4\cdot 1 & 4\cdot 2 & 4\cdot 3 & 4\cdot 4 & 4\cdot 5 & 4\cdot 6 & 4\cdot 7 & 4\cdot 8 \dots \\
\downarrow & \downarrow & \downarrow & \downarrow & \downarrow & \downarrow & \downarrow & \downarrow
\end{array}
$$

Multiples of 4:   4,   8,   12,   16,   20,   24,   28,   32, ...

E X A M P L E   1

(a) List the first six multiples of $8x$ and $12x$.
(b) Which of these multiples are common to both $8x$ and $12x$?

SOLUTION

(a)

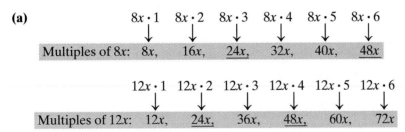

$$
\begin{array}{cccccc}
8x\cdot 1 & 8x\cdot 2 & 8x\cdot 3 & 8x\cdot 4 & 8x\cdot 5 & 8x\cdot 6 \\
\downarrow & \downarrow & \downarrow & \downarrow & \downarrow & \downarrow
\end{array}
$$

Multiples of $8x$: $8x$, $16x$, $\underline{24x}$, $32x$, $40x$, $\underline{48x}$

$$
\begin{array}{cccccc}
12x\cdot 1 & 12x\cdot 2 & 12x\cdot 3 & 12x\cdot 4 & 12x\cdot 5 & 12x\cdot 6 \\
\downarrow & \downarrow & \downarrow & \downarrow & \downarrow & \downarrow
\end{array}
$$

Multiples of $12x$: $12x$, $\underline{24x}$, $36x$, $\underline{48x}$, $60x$, $72x$

(b) The multiples common to both $8x$ and $12x$ are **$24x$** and **$48x$**.

P R A C T I C E   P R O B L E M   1

(a) List the first five multiples of $12x$ and $20x$.
(b) Which of these multiples are common to both $12x$ and $20x$?

2 Finding the Least Common Multiple of Expressions

A whole number that is a multiple of two different numbers is called a **common multiple** of those two numbers. Therefore, in Example 1, we call $24x$ and $48x$ common multiples of $8x$ and $12x$. The number $24x$ is the *smallest common multiple* and is called the *least common multiple*, or *LCM*, of $8x$ and $12x$.

E X A M P L E   2   Find the LCM of 10 and 15.

SOLUTION   First, we list some multiples of 10:   10, 20, 30, 40, 50, 60

Next, we list some multiples of 15:   15, 30, 45, 60

We see that both 30 and 60 are common multiples. Since 30 is the smallest common multiple, we call 30 the least common multiple (LCM).

PRACTICE PROBLEM 2

Find the LCM of 4 and 5.

Suppose that we must find the LCM of 14, 32, 78, and 210. Listing the multiples of each number to find the LCM would be time consuming. A quicker, more efficient way to find LCMs uses prime factorizations. With this method we **build the LCM** using the prime factors of each number. In Example 2 we listed some multiples for 10 and 15 to find the LCM 30. Now let's see how to find the LCM of 10 and 15 using prime factors.

$$10 = 2 \cdot 5, \quad 15 = 3 \cdot 5, \quad \text{LCM} \longrightarrow 30 = 2 \cdot \underline{3 \cdot 5}$$

factors of 10

factors of 15

Notice that the LCM has all the factors of 10 and all the factors of 15 in its prime factorization. Thus the LCM of 10 and 15 must satisfy two requirements:

1. The LCM must have all factors of 10 in its prime factorization: a 2 and a 5.
2. The LCM must have all factors of 15 in its prime factorization: a 3 and a 5.

> To build an LCM we write a prime factorization that satisfies the first requirement, *then add to this factorization* to satisfy the second requirement.

1. The LCM must have a 2 and a 5 as factors. $\boxed{\text{LCM} = 2 \cdot 5 \cdot \,?}$

2. The LCM must have a 3 and a 5 as factors. $\boxed{\text{LCM} = 2 \cdot 5 \cdot 3}$

5 is already a factor, so we just multiply by a 3.

Now, we multiply factors to find the value of the LCM: $2 \cdot 5 \cdot 3 = 30$

The LCM of 10 and 15 is 30.

When we build the LCM, we must use the minimum numbers of factors necessary to satisfy all requirements. Otherwise, we do not create the *smallest* common multiple (LCM). For example, in the above illustration, to satisfy the second requirement—the LCM must have a 3 and a 5 in its prime factorization—we only added a 3 to the existing prime factorization, not a 3 and a 5. There was a 5 in the factorization already, so we did not need to add another one. *If we add an extra 5, we build a multiple of 10 and 15 that is not the smallest common multiple:*

$2 \cdot 5 \cdot 3 = 30$ $2 \cdot 5 \cdot 3 \cdot \underline{5} = 150$

This factorization includes the   This factorization
minimum number of factors needed.   includes an *extra 5*.

Although 30 and 150 are common multiples of 10 and 15, the smallest common multiple or LCM is 30.

PROCEDURE TO FIND THE LCM

1. Factor each number to a product of prime factors.
2. List the requirements for the factorization of the LCM.
3. Build the LCM using the minimum number of factors.

EXAMPLE 3   Find the LCM of 18, 42, and 45.

SOLUTION

Factor each number	→	List requirements for factorization of LCM	→	Build the LCM

$18 = 2 \cdot 3 \cdot 3$ → must have a 2 and a pair of 3's → $\boxed{LCM = 2 \cdot 3 \cdot 3 \cdot ?}$

$42 = 2 \cdot 3 \cdot 7$ → must have a 2, a 3, and a 7 → $\boxed{LCM = 2 \cdot 3 \cdot 3 \cdot 7 \cdot ?}$

2 and a pair of 3's are already factors, so we just multiply by a 7.

$45 = 3 \cdot 3 \cdot 5$ → must have a pair of 3's and a 5 → $\boxed{LCM = 2 \cdot 3 \cdot 3 \cdot 7 \cdot 5}$

A pair of 3's already exist, so we just multiply by a 5.

The LCM of 18, 42, and 45 is $2 \cdot 3 \cdot 3 \cdot 7 \cdot 5 =$ **630**.

PRACTICE PROBLEM 3

Find the LCM of 28, 36, and 70.

EXAMPLE 4   Find the LCM of $2x$, x^2, and $6x$.

SOLUTION

Factor each expression	→	List requirements for factorization of LCM	→	Build the LCM

$2x = 2 \cdot x$ → must have a 2 and an x → $\boxed{LCM = 2 \cdot x \cdot ?}$

$x^2 = x \cdot x$ → must have a pair of x's   → $\boxed{LCM = 2 \cdot x \cdot x \cdot ?}$

One x is already a factor, so we just multiply by another x.

$6x = 2 \cdot 3 \cdot x$ → must have a 2, a 3, and an x → $\boxed{LCM = 2 \cdot x \cdot x \cdot 3}$

A 2 and an x are already factors, so we just multiply by a 3.

The LCM of $2x$, x^2, and $6x$ is $2 \cdot x \cdot x \cdot 3 =$ **$6x^2$**.

PRACTICE PROBLEM 4

Find the LCM of $4x$, x^2, and $10x$.

1. (a) List the first four multiples of 6 and 8.
   (b) Which of these multiples are common to both 6 and 8?

2. (a) List the first five multiples of 4 and 5.
   (b) Which of these multiples are common to both 4 and 5?

3. (a) List the first five multiples of 2 and 5.
   (b) Which of these multiples are common to both 2 and 5?

4. (a) List the first six multiples of 4 and 6.
   (b) Which of these multiples are common to both 4 and 6?

5. (a) List the first four multiples of $12x$ and $18x$.
   (b) Which of these multiples are common to both $12x$ and $18x$?

6. (a) List the first four multiples of $15x$ and $5x$.
   (b) Which of these multiples are common to both $15x$ and $5x$?

Find the LCM of each group of expressions.

7. 9 and 18

8. 12 and 24

9. 16 and 28

10. 18 and 30

11. 15 and 21

12. 12 and 20

13. 40 and 60

14. 30 and 45

15. 5, 8, and 12

16. 6, 14, and 26

17. 7, 14, and 35

18. 8, 12, and 42

19. $4x$ and $7x$

20. $8x$ and $26x$

21. $21a$ and $81a$

22. $15a$ and $35a$

23. $18x$ and $45x^2$

24. $15x$ and $63x^2$

25. $22x^2$ and $4x^3$

26. $8x^2$ and $2x^3$

27. $12x$, 26, and $2x^2$

28. $14x$, 36, and $7x^2$

There are two factors missing in the LCM for the numbers listed. State the two missing factors.

29. $525 = 3 \cdot 5 \cdot 5 \cdot 7$ $\boxed{\text{LCM} = 2 \cdot 3 \cdot 5 \cdot 5 \cdot 7 \cdot ? \cdot ?}$
   $90 = 2 \cdot 3 \cdot 3 \cdot 5$
   $28 = 2 \cdot 2 \cdot 3$

30. $220 = 2 \cdot 2 \cdot 5 \cdot 11$ $\boxed{\text{LCM} = 2 \cdot 2 \cdot 3 \cdot 3 \cdot 7 \cdot 11 \cdot ? \cdot ?}$
   $189 = 3 \cdot 3 \cdot 3 \cdot 7$
   $385 = 5 \cdot 7 \cdot 11$

31. $10x^2 = 2 \cdot 5 \cdot x \cdot x$ $\boxed{\text{LCM} = 2 \cdot 3 \cdot 3 \cdot 5 \cdot 7 \cdot x \cdot ? \cdot ?}$
   $18x = 2 \cdot 3 \cdot 3 \cdot x$
   $49x = 7 \cdot 7 \cdot x$

32. $15y^2 = 3 \cdot 5 \cdot y \cdot y$ $\boxed{\text{LCM} = 3 \cdot 3 \cdot 5 \cdot y \cdot ? \cdot ?}$
   $25y = 5 \cdot 5 \cdot y$
   $9y^2 = 3 \cdot 3 \cdot y$

CALCULATOR EXERCISES

You can use your calculator to help find the prime factors and LCMs.

33. (a) Write the prime factors for 660 and 140.
    (b) Find the LCM of 660 and 140.

34. (a) Write the prime factors for 735 and 455.
    (b) Find the LCM of 735 and 455.

 VERBAL AND WRITING SKILLS

35. Explain why 12 is a multiple of 3 and not a multiple of 5.

36. Explain the relationship between a multiple and a LCM.

ONE STEP FURTHER

Find the LCM of each group of expressions.

37. $2x^3, 8xy^2$, and $10x^2y$ **38.** $4y^2, 2xy$, and $9x^3y$ **39.** $2z^2, 5xyz$, and $15xy$ **40.** $3x, 9xy^2$, and $18xyz$

CUMULATIVE REVIEW

1. Change $\dfrac{15}{2}$ to a mixed number.

2. Change $3\dfrac{2}{5}$ to an improper fraction.

3. Evaluate: $2 + 6(-1) \div 3$.

❶ Add and subtract fractional expressions with common denominators.

❷ Add and subtract fractional expressions without a common denominator.

Math Pro   Video 5.3   SSM

Addition and Subtraction of Fractional Expressions

❶ Adding and Subtracting Fractional Expressions with Common Denominators

What does it mean to add or subtract fractions? Well, the idea is similar to *counting like items* or *adding and subtracting like terms*. For example, if we have 3 floppy disks and we get 2 more, we end up with 5 floppy disks: $3 + 2 = 5$. Similarly, we can add $3x + 2x = 5x$ because $3x$ and $2x$ are like terms. In the case of fractions, instead of adding like items or terms, we add *like parts of a whole*. By this we mean fractions with the same denominators: $\frac{3}{6} + \frac{2}{6} = \frac{5}{6}$. This is because when the denominators of fractions are the same, we are comparing *like parts* of a whole. The following situation and illustration can help us understand this idea.

Suppose that a large piece of wood is cut into 6 equal-sized parts and we take 3 of these parts, then 2 more. We add to find the total amount of wood we have.

Situation	You have **3** out of 6	followed by	an additional **2** out of 6.	You end up with **5** out of 6.
Math symbols	$\frac{3}{6}$	$+$	$\frac{2}{6}$	$=\qquad\frac{5}{6}$
Picture		$+$		$=$

When fractions have the *same denominator*, we say that these fractions have a **common denominator**. From the previous illustration we observe that to add fractions that have a common denominator, we add the numerators and write the sum over the common denominator. A similar rule is followed for subtraction, except that the numerators are subtracted.

PROCEDURE TO ADD OR SUBTRACT FRACTIONAL EXPRESSIONS WITH COMMON DENOMINATORS

1. The fractions added or subtracted must have a common denominator (denominators that are the same).
2. Add or subtract the numerators only.
3. The denominator stays the same.

EXAMPLE 1   Subtract: $\frac{7}{15} - \frac{3}{15}$.

SOLUTION

$$\frac{7}{15} - \frac{3}{15} = \frac{7-3}{15} \qquad \text{Subtract numerators.}$$
$$\text{The denominator stays the same.}$$

$$= \frac{\mathbf{4}}{\mathbf{15}}$$

PRACTICE PROBLEM 1   Add: $\frac{3}{13} + \frac{7}{13}$.

It is important to remember that to add or subtract fractions, the denominators must be the same (common denominators). *Also, we do not add or subtract the denominators—only the numerators.*

EXAMPLE 2   Add: $\dfrac{-11}{20} + \left(\dfrac{-13}{20}\right)$.

SOLUTION

$$\dfrac{-11}{20} + \dfrac{-13}{20} = \dfrac{-11 + (-13)}{20} \qquad \text{Add numerators.}$$

$$\text{The denominator stays the same.}$$

$$= \dfrac{-24}{20}$$

$$= \dfrac{\cancel{4}\,(-6)}{\cancel{4}\,(5)} \qquad \text{Factor and reduce.}$$

$$= \dfrac{-6}{5} \quad \text{or} \quad -1\dfrac{1}{5}$$

Note: The answer may be written as either an improper fraction or a mixed number. In either case, the answer must be reduced.

PRACTICE PROBLEM 2

Add: $\dfrac{-7}{6} + \left(\dfrac{-21}{6}\right)$.

EXAMPLE 3   Perform the indicated operation.

(a) $\dfrac{6}{y} - \dfrac{2}{y}$

(b) $\dfrac{x}{5} + \dfrac{4}{5}$

SOLUTION

(a) $\dfrac{6}{y} - \dfrac{2}{y} = \dfrac{6 - 2}{y} = \dfrac{4}{y}$

(b) $\dfrac{x}{5} + \dfrac{4}{5} = \dfrac{x + 4}{5}$

Note: The answer to part (b), $\dfrac{x + 4}{5}$, is simplified. We cannot add x and 4. They are not like terms.

PRACTICE PROBLEM 3

Perform the indicated operation.

(a) $\dfrac{8}{x} - \dfrac{3}{x}$

(b) $\dfrac{y}{9} + \dfrac{5}{9}$

❷ Adding and Subtracting Fractional Expressions without a Common Denominator

As we stated earlier, we can compare, add, and subtract fractions if the fractions have common denominators (denominators that are the same). Now, what do we do when the denominators are not the same? The following example will help us determine the answer to this question.

Suppose that we have 2 equal-sized blocks of molding clay. One block of clay is cut into 2 equal-sized parts and another is cut into 3 equal-sized parts. If we take 1 of the 2 parts, then 1 of the 3 parts, we must add to find the total amount of clay we have.

$$\frac{1}{2} + \frac{1}{3} = \,?$$

How much molding clay do we have?

Since we are comparing pieces of molding clay that are different sizes, we are not adding like parts of a whole. The total amount of clay could be determined more easily if each block of clay had been cut into 6 equal-sized parts.

$$\frac{1}{2} + \frac{1}{3} = \,?$$
$$\downarrow \quad \downarrow \quad \downarrow$$
$$\frac{3}{6} + \frac{2}{6} = \frac{5}{6}$$

As we can see, cutting the blocks of clay into the same number of pieces makes it possible to work with fractions that have a common denominator. We know how to compare, add, and subtract these fractions. *We use a similar idea to add and subtract fractions with different denominators.*

> **PROCEDURE TO ADD OR SUBTRACT FRACTIONAL EXPRESSIONS WITH DIFFERENT DENOMINATORS**
>
> 1. Find the least common denominator (LCD).
> 2. Write equivalent fractions that have the LCD as the denominator.
> 3. Add or subtract the fractions with common denominators.
> 4. Simplify the answer if necessary.

How do we find a least common denominator? Let's look at the fractions $\frac{1}{2}$ and $\frac{1}{3}$ from the earlier illustration. Notice that the least common denominator 6 is also the least common multiple of 2 and 3. In fact, the least common denominator of two fractions is the least common multiple (LCM) of the two denominators. Since we are working with denominators, we call the LCM the least common denominator, or LCD.

> The *least common multiple* (LCM) and the *least common denominator* (LCD) are two different names for the *same expression*. When we are working with denominators of fractions, we refer to this expression as the LCD.

E X A M P L E 4   Find the least common denominator (LCD) of the fractions.

(a) $\dfrac{1}{5}, \dfrac{1}{3}$

(b) $\dfrac{2}{7}, \dfrac{5}{14}$

SOLUTION

(a) $\dfrac{1}{5}, \dfrac{1}{3}$ The LCD of 5 and 3 is **15**.

(b) $\dfrac{2}{7}, \dfrac{5}{14}$ The LCD of 7 and 14 is **14**.

P R A C T I C E   P R O B L E M 4

Find the least common denominator of the fractions.

(a) $\dfrac{1}{4}, \dfrac{1}{5}$

(b) $\dfrac{1}{6}, \dfrac{7}{12}$

How do we write equivalent fractions that have the LCD as the denominator? In Chapter 4 we learned that to find equivalent fractions we multiply the numerator and denominator by the same nonzero number. Therefore, we must determine the value of the nonzero number that when multiplied by the denominator yields the LCD.

E X A M P L E 5   The LCD of the following fractions is 40. Write equivalent fractions that have the LCD as the denominator.

(a) $\dfrac{1}{5}$

(b) $\dfrac{3}{8}$

SOLUTION

(a) $\dfrac{1}{5} = \dfrac{?}{40}$ What number multiplied by the denominator, 5, yields the LCD 40?   8, since 5(8) = 40.

$\dfrac{1\cdot 8}{5\cdot 8} = \dfrac{8}{40}$ We multiply the numerator and denominator of $\dfrac{1}{5}$ by 8.

$\dfrac{1}{5} = \dfrac{8}{40}$

(b) $\dfrac{3}{8} = \dfrac{?}{40}$ What number multiplied by the denominator, 8, yields the LCD 40?   5, since 5(8) = 40.

$\dfrac{3\cdot 5}{8\cdot 5} = \dfrac{\mathbf{15}}{\mathbf{40}}$ We multiply the numerator and denominator of $\dfrac{3}{8}$ by 5.

$\dfrac{3}{8} = \dfrac{15}{40}$

Note: We do not reduce our answers. Why?

P R A C T I C E   P R O B L E M 5

The LCD of the following fractions is 10. Write equivalent fractions that have the LCD as the denominator.

(a) $\dfrac{3}{5}$

(b) $\dfrac{1}{2}$

Once we find the LCD and write equivalent fractions with the LCD as the denominator, we simply add or subtract the fractions with the common denominators. We summarize the process:

1. Find the LCD.
2. Write equivalent fractions.
3. Add or subtract fractions.
4. Simplify if necessary.

EXAMPLE 6   Perform the indicated operation .

(a) $\dfrac{-5}{7} + \dfrac{3}{4}$

(b) $\dfrac{11}{12} - \dfrac{3}{20}$

SOLUTION

(a) $\dfrac{-5}{7} + \dfrac{3}{4}$

Step 1: Find the LCD.

$$\text{LCD} = 28$$

Step 2: Write equivalent fractions.

$$\dfrac{-5 \cdot 4}{7 \cdot 4} = \boxed{\dfrac{-20}{28}} \qquad \dfrac{3 \cdot 7}{4 \cdot 7} = \boxed{\dfrac{21}{28}}$$

Step 3: Add fractions with common denominators.

$$\dfrac{-5}{7} + \dfrac{3}{4} = \boxed{\dfrac{-20}{28} + \dfrac{21}{28}} = \dfrac{1}{28}$$

$$\dfrac{-5}{7} + \dfrac{3}{4} = \mathbf{\dfrac{1}{28}}$$

(b) $\dfrac{11}{12} - \dfrac{3}{20}$

Step 1: Find the LCD.

$$\text{LCD} = 60$$

Step 2: Write equivalent fractions.

$$\dfrac{11 \cdot 5}{12 \cdot 5} = \boxed{\dfrac{55}{60}} \qquad \dfrac{3 \cdot 3}{20 \cdot 3} = \boxed{\dfrac{9}{60}}$$

Step 3: Subtract fractions with common denominators.

$$\dfrac{11}{12} - \dfrac{3}{20} = \boxed{\dfrac{55}{60} - \dfrac{9}{60}} = \dfrac{46}{60}$$

Step 4: Simplify.

$$\dfrac{46}{60} = \dfrac{\cancel{2} \cdot 23}{\cancel{2} \cdot 2 \cdot 3 \cdot 5} = \dfrac{23}{30}$$

$$\dfrac{11}{12} - \dfrac{3}{20} = \mathbf{\dfrac{23}{30}}$$

PRACTICE PROBLEM 6

Perform the indicated operation.

(a) $\dfrac{-3}{8} + \dfrac{7}{9}$

(b) $\dfrac{13}{30} - \dfrac{2}{15}$

E X A M P L E   7   Perform the indicated operation.

(a) $\dfrac{6}{x} + \dfrac{5}{3x}$

(b) $\dfrac{4}{y} - \dfrac{2}{x}$

SOLUTION

(a) $\dfrac{6}{x} + \dfrac{5}{3x}$

Step 1: Find the LCD.

$$\text{LCD} = 3x$$

Step 2: Write an equivalent fraction.

$$\frac{6 \cdot 3}{x \cdot 3} = \boxed{\frac{18}{3x}}$$

Step 3: Add fractions with common denominators.

$$\frac{6}{x} + \frac{5}{3x} = \boxed{\frac{18}{3x} + \frac{5}{3x}} = \frac{18 + 5}{3x} = \frac{23}{3x}$$

$$\frac{6}{x} + \frac{5}{3x} = \mathbf{\frac{23}{3x}}$$

Note: We did not need to write $\dfrac{5}{3x}$ as an equivalent fraction with the common denominator. Why?

(b) $\dfrac{4}{y} - \dfrac{2}{x}$

Step 1: Find the LCD.

$$\text{LCD} = xy$$

Step 2: Write equivalent fractions.

$$\frac{4 \cdot x}{y \cdot x} = \boxed{\frac{4x}{xy}} \qquad \frac{2 \cdot y}{x \cdot y} = \boxed{\frac{2y}{xy}}$$

Step 3: Subtract fractions with common denominators.

$$\frac{4}{y} - \frac{2}{x} = \boxed{\frac{4x}{xy} - \frac{2y}{xy}} = \frac{4x - 2y}{xy}$$

$$\frac{4}{y} - \frac{2}{x} = \mathbf{\frac{4x - 2y}{xy}}$$

Note: We cannot subtract $4x - 2y$ since $4x$ and $2y$ are not like terms. Therefore, we leave the numerator as the expression $4x - 2y$.

P R A C T I C E   P R O B L E M   7

Perform the indicated operation.

(a) $\dfrac{8}{x} + \dfrac{2}{4x}$

(b) $\dfrac{7}{y} - \dfrac{4}{x}$

E X A M P L E   8   Perform the indicated operation: $\dfrac{7x}{16} + \dfrac{3x}{32}$.

SOLUTION

$$\frac{7x}{16} + \frac{3x}{32}$$

Step 1: Find the LCD.

$$\text{LCD} = 32$$

Step 2: Write an equivalent fraction.

$$\frac{7x \cdot 2}{16 \cdot 2} = \boxed{\frac{14x}{32}}$$

Step 3: Add fractions with common denominators.

$$\frac{7x}{16} + \frac{3x}{32} = \boxed{\frac{14x}{32} + \frac{3x}{32}} = \frac{14x + 3x}{32} = \mathbf{\frac{17x}{32}}$$

P R A C T I C E   P R O B L E M   8

Perform the indicated operation: $\dfrac{8x}{15} + \dfrac{9x}{24}$.

EXERCISES 5.3

Perform the indicated operations. Be sure to simplify your answer.

1. $\dfrac{4}{21} + \dfrac{7}{21}$
2. $\dfrac{5}{35} + \dfrac{3}{35}$
3. $\dfrac{6}{17} - \dfrac{3}{17}$
4. $\dfrac{8}{43} - \dfrac{2}{43}$

5. $\dfrac{-13}{28} + \left(\dfrac{-11}{28}\right)$
6. $\dfrac{-17}{74} + \left(\dfrac{-41}{74}\right)$
7. $\dfrac{-31}{51} + \dfrac{11}{51}$
8. $\dfrac{-27}{43} + \dfrac{15}{43}$

9. $\dfrac{7}{x} - \dfrac{5}{x}$
10. $\dfrac{12}{x} - \dfrac{4}{x}$
11. $\dfrac{31}{a} + \dfrac{8}{a}$
12. $\dfrac{22}{a} + \dfrac{13}{a}$

13. $\dfrac{x}{7} - \dfrac{5}{7}$
14. $\dfrac{x}{12} - \dfrac{11}{12}$
15. $\dfrac{y}{9} + \dfrac{14}{9}$
16. $\dfrac{y}{5} + \dfrac{42}{5}$

Find the least common denominator (LCD) of the fractions.

17. $\dfrac{1}{4}, \dfrac{1}{7}$
18. $\dfrac{1}{3}, \dfrac{1}{8}$
19. $\dfrac{1}{12}, \dfrac{1}{14}$
20. $\dfrac{1}{21}, \dfrac{1}{15}$

21. $\dfrac{3}{5}, \dfrac{7}{25}$
22. $\dfrac{7}{8}, \dfrac{9}{32}$
23. $\dfrac{4}{9}, \dfrac{6}{21}$
24. $\dfrac{8}{5}, \dfrac{3}{45}$

The LCD of the following fractions is 60. Write equivalent fractions that have the LCD 60 as the denominator.

25. $\dfrac{1}{4}$
26. $\dfrac{2}{15}$
27. $\dfrac{5}{6}$
28. $\dfrac{3}{10}$

The LCD of the following fractions is 72. Write equivalent fractions that have the LCD 72 as the denominator.

29. $\dfrac{1}{6}$
30. $\dfrac{2}{8}$
31. $\dfrac{5}{9}$
32. $\dfrac{3}{24}$

Perform the indicated operations. Be sure to simplify your answer.

33. $\dfrac{3}{8} + \dfrac{4}{7}$
34. $\dfrac{7}{4} + \dfrac{5}{9}$
35. $\dfrac{5}{6} + \dfrac{4}{5}$
36. $\dfrac{3}{7} + \dfrac{7}{2}$

37. $\dfrac{11}{18} - \dfrac{6}{45}$

38. $\dfrac{21}{12} - \dfrac{8}{18}$

39. $\dfrac{16}{24} - \dfrac{5}{27}$

40. $\dfrac{15}{32} - \dfrac{7}{28}$

41. $\dfrac{-3}{14} + \dfrac{5}{21}$

42. $\dfrac{-5}{16} + \dfrac{7}{24}$

43. $\dfrac{-2}{13} + \dfrac{7}{26}$

44. $\dfrac{-4}{15} + \dfrac{11}{30}$

45. $\dfrac{3}{16} + \left(\dfrac{-9}{20}\right)$

46. $\dfrac{3}{18} + \left(\dfrac{-5}{27}\right)$

47. $\dfrac{-3}{14} + \left(\dfrac{-9}{28}\right)$

48. $\dfrac{-4}{9} + \left(\dfrac{-11}{18}\right)$

49. $\dfrac{7}{10} - \dfrac{13}{100}$

50. $\dfrac{3}{10} - \dfrac{7}{100}$

51. $\dfrac{3}{25} + \dfrac{1}{35}$

52. $\dfrac{12}{35} + \dfrac{1}{10}$

53. $\dfrac{7}{20} + \dfrac{4}{15}$

54. $\dfrac{5}{12} + \dfrac{5}{16}$

55. $\dfrac{9}{48} - \dfrac{5}{32}$

56. $\dfrac{9}{50} - \dfrac{2}{25}$

57. $\dfrac{5}{2x} + \dfrac{8}{x}$

58. $\dfrac{7}{5x} + \dfrac{3}{x}$

59. $\dfrac{2}{7x} + \dfrac{3}{x}$

60. $\dfrac{2}{5x} + \dfrac{5}{x}$

61. $\dfrac{3}{2x} + \dfrac{5}{6x}$

62. $\dfrac{2}{4x} + \dfrac{5}{8x}$

63. $\dfrac{3}{x} + \dfrac{4}{y}$

64. $\dfrac{5}{x} + \dfrac{3}{y}$

65. $\dfrac{2}{x} - \dfrac{7}{y}$

66. $\dfrac{6}{x} - \dfrac{4}{y}$

67. $\dfrac{9}{y} + \dfrac{1}{x}$

68. $\dfrac{6}{y} + \dfrac{1}{x}$

69. $\dfrac{4x}{15} + \dfrac{3x}{5}$

70. $\dfrac{6x}{12} + \dfrac{5x}{6}$

71. $\dfrac{8x}{10} - \dfrac{7x}{20}$

72. $\dfrac{9x}{14} - \dfrac{3x}{28}$

73. $\dfrac{x}{3} + \dfrac{6x}{12}$

74. $\dfrac{x}{4} + \dfrac{7x}{12}$

VERBAL AND WRITING SKILLS

Fill in the blanks.

75. When we add two fractions with the same denominator, we add the _____, and the _____ stays the same.

76. When we add two fractions with different denominators, we must first find the _____.

ONE STEP FURTHER

Perform the indicated operations.

77. $\dfrac{5}{30} + \dfrac{3}{40} + \dfrac{1}{8}$

78. $\dfrac{1}{12} + \dfrac{3}{14} + \dfrac{4}{21}$

79. $\dfrac{1}{3} + \dfrac{1}{12} - \dfrac{1}{6}$

80. $\dfrac{1}{5} + \dfrac{2}{3} - \dfrac{11}{15}$

Find the value of x.

81. $\dfrac{x}{7} + \dfrac{1}{7} = \dfrac{3}{7}$

82. $\dfrac{x}{9} + \dfrac{2}{9} = \dfrac{5}{9}$

83. $\dfrac{x}{4} - \dfrac{1}{4} = \dfrac{1}{4}$

84. $\dfrac{x}{7} - \dfrac{4}{7} = \dfrac{2}{7}$

CUMULATIVE REVIEW

For the following equations:

    (a) *Write the equivalent addition or subtraction problem.*

    (b) *Solve.*

1. $x - 3 = 70$

2. $x + 8 = 130$

3. $x - 9 = 29$

4. $x + 5 = 160$

After studying this section, you will be able to:

❶ Add and subtract mixed numbers.

❷ Multiply and divide mixed numbers.

❸ Solve applied problems involving mixed numbers.

Math Pro Video 5.4 SSM

Operations with Mixed Numbers

❶ **Adding and Subtracting Mixed Numbers**

We add and subtract mixed numbers in a manner similar to the one used for proper fractions. The only difference is that we work with the whole number and the fractional parts separately.

> **ADDING AND SUBTRACTING MIXED NUMBERS**
> We add or subtract the fractions first, then the whole number.

E X A M P L E   1   Add: $4\frac{1}{8} + 3\frac{3}{8}$.

SOLUTION

$$
\begin{array}{r}
4\,\boxed{\dfrac{1}{8}} \\[2mm]
+3\,\boxed{\dfrac{3}{8}} \\[2mm]
\hline
7\,\boxed{\dfrac{4}{8}} \text{ or } 7\frac{1}{2}
\end{array}
$$

Add the whole numbers: $4 + 3 = 7$

Add the fractions $\frac{1}{8} + \frac{3}{8} = \frac{4}{8}$.

P R A C T I C E   P R O B L E M   1

Add: $5\frac{2}{9} + 2\frac{5}{9}$.

If the fractional parts of the mixed numbers do not have common denominators, we must find the LCD and build equivalent fractions to obtain common denominators before adding.

E X A M P L E   2   Add: $4\frac{2}{3} + 2\frac{1}{4}$.

SOLUTION   The LCD of $\frac{2}{3}$ and $\frac{1}{4}$ is 12, so we build equivalent fractions.

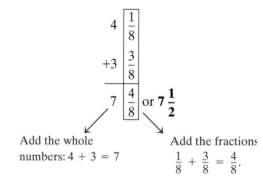

$$
\begin{array}{r}
4\,\boxed{\dfrac{2}{3} \cdot \dfrac{4}{4}} = 4\,\boxed{\dfrac{8}{12}} \\[3mm]
+2\,\boxed{\dfrac{1}{4} \cdot \dfrac{3}{3}} = +2\,\boxed{\dfrac{3}{12}} \\[3mm]
\hline
6\,\dfrac{11}{12}
\end{array}
$$

Add the fractions $\frac{8}{12} + \frac{3}{12} = \frac{11}{12}$.

Add the whole numbers.

PRACTICE PROBLEM 2

Add: $5\dfrac{1}{3} + 6\dfrac{3}{5}$.

EXAMPLE 3   Add: $2\dfrac{5}{7} + 6\dfrac{2}{3}$.

SOLUTION   The LCD of $\dfrac{5}{7}$ and $\dfrac{2}{3}$ is 21.

$$2\left|\dfrac{5}{7}\cdot\dfrac{3}{3}\right| = 2\left|\dfrac{15}{21}\right| \quad \leftarrow \text{Add the fractions:}$$
$$+6\left|\dfrac{2}{3}\cdot\dfrac{7}{7}\right| = +6\left|\dfrac{14}{21}\right| \quad \leftarrow \dfrac{15}{21} + \dfrac{14}{21} = \dfrac{29}{21}.$$

Add the whole numbers.
$$8\dfrac{29}{21} = 8 + 1\dfrac{8}{21} \qquad \text{Since } \dfrac{29}{21} = 1\dfrac{8}{21}$$

$$8\dfrac{29}{21} = 9\dfrac{8}{21} \qquad \text{Add: } 8 + 1 = 9$$

Thus, $2\dfrac{5}{7} + 6\dfrac{2}{3} = \mathbf{9\dfrac{8}{21}}$.

PRACTICE PROBLEM 3

Add: $7\dfrac{3}{4} + 2\dfrac{4}{5}$.

In Example 3 we simplified $8\dfrac{29}{21}$ to $9\dfrac{8}{21}$ using a process similar to *carrying* with whole numbers.

Change to a mixed number
$$8\dfrac{29}{21} = 8 + 1\dfrac{8}{21} = (8 + 1) + \dfrac{8}{21} = 9\dfrac{8}{21}$$
Carry 1 to the whole-number part.

When subtracting mixed numbers it is sometimes necessary to *borrow*. The process used to borrow with mixed numbers is the opposite of carrying with mixed numbers.

Write 9 as the sum 8 + 1. Borrow 1 from the whole-number part and add to the fraction part.
$$9\dfrac{8}{21} = (8 + 1) + \dfrac{8}{21} = 8 + 1\dfrac{8}{21} = 8\dfrac{29}{21}$$
Change $1\dfrac{8}{21}$ to the improper fraction $\dfrac{29}{21}$.

Thus $9\dfrac{8}{21}$ is written as $8\dfrac{29}{21}$ when borrowing is necessary.

E X A M P L E   4   Subtract: $7\frac{4}{15} - 2\frac{7}{15}$.

SOLUTION

We cannot subtract $4 - 7$, so we must borrow:

$7\frac{4}{15} = \mathbf{6 + 1}\frac{4}{15}$ We write 7 as the sum $6 + 1$ (borrowing from 7).

$= 6 + \frac{19}{15}$ We change $1\frac{4}{15}$ to $\frac{19}{15}$.

$$7\frac{4}{15} = 6\frac{19}{15} \longleftarrow$$

$$\underline{-2\frac{7}{15} = -2\frac{7}{15}}$$

$$4\frac{12}{15} \longrightarrow \text{Subtract fractions: } \frac{19}{15} - \frac{7}{15} = \frac{12}{15}.$$

$$\longrightarrow \text{Subtract whole numbers.}$$

We simplify $4\frac{12}{15} = 4\frac{4}{5}$. Thus, $7\frac{4}{15} - 2\frac{7}{15} = \mathbf{4\frac{4}{5}}$.

P R A C T I C E   P R O B L E M   4

Subtract: $5\frac{3}{12} - 3\frac{5}{12}$.

E X A M P L E   5   Subtract: $5\frac{2}{5} - 3\frac{7}{8}$.

SOLUTION   The LCD of $\frac{2}{5}$ and $\frac{7}{8}$ is 40.

We cannot subtract $16 - 35$, so we borrow as follows:

$5\frac{16}{40} = \mathbf{4 + 1}\frac{16}{40}$ We write 5 as the sum $4 + 1$ (borrowing from 5).

$5\frac{16}{40} = 4\frac{56}{40}$ We change $1\frac{16}{40}$ to $\frac{56}{40}$.

$$5\frac{2 \cdot 8}{5 \cdot 8} = 5\frac{16}{40}$$

$$\underline{-3\frac{7 \cdot 5}{8 \cdot 5} = -3\frac{35}{40}}$$

$$5\frac{16}{40} = 4\frac{56}{40} \longleftarrow$$

$$\underline{-3\frac{35}{40} = -3\frac{35}{40}}$$

$$1\frac{21}{40} \longrightarrow \text{Subtract fractions: } \frac{56}{40} - \frac{35}{40} = \frac{21}{40}.$$

$$\longrightarrow \text{Subtract whole numbers.}$$

Thus, $5\frac{2}{5} - 3\frac{7}{8} = \mathbf{1\frac{21}{40}}$.

P R A C T I C E   P R O B L E M   5

Subtract: $6\frac{1}{7} - 2\frac{3}{4}$.

TO THINK ABOUT

We can change mixed numbers to improper fractions, then add or subtract. To illustrate, we work Example 5 this way.

Change to an improper fraction.   Build an equivalent fraction.

$$5\frac{2}{5} = \frac{27}{5} = \frac{216}{40}$$

$$-3\frac{7}{8} = -\frac{31}{8} = -\frac{155}{40}$$

$$\frac{61}{40} \text{ or } 1\frac{21}{40} \quad \text{Subtract improper fractions.}$$

As you can see, the result is the same as the result obtained in Example 5. Which method should you use? Well, there are advantages to both methods. We do not have to carry or borrow when we change to improper fractions. But if the numbers are large, changing to improper fractions can be more difficult.

EXERCISE

1. Add the following numbers using both methods. Which way is easier? Why?

$$25\frac{9}{12} + 32\frac{15}{24}$$

E X A M P L E   6   Subtract: $8 - 3\frac{1}{4}$.

SOLUTION

$$8 = 7\frac{4}{4}$$

$$-3\frac{1}{4} = -3\frac{1}{4}$$

$$4\frac{3}{4}$$

Note: when we borrowed, $8 = 7 + 1$, we changed the 1 to $\frac{4}{4}$ because we wanted a fraction that had the same denominator as $\frac{1}{4}$.

P R A C T I C E   P R O B L E M   6

Subtract: $9 - 4\frac{1}{3}$.

② Multiplying and Dividing Mixed Numbers

In Chapter 4 we learned that the mixed number $2\frac{1}{4}$ means $\left(2 + \frac{1}{4}\right)$. To multiply $3 \cdot 2\frac{1}{4}$ we have $3\left(2 + \frac{1}{4}\right)$, which requires use of the distributive property:

$$3 \cdot 2\frac{1}{4} = 3\left(2 + \frac{1}{4}\right) = 3 \cdot \left(2 + \frac{1}{4}\right) = 3 \cdot 2 + 3 \cdot \frac{1}{4} = 6 + \frac{3}{4} = 6\frac{3}{4}$$

Changing the mixed number to an improper fraction simplifies the calculations.

$$3 \cdot 2\frac{1}{4} = \frac{3}{1} \cdot \frac{9}{4} = \frac{27}{4} = 6\frac{3}{4}$$

> Change mixed numbers to improper fractions before multiplying or dividing.

EXAMPLE 7   Multiply: $5\frac{5}{12} \cdot 3\frac{11}{15}$.

SOLUTION

We change to improper fractions, then multiply.

$$5\frac{5}{12} \cdot 3\frac{11}{15} = \frac{65}{12} \cdot \frac{56}{15} = \frac{\cancel{5} \cdot 13 \cdot \cancel{4} \cdot 14}{3 \cdot \cancel{4} \cdot \cancel{5} \cdot 3} = \frac{182}{9} \quad \text{or} \quad \mathbf{20\frac{2}{9}}$$

PRACTICE PROBLEM 7

Multiply: $4\frac{1}{3} \cdot 2\frac{1}{4}$

EXAMPLE 8   Divide: $2\frac{1}{4} \div 5$.

SOLUTION

$$2\frac{1}{4} \div 5 = \frac{9}{4} \div 5 \qquad \text{Change to improper fractions.}$$

$$= \frac{9}{4} \cdot \frac{1}{5} \qquad \text{Invert and multiply.}$$

$$= \frac{9}{20}$$

PRACTICE PROBLEM 8

Divide: $2 \div \frac{1}{4}$.

③ Solving Applied Problems Involving Mixed Numbers

When solving applications involving fractions it is helpful to draw pictures or diagrams.

EXAMPLE 9   A plumber has a pipe $5\frac{3}{4}$ feet long. He needs $\frac{1}{3}$ of the length of the pipe for a repair job. What length must he cut off the pipe to get the desired size?

SOLUTION   We draw a picture.

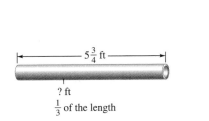

We can see that we must multiply to find the length the plumber must cut off.

$$5\frac{3}{4} \cdot \frac{1}{3} = \frac{23}{4} \cdot \frac{1}{3} = \frac{23}{12} \quad \text{or} \quad 1\frac{11}{12}$$

The plumber must cut off $\mathbf{1\frac{11}{12}}$ **feet** of pipe.

PRACTICE PROBLEM 9

A recipe uses $4\frac{1}{2}$ tablespoons of brown sugar. If Sara only has a $\frac{1}{2}$-tablespoon measuring utensil, how many times must she fill this utensil to get the desired amount of sugar?

Remember when *adding* and *subtracting* mixed numbers we normally *don't* change to improper fractions. However, when *multiplying* and *dividing* mixed numbers we *must* change to improper fractions to avoid using the distributive property.

Add or subtract. Simplify all answers. Express as a mixed number.

1. $12\dfrac{3}{7} + 15\dfrac{2}{7}$

2. $8\dfrac{1}{5} + 3\dfrac{3}{5}$

3. $5\dfrac{5}{8} + 11\dfrac{1}{8}$

4. $7\dfrac{1}{6} + 2\dfrac{1}{6}$

5. $11\dfrac{2}{3} + 7\dfrac{1}{4}$

6. $22\dfrac{3}{5} + 16\dfrac{1}{10}$

7. $7\dfrac{4}{5} - 2\dfrac{1}{10}$

8. $6\dfrac{3}{8} - 2\dfrac{1}{16}$

9. $9\dfrac{2}{3} - 6\dfrac{1}{6}$

10. $15\dfrac{3}{4} - 13\dfrac{1}{6}$

11. $1\dfrac{2}{3} + \dfrac{5}{18}$

12. $1\dfrac{1}{6} + \dfrac{3}{8}$

13. $25\dfrac{2}{3} - 6\dfrac{1}{7}$

14. $45\dfrac{3}{8} - 26\dfrac{1}{16}$

15. $13\dfrac{1}{2} + 7\dfrac{4}{5}$

16. $14\dfrac{7}{9} + 6\dfrac{1}{3}$

17. $11\dfrac{1}{5} - 6\dfrac{11}{25}$

18. $25\dfrac{2}{7} - 16\dfrac{5}{14}$

19. $10\dfrac{5}{12} - 3\dfrac{9}{10}$

20. $12\dfrac{4}{9} - 7\dfrac{5}{6}$

21. $6\dfrac{5}{6} + 4\dfrac{3}{8}$

22. $5\dfrac{4}{5} + 10\dfrac{3}{10}$

23. $8\dfrac{1}{4} + 3\dfrac{5}{6}$

24. $7\dfrac{3}{4} + 6\dfrac{2}{5}$

25. $9 - 2\dfrac{1}{4}$

26. $4 - 2\dfrac{2}{7}$

27. $32 - 1\dfrac{2}{9}$

28. $24 - 3\dfrac{4}{11}$

Multiply or divide and simplify your answer.

29. $2\dfrac{1}{5} \cdot 1\dfrac{2}{3}$

30. $1\dfrac{1}{4} \cdot 2\dfrac{2}{7}$

31. $4\dfrac{1}{3} \cdot 2\dfrac{1}{4}$

32. $6\dfrac{1}{3} \cdot 2\dfrac{1}{4}$

33. $2\dfrac{3}{4} \div 1\dfrac{7}{10}$ **34.** $12\dfrac{1}{2} \div 5\dfrac{5}{6}$ **35.** $4\dfrac{1}{2} \div 2\dfrac{1}{4}$ **36.** $8\dfrac{1}{4} \div 2\dfrac{3}{4}$

37. $1\dfrac{1}{4} \cdot 3\dfrac{2}{3}$ **38.** $2\dfrac{3}{5} \cdot 1\dfrac{4}{7}$ **39.** $\dfrac{3}{4} \cdot 9\dfrac{5}{7}$ **40.** $\dfrac{8}{11} \cdot 4\dfrac{3}{4}$

41. $3\dfrac{1}{4} \div \dfrac{3}{8}$ **42.** $6\dfrac{1}{2} \div \dfrac{3}{4}$ **43.** $\dfrac{1}{4} \div 1\dfrac{7}{8}$ **44.** $\dfrac{1}{2} \div 2\dfrac{5}{8}$

45. $5 \div \dfrac{1}{4}$ **46.** $6 \div \dfrac{1}{2}$ **47.** $7\dfrac{1}{2} \div \dfrac{6}{6}$ **48.** $8\dfrac{2}{9} \div \dfrac{4}{4}$

Solve the following applied problems.

49. To put up the tents for a camping trip, Andy needs several pieces of rope each $6\dfrac{1}{2}$ feet long. If Andy has a rope that is 26 feet long, how many pieces can he cut for his tents?

50. A dress requires $2\dfrac{1}{2}$ yards of material. A bolt of material contains 25 yards. How many dresses can be made from 1 bolt?

Use the following recipe to answer questions 51–52.

Ingredients for 1 Batch of Chocolate Chip Cookies	
$2\dfrac{1}{4}$ cups flour	$\dfrac{3}{4}$ cup granulated sugar
1 teaspoon baking soda	$\dfrac{2}{3}$ cup brown sugar
$\dfrac{3}{4}$ teaspoon salt	1 teaspoon vanilla
1 cup margarine	$2\dfrac{1}{2}$ cups chocolate chips

51. To double the recipe, how much flour do you need?

52. To triple the recipe, how much brown sugar do you need?

53. To make four times the recipe, how many cups of chocolate chips do you need?

54. To make $\dfrac{1}{2}$ of the recipe, how much granulated sugar do you need?

55. Jeff purchased shares of stock at $\$9\dfrac{3}{8}$ per share and sold it at $\$11\dfrac{1}{8}$ per share. How much money per share did he make?

56. If Miami received $2\frac{1}{2}$ inches of rain on Monday and $1\frac{1}{3}$ inches on Tuesday, what was the total rainfall for the 2 days?

57. Marcy multiplied two mixed numbers and got the following results.

$$2\frac{2}{3}\cdot 3\frac{4}{5} \rightarrow 2\cdot 3 = 6 \quad \text{and} \quad \frac{2}{3}\cdot\frac{4}{5} = \frac{8}{15}$$

$$2\frac{2}{3}\cdot 3\frac{4}{5} = 6\frac{8}{15} \qquad \text{This answer is wrong.}$$

(a) What did Marcy do wrong?

(b) What is the correct answer?

58. Lester divided two mixed numbers and got the following results.

$$8\frac{1}{2} \div 4\frac{4}{7} \rightarrow 8 \div 4 = 2 \quad \text{and} \quad \frac{1}{2}\cdot\frac{4}{7} = \frac{2}{7}$$

$$8\frac{1}{2} \div 4\frac{4}{7} = 2\frac{2}{7} \qquad \text{This answer is wrong.}$$

(a) What did Lester do wrong?

(b) What is the correct answer?

CUMULATIVE REVIEW

Perform the indicated operations.

1. $2 + 9\cdot 8$

2. $12 - 4 \div 2$

3. $\dfrac{5 + 7}{2\cdot 3}$

4. $\dfrac{11 - 5}{2}$

Order of Operations and Complex Fractions

After studying this section, you will be able to:

❶ Follow the order of operations with fractions.

❷ Simplify complex fractions.

Math Pro   Video 5.5   SSM

❶ Following the Order of Operations with Fractions

Recall that when we work a problem with more than one operation, we must follow the **order of operations**.

> If a fraction has operations written in the numerator or in the denominator or both, these operations must be done first. We perform operations in the following order:
>
> 1. Perform operations inside parentheses.
> 2. Simplify exponents.
> 3. Do multiplication and division in order from left to right.
> 4. Do addition and subtraction last, working from left to right.

E X A M P L E   1

Simplify: $\dfrac{6}{7} - \dfrac{4}{7} \cdot \dfrac{1}{3}$.

SOLUTION

$$\frac{6}{7} - \frac{4}{7} \cdot \frac{1}{3} = \frac{6}{7} - \frac{4 \cdot 1}{7 \cdot 3} \qquad \text{First we multiply.}$$

$$= \frac{6}{7} - \frac{4}{21} \qquad \text{Then we find the LCD, which is 21.}$$

$$= \frac{6 \cdot \mathbf{3}}{7 \cdot \mathbf{3}} - \frac{4}{21} \qquad \text{Next we build a fraction with the LCD 21.}$$

$$= \frac{18}{21} - \frac{4}{21} \qquad \text{Now we subtract.}$$

$$= \frac{14}{21}$$

$$= \frac{\mathbf{2}}{\mathbf{3}} \qquad \text{Last we reduce.}$$

Do not add or subtract before multiplying even though adding and subtraction comes first in the problem, or appears to be easier. *We must be careful not to make this error.*

P R A C T I C E   P R O B L E M   1

Simplify: $\dfrac{5}{6} + \dfrac{1}{6} \cdot \dfrac{3}{4}$.

❷ Simplifying Complex Fractions

When the numerator and/or the denominator is a fraction, we have a **complex fraction**.

COMPLEX FRACTION
A fraction that contains at least one fraction in the numerator or in the denominator is a complex fraction.

These three fractions are complex fractions:

$$\frac{\dfrac{2}{x}}{\dfrac{12}{x}} \qquad \frac{\dfrac{3}{7}+2}{\dfrac{1}{8}} \qquad \left.\frac{2-\dfrac{1}{x}}{\dfrac{1}{6}+\dfrac{2}{9}}\right\} \begin{array}{l} \text{top fraction} \\ \rightarrow \text{main fraction bar} \\ \text{bottom fraction} \end{array}$$

Now, since the main fraction bar indicates division, we can divide the top fraction by the bottom fraction to simplify.

EXAMPLE 2 Simplify: $\dfrac{\dfrac{3}{4}}{\dfrac{9}{16}}$.

SOLUTION We divide the top fraction by the bottom fraction to simplify.

$$\frac{\dfrac{3}{4}}{\dfrac{9}{16}} = \frac{3}{4} \div \frac{9}{16} = \frac{3}{4} \cdot \frac{16}{9} = \frac{\cancel{3} \cdot \cancel{4} \cdot 4}{\cancel{4} \cdot \cancel{3} \cdot 3} = \frac{4}{3}$$

PRACTICE PROBLEM 2

Simplify: $\dfrac{\dfrac{2}{5}}{\dfrac{16}{15}}$.

EXAMPLE 3 Simplify: $\dfrac{\dfrac{x^2}{8}}{\dfrac{x}{4}}$.

SOLUTION

$$\frac{\dfrac{x^2}{8}}{\dfrac{x}{4}} = \frac{x^2}{8} \div \boxed{\frac{x}{4}} = \frac{x^2}{8} \cdot \boxed{\frac{4}{x}} = \frac{x^2 \cdot \cancel{4}}{2 \cdot \cancel{4} \cdot x} = \frac{\cancel{x} \cdot x}{2 \cdot \cancel{x}} = \frac{x}{2}$$

PRACTICE PROBLEM 3

Simplify: $\dfrac{\dfrac{x^2}{5}}{\dfrac{x}{10}}$.

Although we usually do not write parentheses in the numerator and denominator of a complex fraction, it is understood they exist. Thus we must perform operations above, then below the main fraction bar before we divide.

EXAMPLE 4   Simplify: $\dfrac{\dfrac{2}{3} + \dfrac{1}{6}}{\dfrac{3}{4} - \dfrac{1}{2}}$.

SOLUTION   We add the top fractions and we subtract the bottom fractions.

$$\frac{\dfrac{2}{3} + \dfrac{1}{6}}{\dfrac{3}{4} - \dfrac{1}{2}} = \frac{\boxed{\dfrac{2\cdot 2}{3\cdot 2}} + \dfrac{1}{6}}{\dfrac{3}{4} - \boxed{\dfrac{1\cdot 2}{2\cdot 2}}} = \frac{\boxed{\dfrac{4}{6}} + \dfrac{1}{6}}{\dfrac{3}{4} - \boxed{\dfrac{2}{4}}} = \frac{\dfrac{5}{6}}{\dfrac{1}{4}}$$

Add top fractions.

Subtract bottom fractions.

Now we divide the top fraction by the bottom fraction.

$$\frac{5}{6} \div \frac{1}{4} = \frac{5}{6} \cdot \frac{4}{1} = \frac{5 \cdot \cancel{2} \cdot 2}{3 \cdot \cancel{2}} = \frac{10}{3}$$

$$\frac{\dfrac{2}{3} + \dfrac{1}{6}}{\dfrac{3}{4} - \dfrac{1}{2}} = \mathbf{\frac{10}{3}}$$

PRACTICE PROBLEM 4

Simplify: $\dfrac{\dfrac{3}{5} + \dfrac{1}{2}}{\dfrac{5}{6} - \dfrac{1}{3}}$.

TO THINK ABOUT

ALTERNATIVE METHOD

We can simplify complex fractions using an alternative method. This method requires that we:

1. Find the LCD of all the denominators in the top and bottom fractions.
2. Multiply the top fractions and bottom fractions by this LCD.

We rework the problem in Example 4 using the alternative method.

$$\frac{\dfrac{2}{3} + \dfrac{1}{6}}{\dfrac{3}{4} - \dfrac{1}{2}}$$

1. The LCD of all the denominators 3, 6, 4, and 2, is **12**.

$$= \frac{\mathbf{12} \cdot \left(\dfrac{2}{3} + \dfrac{1}{6} \right)}{\mathbf{12} \cdot \left(\dfrac{3}{4} - \dfrac{1}{2} \right)}$$

2. We use the distributive property and multiply top and bottom fractions by the LCD 12.

$$= \frac{\dfrac{\mathbf{12} \cdot 2}{3} + \dfrac{\mathbf{12} \cdot 1}{6}}{\dfrac{\mathbf{12} \cdot 3}{4} - \dfrac{\mathbf{12} \cdot 1}{2}}$$

We must be sure to multiply all fractions in the top and bottom by 12.

$$= \frac{8 + 2}{9 - 6}$$

After simplifying the fractions, we see that we end up with whole numbers in the numerator and denominator.

$$= \frac{10}{3}$$

As you can see, this method eliminates having to add or subtract fractions and for some students is easier than the previous method. Try using both methods for the first few homework exercises to determine which you prefer.

EXERCISES 5.5

Simplify.

1. $\dfrac{3}{5} - \dfrac{1}{3} \div \dfrac{5}{6}$

2. $\dfrac{1}{2} + \dfrac{3}{8} \div \dfrac{3}{4}$

3. $\dfrac{3}{4} + \dfrac{1}{4} \cdot \dfrac{3}{5}$

4. $\dfrac{4}{5} + \dfrac{1}{5} \cdot \dfrac{2}{3}$

5. $\dfrac{5}{7} \cdot \dfrac{1}{3} \div \dfrac{2}{7}$

6. $\dfrac{2}{7} \cdot \dfrac{3}{4} \div \dfrac{1}{2}$

7. $\dfrac{4}{9} - \dfrac{2}{6} + \dfrac{1}{3}$

8. $\dfrac{5}{6} + \dfrac{1}{12} - \dfrac{1}{4}$

9. $\dfrac{5}{6} \cdot \dfrac{1}{2} + \dfrac{2}{3} \div \dfrac{4}{3}$

10. $\dfrac{3}{5} \cdot \dfrac{1}{2} + \dfrac{1}{5} \div \dfrac{2}{3}$

11. $\dfrac{2}{9} \cdot \dfrac{1}{4} + \left(\dfrac{2}{3} \div \dfrac{6}{7} \right)$

12. $\left(\dfrac{3}{4} \cdot \dfrac{1}{6} \right) + \dfrac{1}{2} \div \dfrac{4}{5}$

13. $\dfrac{3}{4} \cdot \dfrac{1}{4} + \left(\dfrac{3}{4} \right)^2$

14. $\left(\dfrac{2}{5} \right)^2 + \dfrac{3}{5} \cdot \dfrac{1}{2}$

15. $\left(-\dfrac{2}{5} \right) \cdot \left(\dfrac{1}{4} \right)^2$

16. $\left(\dfrac{4}{3} \right)^2 \cdot \left(-\dfrac{1}{2} \right)$

Evaluate for the value given.

17. $x - \dfrac{2}{5} \div \dfrac{4}{15}$, for $x = \dfrac{7}{2}$

18. $x - \dfrac{5}{6} \div \dfrac{25}{12}$, for $x = \dfrac{7}{5}$

19. $-\dfrac{3}{8} \cdot \dfrac{16}{21} + x$, for $x = -\dfrac{4}{7}$

20. $-\dfrac{4}{9} \cdot \dfrac{18}{24} + x$, for $x = -\dfrac{1}{3}$

Simplify.

21. $\dfrac{\frac{2}{3}}{\frac{8}{9}}$

22. $\dfrac{\frac{4}{5}}{\frac{2}{3}}$

23. $\dfrac{\frac{4}{7}}{\frac{12}{21}}$

24. $\dfrac{\frac{5}{8}}{\frac{15}{24}}$

25. $\dfrac{\frac{6}{7}}{\frac{9}{14}}$

26. $\dfrac{\frac{8}{9}}{\frac{4}{27}}$

27. $\dfrac{\frac{x^2}{3}}{\frac{x}{6}}$

28. $\dfrac{\frac{x^2}{5}}{\frac{x}{10}}$

29. $\dfrac{\dfrac{x}{4}}{\dfrac{x^2}{12}}$

30. $\dfrac{\dfrac{x}{6}}{\dfrac{x^2}{18}}$

31. $\dfrac{\dfrac{x}{8}}{\dfrac{x^2}{16}}$

32. $\dfrac{\dfrac{x}{7}}{\dfrac{x^2}{28}}$

33. $\dfrac{\dfrac{1}{2}+\dfrac{3}{4}}{\dfrac{4}{5}+\dfrac{1}{10}}$

34. $\dfrac{\dfrac{3}{7}+\dfrac{1}{14}}{\dfrac{2}{3}+\dfrac{1}{6}}$

35. $\dfrac{\dfrac{1}{9}+\dfrac{2}{3}}{\dfrac{3}{2}+\dfrac{1}{3}}$

36. $\dfrac{\dfrac{5}{6}+\dfrac{1}{3}}{\dfrac{3}{8}+\dfrac{1}{2}}$

37. $\dfrac{\dfrac{4}{25}-\dfrac{3}{50}}{\dfrac{3}{10}+\dfrac{5}{20}}$

38. $\dfrac{\dfrac{5}{12}-\dfrac{7}{24}}{\dfrac{1}{2}+\dfrac{1}{8}}$

39. $\dfrac{\dfrac{3}{7}-\dfrac{5}{21}}{\dfrac{2}{9}-\dfrac{1}{18}}$

40. $\dfrac{\dfrac{2}{9}-\dfrac{1}{27}}{\dfrac{4}{5}-\dfrac{3}{10}}$

VERBAL AND WRITING SKILLS

41. When we perform a series of calculations, do we add first or multiply?

42. When we perform a series of calculations, do we divide first or subtract?

ONE STEP FURTHER

Simplify.

43. $\dfrac{\dfrac{25xy^2}{49}}{\dfrac{15x^2y}{14}}$

44. $\dfrac{\dfrac{36x^2y^2}{45}}{\dfrac{12xy}{30}}$

45. $\dfrac{\dfrac{1}{x}-\dfrac{1}{y}}{\dfrac{1}{x}+\dfrac{1}{y}}$

46. $\dfrac{\dfrac{1}{a}+\dfrac{1}{b}}{\dfrac{1}{a}-\dfrac{1}{b}}$

CUMULATIVE REVIEW

Perform the indicated operations.

1. $(-2)(3)(-1)(-5)$

2. $(-1)(-5)(-1)(-3)$

3. $-25 \div 5$

4. $-36 \div (-6)$

Multiplication and Division Properties of Exponents

After studying this section, you will be able to:

① Divide expressions in exponent form.

② Raise a power to a power.

Math Pro   Video 5.6   SSM

① Dividing Expressions in Exponent Form

Frequently, we must divide variable expressions such as $x^6 \div x^4$. We can rewrite the expression as the fraction $\dfrac{x^6}{x^4}$ and simplify using repeated multiplication.

$$\frac{x^6}{x^4} = \frac{x \cdot x \cdot x \cdot x \cdot x \cdot x}{x \cdot x \cdot x \cdot x} = \frac{\cancel{x} \cdot \cancel{x} \cdot \cancel{x} \cdot \cancel{x} \cdot x \cdot x}{\cancel{x} \cdot \cancel{x} \cdot \cancel{x} \cdot \cancel{x}} = x^2$$

When exponents are large, this process can be time consuming. Let's examine some divisions and look for a pattern to discover a division rule. Notice in the previous division that there are 6 factors in the numerator and 4 factors in the denominator. After we simplify we have $6 - 4 = 2$ factors left in the numerator. Thus we can write

$$\begin{array}{l} 6 \text{ factors} \rightarrow \\ 4 \text{ factors} \rightarrow \end{array} \frac{x^6}{x^4} = \frac{x^{6-4}}{1} = x^2 \qquad \text{2 factors left in the numerator.}$$

Let's consider another divison problem.

$$\frac{2^3}{2^4} = \frac{2 \cdot 2 \cdot 2}{2 \cdot 2 \cdot 2 \cdot 2} = \frac{\cancel{2} \cdot \cancel{2} \cdot \cancel{2}}{\cancel{2} \cdot \cancel{2} \cdot \cancel{2} \cdot 2} = \frac{1}{2}$$

$$\begin{array}{l} 3 \text{ factors} \rightarrow \\ 4 \text{ factors} \rightarrow \end{array} \frac{2^3}{2^4} = \frac{1}{2^{4-3}} = \frac{1}{2^1} \quad \text{or} \quad \frac{1}{2} \qquad \text{1 factor left in the denominator.}$$

We see that we *subtract exponents* to divide these expressions.

DIVIDING EXPRESSIONS IN EXPONENT FORM

If the bases in the numerator and denominator of the fractional expression are the same, then

$$\frac{x^a}{x^b} = x^{a-b} \text{ if the } \textit{larger exponent} \text{ is in the } \textit{numerator} \text{ and } x \neq 0.$$

$$\frac{x^a}{x^b} = \frac{1}{x^{b-a}} \text{ if the } \textit{larger exponent} \text{ is in the } \textit{denominator} \text{ and } x \neq 0.$$

Since division by zero is undefined, in all problems in this book we assume that the denominator of any variable expression is not zero.

E X A M P L E   1   Simplify. Leave your answer in exponent form.

(a) $\dfrac{n^9}{n^6}$ **(b)** $\dfrac{5^8}{5^9}$ **(c)** $\dfrac{2^7}{3^4}$

SOLUTION

(a) $\dfrac{n^9}{n^6} = n^{9-6}$ There are more factors in the numerator.
The leftover factors are in the numerator.

$$\dfrac{n^9}{n^6} = \boldsymbol{n^3} \longleftarrow$$

(b) $\dfrac{5^8}{5^9} = \dfrac{1}{5^{9-8}}$ There are more factors in the denominator.
    The leftover factors are in the denominator.

$\dfrac{5^8}{5^9} = \dfrac{1}{5^1}$ or $\dfrac{1}{5}$ ⟵

(c) $\dfrac{2^7}{3^4}$ We cannot divide using the rule for exponents. The bases are not the same.

PRACTICE PROBLEM 1

Simplify. Leave your answer in exponent form.

(a) $\dfrac{4^{11}}{4^7}$ **(b)** $\dfrac{6^9}{8^{14}}$ **(c)** $\dfrac{y^5}{y^9}$

Remember, if you forget the rule, you can always check your answer by writing the problem as repeated multiplication and simplifying.

EXAMPLE 2   Simplify: $\dfrac{16x^6}{20x^8}$.

SOLUTION

$$\dfrac{16x^6}{20x^8} = \dfrac{\cancel{4} \cdot 2 \cdot 2 \cdot x^6}{\cancel{4} \cdot 5 \cdot x^8}$$

$$= \dfrac{4}{5 \cdot x^{8-6}}$$

$$= \dfrac{4}{5x^2}$$ The leftover x factors are in the denominator.

PRACTICE PROBLEM 2   Simplify: $\dfrac{25y^5}{45y^8}$.

TO THINK ABOUT
....

What can we do if we forget algebraic rules? Do I add, subtract, or multiply exponents? If we start with a simple problem, then *think* about what it is we are actually trying to do, we can often determine the rules by observing our calculations.

$$6x^5 + 3x^5 = ?$$

We have six x^5's and add three more x^5's; we have nine x^5's.

$$6x^5 + 3x^5 = 9x^5$$

We add coefficients; the variable stays the same.

$$\left(6x^2\right)\left(3x^4\right) = ?$$

$$6 \cdot x \cdot x \cdot 3 \cdot x \cdot x \cdot x \cdot x = 6 \cdot 3 \cdot x \cdot x \cdot x \cdot x \cdot x \cdot x$$
$$= 18 \cdot x^6$$
$$\left(6x^2\right)\left(3x^4\right) = 18x^6$$

We multiply coefficients, then add exponents of like bases.

$$\dfrac{6x^2}{3x} = ? \qquad \dfrac{6x^2}{3x} = \dfrac{2 \cdot 3 \cdot x \cdot x}{3 \cdot x} = 2x$$

We simplify coefficients, then subtract exponents of like bases.

② Raising a Power to a Power

If we have $\left(x^3\right)^4$, we say that we are *raising a power to a power*. A problem such as $\left(x^3\right)^4$ could be done by first writing $\left(x^3\right)^4$ as a product, then simplifying.

$$\left(x^3\right)^4 = x^3 \cdot x^3 \cdot x^3 \cdot x^3 \qquad \text{By definition of raising a value to the fourth power.}$$

$$\left(x^3\right)^4 = x^{12} \qquad \text{Add exponents } 3 + 3 + 3 + 3 \text{ to simplify.}$$

EXAMPLE 3   Write $\left(2^3\right)^3$ as a product, then simplify. Leave your answer in exponent form.

SOLUTION

$$\left(2^3\right)^3 = 2^3 \cdot 2^3 \cdot 2^3 = 2^{3+3+3} = \mathbf{2^9}$$

PRACTICE PROBLEM 3

Write $\left(4^2\right)^3$ as a product, then simplify. Leave your answer in exponent form.

Since repeated addition can be written as multiplication, we can state this process by the following rule:

RAISING A POWER TO A POWER
To raise a power to a power, keep the same base and multiply the exponents.

$$\left(x^a\right)^b = x^{ab}$$

EXAMPLE 4   Use the rule for raising a power to a power and leave your answer in exponent form.

(a) $\left(3^3\right)^3$ 　　　　　　　　　　　　**(b)** $\left(x^2\right)^0$

SOLUTION

(a) $\left(3^3\right)^3 = 3^{(3)(3)} = \mathbf{3^9}$
The base does not change when raising a power to a power.

(b) $\left(x^2\right)^0 = x^{(2)(0)} = x^{(0)} = \mathbf{1}$

Recall from Chapter 2 that an expression raised to the zero power is equal to 1. As you can see, this is also true when raising a power to a zero power.

PRACTICE PROBLEM 4

Use the rule for raising a power to a power and leave your answer in exponent form.

(a) $\left(3^3\right)^4$ 　　　　　　　　　　　　**(b)** $\left(n^0\right)^7$

Now we introduce two similar rules involving products and quotients that are very useful. We'll illustrate each with an example.

ADDITIONAL POWER RULES

$$(xy)^a = x^a y^a$$

If the product in parentheses is raised to a power, the parentheses indicate that *each factor* within the parentheses must be raised to that power.

$$\left(\frac{x}{y}\right)^a = \frac{x^a}{y^a} \qquad \text{if } y \neq 0$$

If a fraction is raised to a power, the parentheses indicate that the numerator and denominator are *each* raised to that power.

EXAMPLE 5   Simplify.

(a) $\left(5y^4\right)^2$ **(b)** $(ab)^5$

SOLUTION

(a) $\left(5y^4\right)^2 = (5)^2 \cdot \left(y^4\right)^2$ We must *raise both* the 5 and the y^4 to the power 2.

$\qquad\qquad = 25 \cdot y^{4 \cdot 2}$ We *multiply* exponents.

$\qquad\qquad = \mathbf{25y^8}$

(b) $(ab)^5 = a^5 \cdot b^5 = \boldsymbol{a^5 b^5}$ Raise *each factor* to the power 5.

PRACTICE PROBLEM 5

Simplify.

(a) $(xy)^5$ **(b)** $\left(4a^3\right)^4$

EXAMPLE 6   Simplify: $\left(\dfrac{2}{x}\right)^3$.

SOLUTION   We must remember to raise both the numerator and the denominator to the power.

$$\left(\frac{2}{x}\right)^3 = \frac{2^3}{x^3} = \frac{\mathbf{8}}{\mathbf{x^3}}$$

PRACTICE PROBLEM 6

Simplify: $\left(\dfrac{x}{3}\right)^3$.

DEVELOPING YOUR STUDY SKILLS

KEEP TRYING

Do you wish you could improve your math grade? Are you frustrated and starting to become discouraged? Don't give up! Take note of the following suggestions, they will help make a difference.

- *Be patient.* Would you expect to learn how to play a piano easily, without a lot of effort? Of course, those who have had experience with various instruments earlier in life might learn easier and faster than someone who has not. Developing the skills to do math is like learning to play a musical instrument. *Learning mathematics is a process that takes time and effort.* Those who have had more experience working with mathematics may learn faster and easier, but for many students this is catch-up time.
- *Increase your study time.* It is not unusual to study mathematics for 8 to 12 hours a week. Perfecting math skills requires the same intensity as preparing to play a sport. Baseball players practice many hours each day to perfect their swing or curveball. *Increasing your study time will help you improve your understanding of math.* Yes, it can be slow moving at first, but eventually, as your skills develop, you will find that math will become easier and you will understand concepts faster.
- *Seek help.* Make an appointment with your instructor for help, or use any tutorial services that are available. Too often, students *give up* and skip topics they don't understand. Missing a few topics can make it difficult to understand new topics because math builds on previous topics and skills.
- *Be positive.* Don't let past frustrations stand in your way. Start fresh with a positive attitude, then work hard and practice the study skills in this book. Then as you become more successful, your confidence in your mathematical ability will grow.

EXERCISE

1. Reexamine your time management schedule and insert a few more hours each week to study math. If you have not been using a time management schedule, refer to Developing Your Study Skills in Section 2.2 for assistance.

EXERCISES 5.6

Simplify. In this exercise set assume that all variables in any denominator are nonzero. Leave your answers in exponent form.

1. $\dfrac{x^3}{x^8}$

2. $\dfrac{z^2}{z^4}$

3. $\dfrac{7^4}{7^3}$

4. $\dfrac{9^9}{9^8}$

5. $\dfrac{2^3}{2^7}$

6. $\dfrac{3^5}{3^9}$

7. $\dfrac{x^7}{x^4}$

8. $\dfrac{a^8}{a^3}$

9. $\dfrac{9^3}{8^8}$

10. $\dfrac{6^6}{7^8}$

11. $\dfrac{z^8}{y^4}$

12. $\dfrac{z^6}{x^8}$

13. $\dfrac{3^2}{3^6}$

14. $\dfrac{8^4}{8^3}$

15. $\dfrac{z^8}{z^3}$

16. $\dfrac{a^5}{a^2}$

17. $\dfrac{y^3 z^4}{y^5 z^7}$

18. $\dfrac{a^3 b^5}{a^7 b^7}$

19. $\dfrac{m^9 3^6}{m^7 3^7}$

20. $\dfrac{5^6 r^3}{5^2 r^6}$

21. $\dfrac{a^5 7^4}{a^3 7^7}$

22. $\dfrac{p^3 z^9}{p^9 z^2}$

23. $\dfrac{b^9 9^9}{b^7 9^{11}}$

24. $\dfrac{7^7 r^2}{7^2 r^6}$

Simplify.

25. $\dfrac{25y^3}{35y}$

26. $\dfrac{24x^5}{36x}$

27. $\dfrac{9a^4}{27a^3}$

28. $\dfrac{7m^5}{21m^4}$

29. $\dfrac{56x^9}{64x^3}$

30. $\dfrac{32y^7}{48y^5}$

Multiply and write in exponent form.

31. (a) $z^2 \cdot z^2 \cdot z^2$ **(b)** $\left(z^3\right)^3$

32. (a) $a^3 \cdot a^3 \cdot a^3$ **(b)** $\left(a^3\right)^3$

33. (a) $x^2 \cdot x^2$ **(b)** $\left(x^2\right)^2$

34. (a) $y^6 \cdot y^6$ **(b)** $\left(y^6\right)^2$

35. (a) $b^4 \cdot b^4 \cdot b^4$ **(b)** $\left(b^4\right)^3$ **36. (a)** $x^5 \cdot x^5 \cdot x^5$ **(b)** $\left(x^5\right)^3$

Simplify. Leave your answer in exponent form.

37. $\left(z^6\right)^4$ **38.** $\left(x^7\right)^4$ **39.** $\left(3^3\right)^2$ **40.** $\left(2^3\right)^2$

41. $\left(b^1\right)^6$ **42.** $\left(x^3\right)^1$ **43.** $\left(x^0\right)^4$ **44.** $\left(5^5\right)^0$

45. $\left(y^2\right)^3$ **46.** $\left(6^2\right)^3$ **47.** $\left(2^4\right)^5$ **48.** $\left(x^3\right)^9$

49. $\left(x^2\right)^0$ **50.** $\left(y^0\right)^4$ **51.** $\left(7^3\right)^9$ **52.** $\left(8^2\right)^3$

53. $(xy)^2$ **54.** $(ab)^3$ **55.** $(5y)^6$ **56.** $(2x)^4$

57. $\left(3x^2\right)^8$ **58.** $\left(4y^3\right)^5$ **59.** $\left(6x^7\right)^9$ **60.** $\left(5x^4\right)^7$

Simplify.

61. $\left(\dfrac{3}{x}\right)^3$ **62.** $\left(\dfrac{4}{y}\right)^2$ **63.** $\left(\dfrac{a}{b}\right)^4$ **64.** $\left(\dfrac{x}{y}\right)^5$

65. $\left(\dfrac{x}{6}\right)^2$ **66.** $\left(\dfrac{y}{5}\right)^3$ **67.** $\left(\dfrac{3}{4}\right)^2$ **68.** $\left(\dfrac{1}{2}\right)^3$

Simplify. Leave your answer in exponent form.

69. (a) $15x^3 + 5x^3$ **(b)** $(15x^3)(5x^3)$ **(c)** $\left(x^3\right)^3$ **(d)** $\dfrac{15x^3}{5x^5}$

70. (a) $24x^5 + 6x^5$ **(b)** $(24x^5)(6x^5)$ **(c)** $\left(x^5\right)^5$ **(d)** $\dfrac{24x^5}{6x^3}$

71. (a) $3x^3 + 9x^3$ **(b)** $(3x^3)(9x^4)$ **(c)** $\left(3x^2\right)^4$ **(d)** $\dfrac{3x^3}{9x^4}$

72. (a) $7x^6 + 14x^6$ **(b)** $(7x^6)(14x)$ **(c)** $\left(7x^6\right)^2$ **(d)** $\dfrac{7x^6}{14x^2}$

✎ **VERBAL AND WRITING SKILLS**

State the rule for simplifying each of the following.

73. (a) $15x^3 + 5x^3$

(b) $(15x^3)(5x^3)$

(c) $\dfrac{15x^3}{5x}$

74. (a) $14x^2 + 6x^2$

(b) $(14x^2)(6x^2)$

(c) $\dfrac{14x^2}{6x}$

ONE STEP FURTHER

Simplify. Assume that all variables in any denominator are nonzero.

75. $\dfrac{25x^2 y^3 z^4}{135x^7 y}$

76. $\dfrac{40a^9 b^2 c^4}{88a^3 b}$

77. $\dfrac{156a^0 b^8}{144b^6 c^9}$

78. $\dfrac{256x^0 y^{15}}{300y^9 z^8}$

CUMULATIVE REVIEW

Translate each phrase using symbols.

1. Seven times a number.

2. Seven plus a number.

3. A number decreased by 7.

S E C T I O N   5 . 7

Applied Problems Involving Fractions

After studying this section, you will be able to:

❶ Solve applied problems involving fractions.

Math Pro   Video 5.7   SSM

❶ Solving Applied Problems Involving Fractions

E X A M P L E 1   Marian planted a rectangular rose garden in the center of her 26-foot by 20-foot backyard. There is a sidewalk around the garden that is $3\frac{1}{2}$ feet wide. The garden and sidewalk take up the entire 20-foot by 26-foot yard.

(a) What are the dimensions of the rose garden?
(b) How much will it cost to put a fence around the rose garden if the fencing costs $\$2\frac{1}{2}$ per linear foot?

SOLUTION

   1. *Understand the problem.* We draw a picture.

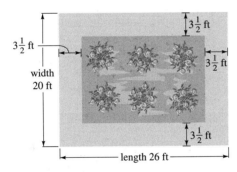

MATHEMATICS BLUEPRINT FOR PROBLEM SOLVING			
GATHER THE FACTS	WHAT AM I ASKED TO DO?	HOW DO I PROCEED?	KEY POINTS TO REMEMBER
Backyard: 26 ft by 20 ft *Garden:* in the center of yard. *Sidewalk:* $3\frac{1}{2}$ ft wide and surrounds garden *Fencing:* costs $\$2\frac{1}{2}$ per foot	(a) Find the dimensions of the rose garden. (b) Calculate the cost of a fence around the rose garden.	*Length of garden:* Find the sum of the widths of the sidewalk. Subtract this amount from the *length* of yard. *Width of garden:* Find the sum of the widths of the sidewalk. Subtract this from the *width* of the yard. Find the perimeter of garden: $P = 2L + 2W$. Multiply the perimeter by $\$2\frac{1}{2}$ to find the cost of the fence.	For the *length* of the garden we find the sum of widths of the sidewalk on the left and right sides of the garden, and for the *width* of the garden the sum above and below the garden.

2. *State and solve the answer.* First, we find the length of the rose garden:

| length of garden | = | length of entire yard | − | width of sidewalk (left side + right side) |

$$L = 26 \text{ ft} - \left(3\frac{1}{2} + 3\frac{1}{2}\right)$$

$$= 26 \text{ ft} - 7 \text{ ft} = \textbf{19 ft}$$

Next, we find the width of the rose garden:

| width of garden | = | width of entire yard | − | width of sidewalk (above + below garden) |

$$W = 20 \text{ ft} - \left(3\frac{1}{2} + 3\frac{1}{2}\right)$$

$$= 20 \text{ ft} - 7 \text{ ft} = \textbf{13 ft}$$

(a) The dimensions of the garden are **19 feet by 13 feet**.
Now we find the perimeter of the garden.

$$P = 2L + 2W = 2(19 \text{ ft}) + 2(13 \text{ ft})$$

$$= 38 \text{ ft} + 26 \text{ ft} = \textbf{64 ft} \quad \textbf{perimeter = 64 ft}$$

We multiply $\$2\frac{1}{2}$ times 64 feet to find the total cost of fencing:

$$2\frac{1}{2} \cdot 64 = \frac{5}{2} \cdot \frac{64}{1} = \frac{5 \cdot 2 \cdot 32}{2 \cdot 1} = 160$$

(b) It will cost **$160** to put a fence around the rose garden.

3. *Check.* We can place the answers on the diagram to check.

Width: $3\frac{1}{2}$ ft $+$ 13 ft $+3\frac{1}{2}$ ft $=$ 20 ft ✓

Length: $3\frac{1}{2}$ ft $+$ 19 ft $+3\frac{1}{2}$ ft $=$ 26 ft ✓

Round the price, $\$2\frac{1}{2} \rightarrow \$3$; round the perimeter, $64 \rightarrow 60$; then multiply to estimate the cost of the fence: $\$3 \times 60 = \$180$, which is close to our answer. ✓

PRACTICE PROBLEM 1

Larry wants to place an Oriental rug in the center of his living room floor, which measures 22 feet by 15 feet. He wants to center the rug so that there is $2\frac{1}{2}$ feet of wood flooring showing on each side of the rug.

(a) What are the dimensions of the rug Larry must buy?
(b) If the brand of rug Larry wants sells for $5 per square foot, how much will the rug cost?

For some applied problems we must work the first part of the problem to determine what operations are needed to answer the question. For applied problems involving fractions, sometimes we must also round fractions to the nearest whole number to help determine what operations to do. For these types of problems the Mathematics Blueprint for Problems Solving may not be appropriate.

EXAMPLE 2   Jason is planning to build a fence on his farm. He determines that he must make 115 wooden fence posts that are each $3\frac{3}{4}$ feet in length. The wood to make the fence posts is sold in 20-foot lengths. How many 20-foot pieces of wood must Jason purchase so that he can make 115 fence posts?

SOLUTION

1. *Understand the problem.* We must first determine how many posts can be cut from one 20-foot piece of wood. Then we can find how many of these 20-foot pieces of wood are needed to make 115 posts. We draw a picture.

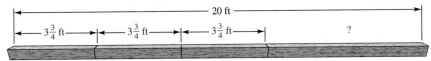

We round the $3\frac{3}{4}$ to 4 to help determine which operation to use. We divide to find how many 4 feet are in 20 feet, so we must also divide to find how many $3\frac{3}{4}$ are in 20 feet.

2. *Solve and state the answer.*

$$20 \div 3\frac{3}{4} = 20 \div \frac{15}{4} = 20 \cdot \frac{4}{15} = \frac{4 \cdot \cancel{5} \cdot 4}{\cancel{5} \cdot 3} = \frac{16}{3} \quad \text{or} \quad 5\frac{1}{3}$$

Five posts can be cut from each 20-foot piece of wood, with some wood left over.

Now we must find how many of the 20-foot pieces are needed.

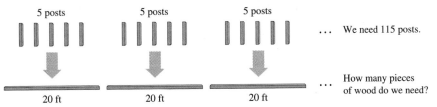

We must find how many groups of 5 are in 115. We divide $115 \div 5$:

$$115 \div 5 = \frac{115}{5} = \frac{\cancel{5} \cdot 23}{\cancel{5}} = 23 \quad \text{Jason must purchase } \mathbf{23} \text{ pieces of wood.}$$

3. *Check.* We can estimate our answer by rounding the fraction $3\frac{3}{4}$ to 4 and reworking the problem.

20-foot board ÷ 4 feet per board = 5 boards per piece of wood
115 posts ÷ 5 = 23 ✓

PRACTICE PROBLEM 2

Nancy wishes to make two bookcases, each with 4 shelves. Each shelf is $3\frac{1}{8}$ feet long. The wood for the shelves is sold in 10-foot boards. How many boards does Nancy need to buy for the shelves?

EXERCISES 5.7

Use the Mathematics Blueprint for Problem Solving to help you organize your work.

1. Joan is trying to lose weight. The first week she lost $2\frac{3}{4}$ pounds, the second week she lost $2\frac{1}{4}$ pounds, the third week she lost $3\frac{1}{8}$ pounds, and the fourth week she lost $1\frac{7}{8}$ pounds. How much did she lose in the four weeks of her diet?

2. In assembling her bookshelves, Keri finds that she needs a bolt that will reach through the $\frac{5}{8}$-inch wood, a $\frac{1}{8}$-inch-thick washer, and a $\frac{1}{4}$-inch-thick nut. How long must the bolt be?

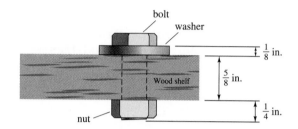

3. John had $15\frac{2}{3}$ gallons of gasoline in his car before traveling to Los Angeles. When he arrived, he only had $9\frac{1}{2}$ gallons left.

(a) How much gasoline did he use to travel to Los Angeles?

(b) If his car get 24 miles per gallon, how many miles did John travel to Los Angeles?

4. Each tile that Amy wants covers $3\frac{1}{2}$ square inches of space. She wants to cover 245 square inches in her bathroom with the tile. How many tiles will she need to purchase?

5. Jeff traveled the Interstate Highway for 310 miles in $5\frac{1}{3}$ hours. What was his average speed?

6. A tank that holds $52\frac{4}{7}$ gallons is used to fill 11 containers of equal size. How much does each container hold?

Use the following recipe to answer questions 7 and 8.

7. To make 6 servings, how much of each ingredient would you need?

8. To make 12 servings, how much of each ingredient would you need?

Cereal Preparation Directions

INGREDIENT	SERVINGS	
	1	2
Water	$1\frac{1}{4}$ cups	$2\frac{1}{2}$ cups
Salt	$\frac{1}{8}$ tsp.	$\frac{1}{4}$ tsp.
Cereal	$\frac{1}{4}$ cup	$\frac{1}{2}$ cup

Use the following chart to answer questions 9 and 10.

9. To make 5 long skirts in size 8, how much 45-inch-wide material would you need?

10. To make a long skirt and a bodice in a size 10, how much material would it take if the material were 60 inches wide?

Pattern Directions

SKIRT PATTERN						
Sizes	4	6	8	10	12	14
Sizes - European	30	32	34	36	38	40
Long Skirt 45"	$2\frac{3}{4}$	$2\frac{3}{4}$	$2\frac{3}{4}$	$2\frac{3}{4}$	$2\frac{3}{4}$	$2\frac{3}{4}$ Yd
60"	$2\frac{1}{4}$	$2\frac{1}{4}$	$2\frac{3}{8}$	$2\frac{1}{2}$	$2\frac{1}{2}$	$2\frac{3}{4}$ Yd
Short Skirt 45"	$1\frac{3}{4}$	$1\frac{3}{4}$	$1\frac{3}{4}$	$1\frac{3}{4}$	$1\frac{3}{4}$	$1\frac{3}{4}$ Yd
60"	1	1	1	1	1	$1\frac{1}{4}$ Yd
Bodice A or B 45"	$\frac{1}{2}$	$\frac{1}{2}$	$\frac{1}{2}$	$\frac{5}{8}$	$\frac{5}{8}$	$\frac{3}{4}$ Yd
60"	$\frac{1}{2}$	$\frac{1}{2}$	$\frac{1}{2}$	$\frac{1}{2}$	$\frac{1}{2}$	$\frac{1}{2}$ Yd

11. In January 1998 an unmanned Athena rocket blasted off to the moon carrying the Lunar Prospector probe. The Prospector looked for water that could one day be used by human settlers. The 4-foot, 650-pound Prospector, took $4\frac{1}{2}$ days to travel 240,000 miles.

    (a) How many miles per day did the spacecraft travel?

    (b) How many miles per hour did the spacecraft travel?

12. The night of the *Titanic* cruise ship disaster, the captain decided to run his ship at $22\frac{1}{2}$ knots (nautical miles per hour). The *Titanic* traveled at that speed for $4\frac{3}{4}$ hours before it met its tragic demise. How far did the *Titanic* travel at this excessive speed before the disaster?

13. Monica planted a rectangular flower bed in the center of her 40-foot by 25-foot front lawn. There is a $4\frac{3}{4}$-foot wide grass area around the entire flower bed. The grass and the flower bed take up the entire 40-foot by 25-foot area.

    (a) What are the dimensions of the flower bed?

    (b) How much will it cost to put a fence around the flower bed if the fencing costs $2\frac{1}{4}$ per linear foot?

14. Howard put a rectangular pool in his yard, which is $45\frac{1}{2}$ feet by 20 feet. There is a grass area 5 feet wide around the entire perimeter of the pool. The pool and grass take up the entire $45\frac{1}{2}$-foot by 20-foot yard.

    (a) What are the dimensions of the pool?

    (b) How much will it cost to put tile around the edge of his pool if the tile costs $3\frac{1}{2}$ per linear foot?

15. Brenda wishes to build 3 bookcases, each with 4 shelves. Each shelf is $3\frac{3}{4}$ feet long. The wood for the shelves is sold in 8-foot boards. How many boards does Brenda need to buy for the shelves?

16. Julie wishes to build 2 bookcases, each with 5 shelves. Each shelf is $2\frac{3}{4}$ feet long. The wood for the shelves is sold in 6-foot boards. How many boards does Julie need to buy for the shelves?

Use the following bar graph to answer questions 17–20.

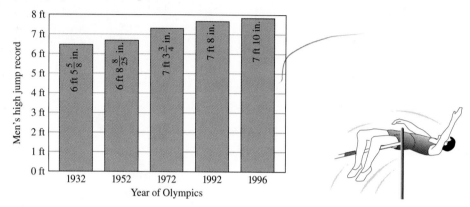

17. How much higher was the Olympic record for the men's high jump in 1996 than in 1992?

18. How much higher was the Olympic record for the men's high jump in 1992 than in 1972?

19. How much higher was the Olympic record for the men's high jump in 1972 than in 1932?

20. How much higher was the Olympic record for the men's high jump in 1992 than in 1952?

CUMULATIVE REVIEW

Write the equivalent equation and solve.

1. $3x = 12$

2. $5x = 45$

3. $\dfrac{x}{7} = 6$

4. $\dfrac{x}{8} = 9$

INVESTING IN THE STOCK MARKET

One way of investing money is through the stock market. Stocks are purchased at the selling price on the day of purchase. Each day it is recorded how much the price goes up or down, + being up and − being down. Therefore, on any given day one can figure out the value of a share of stock. To make the most money, investors hope to buy stocks when prices are down and sell their stocks when prices are high. A person's stock portfolio is a collection of all the stock that he or she owns. The value of the portfolio is the total amount of money that the stock is worth on any given day.

The following table shows the beginning price of one share for three stocks and the daily changes for 6 days.

STOCK	PURCHASE PRICE	CHANGE					
		DAY 1	DAY 2	DAY 3	DAY 4	DAY 5	DAY 6
Newtel Communications	$34\frac{1}{8}$	$-\frac{5}{16}$	$+\frac{1}{4}$	$-\frac{5}{16}$	$-\frac{1}{4}$	$+\frac{5}{8}$	$+\frac{1}{16}$
STR Foods	$21\frac{11}{16}$	$+\frac{1}{8}$	$-\frac{5}{16}$	$-\frac{1}{4}$	$+\frac{7}{8}$	$+\frac{1}{8}$	$-\frac{1}{4}$
Biolab Corp.	$40\frac{1}{2}$	$-\frac{1}{8}$	$-\frac{1}{8}$	$-\frac{5}{16}$	$+\frac{5}{8}$	$-\frac{1}{8}$	$+1\frac{7}{16}$

PROBLEMS FOR INDIVIDUAL INVESTIGATION

Suppose that you purchase 100 shares of Newtel.

1. What is the value of your 100 shares of stock on day 3? On day 6?

2. What is the highest value of your stock in this 6-day period?

3. What is the lowest value of your stock in this 6-day period?

4. If you sold 50 shares on day 6, what would be your profit or loss?

5. On day 4 you buy another 100 shares of Newtel. Was this a good decision? That is, were the shares bought at the lowest price during the week?

6. On day 6, what is the value of your portfolio (original purchase of 100 shares plus the 100 additional shares purchased on day 4, minus the 50 shares sold on day 6.)?

PROBLEMS FOR GROUP INVESTIGATION AND COOPERATIVE STUDY

Suppose that you purchase 100 shares each of Newtel, STR, and Biolab.

1. What is the value of your portfolio on day 6?

2. What was your profit or loss on all your stock purchases at the end of day 6?

3. On day 4 you purchase 100 additional shares of one stock. Which one do you think would be the best choice, and why?

4. You are going to buy 1000 shares of stock on day 6, and you must buy at least 200 shares of each stock. How much of each would you buy? Why?

5. What is the value of the portfolio you purchased in question 4 on day 6?

INTERNET: Go to http://www.prenhall.com/blair to explore this application.

CHAPTER ORGANIZER

TOPIC	PROCEDURE	EXAMPLES
Multiplying fractions	1. Simplify by removing factors of 1 whenever possible. 2. Multiply numerators. 3. Multiply denominators.	Multiply: $-\dfrac{15}{49} \cdot \dfrac{7}{12}$. $-\dfrac{15}{49} \cdot \dfrac{7}{12} = -\dfrac{\cancel{3} \cdot 5}{7 \cdot \cancel{7}} \cdot \dfrac{\cancel{7}}{\cancel{3} \cdot 4} = -\dfrac{5}{28}$
Dividing fractions	To divide two fractions, we invert the second fraction and multiply.	Divide: $12x^2 \div \dfrac{8x}{5}$. $12x^2 \div \dfrac{8x}{5} = 12x^2 \cdot \dfrac{5}{8x}$ $\qquad = \dfrac{3 \cdot \cancel{4} \cdot \cancel{x} \cdot x}{1} \cdot \dfrac{5}{\cancel{4} \cdot 2 \cdot \cancel{x}}$ $\qquad = \dfrac{15x}{2}$
Finding the least common multiple (LCM) and the least common denominator (LCD)	The LCM and LCD are two different names for the same number. 1. Write each denominator as the product of prime factors. 2. List the requirements for the factorization of LCD. 3. Build a LCD that has all the factors of each denominator, using a minimum number of factors.	Find the LCD: $\dfrac{2}{9}, \dfrac{4}{15}, \dfrac{1}{10}$. $9 = 3 \cdot 3 \quad 15 = 3 \cdot 5 \quad 10 = 2 \cdot 5$ $LCD = 2 \cdot 3 \cdot 3 \cdot 5 = 90$
Adding or subtracting fractions with a common denominator	1. Add or subtract the numerators. 2. Keep the common denominator.	Add: $\dfrac{2}{11} + \dfrac{3}{11}$. $\dfrac{2}{11} + \dfrac{3}{11} = \dfrac{5}{11}$ Subtract: $\dfrac{5}{x} - \dfrac{2}{x}$. $\dfrac{5}{x} - \dfrac{2}{x} = \dfrac{3}{x}$
Adding or subtracting fractions without a common denominator	1. Find the LCD of the fractions. 2. Build up each fraction, if needed, to obtain the LCD in the denominator. 3. Follow the steps for adding and subtracting fractions with a common denominator.	Add: $\dfrac{2}{5} + \dfrac{4}{7} + \dfrac{3}{10}$. $LCD = 70$ $\dfrac{2 \cdot 14}{5 \cdot 14} + \dfrac{4 \cdot 10}{7 \cdot 10} + \dfrac{3 \cdot 7}{10 \cdot 7}$ $= \dfrac{28}{70} + \dfrac{40}{70} + \dfrac{21}{70} = \dfrac{89}{70} = 1\dfrac{19}{70}$

TOPIC	PROCEDURE	EXAMPLES
Adding mixed numbers	1. Change fractional parts to equivalent fractions with LCD as a denominator, if needed. 2. Add whole numbers and fractions separately. 3. If improper fractions occur, change to mixed numbers and simplify.	Add: $7\frac{1}{2} + 2\frac{5}{6}$. $7\frac{1 \cdot 3}{2 \cdot 3} = 7\frac{3}{6}$ $+2\frac{5}{6} = 2\frac{5}{6}$ $9\frac{8}{6} = 9\frac{4}{3} = 10\frac{1}{3}$
Subtracting mixed numbers	1. Change fractional parts to equivalent fractions with LCD as a denominator, if needed. 2. If necessary, borrow from the whole number to subtract fractions. 3. Subtract whole numbers and fractions separately.	Subtract: $9\frac{1}{5} - 4\frac{3}{4}$. $9\frac{1 \cdot 4}{5 \cdot 4} = 9\frac{4}{20} = 8\frac{24}{20}$ $-4\frac{3 \cdot 5}{4 \cdot 5} = 4\frac{15}{20} = -4\frac{15}{20}$ $4\frac{9}{20}$
Multiplying mixed and/or whole numbers	1. Change any whole number to a fraction with a denominator of 1. 2. Change any mixed numbers to improper fractions. 3. Use the multiplication rule for fractions.	Multiply: $4 \cdot 2\frac{1}{3}$. $4 \cdot 2\frac{1}{3} = \frac{4}{1} \cdot \frac{7}{3} = \frac{28}{3} = 9\frac{1}{3}$
Dividing mixed and/or whole numbers	1. Change any whole number to a fraction with a denominator of 1. 2. Change any mixed numbers to improper fractions. 3. Use the rule for division of fractions.	Divide: $8\frac{1}{3} \div 5\frac{5}{9}$. $8\frac{1}{3} \div 5\frac{5}{9} = \frac{25}{3} \div \frac{50}{9} = \frac{25}{3} \cdot \frac{9}{50}$ $= \frac{\cancel{5} \cdot \cancel{5}}{\cancel{3}} \cdot \frac{\cancel{3} \cdot 3}{2 \cdot \cancel{5} \cdot \cancel{5}}$ $= \frac{3}{2}$ or $1\frac{1}{2}$
Order of operations with fractions	To simplify fractions, we simplify above, then below the fraction bar. We perform operations in the following order: 1. Perform operations inside parentheses. 2. Simplify exponents. 3. Multiplication and division working left to right. 4. Addition and subtraction working left to right.	Simplify: $\frac{5}{6} + \frac{1}{3} \cdot \left(\frac{3}{4} - \frac{1}{2}\right)$. $\frac{5}{6} + \frac{1}{3} \cdot \left(\frac{3}{4} - \frac{1}{2}\right) = \frac{5}{6} + \frac{1}{3} \cdot \frac{1}{4}$ $= \frac{5}{6} + \frac{1}{12}$ $= \frac{10}{12} + \frac{1}{12} = \frac{11}{12}$

TOPIC	PROCEDURE	EXAMPLES
Simplifying complex fractions	To simplify a complex fraction: 1. Perform all operations in the numerator. 2. Perform all operations in the denominator. 3. Divide the top fraction by the bottom fraction. This is done by inverting the bottom fraction and multiplying it by the top fraction.	Simplify: $\dfrac{\dfrac{3}{4}+\dfrac{1}{2}}{\dfrac{1}{3}-\dfrac{1}{6}}$. $\dfrac{\dfrac{3}{4}+\dfrac{1}{2}}{\dfrac{1}{3}-\dfrac{1}{6}}=\dfrac{\dfrac{3}{4}+\dfrac{2}{4}}{\dfrac{2}{6}-\dfrac{1}{6}}=\dfrac{\dfrac{5}{4}}{\dfrac{1}{6}}$ $=\dfrac{5}{4}\div\dfrac{1}{6}=\dfrac{5}{4}\cdot\dfrac{6}{1}$ $=\dfrac{15}{2}$ or $7\dfrac{1}{2}$
Dividing expressions in exponent form	If the bases are the same and $x \neq 0$, then $\dfrac{x^a}{x^b}=\begin{cases} x^{a-b} & \text{if the larger exponent is in the numerator.} \\ \dfrac{1}{x^{b-a}} & \text{if the larger exponent is in the denominator.} \end{cases}$	Simplify: (a) $\dfrac{x^5}{x^3}$ (b) $\dfrac{3^2}{3^4}$ (a) $\dfrac{x^5}{x^3}=x^2$ (b) $\dfrac{3^2}{3^4}=\dfrac{1}{3^2}=\dfrac{1}{9}$
Raising a power to a power	We keep the same base and multiply exponents. $\left(x^a\right)^b = x^{ab}$ The parentheses indicate that *each factor* within the parentheses must be raised to a power. $(xy)^a = x^a y^a$ $\left(\dfrac{x}{y}\right)^a = \dfrac{x^a}{y^a}$ if $y \neq 0$	Simplify: (a) $\left(x^4\right)^6$ (b) $(4x)^2$ (c) $\left(\dfrac{x}{2}\right)^3$ (a) $\left(x^4\right)^6 = x^{24}$ (b) $(4x)^2 = 16x^2$ (c) $\left(\dfrac{x}{2}\right)^3 = \dfrac{x^3}{8}$

CHAPTER 5 REVIEW

In this exercise set assume that all variables in any denominator are nonzero.

SECTION 5.1

Find the reciprocal.

1. $\dfrac{2}{-9}$

2. 7

3. $\dfrac{-a}{b}$

Multiply or divide.

4. $\dfrac{5}{21} \cdot \dfrac{3}{15}$

5. $\dfrac{-6}{35} \cdot \dfrac{14}{18}$

6. $\dfrac{9x}{15} \cdot \dfrac{21}{18x^2}$

7. $\dfrac{8x^2}{25} \cdot \dfrac{-45}{18x}$

8. $\dfrac{-5x^4}{6} \cdot 12x^3$

9. $\dfrac{9}{14} \div \dfrac{45}{12}$

10. $\dfrac{7}{15} \div \dfrac{-35}{20}$

11. $\dfrac{8}{42} \div \dfrac{-22}{7}$

12. $\dfrac{11x^5}{25} \div \dfrac{3}{5x^2}$

13. $\dfrac{16x^2}{9} \div \dfrac{24}{6x^4}$

14. $5\dfrac{1}{2} \div \dfrac{5}{8}$

15. $3\dfrac{1}{3} \div \dfrac{4}{9}$

16. Sam and Michele Smith have $\dfrac{2}{7}$ of their income withheld for taxes, dues, and medical coverage. What amount is withheld each month for taxes, dues, and medical coverage if their rate of pay is $3500 per month?

17. Les wants to store $\dfrac{1}{2}$ pound of flour in 3 equal-sized containers. How much flour should Les place in each container?

SECTION 5.2

Find the LCM.

18. 7 and 14

19. 10 and 20

20. 18 and 20

21. 42 and 12

22. $4x, 8,$ and $16x$

23. $7x, 14x,$ and 20

24. $18x$ and $45x^2$

25. $20x$ and $25x^2$

SECTION 5.3

Perform the indicated operation.

26. $\dfrac{6}{17} - \dfrac{3}{17}$

27. $\dfrac{-23}{27} + \dfrac{-11}{27}$

28. $\dfrac{7}{x} - \dfrac{5}{x}$

29. $\dfrac{x}{7} - \dfrac{5}{7}$

Perform the indicated operation.

30. $\dfrac{5}{6} + \dfrac{4}{9}$

31. $\dfrac{15}{32} - \dfrac{7}{28}$

32. $\dfrac{-3}{14} + \dfrac{7}{21}$

33. $\dfrac{5}{2x} + \dfrac{8}{3x}$

34. $\dfrac{4x}{15} + \dfrac{3x}{45}$

35. $\dfrac{3x}{14} - \dfrac{5x}{42}$

SECTION 5.4

Perform the indicated operations.

36. $10\dfrac{1}{2} + 3\dfrac{4}{5}$

37. $12\dfrac{7}{9} + 6\dfrac{2}{3}$

38. $11\dfrac{1}{5} - 6\dfrac{11}{25}$

39. $25 - 16\dfrac{5}{14}$

40. $4\dfrac{1}{2} \cdot 2\dfrac{2}{9}$

41. $2\dfrac{3}{4} \div 1\dfrac{3}{7}$

42. $4\dfrac{2}{5} \div 8\dfrac{1}{3}$

43. $12 \div \dfrac{2}{3}$

44. Michelle had purchased stock at $\$7\dfrac{1}{2}$ per share and sold it at $\$9\dfrac{2}{3}$ per share. How much money did she make per share?

45. Mike has a piece of wood $7\dfrac{1}{2}$ feet long. If he needs $\dfrac{1}{4}$ of the length of the wood, what length of wood does he need?

SECTION 5.5

Simplify.

46. $\dfrac{3}{4} + \dfrac{1}{2} \cdot \dfrac{2}{7}$

47. $\left(\dfrac{3}{4}\right)^2 + \dfrac{1}{8} \div \dfrac{1}{2}$

Simplify.

48. $\dfrac{\frac{2}{3}}{\frac{1}{9}}$

49. $\dfrac{\frac{x^2}{2}}{\frac{x}{4}}$

50. $\dfrac{\frac{x}{12}}{\frac{x^2}{20}}$

51. $\dfrac{\frac{1}{2} + \frac{1}{4}}{\frac{2}{3} - \frac{1}{9}}$

52. $\dfrac{\frac{1}{3} + \frac{2}{5}}{\frac{1}{5} + \frac{1}{10}}$

SECTION 5.6

Simplify.

53. $\dfrac{y^5}{y^3}$

54. $\dfrac{3^2}{3^3}$

55. $\dfrac{a^2 b^4}{a^5 b^5}$

56. $\dfrac{x^5 y^3}{x^2 y^9}$

57. $\dfrac{2^3 x^0}{2^6 x^9}$

58. $\dfrac{3^2 y^0}{3^3 y^6}$

59. $\dfrac{20x^5}{35x^9}$

60. $\dfrac{18y^6}{6y^4}$

Simplify.

61. $\left(y^2\right)^3$

62. $\left(2^4\right)^2$

63. $(xy)^3$

64. $\left(3x^2\right)^2$

65. $\left(2a^4\right)^0$

66. $\left(\dfrac{3}{y}\right)^2$

67. $\left(\dfrac{x}{2}\right)^3$

SECTION 5.7

68. Leslie put a rectangular pool in her yard, which is $38\frac{1}{4}$ feet by 25 feet. There is concrete around the entire pool, which is $4\frac{1}{2}$ feet wide. The pool and the conrete area take up the entire $38\frac{1}{4}$-foot by 25-foot yard.

  (a) What are the dimensions of the pool?

  (b) How much will it cost to put tile around the edge of her pool if the tile costs $\$3\frac{1}{4}$ per linear foot?

69. The Pleasantville Country Club maintains the putting greens with a grass height of $\frac{7}{8}$ inch. The grass on the fairways is maintained at $2\frac{1}{2}$ inches. How much lower must the mower blade be lowered by a person mowing the fairways if that person will be using the same mowing machine on the putting green?

CHAPTER 5 TEST

1. Find $\dfrac{2}{3}$ of $\dfrac{9}{10}$.

Multiply and simplify.

2. $\dfrac{-1}{4} \cdot \dfrac{2}{5}$

3. $\dfrac{7x^2}{8} \cdot \dfrac{16x}{14}$

Divide and simplify.

4. $\dfrac{-1}{2} \div \dfrac{1}{4}$

5. $\dfrac{2x^8}{5} \div \dfrac{22x^4}{15}$

6. Anna wishes to build 2 bookcases, each with 5 shelves. Each shelf is $3\dfrac{1}{2}$ feet long. The wood for the shelves is sold in 10-foot boards. How many boards does Anna need to buy for the shelves?

7. List the first 5 multiples of 9.

Find the least common multiple (LCM).

8. 14, 21

9. $5a, 10a^4, 20a^2$

10. 5, 7, 51

11. Find the least common denominator (LCD): $\dfrac{17}{30}, \dfrac{1}{4}$.

Add.

12. $\dfrac{12}{x} + \dfrac{3}{x}$

13. $\dfrac{1}{5a} + \dfrac{3}{4a}$

14. $4\dfrac{5}{6} + 3\dfrac{1}{3}$

1. _____

2. _____

3. _____

4. _____

5. _____

6. _____

7. _____

8. _____

9. _____

10. _____

11. _____

12. _____

13. _____

14. _____

Subtract.

15. $\dfrac{8}{12} - \dfrac{2}{3}$

16. $10\dfrac{5}{8} - 2\dfrac{1}{3}$

17. $\dfrac{10x}{15} - \dfrac{x}{5}$

Multiply.

18. $(-4)5\dfrac{1}{3}$

19. $\dfrac{1}{4} \cdot 3\dfrac{1}{2}$

Divide.

20. $1\dfrac{2}{3} \div 3$

21. $6 \div 2\dfrac{3}{4}$

22. $2\dfrac{1}{2} \div \left(-\dfrac{3}{5}\right)$

Simplify.

23. $\dfrac{2}{3} + \dfrac{1}{2} \cdot \dfrac{1}{4}$

24. $\dfrac{1}{2} + \dfrac{7}{8} \div \dfrac{1}{4}$

25. $\dfrac{\frac{x}{2}}{\frac{x}{4}}$

26. $\dfrac{\frac{1}{3}}{\frac{5}{9}}$

27. $\left(x^3\right)^4$

28. $\left(5x^2\right)^3$

29. $\dfrac{5y^3}{20xy}$

30. Elizabeth bought Disney stock at $\$87\dfrac{1}{2}$. The stock's value went up $\$4\dfrac{3}{4}$ the first week. The second week it went down $\$1\dfrac{1}{4}$.

    (a) What was the value of the stock per share after the second week?

    (b) Was there a profit or loss per share at the end of the second week?

    (c) How much was this profit or loss per share?

15. _____

16. _____

17. _____

18. _____

19. _____

20. _____

21. _____

22. _____

23. _____

24. _____

25. _____

26. _____

27. _____

28. _____

29. _____

30. _____

CUMULATIVE TEST FOR CHAPTERS 1–5

1. _____

2. _____

3. _____

4. _____

5. _____

6. _____

7. _____

8. _____

9. _____

10. _____

11. _____

12. _____

13. _____

14. _____

15. _____

16. _____

1. Write 3401 in expanded notation.

Combine like terms.

2. $2 + x + 8$

3. $5x - 3x + x + 5$

4. $-8r + 3 - 5r - 8$

Translate into an equation and solve.

5. Four added to what number equals 15?

6. Triple a number plus three equals twenty-four.

7. A number minus 8 equals eighteen.

Perform the indicated operations.

8. $(3x^2)(x^3)(x)$

9. $5(x + 3)$

10. $8^2 - 10 + 4$

11. $-4 + 6 - 9$

12. $(-15) \div 3$

13. $(-7)(-1)$

14. $(-10x^2)(5x)$

15. $3 - 12 \div (-2) + 4^2$

16. Identify as prime, composite, or neither:
   (a) 72
   (b) 19

17. Express 48 as a product of prime factors.

18. A student answers 16 questions correctly and 5 questions incorrectly on a test. Write the fraction that describes the number of test questions she answered correctly.

Replace the question mark with the inequality symbol $<$ *or* $>$.

19. 10 m ? 10 cm

20. $\dfrac{3}{10}$? $\dfrac{12}{30}$

21. A car travels 276 miles on 12 gallons of gas. Find the miles per gallon.

22. Find the value of x: $\dfrac{5}{9} = \dfrac{x}{27}$.

23. Add: $\dfrac{1}{6x} + \dfrac{3}{x}$.

24. Subtract: $\dfrac{5x}{2} - \dfrac{x}{10}$.

25. Multiply and simplify: $\dfrac{2x^4}{5x} \cdot \dfrac{10x^2}{4}$.

Divide.

26. $-\dfrac{1}{6} \div -\dfrac{2}{3}$

27. $7\dfrac{1}{2} \div \left(-\dfrac{3}{8}\right)$

Simplify.

28. $\dfrac{1}{2} + \dfrac{3}{5} \cdot 2$

29. $(4x)^3$

30. Frank had $12\dfrac{1}{2}$ gallons of gasoline in his car before leaving for work one morning. When he returned home that evening, he had $9\dfrac{1}{3}$ gallons of gas left. How many gallons of gas did Frank use that day?

17. _____

18. _____

19. _____

20. _____

21. _____

22. _____

23. _____

24. _____

25. _____

26. _____

27. _____

28. _____

29. _____

30. _____

EQUATIONS AND POLYNOMIALS

A ccountants, managers, health care officials, and business owners must be able to determine the profit, cost, and break-even point for their businesses. To find this information, they solve profit, cost, and revenue equations. To find out how to determine these values, turn to Putting Your Skills to Work on page 410.

6.1 Solving Equations Using One Principle of Equality

6.2 Solving Equations Using More Than One Principle of Equality

6.3 Solving Equations Containing Fractions

6.4 Multiplication of Polynomials

6.5 Addition and Subtraction of Polynomials

6.6 Solving Equations Involving Polynomial Expressions

6.7 Applied Problems Involving Polynomials

CHAPTER 6 PRETEST

This test provides a preview of the topics in this chapter. It will help you identify which concepts may require more of your studying time. If you are familiar with the topics in this chapter, take this test now. Check your answers with those in the back of the book. If you are not familiar with the topics in this chapter, begin studying the chapter now.

SECTION 6.1   *Solve for the variable and check.*

1. $x - 9 = 16$ **2.** $x + 8 = -3$ **3.** $1 - 4 + x = 9 - 3$

4. $\dfrac{y}{4} = 3$ **5.** $13y = 52$

SECTION 6.2   *Solve for z and check.*

6. $2z - 6 = 2$ **7.** $-5z - 3 = 7$ **8.** $\dfrac{z}{3} + 5 = 11$

9. $z + 2z = -8 - 10$ **10.** $3(4z + 1) = 27$

SECTION 6.3   *Solve for x and check.*

11. $\dfrac{x}{5} + \dfrac{x}{2} = 7$ **12.** $4x + \dfrac{2}{3} = -\dfrac{10}{3}$ **13.** $\dfrac{x}{3} + x = 4$

14. $\dfrac{3}{x} + 1 = \dfrac{1}{2}$

SECTION 6.4

15. Identify this polynomial as a monomial, binomial, or trinomial: $4x^2 + 3x - 5$.

Multiply.

16. $(-3a)(4a + 6)$ **17.** $2(4x^2 + 3x - 5)$ **18.** $(x + 2)(x - 7)$

SECTION 6.5

19. Simplify: $-(-3x + 2y - 8)$.

20. Add: $(5x^2 + 3x - 1) + (7x^2 - 2x + 3)$.

21. Subtract: $(8y + 4) - (-3y + 1)$.

SECTION 6.6   *Solve for x and check.*

22. $2(4x - 1) + 2x = -42$ **23.** $2(x + 6) = -3(x - 4) + 15$

SECTION 6.7   *If double a number is decreased by 7, the result is 33.*

24. Write an equation to find the number.

25. Solve the equation.

1. _____
2. _____
3. _____
4. _____
5. _____
6. _____
7. _____
8. _____
9. _____
10. _____
11. _____
12. _____
13. _____
14. _____
15. _____
16. _____
17. _____
18. _____
19. _____
20. _____
21. _____
22. _____
23. _____
24. _____
25. _____

364

S E C T I O N   6 . 1

Solving Equations Using One Principle of Equality

After studying this section, you will be able to:

❶ Solve equations using the addition principle.

❷ Solve equations using the multiplication principle.

❸ Solve equations using the division principle.

Math Pro   Video 6.1   SSM

❶ Solving Equations Using the Addition Principle

There are many kinds of equations and many techniques for solving them. But all rest on the same idea. We transform the equation into a simpler, equivalent equation by choosing a logical step. By looking at this second equation, we can sometimes see the solution. For example, in Chapter 1 we wrote equivalent addition problems to solve equations that involved subtraction. Then, from this second equation, the solution was easily seen.

$$x - 2 = 7$$
$$x = 7 + 2$$
$$x = 9$$

We start with this equation, and transform to a simpler, *equivalent* equation. From the second equation we can see the solution.

However, for more complex equations, such as $2x - 4 = x - 8$, we often have to choose many logical steps until the solution is finally apparent. Finding these logical steps often requires that we use the principles of equality described in this chapter.

We observe in our everyday world that if we add the same amount to two equal values, the results are equal. We illustrate this below.

We add 5 pounds on both sides of a seesaw.

We add a 2-pound weight to the center of each 5-pound weight simultaneously.

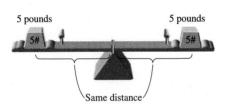

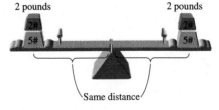

We would expect the seesaw to balance.

The seesaw should still balance.

In mathematics, a similar principle is observed. That is, we can add the same number to both sides of an equation without changing the solution.

$$x - 2 = 7$$
$$x - 2 + 2 = 7 + 2$$
$$x + 0 = 9$$
$$x = 9$$

We saw that the solution to this equation is $x = 9$.

We add 2 to each side of the equation.

We simplify.

The solution remains the same: $x = 9$.

Adding 2 to both sides of the equation did not change the solution. We state this principle of equality.

> **THE ADDITION PRINCIPLE OF EQUALITY**
>
> For real numbers a, b, and c: if $a = b$, then $a + c = b + c$.
>
> If the *same number* is added to both sides of an equation, the results on each side are equal in value.

How do we know what number to add to both sides of the equation? We add the *opposite* of the number we must remove in order to get x alone on one side of the equation: $x = $ *some number.* Why do we add the opposite of this number? Because the sum of opposite numbers equals zero: $-2 + 2 = 0$; $6 + (-6) = 0$.

EXAMPLE 1   Solve: $x - 22 = -14$.

SOLUTION   We must remove -22 from the left side of the equation to get x alone: $x = $ *some number.* So we add its opposite, $+22$, to both sides.

$$x - 22 = -14$$
$$x - 22 + 22 = -14 + 22 \qquad \text{Add } +22 \text{ to both sides, since } -22 + 22 = 0.$$
$$x + 0 = 8 \qquad x - 22 + 22 = x + 0; \text{and } -14 + 22 = 8.$$
$$\boldsymbol{x = 8} \qquad \text{The solution is 8.}$$

Note: We can solve $x - 22 = -14$ by writing the equivalent addition problem, $x = -14 + 22$, then adding to get the solution, $x = 8$. So why do we need the addition principle? Solving complicated equations such as $2x - 3 = 4 + 7x$ is easier with the addition principle. In this section we practice using the addition principle on simple equations so that you can apply the technique on more complicated equations later.

PRACTICE PROBLEM 1
Solve: $x - 19 = -31$.

Some problems require that we write many steps. Writing too many steps can make the problem seem more complicated than it really is. We can eliminate some steps if, when adding a number to both sides of the equation, we place the addition *below* the terms rather than *beside* the terms. Try this with a few problems. You may find it easier and decide to use this format.

EXAMPLE 2   Solve and check your solution: $-15 = x + 9$.

SOLUTION   The variable is on the right side of the equation, so we will simplify the equation to the form *some number* $= x$.

$$\begin{array}{rl} -15 = x + 9 & \quad \textit{Think: } \text{"Add the opposite of 9 to both sides of the} \\ \underline{+ \quad -9 \qquad - 9} & \quad \text{equation."} \\ -24 = x + 0 & \quad \text{Add } -9 \text{ to both sides.} \end{array}$$
$$\boldsymbol{-24 = x}$$

$$\textit{Check: } -15 = x + 9$$
$$-15 \overset{?}{=} -24 + 9$$
$$-15 = -15 \quad \checkmark$$

PRACTICE PROBLEM 2
Solve and check your solution: $-58 = x + 3$.

Often we must simplify each side of the equation separately before we use the addition principle.

EXAMPLE 3   Solve: $2 - 6 = x - 7 + 12$.

SOLUTION   First, we must simplify each side of the equation separately.

$$2 - 6 = x - 7 + 12$$
$$-4 = x + 5 \qquad \text{Simplify.}$$

Then we add −5 to both sides of the equation to get x alone on the right side: *some number* $= x$.

$$-4 = x + 5$$
$$\underline{+ -5 \qquad - 5} \qquad \text{Add } -5 \text{ to both sides.}$$
$$\mathbf{-9 = x} \qquad \text{We usually do not write the step } -9 = x + 0.$$

Although we will not always check the solutions in the examples, it is a good idea for you to check your solutions.

PRACTICE PROBLEM 3

Solve: $5 - 8 = x - 2 + 19$.

The important thing to remember when using logical steps is that an equation is like a balanced scale. Whatever you do to one side of the equation, you must do to the other side of the equation to maintain the balance.

DEVELOPING YOUR STUDY SKILLS

READING THE TEXT

Homework time each day should begin with a careful reading of the section(s) assigned in your textbook. Much time and effort have gone into the selection of a particular text, and your instructor has chosen a book that will help you to become successful in this mathematics class. Expensive textbooks can be a wise investment if you take advantage of them by studying them with care.

Reading a mathematics textbook is unlike reading many other types of books that you may use in your literature, history, psychology, or sociology courses. Mathematics texts are technical books that provide you with exercises to practice on. Reading a mathematics text requires slow and careful reading of each word, which takes time and effort.

- Read your textbook with a paper and pencil in hand.
- Underline new definitions or concepts and write them in your notebook on a separate sheet labeled "Important Facts."
- Whenever you encounter unfamiliar terms, look them up, then note their definitions on your "Important Facts" notebook pages.
- When you come to an example, work through it step by step. Be sure to read each word and to follow directions carefully.
- Be sure that you understand what you are reading. Make a note of any of those things that you do not understand and ask your instructor about them.
- Do not hurry through the material. Learning mathematics takes time.

② Solving Equations Using the Multiplication Principle

Recall from Chapter 4 that whenever the numerator and denominator of a fraction are equal, the fraction can be written as 1. That is, $\dfrac{a}{a} = 1$. We will use this fact to help us solve equations.

EXAMPLE 4   Replace the ? with the number that gives the desired result: $\dfrac{? \cdot x}{-6} = x$.

SOLUTION

$$\frac{? \cdot x}{-6} = x$$

$$\frac{-6 \cdot x}{-6} = x \qquad \text{Since} \quad \frac{\cancel{-6} \cdot x}{\cancel{-6}} = 1 \cdot x = x.$$

PRACTICE PROBLEM 4

Replace the ? with the number that gives the desired result: $\dfrac{? \cdot x}{-8} = x$.

As we saw earlier, to solve an equation we can add the same number to both sides of the equation without changing the solution. What would happen if we multiplied each side of an equation by the same number? Let's return to our example of a balanced seesaw to answer this question. If we doubled the number of weights on each side (we are multiplying each side by 2), the seesaw should still balance.

We state this principle of equality.

THE MULTIPLICATION PRINCIPLE OF EQUALITY

For real numbers a, b, and c, with c not equal to 0:

If $a = b$, then $ca = cb$.

If both sides of an equation are multiplied by the same number, the results on each side are equal in value.

To solve an equation in the form $\dfrac{x}{5} = 30$ we can simply think: "What can we do to the left side of the equation to remove the 5 so that x is alone?" We can multiply the left side by 5. However, whatever we do to one side of the equation, we must do to the other side of the equation.

EXAMPLE 5 Solve: $\dfrac{x}{4} = 28$.

SOLUTION Since we are *dividing* the variable x by 4, we can *multiply* both sides of the equation by 4 so that x is alone on one side of the equation: $x = $ *some number*.

$$\frac{x}{4} = 28 \qquad \text{The variable } x \text{ is } \textit{divided} \text{ by 4.}$$

$$\frac{4 \cdot x}{4} = 28 \cdot 4 \qquad \text{We } \textit{multiply} \text{ both sides of the equation by 4.}$$

$$x = \mathbf{112} \qquad \text{Simplify: } \frac{4x}{4} = x, \text{ and } 28 \cdot 4 = 112.$$

Be sure that you *check* your solution.

Note: We can also solve $\dfrac{x}{4} = 28$ by writing the equivalent multiplication problem, $x = 28 \cdot 4$. We are illustrating how to use the *multiplication principle* to prepare us to solve complicated equations, such as $\dfrac{x}{3} - 7 = 5$, which we study later in the chapter.

PRACTICE PROBLEM 5

Solve: $\dfrac{a}{2} = 17$.

③ Solving Equations Using the Division Principle

To solve an equation in the form $6x = 120$, we divide both sides by 6. The multiplication principle allows us to divide, as well as multiply, since multiplying a number by $\dfrac{1}{6}$ is same as dividing by 6. We illustrate this below.

6 multiplied by $\dfrac{1}{3}$ is equivalent to 6 divided by 3.

$$6 \cdot \frac{1}{3} = \frac{6}{1} \cdot \frac{1}{3} = \frac{6}{3}$$

We state this principle of equality.

THE DIVISION PRINCIPLE OF EQUALITY

For real numbers a, b, and c, with c not equal to 0:

If $a = b$,   then $\dfrac{a}{c} = \dfrac{b}{c}$.

If both sides of an equation are divided by the same nonzero number, the results on each side are equal in value.

To solve an equation in the form $5x = 30$, we can simply think: "What can we do to the left side of the equation to remove the 5 so that x is alone?" We divide by 5.

$$\frac{5x}{?} = 30 \longrightarrow \frac{5x}{5} = \frac{30}{5} \qquad x = 6$$

Remember, whatever we do to one side of the equation, we must do to the other side of the equation.

EXAMPLE 6   Solve: $7x = -147$.

SOLUTION   Our goal is to simplify the equation to the form $x = some$ *number*.

$7x = -147$ The variable, x, is *multiplied* by 7.

$\dfrac{7x}{7} = \dfrac{-147}{7}$ We *divide* both sides by 7.

$x = -21$ $\dfrac{7x}{7} = 1 \cdot x = x$, and $-147 \div 7 = -21$.

PRACTICE PROBLEM 6
Solve: $5m = -155$.

EXAMPLE 7 Solve: $45 - 89 = -11m$.

SOLUTION First, we simplify the equation.

$$45 - 89 = -11m$$
$$45 + (-89) = -11m$$
$$-44 = -11m$$

Then, since we are multiplying the variable by -11, we must divide both sides of the equation by -11.

$$\frac{-44}{-11} = \frac{-11m}{-11}$$
$$\boldsymbol{4 = m}$$

PRACTICE PROBLEM 7
Solve: $65 - 53 = -12y$.

EXERCISES 6.1

Replace the ? with the number that gives the desired result.

1. $4 + \underline{\ ?\ } = 0$ **2.** $6 + \underline{\ ?\ } = 0$ **3.** $-9 + \underline{\ ?\ } = 0$ **4.** $-1 + \underline{\ ?\ } = 0$

5. $17 - \underline{\ ?\ } = 0$ **6.** $43 - \underline{\ ?\ } = 0$ **7.** $-28 + \underline{\ ?\ } = 0$ **8.** $-13 + \underline{\ ?\ } = 0$

Solve and check your solution.

9. $y - 15 = 5$ **10.** $m - 21 = 9$ **11.** $x + 13 = 22$

12. $x + 38 = 4$ **13.** $n - 43 = -74$ **14.** $y - 81 = -12$

15. $y + 44 = -50$ **16.** $x + 1 = -15$ **17.** $1 = x - 13$

18. $5 = x - 7$ **19.** $29 = y + 11$ **20.** $46 = a + 14$

21. $-13 = x + 1$ **22.** $-5 = x + 7$

Simplify, then solve.

23. $-23 + 8 + x = -2 + 13$ **24.** $-33 + 9 + x = -1 + 7$

25. $4 - 9 = a - 1 + 14$ **26.** $3 - 8 = m - 4 + 7$

27. $-45 + 9 + m = -6 + 18$ **28.** $-27 + 7 + n = -5 + 8$

29. $-1 + 4 + x = -5 + 9$ **30.** $-3 + 9 + a = -4 + 7$

31. $3(7 - 11) = y - 5$ **32.** $5(8 - 13) = x - 1$

Solve each of the following.

33. (a) $x - 8 = 22$ **(b)** $x + 8 = 22$ **34. (a)** $a - 3 = 29$ **(b)** $a + 3 = 29$

35. (a) $x + 6 = -11$ **(b)** $x - 6 = -11$ **36. (a)** $y + 2 = -10$ **(b)** $y - 2 = -10$

Replace the ? with the number that gives the result desired.

37. $\dfrac{5x}{?} = x$ **38.** $\dfrac{8x}{?} = x$ **39.** $\dfrac{-2x}{?} = x$ **40.** $\dfrac{-3x}{?} = x$

41. $\dfrac{? \cdot x}{6} = x$ **42.** $\dfrac{? \cdot x}{2} = x$ **43.** $\dfrac{? \cdot x}{-5} = x$ **44.** $\dfrac{? \cdot x}{-9} = x$

Solve and check your solutions.

45. $\dfrac{y}{15} = 12$ **46.** $\dfrac{x}{9} = 16$ **47.** $\dfrac{x}{7} = 31$ **48.** $\dfrac{m}{6} = 10$

49. $\dfrac{m}{13} = -30$ **50.** $\dfrac{a}{8} = -14$ **51.** $\dfrac{x}{14} = -6$ **52.** $\dfrac{m}{19} = -7$

53. $3 = \dfrac{y}{11}$ **54.** $10 = \dfrac{x}{7}$ **55.** $15 = \dfrac{a}{4}$ **56.** $44 = \dfrac{m}{2}$

57. $-1 = \dfrac{m}{30}$ **58.** $-9 = \dfrac{x}{15}$ **59.** $-4 = \dfrac{a}{-20}$ **60.** $-6 = \dfrac{m}{-42}$

61. $8x = 104$ **62.** $9y = 135$ **63.** $-15y = 165$ **64.** $-13x = 156$

65. $-19x = -76$ **66.** $-22y = -132$ **67.** $52 = 5a$ **68.** $45 = 4y$

69. $-2 = 3x$ **70.** $-4 = 5y$

Simplify, then solve. Check your solution.

71. $-26 - 18 = 11a$ **72.** $-44 - 16 = 6y$ **73.** $5x = 3 - 11$ **74.** $-7a = 4 - 13$

75. $-5 - 5 = 11y$ **76.** $-2 - 2 = 5x$

Solve each of the following.

77. (a) $4x = 52$ **(b)** $\dfrac{x}{4} = 52$ **78. (a)** $3y = 39$ **(b)** $\dfrac{y}{3} = 39$

79. (a) $4x = -52$ **(b)** $\dfrac{x}{4} = -52$ **80. (a)** $3y = -39$ **(b)** $\dfrac{y}{3} = -39$

81. (a) $x - 7 = 21$ **(b)** $x + 12 = 33$ **(c)** $5x = 3$ **(d)** $\dfrac{x}{5} = 11$

82. (a) $x - 4 = 19$ **(b)** $x + 6 = 23$ **(c)** $5x = 4$ **(d)** $\dfrac{x}{3} = 12$

83. (a) $x - 7 = 12$ **(b)** $3x = 2$ **(c)** $\dfrac{x}{6} = 9$ **(d)** $x + 11 = 34$

84. (a) $x + 15 = 31$ **(b)** $13x = 7$ **(c)** $\dfrac{x}{2} = 8$ **(d)** $x - 3 = 10$

VERBAL AND WRITING SKILLS

Fill in the blank.

85. The sum of two opposite numbers is equal to _____.

86. A number divided by itself is equal to _____.

87. To solve $x - 6 = 2$, we _____ 6 to both sides of the equation.

88. To solve $x + 6 = 2$, we _____ 6 to both sides of the equation.

ONE STEP FURTHER

Simplify, then solve.

89. $2^2 + (5 - 9) = x + 3^3$

90. $4^2 + (3 - 7) = y + 2^3$

CUMULATIVE REVIEW

Perform the operation indicated.

1. $145 \div 11$

2. $(223)(36)$

3. Round to the nearest hundred: 29,441.

❶ Solve equations using more than one principle of equality.

❷ Simplify and solve equations.

Math Pro Video 6.2 SSM

Solving Equations Using More Than One Principle of Equality

To solve equations such as $4x - 9 = 78$, we use more than one principle of equality. In this section we will see how to use the principles of equality to solve more complex equations.

❶ Solving Equations Using More Than One Principle of Equality

When we solve an equation we must be able to determine which principle to use first. We can use the following sequence of steps to solve an equation such as $3x + 6 = 9$.

PROCEDURE TO SOLVE AN EQUATION IN THE FORM
ax + b = some number

1. First, use the addition principle to get the variable term *ax* alone on one side of the equation:

   ax = some number

2. Then apply the multiplication or division principle to get the variable *x* alone on one side of the equation:

   x = some number

E X A M P L E   1   Solve and check your solution: $3x + 7 = 88$.

SOLUTION   First, we use the addition principle to get the variable term, $3x$, alone ($3x$ = some number).

$$3x + 7 = 88$$
$$3x + 7 + (-7) = 88 + (-7) \qquad \textit{Add } -7 \text{ to both sides of the equation.}$$
$$3x = 81 \qquad \textit{The variable term, } 3x, \text{ is alone.}$$

Then we apply the division principle to get the variable x alone (x = some number).

$$3x = 81$$
$$\frac{3x}{3} = \frac{81}{3} \qquad \textit{Divide by 3 on both sides of the equation.}$$
$$\boldsymbol{x = 27} \qquad \textit{The variable, } x, \text{ is alone.}$$

$$\textit{Check:} \quad 3x + 7 = 88$$
$$3(27) + 7 \overset{?}{=} 88$$
$$81 + 7 \overset{?}{=} 88$$
$$88 = 88 \quad \checkmark$$

Note: We choose to add -7 *beside* rather than *below* the terms.

$$3x + 7 + (-7) = 88 + (-7) \quad \text{or} \quad 3x + 7 = 88$$
$$\underline{\quad -7 \quad -7}$$

As stated in Section 6.1, either technique is fine. We present both so that you can become familiar with each and determine which technique works best for you.

PRACTICE PROBLEM 1
Solve and check your solution: $8x + 9 = 105$.

EXAMPLE 2   Solve: $-6x - 4 = 74$.

SOLUTION   First, we must get the variable term, $-6x$, alone by using the addition principle.

$$-6x - 4 = 74$$
$$-6x - 4 + 4 = 74 + 4 \qquad \textit{Add 4 to both sides of the equation.}$$
$$-6x = 78 \qquad \text{The variable term, } -6x, \text{ is alone.}$$

Then we apply the division principle to get the variable, x, alone.

$$-6x = 78$$
$$\frac{-6x}{-6} = \frac{78}{-6} \qquad \textit{Divide by } -6 \text{ on both sides of the equation.}$$
$$x = -13 \qquad \text{The variable, } x, \text{ is alone.}$$

We leave the check to the student.

PRACTICE PROBLEM 2
Solve: $-5m - 10 = 115$.

CALCULATOR

CHECKING SOLUTIONS TO EQUATIONS

We can use a calculator to check our solutions. To check that -5 is a solution to $3x - 1 = 14$, enter:

Scientific calculator:

3 $\times$ 5 $+/-$ $-$ 1 $=$

Graphing calculator:

3 $\times$ $(-)$ 5 $-$ 1 ENT

The calculator displays

14

TO THINK ABOUT

What would happen if we used the multiplication principle before the addition principle? Would we get the right answer? Yes, but often the process is harder. For example, let's rework Example 2 using the division principle first, and see what happens.

$$-6x - 4 = 74$$
$$\frac{-6x}{-6} - \frac{4}{-6} = \frac{74}{-6} \qquad \text{We must divide } \textit{each term} \text{ on both sides of the equation by } -6.$$

As you can see, we sometimes end up with fractions that are more difficult to work with. Also, students often forget to divide every term. Therefore, the step $\dfrac{4}{-6}$ is often left out. It is best to use the addition principle before the multiplication and division principles.

E X A M P L E   3   Solve: $\dfrac{m}{5} + 9 = -6$.

SOLUTION   We must add -9 to both sides of the equation to get $\dfrac{m}{5}$ alone on one side of the equation.

$$\frac{m}{5} + 9 = -6$$

$$\frac{m}{5} + 9 + (-9) = -6 + (-9) \qquad \textit{Add } -9 \textit{ to both sides of the equation.}$$

$$\frac{m}{5} = -15$$

We must multiply both sides of the equation by 5 to get m alone.

$$\frac{m}{5} = -15$$

$$5 \cdot \frac{m}{5} = 5(-15) \qquad \textit{Multiply by 5 on both sides of the equation.}$$

$$\boldsymbol{m = -75}$$

P R A C T I C E   P R O B L E M   3

Solve: $\dfrac{a}{3} + 7 = 4$.

DEVELOPING YOUR STUDY SKILLS

MATHEMATICS AND CAREERS

Students often question the value of mathematics. They see little real use for it in their everyday lives. However, mathematics is often the key that opens the door to a better-paying job. Studying mathematics sharpens your mind and helps you sort out and solve real-life situations that *do not* require the use of mathematics.

In our present-day technological world, may people use mathematics daily. Many vocational and professional areas—such as the fields of business, statistics, economics, psychology, finance, computer science, chemistry, physics, engineering, electronics, nuclear energy, banking, quality control, and teaching—require a certain level of expertise in mathematics. Those who want to work in these fields must be able to function at a given mathematical level. Those who cannot will not be able to enter this job area.

So, whatever your field, be sure to realize the importance of mastering the basics of this course. It is very likely to help you advance to the career of your choice.

Keep in mind that mathematical thinking does not always require performing math calculations. Organizing and planning skills are enhanced when you study mathematics!

EXERCISE

1. Make a list of the types of jobs you plan to seek when you finish your education. Talk with your instructor, counselor, or job placement director about the level and type of mathematics, as well as the planning and organizational skills, that are required for these jobs.

② Simplifying and Solving Equations

The process of solving an equation is easier if we simplify the equation before we begin to solve it. Many students find that it is helpful to have a written procedure to follow when solving more involved equations.

> **PROCEDURE TO SOLVE EQUATIONS**
>
> 1. *Parentheses.* Remove any parentheses.
> 2. *Simplify each side of the equation.* Collect like terms and simplify numerical work.
> 3. *Isolate the ax term: $ax = $ some number.* Use the addition principle to get all ax terms on one side of the equation and the numerical values on the other side.
> 4. *Isolate the x term: $x = $ some number or some number $= x$.* Use the multiplication or division principle to get x alone on one side of the equation.
> 5. Check your solution.

All variable terms must be on one side of the equation so that we can simplify the equation to the form $x = $ some number.

EXAMPLE 4   Solve: $5x - 3 = 6x + 2$.

SOLUTION   First, we add $-6x$ to both sides of the equation so that all variable terms are on one side of the equation.

$$
\begin{array}{rcr}
5x - 3 = & & 6x + 2 \\
+\ -6x & & -6x \\
\hline
-x\ - 3 = & & 2
\end{array}
$$

Then, we solve the equation.

$$
\begin{array}{rcr}
-x - 3 = & & 2 \\
+ 3 & & +3 \\
\hline
-x\quad\ = & & 5
\end{array}
$$

$$\frac{-1x}{-1} = \frac{5}{-1}$$

$$x = -5$$

Note: Another way to solve $-1x = 5$ is to multiply both sides of the equation by -1.

PRACTICE PROBLEM 4
Solve: $3x - 1 = 4x - 6$.

TO THINK ABOUT
··

When a variable appears on both sides of an equation, does it matter which variable term we remove? No. To see why, let's look at the equation $9x + 1 = 7x - 4$.

Remove $7x$ from the right side:

$$9x + 1 = 7x - 4$$
$$+\underline{-7x \qquad -7x}$$
$$2x + 1 = \qquad -4$$

$$\boxed{2x + 1 = -4}$$

$2x = -5$ Add −1 to both sides.

$x = -\dfrac{5}{2}$ Divide both sides by 2.

Remove $9x$ from the left side:

$$9x + 1 = 7x - 4$$
$$+\underline{-9x \qquad -9x}$$
$$1 = -2x - 4$$

$$\boxed{1 = -2x - 4}$$

$5 = -2x$ Add 4 to both sides.

$-\dfrac{5}{2} = x$ Divide both sides by −2.

As we can see, the answers are the same—the difference being that in one case we have an equation with the variable on the *left side*, and in the other case we have an equation with the variable on the *right side*.

Some students prefer to eliminate the variable term that has the effect of getting a positive coefficient, while others prefer always to have the variable on the left side. This is just a preference; either way is correct.

EXAMPLE 5   Solve: $3(x + 2) = 27$.

SOLUTION   We use the distributive property to remove parentheses and simplify the left side.

$$3(x + 2) = 27$$
$$\mathbf{3}(x) + \mathbf{3}(2) = 27 \qquad \text{Multiply using the distributive property.}$$
$$3x + 6 = 27 \qquad \text{Simplify.}$$
$$+\underline{\quad -6 \quad -6}$$
$$3x \quad = 21$$
$$\frac{3x}{3} = \frac{21}{3}$$
$$x = 7$$

PRACTICE PROBLEM 5
Solve: $5(x + 4) = 40$.

EXERCISES 6.2

Solve for x and check your solution.

1. $3x + 9 = 27$ **2.** $8x + 3 = 19$ **3.** $5x - 10 = 25$ **4.** $6x - 15 = 9$

5. $4x + 10 = 18$ **6.** $2x + 6 = 24$ **7.** $5x - 1 = 16$ **8.** $3x - 2 = 12$

Solve for the variable.

9. $-4y + 7 = 63$ **10.** $-8w + 5 = 69$ **11.** $-6m - 10 = 88$ **12.** $-5y - 9 = 74$

13. $-2x - 10 = 40$ **14.** $-3x - 5 = 16$ **15.** $-5y - 7 = -1$ **16.** $-4x - 9 = -2$

17. $6 = 4 - 2x$ **18.** $8 = 5 - 3x$ **19.** $-3 = 6 - 3y$ **20.** $-2 = 8 - 2y$

21. $\dfrac{x}{8} + 2 = 4$ **22.** $\dfrac{y}{4} + 3 = 7$ **23.** $\dfrac{m}{6} + 7 = 13$ **24.** $\dfrac{n}{5} + 8 = 11$

25. $\dfrac{x}{-2} + 4 = -5$ **26.** $\dfrac{x}{-3} + 5 = -2$ **27.** $\dfrac{x}{-6} - 2 = 4$ **28.** $\dfrac{x}{-5} - 1 = 9$

29. $\dfrac{x}{4} - 3 = -5$ **30.** $\dfrac{x}{6} - 5 = -8$ **31.** $9 = \dfrac{y}{-8} + 2$ **32.** $8 = \dfrac{y}{-4} + 6$

Simplify and solve.

33. $8y + 6 - 2y = 18$ **34.** $7x + 3 - 4x = 12$ **35.** $9x - 2 + 2x = 6$

36. $3m - 1 + 4m = 5$ **37.** $7x + 8 - 8x = 11$ **38.** $4x + 7 - 5x = 12$

Solve for the variable. You may move the variable terms to the right or to the left.

39. $15x = 9x + 30$ **40.** $16x = 7x + 36$ **41.** $2x = -12x + 5$ **42.** $14x = -2x + 11$

43. $8x + 2 = 5x - 4$ **44.** $7x + 6 = 2x - 9$ **45.** $11x + 20 = 12x + 2$ **46.** $6y - 5 = 13y + 9$

47. $4x - 24 = 6x - 8$ **48.** $9x - 4 = 3x - 10$

Remove parentheses. Solve for the variable. Check your solution.

49. $5(x + 2) = 25$

50. $7(x - 3) = 14$

51. $4(x - 1) = 12$

52. $2(x + 2) = 14$

53. $2(y + 1) = -10$

54. $3(x + 1) = -18$

55. $3(m - 4) = 7$

56. $5(y - 2) = 25$

Solve (mixed practice).

57. $2y + 6 = 12$

58. $8x + 4 = 20$

59. $3x - 1 = 14$

60. $7x - 2 = 19$

61. $\dfrac{x}{-2} + 3 = 6$

62. $\dfrac{x}{-3} + 7 = 10$

63. $5x + 5 - 2x = 15$

64. $4x + 3 - 2x = 12$

65. $13x = 8x + 20$

66. $15x = 6x + 30$

67. $5(x + 3) = -10$

68. $4(x + 3) = -16$

 CALCULATOR EXERCISES

Use your calculator to verify the solution to each equation.

69. Is $x = -21$ a solution to
$45(x - 17) + 856 = -854$?

70. Is $x = -15$ a solution to
$22(x - 12) + 345 = -225$?

VERBAL AND WRITING SKILLS

Write in words the question asked by each equation.

71. $1 + 2n = 5$

72. $3x - 2 = 7$

ONE STEP FURTHER

Solve.

73. $2x - 6 + 8x + 13 = -2 - 5x + 6$

74. $9x - 8 - 6x + 19 = -4x - 3 + 2x$

CUMULATIVE REVIEW

Find the least common denominator.

1. $\dfrac{2}{3}, \dfrac{1}{4}, \dfrac{5}{2}$

2. $\dfrac{3}{4}, \dfrac{4}{2}, \dfrac{1}{5}$

3. $\dfrac{1}{2x}, \dfrac{7}{x}$

4. $\dfrac{2}{x}, \dfrac{3}{5x}$

Solving Equations Containing Fractions

After studying this section, you will be able to:

① Solve equations with fractions having numerical denominators.

② Solve equations with fractions having a variable as denominator.

Math Pro   Video 6.3   SSM

① **Solving Equations Containing Fractions Having Numerical Denominators**

Equations containing two or more fractions can be rather difficult to solve. This difficulty is due to the lengthy process required to compute with fractions. To avoid unnecessary work we transform the given equation containing fractions to an equivalent equation that does not contain fractions. This process is often referred to as **clearing the fractions**. How do we do clear the fractions from the equation? We multiply *all terms* on both sides of the equation by the lowest common denominator (LCD) of all the fractions contained in the equation. That is, we multiply both sides of the equation by the LCD, then use the distributive property so that the LCD is multiplied by each term of the equation. To illustrate this process we solve the equation $\frac{x}{4} + \frac{1}{2} = 5$.

$$\frac{x}{4} + \frac{1}{2} = 5 \qquad \text{The LCD is 4.}$$

$$4\left(\frac{x}{4} + \frac{1}{2}\right) = 4 \cdot 5 \qquad \begin{array}{l}\text{Multiply both sides of the}\\ \text{equation by 4.}\end{array}$$

$$4\left(\frac{x}{4}\right) + 4\left(\frac{1}{2}\right) = 4(5) \qquad \text{We multiply each term by 4.}$$

$$x + 2 = 20 \qquad \begin{array}{l}\text{The equivalent equation does not}\\ \text{contain fractions; we cleared the}\\ \text{fractions from the equation.}\end{array}$$

$$x = 18$$

The equation $x + 2 = 20$ is equivalent to $\frac{x}{4} + \frac{1}{2} = 5$ and much easier to work with.

EXAMPLE 1   Solve for x and check your solution: $\frac{x}{3} + \frac{x}{2} = 5$.

SOLUTION   First, we clear the fractions from the equation by multiplying each term by the LCD = 6; then we solve the equation.

$$6\left(\frac{x}{3}\right) + 6\left(\frac{x}{2}\right) = 6(5) \qquad \text{Multiply each term by the LCD to clear the fractions.}$$

$$2x + 3x = 30 \qquad \text{Simplify: } \frac{6x}{3} = 2x \text{ and } \frac{6x}{2} = 3x.$$

$$5x = 30 \qquad \text{Combine like terms.}$$

$$\frac{5x}{5} = \frac{30}{5} \qquad \text{Divide both sides by 5.}$$

$$x = 6$$

Check: $\frac{x}{3} + \frac{x}{2} = 5$

$$\frac{6}{3} + \frac{6}{2} \overset{?}{=} 5$$

$$2 + 3 = 5 \quad \checkmark$$

It is important to *multiply each term on both sides of the equation by the lowest common denominator.* A common mistake made when solving $\frac{x}{3} + \frac{x}{2} = 5$ is to multiply the fractions, $\frac{x}{3}$ and $\frac{x}{2}$, by the LCD but not the 5.

PRACTICE PROBLEM 1

Solve for x and check your solution: $\frac{x}{5} + \frac{x}{2} = 7$.

Let us now write down the steps we have used.

PROCEDURE TO SOLVE AN EQUATION CONTAINING FRACTIONS

1. Determine the LCD of all the denominators.
2. Multiply each term of the equation by the LCD (clear the fraction).
3. Solve the resulting equation.
4. Check your solution.

EXAMPLE 2   Solve for x: $-4x + \frac{3}{2} = \frac{2}{5}$.

SOLUTION

$$10(-4x) + 10\left(\frac{3}{2}\right) = 10\left(\frac{2}{5}\right)$$ Multiply each term by the LCD = 10.

$$-40x + 15 = 4$$ Simplify: $10 \cdot \frac{3}{2} = 15$ and $10 \cdot \frac{2}{5} = 4$.

$$-40x + 15 = \ \ \ 4$$ Solve for x.
$$+ \ \underline{\ \ \ -15 \ \ \ \ -15}$$
$$-40x \ \ \ \ \ = -11$$

$$\frac{-40x}{-40} = \frac{-11}{-40}$$

$$x = \frac{11}{40}$$

We leave the check to the student.

PRACTICE PROBLEM 2

Solve for x: $-5x + \frac{2}{7} = \frac{3}{2}$.

EXAMPLE 3   Solve for x: $\frac{x}{7} + x = 8$.

SOLUTION   Since the x has a denominator of 1, $x = \frac{x}{1}$, the LCD is 7.

$$7\left(\frac{x}{7}\right) + 7(x) = 7(8)$$ Multiply each term by the LCD.

$$1x + 7x = 56$$ Simplify: $\frac{7x}{7} = 1x$.

$$8x = 56$$ Combine like terms.

$$\frac{8x}{8} = \frac{56}{8}$$ Divide both sides by 8.

$$x = 7$$

PRACTICE PROBLEM 3

Solve for x: $\dfrac{x}{4} + x = 5$.

2 Solving Equations Containing Fractions Having a Variable as Denominator

Sometimes the denominator is a variable. Since we cannot divide by 0, we assume that the variable in the denominator is not 0.

EXAMPLE 4 Solve for x: $\dfrac{5}{x} + 1 = \dfrac{1}{3}$.

SOLUTION

$$3x\left(\dfrac{5}{x}\right) + 3x(1) = 3x\left(\dfrac{1}{3}\right)$$

Multiply each term by the LCD $= 3x$.

$$15 + 3x = 1x$$

Simplify: $\dfrac{3x \cdot 5}{x} = 15$ and $\dfrac{3x}{3} = 1x$.

$$\begin{array}{r} 15 + 3x = 1x \\ + \quad -3x \quad -3x \\ \hline 15 \quad = -2x \end{array}$$

Subtract $3x$ from both sides.

$$\dfrac{15}{-2} = \dfrac{-2x}{-2}$$

Divide both sides by -2.

$$\dfrac{15}{-2} = x \quad \text{or} \quad x = -\dfrac{15}{2}$$

PRACTICE PROBLEM 4

Solve for x: $\dfrac{4}{x} + 2 = \dfrac{1}{7}$.

EXERCISES 6.3

Solve for x and check your solution.

1. $\frac{x}{2} + \frac{x}{4} = 12$

2. $\frac{x}{3} + \frac{x}{6} = 9$

3. $\frac{x}{6} + \frac{x}{4} = 5$

4. $\frac{x}{8} + \frac{x}{12} = 5$

5. $\frac{x}{2} - \frac{x}{7} = 10$

6. $\frac{x}{2} - \frac{x}{9} = 7$

7. $3x + \frac{2}{3} = \frac{9}{2}$

8. $2x + \frac{1}{4} = \frac{2}{3}$

9. $5x + \frac{1}{8} = \frac{3}{4}$

10. $3x + \frac{2}{5} = \frac{1}{2}$

11. $5x - \frac{1}{2} = \frac{1}{8}$

12. $3x - \frac{1}{3} = \frac{1}{9}$

13. $-2x + \frac{1}{2} = \frac{3}{7}$

14. $-3x + \frac{1}{4} = \frac{1}{3}$

15. $-4x + \frac{2}{3} = \frac{1}{6}$

16. $-2x + \frac{1}{5} = \frac{1}{10}$

17. $\frac{x}{5} + x = 6$

18. $\frac{x}{3} + x = 4$

19. $\frac{x}{3} + x = 6$

20. $\frac{x}{7} + x = 8$

21. $\frac{x}{2} + x = 3$

22. $\frac{x}{5} - 2x = 8$

23. $\frac{x}{4} - 3x = 3$

24. $\frac{x}{3} - 4x = 5$

25. $\frac{5}{x} + 1 = \frac{1}{6}$

26. $\frac{4}{x} + 1 = \frac{1}{3}$

27. $\frac{3}{x} + 1 = \frac{1}{5}$

28. $\frac{2}{x} + 1 = \frac{1}{3}$

29. $\frac{5}{x} - 2 = \frac{5}{6}$

30. $\frac{6}{x} - 4 = \frac{2}{5}$

VERBAL AND WRITING SKILLS

31. To clear the fractions from the equation $\dfrac{x}{6} + \dfrac{x}{4} = 3$, Sara multiplied each term by 6. The result was an equation that still had a fraction. Why didn't this multiplication clear the fractions?

32. Do you have to clear the fractions from an equation in order to solve it? Explain.

ONE STEP FURTHER

Solve for x.

33. $x + \dfrac{2}{3} + 2 = \dfrac{3}{2} + \dfrac{1}{4}$

34. $2x + \dfrac{3}{4} + 3 = \dfrac{2}{3} + \dfrac{1}{6}$

35. $4 + \dfrac{6}{x} + \dfrac{2}{5} = \dfrac{3}{2x}$

36. $3 + \dfrac{2}{x} + \dfrac{5}{6} = \dfrac{5}{3x}$

CUMULATIVE REVIEW

Combine like terms.

1. (a) $4x + 5 + 7x$

(b) $3y + 2x + 8 + 4y$

Multiply.

2. (a) $(2x)(4x)$

(b) $(5y)(3y^3)(2y)$

3. $(-4x)(2x^2)$

4. $(3y)(-2y)(5y)$

After studying this section, you will be able to:

❶ Identify the types of polynomials.

❷ Multiply mononials times polynomials.

❸ Multiply binomials using FOIL.

Math Pro   Video 6.4   SSM

S E C T I O N   6 . 4

Multiplication of Polynomials

We use the distributive property often in mathematics. In this section you will see how we use this property to multiply variable expressions such as polynomials, monomials, and binomials.

❶ Identifying the Types of Polynomials

The type of expressions we deal with in mathematics are called **polynomials**. Polynomials are variable expressions that contain terms with nonnegative integer exponents. The following three expressions are all polynomials:

$$3xy + 1 \qquad 2a^3 - 3 \qquad 4x + 2y - 9$$

There are special names for polynomials with one, two, or three terms.

A *monomial* has *one* term. $4a$

A *binomial* has *two* terms. $5a^3 + 3b$

A *trinomial* has *three* terms. $3x^2 + 7x - 4$

E X A M P L E   1   Identify each polynomial as a monomial, binomial, or trinomial.

(a) $2x^2 + 5$ **(b)** $8x$ **(c)** $5x^3 + 8x - 1$

SOLUTION

(a) $2x^2 + 5$ binomial   There are 2 terms.

(b) $8x$ monomial   There is 1 term.

(c) $5x^3 + 8x - 1$ trinomial   There are 3 terms.

P R A C T I C E   P R O B L E M   1

Identify each polynomial as a monomial, binomial, or trinomial.

(a) $4x^2$ **(b)** $8x^2 - 9x + 1$ **(c)** $5x^3 + 8x$

Operations involving polynomials often require that we identify the terms of a polynomial. In a variable expression such as a polynomial, the sign in front of the term is considered part of the term. Thus the terms of the polynomial $12ab^2 - 3b^2 + 2a - 5b$ are $+12ab^2, -3b^2, +2a$, and $-5b$.

Polynomial: $12ab^2 - 3b^2 + 2a - 5b$

Terms: $+12ab^2, -3b^2, +2a, -5b$

E X A M P L E   2   Identify the terms of the polynomial: $xy^2 - 7y^2 - 2x + 5y$.

SOLUTION   We include the sign in front of the term as part of the term.

Polynomial: $xy^2 - 7y^2 - 2x + 5y$

Terms: $+xy^2, -7y^2, -2x, +5y$

P R A C T I C E   P R O B L E M   2

Identify the terms of the polynomial: $y^2 - 4x^2 + 5x - 9y$.

2 Multiplying Monomials Times Polynomials

We are familiar with using the distributive property to multiply a monomial times a binomial: $2(x + 3)$. We use the same process to multiply a monomial times a polynomial. That is, we multiply each term of the polynomial by the monomial.

When we use the distributive property with signed numbers, it is important to identify each term carefully when we multiply so that we take the *sign of the term* into consideration when multiplying.

EXAMPLE 3   Multiply: $(-4x)(2x - 6y - 7)$.

SOLUTION   We multiply each term by $-4x$.

$$(-4x)(2x - 6y - 7)$$
$$-8x^2 + 24xy + 28x$$

First term of the product: $(-4x)(2x) = -8x^2$
Second term of the product: $(-4x)(-6y) = +24xy$
Third term of the product: $(-4x)(-7) = +28x$

PRACTICE PROBLEM 3
Multiply: $(-6x)(3x - 8y - 2)$.

Multiplying a polynomial by a negative monomial has the effect of changing the sign of each term of the polynomial.

To illustrate, let's look at the product in Example 3, where we multiplied by the negative monomial, $-4x$.

A *positive* term changes to a *negative* term.

$$(-4x)(2x - 6y - 7) = -8x^2 + 24xy + 28x$$

Negative terms change to *positive* terms.

Therefore, when multiplying by a negative monomial it is a good idea to check the product, verifying that the sign of each term changes.

The distributive property also works if the multiplier (monomial) is on the right. The process is easier, however, if we move the monomial to the left side of the polynomial before we multiply.

EXAMPLE 4   Multiply: $(3x^2 - 6)(-7x^3)$.

SOLUTION   We move the monomial to the left side:

$$(-7x^3)(3x^2 - 6)$$

We multiply each term by $(-7x^3)$.

$$(-7x^3)(3x^2 - 6)$$
$$= -21x^5 + 42x^3$$

$(-7x^3)(3x^2) = -21x^5$
$(-7x^3)(-6) = +42x^3$

Since we are multiplying by a negative monomial, we check the sign of each term in the product.

A *positive* term changes to a *negative* term.

$$(-7x^3)(3x^2 - 6) = -21x^5 + 42x^3 \quad \checkmark$$

A *negative* term changes to a *positive* term.

PRACTICE PROBLEM 4
Multiply: $(2y^2 - 5)(-3y^4)$.

❸ Multiplying Binomials Using FOIL

In this section we use a repeated application of the distributive property to multiply a binomial times a binomial. To illustrate this process, we multiply $(x + 2)(x + 3)$ by multiplying each term of $(x + 2)$ times the binomial $(x + 3)$.

$$\overbrace{x \cdot (x + 3)}$$
$$(x + 2)(x + 3) = (x + 2) \boxed{(x + 3)}$$
$$\underbrace{(+2) \cdot (x + 3)}$$

$$
\begin{aligned}
(x + 2)(x + 3) &= \boldsymbol{x}(x + 3) + \boldsymbol{2}(x + 3) & & \text{Multiply:}\\
&= x \cdot x + x \cdot 3 + 2 \cdot x + 2 \cdot 3 & & x(x + 3) = x \cdot x + x \cdot 3\\
&= x^2 + 3x + 2x + 6 & & (+2)(x + 3) = +2 \cdot x + 2 \cdot 3.\\
&= x^2 + 5x + 6 & & \text{Combine like terms.}
\end{aligned}
$$

Thus $(x + 2)(x + 3) = x^2 + 5x + 6$.

EXAMPLE 5   Use the distributive property to multiply: $(x + 1)(x + 4)$.

SOLUTION

$$\overbrace{x \cdot (x + 4)}$$
$$(x + 1) \boxed{(x + 4)} = x(x + 4) + 1(x + 4)$$
$$\underbrace{(+1) \cdot (x + 4)}$$

$$
\begin{aligned}
(x + 1)(x + 4) &= \boldsymbol{x}(x + 4) + \boldsymbol{1}(x + 4)\\
&= x \cdot x + x \cdot 4 + 1 \cdot x + 1 \cdot 4 & & \text{Use the distributive}\\
&= x^2 + 4x + 1x + 4 & & \text{property again;}\\
& & & \text{multiply } x(x + 4)\\
& & & \text{and } (+1)(x + 4).\\
&= x^2 + 5x + 4 & & \text{Combine like terms.}
\end{aligned}
$$

PRACTICE PROBLEM 5
Use the distributive property to multiply: $(x + 3)(x + 5)$.

The distributive property shows us how the problem can be done and why it can be done. In actual practice there is a somewhat easier approach to obtain the answer. It is often referred to as the **FOIL method**. The letters FOIL stand for:

F   Multiply the *First* terms.
O   Multiply the *Outer* terms.
I   Multiply the *Inner* terms.
L   Multiply the *Last* terms.

$$(x + 1) \qquad\qquad (x + 1)(x + 4) \qquad (x + 1)(x + 4)$$

first term of   last term of   outer terms   inner terms
a binomial   a binomial

The FOIL letters are simply a way to remember the four terms in the final product and how they are obtained. Let's return to Example 5 and rewrite the steps we used to multiply the binomials using the distributive property, then compare these results with the FOIL method.

The distributive property:

$$(x + 1)(x + 4) = x(x + 4) + 1(x + 4)$$
$$= x \cdot x + x \cdot 4 + 1 \cdot x + 1 \cdot 4$$
$$= x^2 + 4x + 1x + 4$$
$$= x^2 + 5x + 4$$

The result is the same.

The FOIL method:

$$(x + 1)(x + 4) \qquad \text{F} \quad \text{Multiply } \textit{first} \text{ terms:} \qquad x \cdot x = \boxed{x^2}$$

$$(x + 1)(x + 4) \qquad \text{O} \quad \text{Multiply } \textit{outer} \text{ terms:} \qquad (+4)x = \boxed{+4x}$$

$$(x + 1)(x + 4) \qquad \text{I} \quad \text{Multiply } \textit{inner} \text{ terms:} \qquad (+1)x = \boxed{+1x}$$

$$(x + 1)(x + 4) \qquad \text{L} \quad \text{Multiply } \textit{last} \text{ terms:} \qquad (+1)(+4) = \boxed{+4}$$

$$(x + 1)(x + 4) = x^2 + 4x + 1x + 4$$
$$= x^2 + 5x + 4$$

Our result is the same as when we used the distributive property. Now let's study the use of the FOIL method in a few examples.

EXAMPLE 6   Multiply: $(x + 5)(x + 3)$.

SOLUTION

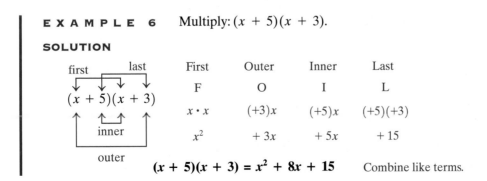

$$(x + 5)(x + 3) = x^2 + 8x + 15 \qquad \text{Combine like terms.}$$

PRACTICE PROBLEM 6

Multiply: $(x + 2)(x + 4)$.

EXAMPLE 7   Multiply: $(y - 1)(y - 7)$.

SOLUTION

	First	Outer	Inner	Last
	F	O	I	L
$(y - 1)(y - 7)$	y^2	$-7y$	$-1y$	$+7$

$$(y - 1)(x - 7) = y^2 - 8y + 7 \qquad \text{Combine like terms.}$$

Note: The products of the outer terms and the inner terms are negative. We must be sure to pay special attention to the signs of the terms when we multiply.

PRACTICE PROBLEM 7

Multiply: $(y - 5)(y - 3)$.

EXAMPLE 8   Multiply: $(2x + 3)(x - 1)$.

SOLUTION

	First	Outer	Inner	Last
	F	O	I	L
$(2x + 3)(x - 1)$	$2x^2$	$-2x$	$+3x$	-3

$$(2x + 3)(x - 1) = 2x^2 + x - 3 \qquad \text{Combine like terms.}$$

Be sure to check the sign of each term.

PRACTICE PROBLEM 8

Multiply: $(3x - 1)(x + 1)$.

EXERCISES 6.4

Identify each polynomial as a monomial, binomial, or trinomial.

1. (a) $5z^3 + 4$ **(b)** 9 **(c)** $2x^7 - 3x^4 - 3$

2. (a) $6a^4$ **(b)** $5y - 9y - 3$ **(c)** $7z - 6$

Identify the terms of each polynomial.

3. $2z^2 + 4z - 2y^4 + 3$ **4.** $4x^5 + 2c^2 - 6c + 4$

5. $6x^6 - 3x^3 - 3y - 7$ **6.** $3a^7 + 4b^4 - 7a - 7$

Multiply.

7. $8(2x^2 + 3x - 5)$ **8.** $4(5y^2 + 7y - 3)$ **9.** $(-5)(2y^2 + 3y - 6)$

10. $(-4)(8y^2 + 4y - 2)$ **11.** $(-2)(5y^2 + 2y - 8)$ **12.** $(-5y)(y - 8)$

13. $(-6x)(x - 2)$ **14.** $(-9a)(a + 1)$ **15.** $(-4a)(2a + 3)$

16. $(-7x)(3x - 2y + 4)$ **17.** $(-2x)(4x - 2y + 5)$ **18.** $(-3a)(2a - 6b - 5)$

19. $(-5a)(2a - 4b - 7)$ **20.** $(-4x)(-3x + 5y - 4)$ **21.** $(-2x)(-4x + 2y - 8)$

22. $(-7x^3)(x - 3)$ **23.** $(-8x^3)(x - 5)$ **24.** $(-6x^5)(x^2 + 4)$

25. $(-4y^2)(y^6 + 9)$ **26.** $(x^3 + 2)(-5x^2)$ **27.** $(x^2 + 1)(-3x^4)$

28. $(y^5 - 3)(-2y^2)$ **29.** $(y^4 - 6)(-4y^5)$ **30.** $(x^3)(x^5 + x^3 + 1)$

31. $(x^4)(x^3 + x^2 + 1)$ **32.** $(x^2)(2x^6 + x^4 + 5)$ **33.** $(x^3)(3x^3 + x^2 + 4)$

Use FOIL to multiply.

34. $(x + 6)(x + 7)$ **35.** $(a + 2)(a + 1)$ **36.** $(x + 3)(x + 9)$ **37.** $(y + 2)(y + 5)$

38. $(a + 6)(a + 2)$ **39.** $(x + 4)(x + 1)$ **40.** $(y + 4)(y - 8)$ **41.** $(a + 7)(a - 4)$

42. $(x + 2)(x - 4)$ **43.** $(x + 3)(x - 5)$ **44.** $(z + 5)(z - 2)$ **45.** $(b + 3)(b - 2)$

46. $(x - 2)(x + 4)$ **47.** $(m - 3)(m + 5)$ **48.** $(y - 2)(y + 7)$ **49.** $(z - 9)(z + 5)$

50. $(2x + 1)(x + 2)$ **51.** $(3x + 1)(x + 2)$ **52.** $(3x + 3)(x + 1)$ **53.** $(4x + 3)(x + 1)$

54. $(2y - 1)(y + 2)$ **55.** $(4y - 2)(y + 1)$ **56.** $(2y + 1)(y - 2)$ **57.** $(4y + 2)(y - 1)$

58. (a) $(z + 5)(z + 1)$ **(b)** $(z - 5)(z + 1)$ **(c)** $(z - 5)(z - 1)$ **(d)** $(z + 5)(z - 1)$

59. (a) $(x + 4)(x + 1)$ **(b)** $(x - 4)(x + 1)$ **(c)** $(x - 4)(x - 1)$ **(d)** $(x + 4)(x - 1)$

 VERBAL AND WRITING SKILLS

60. Erin multiplied $(-4)(x^2 + 2x + 1)$ and obtained this result: $-4x^2 - 8x + 4$. What is wrong with this multiplication?

61. Write in words the multiplication that the word FOIL represents.

ONE STEP FURTHER

62. If $a(2x - 3) = -14x + 21$, what is the value of a?

63. If $b(-3x + 4) = 15x - 20$, what is the value of b?

Use the distributive property to simplify.

64. $-3ab(-4a + 6b - 9)$

65. $-2xy(-3x + 5y - 2)$

66. $(x + 1)(x^2 + 3x + 1)$

67. $(x + 2)(x^2 + 2x + 2)$

CUMULATIVE REVIEW

Perform the indicated operations.

1. $-2 - (-5) + (-7)$

2. $-2x^2 + 7x^2 + (-3x^2)$

3. $-3x^2 + 5x^2 + (-6x^2)$

392

Addition and Subtraction of Polynomials

After studying this section, you will be able to:

❶ Simplify polynomials with grouping symbols.

❷ Add and subtract polynomials.

Math Pro   Video 6.5   SSM

When we write operations with polynomials, we often place grouping symbols such as parentheses, brackets, and braces around each polynomial to distinguish between the polynomials. In this section we see how grouping symbols are used when we add or subtract polynomials.

❶ Simplifying Polynomials with Grouping Symbols

To simplify expressions such as $(-1)(2x + 6)$, we often say that we remove parentheses. Of course we didn't just take them away; we used the distributive property to multiply each term inside the parentheses by -1. Therefore, for the remainder of this book we say "remove parentheses" as a shorthand direction for using the distributive property to multiply.

A negative sign in front of parentheses is equivalent to having a coefficient of negative 1. You can write the -1 and then multiply by -1 using the distributive property.

$$-(x + 4) = -1(x + 4) = -x - 4$$

Similarly, a positive sign in front of parentheses can be viewed as multiplication by $+1$.

$$+(2x - 3) = +1(2x - 3) = 2x - 3$$

If a grouping symbol has a positive or negative sign in front, we mentally multiply by $+1$ or -1. Recall that when we multiply a polynomial by a negative number, the sign of each term of the polynomial changes. Thus, instead of actually multiplying by negative 1, we can simply change the sign of each term of the polynomial, then remove the parentheses.

> Multiplying a polynomial by $(+1)$ has the effect of removing the parentheses and *leaving the terms unchanged*.
>
> Multiplying a polynomial by (-1) has the effect of removing the parentheses, but *the sign of each term is changed*.

E X A M P L E 1   Simplify.

(a) $+(x - 4)$ **(b)** $-(-2a + 5b - 7c)$

SOLUTION

(a) $+(x - 4)$ $=$ $(+1)(x - 4) = x - 4$
              ↓   ↓
         "+" in front   We multiply each term by $(+1)$:
         of parentheses $(+1)(x) = x$; $(+1)(-4) = -4$.

Since multiplication by $+1$ has the same effect as removing parentheses and leaving the terms unchanged, we often skip the multiplication step.

(b) $-(-2a + 5b - 7c) = (-1)(-2a + 5b - 7c) = 2a - 5b + 7c$
              ↓   ↓
         "−" in front   We multiply each term by (-1):
         of parentheses $(-1)(-2a) = 2a$; $(-1)(5b) = -5b$; $(-1)(-7c) = 7c$.

Since multiplication by -1 has the same effect as removing parentheses and changing the sign of each term, we often skip the multiplication step.

PRACTICE PROBLEM 1

Simplify.

(a) $+(a - 2)$ **(b)** $-(-4x + 6y - 2n)$.

❷ Adding and Subtracting Polynomials

We write the sum of the polynomials $8x^2 + 2x - 4$ and $6x^2 - 2$ as follows: $(8x^2 + 2x - 4) + (6x^2 - 2)$. Since the "+" in front of the parentheses can be considered as multiplication by $+1$, to add polynomials we can remove the parentheses and combine like terms.

$$\begin{aligned}
&\overset{\substack{\text{"+" in front} \\ \text{of parentheses} \\ \downarrow}}{} \qquad\qquad \overset{\substack{\text{We remove parentheses and} \\ \text{leave the terms unchanged.} \\ \downarrow}}{} \\
(8x^2 + 2x - 4) + (6x^2 - 2) &= 8x^2 + 2x - 4 + 6x^2 - 2 \\
&= \underline{8x^2} + 2x - 4 \underline{+ 6x^2} - 2 \qquad \text{Now we} \\
&= \underline{14x^2} + 2x - 4 - 2 \qquad\quad \text{combine} \\
(8x^2 + 2x - 4) + (6x^2 - 2) &= 14x^2 + 2x - 6 \qquad\qquad \text{like terms.}
\end{aligned}$$

EXAMPLE 2   Perform the indicated operations: $(-3x + 5) + (2x - 6)$.

SOLUTION

$$\begin{aligned}
&(-3x + 5) + (2x - 6) \qquad &\text{A "+" sign is in front of the parentheses.} \\
&= -3x + 5 + 2x - 6 \qquad &\text{We remove parentheses and leave the} \\
& &\textit{terms unchanged.} \\
&= \underline{-3x} + 5 \underline{+ 2x} - 6 \qquad &\text{Now we combine like terms.} \\
&= -1x - 1 \quad \text{or} \quad \boldsymbol{-x - 1}
\end{aligned}$$

PRACTICE PROBLEM 2

Perform the indicated operations: $(8z - 9) + (-7z + 4)$.

We write the difference of the polynomials $3x^2 + 2x$ and $4x^2 - 1$ as follows: $(3x^2 + 2x) - (4x^2 - 1)$. Since the "−" in front of the parentheses can be considered as multiplication by -1, to subtract polynomials we can remove the parentheses and *change the signs of the terms* that follow the "−." Then we combine like terms.

$$\begin{aligned}
&\overset{\substack{\text{"−" in front} \\ \text{of parentheses} \\ \downarrow}}{} \qquad\qquad \overset{\substack{\text{We remove parentheses and change} \\ \text{the signs of terms that follow "−."} \\ \downarrow}}{} \\
(3x^2 + 2x) - (4x^2 - 1) &= 3x^2 + 2x - 4x^2 + 1 \\
&= \underline{3x^2} + 2x \underline{- 4x^2} + 1 \qquad \text{Now we combine} \\
& \qquad\qquad\qquad\qquad\qquad\quad \text{like terms.} \\
&= \underline{-1x^2} + 2x + 1 \quad \text{or} \quad -x^2 + 2x + 1
\end{aligned}$$

EXAMPLE 3 Perform the indicated operations: $(2x + 3) - (5x - 6)$.

SOLUTION

A "−" sign is in front of the parentheses.

$(2x + 3) - (5x - 6)$

We remove parentheses and *change the signs of terms* that follow "−".

$= 2x + 3 - 5x + 6$

Now we combine like terms.

$= 2x + 3 - 5x + 6$

$= \mathbf{-3x + 9}$

PRACTICE PROBLEM 3

Perform the indicated operations: $(4y + 7) - (8y - 6)$.

EXAMPLE 4 Perform the indicated operations: $(3x^2 + 5x - 7) - (6x^2 - 8x - 1)$.

SOLUTION

$(3x^2 + 5x - 7) - (6x^2 - 8x - 1)$

$= 3x^2 + 5x - 7 - 6x^2 + 8x + 1$

Remove parentheses and *change the signs of terms* that follow "−."

$= 3x^2 + 5x - 7 - 6x^2 + 8x + 1$

Simplify by combining like terms.

$= \mathbf{-3x^2 + 13x - 6}$

PRACTICE PROBLEM 4

Perform the indicated operations: $(4x^2 + 6x - 8) - (7x^2 - 6x - 5)$.

Sometimes it is necessary to simplify before we add or subtract polynomials.

EXAMPLE 5 Perform the indicated operations: $6x - 3(-4x^2 + 3) - (-2x^2 + x - 5)$.

SOLUTION   First we multiply (-3) times the binomial $(-4x^2 + 3)$.

$6x - 3(-4x^2 + 3) - (-2x^2 + x - 5)$

$= 6x + 12x^2 - 9 - (-2x^2 + x - 5)$

Multiply: $-3(-4x^2) = 12x^2$; $-3(+3) = -9$.

$= 6x + 12x^2 - 9 + 2x^2 - x + 5$

Remove parentheses and change the sign of each term following the "−."

$= \mathbf{5x + 14x^2 - 4}$

Combine like terms.

PRACTICE PROBLEM 5

Perform the indicated operations: $4x - 2(3x^2 + 1) - (-3x^2 + x - 6)$.

EXERCISES 6.5

Simplify.

1. $+(5m - 2)$ **2.** $+(7y - 1)$ **3.** $-(-3x + 6z - 5y)$ **4.** $-(-5a + 3b - 7c)$

5. $+(-9x - 4)$ **6.** $+(-3x - 6)$ **7.** $-(5x + 2y)$ **8.** $-(9x + 6z)$

9. $-(-8x + 9)$ **10.** $-(-7x + 6)$ **11.** $+(-2x - 8y + 3z)$ **12.** $+(-3a - 6b + 5c)$

Perform the indicated operations.

13. $(7y - 3) + (-4y + 6)$ **14.** $(5y - 2) + (-7y + 8)$

15. $(-3a + 5) + (4a - 3)$ **16.** $(-6c + 3) + (2c - 7)$

17. $(2y - 5) + (-8y + 2)$ **18.** $(9z - 3) + (-6z + 9)$

19. $(2x + 8) - (4x - 1)$ **20.** $(3x + 5) - (9x - 1)$

21. $(7x + 3) - (-4x - 6)$ **22.** $(5y + 2) - (-7y - 8)$

23. $(-8a + 5) - (4a - 3)$ **24.** $(-6c + 3) - (2c - 7)$

25. $(3y^2 + 4y - 6) - (4y^2 - 6y - 9)$ **26.** $(5z^2 + 8z - 5) - (6z^2 - 3z - 9)$

27. $(2x^2 + 6x - 5) - (6x^2 - 4x - 8)$ **28.** $(3a^2 + 4a - 7) - (7a^2 - 2a - 5)$

29. $(-6z^2 + 9z - 1) - (3z^2 + 8z - 7)$ **30.** $(-9x^2 + 4x - 9) - (6x^2 + 2x - 8)$

31. $(2a^2 + 9a - 1) - (-5a^2 - 8a - 4)$ **32.** $(7c^2 - 3c + 6) - (-9c^2 + 2c - 8)$

33. $(-6x^2 - 6x - 1) - (3x^2 + 8x + 4)$ **34.** $(-5m^2 - 2m - 9) - (5m^2 + 2m + 7)$

35. $(x^2 - x - 1) + (-x^2 + 4x + 6) - (x^2 + 1)$ **36.** $(x^2 - x - 7) + (-x^2 + 3x + 8) - (x^2 + 7)$

37. $\left(-4x^2 - 1\right) - \left(3x^2 + 8x + 2\right) + \left(-x^2 - 9\right)$

38. $\left(-8x^2 - 9\right) - \left(5x^2 + 6x + 2\right) + \left(-x^2 - 5\right)$

39. $\left(6x + 9\right) - \left(-4x^2 + 7x + 1\right) - \left(x^2 - 7\right)$

40. $\left(2x - 5\right) - \left(-3x^2 + 7x + 1\right) - \left(x^2 - 3\right)$

41. $\left(4x^2 - 5x - 8\right) - \left(3x^2 - 8x - 1\right) + \left(x^2 - 9\right)$

42. $\left(7x^2 - 5x - 6\right) - \left(6x^2 - 9x + 1\right) - \left(x^2 + 7\right)$

43. $\left(4x^2 + 6x - 2\right) - \left(3x^2 + 7x + 1\right) + \left(x^2 - 1\right)$

44. $\left(8x^2 - 3x + 4\right) + \left(-5x^2 + 6x + 3\right) - \left(6x^2 - 1\right)$

45. $\left(-7x^2 + 3x - 4\right) - \left(5x^2 + 7x - 1\right) + \left(x^2 - 8\right)$

46. $\left(-3x^2 + 2x - 4\right) - \left(4x^2 + 6x - 5\right) + \left(x^2 - 7\right)$

47. $5x - 2\left(4x^2 + 3x - 2\right) - \left(2x^2 + 8x - 1\right)$

48. $2x - 3\left(5x^2 + 2x - 6\right) - \left(2x^2 + 8x - 1\right)$

49. $4(-9x - 6) - \left(3x^2 - 7x + 1\right) + 4x$

50. $2(-4x + 2) - \left(2x^2 + 2x - 4\right) + 3x$

 VERBAL AND WRITING SKILLS

51. Explain why we change the signs of all the terms inside the parentheses when we remove the parentheses if there is a negative sign in front of the parentheses.

52. Review the exercises you just completed, then describe the types of mistakes you made on those you did not get correct. Explain what you will do to avoid these types of errors. If you did not make any errors, explain what you do to avoid errors.

ONE STEP FURTHER

Determine the value of a.

53. $(ax + 3) + \left(2x^2 + 5x - 6\right) + (8x - 2) = 2x^2 - 10x - 5$

54. $(ax - 5) + \left(4x^2 + 6x + 9\right) + (3x - 1) = 4x^2 - 2x + 3$

CUMULATIVE REVIEW

Solve for x.

1. $2x + 3 = 10$ **2.** $5x + 2 = 11$ **3.** $-3x - 5 = 14$ **4.** $-2x - 3 = 12$

After studying this section, you will be able to:

1 Solve equations with parentheses and polynomial expressions.

Math Pro   Video 6.6   SSM

Solving Equations Involving Polynomial Expressions

1 Solving Equations with Parentheses and Polynomial Expressions

Sometimes we must perform several steps of simplifying an equation before we can begin the process of solving for the variable. If equations contain parentheses, we remove the parentheses first. Then when we simplify the polynomial expressions, the equations become just like those encountered previously in this chapter. We review the procedure to solve equations presented in Section 6.2.

PROCEDURE TO SOLVE EQUATIONS

1. *Parentheses.* Remove any parentheses.
2. *Simplify each side of the equation.* Collect like terms and simplify numerical work.
3. *Isolate the ax term: ax = some number.* Use the addition principle to get all *ax* terms on one side of the equation and the numerical values on the other side.
4. *Isolate the x term: x = some number or some number = x.* Use the multiplication or division principle to get *x* alone on one side of the equation.
5. Check your solution.

E X A M P L E   1   Solve for x and check: $-3(2x + 1) + 4x = 27$.

SOLUTION   We use the distributive property to remove parentheses; then we simplify and solve.

$$-3(2x + 1) + 4x = 27$$
$$-6x - 3 + 4x = 27 \qquad \text{Multiply: } (-3)(2x) = -6x; (-3)(+1) = -3.$$
$$-2x - 3 = 27 \qquad \text{Combine like terms: } -6x + 4x = -2x.$$
$$\underline{+ + 3 \quad +3} \qquad \text{Add 3 to both sides of the equation.}$$
$$-2x = 30$$
$$\frac{-2x}{-2} = \frac{30}{-2} \qquad \text{Divide by } -2 \text{ on both sides of the equation.}$$
$$x = \mathbf{-15}$$

Check: $-3(2x + 1) + 4x = 27$

$$-3[2(-15) + 1] + 4(-15) \overset{?}{=} 27$$
$$-3[-30 + 1] + (-60) \overset{?}{=} 27$$
$$-3[-29] + (-60) \overset{?}{=} 27$$
$$27 = 27 \quad ✓$$

You should check your solution in the *original* equation. Why?

P R A C T I C E   P R O B L E M   1

Solve for x and check: $-5(3x + 2) - 6x = 32$.

We must pay special attention to the *sign* of each term.

E X A M P L E   2   Solve for y and check: $5(y + 8) = -6(y - 2) + 94$.

SOLUTION

$$5(y + 8) = -6(y - 2) + 94$$

$$5(y) + 5(8) = -6(y) - 6(-2) + 94 \qquad \text{Remove parentheses.}$$

$$5y + 40 = -6y + 12 + 94 \qquad \text{Simplify each side of the equation separately.}$$

$$
\begin{array}{rl}
5y + 40 = -6y + 106 & \text{We add } 6y \text{ to both sides so that} \\
\underline{+6y \qquad\quad +6y} & \text{all } y \text{ terms are on one side of the} \\
11y + 40 = \qquad 106 & \text{equation.}
\end{array}
$$

$$
\begin{array}{rl}
11y + 40 = \ 106 & \text{Now we use the principles of} \\
\underline{+ \quad -40 \ \ -40} & \text{equality to solve for } y. \\
11y \quad\ = \ 66 &
\end{array}
$$

$$\frac{11y}{11} = \frac{66}{11}$$

$$y = \mathbf{6}$$

Check: $5(y + 8) = -6(y - 2) + 94$

$$5(6 + 8) \overset{?}{=} -6(6 - 2) + 94$$

$$5(14) \overset{?}{=} -6(4) + 94$$

$$70 \overset{?}{=} -24 + 94$$

$$70 = 70 \ \checkmark$$

P R A C T I C E   P R O B L E M   2
Solve for x and check: $2(x + 4) = -7(x - 3) + 2$.

E X A M P L E   3   Solve for x: $(2x^2 + 3x + 1) - (2x^2 + 6) = 4x + 1$.

SOLUTION   Since the first polynomial does not have a minus sign in front, we can remove the parentheses. We must change the signs of the terms of the second polynomial since it is preceded by a minus sign.

$$
\begin{array}{ll}
(2x^2 + 3x + 1) - (2x^2 + 6) = 4x + 1 & \text{Change the signs of the terms} \\
2x^2 + 3x + 1 - 2x^2 - 6 = 4x + 1 & \text{that follow the minus sign.} \\
3x - 5 = 4x + 1 & 2x^2 - 2x^2 = 0;\ 1 - 6 = -5. \\
3x - 5 = \quad 4x + 1 & \text{Solve for } x \\
\underline{+ -3x \qquad\quad -3x} & \\
-5 = \quad\ x + 1 & \\
\underline{+ \quad -1 \qquad\ -1} & \\
-6 = \quad\ x & \text{or}\quad \mathbf{x = -6}
\end{array}
$$

We leave the check for the student.

P R A C T I C E   P R O B L E M   3
Solve for x: $(4x^2 + 6x + 3) - (4x^2 + 2) = 3x + 1$.

EXERCISES 6.6

Solve for the variable.

1. $-5(2x + 1) = 25$

2. $-6(2x + 3) = 18$

3. $-4(3x - 1) = 12$

4. $-2(3x - 2) = 15$

5. $-2(4y - 2) = 36$

6. $-5(2y - 1) = 30$

7. $-5(2x + 1) + 3x = 32$

8. $-4(2x + 2) + 4x = 12$

9. $-4(2x + 1) + 4x = -5$

10. $-2(3x + 2) + 2x = -9$

11. $-2(5y - 1) + 7y = -1$

12. $-7(4y - 1) + 15y = -6$

13. $4(-6x + 2) - 6x = 68$

14. $2(-7x + 3) - 2x = 4$

15. $3(y - 4) + 6(y + 1) = 57$

16. $5(y - 2) + 2(y + 4) = 26$

17. $2(x - 1) + 4(x + 2) = 18$

18. $3(x - 4) + 6(x + 1) = 21$

19. $-2(x - 2) + 6(x - 1) = 10$

20. $-3(x - 3) + 4(x - 1) = 16$

21. $6(x - 4) = -9(x + 1) + 10$

22. $4(x - 2) = -2(x + 6) + 31$

23. $2(y + 3) = -6(y - 1) - 7$

24. $3(y + 1) = -2(y - 2) - 6$

25. $6(x - 1) + 2(x + 1) = 10 - 3x$

26. $7(x - 1) + 3(x + 1) = 20 - 4x$

27. $\left(-6x^2 + 4x - 1\right) + \left(6x^2 + 9\right) = 12$

28. $\left(-9y^2 + 8y - 2\right) + \left(9y^2 + 5\right) = 6$

29. $\left(5x^2 + 2x + 3\right) - \left(5x^2 + 9\right) = 4x + 1$

30. $\left(8x^2 + 8x - 2\right) - \left(8x^2 + 5\right) = 6x + 2$

31. $\left(x^2 + 3x - 1\right) - \left(x^2 - 5\right) = 2x + 1$

32. $\left(y^2 + 2y - 5\right) - \left(y^2 - 1\right) = 3y + 2$

ONE STEP FURTHER

33. Is $x = 19$ a solution to:

$$(2x + 9) + (3x - 2) - (5x + 1) = 2(x - 6) - (x - 1)?$$

34. Is $a = 3$ a solution to:

$$(9a + 2) + (4a - 1) - (6a + 3) = 4(a - 1) + 12?$$

CUMULATIVE REVIEW

Translate using symbols.

1. Six more than twice a number.

2. Twelve less than some number.

3. The sum of 4 and x.

4. The sum of 5 and y is multiplied by 2.

1 Solve applied problems involving geometric figures.

2 Solve applied problems involving comparison.

Math Pro   Video 6.7   SSM

Applied Problems Involving Polynomials

One of the first steps in solving many real-life applications is writing the equation that represents the situation. In this section we learn how to write the equations needed to solve an applied problem.

1 Solving Applied Problems Involving Geometric Figures

The Mathematics Blueprint for Problem Solving helps you organize information and plan how to solve applied problems.

E X A M P L E   1   Find the length of the following rectangle if the perimeter of the rectangle is 32 feet.

$$L = x + 6$$

$$W = 2$$

SOLUTION

1. *Understand the problem.*

MATHEMATICS BLUEPRINT FOR PROBLEM SOLVING			
GATHER THE FACTS	WHAT AM I ASKED TO DO?	HOW DO I PROCEED?	KEY POINTS TO REMEMBER
$L = x + 6$ $W = 2$ The formula for the perimeter of a rectangle is: $P = 2L + 2W$. $P = 32$ ft	Determine the length of the rectangle.	1. Substitute $(x + 6)$ and 2 in the formula for perimeter. 2. Solve for x. 3. Substitute the value for x in the expression $L = x + 6$ to find the length. 4. Check my answer.	1. Place the unit "ft" in the answer. 2. After I find x, I must use this value to find L in order to answer the question.

2. *Solve and state the answer.* For the perimeter:

$$P = \quad 2L \quad + 2W$$

$$32 = 2(x + 6) + 2(2) \qquad \text{Substitute.}$$

$$32 = 2x + 12 + 4 \qquad \text{Multiply.}$$

$$32 = 2x + 16 \qquad \text{Simplify.}$$

$$32 = 2x + 16$$

$$\underline{+ -16 \qquad\quad - 16} \qquad \text{Solve the equation.}$$

$$16 = 2x$$

$$\frac{16}{2} = \frac{2x}{2}$$

$$8 = x$$

We substitute this amount into the expression that represents the length.

$$L = x + 6$$
$$\downarrow$$
$$L = 8 + 6 = 14$$

The length of the rectangle is **14 feet**.

    3. *Check.* Since the expression for the length is $L = x + 6$ and $W = 2$ ft, we should expect the length, 14, to be larger than the width, and it is. Now we evaluate the formula for perimeter with our answer to check our calculations:

$$P = 2L + 2W$$
$$32 \overset{?}{=} 2(14) + 2(2)$$
$$32 \overset{?}{=} 28 + 4$$
$$32 = 32 \ \checkmark$$

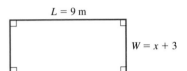

$L = x + 6$ or 14 ft

$W = 2$ ft

PRACTICE PROBLEM 1

Find the width of the rectangle illustrated in the margin if the perimeter of the rectangle is 26 meters.

$L = 9$ m

$W = x + 3$

② Solving Applied Problems Involving Comparison

Often, real-life applications involve comparing two or more quantities. When this is the case, we describe one quantity *in terms of another.* When more than one quantity is compared, it is helpful to let a variable represent the quantity *to which things are being compared.* For example, if John is 4 inches taller than Ed, and Chris is 2 inches shorter than Ed, we are *comparing* all the heights *to Ed's height.* Thus, we let the variable x represent Ed's height, and we describe John's and Chris's heights in terms of Ed's.

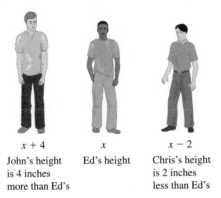

$x + 4$	x	$x - 2$
John's height is 4 inches more than Ed's	Ed's height	Chris's height is 2 inches less than Ed's

Let x = Ed's height, then $(x + 4)$ = John's height, and $(x - 2)$ = Chris's height. $x, (x + 4)$, and $(x - 2)$ are called *variable expressions.* When we write these expressions, we are *defining the variable expressions.* This is an important step; it helps us create the equation needed to find the solution to the problem.

EXAMPLE 2 Define the variable expressions for the length of each side of the triangle: The second side of a triangle is 2 inches longer than the first; the third side is 8 inches shorter than three times the length of the first side.

SOLUTION Since we are *comparing* all sides *to the first side,* we let the *variable represent* the length of the *first side.* We may choose any variable, so we use the letter f.

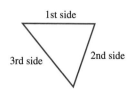

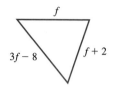

Let f = the length of the first side.

The second side is *2 inches longer than the first side.*

$$\downarrow \quad \downarrow \quad \downarrow$$

second side = $f + 2$

$(f + 2)$ = the length of the second side.

The third side is *8 inches shorter than 3 times the first side.*

$$\downarrow \quad \downarrow \quad \downarrow$$

third side = $3f - 8$

$(3f - 8)$ = the length of the third side.

We define our variable expressions as follows:

$$f = \textbf{the length of the first side}$$
$$(f + 2) = \textbf{the length of the second side}$$
$$(3f - 8) = \textbf{the length of the third side}$$

PRACTICE PROBLEM 2

Define the variable expressions for the length of each side of the triangle: The second side of a triangle is 3 inches longer than the first; the third side is 12 inches shorter than three times the length of the first side.

The following six-step process will help you organize the Mathematics Blueprint and work through the word problem.

SOLVING WORD PROBLEMS

1. *Read the problem* carefully to get an overview.
2. Write down *formulas* or *draw pictures* if possible.
3. *Define the variable expressions.*
4. *Write an equation* using the variable expressions selected.
5. *Solve the equation* and determine the values asked for in the problem.
6. *Check your answer.* Ask if the answers obtained are reasonable.

EXAMPLE 3   For the following word problem:

(a) Define the variable expressions.
(b) Write an equation.
(c) Solve the equation and determine the value asked for.
(d) Check your answer.

Linda is a store manager. The assistant manager earns $8500 less annually than does Linda. The sum of Linda's annual salary and the assistant manager's annual salary is $72,000. How much does each earn annually?

SOLUTION   1. *Understand the problem.* To help us understand the problem, we use the Mathematics Blueprint.

MATHEMATICS BLUEPRINT FOR PROBLEM SOLVING			
GATHER THE FACTS	WHAT AM I ASKED TO DO?	HOW DO I PROCEED?	KEY POINTS TO REMEMBER
Assistant manager earns $8500 less than Linda. The sum of the two salaries is $72,000.	Find Linda's and the assistant manager's salaries.	(a) Define the variable expression for both Linda and the assistant manager. (b) Form an equation by setting the sum of expressions equal to $72,000. (c) Solve the equation and find both salaries. (d) Check my answer.	I must find both Linda's and the assistant manager's salaries.

2. *Solve and state the answer.*

(a) Since we are comparing the assistant manager's salary to Linda's, we let the variable represent Linda's salary.

Let L = Linda's salary

$(L - 8500)$ = the assistant manager's salary

(b)

Linda's salary	+	assistant manager's salary	=	total annual salary for both people
L	+	$(L - 8500)$	=	72,000

The equation is $L + (L - 8500) = 72,000$.

(c)
$$L + (L - 8500) = 72,000$$
$$2L - 8500 = 72,000$$
$$\underline{+\ 8500 \qquad +8,500}$$
$$2L = 80,500$$
$$\frac{2L}{2} = \frac{80,500}{2}$$
$$L = 40,250$$

L = Linda's salary; Linda earns $40,250 annually. $L - 8500$ = the assistant manager's salary ($8500 less than Linda).

$$L - 8500$$
$$\downarrow$$
$$\$40,250 - \$8500 = \$31,750$$

The assistant manager earns $31,750 annually and Linda earns $40,250 annually.

(d) 3. *Check:* Is the sum of their salaries equal to $72,000?

$$\$31,750 + \$40,250 \overset{?}{=} \$72,000$$
$$\$72,000 = \$72,000 \quad \checkmark$$

PRACTICE PROBLEM 3

For the following word problem:

(a) Define the variable expressions.
(b) Write an equation.
(c) Solve the equation and determine the value asked for.
(d) Check your answer.

Jason is a foreman for a construction company. The apprentice for the company earns $7400 less annually than does Jason. The sum of the Jason's annual salary and the apprentice's annual salary is $83,000. How much does each earn annually?

DEVELOPING YOUR STUDY SKILLS

REAL-LIFE MATHEMATICS APPLICATIONS

Applications or word problems are the very life of mathematics! They are the reason for doing mathematics, because they teach you how to put into use the mathematical skills you have developed. Learning mathematics without ever doing word problems is similar to learning all the skills of a sport without ever playing a game or learning all the notes on an instrument without ever playing a song.

The key to success is practice. Make yourself do as many problems as you can. You may not be able to do them all correctly at first, but keep trying. Do not give up whenever you reach a difficult one. If you cannot solve it, just try another one. Then come back and try it again later. Ask for help from your teacher or the tutoring lab. Ask other classmates how they solved the problem.

A misconception among students when they begin studying word problems is that each problem is different. At first the problems may seem this way, but as you practice more and more, you will begin to see the similarities, the different "types." You will see patterns in solving problems, which will enable you to solve problems of a given type more easily.

EXERCISE

1. Spend this week thinking about situations that you have encountered that require the use of mathematics. Write out the facts for these situations, then form a word problem. Share this word problem with other students in your class.

EXERCISES 6.7

For each problem:

   (a) *Write an equation.*   **(b)** *Solve the equation.*

1. If two times a number is increased by five, the result is fifteen. What is the number?

2. If three times a number is increased by one, the result is nineteen. What is the number?

3. If triple a number is decreased by four, the result is five. What is the number?

4. If double a number is decreased by six, the result is eight. What is the number?

5. If the sum of five and a number is multiplied by 2, the result is 12. What is the number?

6. If the sum of eight and a number is multiplied by 6, the result is 54. What is the number?

7. Find the length of the following rectangle if the perimeter is 60 meters.

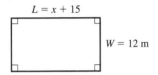

$L = x + 15$

$W = 12$ m

8. Find the length of the following rectangle if the perimeter is 66 m.

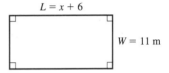

$L = x + 6$

$W = 11$ m

9. Find the length of each side of the following triangle if the perimeter is 12 centimeters.

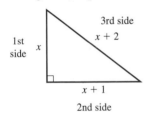

1st side x

3rd side $x + 2$

$x + 1$

2nd side

10. Find the length of each side of the following triangle if the perimeter is 24 decimeters.

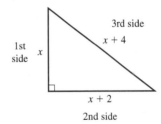

1st side x

3rd side $x + 4$

$x + 2$

2nd side

11. Find the width of the following rectangle if the area is 90 square inches.

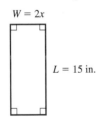

$W = 2x$

$L = 15$ in.

12. Find the length of the following rectangle if the area is 216 square inches.

$L = 3x$

$W = 6$ in.

13. Wendy is 2 inches taller than Janet, and Marci is 3 inches shorter than Janet. Write variable expressions for the heights of Janet, Wendy, and Marci.

14. Sion is 3 inches taller Damien, and Brad is 4 inches shorter than Damien. Write variable expressions for the height of Sion, Damien, and Brad.

15. The second side of a triangle is 4 inches longer than the first. The third side is 10 inches shorter than two times the first. Write a variable expression for the length of each side of the triangle.

16. The second side of a triangle is 3 inches longer than the first. The third side is 7 inches shorter than three times the first. Write a variable expression for the length of each side of the triangle.

17. Jerome is a store supervisor. The cashier earns $7500 less annually than does Jerome. The sum of Jerome's annual salary and the cashier's annual salary is $62,000. How much does each earn?

(a) Define the variable expressions.

(b) Write an equation.

(c) Solve the equation and determine the values asked for.

(d) Check your answer.

18. Lena is a sales supervisor. The salesclerk earns $6200 less annually than does Lena. The sum of Lena's annual salary and the clerk's annual salary is $58,000. How much does each earn?

(a) Define the variable expressions.

(b) Write an equation.

(c) Solve the equation and determine the values asked for.

(d) Check your answer.

19. In two days Juan drove 385 miles to a friend's house. He drove 85 more miles the first day than the second day. How far did he drive each day?

(a) Define the variable expressions.

(b) Write an equation.

(c) Solve the equation and determine the values asked for.

20. Andrew walked 4 miles less than Dave last week in a Boys and Girls Club "walk for the homeless" program. The two boys together walked 34 miles. How many miles did each boy walk?

(a) Define the variable expressions.

(b) Write an equation.

(c) Solve the equation and determine the values asked for.

21. The total flying time for two flights is 15 hours. The flight time of the first flight is half of the second. How long is each flight?

(a) Define the variable expressions.

(b) Write an equation.

(c) Solve the equation and determine the values asked for.

22. A triangle has a perimeter of 176 feet. The second side is 25 feet longer than the first. The third side is 5 feet shorter than the first. Find the length of each side.

(a) Define the variable expressions.

(b) Write an equation.

(c) Solve the equation and determine the values asked for.

23. A triangle has a perimeter of 120 meters. The length of the second side is double the first side. The length of the third side is 12 meters longer than the first side. Find the length of each side.

    (a) Define the variable expressions.

    (b) Write an equation.

    (c) Solve the equation and determine the values asked for.

24. The perimeter of a rectangle is 48 feet. The length is 4 feet less than triple the width. What are the dimensions of the rectangle?

    (a) Define the variable expressions.

    (b) Write an equation.

    (c) Solve the equation and determine the values asked for.

25. The perimeter of a rectangle is 68 meters. The length is 2 meters less than triple the width. What are the dimensions of the rectangle?

    (a) Define the variable expressions.

    (b) Write the equation.

    (c) Solve the equation and determine the values asked for.

26. Last year a total of 395 students took English. 95 more students took it in the spring than in the fall. 75 fewer students took it in the summer than in the fall. How many students took it during each semester?

    (a) Define the variable expressions.

    (b) Write the equation.

    (c) Solve the equation and determine the values asked for.

27. A small community college has a total of 1704 students. 115 more students live on campus than live in nearby off-campus housing. 55 fewer students live at home and commute than live in nearby off-campus housing. How many students are there in each of the three categories?

    (a) Define the variable expressions.

    (b) Write the equation.

    (c) Solve the equation and determine the values asked for.

ONE STEP FURTHER

28. If the perimeter and area of a square are equal in value, what is the length of the side for this square?

29. A triangle with all sides of equal length is called an *equilateral triangle*. If the perimeter of an equilateral triangle is equal to the length of a side squared, what is the length (in feet) of each side of the triangle.

CUMULATIVE REVIEW

Perform the indicated operations.

    1. $3044 \div 21$

    2. 2459×120

    3. Find the area of a parallelogram with a 3-foot base and a height of 4 feet.

    4. Change to a mixed number: $\dfrac{44}{7}$.

PROFIT, COST, AND REVENUE IN BUSINESS

The concepts of cost, revenue, and profit are important parts of running a business. Since cost, revenue, and profit can change depending on the number of items produced and sold, we use linear equations to represent them. By allowing a variable (x) to stand for the number of items produced or sold, these equations can represent what happens for any number of items. The following equations are used to calculate cost, revenue, profit, and break-even points.

Items, x	The number of items produced or sold.
Cost, C	The amount of money it cost to produce x items.
Revenue, R	The amount of money received for selling x items.
Profit, P	The amount of money left after costs.
Profit equation: $P = R - C$	Profit = Revenue − Cost: the equation used to find profit.
Break-even equation: $R = C$	The number of items that must be sold to cover the cost of producing the items. Profit begins after you reach this break-even point.

A business takes these general equations and, based on data collected for their business, develops cost and revenue equations for their business. These equations can then be thought of as formulas for finding their cost and revenue. For example, Leon Manufacturing has determined the following for a Model A door hinge:

- Their *cost* for producing x door hinges is equal to two times the number of hinges produced plus fifty.
- The *revenue* they receive from selling x door hinges is equal to three times the number of door hinges produced minus 5.

The manufacturing company then used this information to write the following cost, revenue, and profit equations.

Cost, Revenue, and Profit Equations for Leon Manufacturing

Door Hinges: Model A

$C = 2x + 50$
$R = 3x - 5$
$P = (3x - 5) - (2x + 50)$

PROBLEMS FOR INDIVIDUAL INVESTIGATION

1. How many Model A door hinges can Leon Manufacturing produce for a cost of $750?

2. If the cost for production is $750, how much is the revenue?

3. How much profit will Leon Manufacturing make from selling all the door hinges produced with the cost at $750?

4. Since $R = C$ represents the break-even point, the break-even equation is $3x - 5 = 2x + 50$. How many door hinges does Leon Manufacturing have to sell to cover the cost of production?

5. How many door hinges does Leon Manufacturing have to sell to make a $500 profit?

PROBLEMS FOR GROUP INVESTIGATION AND COOPERATIVE STUDY

Leon Manufacturing decided to change the type of door hinge they produce to a new model, Model B. Although the cost will be $100 more to produce the same amount of Model B hinges as Model A, it is predicted that the revenue for Model B will double.

1. What is the cost equation for Model B?

2. What is the revenue equation for Model B?

3. What is the profit equation for Model B?

4. (a) What is the break-even point for Model B?
 (b) Is the break-even point for Model B more or less than Model A?
 (c) Explain what it means to have break-even points that are different.

5. Fill in the table.

Model A		Model B	
x hinges	Profit	x hinges	Profit
15		15	
30		30	
40		40	
45		45	
55		55	

Based on the information in the table, answer questions 6 and 7.

6. (a) What does it mean if the profit is negative?

INTERNET: Go to http://www.prenhall.com/blair to explore this application.

(b) If Leon Manufacturing receives an order for 45 door hinges, which model would it be more profitable for them to produce? Why?

7. (a) When is the profit equal to zero for Model A; Model B?

(b) Is there a relationship between a zero profit and the break-even point? Explain your answer.

CHAPTER ORGANIZER

TOPIC	PROCEDURE	EXAMPLES
Solving equations using the addition principle	1. Add the appropriate value to both sides of the equation so that the variable is one side and a number is on the other side of the equal sign: $x = some\ number$. 2. Check by substituting your answer back into the original equation.	Solve for x: $x + 13 = 41$. $x + 13 = 41$ $+\quad -13\quad -13$ $x + 0 = 28$ $x = 28$ $Check$: $x + 13 = 41$ $28 + 13 \overset{?}{=} 41$ $41 = 41$ ✓
Solving equations using the multiplication principle	1. If the variable is *divided* by a number, we *multiply* both sides of the equation by this number so that the variable is on one side and a number is on the other side of the the equal sign: $x = some\ number$. 2. Check by substituting your answer back into the original equation.	Solve for x: $\dfrac{x}{-2} = 16$. $\dfrac{x}{-2} = 16$ The variable is divided by -2. $(-2) \cdot \dfrac{x}{-2} = 16 \cdot (-2)$ We multiply both sides by -2. $x = -32$ $Check$: $\dfrac{x}{-2} = 16$ $\dfrac{-32}{-2} \overset{?}{=} 16$ $16 = 16$ ✓

TOPIC	PROCEDURE	EXAMPLES
Solving equations using the division principle	1. If the variable is *multiplied* by a number, we *divide* both sides of the equation by this number so that the variable is on one side and a number is on the other side of the equal sign: $x = $ *some number*. 2. Check by substituting your answer back into the original equation.	Solve for x: $-36 = 12x$. $-36 = 12x$ The variable is multiplied by 12. $\dfrac{-36}{12} = \dfrac{12x}{12}$ We divide both sides by 12. $-3 = x$ *Check:* $-36 = 12x$ $-36 \stackrel{?}{=} 12(-3)$ $-36 = -36$ ✓
Solving equations using more than one principle of equality	1. Remove any parentheses. 2. Collect like terms and simplify numerical work. 3. Use the addition principle to get all ax terms on one side of the equal sign and numerical values on the other side. 4. Use the multiplication or division principle to get x alone on one side of the equal sign. 5. Check your solution.	Solve for x: $4x - 7 = 2 + 6x + 9$. $4x - 7 = 2 + 6x + 9$ $4x - 7 = 11 + 6x$ $\begin{aligned} 4x - 7 &= 11 + 6x \\ +\;-4x & -4x \\ \hline -7 &= 11 + 2x \end{aligned}$ $\begin{aligned} -7 &= 11 + 2x \\ +\;-11 & -11 \\ \hline -18 &= 2x \end{aligned}$ $\dfrac{-18}{2} = \dfrac{2x}{2}$ $-9 = x$ *Check:* $4x - 7 = 2 + 6x + 9$ $4(-9) - 7 \stackrel{?}{=} 2 + 6(-9) + 9$ $-36 - 7 \stackrel{?}{=} 2 + (-54) + 9$ $-43 = -43$ ✓
Solving equations involving two or more fractions	1. Determine the LCD of all the denominators. 2. Multiply each term of the equation by the LCD. 3. Solve the resulting equation. 4. Check your solution.	Solve for x: $4x + \dfrac{1}{2} = \dfrac{2}{5}$. $4x + \dfrac{1}{2} = \dfrac{2}{5}$ The LCD is 10. $(10)4x + (10) \cdot \dfrac{1}{2} = (10) \cdot \dfrac{2}{5}$ $40x + 5 = 4$ $\begin{aligned} 40x + 5 &= 4 \\ + -5 & -5 \\ \hline 40x &= -1 \end{aligned}$ $x = -\dfrac{1}{40}$

TOPIC	PROCEDURE	EXAMPLES
Multiplying a monomial by a polynomial	Use the distributive property and multiply each term of the polynomial by the monomial.	Multiply: $(-3x)(5x - 3y + 8)$. $(-3x)(5x - 3y + 8) =$ $\qquad\qquad -15x^2 + 9xy - 24x$
Multiplying binomials using FOIL	1. Multiply the *F*irst terms: F 2. Multiply the *O*uter terms: O 3. Multiply the *I*nner terms: I 4. Multiply the *L*ast terms: L 5. Combine like terms.	Multiply using Foil: $(x - 3)(x + 4)$. $\qquad$ first $\qquad$ last $(x - 3)(x + 4)$ $\quad$ F $\quad$ O $\quad$ I $\quad$ L $\qquad\qquad\qquad$ $x^2 + 4x - 3x - 12$ $\qquad$ inner $\qquad$ $x^2 + x - 12$ Combine $\qquad\qquad\qquad\qquad\qquad\qquad$ like terms. $\qquad$ outer
Adding polynomials	To add two polynomials, we add like terms.	$(-6x^2 + 9x - 7) + (2x^2 - 5x - 4)$ $= -6x^2 + 2x^2 + 9x - 5x - 7 - 4$ $= -4x^2 + 4x - 11$
Subtracting polynomials	To subtract two polynomials: 1. Change all signs of the second polynomial. 2. Combine like terms.	$(5x^2 - 7x - 9) - (3x^2 - 4x - 2)$ $= 5x^2 - 7x - 9 - 3x^2 + 4x + 2$ $= 2x^2 - 3x - 7$
Solving equations involving polynomials	Simplify the equation, then solve for the variable.	Solve: $-2(9x - 1) = 4(8x + 3) + 15$ $-18x + 2 = 32x + 12 + 15$ $\qquad\qquad\qquad\qquad$ Simplify. $-18x + 2 = 32x + 27$ $\qquad\quad 2 = 32x + 18x + 27$ $\qquad\qquad\qquad\qquad$ Add $18x$ to both $\qquad\qquad\qquad\qquad$ sides of equation. $\quad 2 - 27 = 50x$ $\quad$ Subtract 27 from $\qquad\qquad\qquad\qquad$ both sides of $\qquad\qquad\qquad\qquad$ equation. $\qquad -25 = 50x$ $\qquad -\dfrac{1}{2} = x$

CHAPTER 6 REVIEW

SECTION 6.1

Solve and check your solution.

1. $y - 12 = 33$

2. $a - 25 = -42$

3. $x + 27 = -34$

4. $-13 = y + 7$

Simplify, then solve.

5. $2 - 9 = x - 7 + 6$

6. $4(8 - 11) = x - 7$

Solve and check your solution.

7. $3 = \dfrac{y}{11}$

8. $-5 = \dfrac{y}{-2}$

9. $5x = -110$

10. $-11y = 9$

Simplify, then solve. Check your solution.

11. $-16 - 10 = 13y$

12. $-7x = 3 - 20$

SECTION 6.2

Solve for the variable and check your solution.

13. $6x - 8 = 34$

14. $-2y + 6 = 58$

15. $46 = 10 - 4x$

16. $\dfrac{y}{7} - 2 = 9$

17. $8x - 7 - 5x = 15$

18. $-6x = 9x + 36$

19. $5x = 3x + 30$

20. $4(x - 8) = -8$

SECTION 6.3

Solve.

21. $\dfrac{x}{3} + \dfrac{x}{4} = 7$

22. $2x - \dfrac{3}{4} = \dfrac{1}{2}$

23. $y + \dfrac{y}{9} = -10$

24. $2x - \dfrac{3}{4} = \dfrac{1}{3}$

SECTION 6.4

Identify each polynomial as a monomial, binomial, or trinomial.

25. (a) $3a^4 + 6$

(b) $5x^2$

(c) $5z^2 - 8z + 4$

Identify the terms of each polynomial.

26. $2x^2 + 5x - 3z^3 + 4$

27. $8a^5 - 7b^3 - 5b - 4$

Multiply.

28. $(-4)(6x^2 - 8x + 5)$

29. $(-2y)(y - 6)$

30. $(3x)(9x - 3y + 2)$

31. $(-5n)(-4n - 9m - 7)$

32. $(4x^2)(x^4 - 4)$

33. $(x^4)(x^5 - 2x - 3)$

34. $(z - 4)(5z)$ **35.** $(y + 10)(-6y)$

Use the FOIL method to multiply.

36. $(x + 2)(x + 4)$ **37.** $(y + 4)(y - 7)$ **38.** $(x - 2)(3x + 4)$ **39.** $(x - 3)(5x - 6)$

SECTION 6.5

Simplify.

40. $+(3a - 4)$ **41.** $-(-3y + 2z - 3)$

Perform the indicated operations.

42. $(-3x + 9) + (5x - 2)$ **43.** $(4x + 8) - (8x - 2)$

44. $(9a^2 - 3a + 5) - (-4a^2 - 6a - 1)$ **45.** $(-4x^2 - 3) - (3x^2 + 7x + 1) + (-x^2 - 4)$

SECTION 6.6

Solve.

46. $-3(x + 5) = 21$ **47.** $4(2x + 9) + 5x = -3$

48. $7(x + 1) + 3(x + 1) = -10$ **49.** $-3(x - 7) = 5(x + 6) - 10$

50. $(9y^2 + 8y - 2) - (9y^2 + 5y - 1) = 14$ **51.** $(5x^2 + x - 2) - (5x^2 - 5) = 6x + 9$

SECTION 6.7

For each of the following problems:

     (a) *Write an equation.*   **(b)** *Solve the equation.*

52. Two times a number increased by four is sixteen. What is the number?

53. Find the width of the following rectangle if the perimeter is 54 feet.

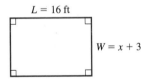

$L = 16$ ft

$W = x + 3$

For each of the following problems:

     (a) *Define the variable expressions.*   **(b)** *Write an equation.*   **(c)** *Solve the equation.*   **(d)** *Check your answer.*

54. Jamie is a store supervisor. The cashier earns $8500 less annually than does Jamie. The sum of Jamie's annual salary and the cashier's annual salary is $59,000. How much does each earn?

55. Last year a total of 491 students took Spanish. 85 more students took it in the spring than in the fall. 65 fewer students took it in the summer than in the fall. How many students took it during each semester?

CHAPTER 6 TEST

Solve for the variable and check.

1. _____

2. _____

3. _____

4. _____

5. _____

6. _____

7. _____

8. _____

9. _____

10. _____

11. _____

12. _____

13. _____

14. _____

15. _____

16. _____

17. _____

18. _____

19. _____

20. _____

21. _____

22. _____

23. _____

1. $x - 6 = 18$

2. $x + 5 = -12$

3. $2 - 6 + x = 4 - 8$

4. $\dfrac{y}{3} = 9$

5. $-5 = \dfrac{y}{2}$

6. $7y = 28$

7. $6(2z - 7) = -18$

8. $4z - 2 = -14$

9. $\dfrac{z}{4} + 8 = 12$

10. $3z + z = -12 - 8$

11. $\dfrac{x}{4} + \dfrac{x}{3} = 7$

12. $6x + \dfrac{1}{2} = -\dfrac{23}{2}$

13. $\dfrac{y}{2} + y = 9$

14. $\dfrac{9}{z} + 2 = \dfrac{11}{4}$

15. Determine if this polynomial is a monomial, binomial, or trinomial: $8xy^2 + 10x$.

Multiply.

16. $(-7b)(2b - 4)$

17. $3(6x^2 - x + 1)$

18. $(-2x^3)(4x^2 - 3)$

19. $(x + 5)(x + 9)$

20. $(x + 3)(x - 2)$

21. Simplify: $-(4x - 2y - 6)$.

Add.

22. $(-5x + 3) + (-2x + 4)$

23. $(4x^2 + 8x - 3) + (9x^2 - 10x + 1)$

Subtract.

24. $(4y + 5) - (2y - 3)$

25. $(-7p - 2) - (3p + 4)$

Solve for the variable and check.

26. $3(2x + 6) + 3x = -27$

27. $4(x - 1) = -6(x + 2) + 48$

28. $(3x^2 - 2x + 1) + (-3x^2 - 10) = 5x + 5$

Use the following information for 29–31:

The first side of a triangle is 2 feet longer than the second. The third side is triple the second. The perimeter of the triangle is 42 feet. Find the length of each side of the triangle.

29. Define the variable expressions.

30. Write an equation.

31. Solve the equation and determine the length of each side of the triangle.

Use the following information for 32–34:

Anna is a store supervisor. The sales clerk earns $4000 per year less than Anna. The sum of Anna's annual salary and the clerk's annual salary is $61,200.

32. Define the variable expressions.

33. Write an equation.

34. Solve the equation and determine each person's annual salary.

24. _____

25. _____

26. _____

27. _____

28. _____

29. _____

30. _____

31. _____

32. _____

33. _____

34. _____

CUMULATIVE TEST FOR CHAPTERS 1–6

1. _____

2. _____

3. _____

4. _____

5. _____

6. _____

7. _____

8. _____

9. _____

10. _____

11. _____

12. _____

13. _____

14. _____

15. _____

16. _____

17. _____

18. _____

19. _____

20. _____

This graph shows the number of customers that entered Videos R Us each day last week.

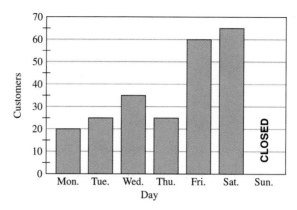

1. How many customers went to Videos R Us on Wednesday?

2. On which day of the week did the most customers enter the store?

3. How many more customers went on Saturday than on Monday?

4. Identify the factors and product:
   (a) $7x = -21$ **(b)** $xy = z$

Evaluate.

5. $5 - 12(9 - 10)$ **6.** $7 - 24 \div 6(-2)^2 - 3$

7. $-6 - (-1)$ **8.** $-9 - 7 + (-10)$

9. $9(0)(1)$

Replace the ? with the appropriate symbol: =, <, or >.

10. $0 \ ? \ -72$ **11.** $-12 \ ? \ -1.2$

12. $|-6| \ ? \ |3|$ **13.** $|8| \ ? \ |-8|$

14. $\dfrac{5}{9} \ ? \ \dfrac{20}{45}$ **15.** The opposite of 9 is ____.

16. The opposite of -12 is ____.   **17.** The absolute value of 9 is ____.

18. The absolute value of -12 is ____.   **19.** Write $\dfrac{15}{2}$ as a mixed number.

20. Change $6\dfrac{3}{4}$ to an improper fraction.

21. A class has 25 freshman and 40 juniors. Write the ratio of freshman to juniors as a fraction in simplest form.

_____

Use the following information for 22–24.

A recipe for chicken with rice calls for $\frac{3}{4}$ pound of boneless chicken. When prepared, it serves 4 people.

_____

22. How much chicken is in each serving of chicken with rice?

_____

23. To prepare this recipe for 8 people, how many pounds of chicken are necessary?

_____

24. To prepare this recipe for 2 people, how many pounds of chicken are necessary?

Use the following information for 25–27.

The second side of a triangle is 3 inches shorter than the first. The third side is double the first. The perimeter of the triangle is 29 inches.

_____

25. Define the variable expressions.

_____

26. Write an equation.

_____

27. Solve the equation and determine the length of each side of the triangle.

Use the following information for 28–30.

Juan is a sales supervisor. The cashier earns $5500 less per year than Juan. The sum of Juan's annual salary and the cashier's annual salary is $52,000.

_____

28. Define the variable expressions.

_____

29. Write an equation.

30. Solve the equation and determine each person's annual salary.

_____

Solve for the variable and check.

31. $y + 10 = -2$

_____

32. $-3x + 1 = -11$

_____

33. $\frac{x}{3} + \frac{x}{2} = 5$

_____

DECIMAL EXPRESSIONS

A checking account is a convenient way to pay bills and make purchases. It can also save you time and money. Yet many people either do not have a checking account or have closed their account because they do not know how to balance their checkbook. Others have had to pay fees for checks written with insufficient funds simply because they do not know how to record and verify their banking transactions correctly. To see how to maintain bank records and balance your checkbook, turn to Putting Your Skills to Work on page 435.

7.1 Understanding Decimal Fractions
7.2 Addition and Subtraction of Decimals
7.3 Multiplication of Decimals and Applied Problems
7.4 Division of Decimals and Scientific Notation
7.5 Solving Equations and Applied Problems Involving Decimals
7.6 Percents
7.7 Solving Percent Problems Using Equations
7.8 Applied Problems Involving Percent

CHAPTER 7 PRETEST

This test provides a preview of the topics in this chapter. It will help you identify which concepts may require more of your studying time. If you are familiar with the topics in this chapter, take this test now. Check your answers with those in the back of the book. If you are not familiar with the topics in this chapter, begin studying the chapter now.

SECTION 7.1

1. Write the word name: 18.071.

2. Write in fractional notation: 4.542. (Do not simplify.)

3. Write as a decimal: $\dfrac{7}{100}$.

4. Replace ? with $<$ or $>$: 0.31 ? 0.312.

5. Round 635.073 to the nearest tenth.

SECTION 7.2

6. Add: $3.12 + 0.48$.

Subtract.

7. $27.16 - 15.82$

8. $14.1 - (-2.5)$

9. On Monday, Joe drove 15.12 miles to work. After work, he drove 1.27 miles to the store, then he drove 16.3 miles home. Estimate the total miles Joe drove on Monday.

10. Christine's bill for dinner, including tax, was $13.57. She gave the waiter $15. How much change should she receive?

SECTION 7.3

Multiply.

11. $(4.12)(0.3)$

12. $(0.536)(10^2)$

13. Evaluate: $-2.4 + 3.16 \times 10^2 - 0.5$.

14. Acme Truck Rentals charges $29.95 per day plus 30 cents per mile to rent a 15-foot truck. How much will it cost to rent a 15-foot truck for two days and drive 90 miles?

1. _____

2. _____

3. _____

4. _____

5. _____

6. _____

7. _____

8. _____

9. _____

10. _____

11. _____

12. _____

13. _____

14. _____

SECTION 7.4

Divide.

15. $8.5 \div 4$

16. $8.2 \div 1.25$

17. Write $\dfrac{29}{4}$ as a decimal.

18. Write 0.000267 in scientific notation.

19. A chemist mixed up 27.5 quarts of fluid. She wishes to pour it in several equal-sized containers that hold 2.5 quarts.

  (a) How many containers will she need?

  (b) If the fluid cost $4.26 per quart, how much will 1 full container of fluid cost?

SECTION 7.5

20. Solve for x: $0.1x + 0.3x = 2.0$.

SECTION 7.6

21. 43 of 50 prealgebra students plan to take algebra next semester. What percentage of prealgebra students plan to take algebra next semester?

22. (a) Write $\dfrac{3}{8}$ as a percent.   **(b)** Write 35.8% as a fraction.

  (c) Write 4.01% as a decimal

SECTION 7.7

Round answers to the nearest hundredth, if necessary.

23. (a) What is 18% of 53?   **(b)** 19 is 72% of what number?
  (c) What percent of 312 is 35?

24. Sue left a $5.25 tip for her dinner, which cost $30.55. What percent of the total bill did Sue leave for a tip? Round your answer to the nearest hundredth of a percent.

SECTION 7.8

25. Marco is paid 15% commission each week based on the total dollar amount of sales. Last week his total sales were $2560. How much did Marco earn in commission?

15. _____

16. _____

17. _____

18. _____

19. _____

20. _____

21. _____

22. _____

23. _____

24. _____

25. _____

1 Write word names for decimal fractions.

2 Convert between decimals and fractions.

3 Compare and order decimals.

4 Round decimals.

Math Pro   Video 7.1   SSM

S E C T I O N   7 . 1

Understanding Decimal Fractions

1 Writing Word Names for Decimal Fractions

> A decimal fraction is a fraction whose denominator is a power of 10: 10, 100, 1000, and so on.

$\frac{9}{10}$ is a decimal fraction. $\frac{41}{100}$ is a decimal fraction.

We can represent the shaded part of a whole as a decimal fraction in different ways (forms): using word names, as a fraction, or in decimal form.

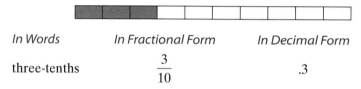

In Words	*In Fractional Form*	*In Decimal Form*
three-tenths	$\frac{3}{10}$	.3

All mean the same quantity, namely 3 out of 10 equal parts of a whole. The "." in the decimal form .3 is called a **decimal point**. Usually a zero is placed in front of the decimal point to make sure that we don't miss seeing the decimal point.

$$.3 \text{ has the same value as } 0.3$$

↑   ↑

decimal point   extra zero used only for clarity

Since situations may require that we use different forms of decimal fractions, it is important that we understand the meaning of decimal fractions as well as how to write decimal fractions in each form. A place-value chart is helpful.

HUNDREDS	TENS	ONES	DECIMAL POINT	TENTHS	HUNDRED**THS**	THOUSAND**THS**	TEN THOUSAND**THS**
100	10	1	. "and"	$\frac{1}{10}$	$\frac{1}{100}$	$\frac{1}{1000}$	$\frac{1}{10,000}$

This place-value chart is an extension of the one we used in Chapter 1 to name whole numbers. From the chart we see that:

1. The names for the place values to the right of the decimal point end with "*ths*" compared to those to the left of the decimal point. Therefore, we must take care when stating the name of the place value.

   "3 hundreds" represents the number 300.

   "3 hundred*ths*" represents the number $\frac{3}{100}$.

2. The word name for a decimal point is *and*.

We can use the place-value chart to help us write a word name for a decimal fraction.

EXAMPLE 1   Write a word name for the decimal 0.561.

SOLUTION

⟶ We do not include 0 as part of word name.

0.<u>561</u>
↓

five hundred sixty-one <u>thousandths</u>

The *last word* names the place value of the *last digit*: 1 is three places to the right of the decimal point—the *thousandths* place.

PRACTICE PROBLEM 1

Write a word name for the decimal 5.32.

Decimal notation is used when writing a check. Often, we write the amount that is less than 1 dollar, such as 35 cents, as $\dfrac{35}{100}$ dollar.

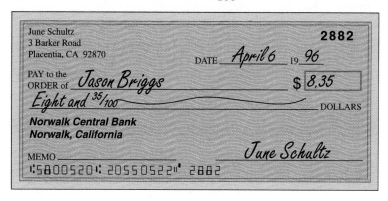

EXAMPLE 2

Write the word name for a check written to Shandell Strong for $126.87.
_____ Dollars

SOLUTION   **One hundred twenty-six and 87/100**

PRACTICE PROBLEM 2

Write the word name for a check written to Wendy King for $245.09.
_____ Dollars

② Converting Between Decimals and Fractions

Decimal fractions can be written with numerals in two ways: decimal notation or fractional notation.

DECIMAL NOTATION	FRACTIONAL NOTATION
0.7	$\dfrac{7}{10}$
3.49	$3\dfrac{49}{100}$
0.131	$\dfrac{131}{1000}$

From the table, note the following:

0.7 $\qquad$ $\dfrac{7}{10}$ $\qquad$ 0.131 $\qquad$ $\dfrac{131}{1000}$

↓	↘	↓	↘
1 decimal place	1 zero	3 decimal places	3 zeros

PROCEDURE TO CHANGE FROM DECIMAL TO FRACTIONAL NOTATION

1. Write the whole number (if any). $\qquad\qquad$ $9.7653 \rightarrow 9$
$\qquad\qquad\qquad\qquad\qquad\qquad\qquad\qquad\qquad\qquad$ ↓
2. Count the number of decimal places. $\qquad$ 4 decimal places

3. Write the decimal part over a denominator that has a 1 and the same number of zeros as the number of decimal places found in step 2. $\qquad$ $9\dfrac{7653}{10,000}$
$\qquad\qquad\qquad\qquad\qquad\qquad\qquad\qquad\qquad\qquad\qquad$ ↓
$\qquad\qquad\qquad\qquad\qquad\qquad\qquad\qquad\qquad\qquad$ 4 zeros

---

E X A M P L E   3   Write 0.86132 using fractional notation. Do not simplify.

SOLUTION

We do not need to write 0 as part of the fraction.
$\qquad$ ↓

$0.86132 \quad = \quad \dfrac{86,132}{100,000}$

↓	↓
5 decimal places	5 zeros

P R A C T I C E   P R O B L E M   3

Write 8.723 using fractional notation. Do not simplify.

If the fractional notation has a denominator that is a power of 10 (10, 100, 1000, and so on), we use a similar procedure to change this type of fraction to a decimal.

PROCEDURE TO CHANGE FRACTIONAL NOTATION TO DECIMAL NOTATION WHEN THE DENOMINATOR IS A POWER OF 10

1. Count the number of zeros in the denominator. $\qquad\qquad$ $9\dfrac{7653}{10,000}$
$\qquad\qquad\qquad\qquad\qquad\qquad\qquad\qquad\qquad\qquad$ ↓
$\qquad\qquad\qquad\qquad\qquad\qquad\qquad\qquad\qquad$ 4 zeros

2. In the numerator, move the decimal point as many places to the left as the number of zeros in step 1 and delete the denominator. $\quad$ ←4 places
$\qquad\qquad\qquad\qquad\qquad\qquad\qquad\qquad\qquad$ $9\dfrac{7653}{10,000} \rightarrow 9.7653$

EXAMPLE 4   Write $7\dfrac{56}{1000}$ as a decimal.

SOLUTION

Move decimal point
3 places to the left

$$7\dfrac{56}{1000} = 7\dfrac{056}{1000} = 7.056$$

3 zeros

Note: We had to insert a 0 before 56 so we could move the decimal point digit 3 places to the left.

PRACTICE PROBLEM 4

Write $\dfrac{17}{1000}$ as a decimal.

③ Comparing and Ordering Decimals

In Chapter 1 we studied the inequality symbols "<" and ">." Recall that

$a < b$ is read "a is less than b."

$a > b$ is read "a is greater than b."

To compare and order decimals using inequality symbols, we compare each digit.

PROCEDURE TO COMPARE TWO POSITIVE NUMBERS IN DECIMAL NOTATION

1. Start at the left and compare corresponding digits. If the digits are the same, move one place to the right.
2. When two digits are different, the larger number is the one with the larger digit.

It is easier to compare two decimals if the decimal parts of each have the same number of digits. Whenever necessary, extra zeros can be written to the right of the last digit—that is, to the right of the last digit after the decimal point—without changing the value of the decimal. To see why, let's look at $\dfrac{3}{10}, \dfrac{30}{100}, \dfrac{300}{1000}$.

$$\dfrac{300}{1000} = \dfrac{30}{100} = \dfrac{3}{10}$$

$\dfrac{3}{10}$ is the reduced form of the fractions $\dfrac{300}{1000}$ and $\dfrac{30}{100}$.

$$0.300 = 0.30 = 0.3$$

We can think of 0.3 as the reduced form of 0.300 and 0.30.

E X A M P L E 5 Replace the ? with < or >: 0.24 ? 0.244.

SOLUTION

0.24 ? 0.244

0.24**0** ? 0.244 Add a zero to 0.24 so that both decimal parts have the same number of digits.

0.2<u>4</u>0 ? 0.2<u>4</u>4 The tenths and hundredths digits are equal.

0.24<u>0</u> ? 0.24<u>4</u> The thousandths digits differ.

Since 0 < 4, **0.240 < 0.244.**

P R A C T I C E P R O B L E M 5
Replace the ? with < or >: 0.77 ? 0.771.

---

4 Rounding Decimals

Just as with whole numbers, we must sometimes round decimals. The rule for rounding is similar to the one we used in Chapter 1 for whole numbers.

> **PROCEDURE TO ROUND DECIMALS**
>
> 1. Identify the round-off place digit.
> 2. If the digit to the *right* of the round-off place digit is:
>   (a) *Less than 5*, do not change the round-off place digit.
>   (b) *5 or more*, increase the round-off place digit by 1.
> 3. In either case, drop all digits to the right of the round-off place digit.

Thus, when rounding decimals we either *increase* the round-off place digit by 1, or *leave it the same.* We always drop all digits to the right of the round-off place digit.

E X A M P L E 6 Round 237.8435 to the nearest hundredth.
SOLUTION

The round-off place digit is the *hundredths place.*
$$\downarrow$$
237.8435

⌐— The digit to the *right* of the round-off place digit is *less than 5*.

Do not change the round-off place digit.
$$\downarrow$$
237.84__

⌐— *Drop* all digit to the *right* of the round-off place digit.

237.8435 rounded to the nearest hundredth is **237.84**.

P R A C T I C E P R O B L E M 6
Round 369.2649 to the nearest hundredth.

Remember that rounding up to the next digit in a position may result in several digits being changed.

EXAMPLE 7   Round to the nearest hundredth: Alex and Lisa used 204.9954 kilowatthours of electricity in their house in June.

SOLUTION

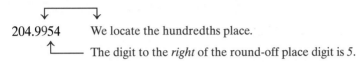

204.9954   We locate the hundredths place.

———— The digit to the *right* of the round-off place digit is *5*.

Since the digit to the right of 9 is 5, we increase the 9 to 10 by changing 9 to 0. Then we must increase the 9 in the tenths place by 1, followed by increasing 4 to 5. Why?

204.9954
↓↓↓

The result is **205.00** kilowatthours.

PRACTICE PROBLEM 7

Round to the nearest tenth: Last month the college auditorium used 16,499.952 kilowatthours of electricity.

DEVELOPING YOUR STUDY SKILLS

EVALUATING YOUR STUDY SKILLS

At this time of the semester it is a good idea to reevaluate your study habits and determine what you can do to maintain or improve your performance in your math class.

1. Starting in Chapter 1, reread all the Developing Your Study Skills.
2. Make a list of learning activities that you are either not doing or can spend more time doing.
3. Based on this list, write out a study plan that you will follow for the remainder of the semester.
4. Discuss this plan with your instructor or counselor to see if they have any other suggestions.
5. Follow this plan for the remainder of the semester!

Learning how to learn and practicing these techniques will help you improve your performance in all your classes as well as in your place of employment.

EXERCISES 7.1

Write a word name for each decimal.

1. 5.32

2. 11.78

3. 0.428

4. 0.983

5. 8.5

6. 14.3

Write in decimal notation.

7. Three hundred twenty-four thousandths

8. One hundred twenty-six thousandths

9. Fifteen and three hundred forty-six ten thousandths.

10. Twenty-four and one hundred seventy-six ten thousandths

Write the word name for the amount on the check.

11.

PAY to the ORDER of *Orange Coast College* $ 25.54

_____ DOLLARS

12.

PAY to the ORDER of *Rita Smith* $ 8.75

_____ DOLLARS

13.

PAY to the ORDER of *The Gas Company* $ 143.56

_____ DOLLARS

14.

PAY to the ORDER of *Pep Boys Auto Parts* $ 146.32

_____ DOLLARS

Write in fractional notation. Do not simplify.

15. 0.7

16. 0.3

17. 3.64

18. 6.23

19. 0.1743

20. 0.8436

21. 100.011

22. 20.022

Write each fraction as a decimal.

23. $\dfrac{1}{10}$

24. $\dfrac{7}{10}$

25. $12\dfrac{37}{1000}$

26. $63\dfrac{31}{1000}$

27. $\dfrac{2}{100}$

28. $\dfrac{8}{100}$

29. $\dfrac{1}{1000}$

30. $\dfrac{2}{1000}$

Replace the ? with < or >.

31. 0.426 ? 0.429

32. 0.526 ? 0.521

33. 0.63 ? 0.62

34. 0.93 ? 0.94

35. 0.36 ? 0.366

36. 0.12 ? 0.127

37. 0.7431 ? 0.743

38. 0.6362 ? 0.636

Round to the nearest hundredth.

39. 523.7235   **40.** 124.6345   **41.** 43.995   **42.** 76.996

Round to the nearest tenth.

43. 9.0746   **44.** 21.087   **45.** 462.951   **46.** 125.972

Round to the nearest thousandth.

47. 312.95166   **48.** 63.44461

Round to the nearest ten thousandth.

49. 0.063148   **50.** 0.047357

51. In the 1960s, the average person in the United States ate 10.62 pounds of chocolate per year. In the 1990s, the average person ate 21.78 pounds of chocolate per year. Round these values to the nearest tenth.

52. The average rainfall in Santiago Peak, California is 33.46 inches per year. From July 1, 1997 to May 29, 1998 Santiago Peak had an increase in rainfall due to an El Niño condition, receiving 105.12 inches of rain. Round the average rainfall, and the rainfall during the El Niño season to the nearest tenth.
Source: Orange County Register

The menu in Emerald's Dinner House reads as follows:

53. Round the price of the New York steak to the nearest dollar.

54. Round the price of the prime rib to the nearest dollar.

> **Emerald's Dinner House**
>
> Grilled chicken breast . . . . $ 12.95
> New York steak . . . . $ 15.25
> Stuffed pork chops . . . . . . $ 13.75
> Fish of the day . . . . $ 14.50
> Lobster . . . . $ 18.25
> Prime rib . . . . $ 15.75

 VERBAL AND WRITING SKILLS

55. 12.97 rounded to the nearest tenth is 13.0. Explain how rounding 9 up is similar to carrying.

56. Explain why we write a zero before the decimal point, 0.23, when there is no whole-number part to the decimal fraction.

57. Arrange in order from largest to smallest:
$0.0069, 0.73, \frac{7}{10}, 0.007, 0.071$.

58. Arrange in order from smallest to largest:
$0.053, 0.005, 0.52, 0.0059, \frac{5}{100}$.

CUMULATIVE REVIEW

Perform the indicated operation.

1. $3005 - 1578$ **2.** $(356)(28)$ **3.** $3870 \div 45$ **4.** $15{,}708 \div 231$

After studying this section, you will be able to:

❶ Add and subtract decimals.

❷ Estimate sums and differences.

❸ Solve applied problems involving decimals.

Math Pro Video 7.2 SSM

Addition and Subtraction of Decimals

❶ Adding and Subtracting Decimals

We often add or subtract decimals when we work with money. For example, if we buy a drink for $1.25 and a sandwich for $3.55, how much is the total bill? It is $1.25 + $3.55, or $4.80. Now, if we pay with a $5 bill, how much change do we receive? $5.00 − $4.80, or $0.20 (20 cents).

We can relate addition and subtraction of fractions to decimals. For example,

$$2\frac{3}{10} = 2.3 \qquad\qquad 8\frac{5}{10} = 8.5$$
$$+\,5\frac{6}{10} = 5.6 \qquad\quad -\,3\frac{2}{10} = 3.2$$
$$\overline{7\frac{9}{10} = 7.9} \qquad\quad \overline{5\frac{3}{10} = 5.3}$$

We see from the illustration above that adding and subtracting decimals is very much like whole numbers, except that we must consider placement of the decimal point.

PROCEDURE TO ADD OR SUBTRACT DECIMALS

1. Line up the decimal points, then either add or subtract as indicated.
2. In the answer, place the decimal point in line with the decimal points of the numbers in the problem.

CALCULATOR

ADDING AND SUBTRACTING DECIMALS

The calculator can be used to verify your work. To find 13.04 + 2.33, enter:

Scientific calculator:

   13.04 $\boxed{+}$ 2.33 $\boxed{=}$

Graphing calculator:

   13.04 $\boxed{+}$ 2.33 $\boxed{\text{ENT}}$

The calculator displays

   $\boxed{15.37}$

E X A M P L E   1   Add: 40 + 8.77 + 0.9.

SOLUTION

$$\begin{array}{r} \overset{1}{} \\ 40.00 \\ 8.77 \\ +\ 0.90 \\ \hline \mathbf{49.67} \end{array}$$

We add zeros so that each number has the same number of decimal places.

Then we line up decimal points and add.

P R A C T I C E   P R O B L E M   1

Add: 50 + 4.39 + 0.7.

E X A M P L E   2   Subtract: 19.02 − 8.6.

SOLUTION

$$\begin{array}{r} {\scriptstyle 8\ \ 10} \\ 19.\cancel{0}2 \\ -\ \ 8.60 \\ \hline \mathbf{10.42} \end{array}$$

Line up decimal points.

Add a zero.

The decimal point in the difference is in line with the other decimal points.

P R A C T I C E   P R O B L E M   2

Subtract: 26.01 − 5.7.

We use the same rules for signed numbers stated in Chapter 3 when adding or subtracting decimals.

EXAMPLE 3   Perform the operation indicated: $-9.79 - (-0.68)$.

SOLUTION

To subtract signed numbers, we add the opposite of the second number.

$$-9.79 - (-0.68)$$
$$-9.79 + (0.68)$$

Add   the opposite of (-0.68).

Next, to add numbers with different signs, we "keep the sign of the larger absolute value and subtract."

$$\begin{array}{r} -9.79 \\ + 0.68 \\ \hline \mathbf{-9.11} \end{array}$$

The answer is negative since $|-9.79|$ is larger than $|0.68|$.

PRACTICE PROBLEM 3
Perform the operations indicated: $9.02 - (-5.1)$.

We combine like terms with decimal coefficients the same way we do when coefficients are whole numbers—that is, we add coefficients of like terms and the variable part stays the same.

EXAMPLE 4   Combine like terms: $11.2x + 3.6x$.

SOLUTION

$$\begin{array}{r} 11.2x \\ + 3.6x \\ \hline \mathbf{14.8x} \end{array}$$

We line up decimal points and add.

PRACTICE PROBLEM 4
Combine like terms: $4.5y + 7.2y$.

② Estimating Sums and Differences

When we estimate a sum or difference, we can round each decimal to the nearest whole number.

EXAMPLE 5   Julie runs on her treadmill each day. She wants to run approximately 25 miles each week to prepare for a track race. She logs the distance she runs each day on the following chart. Estimate the total number of miles Julie ran this week.

Monday	Tuesday	Wednesday	Thursday	Friday	Saturday	Sunday
2.13 mi	2.79 mi	2.9 mi	3.11 mi	3.8 mi	4.12 mi	4.9 mi

SOLUTION   We round each decimal to the nearest whole number, then add.

$$2 + 3 + 3 + 3 + 4 + 4 + 5 = 24 \text{ miles}$$

Julie ran **approximately 24 miles**.

PRACTICE PROBLEM 5

Allen ate at restaurants for two days while on a business trip. He logged his expenses on the chart located in the margin. Estimate the total cost of meals for the two days.

Tuesday	
Breakfast	$ 7.29
Lunch	8.99
Dinner	19.10

Wednesday	
Breakfast	6.99
Lunch	9.20
Dinner	21.76

③ Solving Applied Problems Involving Decimals

EXAMPLE 6   Use the bar graph to answer the questions.

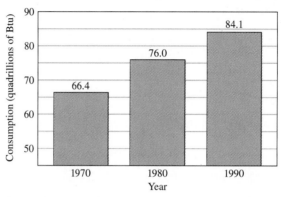

Source: U.S. Department of Energy

(a) How many more Btu were consumed in the United States during 1990 than in 1980?

(b) What was the total consumption of energy in 1990 and 1980?

SOLUTION

(a) We subtract:
$$\begin{array}{ll} 1990\text{:} & 84.1 \\ 1980\text{:} & -76.0 \end{array} \rightarrow \begin{array}{r} \overset{7\ \ 14}{8\cancel{4}.1} \\ -\ 76.0 \\ \hline \end{array}$$
8.1 quadrillion Btu

(b) We add: $84.1 + 76.0 =$ **160.1 quadrillion Btu**

PRACTICE PROBLEM 6

Use the bar graph in Example 6 to answer the questions.

(a) How many more Btu were consumed in the United States during 1980 than in 1970?

(b) What was the total consumption of energy in 1980 and 1970?

BALANCING A CHECKING ACCOUNT

If you have a checking account, you must keep records of the checks you write, ATM withdrawals, deposits, and other transactions on a check register. When the bank sends you a bank statement of their records, you should verify that these bank records match yours. This is called *balancing your checkbook*. Balancing your checkbook allows you to make sure that the balance you think you have in your checkbook is correct. If your checkbook does not balance, you must look for any mistakes. It is best to record every bank transaction on a check register, as you will see in the following problems.

CHECK REGISTER *Jesse Holm*						19 _98_	
CHECK NUMBER	DATE	DESCRIPTION OF TRANSACTION	PAYMENT/ DEBIT (−)	✓	DEPOSIT/ CREDIT (+)	BALANCE $ 1254	32
243	9/2	Manor Apartments	575				
244	9/2	Electric Company	23 41				
245	9/2	Gas Company	15 67				
246	9/3	Jack's Market	125 57				
247	9/4	Clothing Mart	35 85				
	9/5	Deposit			634 51		

Transactions not recorded on check register: On 9/7 Jesse wrote check 248 to the College Bookstore for $168.96, then made an ATM withdrawal of $100. On 9/15 Jesse wrote check 249 to the telephone company for $43.29, check 250 to Sports Emporium for $40, and made a $634.51 deposit. Jesse wrote check 251 to Jack's Market for $78.94, check 252 to Visa Gold Card for $125, and made a $100 ATM withdrawal on 9/20. On 9/25 Jesse wrote the following two checks: 253 to L&R Drugstore for $21.56, 254 to the Computer Store for $231.45. On 10/1 he deposited $423.62 in his account.

PROBLEMS FOR INDIVIDUAL INVESTIGATION

1. Record Jesse Holm's transactions on the check register below.

2. Find Jesse's balance on October 1.

CHECK REGISTER *Jesse Holm*						19 _98_	
CHECK NUMBER	DATE	DESCRIPTION OF TRANSACTION	PAYMENT/ DEBIT (−)	✓	DEPOSIT/ CREDIT (+)	BALANCE $ 1254	32
243	9/2	Manor Apartments	575				
244	9/2	Electric Company	23 41				
245	9/2	Gas Company	15 67				
246	9/3	Jack's Market	125 57				
247	9/4	Clothing Mart	35 85				
	9/5	Deposit			634 51		

3. The bank is changing its checking account plans and offering customers the option of four new plans. Based on Jesse's banking record, which plan is the most economical choice?

Checking Account Options   * Minimum balance = the lowest balance during the month.

	Minimum Balance*	Monthly Service Charge	Free Checks	Per Check Charge	ATM Withdrawal Fee
Plan 1	None	None	5	75¢ after 5 free	$1.50 each
Plan 2	$1,000	None	Yes, with min. balance	35¢ w/o min. balance	$1.50 each
Plan 3	$1,000	$9.00 w/o min. balance	Yes	None	2 free then $1.50 each
Plan 4	$2,500	$12.00 w/o min. balance	Yes	None	None

PROBLEMS FOR GROUP INVESTIGATION AND COOPERATIVE STUDY

1. Fill in Jesse's bank register on the previous page if you have not already done so. Then use the information on his bank statement below to verify that Jesse's checking account balances with the bank (balance his checking account).

Bank Statement: JESSE HOLM   9-1-98 to 9-30-98

Beginning Balance $1254.32
Ending Balance $1265.69

Checks cleared by the bank

#243	$575.00	#245	$15.67	#248	$168.96	#251*	$78.94
#244	$23.41	#246*	$125.57	#249*	$43.29	#253	$21.56

* Indicates that the next check in the sequence is outstanding (hasn't cleared).

Deposits

9/5 $634.51
9/15 $634.51

Other withdrawals

9/7 ATM $100.00   Service charge $5.25
9/20 ATM $100.00

CHECKING RECONCILEMENT				THIS FORM IS PROVIDED TO ASSIST YOU IN BALANCING YOUR CHECKING ACCOUNT

LIST CHECKS OUTSTANDING NOT CHARGED TO YOUR CHECKING ACCOUNT				PERIOD ENDING
CHECK NUMBER	AMOUNT	CHECK NUMBER	AMOUNT	, 19
				1. **SUBTRACT** FROM YOUR CHECK REGISTER ANY CHARGES LISTED ON THE STATEMENT WHICH YOU HAVE NOT PREVIOUSLY DEDUCTED FROM YOUR BALANCE. ALSO **ADD** ANY DIVIDENDS.
				2. **ENTER** CHECK BALANCE SHOWN ON THE STATEMENT HERE. $
				+ $
				3. ENTER DEPOSITS MADE LATER THAN THE ENDING DATE ON THE STATEMENT. + $
				+ $
				TOTAL (2 PLUS 3) $
				4. IN YOUR CHECK REGISTER **CHECK OFF** ALL CHECKS PAID AND IN AREA PROVIDED AT LEFT, **LIST** NUMBERS AND AMOUNTS OF ALL UNPAID CHECKS
		TOTAL		**5.** **SUBTRACT** TOTAL CHECKS OUTSTANDING − $
				6. THIS AMOUNT SHOULD EQUAL YOUR CHECK REGISTER BALANCE $

INTERNET: Go to http://www.prenhall.com/blair to explore this application.

EXERCISES 7.2

Add.

1. $0.34 + 5.23$

2. $0.63 + 2.73$

3. $35.4 + 0.8759$

4. $63.2 + 0.2348$

5. $0.74 + 9.32$

6. $0.23 + 3.46$

7. $73 + 7.54 + 0.483$

8. $59 + 1.27 + 0.345$

9. $73.1 + 0.3169$

10. $74.2 + 0.4524$

11. $25 + 2.73 + 0.423$

12. $0.658 + 23 + 6.24$

Subtract.

13. $34.57 - 1.43$

14. $94.83 - 3.21$

15. $347.69 - 5.87$

16. $938.46 - 2.86$

17. $53.783 - 2.34$

18. $48.575 - 5.44$

19. $616.78 - 3.9$

20. $125.43 - 2.8$

21. $20 - 0.16$

22. $30 - 0.82$

23. $12.1 - 0.23$

24. $13.6 - 0.51$

25. $-91.13 - 14.213$

26. $-88.14 - 16.315$

27. $-8.69 - (-4.12)$

28. $-7.22 - (-2.11)$

Perform the indicated operation.

29. (a) $-3.4 + (-2.1)$ **(b)** $9.7 - (-5.4)$ **(c)** $-9.2 - 4.1$

30. (a) $-1.13 + (-8.84)$ **(b)** $8.31 - (-2.36)$ **(c)** $-4.99 - 1.73$

Combine like terms.

31. $2.3x + 3.9x$

32. $4.6x + 1.7x$

33. $24.8y - 11.3y$

34. $15.6y - 8.2y$

35. $3.5x + 9.1x$

36. $5.5x + 3.2x$

37. $1.4x + 6.2y + 3.5x$

38. $2.6x + 3.1y + 4.2x$

Solve each applied problem.

39. The total daily rainfall from a storm is recorded below. Estimate the total inches of rainfall for the 3-day period.

Sunday	Monday	Tuesday
1.2 in.	1.79 in.	0.98 in.

40. Darlene kept the following log of the miles she drove each day for charity work. Estimate the total miles driven in the 4 days.

Saturday	Sunday	Monday	Tuesday
11.2 mi	9.93 mi	5.12 mi	6.89 mi

41. Ann spent $72.31 on groceries for her family. If she gives the clerk a 100-dollar bill, how much change should she get?

42. Charles checked his odometer before the summer began. It read 2301.22 miles. He traveled 1236.9 miles that summer in his car. What was the odometer reading at the end of the summer?

43. John makes $1763.24 a month. $161.96 is deducted for federal income tax, $61.23 for social security, and $47.82 for state taxes. How much money does he take home each month after the deductions are taken out?

44. Karen has $321.45 in her checking account. She makes deposits of $38.97 and $86.23. She writes checks for $23.10, $45.67, and $8.97. What is the new balance in her checking account?

Use the following information to answer questions 45–50.

Comparison of Auto Insurance Rates in Utah				
Six-month premiums are based on Salt Lake City residents with a clean driving record who drive a 3-year-old car less than three miles per day.				
Insurance Company	Single Male Age 20	Single Female Age 20	Married Male or Female Age 37	Married Male or Female Age 66
Allstate Insurance	$1,054.30	$669.30	$294.30	$294.30
Bear River Mutual	836.40	522.18	293.45	279.73
Nationwide Mutual	899.40	525.60	290.20	262.50
State Farm Mutual	1,199.85	714.10	345.60	295.35
United Services Auto	1,016.43	696.89	333.89	290.32

Source: Salt Lake Tribune.

45. How much more does Allstate charge for a single male, age 20, than for a single female, age 20?

46. How much more does State Farm Mutual charge than Bear River Mutual for a single female, age 20?

47. Which company charges the lowest rate for a married male, age 37?

48. Which company charges the lowest rate for a married female, age 66?

49. Each of the following three people bought insurance from Jerry Denton of Nationwide Mutual on Monday: a married couple, age 37, and their neighbor Arnold, age 20. How much is the total premium for all three people?

50. On Friday an insurance agent for United Services Auto wrote policies for a single female, age 20, and a married couple, age 66. What is the total of the three premiums?

✎ **VERBAL AND WRITING SKILLS**

Fill in the blanks.

51. To add numbers in decimal notation, we _____ the decimal points.

52. When adding decimals, we place the decimal point in the answer _____ with the decimal points in the problem.

53. When subtracting decimals, we place the decimal point in the answer _____ with the decimal point in the problem.

54. When subtracting $73 - 23.4$, we rewrite 73 as _____ so that we can line up _____ .

ONE STEP FURTHER

Evaluate for the value given.

55. $x + 2.3$, for $x = -6.7$

56. $x - (-19.2)$, for $x = -0.09$

Solve.

57. $x - 5.1 = 9.33$

58. $x + 6.7 = -10.88$

CALCULATOR EXERCISES

59. $3425.723 + 181.511 + 26.809$

60. $3456.89 + 259.6 + 2147.813$

61. $6007.03 - 381.95$

62. $7009.4 - 3476.421$

CUMULATIVE REVIEW

Simplify.

1. $(231)(14)$

2. $(-12)(9)$

3. $(19)(-15)$

4. $(-21)(-33)$

After studying this section, you will be able to:

① Multiply decimals.

② Multiply a decimal by a power of 10.

③ Follow the order of operations with decimals.

④ Solve applied problems involving decimals.

Math Pro   Video 7.3   SSM

S E C T I O N   7 . 3

Multiplication of Decimals and Applied Problems

① **Multiplying Decimals**

Just as with addition and subtraction, we can relate multiplication of fractions to decimals. For example,

$$\frac{5}{10} \times \frac{9}{100} = \frac{45}{1000} \qquad \text{Fractional notation}$$

$$\downarrow \qquad \downarrow \qquad \downarrow$$

$$0.5 \times 0.09 = 0.045 \qquad \text{Decimal notation}$$

In both cases we multiply $9 \times 5 = 45$. When we multiply using decimal notation, we must decide where to place the decimal point in the product, 45. We determine this by adding the number of decimal places in each factor.

$$0.5 \quad \times \quad 0.09 \quad = \quad 0.045$$

$$\downarrow \qquad\qquad \downarrow \qquad\qquad \downarrow$$

one	+	*two*	=	*three*
decimal		decimal		decimal
place		places		places

PROCEDURE TO MULTIPLY DECIMALS

1. Multiply the numbers just as you would multiply whole numbers.
2. Find the sum of the decimal places in the factors.
3. Place the decimal point in the product so that the product has the same number of decimal places as the sum in step 2. To do this you may need to write zeros to the left of the number in step 1.

$$0.4 \times 0.06 \rightarrow 4 \times 6 = 24$$

$$\downarrow \qquad \downarrow$$

$$1 \; + \; 2 = 3 \text{ decimal places}$$

$$0.024$$

$$\downarrow$$

Insert   3
zero.   decimal places

E X A M P L E   1   Multiply: 0.08×0.04.

SOLUTION

Multiply just as you
would whole numbers.
$$\downarrow$$

$$0.08 \quad \times \quad 0.04 \rightarrow 8 \times 4 = 32$$

$$0.08 \quad \times \quad 0.04 \qquad\qquad = \textbf{0.0032}$$

2 decimal + 2 decimal = 4 decimal
places   places   places

Note: We had to insert 2 zeros to the left of 32 in order to have 4 decimal places in the product.

P R A C T I C E   P R O B L E M   1

Multiply: 0.05×0.07.

When multiplying larger numbers, it is usually easier to perform the calculation if we multiply vertically, placing the factor with the fewest number of non-zero digits underneath the other factor.

E X A M P L E   2   Multiply: 5.33×7.2.

SOLUTION   We write the multiplication just as we would if there were no decimal points.

$$
\begin{array}{r}
5.33 \\
\times \quad 7.2 \\
\hline
1066 \\
3731 \quad \\
\hline
\mathbf{38.376}
\end{array}
$$

2 decimal places
1 decimal place

We need 3 decimal places.

Note: We do not line up the decimal points when we multiply.

P R A C T I C E   P R O B L E M   2
Multiply: 20.1×4.32.

We use the same rules for multiplying signed numbers stated in Chapter 3.

E X A M P L E   3   Multiply: $(-2)(4.51)$.

SOLUTION   Recall that the product of a negative number and a positive number is a negative number.

$$
\begin{array}{r}
4.51 \\
\times \quad (-2) \\
\hline
\mathbf{-9.02}
\end{array}
$$

2 decimal places
0 decimal places

We need 2 decimal places ($2 + 0 = 2$).

P R A C T I C E   P R O B L E M   3
Multiply: $(-3)(6.22)$.

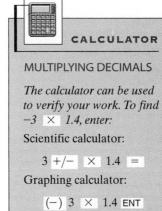

CALCULATOR

MULTIPLYING DECIMALS

The calculator can be used to verify your work. To find
-3 ☒ 1.4, *enter:*

Scientific calculator:

3 +/− ☒ 1.4 =

Graphing calculator:

(−) 3 ☒ 1.4 ENT

The calculator displays

-4.2

❷   Multiplying a Decimal by a Power of 10

Observe the following pattern:

one zero   Decimal point moves *one* place to the right.
$$0.042 \times 10^1 = 0.042 \times 10 = 0.42$$

two zeros   Decimal point moves *two* places to the right.
$$0.042 \times 10^2 = 0.042 \times 100 = 4.2$$

three zeros   Decimal point moves *three* places to the right.
$$0.042 \times 10^3 = 0.042 \times 1000 = 42.$$

PROCEDURE TO MULTIPLY A DECIMAL BY A POWER OF 10
To multiply a decimal by a power of 10, move the decimal point to the *right* the same number of places as the number of zeros in the power of 10. It may be necessary to add zeros at the end of the number.

EXAMPLE 4 Multiply: 0.2345×1000.

SOLUTION

three zeros
↓
$$0.2345 \times 1000 = \mathbf{234.5}$$

Move decimal point *three* places to the right

PRACTICE PROBLEM 4 Multiply: 0.123×100.

If the number that is a power of 10 is in exponent form, move the decimal point to the right the same number of places as the number that is the exponent.

EXAMPLE 5 Multiply: 15×10^4.

SOLUTION Since 10^4 has 4 zeros ($10^4 = 10{,}000$) we must move the decimal point to the right 4 places.

$$(15)(10^4) = (15.0)(10^4) = 150000. \text{ or } \mathbf{150{,}000}$$

Rewrite in decimal notation.   Add zeros and move decimal point.

PRACTICE PROBLEM 5
Multiply: $(0.6944)(10^3)$.

③ Following the Order of Operations with Decimals

The rules for the order of operations that we discussed earlier apply to operations with decimals.

> **ORDER OF OPERATIONS**
>
> 1. Perform operations inside parentheses.
> 2. Simplify any expressions with exponents.
> 3. Multiply and divide from left to right.
> 4. Add and subtract from left to right.

EXAMPLE 6 Evaluate: $-5.1 + 3.671 \times 10^2 - 0.8$.

SOLUTION

$$-5.1 + 3.671 \times 10^2 - 0.8 = -5.1 + \mathbf{367.1} - 0.8$$

Multiply:
$3.671 \times 10^2 = 367.1$.

$$= \mathbf{362.0} - 0.8$$

Add:
$-5.1 + 367.1 = 362.0$.

$$= \mathbf{361.2}$$

Subtract:
$362.0 - 0.8 = 361.2$.

PRACTICE PROBLEM 6
Evaluate: $-11.9 - 6.542 \times 10^2 + 2.7$.

4 Solving Applied Problems Involving Decimals

We use the basic plan of solving applied problems that we discussed in earlier sections. Let us review how we analyze real-life situations.

1. *Understand the problem.*
2. *Solve and state the answer.*
3. *Check.*

E X A M P L E   7   A long-distance phone carrier in New Jersey charges a base fee of $4.95 per month, plus 30 cents per minute for calls outside New Jersey and 10 cents per minute for long-distance calls within the state of New Jersey. On average Natasha's monthly long-distance calls total 45 minutes out of state and 220 minutes within the state. If Natasha's budget for long-distance calls is $45 per month, how much more or less will her average long-distance bill be than her budget of $45?

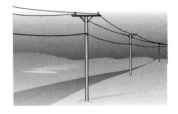

SOLUTION

1. *Understand the problem.* Fill in the Mathematics Blueprint for Problem Solving.

MATHEMATICS BLUEPRINT FOR PROBLEM SOLVING			
GATHER THE FACTS	WHAT AM I ASKED TO DO?	HOW DO PROCEED?	KEY POINTS TO REMEMBER
Base fee: $4.95 Rates, per minute: Outside state 30¢ Inside state 10¢ Budget $45 Total average length of calls: Outside state 45 Inside state 220	Determine how much more or less her average calls will cost than her budget allows.	Add: base fee, $4.95, *plus* calls outside the state, 45 × 30¢, *plus* calls inside the state, 220 × 10¢. Subtract this amount from $45.	Change 30¢ and 10¢ to decimals. Follow the order of operations.

2. *Solve and state the answer.*

base fee	+	charge for calls outside state	+	charge for calls inside state

$$4.95 \;+\; 45 \text{ min} \times 30¢ \text{ per min} \;+\; 220 \text{ min} \times 10¢ \text{ per min}$$
$$= 4.95 \;+\; 45 \times 0.30 \qquad\qquad\quad +\; 220 \times 0.10$$
$$= 4.95 \;+\; 13.5 + 22 \qquad\qquad \text{Multiply: } 45 \times 0.30 = 13.5, \text{and}$$
$$\qquad\qquad\qquad\qquad\qquad\qquad\quad 220 \times 0.10 = 22.$$
$$= 40.45 \qquad\qquad\qquad\qquad\quad \text{Add.}$$

$40.45 is the average cost of Natasha's long-distance calls, and this amount is less than her monthly budget of $45.

We subtract: $45 − $40.45 = $4.55

Her average monthly bill will be **$4.55 less** than her budget of $45.

3. *Check.* Use your calculator to verify your results.

PRACTICE PROBLEM 7

Refer to the information in Example 7 to answer the following. If Natasha averages a total of 55 minutes of long-distance calls outside the state and a total of 185 minutes within the state, will her average monthly bill be more or less than her budget of $45?

DEVELOPING YOUR STUDY SKILLS

PREPARING FOR THE FINAL EXAM

To do well on the final exam, you should begin to prepare many weeks before the final. Cramming for any test, especially the final, often causes anxiety and fatigue and impairs your performance. Complete the following in several 1 to 2-hour study sessions.

- Review all your tests and quizzes.
- Ask your instructor or tutor how to work the problems you missed and still do not understand.
- Complete the Cumulative Test for Chapters 1–7.
- Complete any review sheets that your instructor has given you.
- If you come across a topic that you cannot understand even after seeking assistance, move on to another topic. When you finish reviewing all the topics for the final, return to this topic and try again.
- Find a few students in your class to study with. When you study in groups, you can help each other. Discussing mathematics with others in the group (verbalizing) is an important part of the learning cycle.
- Repeat this process after Chapters 8 and 9.

EXERCISES 7.3

Multiply.

1. 0.03×0.07 **2.** 0.09×0.02 **3.** 0.04×0.08 **4.** 0.05×0.06

5. 7.43×8.3 **6.** 2.5×6.34 **7.** 15.2×3.1 **8.** 21.7×2.2

9. 5.23×1.41 **10.** 7.62×2.13 **11.** $(-4.23)(2.7)$ **12.** $(-3.16)(4.1)$

13. $(-25)(-0.613)$ **14.** $(-31)(-0.314)$ **15.** $(12.1)(-2.81)$ **16.** $(-11.3)(4.11)$

17. $(-21.011)(8.1)$ **18.** $(-31.022)(3.22)$

Multiply by powers of 10.

19. 0.1498×100 **20.** 0.1931×100 **21.** 1.23×1000 **22.** 3.45×1000

23. $85.54 \times 10{,}000$ **24.** $96.12 \times 10{,}000$ **25.** 0.3088×10^3 **26.** 0.97371×10^4

27. 24×10^4 **28.** 35×10^3 **29.** 9.3×10^5 **30.** 7.4×10^5

31. 0.2×10^4 **32.** 0.3×10^3

Evaluate.

33. $3.5 + 4.1 \times 0.5$ **34.** $2.1 + 4.2 \times 1.5$ **35.** $21 - 5.2 \times 1.1 + 0.2$

36. $18 + 3.3 \times 2.4 - 0.5$ **37.** $-6 - 0.4 + 1.5 \times 2.3$ **38.** $-9 + 2.7 - 1.6 \times 0.1$

39. $0.43 \times 100 + 3.1$ **40.** $0.76 \times 1000 + 2.3$ **41.** $3.2 + 0.0052 \times 10^2$

42. $9.3 + 0.0433 \times 10^3$

Solve each problem. The Mathematics Blueprint can help you organize your work when solving the following real-life applications.

MATHEMATICS BLUEPRINT FOR PROBLEM SOLVING			
GATHER THE FACTS	WHAT AM I ASKED TO DO?	HOW DO I PROCEED?	KEY POINTS TO REMEMBER

43. While shopping for school clothes, Amy bought 2 pairs of jeans for $24.95 each, 3 shirts for $12.98 each, and a pair of shoes for $54.25. How much money did she spend for her purchases?

44. Sharon earns $8.75 an hour for the first 40 hours and $13.13 for overtime hours (hours worked over 40 hours). If she works 52 hours this week, how much will she earn?

45. A compact car from Day One Rental Service costs $18.95 a day plus 12 cents a mile for all miles driven over 200 miles. How much will it cost Jack to rent a compact car from this company for 3 days if he drives a total of 423 miles?

46. A mouse pad measures 22.4 centimeters by 26.3 centimeters.
   (a) What is the area of the mouse pad in square centimeters?
   (b) What is the perimeter of the mouse pad?

47. The phone company charges 24 cents for each minute of phone calls during business hours and 14 cents a minute for evening calls. How much will a 17-minute phone call cost during:
   (a) Business hours?
   (b) Evening hours?
   (c) How much is saved by placing the 17-minute call in the evening?

48. (a) The round-trip bus fare to work for Jerry cost $1.90. If Jerry works 20 days a month, how much will it cost him to ride the bus to work?
   (b) A bus pass for a month is $32. How much will Jerry save by buying the bus pass?
   (c) If Jerry takes the bus only 15 days a month, is it still cheaper to buy the bus pass?

 CALCULATOR EXERCISES

Multiply.

49. $(-225.7)(128.59)$

50. $(-8.2365)(-178.34)$

51. $4.036 \times 10^3 \times 54.0052$

52. $10.235 \times 10^4 \times 4.205$

 VERBAL AND WRITING SKILLS

Fill in the blank.

53. If one factor has 3 decimal places and the second factor has 2 decimal places, the product has ____ decimal places.

54. If one factor has 5 decimal places and the second factor has 1 decimal place, the product has ____ decimal places.

ONE STEP FURTHER

Use the following price table to answer questions 55–57.

Item	All-Mart	A&E Foods	The Market
1 gal 2% low-fat milk	$2.89	$3.15	$3.35
1 dozen eggs	1.99	1.99	1.94
1 can cream of chicken soup	1.19	1.09	1.13
1 can tuna	0.79	0.89	0.85
16.9-oz package Rice-A-Roni	0.89	0.99	1.59
1 lb bananas	0.59	0.55	0.44
1 lb apples	1.29	0.79	0.79
1 head of lettuce	1.69	1.99	1.39
1 lb chicken breasts	4.49	4.59	4.99

55. What is the cost of buying one of each item from:
   (a) All Mart?
   (b) A&E Foods?
   (c) The Market?

56. Which item has the most difference between the high and low price?

57. If Jon went to the grocery store and purchased 1 gallon of milk, 2 dozen eggs, 3 cans of soup, 4 pounds of bananas, 3 pounds of apples, and 5 pounds of chicken, which store would cost him the least?

CUMULATIVE REVIEW

Divide.

1. $1330 \div 14$

2. $1029 \div 21$

3. $1377 \div 125$

4. $1561 \div 130$

S E C T I O N   7 . 4

Division of Decimals and Scientific Notation

After studying this section, you will be able to:

1 Divide a decimal by a whole number.

2 Divide a decimal by a decimal.

3 Change a fraction to a decimal.

4 Write numbers using scientific notation.

5 Solve applied problems involving decimals.

Math Pro   Video 7.4   SSM

1 Dividing a Decimal by a Whole Number

Just as with addition, subtraction, and multiplication, the only new rule we must learn when dividing decimal numbers concerns the placement of the decimal point. Let's start our discussion by reviewing the process used to divide whole numbers. Recall that $33 \div 6 = 5$ R3 (5 with remainder 3).

$$\begin{array}{r} 5\text{ R3} \\ 6\overline{\smash{)}33} \\ \underline{30} \\ 3 \end{array} \qquad 33 \div 6 = 5\text{ R3}$$

We can express the answer as a decimal instead of with a remainder as follows:

Place the decimal point directly above the decimal point in the dividend.

$$6\overline{\smash{)}33.0}$$

↘ Rewrite 33 using decimal notation.

Now we divide as if there were no decimal point.

$$\begin{array}{r} 5.5 \\ 6\overline{\smash{)}33.0} \\ \underline{30} \\ 30 \\ \underline{30} \\ 0 \end{array} \qquad 33.0 \div 6 = 5.5$$

PROCEDURE TO DIVIDE A DECIMAL BY A WHOLE NUMBER

1. Place the decimal point in the answer directly above the decimal point in the dividend.
2. Divide as if there were no decimal point involved.

Often, we must add extra zeros to the right of the decimal point so that we can continue dividing until there is a zero remainder.

E X A M P L E   1   Divide: $2.3 \div 5$.

SOLUTION

We place the decimal point directly above the decimal point in the dividend.

$$2.3 \div 5 \longrightarrow 5\overline{\smash{)}2.3}$$

Now we divide as if there were no decimal point.

$$\begin{array}{r} 0.46 \\ 5\overline{\smash{)}2.30} \\ \underline{2\ 0} \\ 30 \\ \underline{30} \\ 0 \end{array} \longrightarrow \text{We add a zero so that we can continue to divide.}$$

$$\mathbf{2.3 \div 5 = 0.46}$$

PRACTICE PROBLEM 1

Divide: $1.3 \div 2$.

Sometimes the division problem does not yield a remainder of zero, or we must carry out the division many decimal places before we get a zero remainder. In such cases we may be asked to round the answer to a specified place.

EXAMPLE 2   Divide: $-6.68 \div 13$. Round your answer to the nearest hundredth.

SOLUTION   We must divide one place beyond the hundredths place—that is, to the thousandths place—so we can round to the nearest hundredth.

The answer is negative.

Place a decimal point directly above the one in the dividend.

$$
\begin{array}{r}
-0.513 \\
13\,\overline{)-6.680} \\
\underline{6\;5} \\
18 \\
\underline{13} \\
50 \\
\underline{39} \\
11 \\
\end{array}
$$

Add a zero to divide to the thousandths place.

0.513 rounded to the nearest hundredth is 0.51; $-6.68 \div 13 \approx -0.51$.

     Note: we used the symbol $\approx$ to indicate that our answer is an approximate value.

PRACTICE PROBLEM 2

Divide: $-36.12 \div 14$. Round your answer to the nearest tenth.

❷ Dividing a Decimal by a Decimal

So far we have considered only division by whole numbers. When the *divisor is not a whole number*, we must adjust the placement of the decimal point so that we have an equivalent division with a whole number as the divisor.

EXAMPLE 3   Divide $6.93 \div 2.2$ and show your work.

SOLUTION   Since the divisor, 2.2, is *not* a whole number, let's write the division problem using fraction notation.

$$6.93 \div 2.2 = \frac{6.93}{2.2}$$

Now, if we multiply the numerator and denominator by 10, the divisor becomes a whole number.

$$\frac{(6.93)(10)}{(2.2)(10)} = \frac{69.3}{22} = 69.3 \div 22$$

The divisor is a whole number.

Once the divisor is a whole number, we divide:

$$
\begin{array}{r}
3.15 \\
22\overline{)69.30} \\
\end{array}
\quad\longrightarrow\quad \text{Add a zero so that we can continue the division.}
$$

$$
\begin{array}{r}
3.15 \\
22\,\overline{)\,69.30} \\
\underline{66} \\
33 \\
\underline{22} \\
110 \\
\underline{110} \\
0 \\
\end{array}
$$

6.93 ÷ 2.2 = 3.15

PRACTICE PROBLEM 3
Divide 14.56 ÷ 3.5 and show your work.

Since multiplying by a power of 10 is the same as moving the decimal point to the right, we can rewrite the division in Example 3 by moving the decimal point to the right one place in both the divisor and dividend.

$$6.93 \div 2.2$$
Move the decimal point 1 place to the right.
or
$$69.3 \div 22.0$$
The divisor is a whole number.

We can summarize the division process with the following procedure for dividing with decimals.

PROCEDURE TO DIVIDE WITH DECIMALS

1. If the divisor is a decimal, change it to a whole number by moving the decimal point to the right as many places as necessary.
2. Then move the decimal point in the dividend to the right the *same* number of places.
3. Place the decimal point in the answer directly above the decimal point in the dividend.
4. Divide until the remainder becomes zero, or the remainder repeats itself, or the desired number of decimal places is achieved.

EXAMPLE 4 Divide 0.7 ÷ 1.5 and show your work.

SOLUTION

The divisor is *not* a whole number.

$$1.5\,\overline{)\,0.7}$$

Move the decimal point *1 place* to the right.

Now that the divisor is a whole number, we rewrite the division and divide.

$$
\begin{array}{r}
.466 \\
15\overline{)7.000} \quad \text{Add zeros.} \\
\underline{6\,0} \\
1\,00 \\
\underline{90} \\
100 \\
\underline{90} \\
10
\end{array}
$$

Decimals that have a digit, or a group of digits, that repeats are called **repeating decimals**. We often indicate the repeating pattern with a bar over the repeating group of digits. Thus **0.7 ÷ 1.5 = 0.4$\overline{6}$** because if we continued the division the 6 repeats.

PRACTICE PROBLEM 4
Divide: $1.1 \div 1.8$.

③ Changing a Fraction to a Decimal

Earlier we saw how to change fractions to decimals when the fraction had a power of 10 as a denominator: $\dfrac{2}{10} = 0.2$, $\dfrac{3}{100} = 0.03$, and so on. In this section we see how to change fractions, whose denominators are not a power of 10, to decimals.

The fraction $\dfrac{21}{5}$ can be written as the division $21 \div 5$. Thus, to change a fraction to a decimal, we divide the numerator by the denominator.

Fraction   Division   Decimal
$$\frac{21}{5} \quad = \quad 21 \div 5 \quad = \quad 4.2$$

PROCEDURE TO CONVERT A FRACTION TO AN EQUIVALENT DECIMAL
Divide the denominator into the numerator until

(a) the remainder becomes zero, *or*
(b) the remainder repeats itself, *or*
(c) the desired number of decimal places is achieved.

EXAMPLE 5   Write $5\dfrac{7}{11}$ as a decimal.

SOLUTION

We divide.

$5\dfrac{7}{11}$ means $5 + \boxed{\dfrac{7}{11}}$

$$
\begin{array}{r}
0.6363 \\
11\overline{)7.000} \\
\underline{6\,6} \\
40 \\
\underline{33} \\
70 \\
\underline{66} \\
40 \\
33
\end{array}
$$

We can see that the pattern repeats.

Thus $\dfrac{7}{11} = 0.\overline{63}$ and $5\dfrac{7}{11} = \mathbf{5.\overline{63}}$.

PRACTICE PROBLEM 5

Write $2\dfrac{5}{11}$ as a decimal.

4 **Writing Numbers Using Scientific Notation**

Earlier we saw that when we multiply by a power of 10, we move the decimal point to the *right* the same number of places as the number of zeros in the power. We will see a similar pattern when we divide by a power of 10.

$$8.0 \div 10^1 = \frac{8}{10} = 0.80 \qquad \text{The decimal point moves left 1 place.}$$

$$8.0 \div 10^2 = \frac{8}{100} = 0.080 \qquad \text{The decimal point moves left 2 places.}$$

$$8.0 \div 10^3 = \frac{8}{1000} = 0.0080 \qquad \text{The decimal point moves left 3 places.}$$

The decimal point moves *left* the same number of places as the number of zeros in the power of 10.

EXAMPLE 6   Perform the indicated operation.

(a) $12.3 \div 10^4$ **(b)** 6.2×10^4

SOLUTION

(a) We are *dividing*, so we move the decimal point *left* 4 places.

$$12.3 \div 10^4 = .00123 \qquad \mathbf{12.3 \div 10^4 = 0.00123}$$
4 places

(b) We are *multiplying*, so we move the decimal point *right* 4 places.

$$6.2 \times 10^4 = 62000. \qquad \mathbf{6.2 \times 10^4 = 62{,}000}$$
4 places

PRACTICE PROBLEM 6
Perform the indicated operation.

(a) $35.12 \div 10^3$ **(b)** 54.2×10^3

If you forget which way to move the decimal point, just think: "When we *divide*, the number gets *smaller*, and moving the decimal point *left* gives that result." "When we *multiply* the number gets *larger*, and moving the decimal point *right* gives that result."

Multiplying and dividing by powers of 10 is often used to convert within the metric system, and in the field of science to write very small or very large numbers using *scientific notation*. For example, the distance to the nearest star, Proxima Centauri, can be written more conveniently in scientific notation.

Standard Notation	*Scientific Notation*
24,800,000,000,000 miles	2.48×10^{13} miles

For very small numbers such as 0.00000321, we write 3.21×10^{-6}.

We observe that a number in scientific notation has two parts, a decimal and a power of 10.

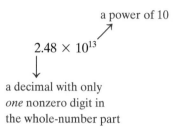

a power of 10

2.48×10^{13}

a decimal with only *one* nonzero digit in the whole-number part

> A positive number is in scientific notation if it is in the form $a \times 10^{n}$, where:
>
> 1. *a* is a number greater than (or equal to) 1 and less than 10.
> 2. *n* (the power) is an integer.

Note: integers are numbers such as $\ldots, -3, -2, -1, 0, 1, 2, 3, \ldots$.

EXAMPLE 7   Write 462 in scientific notation.

SOLUTION   First, we write 462 in the form $a \times 10^{n}$.

$$462 = 4.62 \times 10^{?}$$

The decimal must have only one nonzero digit in the whole-number part.

Now we must determine *what power of 10* we multiply 4.62 by to get 462.

$$4.62 \times 10^{?} = 462$$
$$\mathbf{4.62 \times 10^{2} = 462}$$

We put 2 for the power since we must multiply 4.62 by 10^{2} to get 462.

PRACTICE PROBLEM 7

Write 2119 in scientific notation.

Earlier in this section we saw that 0.00000321 can be written as 3.21×10^{-6}. How is it possible to have a negative exponent? Let's take another look at the powers of 10.

$$10^{2} = 100$$
$$10^{1} = 10$$
$$10^{0} = 1$$
$$10^{-1} = \frac{1}{10}$$
$$10^{-2} = \frac{1}{100}$$

Now, continuing in this pattern, what would you expect the next number on the left to be? What would be the next number on the right? Each number on the right is one-tenth the number above it. We continue in this pattern.

From the pattern we see that

$$\boxed{9 \times 10^{-2}} = 9 \times \frac{1}{100} = \frac{9}{100} \text{ or } \boxed{9 \div 100}$$

$$9 \times 10^{-2} = 9 \div 100$$

Since multiplying by a negative exponent is the same as dividing by a power of 10, we move the decimal point to the *left* when multiplying by 10 to a *negative power*. We use negative exponents to write numbers in scientific notation that are *less* than 1.

EXAMPLE 8   Write 0.00851 in scientific notation.

SOLUTION   First, we write 0.00851 in the form $a \times 10^n$.

$$0.00851 = 8.51 \times 10^?$$ One nonzero digit in the whole-number part.

Now, we must determine what power of 10 we multiply 8.51 by to get 0.00851.

$$8.51 \times 10^? = 0.00851$$
$$8.51 \times 10^{-3} = 0.00851$$ We put -3 for the power since we must move the decimal point *left 3* places to get 0.00851.

$$\mathbf{0.00851 = 8.51 \times 10^{-3}}$$

PRACTICE PROBLEM 8

Write 0.00345 in scientific notation.

Let us summarize the results of Examples 7 and 8 to clarify the differences.

$$4.62 \times 10^2 = 462.$$

This decimal point must move *right* 2 places; we multiply by the *positive power* 2.

$$8.51 \times 10^{-3} = 0.00851.$$

This decimal point must move *left* 3 places; we multiply by the *negative power* –3.

⑤ Solving Applied Problems Involving Decimals

EXAMPLE 9   A chemist mixed up 19.5 quarts of fluid. She wishes to pour it in several equal-sized containers that hold 1.5 quarts each.

(a) How many containers will she need?
(b) If the fluid costs $3.00 per quart, how much will 1 full container of fluid cost?

SOLUTION

(a) Draw a picture.

We divide to find out how many 1.5's are in 19.5: $19.5 \div 1.5 = 13$

The chemist will need **13 containers**.

(b) 1 container holds 1.5 quarts and each quart cost $3.00.
If 1 quart costs 1 × $3.00, then 1.5 quarts cost

$1.5 \times \$3.00 = \$4.50$ 1 full container costs **$4.50**.

PRACTICE PROBLEM 9

A janitor wishes to pour 22.5 liters of cleaning fluid in several equal-sized containers that hold 2.5 liters each.

(a) How many containers will he need?

(b) If the cleaning fluid costs $5.00 per liter, how much will 1 full container of fluid cost?

MONEY EXCHANGE

Each country has its own money system. When you are traveling, you will often need to convert the currency of one country into that of another. The following chart indicates exchange rates for a particular day. These rates can change daily. To change a foreign currency into U.S. dollars, the currency amount is multiplied by the rate shown under *Currency to U.S. Dollars*. To change U.S. dollars to a foreign currency, the U.S. dollar amount is multiplied by the rate shown under *U.S. Dollars to Currency*.

Exchange Rate for October 7, 1998

COUNTRY	CURRENCY TO U.S. DOLLARS	U.S. DOLLARS TO CURRENCY
France (franc)	0.1653	6.0505
Germany (mark)	0.5570	1.7953
Britain (pound)	1.6147	0.6193

PROBLEMS FOR INDIVIDUAL INVESTIGATION

You are on vacation in Europe and must make the following money exchanges.

1. You have $100 to exchange. How much is it worth in francs, marks, and pounds?

2. You have $1000 to exchange. How much is it worth in francs, marks, and pounds?

3. You have 1000 francs to exchange. How much is it worth in U.S. dollars?

4. You have 10 pounds to exchange. How much is it worth in U.S. dollars? Round to the nearest cent.

5. You have 100 marks to exchange. How much is it worth in U.S. dollars?

PROBLEMS FOR GROUP INVESTIGATION AND COOPERATIVE STUDY

From October 7 to October 30 the exchange rates changed. The following chart indicates the exchange rates on October 30.

Exchange Rate for October 30, 1998

COUNTRY	CURRENCY TO U.S. DOLLARS	U.S. DOLLARS TO CURRENCY
France (franc)	0.1679	5.9570
Germany (mark)	0.5639	1.7734
Britain (pound)	1.6054	0.6229

1. Is the U.S. dollar worth more or less on October 30 than it was October 7 in francs, marks, and pounds?

2. On October 30 a product costs 550 marks in Germany. The same product is 200 pounds in Britain. Where should you purchase the product to get the best price in U.S. dollars? Round to the nearest cent.

3. As you travel on your trip, you need to make the following money exchanges. Find the value of each of these exchanges for October 30.
   (a) 100 German marks equals how many pounds?
   (b) 100 pounds equals how many German marks?
   (c) 100 pounds equals how many French francs?
   (d) 100 francs equals how many German marks?

INTERNET: Go to http://www.prenhall.com/blair to explore this application.

EXERCISES 7.4

Divide.

1. $7.44 \div 12$ **2.** $5.12 \div 16$ **3.** $17.28 \div 8$ **4.** $12.6 \div 6$

5. $3.616 \div 64$ **6.** $12.6672 \div 39$ **7.** $82.824 \div 24$ **8.** $44.95 \div 31$

Divide. Round your answer to the nearest hundredth when necessary.

9. $3.25 \div 14$ **10.** $8.23 \div 11$ **11.** $-7.24 \div 8.2$ **12.** $-5.62 \div 9.1$

13. $-20.8 \div (-1.7)$ **14.** $-36.5 \div (-1.6)$ **15.** $6.729 \div 0.27$ **16.** $8.378 \div 0.41$

Divide. If a repeating decimal is obtained, use notation such as $0.\overline{7}$ or $0.\overline{16}$.

17. $5 \div 1.8$ **18.** $14 \div 1.5$ **19.** $0.7 \div 1.1$ **20.** $0.5 \div 3.7$

21. $-100 \div 3.3$ **22.** $-200 \div 3.3$

Write as a decimal.

23. $\dfrac{54}{4}$ **24.** $\dfrac{43}{4}$ **25.** $\dfrac{27}{2}$ **26.** $\dfrac{81}{4}$

Write as a decimal. Round to the nearest hundredth.

27. $\dfrac{11}{6}$ **28.** $\dfrac{15}{7}$ **29.** $\dfrac{16}{3}$ **30.** $\dfrac{23}{7}$

Write as a decimal. Round to the nearest hundredth if necessary.

31. $3\dfrac{2}{5}$ **32.** $5\dfrac{3}{10}$ **33.** $2\dfrac{4}{11}$ **34.** $7\dfrac{2}{15}$

Write as a decimal. Round to the nearest thousandth.

35. $12\dfrac{2}{15}$ **36.** $14\dfrac{3}{16}$

37. A fly can detect motion in $\dfrac{1}{300}$ of a second, whereas the human eye detects motion in $\dfrac{1}{30}$ of a second.

    (a) Write $\dfrac{1}{300}$ of a second as a decimal.

    (b) Write $\dfrac{1}{30}$ of a second as a decimal

38. Erin inherited $\frac{1}{3}$ of her father's estate, and the family's favorite charity inherited $\frac{1}{60}$ of his estate.

   (a) Write $\frac{1}{3}$ as a decimal.

   (b) Write $\frac{1}{60}$ as a decimal.

Perform the indicated operation.

39. (a) $45.6 \div 10^3$ **(b)** 34.7×10^3 **40. (a)** $0.98 \div 10^2$ **(b)** 0.765×10^2

41. (a) $982 \div 10^4$ **(b)** 8.6×10^4 **42. (a)** $1225.6 \div 10^6$ **(b)** 0.1347×10^6

Write in scientific notation.

43. 546   **44.** 631   **45.** 31,235   **46.** 20,231

47. 0.09543   **48.** 0.00478   **49.** 0.00215   **50.** 0.06542

51. 5000   **52.** 1000   **53.** 238,000   **54.** 452,000

Write each number in scientific notation.

55. In one year light will travel 5,878,000,000,000 miles

56. The world forests total 2,700,000,000 acres of wooded area.

57. An electron has a charge of 0.00000000048 electrostatic unit.

58. Yellow light has a wavelength of 0.00000059 meter.

Solve each applied problem.

59. Laura's car travels 901 miles on 34 gallons of gas.

   (a) How many miles per gallon does her car get?

   (b) How many gallons of gasoline can she expect to use driving 1484 miles?

60. Jason's car payment is $303.12 a month.

   (a) If the loan is for 60 months, how much will he pay in total for the car?

   (b) If the original price of the car Jason bought was $14,297.15, how much would he have saved by paying cash?

61. A painter wishes to mix 27 liters of paint in several equal-sized containers that hold 1.5 liters.

   (a) How many containers will he need?

   (b) If the paint cost $5.50 per liter, how much will 1 full container of paint cost?

62. A chemist wishes to mix 65 liters of a mixture in several equal-sized containers that hold 2.5 liters.

   (a) How many containers will he need?

   (b) If the mixture cost $5.70 per liter, how much will 1 full container of this mixture cost?

CALCULATOR EXERCISES

For very large numbers, the display screen on the calculator is not large enough to show all the digits of the number. Thus the number appears on the screen in scientific notation. For example 600,000,000 might be

displayed: | 6.   08 | *indicating* 6×10^8.

Write the number which appears displayed on a calculator.

(a) *Express your answer in scientific notation.*　　　　**(b)** *Multiply to express your answer in standard form.*

63. | 8.23   06 |　　　　　　　　　　　　　　　**64.** | 3.089   04 |

Use your calculator to find each product. Express your answer in scientific notation. Round to the nearest ten thousandth.

65. $20{,}567 \times 98{,}560$　　　　　　　　　　**66.** $70{,}897 \times 65{,}402$

VERBAL AND WRITING SKILLS

Fill in the blank.

67. When we divide $4.62\overline{\smash{)}12.7}$, we rewrite the
equivalent division: _____ , then divide.

68. When we divide $8.23\overline{\smash{)}19.2}$, we rewrite the
equivalent division: _____ , then divide.

ONE STEP FURTHER

To find the average of a list of numbers, use the formula $\text{average} = \dfrac{\text{sum of numbers on the list}}{\text{number of numbers added}}$. *So if you found*

the sum of 255 by adding 5 numbers, to find the average you divide: $\dfrac{255}{5}$. *Use the following information to answer*

questions 69 and 70. Round to the nearest cent.

White-collar Jobs	Hourly Salary
Executives, managers, administrators	$ 39.00
Math and computer scientists	31.85
Registered nurses	25.42
College instructors	35.78

Blue-collar Jobs	Hourly Salary
Stock handlers and baggers	9.19
Waiters and waitresses	5.96
Maids	9.73
Nursing aides	10.51

Source: Federal Bureau of Labor Statistics

69. What is the average hourly salary for the white-collar jobs?

70. What is the average hourly salary for the blue-collar jobs?

CUMULATIVE REVIEW

Solve for x.

1. $12x = 96$　　　　　**2.** $25x = 625$　　　　　**3.** $(23)(45) = x$　　　　　**4.** $x(15) = 225$

458

S E C T I O N   7 . 5

Solving Equations and Applied Problems Involving Decimals

After studying this section, you will be able to:

① Solve equations involving decimals.

② Solve applied problems involving decimals.

Math Pro   Video 7.5   SSM

① Solving Equations Involving Decimals

We use the same rule for solving equations with decimals as that stated in Chapter 6.

> **PROCEDURE TO SOLVE EQUATIONS**
>
> 1. Remove any parentheses.
> 2. Collect like terms and simplify numerical work.
> 3. Add or subtract to get the *ax* term alone on one side of the equation.
> 4. Multiply or divide to get *x* alone on one side of the equation.

E X A M P L E   1   Solve and check your solution: $x - 4.51 = 6.74$.

SOLUTION   To get *x* alone on one side of the equation, we must *add* 4.51 to both sides of the equation.

$$
\begin{array}{rcl}
x - 4.51 & = & 6.74 \\
+\ 4.51 & & +\ 4.51 \\
\hline
x + \ \ 0 & = & 11.25 \\
x & = & \mathbf{11.25}
\end{array}
$$

Check:

$$
\begin{array}{rcl}
x - 4.51 & = & 6.74 \\
11.25 - 4.51 & \overset{?}{=} & 6.74 \\
6.74 & = & 6.74 \checkmark
\end{array}
$$

P R A C T I C E   P R O B L E M   1

Solve and check your solution: $y - 2.23 = 4.69$.

E X A M P L E   2   Solve and check your solution: $-1.2x = 7.8$.

SOLUTION   We must *divide* both sides of the equation by -1.2 to get *x* alone on one side of the equation.

$$
\frac{-1.2x}{-1.2} = \frac{7.8}{-1.2}
$$

$$
x = \mathbf{-6.5}
$$ The answer is negative since a positive number divided by a negative number is negative.

Check:

$$
\begin{array}{rcl}
-1.2x & = & 7.8 \\
(-1.2)(-6.5) & \overset{?}{=} & 7.8 \\
7.8 & = & 7.8 \checkmark
\end{array}
$$

P R A C T I C E   P R O B L E M   2

Solve and check your solution: $5.6x = -25.2$.

EXAMPLE 3   Solve: $2(x + 1.5) = x - 4.62$.

SOLUTION   We must first multiply each term inside the parentheses by 2 to remove the parentheses.

$$2(x + 1.5) = x - 4.62$$

$$2(x) + 2(1.5) = x - 4.62 \qquad \text{Multiply each term by 2.}$$

$$2x + 3 = x - 4.62 \qquad \text{To collect like terms, we must add } -x$$
$$\underline{+ \; -x \qquad\qquad -x} \qquad\quad \text{to both sides of the equation.}$$
$$x + 3 = \qquad -4.62$$

$$x + 3 = \qquad -4.62 \qquad \text{We add } -3 \text{ to both sides of the equation.}$$
$$\underline{+ \qquad - 3 \qquad\quad -3}$$
$$x \qquad = \qquad \mathbf{-7.62}$$

We leave the check to the student.

PRACTICE PROBLEM 3
Solve: $4(x - 2.5) = 3x + 0.6$.

Sometimes it is easier to solve equations that involve decimals if we eliminate the decimal. That is, we rewrite an equivalent equation that does not have decimal terms. We do this by multiplying both sides of the equation by the power of 10 necessary to eliminate the decimal from each term.

EXAMPLE 4   Solve: $0.35x + 0.3 = 1.7$.

SOLUTION   The term 0.35 has the *most* decimal places (2 decimal places), so we must multiply both sides of the equation by 100.

$$0.35x + 0.3 = 1.7$$

$$\mathbf{100}(0.35x + 0.3) = \mathbf{100}(1.7) \qquad \text{Multiply } both \text{ sides of the equation by 100.}$$

$$\mathbf{100}(0.35x) + \mathbf{100}(0.3) = \mathbf{100}(1.7) \qquad \text{Use the distributive property.}$$

$$35x + 30 = 170 \qquad \text{Multiply each term by 100 (moving the decimal point to the right 2 places).}$$

$$35x = 140 \qquad \text{Subtract 30 from both sides of the equation.}$$

$$\frac{35x}{35} = \frac{140}{35}$$

$$x = \mathbf{4}$$

We leave the check to the student.

PRACTICE PROBLEM 4
Solve: $0.3x + 0.15 = 2.7$.

We can shorten this process if we skip the step in which we multiply by a power of 10. Instead, we can just move the decimal point in each term to the right the required number of places. The solution in Example 4 can be shortened as follows.

$$0.35x + 0.3 = 1.7$$
$$35x + 30 = 170$$

Move the decimal point in each term 2 places to the right to multiply each term by 100.

When using this method we must remember to move the decimal point in *each term* the same number of places to the right.

② Solving Applied Problems Involving Decimals

We often deal with decimal equations when we solve problems that involve money. One type of money problem requires that we make a distinction between *how many coins* and the *value of the coins*. For example, if we let the variable d represent *how many dimes* there are, we calculate the *value of the dimes* as follows:

	How Many	Value
Each dime is worth 10 cents or $0.10.	↓	↓
If we had *two* dimes, they would be worth:	$(2)(0.10)$	$= 0.20$
If we had *three* dimes, they would be worth:	$(3)(0.10)$	$= 0.30$
Thus, if we had d dimes, they would be worth:	$(d)(0.10)$	$= 0.10d$

EXAMPLE 5   Andy must fill his vending machine with change. From past experience he knows that the number of dimes he needs to place in the machine for change is twice the number of quarters. The total value of dimes and quarters needed for change is $4.95. How many of each coin must Andy put in the vending machine?

SOLUTION

1. *Understand the problem.*

MATHEMATICS BLUEPRINT FOR PROBLEM SOLVING			
GATHER THE FACTS	WHAT AM I ASKED TO DO?	HOW DO I PROCEED?	KEY POINTS TO REMEMBER
The number of dimes is twice the number of quarters. The total change needed in dimes and quarters is $4.95.	Determine the number of dimes and the number of quarters that must be used for the $4.95 change.	1. Define the variables. 2. Form an equation. 3. Solve the equation.	The value of the dimes is $0.10 \times$ number of dimes. The value of quarters is $0.25 \times$ number of quarters.

2. *Solve and state the answer.* We define our variables. Since we are comparing to quarters, we let our variable represent the number of quarters.

How Many
↓

Q = number of quarters

$2Q$ = number of dimes
(twice the number of quarters)

Value
↓

$(0.25)Q$ = *value* of the quarters

$(0.10)(2Q)$ = *value* of the dimes

We form an equation, then solve the equation.

The *value* of quarters	+	the *value* of dimes	=	the *total value* of coins

$$(0.25)Q \; + \; (0.10)(2Q) \; = \; \$4.95$$

$$0.25Q \; + \; 0.20Q \; = \; 4.95 \qquad \text{Multiply } (0.10)(2Q)$$
$$= (0.10 \times 2)Q = 0.20Q$$

$$25Q \; + \; 20Q \; = \; 495 \qquad \text{Move the decimal point in each term 2 places to the}$$
$$45Q \; = \; 495 \qquad \text{right.}$$

$$\frac{45Q}{45} \; = \; \frac{495}{45}$$

$$Q \; = \; 11 \qquad \text{There are 11 quarters.}$$

$$2Q \; = \; 2(11) = 22 \qquad \text{There are 22 dimes.}$$

Andy must put **11 quarters and 22 dimes** in the vending machine for change.

3. *Check.* Do we have 2 times as many dimes as quarters, as stated in the problem? Yes. 22 dimes is 2×11, or 2 times the number of quarters. Is the value of the coins equal to $4.95? Yes. 22 dimes = $2.20 and 11 quarters = $2.75; $2.20 + $2.75 = $4.95. ✓

PRACTICE PROBLEM 5

Julio must fill his vending machine with change. From past experience he knows that he must put 4 times the number of dimes as quarters in the machine for change. The total value of dimes and quarters needed for change is $3.25. How many of each coin must Julio put in the vending machine?

EXERCISES 7.5

Solve.

1. $x + 3.7 = 9.8$

2. $x + 5.4 = 8.1$

3. $y - 2.5 = 6.95$

4. $y - 5.7 = 3.64$

5. $2.9 + x = 5$

6. $3.7 + x = 8$

7. $x + 2.5 = -9.6$

8. $x + 3.6 = -7.2$

9. $5.6 + x = -4.8$

10. $8.4 + x = -2.9$

11. $3.4x = 10.2$

12. $1.2x = 4.8$

13. $5.1x = 25.5$

14. $1.2x = 7.2$

15. $2x = 11.24$

16. $5x = 45.5$

17. $-5.6x = -19.04$

18. $-3.6x = -10.08$

19. $-5.2x = 26$

20. $-3.5x = 14$

21. $3(x + 1.4) = 6.9$

22. $2(x + 2.2) = 13.2$

23. $2(x - 4) = 26.4$

24. $4(x - 1) = 8.9$

25. $5(x + 2.1) = 21$

26. $3(x + 1.1) = 13.2$

27. $2(x + 3.2) = x + 9.9$

28. $5(x + 1.5) = 4x + 9.6$

Eliminate the decimals by multiplying by a power of 10, then solve.

29. $0.3x + 0.2 = 1.7$

30. $0.4x + 0.3 = 1.9$

31. $0.8x + 0.6 = 5.4$

32. $0.5x + 0.6 = 4.6$

33. $0.12x + 1.1 = 1.22$

34. $0.75x + 1.5 = 2.25$

35. $0.15x + 0.23 = 1.43$

36. $0.52x + 0.31 = 1.35$

Solve each applied problem.

37. Cody must place change in pay phones. From past experience he knows that the number of dimes he must place in a phone is 5 times the number of quarters. The total value of dimes and quarters needed for change is $4.50. How many of each coin must Cody put in each pay phone?

38. Patricia must fill her vending machine with change. From past experience she knows that the number of dimes she must place in the machine for change is 6 times the number of quarters. The total value of dimes and quarters needed for change is $4.25. How many of each coin must Patricia put in the vending machine?

39. Karin has a collection of coins (dimes and nickels) minted in the late 1800s. The dollar value of these coins is $2.20. If the number of dimes in her collection is 5 times the number of nickels, how many of each coin does she have?

40. Suppose that you have $10.50 in dimes and nickels. How many of each coin do you have if the number of dimes you have is twice the number of nickels?

Solve for x.

41. $23.098x = 103.941$

42. $42.0876x = 176.76792$

43. $x + 3.0012 = 21.566$

44. $x + 2.0011 = 33.5401$

VERBAL AND WRITING SKILLS

Translate each question into an equation. Do not solve.

45. What number plus 1.2 equals 2.6?

46. What number minus 5.77 equals 9?

47. 4.2 times what number is equal to 8.992?

48. 75.5 is equal to 5.5 times what number?

ONE STEP FURTHER

49. The average acceleration of an object is given by $a = \dfrac{v}{t}$, where a is the average acceleration, v is the velocity, and t is the time. Find the velocity after 3.5 seconds of an object whose acceleration is 15 feet per second squared.

50. The formula for calculating the temperature in degrees Fahrenheit when you know the temperature in degrees Celsius is $F = 1.8C + 32$. Use this formula to find the Celsius temperature when the Fahrenheit temperature is $-49°$.

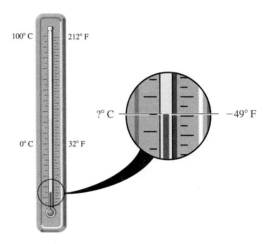

CUMULATIVE REVIEW

Change each fraction to a decimal.

1. $\dfrac{4}{5}$

2. $\dfrac{5}{8}$

3. $\dfrac{1}{3}$

4. $\dfrac{5}{6}$

464

SECTION 7.6

Percents

❶ Understand the meaning of percent.

❷ Change between decimals and percents.

❸ Change between fractions, decimals, and percents.

We use percents in all aspects of life: business, science, sports, and in our everyday life. A suit you want to buy is on sale for 25% off, you receive a 5% increase in pay, and so on. What does this mean? How do we calculate percents? In this section we gain the knowledge to answer these questions.

Math Pro Video 7.6   SSM

❶ Understanding the Meaning of Percent

In previous chapters we used decimals or fractions to describe parts of a whole. Using a percent is another way to describe part of a whole. A percent is a fraction with a denominator of 100.

> It is important to know that 47% means 47 out of 100 parts. It can also be written $\frac{47}{100}$. Understanding the meaning of the notation allows you to work with percents as well as change from one notation to another.

EXAMPLE 1   State using percents: 13 out of 100 radios are defective.

SOLUTION

$$\frac{13}{100} = 13\%$$ **13%** of the radios are defective.

PRACTICE PROBLEM 1

State using percents: 42 out of 100 students in the class voted.

Percents can be larger than 100% or less than 1%. Consider the following situations.

EXAMPLE 2   Last year's attendance at the school's winter formal was 100 students. This year the attendance was 121. Write this year's attendance as a percent of last year's.

SOLUTION

This year's attendance → $\dfrac{121}{100}$ = 121%
Last year's attendance →

This year's attendance at the formal was **121%** of last year's.

PRACTICE PROBLEM 2

Ten years ago a lawn mower cost $100. Now the average price for a lawn mower is $215. Write the present cost as a percent of the cost ten years ago.

EXAMPLE 3   There are 100 milliliters (mL) of solution in a container. Sara takes 0.3 mL of the solution. What percentage of the fluid does Sara take?

SOLUTION Sara takes 0.3 mL out of 100 mL, or

$$\frac{0.3}{100} = \mathbf{0.3}\% \text{ of the solution}$$

PRACTICE PROBLEM 3

There are 100 mL of solution in a container. Julio takes 0.7 mL of the solution. What percentage of the fluid does Julio take?

② Changing between Decimals and Percents

Earlier we saw how to change between fraction and decimal notation. We review below.

Fraction → Decimal	Decimal → Fraction
$\dfrac{5}{8} = 5 \div 8 = 0.625$	$0.625 = \dfrac{625}{1000} = \dfrac{5}{8}$

Now we combine this skill with our knowledge of percents to see how to change between percents and decimals. Observe the pattern in the following illustrations.

We write a decimal as a percent.

Decimal → Percent

$$0.27 = \frac{27}{100} = 27\% \text{ or } 27.0\%$$

0.27 = 27.0%

Decimal point moves 2 places to the *right*.

We reverse the process to write a percent as a decimal.

Decimal ← Percent

0.27 = 27.0%

Decimal point moves 2 places to the *left*.

We summarize below.

PROCEDURE TO CHANGE BETWEEN PERCENTS AND DECIMALS

Writing a decimal as a percent:

1. Move the decimal point 2 places to the *right*.
2. Write the percent symbol at the end of the number.

Decimal → Percent

0.712 = 71.2%

Writing a percent as a decimal:

1. Move the decimal point 2 places to the *left*.
2. Drop the percent symbol.

Decimal ← Percent

0.712 = 71.2%

Using the following chart can be helpful when changing between decimal and percent form since **the decimal point moves the same direction as you move on the chart.**

EXAMPLE 4

(a) Write 3.8% as a decimal.   **(b)** Write 0.09 as a percent.

SOLUTION

(a) Decimal ← Percent   We write the chart.

    ____   3.8%   We move left on the chart, so the decimal
         ←⎵   point moves 2 *places left*.
0.038 = 3.8%   We must place an extra zero to the left of the 3.

(b) Decimal → Percent   We write the chart.

  0.09%   ____   We move right on the chart, so the decimal
  ⎵→   point moves 2 *places right*.
  0.09   =   **9%**

PRACTICE PROBLEM 4

(a) Write 266% as a decimal.   **(b)** Write 0.001 as a percent.

EXAMPLE 5   Complete the table of equivalent notations.

Decimal Form	Percent Form
0.457	
	58.2%
	0.6%
2.996	

SOLUTION

Decimal Form	Percent Form
0.457	45.7%
0.582	58.2%
0.006	0.6%
2.996	299.6%

PRACTICE PROBLEM 5
Complete the table of equivalent notations.

Decimal Form	Percent Form
0.511	
	84.1%
	0.2%
6.776	

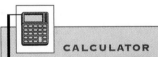

CALCULATOR

PERCENT TO DECIMAL

You can use a calculator to change a percent, 59.3%, to a decimal. Enter:

Scientific calculator with a percent key:

59.3 % =

Scientific calculator without a percent key:

59.3 ÷ 100 =

Graphing calculator:

59.3 ÷ 100 ENT

The calculator displays

0.593

③ Changing between Fractions, Decimals, and Percents

Now that you can change between decimals and percents, you are ready to change between fractions and decimals and percent.

$$\boxed{\text{fraction} \; \overset{\leftarrow}{\underset{\rightarrow}{}} \; \text{decimal} \; \overset{\leftarrow}{\underset{\rightarrow}{}} \; \text{percent}}$$

EXAMPLE 6

(a) Write $\dfrac{211}{500}$ as a percent.

(b) Write 42.2% as a fraction.

SOLUTION

(a) Using a chart often helps:

Fraction → Decimal → Percent		
$\dfrac{211}{500}$ → 0.422	?	Compute: $211 \div 500 = 0.422$
$\dfrac{211}{500}$ → 0.422 → **42.2%**		Move the decimal point 2 places to the right.

(b) We reverse the process to write 42.2% as a fraction.

Fraction ← Decimal ← Percent		
? 0.422 ← 42.2%		Move the decimal point 2 places to the left.
$\dfrac{211}{500}$ ← 0.422 ← 42.2%		$0.422 = \dfrac{422}{1000} = \dfrac{211}{500}$.

PRACTICE PROBLEM 6

(a) Write $\dfrac{7}{40}$ as a percent.

(b) Write 17.5% as a fraction.

Changing some fractions to decimals results in a repeating decimal. For example, $\dfrac{1}{3} = 0.333\ldots$ or $0.\overline{3}$. In such cases, we usually round to the nearest hundredth of a percent.

EXAMPLE 7

Write each fraction as a percent. Round to the nearest hundredth of a percent.

(a) $\dfrac{5}{9}$

(b) $\dfrac{14}{33}$

SOLUTION

(a) Fraction → Decimal → Percent

$$\frac{5}{9} \;=\; 0.5555\,\ldots \approx \mathbf{55.56}\%$$ Round 55.555…% to the nearest hundredth.

Note, if we do not round, $\dfrac{5}{9} = 55.\overline{5}\%$.

(b) Fraction → Decimal → Percent

$$\frac{14}{33} \;=\; 0.424242\,\ldots \approx \mathbf{42.42}\%$$ Round 42.4242…% to the nearest hundredth.

PRACTICE PROBLEM 7

Write each fraction as a percent. Round to the nearest hundredth of a percent.

(a) $\dfrac{2}{3}$ **(b)** $\dfrac{15}{33}$

Certain percents occur very often, especially in money matters. Here are some common equivalents that you may already know. If not, be sure to memorize them.

$$\frac{1}{4} = 0.25 = 25\% \qquad \frac{1}{3} = 0.3\overline{3} = 33\frac{1}{3}\% \qquad \frac{1}{10} = 0.10 = 10\%$$

$$\frac{1}{2} = 0.5 = 50\% \qquad \frac{2}{3} = 0.6\overline{6} = 66\frac{2}{3}\% \qquad \frac{3}{4} = 0.75 = 75\%$$

EXERCISES 7.6

State using percents.

1. 24 out of 100 cars are blue.

2. 84 out of 100 students pass the exam.

3. 31 out of 100 students in the class voted.

4. 71 out of 100 students in the class are women.

5. 7 out of 100 phones are defective.

6. 11 out of 100 new computers are defective.

7. 63 out of 100 power boats had a radar navigation system.

8. 93 out of 100 new cars have compact disc players.

9. Last year's attendance at the medical school was 100 students. This year the attendance is 113. Write this year's attendance as a percent of last year's.

10. Last year's attendance at the College Service Club was 100 students. This year the attendance is 160. Write this year's attendance as a percent of last year's.

11. 4 out of 25 people use electric toothbrushes. What percentage of people use electric toothbrushes?

12. 29 out of 50 people play organized sports. What percentage of people play organized sports?

13. 13 out of 20 people play games on their computers. What percentage of people play games on their computers?

14. 7 out of 10 families own a washer and a dryer. What percentage of families own a washer and a dryer?

15. Write 53.8% as a decimal.

16. Write 24.4% as a decimal.

17. Write 0.0024 as a percent.

18. Write 0.006 as a percent.

19. Write 2.33% as a decimal.

20. Write 75.2% as a decimal.

21. In Alaska, 0.03413 of the state is covered by water. Write the part of the state that is covered by water as a percent.

22. In Florida, 0.07689 of the state is covered by water. Write the part of the state that is covered by water as a percent.

Complete each table of equivalent notations.

23.

Decimal Form	Percent Form
0.576	
	24.9%
	0.3%
1.546	

24.

Decimal Form	Percent Form
0.139	
	57.8%
	0.9%
5.612	

25.

Decimal Form	Percent Form
3.742	
	23.8%
	0.6%
12.882	

26.

Decimal Form	Percent Form
2.869	
	23.8%
	0.1%
13.145	

27. Write $\dfrac{35}{40}$ as a percent.

28. Write $\dfrac{129}{250}$ as a percent.

29. Write 72.8% as a fraction.

30. Write 57.5% as a fraction.

31. (a) Write $\dfrac{14}{40}$ as a percent.

(b) Write 22.3% as a fraction.

32. (a) Write $\dfrac{32}{80}$ as a percent.

(b) Write 72.1% as a fraction.

33. In the 1994 budget of the United States, approximately $4\dfrac{1}{11}\%$ was designated for the Department of Agriculture. Write this percent as a fraction.

34. Arran Copy Center wastes $2\dfrac{1}{2}\%$ of its paper supply due to poor quality of the photocopy produced. Write this percent as a fraction.

35. The brain represents $\dfrac{1}{40}$ of an average person's weight. Express this fraction as a percent.

36. During waking hours a person blinks $\dfrac{9}{2000}$ of the time. Express the fraction as a percent.

Complete each table of equivalent notations.

37.

Fraction Form	Decimal Form	Percent Form
$\dfrac{4}{5}$		
	0.27	
		0.7%
$4\dfrac{1}{3}$		

38.

Fraction Form	Decimal Form	Percent Form
$\dfrac{9}{12}$		
	0.61	
		2.8%
$9\dfrac{5}{8}$		

39.

Fraction Form	Decimal Form	Percent Form
$\dfrac{5}{16}$		
	2.6	
		$\dfrac{1}{4}\%$
$6\dfrac{1}{2}$		

40.

Fraction Form	Decimal Form	Percent Form
$\dfrac{8}{15}$		
	3.5	
		$\dfrac{1}{8}\%$
$5\dfrac{1}{4}$		

Write each fraction as a percent. Round to the nearest hundredth of a percent.

41. $\dfrac{5}{7}$ **42.** $\dfrac{16}{35}$ **43.** $\dfrac{25}{57}$ **44.** $\dfrac{4}{9}$

 CALCULATOR EXERCISES

Use your calculator to write each percent as a decimal.

45. 46.8% **46.** 28.9% **47.** 0.0137% **48.** 0.00914%

 VERBAL AND WRITING SKILLS

Fill in the blanks.

49. To change a percent to a decimal, move the decimal point 2 places to the _____ and drop the _____ .

50. To change a decimal to a percent, move the decimal point 2 places to the _____ and add the _____ to the end of the number.

CUMULATIVE REVIEW

Translate and solve.

1. Three times what number is equal to forty-eight?

2. Twice what number is equal to three hundred thirty?

3. One-fourth of what number is equal to 60?

4. What is one-third of sixty-nine?

Solving Percent Problems Using Equations

After studying this section, you will be able to:

① Translating and Solving Percent Problems

❶ Translate and solve percent problems.

❷ Solve general applied percent problems.

Math Pro   Video 7.7   SSM

In this section we solve problems involving percents. It should be noted that a *percent* is used primarily for comparative and descriptive purposes. *When we perform calculations with percents we must first change the percent to its equivalent decimal or fraction form, then perform the calculations.*

→ The word *of* indicates multiplication.

$$35\% \text{ of } 50$$
$$\downarrow \quad \downarrow \quad \downarrow$$
$$0.35 \times 50 = 17.5$$
↑ —————— We change 35% to decimal form.

We know that the fraction $\dfrac{3}{4} = 75\%$. Now let's look at each of the three parts of this relationship—3, 4, and 75%—and see how we can find any one part if we know the value of the others.

The *part of* the base → $\dfrac{3}{4} = 75\%$ ← *Percent*

↘ The entire *base* amount

To solve applied percent problems, we must understand what each of these three parts means. Therefore, we draw a picture of each situation.

E X A M P L E   1   Translate to an equation and solve.

(a) What is 25% of 40?   **(b)** 10 is 25% of what number?

(c) 10 is what percent of 40?

SOLUTION

(a) We want to find *part of* (25% of) the entire *base* amount of 40.

What is 25% of 40
↓   ↓   ↓   ↓   ↓
n = 25% × 40   Write in symbols.
n = 0.25 × 40   Change 25% to decimal form.
n = 10   Multiply.

10 is 25% of 40.

(b) We want to find the entire *base* amount.

10 is 25% of what number?
↓   ↓   ↓   ↓   ↓
10 = 0.25 × n Change 25% to a decimal.

$\dfrac{10}{0.25} = n$ Solve the equation for n.

40 = n Divide.

10 is 25% of **40.**

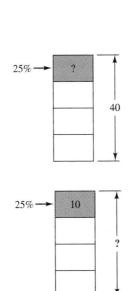

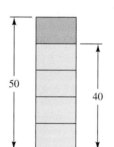

(c) We want to find the *percent* (the part 10 is of 40).

$$10 \text{ is what percent of } 40?$$

$$10 = n \quad \% \times 40$$

$$\frac{10}{40} = n\% \qquad \text{Solve the equation for } n\%.$$

$$0.25 = n\% \qquad \text{The \% symbol reminds us to write 0.25 as a percent.}$$

$$25\% = n \qquad \text{Change 0.25 to a percent.}$$

10 is **25%** of 40.

Note: When forming the equation we use the percent symbol, %, to represent the word *percent*, and remind us that the final answer must be in percent form.

P R A C T I C E   P R O B L E M   1

Translate to an equation and solve.

(a) What is 40% of 90?   **(b)** 36 is 40% of what number?
(c) 36 is what percent of 90?

Sometimes we have more than 100%—this means that the *part* is more than the *base* amount.

E X A M P L E   2   Translate to an equation and solve. 50 is what percent of 40?

S O L U T I O N   We should expect to get more than 100% since 50 is more than the base 40.

$$50 \text{ is what percent of } 40?$$

$$50 = n \quad \% \times 40$$

$$50 = n\% \times 40$$

$$\frac{50}{40} = n\% \qquad \text{Solve for } n\%.$$

$$1.25 = n\% \qquad \text{Divide.}$$

$$125\% = n \qquad \text{Write 1.25 as a percent.}$$

50 is 125% of 40.

P R A C T I C E   P R O B L E M   2

Translate to an equation and solve. 80 is what percent of 20?

E X A M P L E   3   Find 55% of 36.

S O L U T I O N

$$\text{Find } 55\% \text{ of } 36$$

$$n = 55\% \times 36 \qquad \text{We translate } \textit{find} \text{ to } n = \text{ since it has the same meaning as } \textit{what is}.$$

$$= 0.55 \times 36 \qquad \text{Change 55\% to a decimal.}$$

$$= 19.8 \qquad \text{Multiply.}$$

19.8 is 55% of 36.

PRACTICE PROBLEM 3

Find 22% of 60.

TO THINK ABOUT
....

THE PROPORTION METHOD

We can also use proportions to solve percent problems. To use proportions, we must first identify the three parts of a percent that we discussed earlier.

the *part of*
the base $\longrightarrow$ $\dfrac{3}{4} = 75\%$ $\nearrow$ *percent*

$\searrow$ the entire
base amount

Next, we use the methods we learned in Chapter 4 to find the missing part of a proportion. For example, if the *base* is unknown, we would find it as follows:

Percent situation: 3 is 75% of what number?

We write the proportion: $\dfrac{3}{n} = \dfrac{75}{100}$ $\longrightarrow$ We write the
percent as the
fraction with
denominator 100.

The *base* is written in the
denominator of the fraction.

Now we solve the proportion: $100 \times 3 = n \times 75$

$$300 = 75n$$

$$4 = n$$ Divide both sides

Thus 3 is 75% of 4.   by 75.

When we use this method, we must be sure to write the *base* as the denominator of the proportion. If you like this method, there are more examples and explanations in Appendix C.

---

② Solving General Applied Percent Problems

We can solve word problems that involve percents by:

1. Writing a percent statement to represent the situation.
2. Translating the statement to an equation.
3. Solving the equation.

EXAMPLE 4   Marilyn has 850 out of 1000 points possible in her English class. What percent of the total points does Marilyn have?

SOLUTION   We must find the *percent*, so we write the statement that represents the percent situation.

$$850 \text{ is what percent of } 1000?$$
$$\downarrow \quad \downarrow \downarrow \qquad \downarrow \qquad \downarrow \qquad \downarrow$$

Write the statement that represents this situation.

$$850 = n \qquad \% \quad \times 1000? \qquad \text{Form an equation.}$$

$$\frac{850}{1000} = n\% \times \frac{1000}{1000} \qquad \text{Solve the equation.}$$

$$0.85 = n\%$$

$$85\% = n$$

Mary has **85%** of the total points.

There are several ways to write equivalent statements for a percent situation. We could have written, "What percent of 1000 is 850?" Try translating and solving this statement to verify that it is equivalent.

PRACTICE PROBLEM 4

Alisha received 35 out of 50 points on a quiz. What percent of the total points did Alisha earn on the quiz?

EXAMPLE 5   Sean's bill for his dinner at the Spaghetti House was $19.75. How much should he leave for a 15% tip? Round this amount to the nearest cent.

SOLUTION   We must find *part* of (15% of) the *base* amount of $19.75.

$$\text{What is } 15\% \text{ of } \$19.75?$$
$$\downarrow \quad \downarrow \downarrow \quad \downarrow \quad \downarrow$$

Write the statement for this situation.

$$n = 15\% \times \$19.75$$

$$= 0.15 \times \$19.75$$

$$= 2.9625 \qquad \text{Solve the equation.}$$

$$\approx \$2.96 \qquad \text{Round.}$$

The tip is **$2.96**.

Note: Rounding to the nearest cent is the same as rounding to the nearest hundredth. Why?

PRACTICE PROBLEM 5

Francis left a $3.50 tip for her dinner, which cost $18.55. What percent of the total bill did Francis leave for a tip? Round your answer to the nearest hundredth of a percent.

EXERCISES 7.7

Translate to an equation and solve.

1. What is 32% of 84?

2. What is 26% of 72?

3. Find 24% of 145.

4. Find 53% of 210.

5. What is 46% of 60?

6. What is 18% of 66?

Solve each applied problem that involves finding the part of a number.

7. The bill for Marie's dinner was $12.95. How much should she leave for a 15% tip? Round your answer to the nearest cent.

8. The Boyd farm is 250 acres. 70% of it is suitable land for farming. How many acres can be used to farm?

9. 60% of the graduates of Trinity, a two-year college, transfer to a four-year college. If the graduating class at Trinity has 650 students, how many are transferring to a four-year college?

10. H&B Manufacturing claims that no more than 0.5% of its parts are defective. If a client orders 8600 parts from H&B manufacturing, what is the largest number of parts that could be defective?

Translate to an equation and solve. Round to the nearest hundredth of a percent.

11. What percent of 650 is 70?

12. What percent of 350 is 20?

13. 65 is what percent of 850?

14. 400 is what percent of 80?

Solve each applied problem that involves finding the percent.

15. A snack bar has 80 calories. If 15 of those calories are from fat, what percent of the calories are fat?

16. Wesley paid $21 tax when he bought a mountain bike for $300. What percent tax did he pay?

17. In a soccer game Amy made 2 out of 5 shots on goal. What percent of shots did she make?

18. On the Almen High School basketball team, 9 out of the 22 players are over 6 feet tall. What percent of the players are over 6 feet tall?

Translate to an equation and solve.

19. 56 is 70% of what number?

20. 72 is 25% of what number?

21. 24 is 40% of what number?

22. 32 is 64% of what number?

Solve each applied problem that involves finding the base amount.

23. 5% of the employees at the Lido Insurance Insurance Company called in sick with the flu. If 10 employees called in sick, how many employees are there at the company?

24. 25% of the graduating class at Springdale Community College received a scholarship. If 1800 students received a scholarship, how many students are at the college?

25. The new 8-mile nature trail is 125% of the length of the original trail. How long was the original trail?

26. The new 350-seat auditorium is 140% of the original auditorium in seating. How many seats were in the original auditorium?

Translate each of these different types of percent statements to equations, then solve.

27. 15% of 25 is what number?

28. 18% of 97 is what number?

29. What is 27% of 78?

30. What is 32% of 85?

31. 125% of 85 is what number?

32. 118% of 48 is what number?

33. 44 is 50% of what number?

34. 90 is 45% of what number?

35. What percent of 87 is 53.94?

36. What percent of 74 is 39.96?

Write the statement that represents each of these different types of percent situations. Then solve the equation.

37. Robert got 86 out of the 92 questions right on his test. What percent did he get correct? Round your answer to the nearest percent.

38. A basketball player has made 25 of his 30 free throws. What percent of free throws did he make? Round your answer to the nearest percent.

About 208 million tons of residential and commercial trash are generated each year. Use the information in the illustration below to answer questions 39 and 40.

39. How much greater is the percentage of aluminum recycled than the percentage of glass recycled?

40. How much greater is the percentage of yard waste recycled than the percentage of plastics recycled?

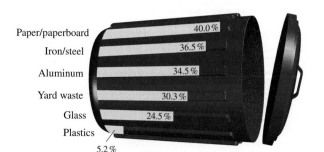

Percent of Residential and Commercial Trash Recycled Annually

Paper/paperboard 40.0%
Iron/steel 36.5%
Aluminum 34.5%
Yard waste 30.3%
Glass 24.5%
Plastics 5.2%

![calculator icon] **CALCULATOR EXERCISES**

41. What is 67.3% of 348.9?

42. What percent of 875 is 625?

43. 368 is 20% of what number?

44. What is 18.9% of $9500?

478

VERBAL AND WRITING SKILLS

Since we use the following percents often, we should memorize them.

25% of a number is the same as $\frac{1}{4}$ of the number.   50% of a number is the same as $\frac{1}{2}$ of the number.

Knowing these facts, explain why it is obvious that there is an error in the following statements.

45. 35 is 50% of 40.

46. 50% of 30 is 40.

47. 25% of 10 is 9.

48. 200 is 25% of 60.

ONE STEP FURTHER

Jeremy has a monthly income of $1250. He allocates it as shown on the circle graph. Use the graph to answer questions 49–52.

49. What percent does he spend on recreation?

50. How much money does he spend on food?

51. How much money does he save each month?

52. After rent, food, and clothing, how much does he have left each month?

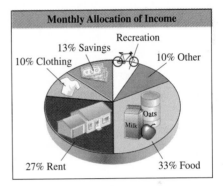

Monthly Allocation of Income

Recreation
13% Savings
10% Clothing
10% Other
Oats
Milk
27% Rent
33% Food

53. Sharon purchased 2 pairs of jeans at $25 each, 3 shirts at $15 each, and a pair of shoes for $45.
   (a) If the sales tax rate is 7%, how much tax will she pay?
   (b) What is the total cost of her purchase, including tax?

CUMULATIVE REVIEW

Solve for x.

1. $2x + 3 = 13$

2. $3x - 1 = 2x + 4$

3. $5x - 3 = 3x + 9$

After studying this section, you will be able to:

1 Solve commission problems.

2 Solve percent increase, decrease, and discount problems.

3 Solve interest problems.

Math Pro   Video 7.8   SSM

S E C T I O N   7 . 8

Applied Problems Involving Percent

1 **Solving Commission Problems**

If you work as a salesperson, your earnings may be in part, or in total, a certain percentage of the sales you make. This type of earnings is called a **commission**. For example, if you work on a 15% commission basis, your commission is 15% of your total sales. We call the percent, 15%, the **commission rate**. We can either write a *percent statement* or use the *formula* given below when solving commission problems.

Your commission is   15% of your total sales.

$$\text{commission} = \text{commission rate} \times \text{total sales}$$

Since the *formula* for a commission problem is equivalent to the *percent statement*, we will only write the formula. Remember, if you forget the formula, you can always write the percent statement.

E X A M P L E   1   Alex is a car salesman and earns a commission rate of 9% of the price of each car he sells. If he earned $3150 commission this month, what were his total sales for the month?

SOLUTION

$$\text{Commission} = \text{commission rate} \times \text{total sales}$$
$$\$3150 = 9\% \qquad \times n$$
$$\$3150 = 0.09 \times n$$
$$\frac{\$3150}{0.09} = \frac{0.09n}{0.09}$$
$$\$35{,}000 = n$$

Alex's total sales were **$35,000**.

Check: We will estimate to check our answer. We round 9% to 10%, then verify that 10% of his total sales ($35,000) equals his commission ($3150). 10% of $35,000 = $3500, which is close to his commission of $3150. ✓

P R A C T I C E   P R O B L E M   1

You must pay your real estate agent a 6% commission on the sale price of your home. If the agent sells your home for $145,000, what amount of commission must you pay the agent?

2 **Solving Percent Increase, Decrease, and Discount Problems**

There are many situations that involve increasing or decreasing an amount by a certain percent. If you receive a 5% raise or buy a CD player for 20% off the original price, you are working with a *percent increase* or a *percent decrease* situation. To find the amount of the increase or decrease, we can either write a percent statement or use a formula.

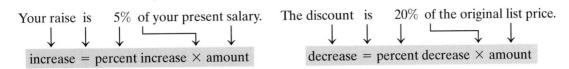

Your raise is   5% of your present salary.   The discount is   20% of the original list price.

increase = percent increase × amount   decrease = percent decrease × amount

EXAMPLE 2   The enrollment at Laird Elementary School was 450 in 1997. In 1998 the enrollment increased by 8%. How many more students enrolled in 1998 than in 1997?

SOLUTION

Percent statement: Increase in enrollment is 8% of 450.

Formula: increase = percent increase × amount

$$increase = 8\% \times 450$$
$$= 0.08 \times 450$$
$$= 36$$

There were **36 more students** in 1998 than 1997.

 Check: We round 8% to 10%. 10% of 450 is 45, which is close to our answer. ✓ Our estimate should be more than our answer. Why?

PRACTICE PROBLEM 2

A dining room set is reduced 30% from the original price of $1640. How much is the dining room set reduced in price?

Suppose that we want to find the new amount after the increase or decrease. We do this as follows:

original amount + increase = new amount
original amount − decrease = new amount

EXAMPLE 3   Arnold earns $26,000 a year and received a 6% raise. How much is his new yearly salary?

SOLUTION   First, we find the amount of the raise:

percent increase × amount = increase (raise)

$$6\% \times 26,000 = 0.06 \times 26,000 = 1560 \text{ raise}$$

Now we find his new yearly salary.

original salary + raise = new salary

$$26,000 + 1560 = 27,560$$

Arnold's new salary is $27,560.

Check: 6% is a little more than $\frac{1}{2}$ of 10%. Use this fact to check the answer.

PRACTICE PROBLEM 3

Rebecca makes $9.45 per hour and received a raise of 3% of her hourly rate. What is her new hourly rate?

EXAMPLE 4   An advertisement states that all items in a department store are reduced 30% off the original list price. What is the sale price of a big-screen television set with a list price of $2700?

SOLUTION   First, we find the amount of the discount.

percent decrease × amount = decrease (discount)

30% × 2700   = 0.30 × 2700 = 810 discount

Next we find the sale price.

original price − discount = sale price

2700 − 810   = 1890

The sale price is $1890.

Check: A 30% discount is about $\frac{1}{3}$ of the price. Use this fact to check the answer.

PRACTICE PROBLEM 4

1500 people attended the Westmont College Jazz Concert last year. This year the number of tickets sold is 14% less than last year. How many people have bought tickets to the concert this year?

❸ Solving Interest Problems

Interest is money paid for the use of money. If you deposit money in a bank, the bank uses that money and pays you interest. If you borrow money, you pay the bank interest for the use of that money. To solve simple interest problems we must be familiar with the terms; *interest, principal, interest rate,* and *time.*

Principal: the amount deposited or borrowed.

Interest rate: the percent used in computing the interest.

Time: the period of time interest is calculated.

Interest: the money earned or paid for the use of money.

The formula used in business to compute simple interest is:

interest = principal × rate × time

$$I = P \times R \times T$$

The interest rate is assumed to be *per year* unless otherwise stated. The time T must be in years.

EXAMPLE 5   Larsen borrowed $9400 from the bank at 13%.

(a) Find the interest on the loan for 1 year.
(b) How much does Larsen pay back to the bank at the end of the year when he pays off the loan?

SOLUTION

(a) $I = P \times R \times T$; P = principal = $9400, R = rate = 13%,
T = time = 1 year.

I = $9400 × 0.13 × 1 = $1222   We change 13% to 0.13, then multiply.

The interest for 1 year is **$1222**.

(b) At the end of the year he must pay back the original amount he borrowed plus the interest.

$$\text{original loan} + \text{interest} = \text{payoff amount}$$
$$\$9400 + \$1222 \quad = \$10{,}622$$

The total amount that Larsen must pay back is **\$10,622**.

PRACTICE PROBLEM 5

Damon put \$2000 in a saving account that pays 6% per year.

(a) How much interest will Damon earn in 1 year?
(b) How much money will be in Damon's saving account at the end of the year?

Our formula is based on a yearly interest rate. Time periods of more than 1 year or a fractional part of a year are sometimes needed.

EXAMPLE 6   Find the interest on a loan of \$2500 that is borrowed at 9% for 3 months.

SOLUTION   We must change 3 months to years since the formula requires that the time be in years: $T = 3 \text{ months} = \dfrac{3}{12} = \dfrac{1}{4}$ year.

$$I = P \times R \times T$$
$$= 2500 \times 0.09 \times \frac{1}{4} = 225 \times \frac{1}{4} = \$56.25 \qquad \text{We divide 225 by 4.}$$

The interest for 3 months is **\$56.25**.

PRACTICE PROBLEM 6

Find the interest on a loan of \$3100 that is borrowed at 9% for 4 months.

You may have had experience with a savings account or a loan that had interest rates that are *compounded quarterly*. If the interest is 5% and it is compounded quarterly, every $\dfrac{1}{4}$ of the year (90 days) the interest is added to the savings or the loan. If it is *compounded semiannually*, the interest is added to the savings or loan every 6 months.

EXAMPLE 7   If Sheena invests \$5000 in a savings account that pays 6% compounded semiannually, how much is in the account at the end of 1 year?

SOLUTION   We find the balance at the end of 1 year as follows:

$$\boxed{\begin{array}{c}\text{balance at the end}\\\text{of first 6 months}\end{array}} + \boxed{\begin{array}{c}\text{interest at the end}\\\text{of second 6 months}\end{array}} = \boxed{\begin{array}{c}\text{balance at the}\\\text{end of the year}\end{array}}$$

In this case $T = 6$ months or $\dfrac{1}{2}$ of a year. $I = P \times R \times T$; $P = \$5000$, $R = 6\%$ or 0.06, $T = \dfrac{1}{2}$ year.

Interest at the end of *first* 6 months $= \$5000 \times 0.06 \times \dfrac{1}{2} = \$150$

Balance at the end of *first* 6 months $= \$5000 + \$150 = \$5150$

We use the balance from the end of the first 6 months as the principal for the second 6 months. Why?

Interest at the end of *second* 6 months $= \$5150 \times 0.06 \times \dfrac{1}{2} = \$154.50$

Balance at the end of *second* 6 months $= \$5150 + \$154.50 = \$5304.50$

There is **$5304.50** in Sheena's account at the end of 1 year.

P R A C T I C E   P R O B L E M   7

If Jerry invests $7000 in a saving account that pays 5% compounded semi-annually, how much is in the account at the end of 1 year?

TO THINK ABOUT
··········

Which type of savings plan earns you more money at the end of the year: one that compounds interest 1 time per year (simple interest), or one that compounds interest semiannually? Let's look at the results from Example 7 to determine the answer to this question.

When the interest is compounded semiannually, the balance at the end of the year is $5304.50. Now let's compute the balance at the end of 1 year using simple interest.

$$I = P \times R \times T$$
$$= \$5000 \times 0.06 \times 1$$
$$= \$300$$

The balance at the end of the year is $5000 + $300 = $5300.

As you can see, when interest is compounded semiannually the savings earn more interest. Why do you think this is true? Will a savings plan that compounds interest quarterly earn more or less than semiannually? Why?

EXERCISES 7.8

Solve. If necessary, round your answer to the nearest hundredth.

COMMISSION

1. Jeff is paid 15% commission each week based on the total dollar amount of sales. Last week his total sales were $2350. How much did Jeff earn in commission?

2. Last week Dave sold $3255 worth of computer software. If he is paid a commission rate of 14% of his sales, how much did he earn?

3. Dalley Dodge pays its sales personnel a commission of 25% of the dealer profit on each car. The dealer made a profit of $13,500 on the cars that Brandon sold last month. What was his commission last month?

4. Leslie works as a phone solicitor and is paid a 10% commission rate based on the dollar amount of sales she makes. If Leslie earned $1150 last month, what were her total sales for the month?

PERCENT INCREASE AND DECREASE

5. A stereo system is reduced 22% from the original price of $1725. How much is the stereo system reduced in price?

6. A computer system is reduced 18% from the original price of $2750. How much is the computer reduced in price?

7. Lisa has a salary of $35,500 a year. If she gets an 8% raise, what is her new salary?

8. Each year employees at Brentwood Electronics get a cost-of-living raise (COLA). This year the raise is 4%. How much will the new salary be for an employee making $36,000 a year?

9. Last year 760 people attended the Outdoor Concert in the Park. If 15% more tickets sold this year than last year, how many people will attend the concert this year?

10. There were 450 seats in the Johnson School of Law's largest auditorium. This year the college remodeled the auditorium, increasing the seating 20%. How many seats does the new auditorium have?

SIMPLE AND COMPOUND INTEREST

11. Chris puts $3000 into a certificate account at an interest rate of 7% per year.

   (a) How much interest will Chris earn in 1 year?

   (b) How much money will Chris have in his account at the end of the year?

12. A farmer borrows $9500 from the bank at an interest rate of 12%.

   (a) Find the interest on the loan for 1 year.

   (b) How much money does the farmer pay back to the bank at the end of the year to pay off the loan?

13. A student gets an emergency student loan of $500 for books and supplies at an interest rate of 8%. How much interest does the student pay if the loan is paid back in 6 months?

14. The financial aid office at a college gives $300 short-term loans to low-income students for money to buy books, supplies, and materials. If the interest rate is 6%, how much interest does a student pay if the loan is paid back in 3 months?

Round your answer to the nearest cent.

15. Jennifer has $900 to spend on furniture. If the sales tax rate is 6%, what amount can she spend on the furniture?

16. Katie makes purchases of $35, $60, and $24. She gets an employee's discount of 25% on everything she buys. How much will she have to pay after her discount is taken off?

17. Jim gets a car loan of $3500 at an interest rate of 8% for 1 year. If he pays back the loan in 12 equal payments, how much is each payment?

18. Carlos gets a business loan of $8500 at an interest rate of 9% for 1 year. If he pays back the loan in 12 equal payments, how much is each payment?

19. Lei weighed 165 pounds two years ago. After careful supervision at a weight-loss center, she reduced her weight by 15%. How much did she weigh after weight loss?

20. Marcia buys a membership in a discount warehouse for $35. She saves 5% on the purchase price of everything. How much will she have to purchase to save the $35 that she pays for membership?

21. If the cost of living goes up 6% in a year, how much would Bob need to make this year to have the equivalent income of his $24,000 salary from last year?

22. In Atlanta, the median home price is $110,200. This is an 8.4% increase from a year ago.
Source: National Association of Realtors

  (a) How much would the same house have cost a year ago?

  (b) If the price increase stays the same, how much will the same house cost a year from now?

23. Personal income growth in 1997 was 5.8% and personal spending growth was 5.4%.
Source: Commerce Department

  (a) If the Jackson family's income was $56,725 last year, how much would it be this year?

  (b) If they spent $55,000 last year, how much would they spend this year?

24. Laird, a high-jumper on the track and field team, hit the bar 58 out of 200 jump attempts. This means that he did not succeed in 29% of his jump attempts. As a result of extra practice, Laird reduced these unsuccessful attempts by 6%. If he attempts another 200 high-jumps, how many times can he expect to hit the bar?

Use the following chart, which gives the prices for custom window shades, for questions 25 and 26.

Window Shade Price Chart

Width to:	24″	30″	36″	42″	48″	54″	60″	66″	72″	84″
Height 36″	$208	235	268	301	331	359	389	421	452	515
to: 42″	223	254	289	327	359	395	428	465	499	569
48″	237	275	313	354	390	427	466	506	546	625
54″	252	291	334	378	421	461	503	549	591	677
60″	270	310	358	401	450	496	540	592	638	734
66″	285	330	381	430	481	531	579	633	684	787
72″	302	348	402	457	512	565	617	675	731	840
78″	315	367	426	483	543	598	655	717	776	894
84″	327	387	446	508	570	629	691	756	817	941
90″	342	403	470	534	601	663	728	798	864	996
96″	355	421	489	558	629	694	763	836	904	1043
108″	387	457	534	610	690	763	838	918	995	1148
120″	417	494	580	665	750	831	914	1000	1084	1252

25. Kristen buys two shades 36 inches wide and 42 inches long. She also buys a shade that is 48 inches wide and 54 inches long.

(a) What is the total price of the three shades?

(b) If Kristen gets a 30% discount on the shades, how much will she pay for the three shades?

26. Pam buys 3 shades 54 inches wide and 60 inches long and 2 shades 72 inches wide and 96 inches long.

(a) What is the total price of the 5 shades?

(b) If Pam gets a 40% discount on the shades, how much will she pay for the 5 shades?

CALCULATOR EXERCISES

Round your answer to the nearest hundredth.

27. Alice traveled 22,437 miles in her car last year. She is a salesperson, and 68% of her mileage is for business purposes.

(a) How many miles did she travel on business last year?

(b) If she can deduct 31 cents per mile for her income tax return, how much can she deduct for business mileage?

28. (a) How much sales tax would you pay on a Mazda 626 that cost $18,456 if the sales tax rate is 4.6%?

(b) What would the total price of the Mazda be?

CUMULATIVE REVIEW

1. Find the area of a parallelogram with a base 6 centimeters and a height 4 centimeters.

2. Find the volume of a rectangular solid with $L = 9$ centimeters, $W = 4$ centimeters, and $H = 7$ centimeters.

CHAPTER ORGANIZER

TOPIC	PROCEDURE	EXAMPLES
Writing a decimal as a fraction	1. Count the number of decimal places. 2. Write the decimal part over a denominator that has a 1 and the same number of zeros as the number of decimal places found in step 1. 3. Reduce if possible.	Write 3.048 as a fraction. $3.048 = 3\dfrac{48}{1000} = 3\dfrac{6}{125}$ $\qquad\quad\downarrow\qquad\qquad\downarrow$ 3 decimal 3 zeros places
Comparing two positive numbers in decimal notation	1. Start at the left and compare corresponding digits. Write in extra zeros if needed. 2. When two digits are different, the larger number is the one with the larger digit.	Which is larger? 0.138 or 0.13 $0.138\quad\text{?}\quad0.130$ $\qquad\uparrow\qquad\qquad\uparrow$ $8 > 0$ So 0.138 > 0.130.

Rounding decimals	1. Identify the round-off place digit.
	2. If the digit to the *right* of the round-off place digit is:
	  (a) *Less than 5*, do not change the round-off place digit.
	  (b) *5 or more*, increase the round-off place digit by 1.
	3. In either case, drop all digits to the right of the round-off place digit.
		Round to the nearest hundredth: 0.8652.
		
		0.87
		
		Round to the nearest thousandth: 0.21648.
		
		0.216

Adding and subtracting decimals	1. Write the numbers vertically and line up the decimal points. Extra zeros may be written to the right of the decimal points if needed.
	2. Add or subtract all the digits with the same place value, starting with the right column, moving to the left. Use carrying or borrowing as needed
	3. Place the decimal point of the sum in line with the decimal points of all the numbers added or subtracted.
		Add: (a) 36.3 + 8.007 + 5.26;
		      (b) −6.8 + 2.6
		
		(a) $\overset{1}{3}6.300$
		        8.007
		  + 5.260
		   49.567
		
		(b) −6.8 Keep the
		  + 2.6 sign of larger
		    −4.2 absolute value
		             and subtract.

Multiplying decimals	1. Multiply the numbers just as you would multiply whole numbers.
	2. Find the sum of the decimal places in the two factors.
	3. Place the decimal point in the product so that the product has the same number of decimal places as the sum in step 2. You may need to insert zeros to the left of the number found in step 1.
		Multiply: (a) 0.0064 × 0.21;
		      (b) 10.2 × (−1.3).
		
		(a) 0.0064 (b) 10.2
		  × 0.21 × (−1.3)
		      64 3 06
		    128 10 2
		 0.001344 −13.26 → The product is negative.

Multiplying a decimal by a power of 10	Move the decimal point to the right the same number of places as there are zeros in the power of 10. (Sometimes it is necessary to write extra zeros before placing the decimal point in the answer.)	Multiply. $0.597 \times 10^4 = 5970$ $0.0082 \times 1000 = 8.2$ $0.075 \times 10^6 = 75{,}000$ $28.93 \times 10^2 = 2893$

TOPIC	PROCEDURE	EXAMPLES
Dividing by a decimal	1. Make the divisor a whole number by moving the decimal point to the right. 2. Move the decimal point in the dividend to the right the same number of places. 3. Place the decimal point of your answer directly above the decimal point in the dividend. 4. Divide as with whole numbers.	Divide: $0.003\overline{)85.8}$ $\begin{array}{r} 28600. \\ 3\overline{)85800.} \\ \underline{6} \\ 25 \\ \underline{24} \\ 18 \\ \underline{18} \\ 0 \end{array}$
Converting a fraction to a decimal	Divide the denominator into the numerator until 1. the remainder is zero, or 2. the decimal repeats, or 3. the desired number of decimal places is achieved.	Find the decimal equivalent: $\dfrac{13}{22}$. $\begin{array}{r} 0.5909 \\ 22\overline{)13.0000} \\ \underline{110} \\ 200 \\ \underline{198} \\ 200 \\ \underline{198} \\ 2 \end{array}$ $\qquad \dfrac{13}{22} = 0.5\overline{90}$ or $0.5909090\ldots$
Writing a number using scientific notation	1. Write the positive number in the form $a \times 10^n$. 2. The power of 10 is: (a) *Positive* if we must move the decimal point to the *right n* places to get the positive number. (b) *Negative* if we must move the decimal point to the *left n* places to get the positive number.	Write in scientific notation. (a) 0.00987 (b) 549 (a) $9.87 \times 10^? = 0.00987$ $9.87 \times 10^{-3} = 0.00987$ Decimal point moves left. (b) $5.49 \times 10^? = 549$ $5.49 \times 10^2 = 549.$ Decimal point moves right.
Solving equations with decimals	1. Remove any parentheses. 2. Collect like terms and simplify. 3. Add or subtract to get the *ax* term alone on one side of the equation. 4. Multiply or divide to get *x* alone on one side of the equation.	Solve: $2x - 5.6 = x - 9.87$. $\begin{array}{rcr} 2x - 5.6 = & & x - 9.87 \\ +\underline{-x} & & \underline{-x} \\ x - 5.6 = & & -9.87 \\ +\underline{5.6} = & & \underline{5.6} \\ x = & & -4.27 \end{array}$
Converting a decimal to a percent	1. Move the decimal point two places to the right. 2. Add the percent sign.	$0.19 = 19\%$ $0.516 = 51.6\%$ $0.04 = 4\%$ $1.53 = 153\%$ $0.006 = 0.6\%$
Changing a percent to a decimal	1. Move the decimal point two places to the left. 2. Drop the percent sign.	$49\% = 0.49$ $2\% = 0.02$ $0.5\% = 0.005$ $196\% = 1.96$

TOPIC	PROCEDURE	EXAMPLES
Changing a fraction to a percent	1. Divide the numerator by the denominator and obtain a decimal. 2. Change the decimal to a percent.	$F \rightarrow D \rightarrow P$ $\dfrac{3}{800} = 0.00375 = 0.375\%$ $F \rightarrow D \rightarrow P$ $\dfrac{312}{200} = 1.56 = 156\%$ $F \rightarrow D \rightarrow P$ $3\dfrac{1}{4} = 3.25 = 325\%$
Changing a percent to a fraction	Remove the % sign by moving the decimal point two places to the left. Then write the decimal as a fraction, and reduce the fraction if possible.	$F \leftarrow D \leftarrow P$ $\dfrac{29}{50} = \dfrac{58}{100} \leftarrow 0.58 \leftarrow 58\%$

Solving percent problems by translating to equations	1. Translate by replacing
	  "of" by $\times$ "find" by $n =$
	  "is" by $=$ "percent" by %
	  "what" by n
	2. Solve the resulting equation.
		(a) What is 3% of 56?
		$\downarrow \downarrow \downarrow \downarrow \downarrow$
		$n = 3\% \times 56$
		$n = (0.03)(56)$
		$n = 1.68$
		
		(b) What percent of 70 is 30?
		$\downarrow \swarrow \quad \downarrow \downarrow \downarrow \downarrow$
		$n\% \quad\quad \times 70 = 30$
		$70n\% = 30$
		$\dfrac{70n\%}{70} = \dfrac{30}{70}$
		$n = 0.4285714\ldots$
		n is approximately 42.86%.

Solving commission problems	Commission = commission rate $\times$ value of sales	A housewares salesperson gets a 16% commission on sales he makes. How much commission does he earn if he sells \$12,000 in housewares? Commission $= (0.16)(12,000)$ $= \$1920$
Solving percent increase, decrease, and discount problems	Increase = percent increase $\times$ amount Decrease = percent decrease $\times$ amount (discount) Original amount + increase = new amount Original amount − decrease = new amount	A \$150 men's suit is discounted 20%. How much is the sale price of the suit? Discount = percent discount $\times$ amount $\$30 = 0.20 \times \$150$ Original amount − discount = sale price $\$150 - \$30 = \$120$
Solving simple interest problems	Interest = principal $\times$ rate $\times$ time $I = P \times R \times T$	Hector borrowed \$3000 for 4 years at a simple interest rate of 12%. How much interest did he owe after 4 years? $I = P \times R \times T$ $I = (3000)(0.12)(4)$ $= (360)(4)$ $= 1440$ Hector owed \$1440 in interest.

CHAPTER 7 REVIEW

SECTION 7.1

Write a word name for each decimal.

1. 6.23

2. 0.679

3. 7.0083

Write the word name for the amount on the check.

4.

PAY to the
ORDER of _____ *UC Regents* _____ $ | 46.85 |

_____ DOLLARS

Write in fractional notation. Do not simplify.

5. 4.268

6. 43.94

Write each fraction as a decimal.

7. $32\dfrac{761}{1000}$

8. $54\dfrac{26}{1000}$

Replace the ? with < or >.

9. 0.523 ? 0.524

10. 0.16 ? 0.168

Round to the nearest hundredth.

11. 842.8569

12. 359.2581

Round to the nearest thousandth.

13. 521.1025

14. 406.7809

SECTION 7.2

Add.

15. **(a)** 0.52 + 8.11

(b) −5.2 + 0.236

(c) 0.588 + 36 + 8.43

16. **(a)** 0.38 + 7.85

(b) −7.6 + 0.2517

(c) 85 + 4.34 + 0.5269

Subtract.

17. **(a)** 25.98 − 2.33

(b) −2.12 − 9.67

18. **(a)** 75.216 − 4.87

(b) −9.355 − 2.48

Perform the operation indicated.

19. **(a)** (−9.2) + (−5.4)

(b) 4.32 − (−6.43)

(c) −7 − (−6.67)

Fill in the blanks.

20. To subtract numbers in decimal notation, we _____ the decimal points.

21. When adding 85 + 36.5, we rewrite 85 as _____ so that we can line up _____.

Combine like terms.

22. $4.6x + 7.2x$

23. $8.6x + 3.9x$

24. Jerry keeps a log of his gasoline receipts for his company.

| June 6, 1998 | $22.98 | June 20, 1998 | $21.35 |
| June 13, 1998 | $25.05 | June 27, 1998 | $24.15 |

Estimate Jerry's total gasoline expenses for the month of June.

25. MaryAnn bought a lamp for $54.56 and a frame for $21.06. If she gave the salesperson a 100-dollar bill, how much change did she get?

SECTION 7.3

Multiply.

26. 0.091×0.06

27. 0.082×0.02

28. 5.68×7.21

29. 2.62×7.33

30. $(-3.01)(-41.25)$

31. $(-5.6)(-9.01)$

Fill in the blank.

32. If one factor has 2 decimal places and the second factor has 3 decimal places, the product has _____ decimal places.

33. If one factor has 3 decimal places and the second factor has 4 decimal places, the product has _____ decimal places.

Multiply.

34. 0.1249×100

35. 3.24×1000

36. 41×10^5

37. 0.3255×10^3

38. 0.5211×10^2

39. 36×10^4

Evaluate.

40. $2.5 + 6.1 \times 0.2$

41. $16 + 2.7 \times 3.1 - 0.8$

42. Eli earns $8.30 per hour for the first 40 hours and $12.45 for overtime hours (hours worked over 40 hours). If he works 49 hours this week, how much will he earn?

43. Mark has a 5-year truck lease that allows him to buy the truck at the end of the lease for $4500 plus 12 cents a mile for each mile the truck is driven over 36,000 miles. If at the end of the lease the mileage on Mark's truck is 44,322.6 miles, how much will it cost Mark to buy the truck?

SECTION 7.4

Divide. Round your answer to the nearest hundredth when necessary.

44. $8.66 \div 12$

45. $12.42 \div 14$

46. $-3.25 \div 5.1$

47. $-8.52 \div -7.2$

Divide. If an infinitely repeating decimal is obtained, use notation such as $0.\overline{8}$ or $0.\overline{16}$.

48. $16.221 \div 0.33$

49. $13.01 \div 0.33$

492

Fill in the blank.

50. When we divide $7.21\overline{)25.9}$, we rewrite the equivalent division: _____ , then divide.

51. Place the decimal point in the quotient: $12\overline{)4.349}$.

Write as a decimal.

52. $\dfrac{86}{8}$

53. $\dfrac{47}{2}$

54. Write $\dfrac{13}{9}$ as a decimal. Round your answer to the nearest hundredth.

Write as a decimal.

55. $5\dfrac{1}{2}$

56. $9\dfrac{1}{5}$

Perform the operation indicated.

57. $31.66 \div 10^3$

58. 24.3×10^2

Write in scientific notation.

59. 348

60. 21,206

61. 0.092

62. 0.1359

63. A mechanic wishes to mix 38.5 liters of antifreeze in several equal-sized containers that hold 3.5 liters.
   (a) How many containers will he need?
   (b) If the antifreeze costs $6.20 per liter, how much will 1 full container of antifreeze cost?

SECTION 7.5

Solve.

64. $x - 2.68 = 8.23$

65. $-1.6x = 3.68$

66. $2x + 2.4 = 8.7$

67. $3x - 2.8 = 2x + 4.2$

68. Arlene must fill a copy machine with change. The number of nickels she needs to put in the machine is 5 times the number of dimes. The total value of nickels and dimes needed for change is $14. How many of each coin must Arlene put in the vending machine?

SECTION 7.6

State using percents.

69. 85 out of 100 citizens voted.

70. Last year's enrollment in a college Psych 100 course was 110. Now the enrollment is 132. Write this year's attendance as a percent of last year's.

71. Write 5.7% as a decimal.

72. Write 0.016 as a percent.

73. Write 124% as a decimal.

Complete each table of equivalent notations.

74.

Decimal Form	Percent Form
0.379	
	42.8%
	0.5%
3.472	

75.

Decimal Form	Percent Form
1.24	
	3.5%
	0.25%
0.567	

76. Write $\dfrac{56}{168}$ as a percent.

77. Write 81.3 as a fraction.

Complete each table of equivalent notations.

78.

Fraction Form	Decimal Form	Percent Form
$\frac{3}{8}$		
	0.72	
		0.5%
$7\frac{2}{5}$		

79.

Fraction Form	Decimal Form	Percent Form
$\frac{3}{4}$		
	0.56	
		0.2%
$3\frac{6}{15}$		

Write each fraction as a percent. Round to the nearest hundredth of a percent.

80. $\dfrac{3}{10}$

81. $\dfrac{4}{15}$

SECTION 7.7

82. 82 is 25% of what number?

83. What is 30% of 90?

84. What percent of 300 is 15?

85. 75 is 25% of what number?

86. What is 45% of 120?

87. What percent of 360 is 18?

88. Natasha bought a new car costing $17,000. How much sales tax did she pay if the tax is 7% of the price of the car?

89. Dean left a $4.25 tip for his dinner, which cost $32.40. What percent of the total bill did Dean leave for a tip? Round your answer to the nearest hundredth of a percent.

SECTION 7.8

90. Jason has a commission rate of 18% based on the total sales of large appliances he sells. This month Jason sold $13,250 worth of large appliances. What is his commission?

91. The population of Mira County increased from 50,000 to 59,500. What was the percent of increase?

92. Find the interest on a loan of $7500 borrowed at 13% for 1 year.

93. The price of a stereo system is reduced 22% from the original price of $1899. How much is the stereo system reduced in price?

494

CHAPTER 7 TEST

1. Write the word name: 207.402.

2. Write the fractional notation: 0.013.

3. Write as a decimal: $\dfrac{51}{100}$.

4. Replace the ? with $<$ or $>$: 0.45 ? 0.412.

5. Round 746.136 to the nearest hundredth.

6. Add: $12.93 + 0.21$.

Subtract

7. $18.81 - 6.17$

8. $(-13.2) - (-7.1)$

Multiply

9. $(8.24)(1.2)$

10. $(4.72)(10^3)$

11. Divide: $15.75 \div 3.5$.

12. Write $\dfrac{19}{3}$ as a decimal.

13. State as a percent: 2 out of 20 computer chips are defective.

Complete the table of equivalent notations.

	Fraction Form	Decimal Form	Percent Form
14.	$\dfrac{76}{95}$	(a) ___	(b) ___
15.	(a) ___	0.10	(b) ___
16.	(a) ___	(b) ___	5%

17. A computer is reduced 24% from the original price of $3300. How much is the computer reduced in price?

18. A chemist mixes together two solutions, labeled solution A and solution B. Solution A contains 20 mL of fluid that is 35% alcohol. Solution B contains 10 mL of fluid that is 55% alcohol. How many milliliters of alcohol are in the mixture?

19. Evaluate: $4.7 + 0.3 \times 21 - 0.3$.

Write in scientific notation.

20. 43,986

21. 0.0561

1. _____

2. _____

3. _____

4. _____

5. _____

6. _____

7. _____

8. _____

9. _____

10. _____

11. _____

12. _____

13. _____

14. _____

15. _____

16. _____

17. _____

18. _____

19. _____

20. _____

21. _____

22. Fred has a 3-year car lease that allows him to buy the car at the end of the lease for $5500 plus 12 cents a mile for each mile the car is driven over 30,000 miles. If at the end of the lease the mileage on Fred's car is 34,100.5 miles, how much will it cost Fred to buy the car?

23. Solve for x: $0.5x + 0.2x = 2.8$.

24. Sam drove 577.2 miles last week. He used 33.1 gallons of gas.
   (a) Find how many miles per gallon he averaged in his car last week. Round to the nearest tenth.
   (b) If the cost of gas was $1.27 per gallon, how much money did it cost to operate his car last week? (Round to the nearest cent.)

25. 12 is 30% of what number?

26. What percent of 250 is 5?

27. What is 20% of 48?

28. Fred has a commission rate of 15% based on the total sales of electronic equipment he sells. This month, Fred sold $10,500 worth of electronic equipment. What is his commission?

29. Last month, Wanda's phone bill was $35.78. This month, her phone bill was $52.91. What was the percent of increase? (Round to the nearest tenth of a percent.)

30. A $600 television set is on sale for $525. What is the percent of decrease?

31. A car stereo is reduced 18% from the original price of $250. How much is the car stereo reduced in price?

32. Find the interest on a loan of $8200 at 11% for 2 years.

CUMULATIVE TEST FOR CHAPTERS 1–7

Combine like terms.

1. $x + 2x + 3xy$

2. $5x - x + 2x$

3. $3 - 6x + 2xy - 6 - 4xy$

4. Find the perimeter of a square with a side 8 feet in length.

5. Find the area of a room that is 11 feet long and 9 feet wide.

6. For the first quarter of 1998, CyberTechnologies had a $10,000 loss. For the second quarter, the company had a $25,000 profit. What was the company's overall profit or loss at the end of the second quarter?

Evaluate $4x - 2$:

7. For $x = -1$.

8. For $x = 10$.

9. For $x = \dfrac{1}{2}$.

10. Melanie swam for 30 minutes and burned 250 calories. How many calories did she burn per minute?

Simplify

11. $\dfrac{18y}{27y^2}$

12. $\left(x^3\right)^2$

13. $(4xy)^3$

14. Find the value of x: $\dfrac{2}{3} = \dfrac{26}{x}$.

15. Multiply: $(-3x)(2x - 5)$.

16. Add: $(-2x + 1) + (5x - 6)$.

17. Subtract: $(3x - 7) - (2x + 5)$.

18. Solve for the variable and check: $2(x + 4) - 3x = 6$.

19. Joe plans to make a 10-minute phone call to a city 50 miles away using a public pay phone. It costs $0.75 for the first 3 minutes and $0.15 for each additional minute. How much will the phone call cost?

20. Round 180.273 to the nearest hundredth.

21. Add: $20.17 + 13.59$.

22. Subtract: $27.01 - 5.3$.

23. Multiply: $(9.3)(10^2)$.

24. Divide: $14.7 \div 4.2$.

25. Fred earned $15 per hour. If he then gets a 15% raise, how much does he earn now?

Complete the table of equivalent notations.

	Fraction Form	Decimal Form	Percent Form
26.	$\dfrac{15}{25}$	**(a)** ___	**(b)** ___
27.	**(a)** ___	0.30	**(b)** ___
28.	**(a)** ___	**(b)** ___	1%

1. _____
2. _____
3. _____
4. _____
5. _____
6. _____
7. _____
8. _____
9. _____
10. _____
11. _____
12. _____
13. _____
14. _____
15. _____
16. _____
17. _____
18. _____
19. _____
20. _____
21. _____
22. _____
23. _____
24. _____
25. _____
26. _____
27. _____
28. _____

GRAPHING AND STATISTICS

CHAPTER **8**

O ften, we must make decisions that require us to predict the outcome based on information we have about the situation. For example, a foreman on a construction site must decide the most economical way to get a job done. The foreman must delegate the work based on the number of workers he has available and their speed and efficiency in completing the job. To find out one way to do this, turn to Putting Your Skills to Work on page 540.

8.1   Interpreting and Constructing Graphs
8.2   Mean and Median
8.3   The Rectangular Coordinate System
8.4   Linear Equations with Two Variables

1. _____

2. _____

3. _____

4. _____

5. _____

6. _____

7–10. _____

11. _____

CHAPTER 8 PRETEST

This test provides a preview of the topics in this chapter. It will help you identify which concepts may require more of your studying time. If you are familiar with the topics in this chapter, take this test now. Check your answers with those in the back of the book. If you are not familiar with the topics in this chapter, begin studying the chapter now.

SECTION 8.1

The double-bar graph indicates the number of people unemployed in Pacerville during each quarter of 1996 and 1997.

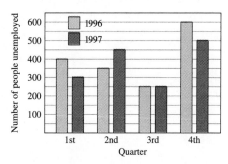

1. How many people were unemployed in Pacerville during the second quarter of 1996?
2. How many people were unemployed in Pacerville during the fourth quarter of 1997?
3. When were the greatest number of people unemployed in Pacerville?
4. How many more people were unemployed during the first quarter of 1996 than the first quarter of 1997?

SECTION 8.2

The table indicates the number of miles Joe ran each day last week.

MON.	TUES.	WED.	THURS.	FRI.	SAT.	SUN.
3	5	2	0	2	4	3

5. Find the mean number of miles that Joe ran per day. Round to the nearest tenth.
6. Find the median number of miles that Joe ran per day.

SECTION 8.3

Plot and label the following ordered pairs on the rectangular coordinate plane.

7. $(4, 2)$

8. $(0, -2)$

9. $(-2, 1)$

10. $(1, -3)$

SECTION 8.4

11. **(a)** Name 3 ordered pairs that are solutions to $y = 2x + 1$
    (b) Plot these ordered pairs on the rectangular coordinate plane and draw a line through coordinate points.

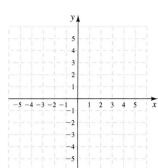

Interpreting and Constructing Graphs

After studying this section, you will be able to:

❶ Read a pictograph.

❷ Read a circle graph with percentage values.

❸ Read and interpret a double-bar graph.

❹ Read and interpret a comparison line graph.

❺ Construct double-bar and line graphs.

Math Pro   Video 8.1   SSM

Statistics is that branch of mathematics that collects and restudies data. Once the data are collected, the data must be organized so that the information is easily readable. As we have seen in earlier chapters, graphs give a visual representation of the data that is easy to read. Their visual nature allows them to communicate information about the complicated relationships among statistical data. For this reason, newspapers often use the types of graphs we will study in this section to help their readers grasp information quickly.

❶ **Reading a Pictograph**

A **pictograph** uses a visually appropriate symbol to represent an amount of items. A pictograph is used in Example 1.

> **E X A M P L E 1**   Consider the following pictograph.
>
>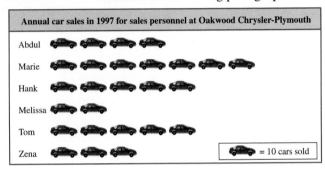
>
> **(a)** How many cars did Melissa sell in 1997?
> **(b)** Who sold the greatest number of cars?
> **(c)** How many more cars did Tom sell than Zena?

SOLUTION

(a) Melissa sold $2 \times 10 =$ **20 cars**.
(b) **Marie** sold the greatest number of cars.
(c) Tom sold $5 \times 10 = 50$ cars. Zena sold $3 \times 10 = 30$ cars. Now $50 - 30 = 20$. Therefore, Tom sold **20 cars more** than Zena.

PRACTICE PROBLEM 1

Consider the following pictograph.

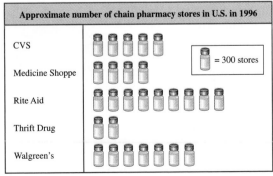

Source: Federal Food and Drug Administration

(a) Approximately how many stores does Walgreen's have?
(b) Approximately how many more stores does Rite Aid have than CVS?
(c) What is the combined number of stores of Thrift Drug and Medicine Shoppe?

2 Reading a Circle Graph with Percentage Values

A **circle graph** indicates how a whole quantity is divided into parts. These graphs help you to visualize the size of the relative proportions of parts. Each piece of the pie or circle is called a **sector**. We sometimes refer to circle graphs as **pie graphs**.

EXAMPLE 2   Together, the Great Lakes form the largest body of fresh water in the world. The total area of these five lakes is about 290,000 square miles, almost all of which is suitable for boating. The percentage of this total area taken up by each of the Great Lakes is shown in the pie graph.

(a) What percentage of the area is taken up by Lake Michigan?

(b) What lake takes up the largest percentage of area?

(c) How many square miles are taken up by Lake Huron and Lake Michigan together?

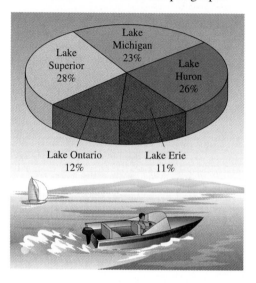

SOLUTION

(a) Lake Michigan takes up **23%** of the area.

(b) **Lake Superior** takes up the largest percentage.

(c) If we add 26 + 23, we get 49. Thus Lake Huron and Lake Michigan together take up 49% of the total area.

49% of 290,000 = (0.49)(290,000) = **142,100 square miles**.

PRACTICE PROBLEM 2

There were 690,000 bachelor's degrees awarded in the United States in 1993 to students majoring in the six most popular subject areas: business, social science, engineering, health science, psychology, and English. This circle graph shows how the degrees in these six subject areas are distributed.

Bachelor's degrees earned in 1993 by field

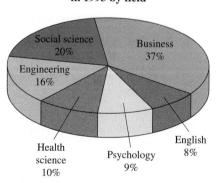

Source: U.S. National Center for Educational Statistics

(a) What percent of the bachelor's degrees represented in this circle graph are in the fields of psychology and business combined?

(b) Of the approximately 690,000 degrees awarded in these six fields, how many are awarded in the field of engineering?

③ Reading and Interpreting a Double-Bar Graph

Double-bar graphs are useful for making comparisons. For example, when a company is analyzing its sales, it may want to compare different years or different quarters. The following double-bar graph illustrates the sales of new cars at a local Ford dealership for two different years, 1995 and 1996. The sales are recorded for each quarter of the year.

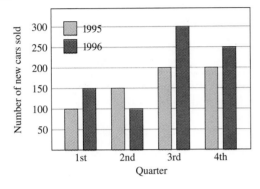

EXAMPLE 3   How many cars were sold in the second quarter of 1995?

SOLUTION   The bar rises to 150 for the second quarter of 1995. Therefore, **150 cars** were sold.

PRACTICE PROBLEM 3

How many cars were sold in the fourth quarter of 1996?

EXAMPLE 4   How many more cars were sold in the third quarter of 1996 than in the third quarter of 1995?

SOLUTION   From the double-bar graph, we see that 300 cars were sold in the third quarter of 1996 and that 200 cars were sold in the third quarter of 1995.

$$300 - 200 = 100$$

Thus **100 more cars** were sold.

PRACTICE PROBLEM 4

How many fewer cars were sold in the second quarter of 1996 than in the second quarter of 1995?

④ Reading and Interpreting a Comparison Line Graph

Two or more sets of data can be compared by using a **comparison line graph**. A comparison line graph shows two or more line graphs together. A different style for each line distinguishes them. Note that using a blue line and a black line in the following graph makes it easy to read.

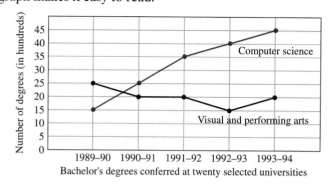

Bachelor's degrees conferred at twenty selected universities

E X A M P L E 5 How many bachelor's degrees in computer science were awarded in the academic year 1991–92?

SOLUTION Because the dot corresponding to 1991–92 is opposite 35 and the scale is in hundreds, we have $35 \times 100 = 3500$. Thus **3500 degrees** were awarded in computer science in 1991–92.

P R A C T I C E P R O B L E M 5

How many bachelor's degrees in visual and performing arts were awarded in the academic year 1992–93?

E X A M P L E 6 In what academic year were more degrees awarded in the visual and performing arts than in computer science?

SOLUTION The only year when more bachelor's degrees were awarded in the visual and performing arts was the academic year **1989–90**.

P R A C T I C E P R O B L E M 6

What was the first academic year in which more degrees were awarded in computer science than in the visual and performing arts?

⑤ Constructing Double-Bar and Line Graphs

There are several ways to construct line and bar graphs. How you design a graph usually depends on the data you must graph and the visual appearance you want. Either way, the intervals on both the horizontal and vertical lines must be equally spaced.

E X A M P L E 7 Construct a comparison line graph of the information given in the table.

CATEGORY	ACTIVITY	HOURS SPENT PER WEEK
Single men	Gym	6
	Outdoor sports	4
	Dating	7
	Reading and TV	3
Single women	Gym	4
	Outdoor sports	2
	Dating	7
	Reading and TV	9

SOLUTION   We plan our graph:

1. First, we draw a vertical and a horizontal line.
2. Since we are comparing *hours per week*, place this label on the vertical line. Mark intervals with equally spaced notches and label 0 to 9.
3. Place the label *Activity* on the horizontal line. Mark intervals with equally spaced notches and label with the name of each activity.
4. Now, we place dots on the graph that correspond to the data given.
5. Choose a different color line for each of the categories we are comparing (men and women), and connect the dots with line segments.

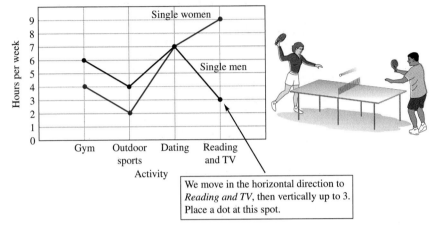

We move in the horizontal direction to *Reading and TV*, then vertically up to 3. Place a dot at this spot.

Note: Instead of colored lines, we could use one solid and one dashed line.

PRACTICE PROBLEM 7

Construct a comparison line graph of the information given in the table.

T-BILL	DATE IN MARCH	RATE %
3-month	7	4.90
	14	5.20
	21	5.00
	28	5.40
6-month	7	5.10
	14	5.40
	21	4.90
	28	5.40

EXAMPLE 8   Construct a double-bar graph of the information given in the table.

The Window Store Profits[a]

	1998	1999
Miniblinds	$12,000	$15,000
Vertical blinds	11,000	16,000
Shutters	16,000	14,000
Drapes	9,000	7,000

[a] Profits rounded to the nearest thousand.

SOLUTION   We plan our graph:

1. First, we draw a vertical and a horizontal line.
2. Since we are comparing the amount of profit per item, place the label *Profit*

on the vertical line. Mark intervals with equally spaced notches and label 7,000 to 17,000.

3. We label the horizontal line with the items that correspond to the dollar amounts.

4. Label a shaded and a nonshaded bar for the years we are displaying (1998 and 1999).

5. Now, draw a bar to the appropriate height for each category of wall coverings.

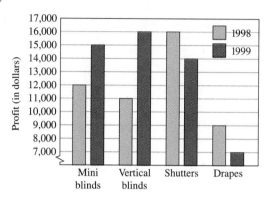

PRACTICE PROBLEM 8

Construct a double-bar graph with the information given.

Number of Men and Women Attending Mira College[a]

YEAR	MEN	WOMEN
1950	1000	700
1960	1100	600
1970	1200	900
1980	1200	1000
1990	1400	1300

[a] Rounded to the nearest hundred.

ANALYZING AND CONSTRUCTING GRAPHS OF PRODUCTION COSTS

How does a company determine how much to sell its products for? The *cost of producing* the products is an important factor in determining the selling price. To find the *cost of production* all expenses are totaled. To make money (profit), a company must sell enough of the products to make a total income (gross income) that is more than the total cost of production.

The accompanying chart lists the monthly production costs of Loren Manufacturing.

1998 Monthly Production Costs

Salaries and benefits	$13,440
Equipment	2,160
Supplies	3,360
Advertising	1,920
Rent	2,640
Utilities	480

PROBLEMS FOR INDIVIDUAL INVESTIGATION

1. What is the total monthly cost of production?

2. Make a bar graph comparing all the expenses.

3. What percentage of the total cost is each expense?

4. Make a circle graph reflecting these percentages. The entire circle graph is 100%.

5. If the company wants to make a 4% profit each month, how much will its gross income have to be?

PROBLEMS FOR GROUP INVESTIGATION AND COOPERATIVE STUDY

For the new year, 1999, salaries and benefits at Loren Manufacturing have been increased by 5%. However, since some of the equipment has been paid for, the equipment expenses are reduced 60%. Also, the company has decided to reduce advertising cost by 21%. Other changes are a rent increase of 5% and a increase in supply costs of 10%.

1. Find the new 1999 cost in each expense category.

2. (a) What is the total 1999 monthly production cost?
   (b) How much more or less are the total costs in 1999 than in 1998?
   (c) What is the percent increase or decrease?

3. Construct a double-bar graph that compares production expenses in each category for 1998 and 1999.

4. (a) For the 1999 production costs, calculate the percentage of the total monthly costs for each expense.
   (b) Draw a circle graph reflecting these percentages. Round answers to the nearest percent so that the total is as close to 100% as possible.

INTERNET: Go to http://www.prenhall.com/blair to explore this application.

EXERCISES 8.1

Number of new homes built in 1997 in each of the five counties	Approximate number of apartments in U.S. in 1995
Essex	Monthly rental under $250
Tarrant	Monthly rental $250-$499
Waverly	Monthly rental $500-$799
Northface	Monthly rental $800-$1249
DuPage	Monthly rental $1250 and up
= 200 homes built	= 800,000 units

(*Source:* U.S. Bureau of the Census.)

Use this pictograph to answer questions 1–6.

Use this pictograph to answer questions 7–10.

1. How many homes were built in Tarrant County in 1997?

2. How many homes were built in Essex County in 1997?

3. In what county were the most homes built?

4. How many more homes were built in Tarrant County than in Waverly County?

5. How many homes were built in Essex County and Northface County combined?

6. How many homes were built in DuPage County and Waverly County combined?

7. How many apartment units were rented for under $250 per month?

8. How many apartment units were rented for $800–$1249 per month?

9. How many more apartment units were available in the $500–$799 range than in the $800–$1249 range?

10. How many more apartment units were available in the $800–$1249 range than in the $1250 and up range?

In 1992 there were approximately 118,000 women physicians in the United States. The circle graph divides them by age. Use the circle graph to answer questions 11–16.

11. What percent of the women physicians were between the ages of 45 and 54?

12. What percent of the women physicians were between the ages of 35 and 44?

13. What percent of the women physicians were under the age of 45?

14. What percent of the women physicians were over the age of 45?

15. How many of the 118,000 women physicians were between the ages of 35 and 44?

16. How many of the 118,000 women physicians were over age 55?

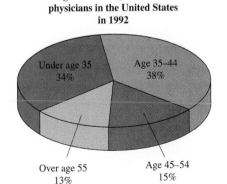

Ages of the 118,000 women physicians in the United States in 1992

Under age 35 34%

Age 35–44 38%

Age 45–54 15%

Over age 55 13%

Researchers estimate that the religious faith distribution of the 5,000,000,000 people in the world in 1993 was approximately that displayed in the circle graph. Use the graph to answer questions 17–22.

17. Approximately how many of the 5,000,000,000 people were Christians?

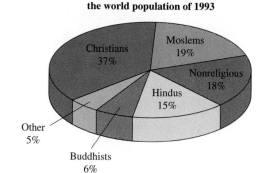

Distribution of religious faith in the world population of 1993

18. Approximately how many of the 5,000,000,000 were Moslems?

19. What percent of the world's population was either Moslem or nonreligious?

20. What percent of the world's population was either Hindu or Buddhist?

21. What percent of the world's population was *not* Moslem?

22. What percent of the world's population was *not* Christian?

The following double-bar graph illustrates the number of students at 10 selected universities at 20-year intervals. The student population is divided into men and women. Use the graph to answer questions 23–30.

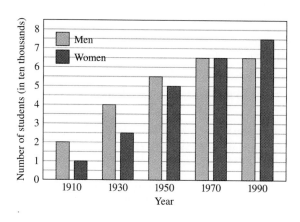

23. How many women students were enrolled in 1950?

24. How many men students were enrolled in 1930?

25. In what year was the number of women students enrolled greater than the number of men students enrolled?

26. In what year was the number of women students enrolled equal to the number of men students enrolled?

27. How many more women students were enrolled in 1950 then in 1930?

28. How many more men students were enrolled in 1970 than in 1950?

29. In what 20-year period did the greatest increase in the enrollment of women take place?

30. In what 20-year period did the greatest increase in the enrollment of men take place?

The following comparison line graph illustrates the number of customers per month coming into the Bay Shore Restaurant and the Lilly Cafe. Use the graph to answer questions 31–36.

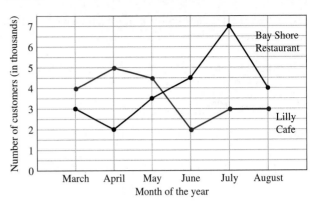

31. In which month did the fewest number of customers come into the Lilly Cafe?

32. In which month did the greatest number of customers come into the Bay Shore Restaurant?

33. (a) Approximately how many customers per month came into the Bay Shore Restaurant during the month of June?
(b) From May to June, did the number of customers increase or decrease?

34. (a) Approximately how many customers per month came into the Lilly Cafe during the month of May?
(b) From March to April, did the number of customers increase or decrease?

35. Between what two months is the *increase* in attendance the largest at the Bay Shore Restaurant?

36. Between what two months did the biggest *decrease* occur at the Lilly Cafe?

The accompanying comparison line graph indicates the rainfall for the last six months of two different years in Springfield. Use the graph to answer problems 37–42.

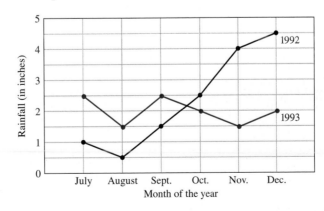

37. In September 1993, how many inches of rain were recorded?

38. In October of 1992, how many inches of rain were recorded?

39. During what months was the rainfall in 1993 less than the rainfall in 1992?

40. During what months was the rainfall in 1993 greater than the rainfall in 1992?

41. How many more inches of rain fell in November of 1992 than in October 1992?

42. How many more inches of rain fell in September of 1992 than in August 1992?

43. Use the table to make a double-bar graph that compares the high and low daily temperatures given in the table.

CITY	HIGH °F	LOW °F
Albany	50	24
Anchorage	30	16
Boise	55	30
Chicago	20	11

44. Use the table to make a double-bar graph that compares the number of grams of fat and cholesterol in the "fast foods" named in the following table.

TYPE OF SANDWICH	FAT (g)	CHOLESTEROL (g)
Wendy's Bacon Cheeseburger	25	65
Burger King BK Broiler	10	50
McDonald's McChicken	20	50
Wendy's Grilled Chicken	7	60
Source: U.S. government and food manufacturers.		

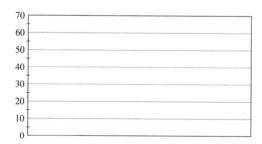

Use the information in the following table for problems 45 and 46.

Profit for Douglas Electronics

QUARTER	1996	1997	1998
1st	$12,000	$10,000	$15,000
2nd	14,000	11,000	13,000
3rd	15,000	13,000	10,000
4th	10,000	15,000	12,000

45. Construct a comparison line graph that compares the profits in 1996 and 1997.

46. Construct a comparison line graph that compares the profits in 1997 and 1998.

Use the information on the following chart for problems 47 and 48.

Dale College Student Enrollment for Fall Quarter

ACADEMIC AREA	FRESHMEN	SOPHOMORES	JUNIORS	SENIORS
English	400	250	150	100
Mathematics	300	500	350	450
History	500	225	275	100
Science	150	325	400	450

47. Construct a comparison line graph that compares the freshman and senior course enrollment.

48. Construct a comparison line graph that compares the sophmore and junior course enrollment.

 CALCULATOR EXERCISES

Use the circle graph for problems 49–52.

49. What percent of the family income was spent for food, medical care, and housing?

50. If the average two-income family earned $52,000 per year, how much was spent on recreation?

51. What percent of the family income was spent for federal, state, and local taxes?

52. If the average two-income family earned $52,000 per year, how much was spent on transportation?

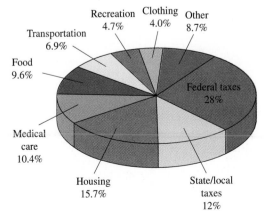

Distribution of spending by "average" two income American family in 1994

Source: Internal Revenue Service

Use the following chart to answer questions 53 and 54.

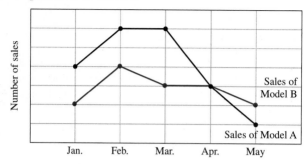

53. Why can you determine that Model A had the largest decrease in sales from March through May, even though there are no numbers on the graph?

54. Why can you determine that Models A and B had the same number of sales in April, even though there are no numbers on the graph?

CUMULATIVE REVIEW

1. Find the area of a rectangle with width 6 inches and length 14 inches.

2. Find the area of a parallelogram with a base of 17 inches and height 12 inches.

3. How many gallons of paint will it take to cover the four sides of a barn? Two sides measure 7 yards by 12 yards and two sides measure 7 yards by 20 yards. Assume that a gallon of paint covers 28 square yards.

4. Find the volume of a box with $L = 5$ inches, $W = 7$ inches, and $H = 6$ inches.

Mean and Median

After studying this section, you will be able to:

① Find the mean of a set of numbers.

② Find the median value of a set of numbers.

Math Pro   Video 8.2   SSM

① Finding the Mean of a Set of Numbers

We often want to know the *middle value* of a group of numbers. In this section we learn that in statistics there is more than one way of describing this middle value: There is the *mean* of the group of numbers, and there is the *median* of the group of numbers. In some situations it's more helpful to look at the mean, and in others it's more helpful to look at the median. We'll learn to tell which situations lend themselves to one or the other.

> The **mean** of a set of values is the sum of the values divided by the number of values. The mean is often called the **average**.

E X A M P L E   1   Find the average or mean test score of a student who has test scores of 71, 83, 87, 99, 80, and 90.

SOLUTION   We take the sum of the six tests and divide the sum by 6.

$$\begin{array}{l}\text{Sum of test scores} \rightarrow \\ \text{Number of tests} \rightarrow\end{array} \frac{71 + 83 + 87 + 99 + 80 + 90}{6} = \frac{510}{6} = 85$$

The mean is 85.

P R A C T I C E   P R O B L E M   1

Find the average or mean of the following test scores: 88, 77, 84, 97, and 89.

The mean value is often rounded to a certain decimal-place accuracy.

E X A M P L E   2   Carl and Wally each kept a log of the miles per gallon achieved by their cars for the last two months. Their results are recorded on the graph. What is the mean miles per gallon figure for the last 8 weeks for Carl? Round your answer to the nearest mile per gallon.

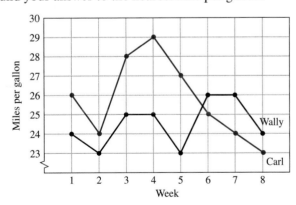

SOLUTION

$$\begin{array}{l}\text{Sum of values} \rightarrow \\ \text{Number of values} \rightarrow\end{array} \frac{26 + 24 + 28 + 29 + 27 + 25 + 24 + 23}{8}$$

$$= \frac{206}{8} \approx 26 \qquad \text{Rounded to the nearest whole number.}$$

The mean miles per gallon rating is **26**.

PRACTICE PROBLEM 2

What is the mean miles per gallon figure for the last 8 weeks for Wally? Round your answer to the nearest mile per gallon.

2 Finding the Median Value of a Set of Numbers

> If a set of numbers is arranged in order from smallest to largest, the **median** is that value that has the same number of values above it as below it.

EXAMPLE 3   The length of telephone calls made by Sara and Brad are indicated on the double-bar graph below. Find the median value for the length of Sara's calls.

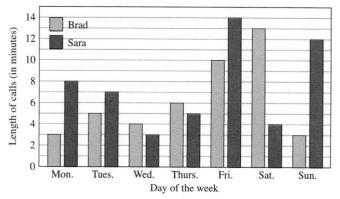

SOLUTION   We arrange the numbers in order from smallest to largest for Sara's calls:

$$\underbrace{3, \quad 4, \quad 5}_{\text{three numbers}} \quad \boxed{7} \quad \underbrace{8, \quad 12, \quad 14}_{\text{three numbers}}$$
$$\uparrow$$
$$\text{middle number}$$

There are three numbers smaller than 7 and three numbers larger than 7. Thus **7** is the median.

PRACTICE PROBLEM 3

Use the double-bar graph in Example 3 to find the median value for the length of Brad's calls.

If a list of numbers contains an even number of different items, then of course there is no one middle number. In this situation we obtain the median by taking the average of the two middle numbers.

EXAMPLE 4   Find the median of the following numbers: 13, 16, 18, 26, 31, 33, 38, and 39.

SOLUTION

$$\underbrace{13, \quad 16, \quad 18}_{\text{three numbers}} \quad \boxed{26, \quad 31} \quad \underbrace{33, \quad 38, \quad 39}_{\text{three numbers}}$$
$$\uparrow$$
$$\text{two middle numbers}$$

The average (mean) of 26 and 31 is $\dfrac{26 + 31}{2} = \dfrac{57}{2} = 28.5$

Thus the median value is **28.5**.

PRACTICE PROBLEM 4

Find the median value of the following numbers: 88, 90, 100, 105, 118, and 126.

TO THINK ABOUT

When would someone want to use the mean, and when would someone want to use the median? Which is more helpful? The mean (average) is used more frequently. It is most helpful when the data are distributed fairly evenly, that is, when no one value is "much larger" or "much smaller" than the rest. For example, a company had employees with annual salaries of $9,000, $11,000, $14,000, $15,000, $17,000, and $20,000. All the salaries fall within a fairly limited range. The mean salary (rounded to the nearest cent)

$$\frac{9000 + 11,000 + 14,000 + 15,000 + 17,000 + 20,000}{6} \approx \$14,333.33$$

gives us a reasonable idea of the "average" salary. However, suppose that the company had six employees with salaries of $9,000, $11,000, $14,000, $15,000, $17,000, and $90,000. Talking about the mean salary, which is $26,000, is deceptive. No one earns a salary very close to the mean salary. The "average" worker in that company does not earn around $26,000. In this case, the median value is more appropriate. Here the median is $14,500. Some problems of this type are included in Exercises 8.2, problems 33 and 34.

Another value that is sometimes used to describe a set of data is the **mode**. The *mode* of a set of data is the number or numbers that occur most often. We will cover this in more detail in exercises 35–40.

EXERCISES 8.2

In problems 1–8, find the mean. Round to the nearest tenth when necessary.

1. A student received grades of 89, 92, 83, 96, and 99 on math quizzes.

2. A student received grades of 77, 88, 90, 92, 83, and 84 on history quizes.

3. The Windy City Passport Photo Center received the following number of telephone calls over the last 6 days: 23, 45, 63, 34, 21, and 42.

4. The local Hertz rental car office received the following number of inquiries over the last 7 days: 34, 57, 61, 22, 43, 80, and 39.

5. The last 5 houses built in town sold for the following prices: $89,000, $93,000, $62,000, $102,000, and $89,000.

6. Luis priced a sofa at 6 local stores. The prices were $499, $359, $600, $450, $529, and $629.

7. Sam watched television last week for the following number of hours per day:

MON.	TUES.	WED.	THURS.	FRI	SAT.	SUN.
8	3	2	5	6.5	7.5	3

8. Steve's car got the following miles per gallon results during the last 6 months:

JAN.	FEB.	MAR.	APR.	MAY	JUNE
23	22	25	28	29	30

The accompanying double-bar graph shows the number of sales made by Alex and Lisa from June through October. Use the graph to answer questions 9 and 10.

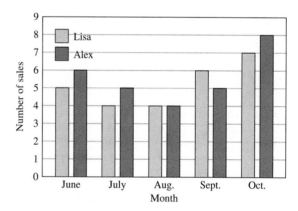

9. Find the average number of sales by Alex for the 5-month period.

10. Find the average number of sales by Lisa for the 5-month period.

11. The captain of the college baseball team achieved the following results:

	GAME 1	GAME 2	GAME 3	GAME 4	GAME 5
Hits	0	2	3	2	2
Times at bat	5	4	6	5	4

Find his batting average by dividing his total number of hits by the total times at bat.

12. The captain of the college bowling team had the following results after practice:

	PRACTICE 1	PRACTICE 2	PRACTICE 3	PRACTICE 4
Score (pins)	541	561	840	422
Number of games	3	3	4	2

Find her bowling average by dividing the total number of pins scored by the total number of games.

13. Frank and Wally traveled to the west coast during the summer. The number of miles they drove and the number of gallons of gas they used are recorded below.

	DAY 1	DAY 2	DAY 3	DAY 4
Miles driven	276	350	391	336
Gallons of gas	12	14	17	14

Find the average miles per gallon achieved by the car on the trip by dividing the total number of miles driven by the total number of gallons used.

14. Cindy and Andrea traveled to Boston this fall. The number of miles they drove and the number of gallons of gas they used are recorded below.

	DAY 1	DAY 2	DAY 3	DAY 4
Miles driven	260	375	408	416
Gallons of gas	10	15	17	16

Find the average miles per gallon achieved by the car on the trip by dividing the total number of miles driven by the total number of gallons used.

In problems 15–24 find the median value.

15. 22, 36, 45, 47, 48, 50, 58

16. 37, 39, 46, 53, 57, 60, 63

17. 1052, 968, 1023, 999, 865, 1152

18. 1400, 1329, 1200, 1386, 1427, 1350

19. 0.52, 0.69, 0.71, 0.34, 0.58

20. 0.26, 0.12, 0.35, 0.43, 0.28

21. The annual salaries of the employees of a local cable television office are $17,000, $11,600, $23,500, $15,700, $26,700, and $31,500.

22. The costs of six cars recently purchased by the Weston Company were $18,270, $11,300, $16,400, $9,100, $12,450, and $13,800.

23. The number of minutes spent on the phone per day by each of 8 San Diego teenagers living on the same block is 40 minutes, 108 minutes, 62 minutes, 12 minutes, 24 minutes and 31 minutes.

24. The ages of 10 people swimming laps at the YMCA pool one morning: 60, 18, 24, 36, 39, 32, 70, 12, 15, and 85.

The comparison line graph indicates the price of the same compact disc sold at music stores and by mail order during a 6-week period. Use the graph to answer questions 25 and 26.

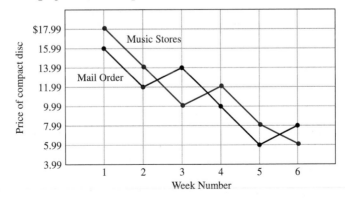

25. Find the median value of the prices at the music stores.

26. Find the median value of the prices by mail order.

Find the median value.

27. The numbers of potential actors who tried out for the school play at Hamilton–Wenham Regional High School over the last 10 years: 36, 48, 44, 64, 60, 71, 22, 36, 53, and 37.

28. The number of injuries during the high school football season for the Badgers over the last 8 years: 10, 17, 14, 29, 30, 19, 25, and 21.

 CALCULATOR EXERCISES

Find the median value.

29. 2576, 8764, 3700, 5000, 7200, 4700, 9365, 1987

30. 15.276, 21.375, 18.90, 29.2, 14.77, 19.02

Find the mean.

31. Oil is imported into the United States in great quantities. The following table was provided by the U.S. Energy Information Administration.

YEAR	AMOUNT OF OIL IMPORTED (MILLIONS OF BARRELS)
1991	2151
1992	2110
1993	2220
1994	2578
1995	2643

(a) Find the mean number of barrels of oil imported into the United States per year.

(b) If 1 barrel contains 42 gallons of oil, find the mean number of gallons of oil imported into the United States per year.

32. The capacity of the United States to produce electricity continues to increase as the demand for electricity continues to grow. The following table was provided by the U.S. Energy Information Administration.

YEAR	SUMMER CAPACITY (MILLIONS OF KILOWATTS)
1990	685.1
1991	690.9
1992	695.4
1993	694.3
1994	703.0

(a) Find the mean capacity of the United States to produce electricity in the summer measured in kilowatts.

(b) A kilowatt is 1000 watts. Find the mean capacity to produce electricity in the summer measured in watts.

 VERBAL AND WRITING SKILLS

33. A local travel office has 10 employees. Their monthly salaries are $1500, $1700, $1650, $1300, $1440, $1580, $1820, $1380, $2900, and $6300.

(a) Find the mean.

(b) Find the median.

(c) Which of these numbers best represents "what the average person earns"? Why?

34. A college track star in California ran the 100-meter event in 8 track meets. Her times were 11.7 seconds, 11.6 seconds, 12.0 seconds, 12.1 seconds, 11.9 seconds, 18 seconds, 11.5 seconds, and 12.4 seconds.

(a) Find the mean.

(b) Find the median.

(c) Which of these numbers best represents "her average running time?" Why?

ONE STEP FURTHER

*Another value that is sometimes used to describe a set of data is the mode. The **mode** of a set of data is the number or numbers that occur most often. For example, if a student had test scores of 89, 94, 96, 89, and 90, we would say that the mode is 89. If two values occur most often, we say that the data have two modes (or are bimodal). For example, if the ages of students in a calculus class were 33, 27, 28, 28, 21, 19, 18, 25, 26, and 33, we would say that the modes were 28 and 33. The mode is not used as frequently in statistics as the mean and the median. We practice finding the mode in problems 35–40.*

Find the mode.

35. 60, 65, 68, 60, 72, 59, 80

36. 86, 84, 82, 87, 84, 88, 90

37. 121, 150, 116, 150, 121, 181, 117, 123

38. 144, 143, 140, 141, 149, 144, 141, 150

39. The last six bicycles sold at the Skol Bike shop cost $249, $649, $439, $259, $269, and $249.

40. The last six color television sets sold at the local Circuit City cost $315, $430, $515, $330, $430, and $615.

CUMULATIVE REVIEW

Evaluate for the value given.

1. $\dfrac{x}{2} + 4$ for $x = 26$

2. $\dfrac{35}{x} - 9$ for $x = 7$

3. $2x + 1$ for $x = 5$.

4. $3x - 7$ for $x = 0$.

After studying this section, you will be able to:

❶ Write data as ordered pairs.

❷ Plot a point given the coordinates.

❸ Name the coordinates of a point.

❹ Graph horizontal and vertical lines.

Math Pro   Video 8.3   SSM

SECTION   8 . 3

The Rectangular Coordinate System

❶   Writing Data as Ordered Pairs

Many things in real life are clearer if we can see a picture of them. Similarly, in mathematics we often find that a drawing is helpful. We can picture relationships by drawing graphs and charts. Consider the following line graph, which shows the number of products produced by a manufacturing company over a 6-year period.

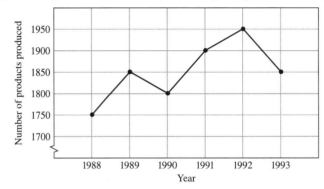

This graph indicates that 1750 products were produced in 1988, 1850 in 1989, 1800 in 1990, 1900 in 1991, 1950 in 1992, and 1850 in 1993. We could also describe the productivity without a graph. Instead, we can use **ordered pairs** of numbers stating the year followed by the number of products produced.

(year,   number of products)
    ↓   ↓
(1988, 1750)
(1989, 1850)
(1990, 1800)
(1991, 1900)
(1992, 1950)
(1993, 1850)

As you can see, *the order in which we list the numbers is important*; otherwise, we could confuse the year with the number of products produced. Since ordered pairs are pairs of numbers represented in a *specific order*, we often use ordered pairs to specify location on a graph. We use a *dot* to represent an ordered pair on a graph.

EXAMPLE   1   Refer to the line graph and list the number of applicants to a certain school in the years 1985, 1987, 1989, and 1991 using ordered pairs: (year, number of applicants).

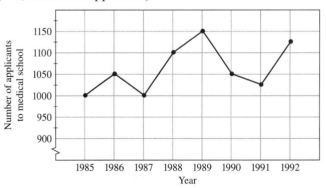

SOLUTION

(year, number of applicants)
  ↓   ↓

(1985, 1000)

(1987, 1000)

(1989, 1150)

(1991, 1025)

PRACTICE PROBLEM 1

Refer to the line graph in Example 1 and list the number of applicants in 1986, 1988, 1990, and 1992 using ordered pairs: (year, number of applicants).

② Plotting a Point Given the Coordinates

We cannot easily use the type of line graph illustrated in Example 1 to display ordered pairs that include negative numbers. Instead, we use a **rectangular coordinate system**. We can think of a rectangular coordinate system as an extension of the line graph—we extend below the horizontal line and to the left of the vertical line to represent negative numbers.

To form this coordinate system we draw two number lines, one horizontally and a second one vertically. We construct the number lines so that the zero point on each number line is at exactly the same place. We refer to this location as the **origin**. Each number line is called an **axis**. The horizontal number line is often called the **x-axis**, and the vertical number line is called the **y-axis**.

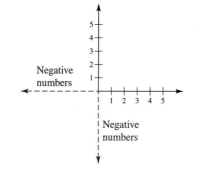

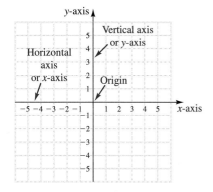

Notice on the vertical number line that the positive numbers are above the origin (positive *y*-direction) and the negative numbers are below the origin (negative *y*-direction).

Since there are two number lines in the rectangular coordinate system, we use this system to display (plot) ordered pairs of numbers. The ordered pair of numbers that represents a point is often referred to as the **coordinates of the point**. The first value is called the **x-coordinate** or **x-value** and the second value the **y-coordinate** or **y-value**.

The *x*-value represents the distance from the origin in the *x*-direction.

The *y*-value represents the distance from the origin in the *y*-direction.

x-value
↑
(x, y)
↓
y-value

E X A M P L E 2   Plot the ordered pair: $(-4, 3)$.

SOLUTION

The first number indicates the x-direction.

$$(-4, 3)$$

The second number indicates the y-direction.

To plot $(-4, 3)$ or $x = -4$, $y = 3$, we start at the origin and move 4 units in the negative x-direction followed by 3 units in the positive y-direction. We end up at the coordinate point $(-4, 3)$ and place a dot there.

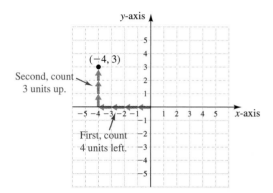

P R A C T I C E P R O B L E M 2

Plot the ordered pair: $(-3, 2)$.

E X A M P L E 3   Plot each ordered pair.

(a) $(-3, 1)$ **(b)** $(1, -3)$ **(c)** $(0, 2)$ **(d)** $\left(5\frac{1}{2}, 1\right)$

SOLUTION

(a) $(-3, 1)$: 3 units left followed by 1 unit up.
(b) $(1, -3)$: 1 unit right followed by 3 units down.
   Notice the difference between parts (a) and (b). We must remember that the first number indicates the x-direction and the second number the y-direction.
(c) $(0, 2)$: start at the origin and move 0 units in the x-direction, followed by 2 units in the positive y-direction. The coordinate point is on the y-axis.
(d) To plot the point $\left(5\frac{1}{2}, 1\right)$ we move $5\frac{1}{2}$ units in the positive x-direction. The measure of $\frac{1}{2}$ unit is located half-way between 5 and 6. Then we move 1 unit up.

P R A C T I C E P R O B L E M 3

Plot each ordered pair.

(a) $(-4, 2)$ **(b)** $(2, -4)$ **(c)** $\left(0, 3\frac{1}{2}\right)$ **(d)** $(1, 0)$

3   Naming the Coordinates of a Point

E X A M P L E 4   Give the coordinates of each point on the graph on the next page.

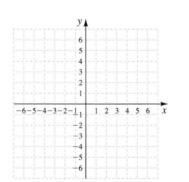

SOLUTION

(a) $S = (-5, 1)$

(b) $T = (0, 3)$

(c) $U = (1, 0)$

(d) $V = (6, 2)$

(e) $W = (-5, -1)$

(f) $X = (6, -2)$

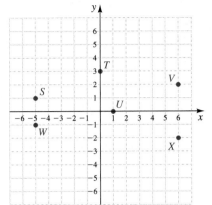

PRACTICE   PROBLEM   4

Give the coordinates of each point.

(a) S

(b) T

(c) U

(d) V

(e) W

(f) X

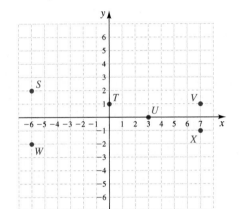

If the x-axis represents the directions east and west, and the y-axis represents north and south, we can indicate direction using ordered pairs.

EXAMPLE   5

(a) State the ordered pair that represents 3 miles west, 1 mile north.
(b) Plot the ordered pair.

SOLUTION

(a) $(-3, 1)$

(b)

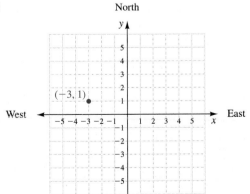

PRACTICE PROBLEM 5

(a) State the ordered pair that represents 5 miles east, 2 miles south.

(b) Plot the ordered pair.

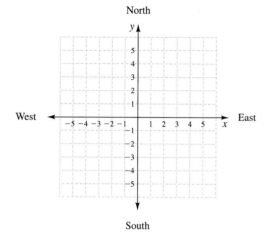

4 Graphing Horizontal and Vertical Lines

EXAMPLE 6

(a) Plot the points corresponding to the ordered pairs, then draw a line connecting the coordinate points: $(2, 1), (2, -3), (2, 0), (2, 3)$.

(b) Plot the points corresponding to the ordered pairs, then draw a line connecting the coordinate points: $(-3, 4), (4, 4), (1, 4), (-5, 4)$.

SOLUTION

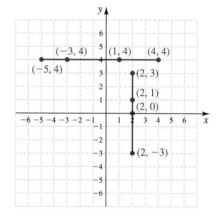

PRACTICE PROBLEM 6

(a) Plot the points corresponding to the ordered pairs, then draw a line connecting the coordinate points: $(1, -2), (1, -3), (1, 0), (1, 2)$.

(b) Plot the points corresponding to the ordered pairs, then draw a line connecting the coordinate points: $(-2, 3), (3, 3), (1, 3), (-6, 3)$.

Let's take another look at the ordered pairs and points on the graph in Example 6. What do you observe about the sets of ordered pairs?

(a) $(2, 1), (2, -3), (2, 0), (2, 3)$ (b) $(-3, 4), (4, 4), (1, 4), (-5, 4)$
         ↓   ↓

The x-values for each ordered pair are the same, $x = 2$.

The y-values for each ordered pair are the same, $y = 4$.

Now let's see how this affects the graphs.

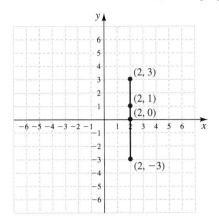

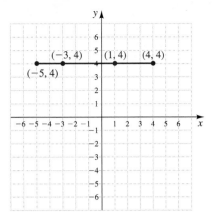

We have a vertical line, since we moved the same distance in the x direction, $x = 2$, for all the ordered pairs.

We have a horizontal line, since we moved the same distance in the y direction, $y = 4$, for all the ordered pairs.

When all x-values of a set of ordered pairs are the same number, $x = a$, the coordinate points on the graph form a *vertical line*.

When all y-values of a set of ordered pairs are the same number, $y = b$, the coordinate points on the graph form a *horizontal line*.

EXERCISES 8.3

Use the following line graph to answer questions 1 and 2.

1. Represent the enrollment in 1987, 1988, 1989 and 1990 using ordered pairs (year, number of students).

2. Represent the enrollment in 1991, 1992, 1993, and 1994 using ordered pairs (year, number of students).

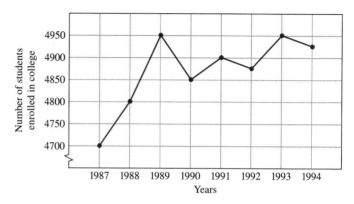

Students graphed their shoe size versus their height on the graph below. Use the line graph to answer questions 3–6.

3. Represent the height and corresponding shoe size for Lena, Janie, and Mark using ordered pairs (height, shoe size).

4. Represent the height and corresponding shoe size for Nho, Kelley, and Jeff using ordered pairs (height, shoe size).

5. Represent the height and corresponding shoe size for Lena, Janie, and Mark using ordered pairs (shoe size, height).

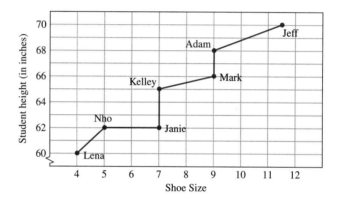

6. Represent the height and corresponding shoe size for Nho, Kelley, and Jeff using ordered pairs (shoe size, height).

Plot and label the ordered pairs on the rectangular coordinate plane below.

7. $(2, 2)$

8. $(3, 3)$

9. $(-1, 4)$

10. $(-2, 3)$

11. $(3, -2)$

12. $(4, -3)$

13. $(-2, -3)$

14. $(-5, -3)$

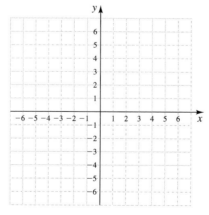

Plot and label the ordered pairs on the rectangular coordinate plane below.

15. $(-1, 2)$ **16.** $(-4, 3)$

17. $(5, -1)$ **18.** $(1, -3)$

19. $(0, -2)$ **20.** $(0, -3)$

21. $(5, 0)$ **22.** $(-1, 0)$

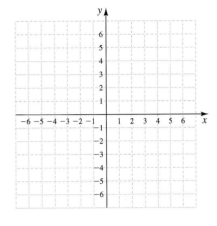

Plot and label the ordered pairs on the rectangular coordinate plane below.

23. $\left(4\frac{1}{2}, 3\right)$ **24.** $\left(2\frac{1}{2}, 2\right)$

25. $\left(-1\frac{1}{2}, -4\right)$ **26.** $\left(-3\frac{1}{2}, -3\right)$

27. $\left(2\frac{1}{2}, -1\right)$ **28.** $\left(1\frac{1}{2}, -3\right)$

29. $\left(-3\frac{1}{2}, 2\right)$ **30.** $\left(-1\frac{1}{2}, 4\right)$

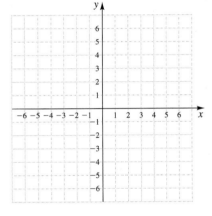

Give the coordinates of the point.

31. K **32.** L

33. M **34.** N

35. O **36.** P

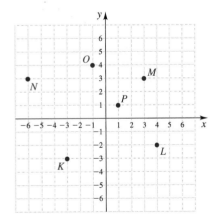

Give the coordinates of the point.

37. *Q*

38. *R*

39. *S*

40. *T*

41. *U*

42. *V*

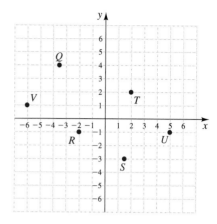

43. (a) State the ordered pair that represents 4 miles west, 2 miles south.

(b) Plot the ordered pair.

North

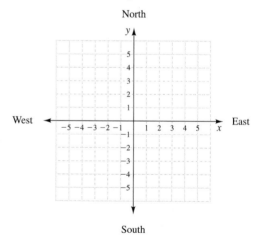

South

44. (a) State the ordered pair that represents 6 miles east, 3 miles south.

(b) Plot the ordered pair.

North

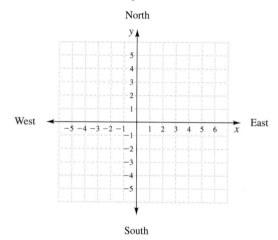

South

45. (a) State the ordered pair that represents 2 miles east, 1 mile north.

(b) Plot the ordered pair.

North

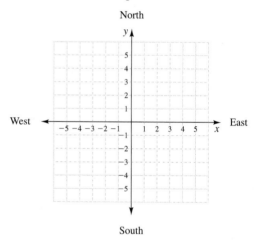

South

46. (a) State the ordered pair that represents 5 miles west, 2 miles north.

(b) Plot the ordered pair.

North

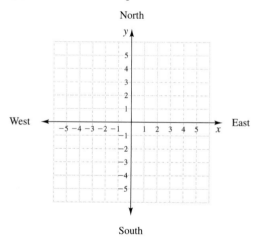

South

530

Plot the points corresponding to the ordered pairs, then draw a line connecting the coordinate points.

47. $(2, 4), (2, -1), (2, -3), (2, 0)$

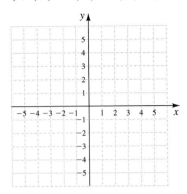

48. $(4, 2), (4, -2), (4, -1), (4, 0)$

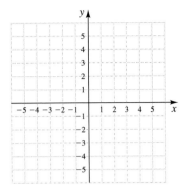

49. $(1, 3), (-5, 3), (0, 3), (-2, 3)$

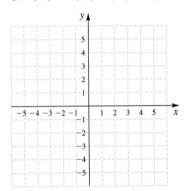

50. $(2, 5), (-3, 5), (3, 5), (-5, 5)$

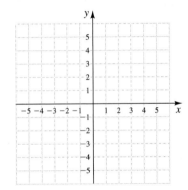

51. $(0, -1), (-2, -1), (-4, -1), (4, -1)$

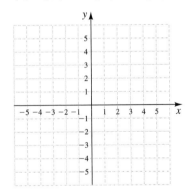

52. $(0, -2), (-3, -2), (3, -2), (2, -2)$

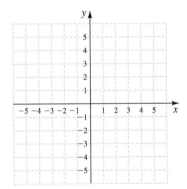

 VERBAL AND WRITING SKILLS

Fill in the blank.

53. To plot a point on a rectangular coordinate plane, we always start at the _____ .

54. The negative x-direction is to the left of 0 (origin). We move in this direction when the _____ is negative.

55. Describe in words the directions you move on the rectangular coordinate plane to plot the point $(2, -1)$.

56. Describe in words the directions you move on the rectangular coordinate plane to plot the point $(-4, -2)$.

57. Why can you tell simply by looking at the ordered pairs $(6, 8), (6, 3),$ and $(6, -1)$ that a line drawn through these points forms a vertical line?

58. Why can you tell simply by looking at the ordered pairs $(1, 5), (7, 5),$ and $(-2, 5)$ that a line drawn through these points forms a horizontal line?

ONE STEP FURTHER

59. (a) Plot $(5, 2)$ and $(5, 6)$, then draw a line connecting the points.

(b) Plot $(2, 2)$ and $(2, 6)$, then draw a line connecting the points.

(c) Draw a line connecting the points $(2, 2), (5, 2)$.

(d) Draw a line connecting the points $(2, 6), (5, 6)$.

(e) Describe the figure.

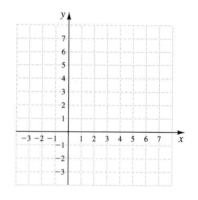

60. (a) Plot $(0, 0)$ and $(4, 0)$, then draw a line connecting the points.

(b) Name the two ordered pairs that you must plot so that you can form a square by drawing lines connecting the appropriate coordinate points.

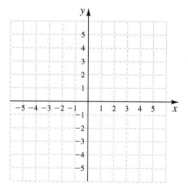

CUMULATIVE REVIEW

Find the value.

1. $4x - 3$ if $x = 2$

2. $5x + 2$ if $x = -3$

3. $2x - 6$ if $x = 1$

4. $3x + 1$ if $x = -2$

SECTION 8.4

Linear Equations with Two Variables

After studying this section, you will be able to:

1 Find solutions to linear equations with two variables

2 Graph linear equations with two variables

3 Graph horizontal and vertical lines.

Math Pro   Video 8.4   SSM

1 Finding Solutions
to Linear Equations with Two Variables

In earlier chapters we found the solutions to linear equations with *one variable*. For example, the solution to $x + 2 = 7$ is $x = 5$. Now, consider the equation $x + y = 7$. This equation has two variables and is called a *linear equation with two variables*. The solution to $x + y = 7$ consists of pairs of numbers, one for x and one for y, whose sum is 7. When we find the solutions to $x + y = 7$, we are answering the question: "The sum of what two numbers equals 7?" Since $5 + 2 = 7$, the pair of numbers $x = 5$ and $y = 2$ or $(5, 2)$ is a solution.

$$x + y = 7 \qquad \text{The sum of what two numbers equals 7?}$$
$$5 + 2 = 7 \qquad x = 5 \text{ and } y = 2 \text{ or } (5, 2) \text{ is a solution.}$$

Now, $3 + 4 = 7$, so the pair of numbers $x = 3$ and $y = 4$ or $(3, 4)$ is also a solution to the equation $x + y = 7$. Can you find another pair of numbers whose sum is 7? Of course you can. In fact, there are infinitely many pairs of numbers with a sum of 7 and thus infinitely many ordered pairs that represent solutions to $x + y = 7$. It is important that you realize that not just any ordered pair of numbers is a solution to $x + y = 7$. Only ordered pairs whose sum is 7 are solutions to $x + y = 7$.

EXAMPLE 1 List three ordered pairs of numbers that are solutions to the equation $x + y = 3$.

SOLUTION There are infinitely many solutions, so answers may vary. An ordered pair represents a solution if, when we substitute the values for x and y in the equation $x + y = 3$, we get a true statement.

We list three ordered pairs.   Check:

$x = -1$ and $y = 4$, or $(-1, 4)$

$$x + y = 3$$
$$\downarrow \quad \downarrow$$
$$-1 + 4 = 3, \qquad 3 = 3 \checkmark$$

$x = 0$ and $y = 3$, or $(0, 3)$

$$x + y = 3$$
$$\downarrow \quad \downarrow$$
$$0 + 3 = 3, \qquad 3 = 3 \checkmark$$

$x = 2$, and $y = 1$, or $(2, 1)$

$$x + y = 3$$
$$\downarrow \quad \downarrow$$
$$2 + 1 = 3, \qquad 3 = 3 \checkmark$$

PRACTICE PROBLEM 1

List four ordered pairs of numbers that are solutions to the equation $x + y = 11$.

EXAMPLE 2 A machine at a manufacturing company can seal 50 jars per minute. The equation that represents the situation is $y = 50x$, where y equals the number of jars sealed and x represents the number of minutes the machine is in operation. Determine how many minutes it takes the machine to seal the following numbers of jars. State your answer as the ordered pair (x, y).

(a) 200 jars   **(b)** 100 jars

SOLUTION

number of
jars sealed ⌐┐   ⌐─ number of minutes

(a) $y = 50x$
                    ↓
          $200 = 50x$ We substitute $y = 200$ in the equation.
          $\dfrac{200}{50} = x$ We solve for x.
          $4 = x$ or $x = 4$ minutes

     Since $x = 4$ when $y = 200$, the ordered pair (x, y) is **(4, 200)**.

number of
jars sealed ⌐┐   ⌐─ number of minutes

(b) $y = 50x$
                    ↓
          $100 = 50x$ We substitute $y = 100$ in the equation.
          $\dfrac{100}{50} = x$ We solve for x.
          $2 = x$ or $x = 2$ minutes

     Since $x = 2$ when $y = 100$, the ordered pair (x, y) is **(2, 100)**.

PRACTICE PROBLEM 2

A machine at a manufacturing company can label 25 bottles per minute. The equation that represents the situation is $y = 25x$, where y equals the number of bottles labeled and x represents the number of minutes the machine is in operation. Determine how many minutes it takes the machine to label the following numbers of bottles. State your answer as the ordered pair (x, y).

(a) 125 bottles   **(b)** 200 bottles

EXAMPLE 3   Fill in the ordered pairs so that they are solutions to the equation $x + 2y = 10$: $(0, \underline{\hspace{0.4cm}}), (\underline{\hspace{0.4cm}}, 1)$.

SOLUTION   We substitute the given value, then solve for the unknown value in the equation $x + 2y = 10$.

$(0, \underline{\hspace{0.4cm}})$ $x = 0$, $y = ?$	$(\underline{\hspace{0.4cm}}, 1)$ $x = ?$ $y = 1$
$x + 2y = 10$	$x + 2y = 10$
↓	↓
$0 + 2y = 10$ Substitute for x.	$x + 2(1) = 10$ Substitute for y.
$2y = 10$	$x + 2 = 10$
$y = 5$ Solve for y.	$x = 8$ Solve for x.
(0, 5)	**(8, 1)**

It is a good idea to check your answers.

Check: $x + 2y = 10$ $x + 2y = 10$
                 ↓   ↓   ↓   ↓
$(0, 5)$ $0 + 2(5) \stackrel{?}{=} 10$ $(8, 1)$ $8 + 2(1) \stackrel{?}{=} 10$
              $10 = 10$ ✓ $10 = 10$ ✓

$(0, 5), (8, 1)$ are solutions to $x + 2y = 10$.

PRACTICE PROBLEM 3

Fill in the ordered pairs so that they are solutions to the equation
$x + 3y = 12$: (__, 0), (__, 2).

We must be careful when we state the ordered pairs: x must be written first, and y second. A common error is to reverse the numbers in the ordered pair. Consider Example 3, which has $(0, 5)$ as a solution. If we reverse the coordinates (in error) and state the solution as $(5, 0)$, our proposed solution does not check in the equation $x + 2y = 10$.

$(5, 0)$ means $x = 5$ and $y = 0$ *Check:* $x + 2y = 10$

$$5 + 2(0) \stackrel{?}{=} 10$$
$$5 + 0 \stackrel{?}{=} 10$$
$$5 \neq 10$$

The ordered pair $(5, 0)$ does not check and therefore is not a solution to the equation $x + 2y = 10$. To help avoid this error, we often organize our work in a chart, as in the following example.

EXAMPLE 4   Fill in the ordered pairs so that they are solutions to the equation $y = x + 4$: $(0, \_\_), (\_\_, 5), (\_\_, 1)$.

SOLUTION   We start by organizing our ordered pairs in a chart. Then we substitute the given value and solve for the unknown value.

(x,	y)
(0	)
(	5)
(	1)

$$y = x + 4 \qquad y = x + 4 \qquad y = x + 4$$
$$y = 0 + 4 \qquad 5 = x + 4 \qquad 1 = x + 4$$
$$y = 4 \qquad\qquad 1 = x \qquad\quad -3 = x$$

We write these values in the appropriate place in the chart.
$(0, 4), (1, 5), (-3, 1)$ are solutions to $y = x + 4$. We leave the *check* to the student.

(x,	y)
(0,	4)
(1,	5)
(-3,	1)

PRACTICE PROBLEM 4

Fill in the ordered pairs so that they are solutions to the equation $y = x + 2$:
$(0, \_\_), (\_\_, 8), (\_\_, 5)$.

If we are not given values for either x or y, we may choose any value for x then solve for y. Or we may choose any value for y then solve for x.

EXAMPLE 5   Find 3 ordered pairs that are solutions to $y = 3x - 1$.

SOLUTION   We choose 3 values for x and write these values in a chart.
We substitute these values in $y = 3x - 1$ and solve for y.

(x,	y)
(0	)
(1	)
(-1	)

$$y = 3x - 1 \qquad y = 3x - 1 \qquad y = 3x - 1$$
$$= 3(0) - 1 \qquad = 3(1) - 1 \qquad = 3(-1) - 1$$
$$= 0 - 1 \qquad\quad = 3 - 1 \qquad\quad = -3 - 1$$
$$y = -1 \qquad\quad y = 2 \qquad\qquad y = -4$$

We write these values in the appropriate place in the chart.

(0, −1), (1, 2), (−1, −4) are solutions to $y = 3x − 1$. Answers may vary since you may start with different values.

(x,	y)
(0,	−1)
(1,	2)
(−1,	−4)

For the equation $y = 3x − 1$, the calculations are simplified if we choose 3 values for x, then solve for y. To verify this, try choosing a few values for y, then solve for x.

PRACTICE PROBLEM 5

Find 3 ordered pairs that are solutions to $y = 4x − 5$.

TO THINK ABOUT

Why can we choose any value for either x or y when we find ordered pairs that are solutions to an equation?

The numbers for x and y that are solutions to the equation come in pairs. When we try to find a pair that fits, we have to start with some number. Usually, it is an x-value that is small and easy to work with. Then we must find the value for y so that the pair of numbers (x, y) is a solution to the given equation.

The following situation might help you understand this idea. The housing director must find a roommate for each student who will live in the dorms. *The director can start with any application*—say applicant 6, or in math symbols, $x = 6$. The director then uses the criteria submitted by all of the applicants and finds that applicant 11 is a perfect fit— $y = 11$. In this case students submit criteria such as *sleeps in late on weekends*, *likes to listen to classical music*, or *comes from South Africa*, and so on. The director uses this criteria to find the **right pair of roommates**, $(6, 11)$, whereas we use an equation to find the **right ordered pair that is a solution** to the equation. In either case, we can choose any applicant or number to start with.

❷ Graphing Linear Equations with Two Variables

We often plot the solutions to a linear equation on a rectangular coordinate plane. We do this by finding 3 or 4 solutions to the equation, then plotting these ordered pairs on a rectangular coordinate plane and connecting the points with a line.

EXAMPLE 6

(a) Name 3 ordered pairs that are solutions to $y = 2x + 4$.

(b) Plot these ordered pairs on a rectangular coordinate plane and connect the coordinate points with a line.

SOLUTION

(a) We choose 3 points for x:

$(x,$	$y)$
$(-1$	$)$
$(0$	$)$
$(1$	$)$

We substitute the values for x and solve for y.

$$
\begin{array}{lll}
y = 2x + 4 & y = 2x + 4 & y = 2x + 4 \\
\quad = 2(-1) + 4 & \quad = 2(0) + 4 & \quad = 2(1) + 4 \\
\quad = -2 + 4 & \quad = 0 + 4 & \quad = 2 + 4 \\
y = 2 & y = 4 & y = 6
\end{array}
$$

We place these values in the chart.

$(x,$	$y)$
$(-1,$	$2)$
$(0,$	$4)$
$(1,$	$6)$

(b) Then we plot these ordered pairs and connect the points.

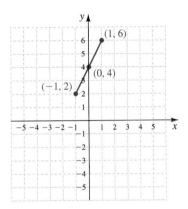

PRACTICE PROBLEM 6

(a) Name 3 ordered pairs that are solutions to $y = 2x + 1$.

(b) Plot these ordered pairs on a rectangular coordinate plane and connect the coordinate points with a line.

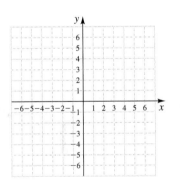

Let's examine the coordinate points in Example 6 (see the illustration below). Notice that when we connect all 3 points they form a straight line. If we plot other solutions to the equation $y = 2x + 4$, we find that these points also lie on this line. For example, $(-3, -2)$, and $(-4, -4)$ are solutions to $y = 2x + 4$ and lie on the line.

In fact, every ordered pair that is a solution to this equation lies on this line. Therefore, to graph the solution set to the equation $y = 2x + 4$ we draw a line through the points and *continue the line beyond the points*, placing an arrow on both ends to indicate that solutions lie beyond the end of the line that we drew.

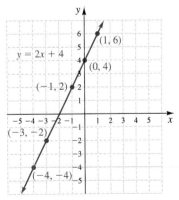

All solutions to $y = 2x + 4$ form a straight line.

> ### PROCEDURE TO GRAPH A LINEAR EQUATION
>
> 1. Look for 3 ordered pairs that are solutions to the equation.
> 2. Plot the points.
> 3. Draw a line through the points.
> 4. Continue this line beyond the points, placing an arrow at both ends.
>
> If the points form a straight line, you do not need to check your ordered pairs for an error.

EXAMPLE 7

(a) Name 3 ordered pairs that are solutions to $y = -2x - 1$.
(b) Plot these ordered pairs on a rectangular coordinate plane and draw a line through the points.

SOLUTION

(a) We choose 3 values for x and find 3 ordered pairs that are solutions to $y = -2x - 1$.

(x,	y)
(2	)
(0	)
(−1	)

$$y = -2x - 1$$
$$= (-2)(2) - 1$$
$$= -4 - 1$$
$$y = -5$$

$$y = -2x - 1$$
$$= (-2)(0) - 1$$
$$= 0 - 1$$
$$y = -1$$

$$y = -2x - 1$$
$$= (-2)(-1) - 1$$
$$= 2 - 1$$
$$y = 1$$

(x,	y)
(2,	**−5**)
(0,	**−1**)
(−1,	**1**)

(b) Then we plot these ordered pairs and draw a straight line through the points. All solutions to $y = -2x - 1$ lie on this straight line. If any of the points we plot do not lie on this line, we made an error and must check our work.

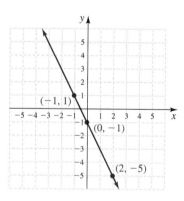

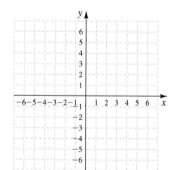

PRACTICE PROBLEM 7

(a) Name 3 ordered pairs that are solutions to $y = -3x - 1$.
(b) Plot these ordered pairs on a coordinate plane and draw a line through the points.

③ Graphing Horizontal and Vertical Lines

We often say "graph the equation" when we mean "plot the set of ordered pair solutions on a graph and connect the points with a line."

EXAMPLE 8   Graph: $y = -1$.

SOLUTION   How do we find solutions and graph $y = -1$? A solution to $y = -1$ is any ordered pair that has the y coordinate of -1. The x-coordinate can be any number, as long as y is -1.

The ordered pairs $(6, -1), (2, -1)$ and $(-3, -1)$ are solutions to $y = -1$ since all the y-values are -1.

We plot these ordered pairs.

Recall that if each y-value of ordered pairs is equal to -1, these coordinate points form a horizontal line at $y = -1$.

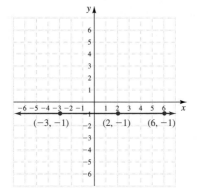

PRACTICE PROBLEM 8

Graph: $y = 4$.

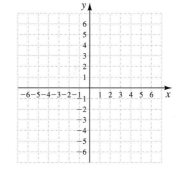

In Example 8 you could write the $y = -1$ as the equation $y + 0x = -1$, and then it would be clear that for any value of x that you substitute you will always obtain $y = -1$.

$$(6, -1) \qquad y + 0x = -1; \quad y + 0(6) = -1; \quad y = -1$$
$$(2, -1) \qquad y + 0x = -1; \quad y + 0(2) = -1; \quad y = -1$$
$$(-3, -1) \quad y + 0x = -1; \quad y + 0(-3) = -1; \quad y = -1$$

EXAMPLE 9   Graph: $x = 5$.

SOLUTION   A solution to $x = 5$ is any ordered pair that has the x-coordinate of 5. The y-coordinate can be any number as long as x is 5.

$(5, -2), (5, 3)$, and $(5, 0)$ are solutions to $x = 5$, since *all the x-values are 5.*

As we saw earlier, when each x-value of ordered pairs is equal to 5, the coordinate points form a vertical line at $x = 5$.

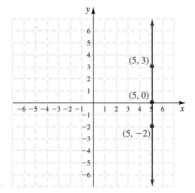

PRACTICE PROBLEM 9

Graph: $x = -4$.

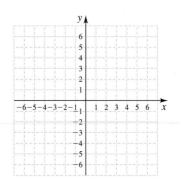

USING DATA AND EQUATIONS TO MAKE PREDICTIONS

Graphs can give a lot of information about the relationship between two things. For example, graphs can show if there is a positive or negative correlation. A *positive correlation* means that if one number goes up, the corresponding number also goes up. A *negative correlation* means that if one number goes up, the corresponding number goes down. We can make predictions about the relationship between data either by reading the graph or by determining an equation from the data, then using the equation to make the prediction. For example, we could determine the most efficient way to complete a job such as painting a house.

PROBLEMS FOR INDIVIDUAL INVESTIGATION

The table lists the price for the corresponding number of items.

NUMBER OF ITEMS	PRICE
1	$3
2	6
4	12
6	18

1. Write the ordered pairs (number of items, price).

2. Plot the points on a graph and draw a line through the points, extending it beyond the endpoints.

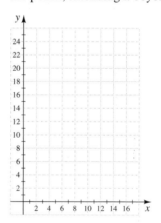

3. Is there a positive or a negative correlation between the number of items and the price of the items?

4. From the graph predict the price of 8 items.

5. Determine an equation from the relationship between coordinate points.

6. From your equation, predict the price of 20 items.

At Elton Electronics the cost of making an electronic component depends on the number made. The following table lists the cost of making the corresponding number of electronic components.

NUMBER OF COMPONENTS	COST
1	$3
2	4
3	5
4	6
6	8

7. Write the ordered pairs (number of components, cost).

8. Plot the points on a graph and draw a line through the points, extending it beyond the endpoints.

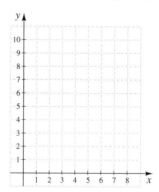

9. From the graph, predict the cost of 8 components.

10. Determine an equation from the relationship between coordinate points.

11. From your equation, predict the cost of 50 components.

PROBLEMS FOR GROUP INVESTIGATION AND COOPERATIVE STUDY

The first table below lists how many jobs can be completed by Sam in a given time. The second table lists how many of the same type of jobs can be completed by Sam and Jesse working together in a given time.

Sam Working Alone

NUMBER OF JOBS	HOURS WORKED
1	3
2	6
3	9
4	12

Sam and Jesse Working Together

NUMBER OF JOBS	HOURS WORKED
1	2
2	4
3	6
4	8

1. List the ordered pairs (number of jobs, hours) for the data given for Sam working alone.

2. Plot the points on a graph and draw a line through the points, extending it beyond the endpoints.

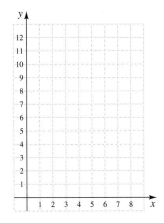

3. List the ordered pairs (number of jobs, hours) for the data given for Sam and Jesse working together.

4. Plot the points on the same graph used for question 2. Draw a line through the points, extending it beyond the endpoints.

5. Determine an equation that shows each relationship.

6. The company needs 12 jobs done.
   (a) How long would it take Sam working alone to get the 12 jobs done?
   (b) How long would it take Sam and Jesse working together to get the jobs done?

7. Sam gets paid $20 per hour for his work. Jesse gets paid $15 per hour for his work. Would it cost less to hire Sam and Jesse or just Sam alone to do the 12 jobs?

8. Does this graph have any real-world meaning for negative values of x; for y? Why?

INTERNET: Go to http://www.prenhall.com/blair to explore this application.

EXERCISES 8.4

1. List 4 ordered pairs that are solutions to the equation $x + y = 4$.

2. List 4 ordered pairs that are solutions to the equation $x + y = 5$.

3. List 4 ordered pairs that are solutions to the equation $x + y = 12$.

4. List 4 ordered pairs that are solutions to the equation $x + y = 9$.

5. A machine at a manufacturing company can label 35 bottles per minute. The equation that represents the situation is $y = 35x$, where y equals the number of bottles labeled and x represents the number of minutes the machine is operating. Determine how many minutes it takes the machine to label the following numbers of bottles. State your answer as the ordered pair (x,y).

   (a) 140 bottles   **(b)** 280 bottles

6. A machine at a manufacturing company can seal 40 jars per minute. The equation that represents the situation is $y = 40x$, where y equals the number of jars sealed and x represents the number of minutes the machine is operating. Determine how many minutes it takes the machine to seal the following numbers of jars. State your answer as the ordered pair (x,y).

   (a) 160 jars   **(b)** 320 jars

7. The secretary for Darwin Electronics can type 80 words per minute. The equation that represents the situation is $y = 80x$, where y equals the number of words typed and x represents the number of minutes the secretary typed. Determine how many minutes it takes the secretary to type the following numbers of words. State your answer as the ordered pair (x,y).

   (a) 240 words   **(b)** 400 words

8. The office manager at A&L Accounting Services can type 60 words per minute. The equation that represents the situation is $y = 60x$, where y equals the number of words typed and x represents the number of minutes the manager typed. Determine how many minutes it takes the manager to type the following numbers of words. State your answer as the ordered pair (x,y).

   (a) 360 words   **(b)** 420 words

Use a chart to organize your work for exercises 9–20.

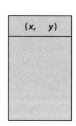

(x,	y)

9. Fill in the ordered pairs so that they are solutions to the equation $x + 2y = 16$:
   $(0, \_), (\_, 0), (\_, 4)$.

10. Fill in the ordered pairs so that they are solutions to the equation $x + 3y = 6$:
    $(0, \_), (\_, 0), (\_, 1)$.

11. Fill in the ordered pairs so that they are solutions to the equation $x + y = 5$:
    $(\_, 2), (0, \_), (1, \_)$.

12. Fill in the ordered pairs so that they are solutions to the equation $x + y = 12$:
    $(\_, 1), (3, \_), (0, \_)$.

13. Fill in the ordered pairs so that they are solutions to the equation $y = x + 2$:
$(-1, \_), (\_, 3), (\_, 0)$.

14. Fill in the ordered pairs so that they are solutions to the equation $y = x + 4$:
$(-1, \_), (\_, 4), (\_, 0)$.

15. Fill in the ordered pairs so that they are solutions to the equation $y = 5x + 3$:
$(0, \_), (-1, \_), (1, \_)$.

16. Fill in the ordered pairs so that they are solutions to the equation $y = 3x + 4$:
$(-2, \_), (-3, \_), (0, \_)$.

17. Find 3 ordered pairs that are solutions to $y = 5x - 3$.

18. Find 3 ordered pairs that are solutions to $y = 6x - 2$.

19. Find 3 ordered pairs that are solutions to $y = x + 6$.

20. Find 3 ordered pairs that are solutions to $y = x + 5$.

Plot 3 ordered-pair solutions of the given equation, then draw a line through the 3 points.

21. $y = 2x + 2$

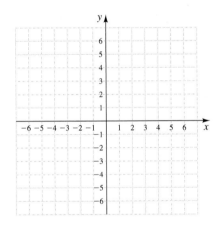

22. $y = 3x + 2$

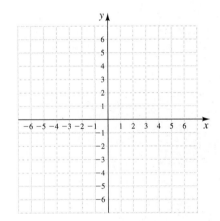

23. $y = -3x + 1$

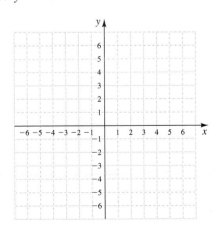

24. $y = -4x - 3$

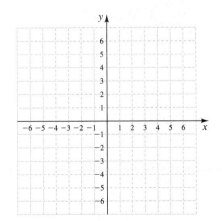

25. $y = 5x - 4$

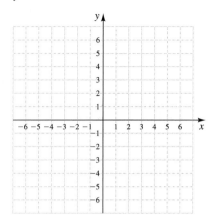

26. $y = 3x - 6$

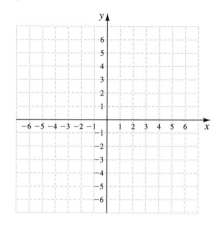

27. $y = 3x - 2$

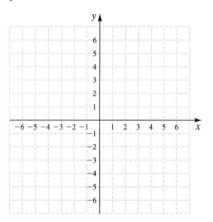

28. $y = 2x - 1$

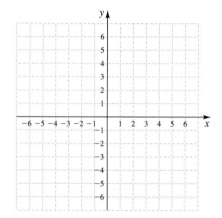

29. $y = -5x - 7$

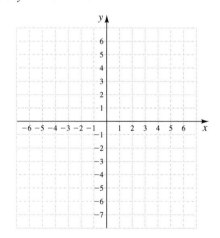

30. $y = -6x - 4$

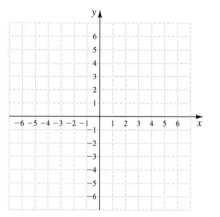

31. $y = 3$

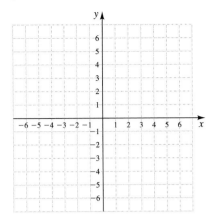

32. $y = -2$

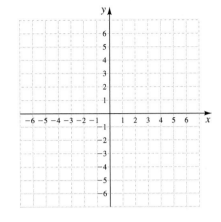

33. $x = 3$

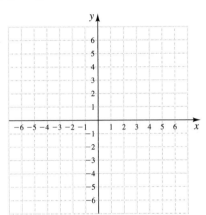

34. $x = -2$

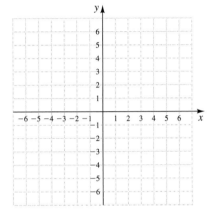

VERBAL AND WRITING SKILLS

35. When graphing the solutions to an equation, what is the advantage of plotting 3 ordered pairs rather than just 2?

36. When solving the equation, if you end up with a large value or a fraction for either x or y, must you use this point for the graph?

37. All of the following ordered pairs are a solution to the equation $y = mx + b$ except one. Which ordered pair is not a solution? Why?

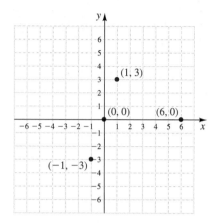

38. The ordered pairs plotted on the rectangular coordinate plane are solutions to the equation $y = mx + b$. Is $(5, 1)$ a solution to this equation? Why or why not?

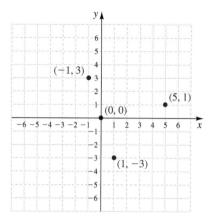

ONE STEP FURTHER

39. Find an equation that has the following solutions; $(2, 4), (3, 6), (4, 8)$.

40. Three solutions to an equation are plotted on the rectangular coordinate plane to the right. Determine 2 more ordered pairs that are solutions to this equation.

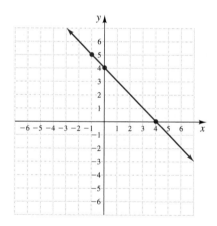

CUMULATIVE REVIEW

1. 36 inches = ____ feet

2. 2 gallons = ____ quarts

3. 32 ounces = ____ pounds

4. 3 hours = ____ minutes

546

CHAPTER ORGANIZER

TOPIC	PROCEDURE	EXAMPLES
Pictographs	The following pictograph illustrates the oil production from a local well over a 5-year period. 1990 [Oil] 1991 [Oil][Oil][Oil] 1992 [Oil][Oil][Oil][Oil][Oil][Oil] 1993 [Oil][Oil][Oil][Oil] 1994 [Oil][Oil] [Oil] = 500 barrels	State the number of barrels produced each year from 1990 to 1994. YEAR · BARRELS PRODUCED 1990 · 500 1991 · 1500 1992 · 3000 1993 · 2000 1994 · 1000
Circle graphs	The following circle graph describes the age of the 200 men and women of the Glover City police force. The percentage of the 200 police officers within a given age range is illustrated. Over Age 50 — 12% Under Age 23 — 10% Age 23–32 — 30% Age 32–50 — 48%	1. What percent of the police force is between 23 and 32 years old? 30% 2. How many men and women in the police force are over 50 years old? 12% of 200 = (0.12)(200) = 24 people
Double-bar graphs	The following double-bar graph illustrates the sales of color television sets by a major store chain for 1989 and 1990 in three regions of the country. □ 1989 ■ 1990 Number of televisions sold (in thousands) West Coast, Mid-west, East Coast	1. How many color television sets where sold by the chain on the East Coast in 1990? 6000 sets 2. How many *more* color television sets were sold in 1990 than in 1989 on the West Coast? 3000 sets were sold in 1990; 2000 sets were sold in 1989. 3000 $\underline{-2000}$ 1000 sets more in 1990

TOPIC	PROCEDURE	EXAMPLES
Comparison line graphs	The following line graph indicates the number of visitors to Wetlands State Park during a 4-month period in 1989 and 1990. 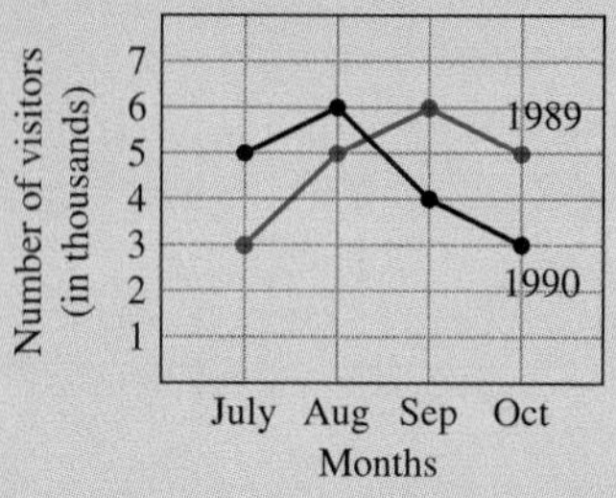	1. How many visitors came to the park in July 1989? 3000 visitors 2. In what months were there more visitors in 1989 than in 1990? September and October 3. The sharpest decrease in attendance took place between what two months? Between August 1990 and September 1990
Constructing comparison line graphs	1. Draw and label a vertical and a horizontal number line. 2. Place dots on the graph that correspond to the first category of data. 3. Connect the dots. 4. Repeat steps 2 and 3 for the second category of data. 5. Label each line.	The number of phone calls received by the receptionist on the day and night shifts of a hospital are listed on the following chart. | DAY OF WEEK	M	T	W	TH	F
---	---	---	---	---	---
Night shift	125	130	120	110	115
Day shift	90	120	80	95	110	 Construct a comparison line graph of the information given on the chart.
Constructing double-bar graphs	1. Draw and label a vertical and a horizontal number line. 2. Label a shaded and a nonshaded bar to represent each category that is being compared. 3. Draw a bar to the appropriate height for the data in each category.	Construct a double-bar graph for the number of calls received by the day and night shifts at the hospital.
Finding the mean	The *mean* of a set of values is the sum of the values divided by the number of values. The mean is often called the *average*.	1. Find the mean of 19, 13, 15, 25, and 18. $$\frac{19 + 13 + 15 + 25 + 18}{5} = \frac{90}{5} = 18$$ The mean is 18.

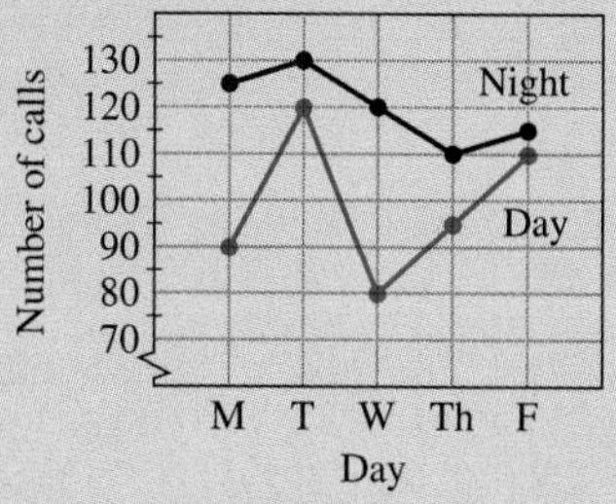

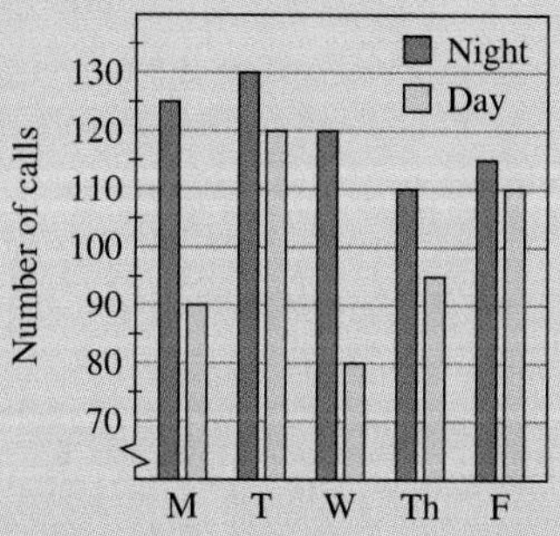

TOPIC	PROCEDURE	EXAMPLES

Finding the median	1. Arrange the numbers in order from smallest to largest.
	2. If there is an odd number of values, the middle value is the median.
	3. If there is an even number of values, the average of the two middle values is the median.
		1. Find the median of 19, 29, 36, 15, and 20.
		First we arrange in order from smallest to largest: 15, 19, 20, 29, 36.
		
		   15, 19 **20** 29, 36
		   two middle two
		  numbers number numbers
		
		The median is 20.
		2. Find the median of 67, 28, 92, 37, 81, and 75.
		First we arrange in order from smallest to largest: 28, 37, 67, 75, 81, 92.
		There is an even number of values.
		
		  28, 37 **67, 75** 81, 92
		       two middle
		       numbers
		
		$$\frac{67 + 75}{2} = \frac{142}{2} = 71$$
		The median is 71.

Plotting points	To plot (x, y): 1. Begin at the origin. 2. If x is positive, move to the right along the x-axis. If x is negative, move to the left along the x-axis. 3. If y is positive, move up. If y is negative, move down. 4. Place a dot at this location and label the point.	To plot $(-2, 3)$:
Graphing straight lines	A linear equation has a graph that is a straight line. To graph such an equation, plot any three points; two points give the line and the third point checks it.	Graph $3x + 2y = 6$. | x	y
---	---
0	3
4	−3
2	0	

CHAPTER 8 REVIEW

SECTION 8.1

Use this pictograph to answer questions 1–4.

1. How many students are enrolled in prealgebra?

2. How many students are enrolled in calculus?

3. How many more students are enrolled in algebra than calculus?

4. What is the combined total enrollment in prealgebra and algebra?

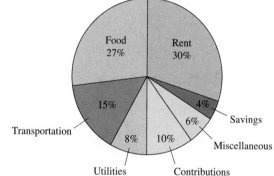

Mathematics class	Current course enrollment figures
Prealgebra	🚶🚶🚶🚶🚶🚶
Algebra	🚶🚶🚶🚶🚶🚶🚶🚶🚶
Calculus	🚶🚶🚶🚶🚶
	🚶 = 50 students

Nancy and Wally Worzowski's family monthly budget of $2400 is displayed in the accompanying circle graph. Use the graph to answer questions 5–12.

5. What percent of the budget is allotted for transportation?

6. What percent of the budget is allotted for savings?

7. What percent of the budget is used up by the food and rent categories?

8. What percent of the budget is used up by the transportation, utilities, and savings categories?

9. Of the total $2400, how much money per month is budgeted for utilities?

10. Of the total $2400, how much money per month is budgeted for transportation?

11. Of the total $2400, how much money per month is budgeted for the rent and savings categories?

12. Of the total $2400, how much money per month is budgeted for the transportation and food categories?

Circle graph: Food 27%, Rent 30%, 4% Savings, 6%, Miscellaneous, 10% Contributions, 8% Utilities, 15% Transportation

The accompanying double-bar graph illustrates the number of customers at Reid's Steak House for each quarter for the years 1993 and 1994. Use the graph to answer questions 13–20.

13. How many customers came to the restaurant in the second quarter of 1993?

14. How many customers came to the restaurant in the third quarter of 1994?

15. When did the restaurant have the greatest number of customers?

16. When did the restaurant have the fewest number of customers?

17. By how much did the number of customers increase from the first quarter of 1993 to the first quarter of 1994?

18. By how much did the number of customers increase from the fourth quarter of 1993 to the fourth quarter of 1994?

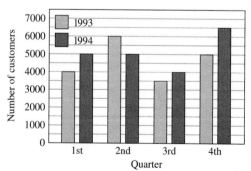

19. During the third quarter (July, August, September) there was road construction in front of the restaurant in both 1993 and 1994. Does the graph indicate the possibility that this might have caused a drop in the number of customers?

20. In the second quarter (April, May, June) the owner spent less on advertising in 1994 than in 1993. Does the graph suggest that this change might have caused a drop in the number of customers?

The accompanying comparison line graph shows the number of ice cream cones purchased at the Junction Ice Cream Stand during a 5-month period in both 1991 and 1992. Use the graph to answer questions 21–28.

21. How many ice cream cones were purchased in July 1992?

22. How many ice cream cones were purchased in August 1991?

23. How many more ice cream cones were purchased in May 1991 than in May 1992?

24. How many more ice cream cones were purchased in August 1992 than in August 1991?

25. How many more ice cream cones were purchased in July 1991 than in June 1991?

26. How many more ice cream cones were purchased in September 1991 than in August 1991?

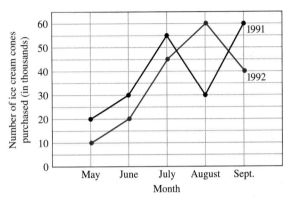

27. At the location of the ice cream stand, August 1991 was cold and rainy. The months of May, June, July, and September of 1991 were warm and sunny. What trend on the graph do you think is dependent on the weather?

28. July and August of 1992 were warm and very sunny, while May and June of 1992 were cloudy at the location of the ice cream stand. What trend on the graph do you think is dependent on the weather?

Use the information in the table to answer questions 29 and 30.

Circus Souvenir Sales

WEEK	HATS	STUFFED ANIMALS	T-SHIRTS	POSTERS
1	55	80	75	65
2	45	60	70	40
3	40	70	80	50

29. Use the graph to construct a comparison line graph that compares the number of sales for week 1 and week 3 at the circus.

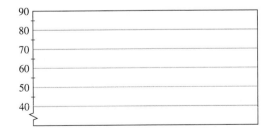

30. Use the graph to construct a comparison line graph that compares the number of sales for week 1 and week 2 at the circus.

551

Use the information in the table to answer questions 31 and 32.

Number of Bus Riders

	MONDAY	TUESDAY	WEDNESDAY	THURSDAY	FRIDAY
Adult Male	100	150	150	150	200
Adult Female	175	150	250	250	300
Children	75	100	150	100	175

31. Construct a double-bar graph that compares the number of female and male adult bus riders.

32. Construct a double bar-graph that compares the number of children and female adult bus riders.

SECTION 8.2

Find the median value.

33. The scores on a recent mathematics exam: 69, 57, 100, 87, 93, 65, 77, 82, and 88.

34. The number of students taking abnormal psychology for the fall semester for the last 9 years at Elmson College: 77, 83, 91, 104, 87, 58, 79, 81, and 88.

35. The number of cups of coffee consumed by each of the students of the 7:00 A.M. Biology III class during the last semester: 38, 19, 22, 4, 0, 1, 5, 9, 18, 36, 43, 27, 21, 19, 25, and 20.

36. The number of deliveries made each day by the Northfield House of Pizza: 21, 16, 0, 3, 19, 24, 13, 18, 9, 31, 36, 25, 28, 14, 15, and 26.

Find the mean value (average).

37. The last 7-day maximum temperature readings in Los Angeles in July were: 86°F, 83°F, 88°F, 95°F, 97°F, 100°F, and 81°F.

38. The amount of groceries purchased by the Michael Stallard family each week for the last 7 weeks was $87, $105, $89, $120, $139, $160, $98.

39. The number of college textbooks purchased by each of 8 men living at Jenkins House during his 4 years of college was 76, 20, 91, 57, 42, 21, 75, and 82.

40. The number of women students enrolled in the school of engineering at Westwood University during each of the last 10 years was 151, 140, 148, 156, 183, 201, 205, 228, 231, and 237.

552

SECTION 8.3

Use the following line graph to answer questions 41 and 42.

41. Represent the number of products sold in the years 1991, 1992, 1993, and 1994 using ordered pairs (year, number sold).

42. Represent the the number of products sold in the years 1995, 1996, 1997, and 1998 using ordered pairs (year, number sold).

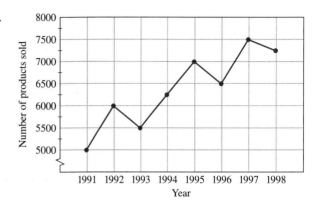

Plot and label the point on the rectangular cooordinate plane.

43. $(3, 2)$

44. $\left(2, 3\frac{1}{2}\right)$

45. $(-2, 0)$

46. $(-3, -1)$

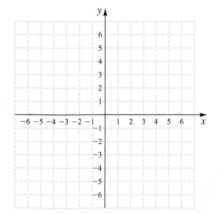

Give the coordinates of each point.

47. R

48. S

49. T

50. U

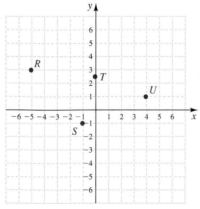

Plot the points corresponding to the ordered pairs, then draw a line connecting the coordinate points.

51. $(2, 0), (2, 3), (2, -1)$

52. $(3, 1), (-4, 1), (0, 1)$

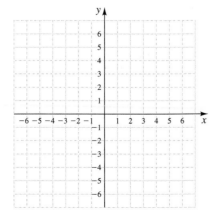

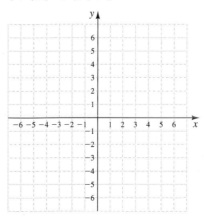

53. List 3 ordered pairs that are solutions to the equation $x + y = 8$.

54. List 3 ordered pairs that are solutions to the equation $x + y = 3$.

55. The secretary for the J&M law offices can type 70 words per minute. The equation that represents the situation is $y = 70x$, where y equals the number of words typed and x represents the number of minutes the secretary typed. Determine how many minutes it takes the secretary to type the following number of words. State your answer as the ordered pair (x, y).

(a) 280 words

(b) 350 words

56. Fill in the ordered pairs so that they are solutions to the equation $y = 2x - 6$:
($\_$, −4), ($\_$, −6), ($\_$, −8).

57. Fill in the ordered pairs so that they are solutions to the equation $y = -6x + 2$:
($\_$, 2), ($\_$, 8), ($\_$, −4).

Graph each equation.

58. $y = 3x - 1$

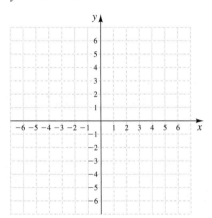

59. $y = -5x - 4$

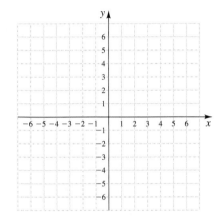

60. $y = 4x - 6$

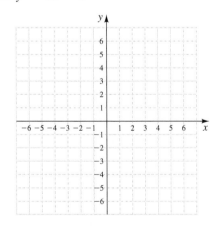

61. $y = -1$

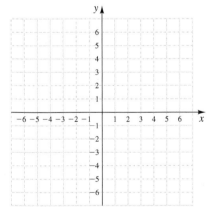

CHAPTER 8 TEST

The ages of 5000 students on campus were recorded. The circle graph depicts the distribution. Use the graph to answer questions 1–4.

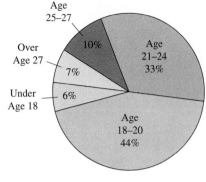

1. What age group has the largest percent of the student body?

2. What percent of the students are between 18 and 24?

3. What percent of the students are 20 or younger?

4. If 5000 students are at the university, how many students are over age 27?

A research study by 10 midwestern universities produced the following line graph. Use the graph to answer questions 5–8.

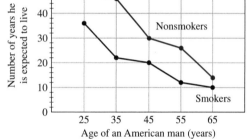

5. Approximately how many more years is a 65-year-old American man expected to live if he smokes?

6. Approximately how many more years is a 35-year-old American man expected to live if he does not smoke?

7. According to this graph, approximately how much longer is a 45-year-old nonsmoker expected to live than a 45-year-old smoker?

8. According to this graph, at what age is the difference between the life expectancy of a smoker and a nonsmoker the greatest?

The following double-bar graph indicates the number of cars sold at Boley's Chrysler during each quarter of 1992 and 1993. Use the graph to answer questions 9–13.

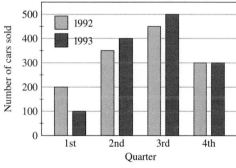

9. How many cars were sold in the first quarter of 1993?

10. How many more cars were sold in the second quarter of 1993 than in the second quarter of 1992?

11. During which quarter and year were the most cars sold?

12. What was the mean number of cars sold in 1992?

13. What was the median number of cars sold in 1993?

1. _____

2. _____

3. _____

4. _____

5. _____

6. _____

7. _____

8. _____

9. _____

10. _____

11. _____

12. _____

13. _____

A student received quiz grades of 89, 76, 85, 91, 83, and 90 on 6 quizzes.

14. Find the student's average (mean) quiz score. Round to the nearest tenth.

15. Find the student's median quiz score.

Plot and label each ordered pair on the rectangular coordinate plane.

16. $(3, 5)$

17. $(0, 0)$

18. $(2, -1)$

19. $(-3, 0)$

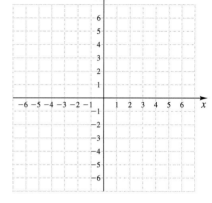

Give the coordinates of each point.

20. A

21. B

22. C

23. D

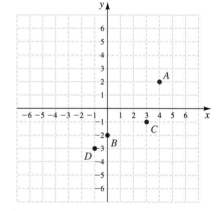

24. Plot the set of ordered pairs, then draw a line connecting the coordinate points: $(1, 2), (-3, 2), (0, 2), (4, 2)$.

14. _____

15. _____

16–19. _____

20. _____

21. _____

22. _____

23. _____

24. _____

556

25. Graph $y = 4x - 2$.

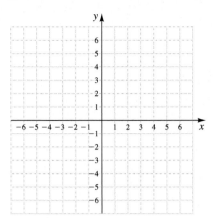

26. Graph $y = 3$.

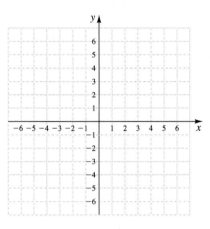

CUMULATIVE TEST FOR CHAPTERS 1-8

1. _____

2. _____

3. _____

4. _____

5. _____

6. _____

7. _____

8. _____

9. _____

10. _____

11. _____

12. _____

13. _____

14. _____

15. _____

16. _____

17. _____

18. _____

19. _____

20. _____

21. _____

22. _____

1. It is reported that as a high school freshman, Michael Jordan was 5 feet 8 inches tall. Two years later, as a junior, he was 6 feet 3 inches tall. How many inches did he grow in those 2 years?

2. Write in exponent form: $5 \cdot 5 \cdot 5 \cdot 5 \cdot 5 \cdot 5$.

Find the value.

3. 4^3

4. $3^2 - 2 + 7$

5. $20 \div 4 + 5(2 + 1)$

6. -2^3

7. Express 98 as a product of prime factors.

8. A car travels 285 miles on 12 gallons of gas. Find the miles per gallon, rounded to the nearest tenth.

9. Multiply and simplify: $\dfrac{xy^2}{18} \cdot \dfrac{9}{x}$.

10. Find the reciprocal of -11.

11. Divide and simplify: $\dfrac{2}{7} \div \left(-\dfrac{8}{14}\right)$.

Solve for x and check in problems 12–14.

12. $10x = 120$

13. $\dfrac{x}{3} = 12$

14. $-5x - 2 = -22$

15. Multiply: $2(2x^3 - x + 6)$.

16. Add: $(3x^2 - 2x - 8) + (x^2 + 4x - 3)$.

17. Write $\dfrac{7}{8}$ as a percent.

18. Write 8% as a decimal.

19. What is 16% of 48?

20. 29 is 18% of what number? Round to the nearest hundredth.

21. What percent of 436 is 37? Round to the nearest tenth of a percent.

22. Michael Jordan scored 26,920 points in 848 games. What was the average number of points he scored per game? Round to the nearest tenth.

The following comparison line graph depicts the annual rainfall in Dixville compared to the annual rainfall in Weston for 5 specific years. Use the graph to answer Exercises 23 to 26.

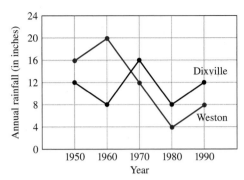

23. How many inches of rain fell in Weston in 1990?

24. In what years was the annual rainfall in Dixville greater than the annual rainfall in Weston?

25. What was the average number of inches of rainfall in Weston from 1950 to 1990?

26. What was the median number of inches of rainfall in Dixville from 1950 to 1990?

27. Graph $y = x + 3$

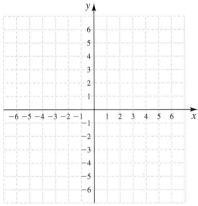

Plot and label the following ordered pairs on the rectangular coordinate plane.

28. $(6, 1)$

29. $(2, 3)$

30. $(0, -2)$

31. $(-4, 3)$

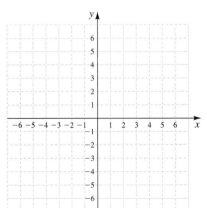

23. _____

24. _____

25. _____

26. _____

27. _____

28–31. _____

MEASUREMENT AND GEOMETRIC FIGURES

When traveling to another country, we often find that we must learn about that country's money system so that we can estimate the cost of our expenses. We also need to become familiar with the metric system since it is used throughout the world. To see the types of situations that require this knowledge, turn to Putting Your Skills to Work on page 578.

9.1   Converting Between U.S. Units; Converting Between Metric Units
9.2   Converting Between the U.S. and Metric Systems (optional)
9.3   Square Roots and Square Root Expressions
9.4   Triangles and the Pythagorean Theorem
9.5   The Circle and Applied Problems
9.6   Volume
9.7   Similar Geometric Figures

CHAPTER 9 PRETEST

This test provides a preview of the topics in this chapter. It will help you identify which concepts may require more of your studying time. If you are familiar with the topics in this chapter, take this test now. Check your answers with those in the back of the book. If you are not familiar with the topics in this chapter, begin studying the chapter now.

SECTION 9.1

1. Convert 370 minutes to hours.

2. A fence is $10\frac{1}{4}$ feet long. How many inches long is the fence?

3. Convert 5000 meters to kilometers.

SECTION 9.2

4. Convert 35 meters to feet.

5. A bolt that is 7 millimeters wide is how many inches wide?

SECTION 9.3

Write the math symbols to represent each statement, then use multiplication facts to solve.

6. Three squared equals what number?

7. What positive number squared equals nine?

8. What positive number multiplied by itself equals four?

Simplify.

9. $\sqrt{36}$

10. $7\sqrt{8} + 2\sqrt{8}$

SECTION 9.4

11. In a triangle with angles A, B, and C: angle A measures $80°$ and angle C measures $20°$. What is the measure of angle B?

12. Find the perimeter of an equilateral triangle with one side 11 inches long.

13. Find the length of the hypotenuse of a right triangle with legs 3 centimeters long and 4 centimeters long.

SECTION 9.5

Use $\pi \approx 3.14$. Round your answers to the nearest hundredth if necessary.

14. Find the circumference of a circle if the diameter is 7 meters.

15. Find the area of a circle whose diameter is 6 inches.

16. Find the area of a circle whose radius is 5.3 centimeters. Round your answer to the nearest tenth.

17. Ellen wants to buy a circular table cloth that is 4 feet in diameter. Find the cost of the table cloth at $15 a square yard.

SECTION 9.6

18. How much air is needed to fully inflate a basketball if the radius of the inner lining is 6 inches? Use $\pi \approx 3.14$.

19. Find the volume of a pyramid with height = 3 meters, length of base = 4 meters, width of base = 5 meters.

SECTION 9.7

20. A flagpole casts a shadow of 30 feet. At the same time, a tree that is 6 feet tall has a shadow of 12 feet. How tall is the flagpole?

1. _____

2. _____

3. _____

4. _____

5. _____

6. _____

7. _____

8. _____

9. _____

10. _____

11. _____

12. _____

13. _____

14. _____

15. _____

16. _____

17. _____

18. _____

19. _____

20. _____

SECTION   9.1

Converting Between U.S. Units;
Converting Between Metric Units

After studying this section, you will be able to:

❶ Use a unit fraction to convert from one U.S. unit to another.

❷ Solve applied problems involving U.S. units.

❸ Convert from one metric unit to another.

❹ Solve applied problems involving metric units.

Math Pro   Video 9.1   SSM

We have worked with units of measurement throughout the book. Now we will see how to change (convert) from one U.S. unit to another. For your review we present the table from Chapter 1 which shows the relationships between units.

Length	Weight
12 inches (in.) = 1 foot (ft)	16 ounces (oz) = 1 pound (lb)
3 feet (ft) = 1 yard (yd)	2000 pounds = 1 ton
5280 feet (ft) = 1 mile (mi)	
Volume	**Time**
2 cups (c) = 1 pint (pt)	60 seconds (sec) = 1 minute (min)
2 pints (pt) = 1 quart (qt)	60 minutes (min) = 1 hour (hr)
4 quarts (qt) = 1 gallon (gal)	24 hours (hr) = 1 day
	7 days = 1 week

❶ Using a Unit Fraction to Convert from One U.S. Unit to Another

For many simple problems such as 24 inches = ? feet, we can easily see how to convert from inches to feet.

$$24 \text{ inches} = 12 \text{ inches} + 12 \text{ inches}$$
$$= 1 \text{ foot} \quad + 1 \text{ foot} \qquad (12 \text{ inches} = 1 \text{ foot})$$
$$24 \text{ inches} = 2 \text{ feet}$$

For more complicated problems, such as changing miles to feet, we need another method to do this conversion so that the process is simple and efficient. The method we use involves multiplying by a unit fraction.

> A **unit fraction** is a fraction that shows the relationship between units and is equal to 1.

For example, since 12 in. = 1 ft, we can say that there are 12 inches per 1 foot or 1 foot per 12 inches. If we read the fraction bar as *per*, we have the following unit fractions that are equal to 1:

$$\text{per} \rightarrow \frac{12 \text{ in.}}{1 \text{ ft}} = \frac{1 \text{ ft}}{12 \text{ in.}} = 1 \qquad \frac{12 \text{ in.}}{1 \text{ ft}} \text{ and } \frac{1 \text{ ft}}{12 \text{ in.}} \text{ are called } \textit{unit fractions.}$$

We can multiply a quantity by a unit fraction since its value is equal to 1, and we know that multiplying by 1 does not change the value of the quantity.

EXAMPLE 1   Convert 35 yards to feet.

SOLUTION   We write the relationship between feet and yards as a unit fraction: 3 ft = 1 yd; $\dfrac{3 \text{ ft}}{1 \text{ yd}}$

$$35 \text{ yd} = \underline{\quad ? \quad} \text{ ft}$$

$$35 \text{ yd} \times \frac{3 \text{ ft}}{1 \text{ yd}} \qquad \text{Multiply by the unit fraction.}$$

$$= 35 \text{ yd} \times \frac{3 \text{ ft}}{1 \text{ yd}} \qquad \text{Divide out the units "yd."}$$

$$= 35 \times 3 \text{ ft} = \mathbf{105 \text{ ft}} \qquad \text{Multiply.}$$

PRACTICE PROBLEM 1

Convert 420 minutes to hours.

How did we know which unit fraction to use in Example 1? We use the unit fraction that relates the units we are working with, in this case it is *feet* and *yards*. Now, to determine which unit to put in the numerator and denominator of the fraction, we must consider what unit we want to end up with. In Example 1 we wanted to end up with feet, so we placed feet in the numerator.

We want to end up with feet, so we place 3 ft in the numerator.

$$35 \text{ yd} \times \frac{3 \text{ ft}}{1 \text{ yd}} = \underline{\quad ? \quad} \text{ ft}$$

The yards divide out, and we end up with feet.

PROCEDURE TO CONVERT FROM ONE UNIT TO ANOTHER

1. Write the relationship between the units.
2. Identify the unit you want to end up with.
3. Write a unit fraction that has the unit you want to end up with in the numerator.
4. Multiply by the unit fraction.

EXAMPLE 2 Convert 560 quarts to gallons.

SOLUTION We write the relationship between quarts and gallons: 4 qt = 1 gal. We want to end up with gallons, so we write *1 gal* in the numerator of the unit fraction: $\dfrac{1 \text{ gal}}{4 \text{ qt}}$.

$$560 \text{ qt} = \underline{\quad ? \quad} \text{ gal}$$

$$560 \text{ qt} \times \frac{1 \text{ gal}}{4 \text{ qt}} \qquad \text{We multiply by the appropriate unit fraction.}$$

$$= 560 \text{ qt} \times \frac{1 \text{ gal}}{4 \text{ qt}} \qquad \text{We divide out the units "qt."}$$

$$= 560 \times \frac{1}{4} \text{ gal} = \frac{560 \text{ gal}}{4} = \mathbf{140 \text{ gal}}$$

PRACTICE PROBLEM 2

Convert 144 ounces to pounds.

② Solving Applied Problems Involving U.S. Units

E X A M P L E 3   A computer printout shows that a particular job took 144 seconds. How many minutes is that? (Express your answer as a decimal.)

SOLUTION

$$144 \text{ \sout{seconds}} \times \frac{1 \text{ minute}}{60 \text{ \sout{seconds}}} = \frac{144}{60} \text{ minutes} = \textbf{2.4 minutes}$$

P R A C T I C E P R O B L E M 3

Joe's time card read "Hours worked today: 7.2." How many minutes are in 7.2 hours?

E X A M P L E 4   The all-night Charlotte garage charges $1.50 per hour for parking both day and night. A businessman left his car there for $2\frac{1}{4}$ days. How much was he charged?

SOLUTION

1. *Understand the problem.* Here it might help to look at a simpler problem. If the businessman had left his car for 2 hours, we would multiply.

$$\text{per} \rightarrow \frac{1.50 \text{ dollars}}{1 \text{ hr}} \times 2 \text{ hr} = 3.00 \text{ dollars or } \$3.00$$

Thus, if the businessman leaves his car for 2 hours, he would be charged $3. We see that we need to multiply by the number of hours the car was in the garage to solve the problem.

Since the original problem gave the time in days, not hours, we will need to **change the days to hours**. Remember the garage charges $1.50 per *hour*.

2. *Solve and state the answer.* Now that we know that the way to solve the problem is to multiply hours, we will begin. To make our calculations easier we will write $2\frac{1}{4}$ as 2.25.

Change days to hours.   Multiply by cost per hour.

$$2.25 \text{ \sout{days}} \times \frac{24 \text{ \sout{hr}}}{1 \text{ \sout{day}}} \times \frac{1.50 \text{ dollars}}{1 \text{ \sout{hr}}} = 81 \text{ dollars or } \$81$$

The businessman was charged $81.

3. *Check.* Is our answer in the desired units? Yes. The answer is in dollars and we would expect it to be in dollars. ✓
    You may want to redo the calculation or use a calculator to check. The check is up to you.

P R A C T I C E P R O B L E M 4

A businesswoman parked her car at a garage for $1\frac{3}{4}$ days. The garage charges $1.50 per hour. How much did she pay to park the car?

TO THINK ABOUT

How did people first come up with the idea of multiplying by a unit fraction? What mathematical principles are involved here? Actually, this is the same as solving a proportion. Consider a situation where we change 34 quarts to 8.5 gallons by multiplying by a unit fraction.

$$34 \text{ qt} \times \frac{1 \text{ gal}}{4 \text{ qt}} = \frac{34}{4} \text{ gal} = 8.5 \text{ gal}$$

What we were actually doing is setting up the proportion, 1 gal is to 4 qt as n gal is to 34 qt, and solving for n.

$$\frac{1 \text{ gal}}{4 \text{ qt}} = \frac{n \text{ gal}}{34 \text{ qt}}$$

We cross-multiply: 34 qt $\times$ 1 gal $=$ 4 qt $\times$ n gal. We divide both sides of the equation by 4 quarts.

$$\frac{34 \text{ qt} \times 1 \text{ gal}}{4 \text{ qt}} = \frac{4 \text{ qt} \times n \text{ gal}}{4 \text{ qt}}$$

$$1 \text{ gal} \times \frac{34}{4} = n \text{ gal}$$

$$\mathbf{8.5 \text{ gal}} = n \text{ gal}$$

Thus the number of gallons is 8.5. Using proportions takes a little longer, so multiplying by a unit fraction is the more popular method.

---

③ Converting from One Metric Unit to Another

In Chapter 1 we were introduced to the metric system. Now we will see how to change from one metric unit to another. We start by reviewing the relationship between metric units. Units that are larger than the *basic unit* use the prefixes *kilo*, meaning 1000; *hecto*, meaning 100; and *deka*, meaning 10. For units smaller than the basic unit we use the prefixes *deci*, meaning $\frac{1}{10}$; *centi*, meaning $\frac{1}{100}$; and *milli*, meaning $\frac{1}{1000}$.

A teaspoon can hold about 5 milliliters.

A 1-liter bottle can hold 1000 milliliters.

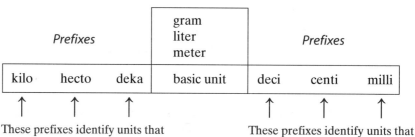

	Prefixes		gram liter meter		*Prefixes*	
kilo	hecto	deka	basic unit	deci	centi	milli

These prefixes identify units that are larger than the basic unit.

These prefixes identify units that are smaller than the basic unit.

We list the relationships between units that are commonly used in the metric system.

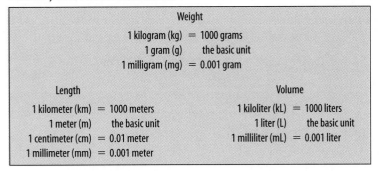

Commonly Used Metric Measurements

Weight

1 kilogram (kg) = 1000 grams
1 gram (g)   the basic unit
1 milligram (mg) = 0.001 gram

Length

1 kilometer (km) = 1000 meters
1 meter (m)   the basic unit
1 centimeter (cm) = 0.01 meter
1 millimeter (mm) = 0.001 meter

Volume

1 kiloliter (kL) = 1000 liters
1 liter (L)   the basic unit
1 milliliter (mL) = 0.001 liter

1 nickel weighs about 5 grams.

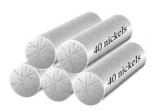

200 nickels weigh about 1000 grams or 1 kilogram (kg).

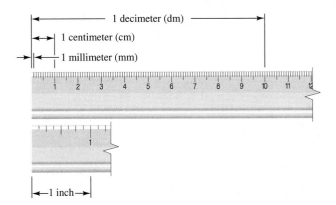

How do we convert from one metric unit to another? For example, how do we change 5 kilometers into an equivalent number of meters?

Recall from Chapter 7 that when we multiply by 10, we move the decimal point one place to the right. When we divide by 10, we move the decimal point one place to the left. Let's see how we use that idea to change from one metric unit to another.

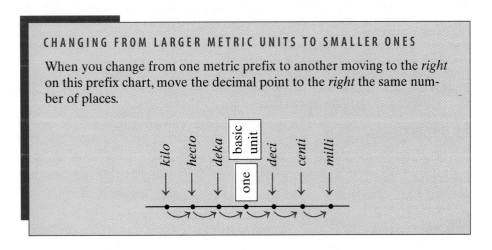

CHANGING FROM LARGER METRIC UNITS TO SMALLER ONES

When you change from one metric prefix to another moving to the *right* on this prefix chart, move the decimal point to the *right* the same number of places.

Thus 1 meter = 100 centimeters because we move two places to the right on the chart of prefixes and we also move the decimal point in 1.00 two places to the right.

EXAMPLE 5

(a) Change 5 kilometers to meters. **(b)** Change 20 liters to centiliters.

SOLUTION

(a) From kilometer to meter we move *three places to the right on the prefix chart*, so we move the decimal point three places to the right.

5 km = 5.000 m (move three places) = **5000 m**

(b) To go from liters to centiliters we move *two places to the right on the prefix chart*. Thus we move the decimal point two places to the right.

20 L = 20.00 cL (move two places) = **2000 cL**

PRACTICE PROBLEM 5

(a) Change 4 meters to centimeters.
(b) Change 30 centigrams to milligrams.

Now let us see how we can change a measurement stated in a smaller unit to an equivalent measurement stated in larger units.

CHANGING FROM SMALLER METRIC UNITS TO LARGER ONES

When you change from one metric prefix to another moving to the *left* on this prefix chart, move the decimal point to the *left* the same number of places.

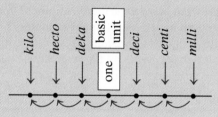

EXAMPLE 6

(a) Change 7 centigrams to grams.
(b) Change 56 millimeters to kilometers.

SOLUTION

(a) To go from centigrams to grams we move *two* places to the left on the prefix chart. Thus we move the decimal point two places to the left.

7 cg = 0.07 grams   Move two places to the left.

= **0.07 grams**

(b) To go from *milli*meters to *kilo*meters, we move *six* places to the left on the prefix chart. Thus we move the decimal point six places to the left.

56 mm = 0.000056. km   Move six places to the left.

= **0.000056 km**

PRACTICE PROBLEM 6

(a) Change 3 *milli*liters to liters.
(b) Change 47 *centi*meters to *kilo*meters.

EXAMPLE 7 Convert.

(a) 426 decimeters to kilometers   **(b)** 9.47 hectometers to meters

SOLUTION

(a) We are converting from a smaller unit, *dm*, to a larger one, *km*. Therefore, there will be fewer kilometers than decimeters (426 will "get smaller"). We move the decimal point four places to the left.

$\quad$ 426 dm = 0.0426 km $\qquad$ Move four places to the left.

$\qquad\quad$ = **0.0426 km**

(b) We are converting from a larger unit, *hm*, to a smaller one, *m*. Therefore, there will be more meters than hectometers (9.47 will "get larger"). We move the decimal point two places to the right.

$\quad$ 9.47 hm = 947. m $\qquad$ Move two places to the right.

$\qquad\quad$ = **947 m**

PRACTICE PROBLEM 7
Convert.

(a) 389 millimeters to dekameters   **(b)** 0.48 hectometer to centimeters

④ Solving Applied Problems Involving Metric Units

EXAMPLE 8   A special cleaning fluid to rinse test tubes in a chemistry lab cost $40.00 per liter. What is the cost per milliliter?

SOLUTION

1. *Understand the problem.* A milliliter is a very small part of a liter, $\frac{1}{1000}$ of a liter. Therefore it costs much less for a milliliter, $\frac{1}{1000}$ of the cost of a liter.

2. *State and solve the problem.* To change from milliliter to liter we move the decimal point 3 places left.

$\quad$ $40.00 per liter = $0.040 per milliliter

Thus **each milliliter costs $0.04**, which is quite expensive.

3. *Check:* $0.04 is much smaller than $40.00, so our answer seems reasonable.

1 liter

1 milliliter

PRACTICE PROBLEM 8
A purified acid costs $110 per liter. What does it cost per milliliter?

EXERCISES 9.1

From memory, write the equivalent values.

1. 1 foot = ____ inches

2. 1 yard = ____ feet

3. ____ pints = 1 quart

4. _____ feet = 1 mile

5. 1 ton = _____ pounds

6. 1 pound = ____ ounces

7. ____ quarts = 1 gallon

8. ____ cups = 1 pint

9. 7 days = ____ week

10. 1 day = ____ hours

11. ____ seconds = 1 minute

12. ____ minutes = 1 hour

Convert.

13. 15 feet = ____ yards

14. 84 inches = ____ feet

15. 10,560 feet = ____ miles

16. 5280 feet = _____ yards

17. 8 pints = ____ quarts

18. 6 miles = _____ feet

19. 12 feet = _____ inches

20. 41 yards = _____ feet

21. 87 inches = _____ feet

22. 192 ounces = ____ pounds

23. 13 tons = _____ pounds

24. 2.25 pounds = ____ ounces

25. 7 gallons = ____ quarts

26. 12 weeks = ____ days

27. 660 minutes = ____ hours

28. 11 days = _____ hours

29. 70 minutes = _____ seconds

30. 18 hours = _____ seconds

31. 6 tons = _____ pounds

32. A stockbroker left his car in an all-night garage for $2\frac{1}{2}$ days. The garage charges $2.25 per hour. How much did he pay for parking?

33. Judy is making a wild mushroom sauce for pasta tonight with a large group of friends. She bought 26 ounces of wild mushrooms at $6.00 per pound. How much did the mushrooms cost?

34. Jan Zelezny of the Czech Republic threw his javelin 318 feet 10 inches to break the world's record in 1993. How many inches did his javelin fly?

35. Mount Whitney in California is approximately 2.745 miles high. How many feet is that? Round your answer to the nearest 10 feet.

The following conversions involve metric units that are very commonly used. You should be able to perform each conversion without any notes and without consulting your text.

36. 37 cm = ____ mm

37. 44 cm = ____ mm

38. 3.6 km = _____ m

39. 7.6 km = _____ m

40. 2.43 kL = _____ mL

41. 1.76 kL = _____ ml

42. 28,156 mL = _____ kL

43. 5261 mL = _____ kL

44. 0.78 g = _____ kg

45. 294 g = _____ kg

46. 5.9 kg = _____ mg

47. 6328 mg = _____ g

48. 14.6 kg = ____ g

49. 0.83 kg = _____ mg

Fill in the blanks with the correct values.

50. 7 mL = _____ L = _____ kL

51. 18 mL = _____ L = _____ kL

52. 522 mg = _____ g = _____ kg

53. 49 mg = _____ g = _____ kg

54. 3607 mL = _____ L = _____ kL

55. 7183 mg = _____ g = _____ kg

56. 35 mm = _____ cm = _____ m

57. 83 mm = _____ cm = _____ m

58. 3582 mm = _____ m = _____ km

59. 7812 mm = _____ m = _____ km

60. 0.32 cm = _____ m = _____ km

61. 0.81 cm = _____ m = _____ km

62. A pharmaceutical firm developed a new vaccine that costs $6.00 per milliliter to produce. How much will it cost the firm to produce 1 liter of the vaccine?

63. A physician in Mexico is striving to perfect a new anticancer drug. At the moment, he is using a chemical solution that costs $95.50 per gram. How much would it cost to buy 6 kilograms of the solution?

64. A very rare essence of an almost extinct flower found in the Amazon jungle of South America is extracted by a biogenetic company trying to copy and synthesize it. The company estimates that if the procedure is successful, the product will cost the company $850 per milliliter to produce. How much will it cost the company to produce 0.4 liter of the engineered essence?

65. The world's longest train run is on the Trans-Siberian line in Russia, from Moscow to Nakhodka on the Sea of Japan. The length of the run, which makes 97 stops, measures 94,380,000 centimeters. The run takes 8 days, 4 hours, and 25 minutes.

(a) How many meters is the run?

(b) How many kilometers is the run?

66. The highest railroad line in the world is a track on the Morococha branch of the Peruvian state railways at La Cima. The track is 4818 meters high.

(a) How many centimeters high is the track?

(b) How many kilometers high is the track?

67. The breakup of the former Soviet Union has delayed the completion of the Rogunskaya dam, which will be 335 meters high.

(a) How many kilometers high is the dam supposed to be?

(b) How many centimeters high is the dam supposed to be?

CALCULATOR EXERCISES

68. A multimillion-dollar shopping complex in Italy is using nothing but marble for the floors and walls of all public areas. The lobby alone is constructed of 267,905,993 pounds of blue-gray marble. How many tons would that be? Round to the nearest ton.

69. A space probe traveled for 3 years to get to the planet Saturn. How many hours did the trip take? (Assume 365 days in 1 year.)

Write the metric prefix that means:

70. Hundred

71. Hundredth

72. Tenth

73. Thousandth

74. Thousand

75. Ten

ONE STEP FURTHER

Metric measurements are also used for computers. A **byte** *is the amount of computer memory needed to store one alphanumeric character. When referring to computers you may hear the following words: kilobytes, megabytes, and gigabytes. The following chart may help you:*

1 gigabyte (GB) = one billion bytes	=	1,000,000,000 bytes
1 megabyte (MB) = one million bytes*	=	1,000,000 bytes
1 kilobyte (KB) or K = one thousand bytes†	=	1000 bytes

* Sometimes in computer science 1 megabyte is considered to be 1,048,576 bytes.
† Sometimes in computer science 1 kilobyte is considered to be 1024 bytes.

76. 1.2 gigabytes = _____ bytes

77. 528 megabytes = _____ bytes

78. 78.9 kilobytes = _____ bytes

79. 24.9 gigabytes = _____ bytes

CUMULATIVE REVIEW

1. 14 out of 70 is what percent?

2. What is 23% of 250?

3. What is 1.7% of $18,900?

4. A salesperson earns a commission of 8%. She sold furniture worth $8960. How much commission will she earn?

572

S E C T I O N   9 . 2

Converting Between the U.S. and Metric Systems (optional)

After studying this section, you will be able to:

❶ Convert units of length, volume, or weight between the U.S. and metric systems.

❷ Convert temperature readings between Fahrenheit and Celsius degrees.

Math Pro   Video 9.2   SSM

❶ Converting Units of Length, Volume, or Weight Between the Metric and U.S. Systems

So far we've seen how to convert units when working *within* either the U.S. or the metric system. Many people, however, work with *both* the metric and U.S. systems. If you study such fields as chemistry, electromechanical technology, business, x-ray technology, nursing, or computers, you will probably need to convert measurements between the two systems. We learn that skill in this section.

To convert between U.S. units and metric units, it is helpful to have equivalent values. The most commonly used equivalents are listed below. Most of these equivalents are approximate.

Equivalent measures

	U.S. TO METRIC	METRIC TO U.S.
Units of length	1 mile = 1.61 kilometers	1 kilometer = 0.62 mile
	1 yard = 0.914 meter	1 meter = 3.28 feet
	1 foot = 0.305 meter	1 meter = 1.09 yards
	1 inch = 2.54 centimeters	1 centimeter = 0.394 inch
Units of volume	1 gallon = 3.79 liters	1 liter = 0.264 gallon
	1 quart = 0.946 liter	1 liter = 1.06 quarts
Units of weight	1 pound = 0.454 kilogram	1 kilogram = 2.2 pounds
	1 ounce = 28.35 grams	1 gram = 0.0353 ounce

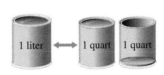

1 liter ⟷ 1 quart 1 quart

Remember that to convert from one unit to the other you multiply by a unit fraction that is equivalent to 1. Create a fraction from the equivalent measures table so that the unit in the numerator is the unit you want to end up with. To change 5 miles to kilometers, we look in the table and find that 1 mile = 1.61 kilometers. We will use the fraction

$$\frac{1.61 \text{ km}}{1 \text{ mi}}$$

We want to have 1.61 kilometers in the numerator.

$$5 \ \cancel{\text{mi}} \times \frac{1.61 \text{ km}}{1 \ \cancel{\text{mi}}} = 5 \times 1.61 \text{ km} = 8.05 \text{ km}$$

Thus 5 miles = 8.05 kilometers.

E X A M P L E   1

(a) Convert 26 m to yd.
(c) Convert 14 gal to L.
(e) Convert 5.6 lb to kg.

(b) Convert 1.9 km to mi.
(d) Convert 2.5 L to qt.
(f) Convert 152 g to oz.

SOLUTION

(a) $26 \ \cancel{\text{m}} \times \dfrac{1.09 \text{ yd}}{1 \ \cancel{\text{m}}} = \mathbf{28.34 \text{ yd}}$ **(b)** $1.9 \ \cancel{\text{km}} \times \dfrac{0.62 \text{ mi}}{1 \ \cancel{\text{km}}} = \mathbf{1.178 \text{ mi}}$

(c) $14 \ \cancel{gal} \times \dfrac{3.79 \ L}{1 \ \cancel{gal}} = \textbf{53.06 L}$ **(d)** $2.5 \ \cancel{L} \times \dfrac{1.06 \ qt}{1 \ \cancel{L}} = \textbf{2.65 qt}$

(e) $5.6 \ \cancel{lb} \times \dfrac{0.454 \ kg}{1 \ \cancel{lb}} = \textbf{2.5424 kg}$ **(f)** $152 \ \cancel{g} \times \dfrac{0.0353 \ oz}{1 \ \cancel{g}} = \textbf{5.3656 oz}$

PRACTICE PROBLEM 1

(a) Convert 17 m to yd.   **(b)** Convert 29.6 km to mi.
(c) Convert 26 gal to L.   **(d)** Convert 6.2 L to qt.
(e) Convert 16 lb to kg.   **(f)** Convert 280 g to oz.

Some conversions require more than one step.

EXAMPLE 2   Convert 235 cm to ft. Round your answer to the nearest hundredth of a foot.

SOLUTION   Our first fraction converts centimeters to inches. Our second fraction converts inches to feet.

$$235 \ \cancel{cm} \times \dfrac{0.394 \ \cancel{in.}}{1 \ \cancel{cm}} \times \dfrac{1 \ ft}{12 \ \cancel{in.}} = \dfrac{92.59}{12} \ ft = 7.72 \ ft$$

(rounded to the nearest hundredth)

PRACTICE PROBLEM 2
Convert 180 cm to ft.

The same rules can be followed for a rate such as 50 miles per hour.

EXAMPLE 3   Convert 100 km/hr to mi/hr.

SOLUTION

$$\dfrac{100 \ \cancel{km}}{hr} \times \dfrac{0.62 \ mi}{1 \ \cancel{km}} = 62 \ mi/hr.$$

Thus 100 km/hr is approximately equal to 62 mi/hr.

PRACTICE PROBLEM 3
Convert 88 km/hr to mi/hr. (Round to the nearest hundredth.)

EXAMPLE 4   A camera film that is 35 mm wide is how many inches wide?

SOLUTION   We first convert from millimeters to centimeters by moving the decimal point in the number 35 one place to the left.

$$35 \ mm = 3.5 \ cm$$

Then we convert to inches using a unit fraction.

$$3.5 \ \cancel{cm} \times \dfrac{0.394 \ in.}{1 \ \cancel{cm}} = \textbf{1.379 in.}$$

PRACTICE PROBLEM 4
The city police use 9-mm automatic pistols. If such a pistol fires a bullet 9 mm wide, how many inches wide is it? (Round to the nearest hundredth.)

TO THINK ABOUT

Suppose we consider a rectangle that measures 2 yd wide by 4 yd long. The area would be 2 yd × 4 yd = 8 yd². How could you change 8 yd² to m²? Suppose that we look at 1 yd². Each side is 1 yd long, which is equivalent to 0.914 m.

Area = 1 yd × 1 yd = 0.914 m × 0.914 m

Area = 1 yd² = 0.8354 m²   rounded to the ten-thousandths place

Thus 1 yd² = 0.8354 m². Therefore,

$$8 \text{ yd}^2 \times \frac{0.8354 \text{ m}^2}{1 \text{ yd}^2} = 6.6832 \text{ m}^2$$

8 yd² = 6.6832 m². Thus, 8 square yards is equivalent to 6.6832 square meters.

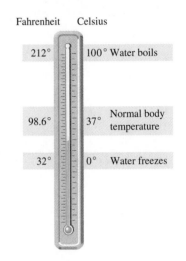

Fahrenheit	Celsius	
212°	100°	Water boils
98.6°	37°	Normal body temperature
32°	0°	Water freezes

Converting Temperature Readings Between Fahrenheit and Celsius Degrees

2

In the metric system temperature is measured on the Celsius scale. Water boils at 100° (100°C) and freezes at 0° (0°C) on the Celsius scale. In the Fahrenheit system water boils at 212° (212°F) and freezes at 32° (32°F).

To convert Celsius to Fahrenheit, we can use the formula

$$F = 1.8 \times C + 32$$

To convert Fahrenheit temperature to Celsius, we can use the formula

$$C = \frac{5 \times F - 160}{9}$$

where F is the number of Fahrenheit degrees and C is the number of Celsius degrees.

EXAMPLE 5   When the temperature is 176°F, what is the Celsius reading?

SOLUTION

$$C = \frac{5 \times F - 160}{9}$$

$$= \frac{5 \times 176 - 160}{9}$$

$$= \frac{880 - 160}{9} = \frac{720}{9} = 80$$

The temperature is **80°C**.

PRACTICE PROBLEM 5

Convert 181°F to Celsius temperature.

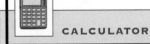

CALCULATOR

CONVERTING TEMPERATURE

You can use your calculator to convert temperature readings between Fahrenheit and Celsius.
To convert 30°C to Fahrenheit temperature, enter:

1.8 ☒ 30 ⊞ 32 ⩵

Display:

86

The temperature is 86°F.
To convert 82.4°F to Celsius temperature, enter:

5 ☒ 82.4 ⊟ 160

⩵ ÷ 9 ⩵

Display:

28

The temperature is 28°C.

E X E R C I S E S   9 . 2

Perform each conversion. Round the answer to the nearest hundredth when necessary.

1. 7 ft to m

2. 11 ft to m

3. 9 in. to cm

4. 13 in. to cm

5. 14 m to yd

6. 18 m to yd

7. 26.5 m to yd

8. 29.3 m to yd

9. 15 km to mi

10. 12 km to mi

11. 24 yd to m

12. 31 yd to m

13. 82 mi to km

14. 68 mi to km

15. 25 m to ft

16. 35 m to ft

17. 17.5 cm to in.

18. 19.6 cm to in.

19. 200 m to yd

20. 350 m to yd

21. 5 gal to L

22. 7 gal to L

23. 19 L to gal

24. 15 L to gal

25. 4.5 L to qt

26. 6.5 L to qt

27. 32 lb to kg

28. 27 lb to kg

29. 7 oz to g

30. 9 oz to g

31. 16 kg to lb

32. 14 kg to lb

33. 126 g to oz

34. 186 g to oz

35. 166 cm to ft

36. 142 cm to ft

37. 16.5 ft to cm

38. 19.5 ft to cm

39. 50 km/hr to mi/hr

40. 60 km/hr to mi/hr

41. 60 mi/hr to km/hr

42. 40 mi/hr to km/hr

43. A wire that is 13 mm wide is how many inches wide?

44. A bolt that is 7 mm wide is how many inches wide?

45. 40°C to Fahrenheit

46. 60°C to Fahrenheit

47. 85°C to Fahrenheit

48. 105°C to Fahrenheit

49. 12°C to Fahrenheit

50. 21°C to Fahrenheit

51. 68°F to Celsius

52. 131°F to Celsius

53. 168°F to Celsius

54. 112°F to Celsius

55. 86°F to Celsius

56. 98°F to Celsius

Solve. Round answers to the nearest hundredth when necessary.

57. Mr. and Mrs. Weston have traveled 67 miles on a boat cruise from Seattle, Washington, to Victoria Island, Vancouver, B.C., Canada. They have 36 kilometers until their rendezvous point with another boat. How many kilometers in total will they have traveled?

58. Marcia is traveling from Ixtapa to Zihuatenejo in Mexico. She drives from the center of town to the cliff that makes the descent down to the beaches. The odometer on her American car shows that the first part of her short trip has taken 4 miles. The sign carved into a wooden post says Zihuatenejo 14 KILOMETERS. How many kilometers in total will she have traveled when she arrives at the beach?

59. Pierre had a Jeep imported into France. During a trip from Paris to Lyon, he used 38 liters of gas. The tank, which he filled before starting the trip, holds 15 gallons of gas. How many liters of gas were left in the tank when he arrived in Lyon, if he had a full tank to start with?

60. A surgeon is irrigating an abdominal cavity after a cancerous growth is removed. There is a supply of 3 gallons of distilled water in the operating room. The surgeon uses a total of 7 liters of the water during the procedure. How many liters of water are left over after the operation?

61. One of the heaviest human males documented in medical records weighed 635 kg in 1978. What would have been his weight in pounds?

62. The average weight for a 7-year-old girl is 22.2 kilograms. What is the average weight in pounds?

63. In the Australian summer, Ayers Rock in Central Australia is visited by tourists who climb beginning at 4 o'clock in the morning, when the temperature is 19° Celsius, because after 7 o'clock in the morning, the temperature can reach 45°C and people could die of dehydration while climbing. What would be equivalent Fahrenheit temperatures?

64. The holiday turkey in Buenos Aires, Argentina, was roasted at 200° Celsius for 4 hours, (20 minutes per pound). What would have been the temperature in Fahrenheit in Joplin, Missouri?

▣ CALCULATOR EXERCISES

65. Metallic tungsten, or wolfram (W), melts at 6188° Fahrenheit. At what Celsius temperature would metallic tungsten melt?

66. On top of Old Smoky, water boils at 205°F. What Celsius temperature is this?

Round your answer to the nearest thousandth.

67. A pathologist found 0.768 ounce of lead in the liver of a child who died of lead poisoning. How many grams of toxic lead were in the child's liver?

68. A prospector in the Yukon, in Alaska, found a gold nugget that weighed 2.552 ounces while panning in a river. How many grams did the nugget weigh?

CUMULATIVE REVIEW

Do the operations in the correct order.

1. $2^3 \times 6 - 4 + 3$

2. $5 + 2 - 3 + 5 \times 3^2$.

3. $2^2 + 3^2 + 4^3 + 2 \times 7$

4. $5^2 + 4^2 + 3^2 + 3 \times 8$

TRAVELING AND CONVERSIONS

Jessica is spending the summer in Germany with a friend from college. However, in Germany the metric system of measurement is used and Jessica is only familiar with the U.S. system. To plan travel, food purchases, and other things, she will need to be able to convert measures of length, weight, and volume into units of measure that she is familiar with.

PROBLEMS FOR INDIVIDUAL INVESTIGATION

1. Jessica is planning a trip to Paris. It is 880 kilometers from where she is staying in Berlin. How many miles is this? Round your answer to the nearest tenth.

2. Jessica's rental car has a gas tank that holds 16 gallons. Gas is sold in liters in Europe. How many liters of gas will her gas tank hold? Round your answer to the nearest tenth.

3. Jessica wants to buy the following items at a grocery store in Berlin.

Meat	5 pounds
Carrots	2 pounds
Apples	3 pounds
Potatoes	10 pounds

   Most food is priced in kilograms. How many kilograms of each item on the list should Jessica buy to get the quantity she wants? Round your answers to the nearest hundredth.

4. The current temperature is 25°C. What is that temperature in Fahrenheit?

5. One dollar equals approximately 1.75 marks. How many marks is $70 worth?

PROBLEMS FOR GROUP INVESTIGATION AND COOPERATIVE STUDY

1. Jessica's car gets 30 miles per gallon on a trip. What would this be in kilometers per liter? Round your answer to the nearest hundredth.

2. Prices at the grocery store are as follows:

Meat	15 marks per kilogram
Apples	3.5 marks per kilogram
Potatoes	12 marks per 10 kilograms

   What are these prices in dollars per pound?

3. Gasoline costs 1.5 marks per liter. What is the price in dollars per gallon?

INTERNET: Go to http://www.prenhall.com/blair to explore this application.

SECTION 9.3

Square Roots and Square Root Expressions

After studying this section, you will be able to:

1 Understand the meaning of square roots.

2 Find the square root of a number that is a perfect square.

3 Approximate the square root of a number that is not a perfect square.

4 Simplify the square root of an expression.

5 Add and subtract square root expressions.

1 Understanding the Meaning of Square Roots

We begin our discussion of square roots by introducing a few new vocabulary words, phrases, and mathematical symbols. When a whole number or fraction is multiplied by itself (squared), the number obtained is called a **perfect square**. The number 9 is a perfect square because 3 squared equals 9.

$$3^2 = 3 \cdot 3 = 9$$

The numbers 15 and 13 are *not* perfect squares. There is no whole number that when squared yields 15 or 13.

Math Pro   Video 9.3   SSM

EXAMPLE 1   Determine if each of the numbers is a perfect square.

(a) 30   **(b)** 64   **(c)** $\dfrac{1}{4}$

SOLUTION

(a) 30 is not a perfect square. There is no whole number that when squared equals 30.

(b) 64 is a perfect square because $8^2 = 8 \cdot 8 = 64$.

(c) $\dfrac{1}{4}$ **is a perfect square** because $\left(\dfrac{1}{2}\right)^2 = \dfrac{1}{2} \cdot \dfrac{1}{2} = \dfrac{1}{4}$.

PRACTICE PROBLEM 1

Determine if each of the numbers is a perfect square.

(a) 81   **(b)** 17   **(c)** $\dfrac{1}{9}$

It is helpful to know the first 15 perfect squares for whole numbers.

$1^2 = 1$	$6^2 = 36$	$11^2 = 121$
$2^2 = 4$	$7^2 = 49$	$12^2 = 144$
$3^2 = 9$	$8^2 = 64$	$13^2 = 169$
$4^2 = 16$	$9^2 = 81$	$14^2 = 196$
$5^2 = 25$	$10^2 = 100$	$15^2 = 225$

Suppose we know the value of the perfect square, say 9, and we want to know what positive number we must multiply by itself (square) to get 9. We write this using mathematical symbols as follows:

What positive number squared equals nine?
$$\downarrow \qquad \downarrow \quad \downarrow$$
$$n^2 \quad = \quad 9$$
$$n \cdot n \quad = \quad 9$$
$$n = 3$$

Note that both $3^2 = 9$ and $(-3)^2 = 9$. We were asked for a positive number; therefore, the answer is 3, not -3. In this section we work with positive numbers only. When we denote a variable, we will also assume that the variable is either positive or zero.

Another way of saying either "What positive number squared equals 9?" or "What positive number multiplied by itself equals 9?" is to ask the question: "What is the positive square root of 9?" We write this as $\sqrt{9}$.

> We use the symbol $\sqrt{}$ to indicate that we want to find the positive square root. The symbol $\sqrt{}$ is called a **radical sign**.

Each of the following can be written using the radical sign.

$$n \cdot n = 36 \rightarrow n = \sqrt{36} \qquad n^2 = 36 \rightarrow n = \sqrt{36}$$
$$\text{What is the positive square root of 36?} \rightarrow n = \sqrt{36}$$

EXAMPLE 2   Write using the radical sign.

(a) $n^2 = 49$

(b) What is the positive square root of 49?

(c) $n \cdot n = 49$

SOLUTION

(a) $n^2 = 49 \qquad n = \sqrt{49}$

(b) What is the positive square root of 49? $\qquad n = \sqrt{49}$

(c) $n \cdot n = 49 \qquad n = \sqrt{49}$

PRACTICE PROBLEM 2
Write using the radical sign.

(a) $n^2 = 81$

(b) What is the positive square root of 81?

(c) $n \cdot n = 81$

Finding the Square Root
② ### of a Number That Is a Perfect Square

Before we continue our discussion on square roots, let us take a closer look at the relationship between squaring and finding the square root. Consider the following:

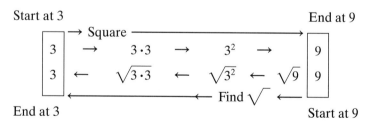

As we can see, squaring 3 gives us 9—finding the square root of 9 gives us 3. Thus, *finding the square root of a number is the inverse of squaring a number.* We find the square root by asking ourselves the question, "What positive number times itself equals the number under the $\sqrt{}$ symbol?"

What positive number times itself equals 9? $\qquad$ 3

$$\sqrt{9} = n \qquad\qquad \sqrt{9} = 3$$

It should be noted that the square roots we are discussing are *principal square roots*, that is, the *positive square roots*. There exists another square root that is negative. For example, (-3) is the *negative square root* of 9 since $(-3)(-3) = 9$, and 3 is the *positive square root* of 9 since $(3)(3) = 9$. In this book our discussion of square roots is limited to positive square roots.

> The square root of a positive number written $\sqrt{a}$ is the number we square to get a.
>   In symbols:
>
>   If $\sqrt{a} = b$, then $b^2 = a$.

EXAMPLE 3   Find the square root.

(a) $\sqrt{36}$ 　　　　　　　　　　　　　　**(b)** $\sqrt{x^2}$, $x > 0$.

SOLUTION

(a) $\sqrt{36} = n$ 　　　What number times itself equals 36?

　　　$\downarrow$

　　$\sqrt{36} = \mathbf{6}$ 　　　6, since $6 \cdot 6 = 36$.

(b) $\sqrt{x^2} = ?$ 　　　What variable times itself equals x^2?

　　　$\downarrow$

　　$\sqrt{x^2} = \mathbf{x}$ 　　　x, since $x \cdot x = x^2$.

PRACTICE PROBLEM 3

Find the square root.

(a) $\sqrt{25}$ 　　　　　　　　　　　　　　**(b)** $\sqrt{y^2}$, $y > 0$

For the remainder of the text we will assume that all variables under the radical sign are greater than zero.

EXAMPLE 4   Find the square root: $\sqrt{\dfrac{36}{49}}$.

SOLUTION

$\sqrt{\dfrac{36}{49}} = n$ 　　　What fraction multiplied times itself equals $\dfrac{36}{49}$?

　　　$\downarrow$

$\sqrt{\dfrac{36}{49}} = \dfrac{\mathbf{6}}{\mathbf{7}}$ 　　　$\dfrac{6}{7}$, since $\left(\dfrac{6}{7}\right)\left(\dfrac{6}{7}\right) = \dfrac{36}{49}$.

PRACTICE PROBLEM 4

Find the square root.

(a) $\sqrt{\dfrac{81}{100}}$ 　　　　　　　　　　　**(b)** $\sqrt{\dfrac{16}{49}}$

EXAMPLE 5   Find the length of the side of a square that has an area of 64 square feet.

SOLUTION   We draw the figure and write the formula for area.

$$A = s^2 \text{ or } s \cdot s$$
$$64 = s^2 \qquad \text{What positive number squared equals 64?}$$
$$\sqrt{64} = s \qquad \text{Write using the radical sign.}$$
$$8 = s \qquad \text{The positive square root of 64 is 8.}$$

The length of the side of the square is **8 feet**.

PRACTICE PROBLEM 5

Find the length of the side of a square that has an area of 49 square feet.

③ Approximating the Square Root of a Number That Is Not a Perfect Square

If a number is not a perfect square, we approximate the square root. This can be done using any calculator that has a square root key. Usually, the key looks like $[\sqrt{}]$ or $[\sqrt{x}]$. To find the square root of 7 on a calculator, enter [7] and press $[\sqrt{}]$ or $[\sqrt{x}]$. You will see displayed 2.645751. (Some calculators require that you enter the $[\sqrt{}]$ key first, then enter the 7.) Your calculator may display fewer or more digits. Remember, no matter how many digits your calculator displays, when we find $\sqrt{7}$—or the square root of any number that is not a perfect square—we have only an approximation.

$$\sqrt{7} \approx 2.645751 \qquad (2.645751)(2.645751) \approx 6.9999984,$$
which is close to, but not equal to, 7.

EXAMPLE 6 Find the approximate value. Round your answer to the nearest thousandth.

(a) $8\sqrt{27}$ 　　　　　　　　**(b)** $6\sqrt{121}$

SOLUTION

(a) The expression $8\sqrt{27}$ means 8 times $\sqrt{27}$.

$$8\sqrt{27} = 8 \cdot \sqrt{27}$$
$$= 8(5.1961524) \qquad \text{Use the calculator to find } \sqrt{27}.$$
$$= 41.569219 \qquad \text{Multiply.}$$
$$\approx \mathbf{41.569} \qquad \text{Round.}$$

Note: We do not round until we complete all calculations.

(b) $6\sqrt{121} = 6 \cdot \sqrt{121}$
$$= 6 \cdot 11 \qquad \text{Use the calculator: } \sqrt{121} = 11.$$
$$= \mathbf{66} \qquad \text{Multiply.}$$

Note: We can use the calculator to find the square root of perfect squares such as 121, although it is advised that you know the first 15 perfect squares.

PRACTICE PROBLEM 6

Find the approximate value. Round your answer to the nearest thousandth.

(a) $4\sqrt{100}$ 　　　　　　　　**(b)** $7\sqrt{23}$

④ Simplifying the Square Root of an Expression

When we do not have a perfect square under the radical we are often asked to *simplify* the radical expressions. Before we begin our discussion on simplifying, we introduce a property that we will use to simplify radical expressions.

> **MULTIPLICATION PROPERTY FOR SQUARE ROOTS**
> For all positive numbers a and b,
> $$\sqrt{a \cdot b} = \sqrt{a} \cdot \sqrt{b} \quad \text{and} \quad \sqrt{a} \cdot \sqrt{b} = \sqrt{ab}$$
> In other words, we can factor the number under the radical sign, then take the square root of each factor separately.

To illustrate this property, notice that the expressions $\sqrt{4 \cdot 9}$ and $\sqrt{4} \cdot \sqrt{9}$ have the same value.

$$\sqrt{4 \cdot 9}$$
$$\downarrow$$
$= \sqrt{36}$ First, we multiply $4 \cdot 9$.

$= 6$ Then we take the square root.

$$\sqrt{4} \cdot \sqrt{9}$$
$$\downarrow \quad \downarrow$$
$2 \quad \cdot \quad 3$ First, we take the square roots of 4 and 9.

6 Then we multiply.

Thus we can factor $\sqrt{36}$ as $\sqrt{4 \cdot 9}$, write it as two separate square roots, $\sqrt{4} \cdot \sqrt{9}$, then find the square root of each factor separately.

$$\sqrt{36} = \sqrt{4 \cdot 9} = \sqrt{4} \cdot \sqrt{9} = 2 \cdot 3 = 6$$

EXAMPLE 7   Find the square root. Do not use a calculator.

(a) $\sqrt{400}$ 　　　　　　　　**(b)** $\sqrt{25x^2}$

SOLUTION

(a) $\sqrt{400} = \sqrt{100 \cdot 4}$ 　　Factor 400 as a product of two perfect squares.

$= \sqrt{100} \cdot \sqrt{4}$ 　　Write as two separate square roots.

$= 10 \cdot 2$ 　　Take the square roots.

$= \mathbf{20}$ 　　Multiply.

(b) $\sqrt{25x^2} = \sqrt{25 \cdot x^2}$ 　　Factor $25x^2$ as a product of two perfect squares.

$= \sqrt{25} \cdot \sqrt{x^2}$ 　　Write as two separate square roots.

$= 5 \cdot x$ 　　Take the square roots.

$= \mathbf{5x}$ 　　Multiply.

PRACTICE PROBLEM 7

Find the square root. Do not use a calculator.

(a) $\sqrt{900}$ 　　　　　　　　**(b)** $\sqrt{49y^2}$

What if we cannot factor a radical expression so that all factors are perfect squares? We *simplify* the expression as much as possible. That is, we factor the radical expression so that as many factors as possible are perfect squares, then find their square roots. We end up with a number left under the radical, but this number will be as small as possible. We simplify a radical expression to find an *exact value*, not an *approximate value*.

EXAMPLE 8 Simplify: $\sqrt{44}$.

SOLUTION

$$\sqrt{44} = \sqrt{4 \cdot 11} \qquad \text{Factor so that as many factors as possible are perfect squares.}$$
$$= \sqrt{4} \cdot \sqrt{11} \qquad \text{Write as two separate square roots.}$$
$$= \mathbf{2} \cdot \sqrt{\mathbf{11}} \qquad \text{Take the square root of all perfect squares.}$$

Note: We cannot simplify $\sqrt{11}$ any further since it is not a perfect square and the number 1 is the only perfect square factor.

PRACTICE PROBLEM 8

Simplify.

(a) $\sqrt{75}$ **(b)** $\sqrt{28}$

Since a perfect square can be written as the product of a whole number times itself, the *factors occur twice*. We can find the square root by removing the factor that occurs twice from under the square root symbol.

$$\sqrt{49} = \sqrt{7 \cdot 7} \quad = \quad 7$$
$$\qquad\qquad \downarrow \qquad\qquad \downarrow$$

The factor 7   The answer is the factor that occurs
occurs twice.   twice under the square root symbol.

We summarize this observation as follows.

REPEATING FACTOR PROPERTY FOR SQUARE ROOTS
For all positive numbers a,

$$\sqrt{a \cdot a} = a$$

In other words, when a factor occurs *twice* under a square root symbol, we can take that factor out from under the square root.

EXAMPLE 9 Simplify: $\sqrt{72y^3}$.

SOLUTION $\sqrt{72y^3}$: We recognize that 9 is a perfect square factor, so we factor: $72 = 9 \cdot 8$.

$$\sqrt{72y^3} = \sqrt{9} \cdot \sqrt{8} \cdot \sqrt{y^3} \qquad\qquad \text{Write as 3 separate square roots.}$$
$$= 3 \cdot \sqrt{8} \cdot \sqrt{y^3} \qquad\qquad \text{Simplify: } \sqrt{9} = 3.$$
$$= 3 \cdot \sqrt{2 \cdot 2 \cdot 2} \cdot \sqrt{y \cdot y \cdot y} \qquad \text{Factor the remaining square root}$$
$$\qquad\qquad\qquad\qquad\qquad\qquad\qquad \text{expressions.}$$
$$= 3 \cdot \sqrt{2 \cdot 2} \cdot \sqrt{2} \cdot \sqrt{y \cdot y} \cdot \sqrt{y}$$
$$\qquad\qquad \downarrow \qquad\qquad\qquad \downarrow$$
$$= 3 \cdot \quad 2 \quad \cdot \sqrt{2} \cdot \quad y \quad \cdot \sqrt{y} \qquad \text{Remove factors that occur twice.}$$
$$= \mathbf{6y}\sqrt{\mathbf{2y}} \qquad\qquad\qquad\qquad \text{Multiply expressions under the}$$
$$\qquad\qquad\qquad\qquad\qquad\qquad\qquad \text{radical separately from those not}$$
$$\qquad\qquad\qquad\qquad\qquad\qquad\qquad \text{under the radical.}$$

PRACTICE PROBLEM 9

Simplify.

(a) $\sqrt{216y^4}$

(b) $\sqrt{160x^3}$

⑤ **Adding and Subtracting Square Root Expressions**

We are familiar with combining like terms by applying the distributive property: $2x + 3x = (2 + 3)x = 5x$. We add and subtract radical expressions in the same way that we combine like terms. **We can add or subtract radical expressions using the distributive property if each term contains the same square root.**

EXAMPLE 10   Simplify each square root, and then combine if possible.

(a) $4\sqrt{3} + 3\sqrt{3}$ **(b)** $3\sqrt{8} + 5\sqrt{18}$ **(c)** $\sqrt{50} - \sqrt{12}$

SOLUTION

(a) Both terms contain the same square root, $\sqrt{3}$, so we can add.

$$4\sqrt{3} + 3\sqrt{3} = (4 + 3)\sqrt{3} = \mathbf{7\sqrt{3}}$$

(b) We simplify each term first, then we add only if each term contains the same square root.

$$3\sqrt{8} + 5\sqrt{18} = 3\sqrt{\mathbf{4 \cdot 2}} + 5\sqrt{\mathbf{9 \cdot 2}} \qquad \text{Factor both 8 and 18.}$$
$$= 3 \cdot \mathbf{2}\sqrt{2} + 5 \cdot \mathbf{3}\sqrt{2} \qquad \text{Simplify } \sqrt{4} = 2 \text{ and } \sqrt{9} = 3.$$
$$\qquad\quad \downarrow \qquad\qquad \downarrow$$
$$= 6\sqrt{2} + \quad 15\sqrt{2} \qquad \text{We can add since each term}$$
$$\text{contains } \sqrt{2}.$$
$$= \mathbf{21\sqrt{2}} \qquad\qquad \text{Add: } (6 + 15)\sqrt{2} = 21\sqrt{2}.$$

Note: We do not add the numbers under the radical: $\sqrt{8} + \sqrt{18} \neq \sqrt{26}$. Verify this using your calculator.

(c) $\sqrt{50} - \sqrt{12} = \sqrt{\mathbf{25 \cdot 2}} - \sqrt{\mathbf{4 \cdot 3}} \qquad$ Factor 50 and 12.
$\quad \sqrt{50} - \sqrt{12} = \mathbf{5\sqrt{2} - 2\sqrt{3}} \qquad\quad \sqrt{25} = 5 \text{ and } \sqrt{4} = 2.$

We cannot subtract, $5\sqrt{2} - 2\sqrt{3}$, since each term does not contain the same square root.

PRACTICE PROBLEM 10

Simplify each square root and then combine if possible.

(a) $7\sqrt{5} - 3\sqrt{5}$ **(b)** $3\sqrt{12} + 2\sqrt{27}$ **(c)** $\sqrt{75} - \sqrt{32}$

EXERCISES 9.3

Determine if each number is a perfect square.

1. 44

2. 25

3. 36

4. 15

5. 81

6. 92

7. $\dfrac{1}{5}$

8. $\dfrac{1}{4}$

Write using the radical sign $(\sqrt{})$.

9. (a) $n^2 = 64$

(b) What is the positive square root of 64?

(c) $n \cdot n = 64$

10. (a) $n^2 = 49$

(b) What is the positive square root of 49?

(c) $n \cdot n = 49$

Find the square root. Assume that all variables represent positive numbers.

11. $\sqrt{49}$

12. $\sqrt{25}$

13. $\sqrt{y^2}$

14. $\sqrt{z^2}$

15. $\sqrt{64}$

16. $\sqrt{121}$

17. $\sqrt{n^2}$

18. $\sqrt{x^2}$

19. $\sqrt{144}$

20. $\sqrt{169}$

21. $\sqrt{b^2}$

22. $\sqrt{c^2}$

23. $\sqrt{225}$

24. $\sqrt{196}$

25. $\sqrt{\dfrac{25}{49}}$

26. $\sqrt{\dfrac{16}{36}}$

27. $\sqrt{\dfrac{9}{81}}$

28. $\sqrt{\dfrac{4}{64}}$

29. $\sqrt{\dfrac{36}{49}}$

30. $\sqrt{\dfrac{25}{81}}$

31. Find the length of the side of a square that has an area of 121 square feet.

32. Find the length of the side of a square that has an area of 16 square feet.

33. Find the length of the side of a square that has an area of 144 square inches.

34. Find the length of the side of a square that has an area of 169 square inches.

Use your calculator to approximate the value. If necessary, round your answer to the nearest thousandth.

35. $\sqrt{31}$

36. $\sqrt{56}$

37. $\sqrt{69}$

38. $\sqrt{26}$

39. $\sqrt{80}$

40. $\sqrt{39}$

41. $\sqrt{90}$

42. $\sqrt{14}$

43. $4\sqrt{12}$

44. $2\sqrt{18}$

45. $5\sqrt{31}$

46. $4\sqrt{26}$

47. $6\sqrt{17}$

48. $3\sqrt{61}$

Find the following square roots. Do not use a calculator. You may assume that all variables represent positive numbers.

49. $\sqrt{64z^2}$ **50.** $\sqrt{81a^2}$ **51.** $\sqrt{36x^2}$ **52.** $\sqrt{9x^2}$

53. $\sqrt{49x^2}$ **54.** $\sqrt{64y^2}$

Simplify each expression by taking as much out from under the radical as possible. Do not use a calculator. You may assume that all variables represent positive numbers.

55. $\sqrt{45}$ **56.** $\sqrt{90}$ **57.** $\sqrt{98}$ **58.** $\sqrt{162}$

59. $\sqrt{72}$ **60.** $\sqrt{52}$ **61.** $\sqrt{60x^4}$ **62.** $\sqrt{32a^6}$

63. $\sqrt{120y^5}$ **64.** $\sqrt{168x^7}$ **65.** $\sqrt{125z^7}$ **66.** $\sqrt{128y^5}$

Simplify the square roots, then combine if possible.

67. $3\sqrt{7} + 2\sqrt{7}$ **68.** $5\sqrt{3} + 4\sqrt{3}$ **69.** $2\sqrt{5} + 4\sqrt{5}$

70. $7\sqrt{2} + 3\sqrt{2}$ **71.** $4\sqrt{8} - 2\sqrt{18}$ **72.** $5\sqrt{12} - 2\sqrt{27}$

73. $4\sqrt{5} - 2\sqrt{20}$ **74.** $2\sqrt{27} - 4\sqrt{3}$ **75.** $6\sqrt{50} + 4\sqrt{12}$

76. $5\sqrt{8} + 3\sqrt{27}$

 VERBAL AND WRITING SKILLS

77. Write as a question: $\sqrt{49}$.

78. (a) Find $\sqrt{16}$, then find $\sqrt{9}$. Now add the results: $\sqrt{16} + \sqrt{9} =$ ____.

   (b) Find $16 + 9$, then find the square root of this sum: $\sqrt{16 + 9} =$ ____.

   (c) Based on these results, what can you say about the following?

   $\sqrt{16} + \sqrt{9}$ (= or ≠) $\sqrt{16 + 9}$

CUMULATIVE REVIEW

Find the perimeter of:

1. A triangle with sides 3 feet, 4 feet, and 5 feet.

2. A rectangle with length 4 meters and width 7 meters.

Find the area of:

3. A rectangle with length 8 centimeters and width 4 centimeters.

4. A parallelogram with base 9 centimeters and height 4 centimeters.

❶ Understand angles in a quadrilateral or a triangle.

❷ Find the perimeter and the area of a triangle.

❸ Use the Pythagorean theorem.

❹ Solve applied problems involving right triangles.

Math Pro   Video 9.4   SSM

S E C T I O N   9 . 4

Triangles and the Pythagorean Theorem

❶ Understanding Angles in a Quadrilateral or a Triangle

A **line** ⟷ extends indefinitely, but a portion of a line, called a **line segment**, has a beginning and an end. An **angle** is formed whenever two line segments meet. The two line segments are called the **sides** of the angle. The point at which they meet is called the **vertex** of the angle.

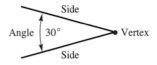

The *amount of opening* of an angle can be measured. Angles are commonly measured in degrees. In this sketch the angle measures 30 degrees, or 30°. The symbol ° indicates degrees. If you fix one side of an angle and keep moving the other side, the angle measure will get larger and larger until eventually you have gone around in one complete revolution.

One complete revolution is 360°.

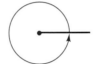

One-half of a revolution is 180°.

One-fourth of a revolution is 90°.

We call two lines **perpendicular** when they meet at an angle of 90°. A 90° angle is called a **right angle**. A 90° angle is often indicated by a small ☐ at the vertex. Thus when you see ⌐ you know that the angle is 90° and also that the sides are perpendicular to each other.

We often label each vertex of a rectangle and a square with the symbol ☐ to indicate that each angle measures 90°. When you see these symbols, you know that the quadrilateral is a rectangle since adjacent sides of a rectangle form right angles.

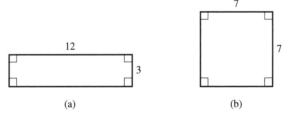

(a)   (b)

Rectangle (b) is also a square. Recall that a square is a rectangle all of whose sides are equal.

A triangle is a three-sided figure with three angles. The prefix *tri* means *three*. Some triangles are shown below.

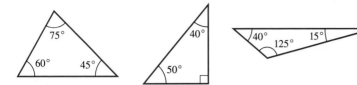

We begin our study of triangles by looking at the angles. Although the size of the angles in triangles may be different, the sum of the angle measures of any triangle is always 180°.

> The sum of the measures of the angles in a triangle is 180°.

We can use this fact to find the measure of an unknown angle in a triangle if we know the measures of the other two angles.

E X A M P L E   1   In the triangle to the right, angle *A* measures 35° and angle *B* measures 95°. Find the measure of angle *C*.

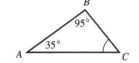

SOLUTION   We will use the fact that the sum of the measures of the angles of a triangle is 180°.

$$35 + 95 + x = 180$$
$$130 + x = 180$$
$$x = 50$$

The measure of angle *C* must equal **50°**.

P R A C T I C E   P R O B L E M   1
In a triangle, angle *B* measures 125° and angle *C* measures 15°. What is the measure of angle *A*?

② Finding the Perimeter and the Area of a Triangle

Recall that the perimeter of any figure is the sum of the lengths of its sides. Thus the perimeter of a triangle is the sum of the lengths of its three sides.

Some triangles have special names. A triangle with two equal sides is called an **isosceles triangle**.

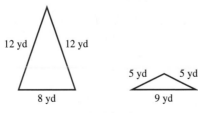

Isosceles Triangles

A triangle with three equal sides is called an **equilateral triangle**. All angles in an equilateral triangle are exactly 60°.

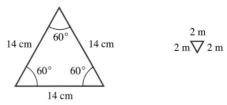

EXAMPLE 2 Find the perimeter of an equilateral triangle if one of its sides is 12 inches.

SOLUTION We know that all the sides of an equilateral triangle are equal. Thus every side measures 12 inches. To find the perimeter, we can add 12 inches three times or we can multiply 3 times 12 inches.

$$P = 12 \text{ in.} + 12 \text{ in.} + 12 \text{ in.} \quad \text{or} \quad P = (3)(12 \text{ in.})$$
$$= 36 \text{ in.} \qquad\qquad\qquad\qquad = 36 \text{ in.}$$

The perimeter of the triangle is **36 in.**

PRACTICE PROBLEM 2

The perimeter of an equilateral triangle is 30 centimeters. Find the length of each side.

A triangle with one 90° angle is called a **right triangle**. The **height** of any triangle is the length of a line segment drawn from a vertex perpendicular to the other side or an extension of the other side. The height of any figure is perpendicular to the **base**. When you see a small □ in a drawing, you know that the line is the height because the line forms a 90° angle. The height may be one of the sides in a right triangle. The height may reach to an extension of one side if one angle of the triangle is greater than 90°.

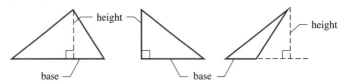

To find the **area** of a triangle, we need to be able to identify the height and the base of the triangle. The area of any triangle is half of the product of the base times the height of the triangle. The height is measured from the vertex above the base to that base.

The **area** of a triangle is the base times the height divided by 2.

$$A = \frac{bh}{2}$$

$h = $ height
$b = $ base

EXAMPLE 3 Find the area of a triangle whose base is 18 millimeters and whose height is 15 millimeters.

SOLUTION

$$A = \frac{bh}{2} = \frac{(18 \text{ mm})(15 \text{ mm})}{2} = \frac{270 \text{ mm}^2}{2} = \mathbf{135 \text{ mm}^2}$$

PRACTICE PROBLEM 3

Find the area of a triangle whose base is 5 centimeters and whose height is 16 centimeters.

In some geometric shapes a triangle is combined with rectangles, squares, and parallelograms.

EXAMPLE 4   Find the area of the side of the house shown in the margin.

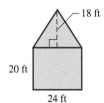

18 ft

20 ft

24 ft

SOLUTION   Because opposite sides of a rectangle are equal, the triangle has a base of 24 feet. Thus

$$A = \frac{bh}{2} = \frac{(24 \text{ ft})(18 \text{ ft})}{2} = \frac{432 \text{ ft}^2}{2} = 216 \text{ ft}^2$$

The area of the rectangle is $A = LW = (24 \text{ ft})(20 \text{ ft}) = 480 \text{ ft}^2$.

The sum of the two areas is $216 \text{ ft}^2 + 480 \text{ ft}^2 = \textbf{696 ft}^2$

18 ft

24 ft

PRACTICE PROBLEM 4

Find the area.

20 ft

24 ft

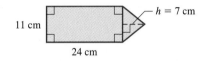

11 cm

24 cm

h = 7 cm

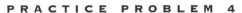

TO THINK ABOUT

Where does the 2 come from in the formula $A = \dfrac{bh}{2}$? Why does this formula for the area of a triangle work? Suppose that we construct a triangle with base b and height h.

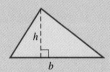

h

b

Now let us make an exact copy of the triangle and turn the copy around to the right exactly 180°. Carefully place the two triangles together. We now have a parallelogram with base b and height h. Recall that the area of a parallelogram is $A = bh$.

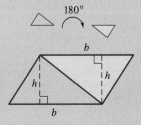

180°

b

h

h

b

Because the parallelogram has area $A = bh$ and is made up of two triangles of identical shape and area, the area of one of the triangles would be the area of the parallelogram divided by 2. Thus the area of a triangle is $A = \dfrac{bh}{2}$.

③ Using the Pythagorean Theorem

The Pythagorean theorem is a mathematical idea formulated long ago. It is as useful today as when it was discovered. The Pythagoreans, followers of the Greek philosopher and mathematician Pythagoras, lived in Italy about 2500 years ago. They studied various mathematical properties. They discovered that for any right triangle, *the square of the longest side is equal to the sum of the squares of the other two sides.* This relationship is known as the **Pythagorean Theorem**.

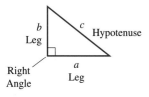

The square of the longest side is equal to the sum of the squares of the other two sides.

$$c^2 \quad = \quad a^2 + b^2$$

The shorter sides, a and b, are referred to as the **legs**. The longest side, c, which is opposite the right angle, is called the **hypotenuse** of the right triangle.

> **PYTHAGOREAN THEOREM**
> In any right triangle, if c is the length of the hypotenuse and a and b are the lengths of the two legs, then
>
> $$c^2 = a^2 + b^2$$
>
> In other words,
>
> $$(\text{hypotenuse})^2 = (\text{leg})^2 + (\text{other leg})^2$$

EXAMPLE 5   Find the length of the hypotenuse of a right triangle with legs of 9 meters and 12 meters.

SOLUTION   We draw the figure and write the formula.

$$(\text{hypotenuse})^2 = (\text{leg})^2 + (\text{other leg})^2$$

$$
\begin{aligned}
c^2 &= a^2 + b^2 \\
&= 9^2 + 12^2 \qquad \text{We evaluate the formula.} \\
&= 81 + 144 \\
c^2 &= 225 \qquad\qquad \text{Therefore} \quad c = \sqrt{225} = 15
\end{aligned}
$$

Thus the length of the hypotenuse of the triangle is **15 meters**.

Recall that the expressions $c^2 = 225$ and $c \cdot c = 225$ can both be written as $c = \sqrt{225}$. If we use a calculator to find square root, we should write the expression using the $\sqrt{}$ symbol. Why?

PRACTICE PROBLEM 5

Find the length of the hypotenuse of a right triangle with legs of 4 meters and 3 meters.

4 Solving Applied Problems Involving Right Triangles

It is important to remember that the hypotenuse, c, is the side across from the right angle.

EXAMPLE 6   A 25-foot ladder is placed against the side of a building. The top of the ladder is 22 feet from the ground. What is the distance from the base of the ladder to the building? Round to the nearest tenth if necessary.

SOLUTION   We draw a picture of the situation.

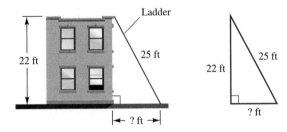

We want to find the length of one of the legs of a right triangle. We write the formula and evaluate.

$$c^2 = a^2 + b^2$$
$$\downarrow\downarrow$$
$$25^2 = a^2 + 22^2$$
$$625 = a^2 + 484 \qquad \text{Now we solve for } a.$$
$$625 - 484 = a^2$$
$$141 = a^2$$
$$\sqrt{141} = a$$
$$11.874 \approx a$$

The ladder is **approximately 11.9 feet** from the base of the building.

PRACTICE PROBLEM 6

A ladder is placed against the side of a building. The top of the ladder is 15 feet from the ground. The base of the ladder is 12 feet from the building. What is the length of the ladder?

EXERCISES 9.4

Find the missing angle in each triangle.

1. Two angles are 30° and 90°.

2. Two angles are 90° and 45°.

3. Two angles are 130° and 20°.

4. Two angles are 16° and 18°.

5. Two angles are both 45°.

6. Two angles are both 60°.

Find the perimeter of each triangle.

7. A triangle whose sides are 36 m, 27 m, and 41 m.

8. A triangle whose sides are 71 m, 65 m, and 82 m.

9. An isosceles triangle whose sides are 50 in., 40 in., and 40 in.

10. An isosceles triangle whose sides are 36 in., 36 in., and 29 in.

11. An equilateral triangle whose side measures 3.5 mi.

12. An equilateral triangle whose side measures 4.6 mi.

Find the area of each triangle.

13.
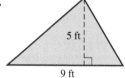
5 ft
9 ft

14.

6 ft
8 ft

15.
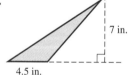
7 in.
4.5 in.

16.
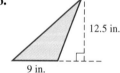
12.5 in.
9 in.

17. The base is 17.5 cm and the height is 9.5 cm.

18. The base is 3.6 cm and the height is 11.2 cm.

19. The base is 6.7 m and the height is 4.2 m.

20. The base is 8.5 m and the height is 3.6 m.

21. Find the area.

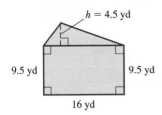

22. Find the area of the shaded region.

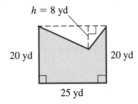

Find the unknown side of each right triangle. Use a calculator or square root table when necessary and round your answer to the nearest thousandth.

23.

24.

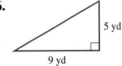

25.

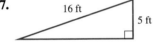

26.

27.

28.

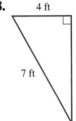

Find the unknown side of each right triangle using the information given. Round your answer to the nearest thousandth.

29. leg = 8 km, hypotenuse = 13 km

30. leg = 5 km, hypotenuse = 11 km

31. leg = 11 m, leg = 3 m

32. leg = 5 m, leg = 4 m

33. leg = 5 m, leg = 5 m

34. leg = 6 m, leg = 6 m

35. hypotenuse = 5 ft, leg = 4 ft

36. hypotenuse = 10 ft, leg = 6 ft

37. leg = 9 in., leg = 12 in.

38. leg = 12 in., leg = 16 in.

39. hypotenuse = 13 yd, leg = 11 yd

40. hypotenuse = 14 yd, leg = 10 yd

Solve each applied problem. Round your answer to the nearest tenth.

41. Find the length of the guy wire supporting the telephone pole.

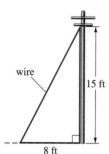

wire

15 ft

8 ft

42. Find the length of this ramp to a loading dock.

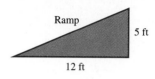

Ramp

5 ft

12 ft

43. Juan runs out of gas in Los Lunas, New Mexico. He walks 4 miles west and then 3 miles south looking for a gas station. How far is he from his starting point?

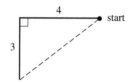

4

start

3

44. A construction project requires a stainless steel plate with holes drilled as shown. Find the distance between the centers of the holes in this triangular plate.

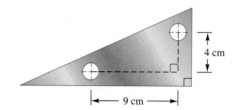

4 cm

9 cm

45. A 20-ft ladder is placed against a college classroom building at a point 18 ft above the ground. What is the distance from the base of the ladder to the building?

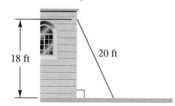

18 ft

20 ft

46. Barbara is flying her dragon kite on 32 yards of string. The kite is directly above the edge of a pond. The edge of the pond is 30 yards from where the kite is tied to the ground. How far is the kite above the pond?

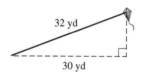

32 yd

30 yd

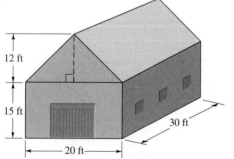

CALCULATOR EXERCISES

Find the total area of all the vertical sides (front, and back, and sides) of each building.

47.

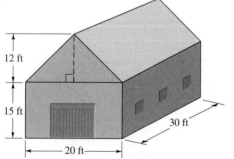

12 ft

15 ft

20 ft

30 ft

48.

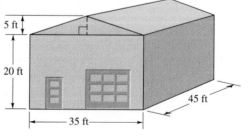

5 ft

20 ft

35 ft

45 ft

Use your calculator to find the exact value of the unknown side of each right triangle.

49. The two legs of the triangle are 48 yd and 20 yd.

50. The hypotenuse of the triangle is 65 yd and one leg is 33 yd.

 VERBAL AND WRITING SKILLS

51. A 90° angle is called a _____ angle.

52. The sum of the angle measures of a triangle is _____ .

53. Explain in your own words how you would find the measure of an unknown angle in a triangle if you knew the measures of the other two angles.

54. If you were told that a triangle was an isosceles triangle, what could you conclude about the sides of that triangle?

55. If you were told that a triangle was an equilateral triangle, what could you conclude about the sides of that triangle?

56. How do you find the area of a triangle?

ONE STEP FURTHER

The top surface of the wings of a test plane must be coated with a special lacquer that costs $90 per square yard. Find the cost to coat the shaded wing surface of each plane.

57.

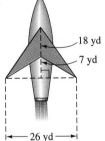

18 yd
7 yd
26 yd

58.

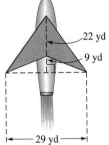

22 yd
9 yd
29 yd

CUMULATIVE REVIEW

Solve for n. Round to the nearest hundredth.

1. $\dfrac{5}{n} = \dfrac{7.5}{18}$

2. $\dfrac{n}{29} = \dfrac{7}{3}$

3. On the cruise ship H.M.S. *Salinora*, all restaurants on board must keep the ratio of waitstaff to patrons at 4 to 15. How many waitstaff should be serving if there are 2685 patrons dining at one time?

4. Recently, an airline found that after the transatlantic flight to Frankfurt, 68 people out of 300 passengers kept their in-flight magazines after being encouraged to take the magazines with them to read at their leisure. On a similar flight carrying 425 people, how many in-flight magazines would the airline expect to be taken? Round to the nearest whole number.

❶ Find the circumference of a circle.

❷ Find the area of a circle.

❸ Solve area problems containing circles and other geometric shapes.

Math Pro   Video 9.5   SSM

SECTION 9.5

The Circle and Applied Problems

Every point on the rim of a circle is the same distance from the center of the circle, so the circle looks the same all around. In this section we learn how to calculate the distance around a circle as well as the area of a circle.

❶ Finding the Circumference of a Circle

A **circle** is a figure for which all points are at an equal distance from a given point. This given point is called the **center** of the circle.

The **radius**, r, is a line segment from the center point on the circle.

The **diameter**, d, is a line segment across the circle that passes through the center.

Clearly, we can see the following relationships between the diameter and radius.

The diameter is two times the radius   or $d = 2r$.

The radius is one-half the diameter   or $r = \dfrac{d}{2}$.

> The distance around the rim of a circle is called the **circumference**, C.

The measure of the circumference is similar to the perimeter of a polygon in that they both *measure the distance around the outer edge of the figure.*

Place a string around the edge of a circle.

Then lay the string out and measure it.

The length of the string is equal to the circumference.

When we divide the circumference by the diameter, we get a special number called **pi**, denoted by the Greek lowercase letter π: $\dfrac{C}{d} = \pi$. π is approximately 3.14159265359. We can approximate π to any number of digits. For all work in this book we use the following:

> π is approximately 3.14 rounded to the nearest hundredth.

If we multiply both sides of $\dfrac{C}{d} = \pi$ by d, we get a formula for the circumference of a circle.

The circumference of a circle is equal to π times the diameter.

$$C = \pi d$$

Since $d = 2r$, an alternative formula for circumference is $C = \pi \cdot 2r$, or

$$C = 2\pi r$$

When we approximate pi with 3.14 in the following examples, the answers are approximate values.

EXAMPLE 1   Find the circumference of a circle if the diameter is 9.2 meters. Use $\pi \approx 3.14$.

SOLUTION   Since we are given the diameter, we use the formula for circumference that includes the diameter.

$C = \pi d$ Write the formula.

$= (3.14)(9.2\text{ m})$ Then substitute the values given.

$C \approx \mathbf{28.888\ m}$

9.2 m

PRACTICE PROBLEM 1

Find the circumference of a circle if the diameter is 5.1 meters. Use $\pi \approx 3.14$.

When solving problems involving circles, be careful. Ask yourself, "Is the radius given or the diameter?" Then choose the appropriate formula for your calculations.

When a circular object rolls through 1 complete revolution, it rolls along the rim of a circle. Thus 1 complete revolution equals the distance around the rim of the circular object, and this distance is called the *circumference*.

Start   End

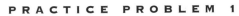

|← 1 revolution →|

EXAMPLE 2   The larger of 2 bicycles has a 26-inch wheel diameter and the smaller bicycle has a 24-inch wheel diameter.

(a) If the wheels on the larger bicycle complete 12 revolutions, what distance does the larger bicycle travel?
(b) How many complete revolutions does each wheel on the smaller bicycle complete to travel the same distance as the larger bicycle?

SOLUTION

1. *Understand the problem for part (a).* We draw a picture of the situation:

12 revolutions of the 26 inch wheel

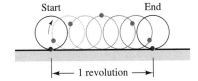

What is the distance traveled?

Smaller bicycle

24 in. diameter

Larger bicycle

26 in. diameter

The distance traveled by the larger bicycle during 1 revolution is equal to the measure of the circumference. We must multiply the circumference by 12 to find 12 revolutions.

2. *Solve and state the answer to part (a).*

12 times the circumference: $c = \pi d$

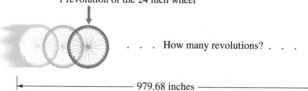

$(12)(\pi d)$

$= (12)(3.14)(26 \text{ in.})$

$= (12)(81.64 \text{ in.}) \approx 979.68 \text{ in.}$

The larger bicycle travels approximately 979.68 inches.

1. *Understand the problem for part (b).* We draw a picture of the situation:

1 revolution of the 24 inch wheel

. . . How many revolutions? . . .

979.68 inches

First, we find the distance traveled for 1 revolution (the circumference), then we divide 979.68 by the circumference to find the total number of revolutions that can be made in this distance.

2. *Solve and state the answer to part (b).*

$C = \pi d$

$C = (3.14)(24 \text{ in.}) \approx 75.36 \text{ in. per revolution}$

We divide:

$979.68 \div 75.36 = 13.$

The smaller bicycle must complete 13 revolutions to travel 979.68 inches.

3. *Check.* You may use your calculator to check.

PRACTICE PROBLEM 2

3.75 ft radius

1.25 ft radius

A pulley has a large wheel that has a 3.75-foot radius and a smaller wheel that has a 1.25-foot radius. If the larger wheel makes 2 complete revolutions, how many revolutions does the small wheel make?

② Finding the Area of a Circle

The area of a circle is similar to the area of a polygon in that they both measure the surface inside the figure.

The area of a circle is the product of π times the radius squared.

$A = \pi r^2$

EXAMPLE 3   Find the area of a circle whose radius is 2.1 meters. Use $\pi \approx 3.14$. Round your answer to the nearest tenth.

SOLUTION

$$A = \pi r^2$$
$$= (3.14)(2.1 \text{ m})^2 \quad \text{We must square the radius first. Then multiply by 3.14.}$$
$$= (3.14)(4.41 \text{ m}^2)$$
$$A \approx 13.8474 \text{ m}^2$$

The area of the circle is approximately 13.8 square meters.

PRACTICE PROBLEM 3

Find the area of a circle whose radius is 6.1 centimeters. Use $\pi \approx 3.14$. Round your answer to the nearest tenth.

③ Solving Area Problems Containing Circles and Other Geometric Shapes

The formula for finding the area uses the length of the radius. If we are given a diameter, we can use the formula, $r = \dfrac{d}{2}$.

EXAMPLE 4   Lester wants to buy a circular braided rug that is 8 feet in diameter. Find the cost of the rug at $35 per square yard.

SOLUTION

1. *Understand the problem.* We are given the diameter in feet. We will need to find the radius. The cost of the rug is in *square yards*. We will need to change square feet to square yards.

2. *Solve and state the answer.* The diameter is 8 feet, so the radius is 4 feet $\left(r = \dfrac{8}{2} = 4 \right)$. Find the area.

$$A = \pi r^2 = (3.14)(4 \text{ ft})^2$$
$$= (3.14)(16 \text{ ft}^2) \quad \text{We square the radius first, then we multiply.}$$
$$A \approx 50.24 \text{ ft}^2 \quad \text{Multiply.}$$

Change square feet to square yards. Since 1 yd = 3 ft, $(1 \text{ yd})^2 = (3 \text{ ft})^2$. That is, **1 yd^2 = 9 ft^2**.

$$50.24 \text{ ft}^2 \times \frac{1 \text{ yd}^2}{9 \text{ ft}^2} = 5.58 \text{ yd}^2 \quad \text{(rounded to the nearest hundredth)}$$

Find the cost.

$$\frac{\$35}{1 \text{ yd}^2} \times 5.58 \text{ yd}^2 = \mathbf{\$195.30}$$

3. *Check.* You may use your calculator to check.

PRACTICE PROBLEM 4

Mary Beth wants to buy a circular crocheted tablecloth that is 60 inches in diameter. Find the cost of the tablecloth at $8 per square foot.

EXERCISES 9.5

In problems 1–20 use π ≈ 3.14. Round each answer to the nearest tenth.

Find the length of the radius of a circle if the diameter has the value given.

1. $d = 45$ yd   **2.** $d = 65$ yd   **3.** $d = 3.8$ cm   **4.** $d = 5.2$ cm

Find the circumference of each circle.

5. Diameter = 32 cm   **6.** Diameter = 22 cm

7. Radius = 11 in.   **8.** Radius = 15 in.

A bicycle wheel makes 1 revolution. Determine how far the bicycle travels in inches.

9. The diameter of the wheel is 26 in.   **10.** The diameter of the wheel is 28 in.

A bicycle wheel makes 5 revolutions. Determine how far the bicycle travels in feet.

11. The diameter of the wheel is 32 in.   **12.** The diameter of the wheel is 24 in.

Find the area of each circle.

13. Radius = 5 yd   **14.** Radius = 7 yd

15. Diameter = 32 cm   **16.** Diameter = 44 cm

A water sprinkler sends water out in a circular pattern. Determine how large an area is watered.

17. The radius of the watered area is 8 ft.   **18.** The radius of the watered area is 12 ft.

A radio station sends out radio waves in all directions from a tower at the center of the circle of broadcast range. Determine how large an area is reached.

19. The diameter is 120 mi.   **20.** The diameter is 90 mi.

Use π ≈ 3.14 in problems 21–32. Round answers to the nearest hundredth.

A semicircle is one-half of a circle and has an area that is one-half of the area of a circle. Find the cost of fertilizing a playing field at $0.20 per square yard for the dimensions stated.

21. The rectangular part of the field is 120 yd long and the diameter of each semicircle is 40 yd.

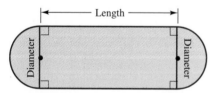

22. The rectangular part of the field is 110 yd long and the diameter of each semicircle is 50 yd.

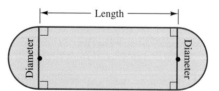

23. A porthole window on a freighter ship has a diameter of 2 ft. What is the length of the insulating strip that encircles the window and keeps out wind and moisture?

24. A manhole cover has a diameter of 3 ft. What is the length of the brass grip-strip that encircles the cover, making it easier to manage?

25. Elena's car has tires with a radius of 14 in. How many feet does her car travel if the wheel makes 35 revolutions?

26. Jimmy's truck has tires with a radius of 30 in. How many feet does his truck travel if the wheel makes 9 revolutions?

27. Carlotta's car has tires with a radius of 16 in. Carlotta moved her car forward 20,096 in. How many complete revolutions did the wheels turn?

28. Mickey's car has tires with a radius of 15 in. He backed his car up a distance of 9891 in. How many complete revolutions did the wheels turn backing up?

29. Find the area of a circular Coca-Cola sign with a diameter of 64 in.

30. Find the area of a circular flower bed with a diameter of 14 ft.

31. Sarah bought a circular marble tabletop to use as her dining room table. The marble is 6 ft in diameter. Find the cost of the tabletop at $72 per square yard of marble.

32. Tom made a base for a circular patio by pouring concrete into a circular space 10 ft in diameter. Find the cost at $18 per square yard.

 CALCULATOR EXERCISES

Use $\pi \approx 3.14159$ in all calculations for problems 33 and 34. Round to the nearest hundred-thousandths.

33. Find the circumference of the circle with a 0.223-m diameter.

34. Find the area of the circle whose radius is 1.39 cm.

 VERBAL AND WRITING SKILLS

Fill in the blank.

35. The distance around a circle is called the _____ .

36. The radius is a line segment from the _____ to a point on the circle.

37. The diameter is two times the _____ of the circle.

38. Explain in your own words how to find the area of a circle if you are given the diameter.

ONE STEP FURTHER

39. A 15-in.-diameter pizza costs $6. A 12-in.-diameter pizza costs $4. The 12-in.-diameter pizza is cut into six slices. The 15-in.-diameter pizza is cut into eight slices.

    (a) What is the cost per slice of the 15-in.-diameter pizza? How many square inches of pizza are in one slice?

    (b) What is the cost per slice of the 12-in.-diameter pizza? How many square inches of pizza are in one slice?

    (c) If you want more value for your money, which slice of pizza should you buy?

40. A 14-in.-diameter pizza costs $5.50. It is cut into eight pieces. A 12.5 in. by 12.5 in. square pizza costs $6. It is cut into nine pieces.

(a) What is the cost of one piece of the 14-in.-diameter pizza? How many square inches of pizza are in one piece?

(b) What is the cost of one piece of the 12.5 in. by 12.5 in. square pizza? How many square inches of pizza are in one piece?

(c) If you want more value for your money, which piece of pizza should you buy?

CUMULATIVE REVIEW

1. Find 16% of 87.

2. What is 0.5% of 60?

Find the volume of each rectangular box.

3. $L = 11$ in., $W = 5$ in., $H = 6$ in.

4. $L = 8$ in., $W = 4$ in., $H = 5$ in.

SECTION 9.6

Volume

❶ Find the volume of a cylinder and a sphere.

❷ Find the volume of a cone.

❸ Find the volume of a pyramid.

Math Pro   Video 9.6   SSM

❶ Finding the Volume of a Cylinder and a Sphere

In Chapter 2 we learned that we find the **volume** to measure the space enclosed in a rectangular three-dimensional figure such as a box. The volume of a cylinder or a sphere is similar to a rectangular solid in that they both measure the *amount of space enclosed in the three-dimensional figure.* Recall that volume is measured in cubic units such as cubic meters (abbreviated m³) or cubic feet (abbreviated ft³).

Cylinders are the shapes we observe when we see a can or a tube.

Spheres are the shapes we observe when we see a ball.

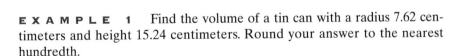

The *volume of a cylinder* is the product of the area of its circular base, πr^2, and the height, h.

$$V = \pi r^2 h$$

The *volume of a sphere* is the product of 4 times π times the radius cubed divided by 3.

$$V = \frac{4\pi r^3}{3}$$

EXAMPLE 1 Find the volume of a tin can with a radius 7.62 centimeters and height 15.24 centimeters. Round your answer to the nearest hundredth.

SOLUTION

$$V = \pi r^2 h = (3.14)(7.62 \text{ cm})^2 (15.24 \text{ cm}) \approx 2778.59 \text{ cm}^3$$

The volume of the tin can is approximately 2778.59 cubic centimeters.

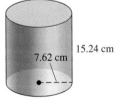

15.24 cm
7.62 cm

PRACTICE PROBLEM 1

Find the volume of a tin can with a radius 10.24 centimeters and height 17.78 centimeters. Round your answer to the nearest hundredth.

EXAMPLE 2 How much air is needed to fully inflate a soccer ball if the radius of the inner lining is 5 inches? Round your answer to the nearest hundredth.

SOLUTION

$$V = \frac{4\pi r^3}{3} = \frac{4(3.14)(5 \text{ in.})^3}{3} \approx 523.33 \text{ in}^3$$

Approximately 523.33 cubic inches of air is needed to fully inflate the ball.

PRACTICE PROBLEM 2

How much air is needed to fully inflate a beach ball if the radius of the inner lining is 7 inches? Round your answer to the nearest hundredth.

EXAMPLE 3   An insulated thermos has a large radius R to the outer rim that is 15 centimeters. The small radius r is 13 centimeters and measures to the edge of the insulation. The height of the thermos is 27 centimeters, and the top and bottom of the thermos do not have a layer of insulation.

(a) What volume of coffee can the thermos hold?
(b) What is the volume of the insulation?

SOLUTION

   1. *Understand the problem for part (a).* We find the volume of the inner shaded region (V_r) to determine how much coffee the thermos will hold.

   2. *Solve and state the answer to part (a).*

$$r = 13 \text{ cm} \qquad V_r = \pi r^2 h$$
$$h = 27 \text{ cm} \qquad\quad = (3.14)(13 \text{ cm})^2(27 \text{ cm})$$
$$V_r \approx 14{,}327.82 \text{ cm}^3$$

The thermos can hold 14,327.82 cubic centimeters of coffee.

   1. *Understand the problem for part (b).* To determine the volume of the insulated region (blue shaded region), we find the volume of the entire cylinder (V_R) minus the volume of the inner region (V_r).

   2. *Solve and state the answer to part (b).*

Volume of insulated region $= V_R - V_r$

$$R = 15 \text{ cm} \qquad V_R = \pi r^2 h$$
$$h = 27 \text{ cm} \qquad\quad = (3.14)(15 \text{ cm})^2(27 \text{ cm})$$
$$V_R \approx \mathbf{19{,}075.5 \text{ cm}^3}$$

$$\begin{array}{ccc} V_R & - & V_r \\ \downarrow & & \downarrow \end{array}$$
$$19{,}075.5 - 14{,}327.82 = 4747.68 \text{ cm}^3$$

The volume of the insulation is 4747.68 cubic centimeters.

PRACTICE PROBLEM 3

If $R = 12$ centimeters and $r = 10$ centimeters for the thermos in Example 3, answer the following.

(a) What volume of coffee can the thermos hold?
(b) What is the volume of the insulation?

 Finding the Volume of a Cone

We see the shape of a cone when we look at the sharpened end of a pencil or at an ice cream cone. To find the volume of a cone we use the following formula.

The **volume of a cone** is π times the radius squared times the height divided by 3.

$$V = \frac{\pi r^2 h}{3}$$

E X A M P L E   4   Find the volume of a cone of radius 7 meters and height 9 meters.

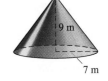

SOLUTION

$$V = \frac{\pi r^2 h}{3}$$

$$= \frac{(3.14)(7\text{ m})^2(9\text{ m})}{3}$$

$$= \frac{(3.14)(7\text{ m})(7\text{ m})(9\text{ m})}{3}$$

$$= (3.14)(49)(3)\text{m}^3 = (153.86)(3)\text{m}^3 = 461.58\text{ m}^3$$

$V \approx \mathbf{461.6\ m^3}$ rounded to the nearest tenth

PRACTICE PROBLEM 4

Find the volume of a cone of radius 5 meters and height 12 meters. Round to the nearest tenth.

③ Finding the Volume of a Pyramid

You have seen pictures of the great pyramids of Egypt. These amazing stone structures are over 4000 years old.

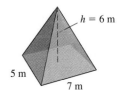

The **volume of a pyramid** is obtained by multiplying the area B of the base of the pyramid by the height h and dividing by 3.

$$V = \frac{Bh}{3}$$

E X A M P L E   5   Find the volume of a pyramid with height = 6 meters, length of base = 7 meters, width of base = 5 meters.

SOLUTION   The base is a rectangle.

   Area of base $= (7\text{ m})(5\text{ m}) = 35\text{ m}^2$

Substituting the area of the base, 35 m², and the height of 6 m, we have

$$V = \frac{Bh}{3} = \frac{(35\text{ m}^2)(6\text{ m})}{3}$$

$$= (35)(2)\text{ m}^3$$

$$V = \mathbf{70\ m^3}$$

PRACTICE PROBLEM 5

Find the volume of a pyramid having the dimensions given.

(a) Height 10 meters, base width 6 meters, base length 6 meters.
(b) Height 15 meters, base width 7 meters, base length 8 meters.

EXERCISES 9.6

Find each volume. Use π ≈ 3.14. Round each answer to the nearest tenth when necessary.

1. A cylinder with radius 2 m and height 7 m

2. A cylinder with radius 3 m and height 8 m

3. A sphere with radius 4 m

4. A sphere with radius 5 m

Problems 5 and 6 involve hemispheres. A hemisphere is exactly one-half of a sphere.

5. Find the volume of a hemisphere with radius = 7 m.

6. Find the volume of a hemisphere with radius = 6 m.

Find each volume. Use π ≈ 3.14. Round each answer to the nearest tenth.

7. A cone with a height of 12 cm and a radius of 9 cm

8. A cone with a height of 15 cm and a radius of 8 cm

9. A cone with a height of 10 ft and a radius of 5 ft

10. A cone with a height of 12 ft and a radius of 6 ft

11. A pyramid with a height of 7 m and a square base of 3 m on a side

12. A pyramid with a height of 10 m and a square base of 7 m on a side

13. A pyramid with a height of 5 m and a rectangular base measuring 6 m by 12 m

14. A pyramid with a height of 10 m and a rectangular base measuring 8 m by 14 m

A collar of Styrofoam is made to insulate a pipe. Find the volume of the blue shaded region (which represents the collar). The large radius R is to the outer rim. The small radius r is to the edge of the insulation.

15. r = 3 in.
R = 5 in.
h = 20 in.

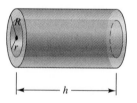

16. r = 4 in.
R = 6 in.
h = 25 in.

17. Find the volume of Sandy's snow cone before it melts! The top of the cone is one-half of a sphere with a diameter of 2 in. The base is the shape of a cone with a height of 6 in. If the snow cone cost $0.09 per cubic inch, how much will Sandy's snow cone cost? Round to the nearest hundredth.

Find the volume of each of the following shapes, which consist of a rectangular box with a cylinder attached to the top.

18. A dentist places a gold filling in a tooth in the shape of a cylinder with a hemispherical top. The radius *r* of the filling is 1 mm. The height is 2 mm. If dental gold costs $95 per cubic millimeter, how much did the gold cost for the filling?

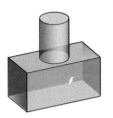

19. The box has $L = 4$ ft, $W = 3$ ft, and $H = 2$ ft. The cylinder has a diameter of 2 ft and height of 2 ft.

20. The box has $L = 10$ in., $W = 5$ in., and $H = 2$ in. The cylinder has a diameter of 4 in. and a height of 4 in.

21. The nose cone of a passenger jet is used to receive and send radar. It is made of a special aluminum alloy that costs $4 per cm³. The cone has a radius of 5 cm and a height of 9 cm. What is the cost of the aluminum needed to make this *solid* nose cone?

22. A special stainless-steel cone sits on top of a cable television antenna. The cost of the stainless steel is $3 per cm³. The cone has a radius of 6 cm and a height of 10 cm. What is the cost of the stainless steel needed to make this *solid* steel cone?

 CALCULATOR EXERCISES

23. Jupiter has a radius of approximately 45,000 mi. Assuming that it is a sphere, what is its volume?

24. A tennis ball has a diameter of 2.5 in. A baseball has a diameter of 2.9 in. What is the difference in *volume* between the baseball and the tennis ball? (Round your answer to the nearest tenth.)

For questions 25–28 use $\pi \approx 3.14$ and round your answer to nearest tenth.

25. Find the volume of a sphere with radius 5.21 m.

26. Find the volume of a cone of radius 21.12 mm and height 32 mm.

27. Find the volume of a pyramid with height = 9.212 ft, length of rectangular base = 6.22 ft, and width of rectangular base = 5.01 ft.

28. Find the volume of a hemisphere with radius = 8.583 ft.

Match the formula for volume with the appropriate figure.

29. $V = LWH$

30. $V = \pi r^2 h$

31. $V = \dfrac{4\pi r^3}{3}$

32. $V = \dfrac{\pi r^2 h}{3}$

33. $V = \dfrac{Bh}{3}$

34. $V = s^3$

ONE STEP FURTHER

Use the formulas for surface area for questions 35 and 36. Use $\pi \approx 3.14$.

   Surface area of a cylinder: $SA = 2\pi r^2 + 2\pi rh$
   Surface area of a sphere: $SA = 4\pi r^2$

35. How much material is needed to make a beach ball that has a diameter of 14 in.?

36. What is the surface area of a cylinder-shaped tank that has a diameter of 10 m and and height of 20 m?

CUMULATIVE REVIEW

Solve each proportion.

1. $\dfrac{21}{40} = \dfrac{x}{120}$

2. $\dfrac{18}{x} = \dfrac{12}{10}$

3. Multiply: $2\dfrac{1}{4} \times 3\dfrac{3}{4}$.

4. Divide: $7\dfrac{1}{2} \div 4\dfrac{1}{5}$.

S E C T I O N   9 . 7

Similar Geometric Figures

After studying this section, you will be able to:

❶ **Finding the Corresponding Parts of Similar Triangles**

In English, *similar* means that the things being compared are, in general, alike. But in mathematics, *similar* means that the things being compared are alike in a special way—they are *alike in shape*, even though they may be different in size. So photographs that are enlarged produce images *similar* to the original; a floor plan of a building is *similar* to the actual building; a model car is *similar* to the actual vehicle.

Two triangles with the same shape but not necessarily the same size are called **similar triangles**. Here are two pairs of similar triangles:

❶ Find the corresponding parts of similar triangles.

❷ Find the corresponding parts of similar geometric figures.

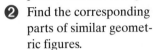

Math Pro   Video 9.7   SSM

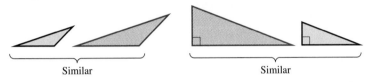

Similar   Similar

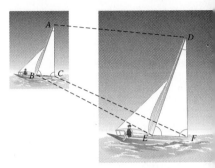

The two triangles below are similar. The smallest angle in the first triangle is angle *A*. The smallest angle in the second triangle is angle *D*. Both angles measure 36°. We say that angle *A* and angle *D* are corresponding angles in these similar triangles.

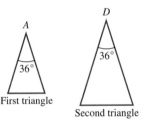

First triangle

Second triangle

> The **corresponding angles** of similar triangles are equal.

The two triangles below are similar. Notice the corresponding sides.

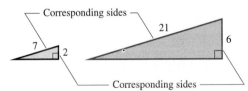

We see that the ratio of 7 to 21 is the same as the ratio of 2 to 6.

$$\frac{7}{21} = \frac{2}{6} \quad \text{is obviously true since} \quad \frac{1}{3} = \frac{1}{3}$$

> The lengths of **corresponding sides** of similar triangles have the same ratio.

We can use the fact that corresponding sides of similar triangles have the same ratio to find the missing lengths of sides of triangles.

E X A M P L E   1   The two triangles below are similar. Find the length of side n. Round to the nearest tenth.

SOLUTION   The ratio of 12 to 19 is the same as the ratio of 5 to n.

$$\frac{12}{19} = \frac{5}{n}$$

$12n = (19)(5)$ Cross-multiply.

$12n = 95$ Simplify.

$$\frac{12n}{12} = \frac{95}{12}$$ Divide each side by 12.

$n = 7.91\overline{6}$

$ = 7.9$ Round to the nearest tenth.

The length of side n is approximately 7.9.

P R A C T I C E   P R O B L E M   1

The two triangles in the margin are similar. Find the length of side n. Round to the nearest tenth.

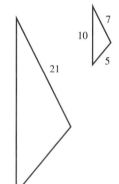

Similar triangles are not always oriented the same way. You may find it helpful to rotate one of the triangles so that the similarity is more apparent.

> The **perimeters** of similar triangles have the same ratios as the corresponding sides.

E X A M P L E   2   Two triangles are similar. The smaller triangle has sides 5 yards, 7 yards, and 10 yards. The 7-yard side on the smaller triangle corresponds to a side of 21 yards on the larger triangle. What is the perimeter of the larger triangle?

SOLUTION

1. *Understand the problem. What are we asked to find?* We are asked to find the perimeter of the larger triangle. But we are not given the lengths of all the sides.

What are we given? We are given the lengths of all the sides of a smaller, similar triangle. We are also told that the 7-yard side in the smaller triangle corresponds to the 21-yard side of the larger triangle.

What do we know? We know that the perimeters of similar triangles have the same ratios as the corresponding sides.

We start by drawing a picture and putting all the pieces together to solve the problem.

2. *Solve and state the answer.* We begin by finding the perimeter of the smaller triangle.

$$5\,\text{yd} + 7\,\text{yd} + 10\,\text{yd} = 22\,\text{yd}$$

We can now write equal ratios. Let $p =$ the unknown perimeter. We will set up our ratios as: $\dfrac{\text{smaller triangle}}{\text{larger triangle}}$

Remember, once you write the first ratio, be sure to write the terms of the second ratio in the same order. We will write

perimeter of the smaller triangle → $\dfrac{22}{p} = \dfrac{7}{21}$ ← side of the smaller triangle
perimeter of the larger triangle → $\phantom{\dfrac{22}{p}}$ ← side of the larger triangle

$$7p = (21)(22)$$
$$7p = 462$$
$$\frac{7p}{7} = \frac{462}{7}$$
$$p = 66$$

The perimeter of the larger triangle is 66 yards.

3. *Check.* Are the ratios the same? Simplify each ratio to check. That is,

$$\frac{22}{66} \overset{?}{=} \frac{7}{21} \qquad \frac{2 \cdot \cancel{11}}{2 \cdot 3 \cdot \cancel{11}} \overset{?}{=} \frac{1 \cdot \cancel{7}}{3 \cdot \cancel{7}} \qquad \text{Simplify each ratio.}$$

$$\frac{1}{3} = \frac{1}{3} \quad \checkmark$$

PRACTICE PROBLEM 2

Two triangles are similar. The smaller triangle has sides of 5 yards, 12 yards, and 13 yards. The 12-yard side of the smaller triangle corresponds to a side of 40 yards on the larger triangle. What is the perimeter of the larger triangle?

Similar triangles can be used to find distances or lengths that are difficult to measure.

EXAMPLE 3   A flagpole casts a shadow of 36 feet. At the same time a tree that is 3 feet tall has a shadow of 5 feet. How tall is the flagpole?

SOLUTION

1. *Understand the problem.* The shadows cast by the sun shining on vertical objects at the same time of day form similar triangles. We draw a picture.

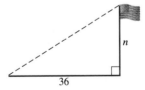

2. *Solve and state the answer.* Let $n =$ the height of the flagpole. Thus we can say that n is to 3 as 36 is to 5.

$$\frac{n}{3} = \frac{36}{5}$$
$$5n = (3)(36)$$
$$5n = 108$$
$$\frac{5n}{5} = \frac{108}{5}$$
$$n = 21.6 \qquad \textbf{The flagpole is 21.6 feet tall.}$$

3. *Check.* The check is up to you.

PRACTICE PROBLEM 3

How tall (*h*) is the building in the margin if the two triangles are similar?

② Finding the Corresponding Parts of Similar Geometric Figures

Geometric figures such as rectangles can also be similar figures.

The *corresponding sides of similar geometric figures* have the same ratio.

EXAMPLE 4   The two rectangles shown here are similar because the corresponding sides of the two rectangles have the same ratio. Find the width of the larger rectangle.

SOLUTION   Let *W* = the width of the larger rectangle.

$$\frac{W}{1.6} = \frac{9}{2}$$

$$2W = (1.6)(9)$$

$$2W = 14.4$$

$$\frac{2W}{2} = \frac{14.4}{2}$$

$$W = 7.2$$

The width of the larger rectangle is **7.2 meters**.

PRACTICE PROBLEM 4

The two rectangles in the margin are similar. Find the width of the larger rectangle.

EXERCISES 9.7

For each pair of similar triangles, find the missing side n. Round your answer to the nearest tenth when necessary.

1.
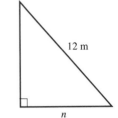
3 m
2 m
12 m
n

2.
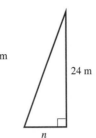
8 m
3 m
24 m
n

3.
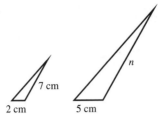
7 cm
2 cm
5 cm
n

4.
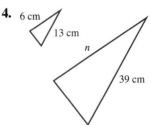
6 cm
13 cm
n
39 cm

5.
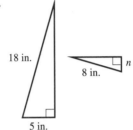
18 in.
5 in.
8 in.
n

6.
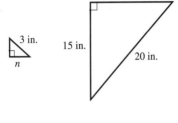
3 in.
n
15 in.
20 in.

Solve each applied problem.

7. A sculptor is designing her new triangular masterpiece. Her scale drawing shows that the shortest side of a triangular piece to be made measures 8 cm. The longest side of the drawing measures 25 cm. The longest side of the actual part to be sculpted must be 10.5 m long. How long will the shortest side of the actual part be? Round to the nearest tenth.

8. The zoo has hired a landscape architect to design the triangular lobby of the children's petting zoo. His scale drawing shows that the longest side of the lobby is 9 cm. The shortest side of the lobby is 5 cm. The longest side of the actual lobby will be 30 m. How long will the shortest side of the actual lobby be? Round to the nearest tenth.

9. Janice took a great photo of the entire family at this year's reunion. She takes it to a professional photography studio and asks that the 3 in. by 5 in. photo be blown up to a poster size, which is 3.5 ft tall. What is the smaller dimension (width) of the poster?

10. Jeff and Shelley are planning a new kitchen. The old kitchen measured 9 ft by 12 ft. The new kitchen is similar in shape, but the longest dimension is 19 ft. What is the smaller dimension (width) of the new kitchen?

A flagpole casts a shadow. At the same time a small tree casts a shadow. Use the sketch to find the height n of each flagpole.

11.
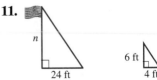
n
24 ft
6 ft
4 ft

12.
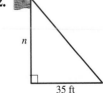
n
35 ft
8 ft
7 ft

13. Lola is standing outside of the shopping mall. She is 5.5 ft tall and her shadow measures 6.5 ft long. The outside of the department store casts a shadow of 96 ft. How tall is the store? Round to the nearest foot.

14. Thomas is rock climbing in Utah. He is 6 ft tall and his shadow measures 8 ft long. The rock he wants to climb casts a shadow of 610 ft. How tall is the rock he is about to climb?

Each pair of figures is similar. Find the missing side. Round to the nearest tenth when necessary.

15.

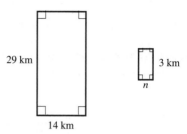

16.

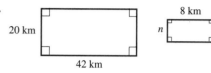

17.

18.

Refer to Example 2 to complete the following exercises.

19. Perimeter = 100 m

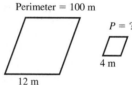

20. Perimeter = 160 m

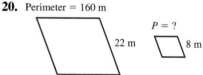

ONE STEP FURTHER

How would you find the relationship between the areas of two similar geometric figures? Consider the following two similar rectangles.

The area of the smaller rectangle is $(3 \text{ m})(7 \text{ m}) = 21 \text{ m}^2$. The area of the larger rectangle is $(9 \text{ m})(21 \text{ m}) = 189 \text{ m}^2$. How could you have predicted this result?

The ratio of small width to large width is $\frac{3}{9} = \frac{1}{3}$. The small rectangle has sides that are $\frac{1}{3}$ as large as the large rectangle. The ratio of the area of the small rectangle to the area of the large rectangle is $\frac{21}{189} = \frac{1}{9}$.

Note that $\left(\frac{1}{3}\right)^2 = \frac{1}{9}$.

Thus we can develop the following principle: *The areas of two similar figures are in the same ratio as the square of the ratio of two corresponding sides.*

Each pair of geometric figures is similar. Find the unknown area. Round to the nearest tenth.

21.

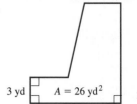

3 yd $A = 26 \text{ yd}^2$ 2 yd $A = ?$

22.

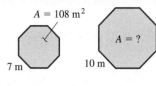

$A = 108 \text{ m}^2$ $A = ?$

7 m   10 m

CUMULATIVE REVIEW

Perform the calculation. Use the correct order of operations.

1. $2 \times 3^2 + 4 - 2 \times 5$

2. $300 \div (85 - 65) + 2^4$

3. $(5)(9) - (21 + 3) \div 8$

4. $108 \div 6 + 51 \div 3 + 3^3$

CHAPTER ORGANIZER

TOPIC	PROCEDURE	EXAMPLES
Changing from one U.S. unit to another	1. Find the equality statement that relates what you want to find and what you know. 2. Form a unit fraction. The numerator will contain the units you want to end up with. 3. Multiply by the unit fraction and simplify.	Convert 210 inches to feet. 1. Use 12 inches = 1 foot. 2. Unit fraction $= \dfrac{1 \text{ foot}}{12 \text{ inches}}$ 3. $210 \text{ in.} \times \dfrac{1 \text{ ft}}{12 \text{ in.}} = \dfrac{210}{12} \text{ ft}$ $= 17.5 \text{ ft}$
Changing from U.S. units to metric units	1. From the list of equivalent measures, pick an equality statement that begins with the unit you now have. $\begin{aligned} 1 \text{ mi} &= 1.61 \text{ km} & 1 \text{ gal} &= 3.79 \text{ L} \\ 1 \text{ yd} &= 0.914 \text{ m} & 1 \text{ qt} &= 0.946 \text{ L} \\ 1 \text{ ft} &= 0.305 \text{ m} & 1 \text{ lb} &= 0.454 \text{ kg} \\ 1 \text{ in.} &= 2.54 \text{ cm} & 1 \text{ oz} &= 28.35 \text{ g} \end{aligned}$ 2. Multiply by a unit fraction.	Convert 7 gallons to liters 1. 1 gal = 3.79 L 2. $7 \text{ gal} \times \dfrac{3.79 \text{ L}}{1 \text{ gal}} = 26.53 \text{ L}$ Convert 18 pounds to kilograms. 1. 1 lb = 0.454 kg 2. $18 \text{ lb} \times \dfrac{0.454 \text{ kg}}{1 \text{ lb}} = 8.172 \text{ kg}$
Changing from metric units to U.S. units	1. From the list of approximate equivalent measures, pick an equality statement that begins with the unit you now have and ends with the unit you want. $\begin{aligned} 1 \text{ km} &= 0.62 \text{ mi} & 1 \text{ L} &= 0.264 \text{ gal} \\ 1 \text{ m} &= 3.28 \text{ ft} & 1 \text{ L} &= 1.06 \text{ qt} \\ 1 \text{ m} &= 1.09 \text{ yd} & 1 \text{ kg} &= 2.2 \text{ lb} \\ 1 \text{ cm} &= 0.394 \text{ in.} & 1 \text{ g} &= 0.0353 \text{ oz} \end{aligned}$ 2. Multiply by a unit fraction.	Convert 605 grams to ounces. 1. 1 g = 0.0353 oz 2. $605 \text{ g} \times \dfrac{0.0353 \text{ oz}}{1 \text{ g}} = 21.3565 \text{ oz}$ Convert 80 km/hr to mi/hr. 1. 1 km = 0.62 mi 2. $80 \dfrac{\text{km}}{\text{hr}} \times \dfrac{0.62 \text{ mi}}{1 \text{ km}} = 49.6 \text{ mi/hr}$
Changing from Celsius to Fahrenheit temperature	1. To convert Celsius to Fahrenheit, we use the formula $$F = 1.8 \times C + 32$$ 2. We replace C by the Celsius temperature. 3. We calculate to find the Fahrenheit temperature.	Convert 65°C to Fahrenheit. 1. $F = 1.8 \times C + 32$ 2. $F = 1.8 \times 65 + 32$ 3. $F = 117 + 32 = 149$ The temperature is 149°F.
Changing from Fahrenheit to Celsius temperature	1. To convert Fahrenheit to Celsius, we use the formula $$C = \dfrac{5 \times F - 160}{9}$$ 2. We replace F by the Fahrenheit temperature. 3. We calculate to find the Celsius temperature.	Convert 50°F to Celsius. 1. $C = \dfrac{5 \times F - 160}{9}$ 2. $C = \dfrac{5 \times 50 - 160}{9}$ 3. $C = \dfrac{250 - 160}{9} = \dfrac{90}{9} = 10$ The temperature is 10°C.

TOPIC	PROCEDURE	EXAMPLES
Evaluating square roots of numbers that are perfect squares	The square root of a number is one of two identical factors of that number.	$\sqrt{1} = 1$ because $(1)(1) = 1$ $\sqrt{4} = 2$ because $(2)(2) = 4$ $\sqrt{100} = 10$ because $(10)(10) = 100$ $\sqrt{169} = 13$ because $(13)(13) = 169$
Approximating the square root of a number that is not a perfect square	If a calculator with a square root key is available, enter the number and then press the $\boxed{\sqrt{x}}$ or $\boxed{\sqrt{}}$ key. Some calculators require that you enter the $\boxed{\sqrt{}}$ key followed by the number. The approximate value will be displayed.	1. Find on a calculator. (a) $\sqrt{13}$ (b) $\sqrt{182}$ Round to the nearest thousandth. (a) 13 $\boxed{\sqrt{x}}$ 3.60555127 rounds to 3.606 (b) 182 $\boxed{\sqrt{x}}$ 13.49073756 rounds to 13.491
Simplifying radicals	$\sqrt{ab} = \sqrt{a}\,\sqrt{b}$ Radicals are simplified by taking the square roots of all perfect squares and leaving other factors under the radical sign. $\sqrt{a \cdot a} = a$ For large numbers or variable expressions, we can factor the expression and remove factors that occur twice. In this text we assume that all variables under a radical sign are positive numbers.	1. $\sqrt{12x^2} = \sqrt{3} \cdot \sqrt{4} \cdot \sqrt{x^2}$ $\qquad \downarrow \quad \downarrow \quad \downarrow$ $\qquad \sqrt{3} \cdot \quad 2 \ \cdot \ x$ $\qquad = 2x\sqrt{3}$ 2. $\sqrt{32y^3}$ $= \sqrt{4} \ \cdot \ \sqrt{8} \ \cdot \ \sqrt{y^3}$ $= 2 \cdot \sqrt{2 \cdot 2} \cdot \sqrt{2} \cdot \sqrt{y \cdot y}\,\sqrt{y}$ $= 2 \cdot \ 2\sqrt{2} \ \cdot \ y \ \sqrt{y}$ $= 4y\sqrt{2y}$
Adding and subtracting radicals	Simplify all radicals and then add or subtract each term that contains the same radical.	1. $\sqrt{12} + \sqrt{27} = \sqrt{4}\,\sqrt{3} + \sqrt{9}\,\sqrt{3}$ $\qquad\qquad\qquad = 2\sqrt{3} + 3\sqrt{3}$ $\qquad\qquad\qquad = 5\sqrt{3}$ 2. $\sqrt{50} - \sqrt{45} = \sqrt{25}\,\sqrt{2} - \sqrt{9}\,\sqrt{5}$ $\qquad\qquad\qquad = 5\sqrt{2} - 3\sqrt{5}$ We cannot subtract; the terms contain different radicals.
Pythagorean theorem	In any right triangle with hypotenuse of length c and legs of lengths a and b, $c^2 = a^2 + b^2$	Find c to the nearest tenth if $a = 7$ and $b = 9$. $c^2 = 7^2 + 9^2 = 49 + 81 = 130$ $c = \sqrt{130} \approx 11.4$
Area of a triangle	$A = \dfrac{bh}{2}$ $b = $ base $h = $ height	Find the area of the triangle whose base is 1.5 m and whose height is 3 m. $A = \dfrac{bh}{2} = \dfrac{(1.5\text{ m})(3\text{ m})}{2} = \dfrac{4.5\text{ m}^2}{2}$ $= 2.25\text{ m}^2$

TOPIC	PROCEDURE	EXAMPLES
Circumference of a circle	$C = \pi d$	Find the circumference of a circle with a diameter of 12 ft. $C = \pi d = (3.14)(12 \text{ ft}) = 37.68 \text{ ft}$ $\approx 37.7 \text{ ft}$ (rounded to nearest tenth)
Area of a circle	$A = \pi r^2$ 1. Square the radius first. 2. Then multiply the result by 3.14.	Find the area of a circle with radius 7 ft. $A = \pi r^2 = (3.14)(7 \text{ ft})^2$ $= (3.14)(49 \text{ ft}^2)$ $= 153.86 \text{ ft}^2$ $\approx 153.9 \text{ ft}^2$ (rounded to nearest tenth)
Volume of a cylinder	$r = \text{radius} \quad h = \text{height} \quad V = \pi r^2 h$ 1. Square the radius first. 2. Then multiply the result by 3.14 and by the height.	Find the volume of a cylinder with a radius of 7 m and a height of 3 m. $V = \pi r^2 h = (3.14)(7 \text{ m})^2(3 \text{ m})$ $= (3.14)(49)(3) \text{ m}^3$ $= (153.86)(3) \text{ m}^3 = 461.58 \text{ m}^3$ $\approx 461.6 \text{ m}^3$ (rounded to nearest tenth)
Volume of a sphere	$V = \dfrac{4\pi r^3}{3}$ $r = \text{radius}$	Find the volume of a sphere of radius 3 m. $V = \dfrac{4\pi r^3}{3} = \dfrac{(4)(3.14)(3 \text{ m})^3}{3}$ $= \dfrac{(4)(3.14)(27) \text{ m}^3}{3}$ $= (4)(3.14)(9) \text{ m}^3$ $= (12.56)(9) \text{ m}^3 = 113.04 \text{ m}^3$ $\approx 113.0 \text{ m}^3$ (rounded to nearest tenth)
Volume of a cone	$V = \dfrac{\pi r^2 h}{3}$ $r = \text{radius}$ $h = \text{height}$	Find the volume of a cone of height 9 m and radius 7 m. $V = \dfrac{\pi r^2 h}{3} = \dfrac{(3.14)(7 \text{ m})^2(9 \text{ m})}{3}$ $= \dfrac{(3.14)(7^2)(9) \text{ m}^3}{3}$ $= (3.14)(49)(3) \text{ m}^3$ $= (153.86)(3) \text{ m}^3 = 461.58 \text{ m}^3$ $\approx 461.6 \text{ m}^3$ (rounded to nearest tenth)

Note, all examples on this page have answers rounded to the nearest tenth.

TOPIC	PROCEDURE	EXAMPLES
Volume of a pyramid	$V = \dfrac{Bh}{3}$ B = area of the base h = height 1. Find the area of the base 2. Multiply this area by the height and divide the result by 3.	Find the volume of a pyramid whose height is 6 m and whose rectangular base is 10 m by 12 m. 1. $B = (12\text{ m})(10\text{ m}) = 120\text{ m}^2$ 2. $V = \dfrac{(120)(\overset{2}{\cancel{6}})\text{ m}^3}{\underset{1}{\cancel{3}}} = (120)(2)\text{ m}^3$ $= 240\text{ m}^3$
Similar figures, corresponding sides	The corresponding sides of similar figures have the same ratio.	Find n in the following similar figures. $\dfrac{n}{4} = \dfrac{9}{3}$ $3n = 36$ $n = 12\text{ m}$

CHAPTER 9 REVIEW

SECTION 9.1

Convert. When necessary, express your answer as a decimal. Round to the nearest hundredth.

1. 27 ft = _____ yd

2. 2160 sec = _____ min

3. 90 in. = _____ ft

4. 15,840 ft = _____ mi

5. 4 tons = _____ lb

6. 15 gal = _____ qt

7. 92 oz = _____ lb

8. 31 pt = _____ qt

Convert. Do not round your answer.

9. 59 mL = _____ L

10. 56 cm = _____ mm

11. 2598 mm = _____ cm

12. 778 mg = _____ g

13. 9.2 m = _____ cm

14. 7 km = _____ m

15. 17 kL = _____ L

16. 473 m = _____ km

17. 196 kg = _____ g

18. 721 kg = _____ g

Solve each problem. Round your answer to the nearest hundredth when necessary.

19. A chemist has a solution of 4 L. She needs to place it into 24 equal-sized smaller jars. How many milliliters will be contained in each jar?

20. Find the perimeter of the triangle.

   (a) Express your answer in feet.

   (b) Express your answer in inches.

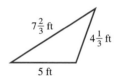

SECTION 9.2

Perform each conversion. Round your answer to the nearest hundredth.

21. 42 kg = _____ lb

22. 20 lb = _____ kg

23. 15 ft = _____ m

24. 1.8 ft = _____ cm

25. 13 oz = _____ g

26. 15°C = _____ °F

27. 14 cm = _____ in.

28. 32°F = _____ °C

29. Keshia traveled at 90 km/hr for 3 hr. She needs to travel 200 mi. How much farther does she need to travel?

30. The unit price on a box of cereal is $0.14 per ounce. The net weight is 450 g. How much does the cereal cost?

SECTION 9.3

Find each of the following square roots without using a calculator.

31. $\sqrt{64}$

32. $\sqrt{y^2}$

Approximate using a calculator or square root table. Round your answer to the nearest thousandth when necessary.

33. $\sqrt{45}$

34. $4\sqrt{5}$

Simplify each expression taking as much out from under the radical as possible. Do not use a calculator. You may assume that all variables represent positive numbers.

35. $\sqrt{200}$ **36.** $\sqrt{300}$ **37.** $\sqrt{45x^3}$ **38.** $\sqrt{32y^5}$

Simplify, then add or subtract if possible.

39. $3\sqrt{8} + 5\sqrt{18}$ **40.** $3\sqrt{27} - 2\sqrt{12}$ **41.** $2\sqrt{20} + \sqrt{54}$

SECTION 9.4

42. The sides of an equilateral triangle are 10 ft. Find the perimeter of the triangle.

43. Two known angles measure 62° and 78°. Find the measure of the third angle in the triangle.

Find the area of the triangle. Round your answer to the nearest tenth if necessary.

44. Base = 9.6 m, height = 5.1 m

Find the unknown side of the right triangle. If the answer cannot be obtained exactly, use a square root table or a calculator with a square root key. Round your answer to the nearest hundredth when necessary.

45.

46.

47. Find the width of a door if it is 6 ft tall and the diagonal line measures 7 ft.

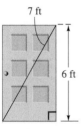

SECTION 9.5

For problems 48–51, use $\pi \approx 3.14$. Round your answer to the nearest tenth if necessary.

48. Find the circumference of a circle with diameter 12 in.

49. Find the circumference of a circle with radius 7 in.

Find the area of each circle.

50. Radius = 6 m

51. Diameter = 16 ft

Find the area of the shaded region. Round your answer to the nearest tenth if necessary.

52.

Find the volume of the object. Use $\pi \approx 3.14$ and round your answer to the nearest tenth if necessary.

53. A sphere with radius 1.2 ft.

54. Find the volume of a concrete connector for the city sewer system. A diagram of the connector is shown. It is shaped like a box with a hole of diameter 2 m. If it is formed using concrete that costs $1.20 per cubic meter, how much will the necessary concrete cost?

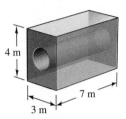

4 m

7 m

3 m

55. Find the volume of the shape of a cylinder with a hemispherical top. The radius r is 2 cm, the height is 9 cm.

h

56. Find the volume of a pyramid that is 18 m high and whose rectangular base measures 16 m by 18 m.

57. A chemical has polluted a volume of ground in a cone shape. The depth of the cone is 30 yd. The radius of the cone is 17 yd. Find the volume of polluted ground.

SECTION 9.7

Find n in the set of similar triangles.

58.

45 m

3 m

2 m

n

Find the perimeter of the larger of the two similar figures.

59.

5 cm   18 cm   5 cm

26 cm

108 cm

CHAPTER 9 TEST

1. Convert 145 ounces to pounds.

2. Convert 162 grams to kilograms.

3. Sharon purchased 2 kilograms of hamburger meat.
   (a) How many grams did she purchase?
   (b) How many pounds did she purchase?

4. A recipe says to roast a turkey at 200° Celsius for 4 hours. What temperature in Fahrenheit should be used?

5. Fred is traveling through Europe. After driving 4 miles, he sees a sign that says 14 kilometers to his destination. How many kilometers in total will he have traveled when he arrives at his destination?

Determine if the following are perfect squares.

6. 81

7. 24

8. $\dfrac{1}{8}$

9. Find the length of the side of a square that has an area of 64 square feet.

Simplify.

10. $\sqrt{y^2}$

11. $\sqrt{49}$

12. $\sqrt{\dfrac{9}{16}}$

Find the approximate value. Round your answer to the nearest hundredth.

13. $\sqrt{6}$

14. $2\sqrt{7}$

Add. Find the exact value; do not approximate.

15. $8\sqrt{3} + \sqrt{3}$

16. In a triangle, angle B measures 50° and angle C measures 70°. What is the measure of angle A?

17. The perimeter of an equilateral triangle is 42 centimeters. Find the length of each side.

18. Find the area of a triangle whose base is 10 centimeters and whose height is 12 centimeters

19. Find the area of the side of this house.

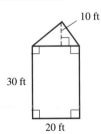

10 ft

30 ft

20 ft

1. _____

2. _____

3. _____

4. _____

5. _____

6. _____

7. _____

8. _____

9. _____

10. _____

11. _____

12. _____

13. _____

14. _____

15. _____

16. _____

17. _____

18. _____

19. _____

20. _____

21. _____

22. _____

23. _____

24. _____

25. _____

26. _____

27. _____

28. _____

29. _____

30. _____

31. _____

32. _____

33. _____

20. Find the length of the hypotenuse of a right triangle with legs 10 inches and 4 inches. Round to the nearest hundredth.

21. Find the length of one leg of a right triangle with one leg of 3 centimeters and the hypotenuse of 7 centimeters. Round to the nearest hundredth.

22. A ladder is placed against the side of a building. The top of the ladder is 14 feet from the ground. The base of the ladder is 11 feet from the building. What is the approximate length of the ladder? Round to the nearest tenth.

For problems 23–29, use $\pi \approx 3.14$.

23. Find the circumference of a circle if the diameter is 5.15 meters.

24. A pulley has a large wheel that has a 2.1-foot radius and a smaller wheel that has a 0.75-foot radius. If the larger wheel makes 2 complete revolutions, how many revolutions does the small wheel make? Round your answer to the nearest tenth.

25. Find the area of a circle whose radius is 1.2 centimeters. Round your answer to the nearest tenth.

26. Joe wants to buy a circular tablecloth that is 6 feet in diameter. Find the cost of the tablecloth at $20 a square yard.

27. Find the volume of a tin can with a radius of 3.4 centimeters and a height of 7.1 centimeters. Round your answer to the nearest hundredth.

28. How much air is needed to fully inflate a beach ball if the radius of the inner lining is 8 inches? Round your answer to the nearest hundredth.

29. Find the volume of a cone with radius 8 centimeters and height 12 centimeters.

30. Find the volume of a pyramid with height 12 meters, length of base 10 meters, and width of base 7 meters.

31. These two triangles are similar. Find the length of side *n*. Round to the nearest tenth.

32. A flagpole casts a shadow of 2.5 feet. At the same time, a tree that is 20 feet tall has a shadow of 2 feet. How tall is the flagpole?

33. The two rectangles shown here are similar. Find the width of the larger rectangle.

1 cm ▭
4.5 cm

W [_____]
36 cm

CUMULATIVE TEST FOR CHAPTERS 1-9

Replace the question mark with the inequality symbol $<$ or $>$.

1. 0 ? -5

2. 0.5 ? 0.15

3. -8 ? -18

4. $\dfrac{5}{8}$? $\dfrac{2}{3}$

5. $\dfrac{7}{8}$? 0.9

6. $|-6|$? $|2|$

7. Jane has a rectangular shaped garden that she would like to fence. The garden has a width of 9 feet and a length of 12 feet. Fencing material costs $2.50 per linear foot.

(a) How much fencing material does Jane need?

(b) How much will it cost to fence Jane's garden?

(c) What is the area of Jane's garden?

Evaluate.

8. $12 - 5^2 + 3$

9. $4(8 - 12)$

10. $-15(-3)$

11. $\dfrac{1}{3}\left(\dfrac{2}{5}\right) + \dfrac{1}{2}$

12. $\dfrac{-3}{4} \div \dfrac{7}{8}$

13. Combine like terms: $2x + 3x^2 - 8 + x - 17$.

Solve for the variable and check.

14. $x - 20 = 18$

15. $x + 3 = -12$

16. $3 - 8 + x = 14 - 9$

17. $\dfrac{x}{6} = 4$

18. $5x - 9(2) = 48$

Simplify.

19. $4(x + 3)$

20. $(x + 2)(x - 1)$

21. $(4x + 3y - 9) + (-3x + 7y + 6)$

22. A television is reduced 17% from the original price of $250. What is the reduced price of the television set?

23. Joe earned $23,450 last year. This year he will earn $26,275. What was the percent of increase in Joe's earnings? Round your answer to the nearest hundredth.

A student received test grades of 76, 85, 73, 75, and 81 on 5 tests.

24. Find the student's average test score. Round to the nearest tenth.

25. Find the student's median test score.

1. _____

2. _____

3. _____

4. _____

5. _____

6. _____

7. _____

8. _____

9. _____

10. _____

11. _____

12. _____

13. _____

14. _____

15. _____

16. _____

17. _____

18. _____

19. _____

20. _____

21. _____

22. _____

23. _____

24. _____

25. _____

Plot each ordered pair on the rectangular coordinate plane.

26. $(4, 2)$

27. $(-2, 1)$

28. $(-4, -5)$

29. $(5, -2)$

30. $(0, 0)$

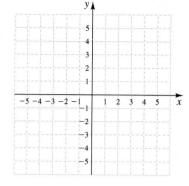

31. (a) Name 3 ordered pairs that are solutions to $y = 2x + 1$.

(b) Graph these ordered pairs on the rectangular coordinate plane and draw a line through the points.

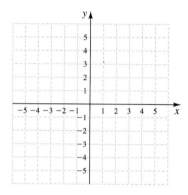

32. Which one of the following metric units would be the most reasonable measurement for the height of an adult male: grams, kilometers, centimeters, or liters?

33. Is 125 a perfect square?

Simplify.

34. $\sqrt{144}$

35. $\sqrt{x^2}$

36. $\sqrt{\dfrac{81}{100}}$

37. Find the approximate value and round to the nearest tenth: $5\sqrt{10}$.

38. Find the exact answer. Do not approximate. Add: $10\sqrt{7} + 3\sqrt{7}$.

39. The perimeter of an equilateral triangle is 51 feet. Find the length of each side.

For problems 40 and 41, use $\pi \approx 3.14$.

40. The diameter of a globe is 14 inches. Find the circumference.

41. Find the volume of a tin can with a radius of 2.7 centimeters and a height of 6.3 centimeters. Round your answer to the nearest hundredth.

Left margin answer blanks:

31.

32.

33.

34.

35.

36.

37.

38.

39.

40.

41.

SQUARE ROOT TABLE

Square root values are rounded to the nearest thousandth unless the answer ends in .000.

n	$\sqrt{n}$	n	$\sqrt{n}$	n	$\sqrt{n}$	n	$\sqrt{n}$	n	$\sqrt{n}$
1	1.000	41	6.403	81	9.000	121	11.000	161	12.689
2	1.414	42	6.481	82	9.055	122	11.045	162	12.728
3	1.732	43	6.557	83	9.110	123	11.091	163	12.767
4	2.000	44	6.633	84	9.165	124	11.136	164	12.806
5	2.236	45	6.708	85	9.220	125	11.180	165	12.845
6	2.449	46	6.782	86	9.274	126	11.225	166	12.884
7	2.646	47	6.856	87	9.327	127	11.269	167	12.923
8	2.828	48	6.928	88	9.381	128	11.314	168	12.961
9	3.000	49	7.000	89	9.434	129	11.358	169	13.000
10	3.162	50	7.071	90	9.487	130	11.402	170	13.038
11	3.317	51	7.141	91	9.539	131	11.446	171	13.077
12	3.464	52	7.211	92	9.592	132	11.489	172	13.115
13	3.606	53	7.280	93	9.644	133	11.533	173	13.153
14	3.742	54	7.348	94	9.695	134	11.576	174	13.191
15	3.873	55	7.416	95	9.747	135	11.619	175	13.229
16	4.000	56	7.483	96	9.798	136	11.662	176	13.266
17	4.123	57	7.550	97	9.849	137	11.705	177	13.304
18	4.243	58	7.616	98	9.899	138	11.747	178	13.342
19	4.359	59	7.681	99	9.950	139	11.790	179	13.379
20	4.472	60	7.746	100	10.000	140	11.832	180	13.416
21	4.583	61	7.810	101	10.050	141	11.874	181	13.454
22	4.690	62	7.874	102	10.100	142	11.916	182	13.491
23	4.796	63	7.937	103	10.149	143	11.958	183	13.528
24	4.899	64	8.000	104	10.198	144	12.000	184	13.565
25	5.000	65	8.062	105	10.247	145	12.042	185	13.601
26	5.099	66	8.124	106	10.296	146	12.083	186	13.638
27	5.196	67	8.185	107	10.344	147	12.124	187	13.675
28	5.292	68	8.246	108	10.392	148	12.166	188	13.711
29	5.385	69	8.307	109	10.440	149	12.207	189	13.748
30	5.477	70	8.367	110	10.488	150	12.247	190	13.784
31	5.568	71	8.426	111	10.536	151	12.288	191	13.820
32	5.657	72	8.485	112	10.583	152	12.329	192	13.856
33	5.745	73	8.544	113	10.630	153	12.369	193	13.892
34	5.831	74	8.602	114	10.677	154	12.410	194	13.928
35	5.916	75	8.660	115	10.724	155	12.450	195	13.964
36	6.000	76	8.718	116	10.770	156	12.490	196	14.000
37	6.083	77	8.775	117	10.817	157	12.530	197	14.036
38	6.164	78	8.832	118	10.863	158	12.570	198	14.071
39	6.245	79	8.888	119	10.909	159	12.610	199	14.107
40	6.325	80	8.944	120	10.954	160	12.649	200	14.142

SCIENTIFIC CALCULATORS

This book does *not require* the use of a calculator. However, you may want to consider the purchase of an inexpensive scientific calculator. It is wise to ask your instructor for advice before you purchase any calculator for this course. It should be stressed that students are asked to avoid using a calculator for any of the exercises in which the calculations can readily be done by hand. The only problems in the text that really demand the use of a scientific calculator are labeled "Calculator Exercises." Dependence on the use of the scientific calculator for regular exercises in the text will only hurt the student in the long run.

The Two Types of Logic Used in Scientific Calculators

Two major styles of scientific calculators are popular today. The most common type employs a type of logic known as **algebraic logic**. Calculators manufactured by Casio, Sharp, Texas Instruments, and many other companies employ this type of logic. An example of calculation on such a calculator would be the following. To add 14 + 26 on an algebraic logic calculator, the sequence of buttons is

14 $\boxed{+}$ 26 $\boxed{=}$

The second type of scientific calculator requires the entry of data in **reverse Polish notation** (RPN). Most calculators manufactured by Hewlett-Packard, and a few other specialized calculators, are made to use RPN. To add 14 + 26 on a RPN calculator, the sequence of buttons is

14 $\boxed{\text{enter}}$ 26 $\boxed{+}$

Graphing scientific calculators such as the TI-82 and TI-83 have a large liquid crystal display for viewing graphs. To perform the calculation on most graphing calculators, the sequence of buttons is

14 $\boxed{+}$ 26 $\boxed{\text{enter}}$

Mathematicians and scientists do not agree on which type of scientific calculator is superior. However, the clear majority of college students own calculators that employ algebraic logic. Therefore, this section of the text is explained with reference to the sequence of steps employed by an algebraic logic calculator. If you already own or intend to purchase a scientific calculator that uses RPN or a graphing calculator, you are encouraged to study the instruction booklet that comes with the calculator and practice the problems shown in the booklet. After this practice you will be able to solve the calculator problems discussed in this section.

Performing Simple Calculations

The following example illustrates the use of the scientific calculator in doing basic arithmetic calculations.

EXAMPLE 1   Add: 156 + 298.

SOLUTION   We first key in the number 156, then press the $+$ key, then enter the number 298, and finally, press the $=$ key.

156 $+$ 298 $=$ **454**

PRACTICE PROBLEM 1
Add: 3792 + 5896.

EXAMPLE 2   Subtract: 1508 − 963.

SOLUTION   We first enter the number 1508, then press the $-$ key, then enter the number 963, and finally, press the $=$ key.

1508 $-$ 963 $=$ **545**

PRACTICE PROBLEM 2
Subtract: 7930 − 5096.

EXAMPLE 3   Multiply: 196 × 358.

SOLUTION

196 $\times$ 358 $=$ **70168**

PRACTICE PROBLEM 3
Multiply: 896 × 273.

EXAMPLE 4   Divide: 2054 ÷ 13.

SOLUTION

2054 $\div$ 13 $=$ **158**

PRACTICE PROBLEM 4
Divide: 2352 ÷ 16.

Decimal Problems

Problems involving decimals can be done readily on a calculator. Entering numbers with a decimal point is done by pressing the $\cdot$ key at the appropriate time.

EXAMPLE 5   Calculate: 4.56 × 283.

SOLUTION   To enter 4.56, we press the 4 key, the decimal point key, then the 5 key, and finally, the 6 key.

4.56 $\times$ 283 $=$ **1290.48**

The answer is 1290.48. Observe how your calculator displays the decimal point.

P R A C T I C E   P R O B L E M   5
Calculate: 72.8×197.

E X A M P L E   6   Add: $128.6 + 343.7 + 103.4 + 207.5$.

SOLUTION

128.6 $\boxed{+}$ 343.7 $\boxed{+}$ 103.4 $\boxed{+}$ 207.5 $\boxed{=}$ **783.2**

The answer is 783.2. Observe how your calculator displays the answer.

P R A C T I C E   P R O B L E M   6
Add: $52.98 + 31.74 + 40.37 + 99.82$.

Combined Operations

You must use extra caution concerning the order of mathematical operations when you are using a calculator to do a problem that involves two or more different operations.

Any scientific calculator with algebraic logic uses a priority system that has a clearly defined order of operations. It is the same order that we use in performing arithmetic operations by hand. In either situation, calculations are performed in the following order:

1. Calculations within parentheses are completed.
2. Numbers are raised to a power or a square root is calculated.
3. Multiplication and division operations are performed from left to right.
4. Addition and subtraction operations are performed from left to right.

This order is followed carefully on *scientific calculators*. Small inexpensive calculators that do not have scientific functions often do not follow this order of operations.

The number of digits displayed in the answer varies from calculator to calculator. In the following examples, your calculator may display more or fewer digits than the answer we have listed.

E X A M P L E   7   Evaluate: $5.3 \times 1.62 + 1.78 \div 3.51$.

SOLUTION   This problem requires that we multiply 5.3 by 1.62 and divide 1.78 by 3.51 first and then add the two results. If the numbers are entered directly into the calculator exactly as the problem is written, the calculator will perform the calculations in the correct order.

5.3 $\boxed{\times}$ 1.62 $\boxed{+}$ 1.78 $\boxed{\div}$ 3.51 $\boxed{=}$ **9.09312251**

P R A C T I C E   P R O B L E M   7
Evaluate: $0.0618 \times 19.22 - 59.38 \div 166.3$.

The Use of Parentheses

To perform some calculations on the calculator, the use of parentheses is helpful. These parentheses may or may not appear in the original problem.

E X A M P L E   8   Evaluate: $5 \times (2.123 + 5.786 - 12.063)$.

SOLUTION   The problem requires that the numbers in the parentheses be combined first. This will be accomplished by entering the parentheses on the calculator.

5 $\boxed{\times}$ $\boxed{(}$ 2.123 $\boxed{+}$ 5.786 $\boxed{-}$ 12.063 $\boxed{)}$ $\boxed{=}$ **−20.77**

Note: The result is a negative number.

PRACTICE PROBLEM 8
Evaluate: $3.152 \times (0.1628 + 3.715 - 4.985)$.

Negative Numbers

To enter a negative number, enter the number followed by the $\boxed{+/-}$ button.

E X A M P L E   9   Evaluate: $(-8.634)(5.821) + (1.634)(-16.082)$.

SOLUTION   The products will be evaluated first by the calculator. Therefore, parentheses are not needed as we enter the data.

8.634 $\boxed{+/-}$ $\boxed{\times}$ 5.821 $\boxed{+}$ 1.634 $\boxed{\times}$ 16.082 $\boxed{+/-}$ $\boxed{=}$ **−76.536502**

Note: The result is negative.

PRACTICE PROBLEM 9
Evaluate: $(0.5618)(-98.3) - (76.31)(-2.98)$.

Scientific Notation

If you wish to enter a number in scientific notation, you should use the special scientific notation button. On most calculators it is denoted as $\boxed{\text{EXP}}$ or $\boxed{\text{EE}}$.

E X A M P L E   1 0   Multiply: $(9.32 \times 10^6)(3.52 \times 10^8)$.

SOLUTION

9.32 $\boxed{\text{EXP}}$ 6 $\boxed{\times}$ 3.52 $\boxed{\text{EXP}}$ 8 $\boxed{=}$ **3.28064 15**

This notation means that the answer is 3.28064×10^{15}.

PRACTICE PROBLEM 1 0
Divide: $(3.76 \times 10^{15}) \div (7.76 \times 10^7)$.

Raising a Number to a Power

All scientific calculators have a key for finding powers of numbers. It is usually labeled $\boxed{y^x}$. (On a few calculators the notation is $\boxed{x^y}$ or sometimes $\boxed{\wedge}$.) To raise a number to a power, first you enter the base, then push the $\boxed{y^x}$ key. Then you enter the exponent, then finally, the $\boxed{=}$ button.

E X A M P L E 1 1   Evaluate: $(2.16)^9$.

SOLUTION

2.16 $\boxed{y^x}$ 9 $\boxed{=}$ **1023.490369**

P R A C T I C E P R O B L E M 1 1
Evaluate: $(6.238)^6$.

There is a special key to square a number. It is usually labeled $\boxed{x^2}$.

E X A M P L E 1 2   Evaluate: $(76.04)^2$.

SOLUTION

76.04 $\boxed{x^2}$ **5782.0816**

P R A C T I C E P R O B L E M 1 2
Evaluate: $(132.56)^2$.

Finding Square Roots of Numbers

There is usually a key to approximate square roots on a scientific calculator labeled $\boxed{\sqrt{}}$. In this example we need to use parentheses.

E X A M P L E 1 3   Evaluate: $\sqrt{5618 + 2734 + 3913}$.

SOLUTION

$\boxed{(}$ 5618 $\boxed{+}$ 2734 $\boxed{+}$ 3913 $\boxed{)}$ $\boxed{\sqrt{}}$ **110.747605**

P R A C T I C E P R O B L E M 1 3
Evaluate: $\sqrt{0.0782 - 0.0132 + 0.1364}$.

Note: On some scientific calculators the square root key must be entered first, followed by the left parenthesis.

SCIENTIFIC CALCULATOR EXERCISES

Use your calculator to complete each problem. Your answers may vary slightly because of the characteristics of individual calculators.

Complete the following table.

TO DO THIS OPERATION	USE THESE KEYSTROKES	RECORD YOUR ANSWER HERE
1. 8963 + 2784	8963 ☐+☐ 2784 ☐=☐	
2. 15,308 − 7980	15308 ☐−☐ 7980 ☐=☐	
3. 2631 × 134	2631 ☐×☐ 134 ☐=☐	
4. 70,221 ÷ 89	70221 ☐÷☐ 89 ☐=☐	
5. 5.325 − 4.031	5.325 ☐−☐ 4.031 ☐=☐	
6. 184.68 + 73.98	184.68 ☐+☐ 73.98 ☐=☐	
7. 2004.06 ÷ 7.89	2004.06 ☐÷☐ 7.89 ☐=☐	
8. 1.34 × 0.763	1.34 ☐×☐ 0.763 ☐=☐	

Write down the answer and then show what problem you have solved.

9. 123.45 ☐+☐ 45.9876 ☐+☐ 8765.3 ☐=☐   **10.** 0.0897 ☐×☐ 234.56 ☐×☐ 2.5428 ☐=☐

11. 34 ☐÷☐ 8 ☐+☐ 12.56 ☐=☐   **12.** 458 ☐÷☐ 4 ☐−☐ 16.897 ☐=☐

Perform each calculation using your calculator.

13. 9.467 + 0.563   **14.** 0.347 + 23.457   **15.** 34.89 + 39.6 + 214.897

16. 12.567 + 48.31 + 189.38   **17.** 412,899 − 34,675   **18.** 87,456 − 2876

19. 3,567,089 − 2,876,805   **20.** 8,345,802 − 4,985,004   **21.** 234 × 4.567

22. 1.9876 × 347   **23.** 0.456 × 3.48   **24.** 67,876 × 0.0946

25. 3458 ÷ 2.5   **26.** 9764 ÷ 8   **27.** 12.107524 ÷ 15.86

28. 16.06513 ÷ 17.98

Perform each calculation using your calculator.

29. 1.98	**30.** 8.92	**31.** $ 103.91	**32.** $3986.21
6.34	9.31	$2653.82	$4502.89
+7.71	+7.79	+$9804.61	+$ 989.30

33. $368,781.5$
$-283,617.8$

34. $571,809.6$
$-539,376.8$

35. $\$1,393,271.86$
$-\$1,289,663.21$

36. $\$8,571,300.76$
$-\$4,098,789.39$

37. 345.34
$\times 45.7$

38. 8954.34
$\times 425.4$

39. 0.6314
$\times 3.96$

40. 0.0789
$\times 12.38$

41. $40.36\overline{)36,202.92}$

42. $52.98\overline{)172,608.84}$

43. $0.7613\overline{)17.12925}$

44. $0.9854\overline{)3.59671}$

Perform the operations in the proper order using your calculator.

45. $4.567 + 87.89 - 2.45 \times 3.3$

46. $4.891 + 234.5 - 0.98 \times 23.4$

47. $7 \div 8 + 3.56$

48. $9 \div 4.5 + 0.6754$

49. $(9.34)(0.345) + 98.345$

50. $(0.628)(398) + 34.4581$

51. $\dfrac{(95.34)(0.9874)}{381.36}$

52. $\dfrac{(0.8759)(45.87)}{183.48}$

53. $2.56 + 8.98 \times 3.14$

54. $1.62 + 3.81 - 5.23 \times 6.18$

55. $(-4.23)(1.863) - 5.998$

56. $12.34 - (26.314)(-1.856)$

57. $5.62(5 \times 3.16 - 18.12)$

58. $9.356(4.8 - 7.2 - 15.94)$

59. $(3.42 \times 10^8)(0.97 \times 10^{10})$

60. $(6.27 \times 10^{20})(1.35 \times 10^3)$

61. $\dfrac{(2.16 \times 10^3)(1.37 \times 10^{14})}{6.39 \times 10^5}$

62. $\dfrac{(3.84 \times 10^{12})(1.62 \times 10^5)}{7.78 \times 10^8}$

63. $\dfrac{2.3 + 5.8 - 2.6 - 3.9}{5.3 - 8.2}$

64. $\dfrac{(2.6)(-3.2) + (5.8)(-0.9)}{2.614 + 5.832}$

65. $\sqrt{253.12}$

66. $\sqrt{0.0713}$

67. $\sqrt{5.6213 - 3.7214}$

68. $\sqrt{3417.2 - 2216.3}$

69. $(1.78)^3 + 6.342$

70. $(2.26)^8 - 3.1413$

71. $\sqrt{(6.13)^2 + (5.28)^2}$

72. $\sqrt{(0.3614)^2 + (0.9217)^2}$

73. $\sqrt{56 + 83} - \sqrt{12}$

74. $\sqrt{98 + 33} - \sqrt{17}$

Find an approximate value. Round your answer to five decimal places.

75. $\dfrac{7}{18} + \dfrac{9}{13}$

76. $\dfrac{5}{22} + \dfrac{1}{31}$

77. $\dfrac{7}{8} + \dfrac{3}{11}$

78. $\dfrac{9}{14} + \dfrac{5}{19}$

A-8

SOLVING PERCENT PROBLEMS USING A PROPORTION

Identifying the Parts of the Percent Proportion

In Section 7.7 we showed you how to use an equation to solve a percent problem. Some students find it easier to use proportions to solve percent problems. We will show you how to use proportions in this section. The two methods work equally well. Using percent proportions allows you to see another of the many uses of proportions that we studied in Chapter 4.

Consider the following relationship:

$$\frac{19}{25} = 76\%$$

This can be written as

$$\frac{19}{25} = \frac{76}{100}$$

In general, we can write this relationship using the percent proportion

$$\frac{\text{amount}}{\text{base}} = \frac{\text{percent number}}{100}$$

To use this equation effectively, we need to find the amount, base, and percent number in a word problem. The easiest of these three parts to find is the percent number. We use the letter p (a variable) to represent the percent number.

EXAMPLE 1 Identify the percent number p.

(a) Find 16% of 370.   **(b)** 28% of what is 25?

(c) What percent of 18 is 4.5?

SOLUTION

(a) Find 16% of 370.   **(b)** 28% of what is 25?
   The value of p is **16**.   The value of p is **28**.

(c) What percent of 18 is 4.5?
   $\underbrace{\text{What percent}}_{p}$

   We let p represent the unknown percent number.

PRACTICE PROBLEM 1

Identify the percent number p.

(a) Find 83% of 460.   **(b)** 18% of what number is 90?

(c) What percent of 64 is 8?

We use the letter *b* to represent the base number. The base is the entire quantity or the total involved. The number that is the base usually appears after the word *of*. The amount, which we represent by the letter *a*, is the part being compared to the whole.

E X A M P L E   2   Identify the base *b* and the amount *a*.

(a) 20% of 320 is 64.   **(b)** 12 is 60% of what?

SOLUTION

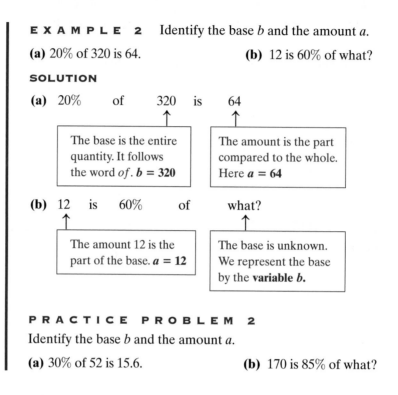

(a)   20%   of   320   is   64

The base is the entire quantity. It follows the word *of*. **b = 320**

The amount is the part compared to the whole. Here **a = 64**

(b)   12   is   60%   of   what?

The amount 12 is the part of the base. **a = 12**

The base is unknown. We represent the base by the **variable b.**

P R A C T I C E   P R O B L E M   2
Identify the base *b* and the amount *a*.

(a) 30% of 52 is 15.6.   **(b)** 170 is 85% of what?

When identifying *p, b,* and *a* in a problem, it is easiest to identify *p* and *b* first. The remaining quantity or variable is *a*.

E X A M P L E   3   Find *p, b,* and *a*.

(a) What is 52% of 300?   **(b)** What percent of 30 is 18?

SOLUTION

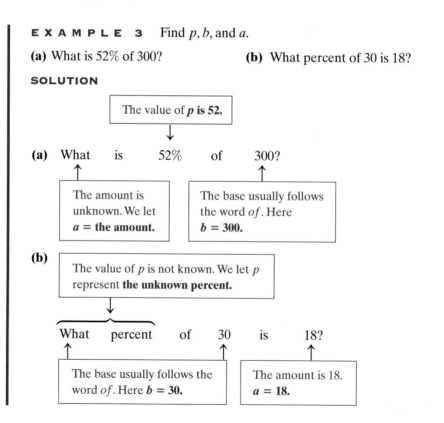

The value of ***p* is 52.**

(a)   What   is   52%   of   300?

The amount is unknown. We let ***a* = the amount.**

The base usually follows the word *of*. Here **b = 300.**

(b)

The value of *p* is not known. We let *p* represent **the unknown percent.**

What   percent   of   30   is   18?

The base usually follows the word *of*. Here **b = 30.**

The amount is 18. **a = 18.**

PRACTICE PROBLEM 3

Find p, b, and a.

(a) What is 18% of 240?　　　　　　　**(b)** What percent of 64 is 4?

Using the Percent Proportion to Solve Percent Problems

When we solve the percent proportion we will have enough information to state the numerical value for two of the three variables a, b, p in the equation

$$\frac{a}{b} = \frac{p}{100}$$

We first identify those two values, then substitute those values into the equation. Then we use the skills that we acquired for solving proportions in Chapter 4 to find the value we do not know.

EXAMPLE 4　　What is 16% of 380?

SOLUTION　　The percent $p = 16$. The base $b = 380$. The amount is unknown. We use the variable a. Thus

$$\frac{a}{b} = \frac{p}{100} \qquad \text{becomes} \qquad \frac{a}{380} = \frac{16}{100}$$

If we reduce the fraction on the right-hand side, we have

$$\frac{a}{380} = \frac{4}{25}$$

$25a = (380)(4)$　　　Cross-multiply.

$25a = 1520$　　　　　Simplify.

$$\frac{25a}{25} = \frac{1520}{25} \qquad \text{Divide each side by 25.}$$

$a = 60.8$　　　　　Divide $1520 \div 25$.

Thus 16% of 380 is **60.8**.

PRACTICE PROBLEM 4

What is 32% of 550?

EXAMPLE 5　　Find 260% of 40.

SOLUTION　　The percent $p = 260$. The number that is the base usually appears after the word *of*. The base $b = 40$. The amount is unknown. We use the variable a. Thus

$$\frac{a}{b} = \frac{p}{100} \qquad \text{becomes} \qquad \frac{a}{40} = \frac{260}{100}$$

If we reduce the fraction on the right-hand side, we have

$$\frac{a}{40} = \frac{13}{5}$$

$5a = (40)(13)$　　　Cross-multiply.

$5a = 520$　　　　　Simplify.

$$\frac{5a}{5} = \frac{520}{5} \qquad \text{Divide each side of the equation by 5.}$$

$a = 104$

Thus 260% of 40 is **104**.

PRACTICE PROBLEM 5
Find 340% of 70.

EXAMPLE 6   85% of what is 221?

SOLUTION   The percent $p = 85$. The base is unknown. We use the variable b. The amount a is 221. Thus

$$\frac{a}{b} = \frac{p}{100} \quad \text{becomes} \quad \frac{221}{b} = \frac{85}{100}$$

If we reduce the fraction on the right-hand side, we have

$$\frac{221}{b} = \frac{17}{20}$$

$$(20)(221) = 17b \quad \text{Cross-multiply.}$$
$$4420 = 17b \quad \text{Simplify.}$$
$$\frac{4420}{17} = \frac{17b}{17} \quad \text{Divide each side by 17.}$$
$$260 = b \quad \text{Divide } 4420 \div 17.$$

Thus 85% of **260** is 221.

PRACTICE PROBLEM 6
68% of what is 476?

EXAMPLE 7   53 is 0.2% of what?

SOLUTION   The percent $p = 0.2$. The base is unknown. We use the variable b. The amount $a = 53$. Thus

$$\frac{a}{b} = \frac{p}{100} \quad \text{becomes} \quad \frac{53}{b} = \frac{0.2}{100}$$

When we cross-multiply, we obtain

$$(100)(53) = 0.2b$$
$$5300 = 0.2b$$
$$\frac{5300}{0.2} = \frac{0.2b}{0.2}$$
$$26{,}500 = b$$

Thus 53 is 0.2% of **26,500**.

PRACTICE PROBLEM 7
216 is 0.3% of what?

EXAMPLE 8   What percent of 4000 is 160?

SOLUTION   The percent is unknown. We use the variable p. The base $b = 4000$. The amount $a = 160$. Thus

$$\frac{a}{b} = \frac{p}{100} \quad \text{becomes} \quad \frac{160}{4000} = \frac{p}{100}$$

If we reduce the fraction on the left-hand side, we have

$$\frac{1}{25} = \frac{p}{100}$$

$100 = 25p$ Cross-multiply.

$$\frac{100}{25} = \frac{25p}{25}$$ Divide each side by 25.

$4 = p$ Divide $100 \div 25$.

Thus **4%** of 4000 is 160.

P R A C T I C E   P R O B L E M   8

What percent of 3500 is 105?

E X A M P L E   9   19 is what percent of 95?

SOLUTION   The percent is unknown. We use the variable p. The base $b = 95$. The amount $a = 19$. Thus

$$\frac{a}{b} = \frac{p}{100} \quad \text{becomes} \quad \frac{19}{95} = \frac{p}{100}$$

Cross-multiplying, we have

$$(100)(19) = 95p$$
$$1900 = 95p$$
$$\frac{1900}{95} = \frac{95p}{95} \quad \text{Divide each side of the equation by 95.}$$
$$20 = p$$

Thus 19 is **20%** of 95.

P R A C T I C E   P R O B L E M   9

42 is what percent of 140?

EXERCISES

Identify p, b, and a. Do not solve for the unknown.

	p	b	a

1. 75% of 660 is 495.

2. 65% of 820 is 532.

3. What is 42% of 400?

4. What is 56% of 600?

5. 49% of what is 2450?

6. 38% of what is 2280?

7. 30 is what percent of 50?

8. 50 is what percent of 250?

9. What percent of 25 is 10?

10. What percent of 24 is 6?

11. 400 is 160% of what?

12. 900 is 225% of what?

Solve by using the percent proportion

$$\frac{a}{b} = \frac{p}{100}$$

In problems 13–18, the amount "a" is not known.

13. 24% of 300 is what?   **14.** 54% of 500 is what?   **15.** Find 250% of 30.

16. Find 320% of 60.   **17.** 0.7% of 8000 is what?   **18.** 0.8% of 9000 is what?

In problems 19–24, the base "b" is not known.

19. 45 is 60% of what?   **20.** 96 is 80% of what?   **21.** 150% of what is 90?

22. 125% of what is 75?   **23.** 3000 is 0.5% of what?   **24.** 6000 is 0.4% of what?

In problems 25–30, the percent "p" is not known.

25. 70 is what percent of 280?   **26.** 90 is what percent of 450?

27. What percent of 140 is 3.5?

28. What percent of 170 is 3.4?

29. What percent of $5000 is $90?

30. What percent of $4000 is $64?

MIXED PRACTICE

31. 26% of 350 is what?

32. 56% of 650 is what?

33. 180% of what is 540?

34. 160% of what is 320?

35. 82 is what percent of 500?

36. 75 is what percent of 600?

37. Find 0.7% of 520.

38. Find 0.4% of 650.

39. What percent of 66 is 16.5?

40. What percent of 49 is 34.3?

41. 68 is 40% of what?

42. 52 is 40% of what?

43. 94.6 is what percent of 220?

44. 85.8 is what percent of 260?

45. What is 12.5% of 380?

46. What is 20.5% of 320?

47. Find 0.05% of 5600.

48. Find 0.04% of 8700.

CALCULATOR EXERCISES

Solve each percent problem. Round to the nearest hundredth.

49. What percent of 4550 is 720?

50. What is 16.25% of 65,250?

51. 4.25% of 256.75 is what number?

52. 2760 is 5.5% of what number?

53. What is $19\frac{1}{4}$% of 798?

54. $140\frac{1}{2}$% of what number is 10,397?

55. Find 18% of 20% of $3300.
(*Hint:* First find 20% of $3300.)

56. Find 42% of 16% of $5500.
(*Hint:* First find 16% of $5500.)

CONGRUENT TRIANGLES

Two objects are **congruent** if they can be made to coincide by placing one on top of the other. This can be done either directly or by flipping one of the objects over. In other words, two objects are congruent if they have the same size and the same shape.

Two triangles are congruent if the measures of corresponding angles are equal and the lengths of corresponding sides are equal.

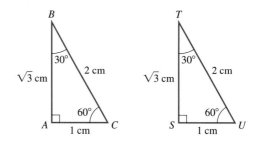

Triangle *ABC* is congruent to triangle *STU*.

We use the symbol $\angle$ to indicate that we are referring to the measure of an angle. In the triangle above, $\angle A = 90°$, $\angle B = 30°$, and $\angle C = 60°$.

Two triangles are congruent when one of the following can be shown: congruence by side–side–side (SSS), side–angle–side (SAS), or angle–side–angle (ASA).

CONGRUENCE BY SIDE–SIDE–SIDE (SSS)
If the three sides of one triangle have the same length as the corresponding three sides of a second triangle, the two triangles are congruent.

E X A M P L E   1   Explain why the following two triangles are congruent.

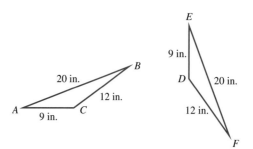

SOLUTION

Side BC and side DF = 12 in.
Side AC and side DE = 9 in.
Side AB and side EF = 20 in.

All three sides of triangle ABC have the same length as the corresponding sides of triangle DEF. The triangles satisfy SSS and are congruent.

    Note: If we rotate the triangle ABC downward at B, we can lay this triangle on top of triangle DEF since both triangles have the same size and shape.

PRACTICE PROBLEM 1

Explain why the following two triangles are congruent.

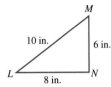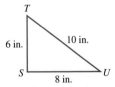

CONGRUENCE BY SIDE–ANGLE–SIDE (SAS)
Two triangles are congruent if:

1. Two sides of one triangle are equal in length to two sides of a second triangle, and
2. The measure of the angle formed by these two sides in each triangle is equal.

EXAMPLE 2   Explain why the following two triangles are congruent.

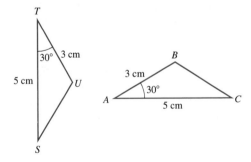

SOLUTION

Side ST = side AC = 5 cm.
Side TU = side AB = 3 cm.
$\angle T = \angle A = 30°$.

Two sides of triangles STU and ABC are equal in length, and the angles formed by these sides are also equal. The triangles are congruent by SAS.

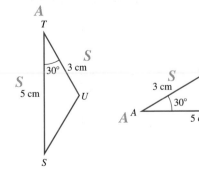

PRACTICE PROBLEM 2

Explain why the following two triangles are congruent.

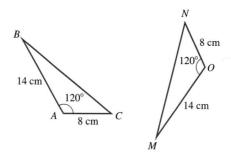

CONGRUENCE BY ANGLE–SIDE–ANGLE (ASA)

Two triangles are congruent if:

1. Two angles of one triangle are equal in measure to two angles of a second triangle, and
2. The side between these two angles in each triangle is equal in length.

EXAMPLE 3 Explain why the following two triangles are congruent.

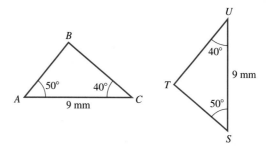

SOLUTION

$\angle A = \angle S = 50°$.

$\angle C = \angle U = 40°$.

Side AC = side SU = 9 mm.

Two angles of triangle ABC equal two angles of triangle STU, and the sides between these angles are equal in length. The triangles are congruent by ASA.

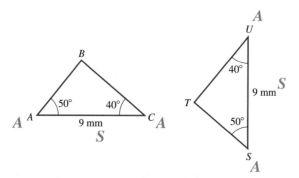

PRACTICE PROBLEM 3

Explain why the following two triangles are congruent.

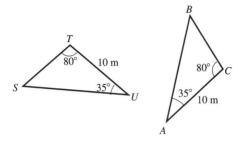

EXAMPLE 4   Determine whether each set of triangles is congruent. If so, state by what rule it is congruent.

(a)

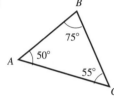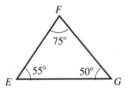

(b)

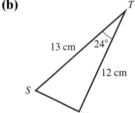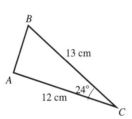

SOLUTION

(a)

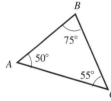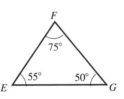

We must determine if one of the rules for congruence is satisfied to determine whether the triangles are congruent. The only information given is that all the corresponding angles of each triangle are equal. There is not enough information given to determine SSS, ASA, or SAS, so the triangles are not necessarily congruent. We must know the length of at least one side of a pair of triangles in order to prove congruency.

(b)

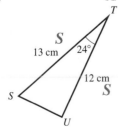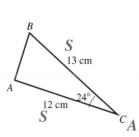

We see from the information given that the triangles are congruent by the SAS rule.

PRACTICE PROBLEM 4

Determine whether each set of triangles is congruent. If so, state by what rule it is congruent.

(a)

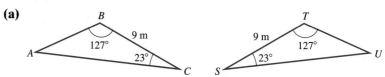

(b)

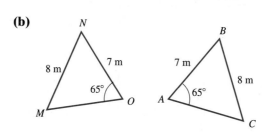

In Chapter 9 we learned that if corresponding angles of triangles are equal, the triangles are similar. In Example 4(a), triangles *ABC* and *EFG* are similar triangles. Why?

The difference between similar triangles and congruent triangles is as follows:

> The corresponding *sides* and *angles* of congruent triangles must be *equal*.
> The corresponding *angles* of similar triangles must be *equal*, but corresponding *sides* are *not necessarily equal*.

Can you draw two triangles that are both congruent and similar?

EXERCISES

The following pairs of triangles are congruent by SAS. Find the length of side EF.

1.

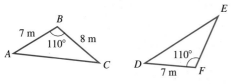

2.

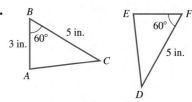

The following pairs of triangles are congruent by ASA. Name the angle in triangle DEF that is 35°.

3.

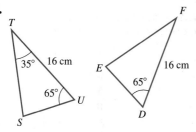

4.

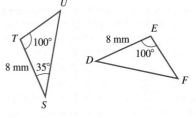

The following pairs of triangles are congruent by SSS. Find the length of the missing side.

5.

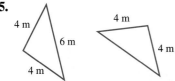

6.

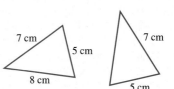

Explain why each set of triangles is congruent.

7.

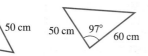

8.

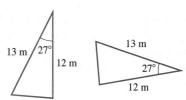

9.

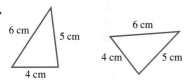

10.

11.

12.

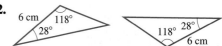

13.

14.

Determine whether each set of triangles is congruent. If so, state by what rule it is congruent.

15.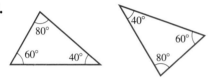

16.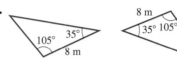

17.

18.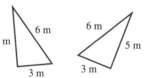

19.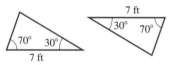

20.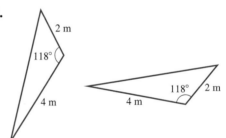

21.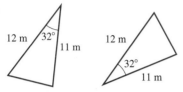

22.

23.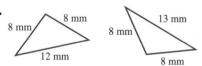

24.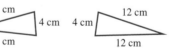

25.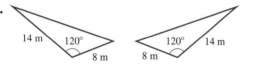

26.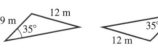

27.

 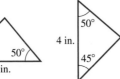

28.

29.

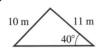

30.

CHAPTER 1

1.1 PRACTICE PROBLEMS

1. **(a)** hundred thousands **(b)** ten thousands
   (c) hundreds

2. $2,507,235 = 2,000,000 + 500,000 + 7000 + 200$
   $+ 30 + 5$

3. $\$582 = 500 + 80 + 2$
   = 5 hundred-dollar bills, 8 ten-dollar bills, and
   2 one-dollar bills

4. **(a)** $3 > 2$ **(b)** $6 < 8$
   (c) $1 < 7$ **(d)** $7 > 1$

5. **(a)** $7 > 2$ **(b)** $3 < 4$

6. 34,627   Locate the hundreds round-off place.

   34,6$\underline{2}$7   The digit to the right of the round-off
   $\downarrow$ place is less than 5.
   34,600   Do not change the round-off place digit.
   Replace all digits to the right of the
   round-off place with zeros.

7. $1,335,627 = 1,340,000$ to the nearest ten thousand.
   The digit to the right of the ten thousands digit was 5,
   so we increased the round-off place digit by 1.

8. Approximately one hour

1.2 PRACTICE PROBLEMS

1. **(a)** Five added to some number
     $\downarrow$ $\downarrow$ $\downarrow$
     5   + x
         $5 + x$

   (b) Four more than 5
     $\downarrow$ $\downarrow$ $\downarrow$
     4   +   5
       $4 + 5$

2. $6 + 0 = 6$
   $5 + 1 = 6$
   $4 + 2 = 6$
   $3 + 3 = 6$
   $2 + 4 = 6$
   $1 + 5 = 6$
   $0 + 6 = 6$

   We need to learn the three addition facts for the
   number 6. The remaining facts are a repeat of these.

3. **(a)** $x + 3 = 3 + x$
   (b) $9 + w = w + 9$
   (c) $4 + 0 = 0 + 4$

4. If $5663 + 412 = 6075$, then $412 + 5663 = 6075$.

5. $6 + 3 + x = 9 + x$ or $x + 9$

6. $(w + 1) + 4 = w + (1 + 4) = w + 5$

7. **(a)** $(2 + x) + 8 = (x + 2) + 8$
                  $= x + (2 + 8)$
                  $= x + 10$

   (b) $(4 + x + 3) + 1 = (x + 4 + 3) + 1$
                        $= (x + 7) + 1$
                        $= x + (7 + 1)$
                        $= x + 8$

8. $\overset{1}{2}47$
   $+\ \ 38$
      285

9. We add the responses for milk, 286; orange juice, 475;
   and other, 91.
      $\overset{21}{2}86$
      475
   $+\ \ 91$
      852

10. **(a)** Four n's $= 4n$ **(b)** $y + y + y = 3y$
    (c) Eight $= 8$ **(d)** One $y = 1y$ or y

11. There are no terms like $5y$; none have the exact
    variable part y. There are no terms like $6n$; none have
    the exact variable part n. The like terms are $2mn$ and
    $4mn$. The variable parts, mn, are the same.

12. **(a)** $2ab + 4a + 3ab = 2ab + 3ab + 4a$
                       $= (2 + 3)ab + 4a$
                       $= 5ab + 4a$

    (b) $4y + 5x + y + x = 4y + 5x + 1y + 1x$
                         $= 4y + 1y + 5x + 1x$
                         $= (4 + 1)y + (5 + 1)x$
                         $= 5y + 6x$

    (c) $7x + 3y + 3z = 7x + 3y + 3z$

    Terms cannot be combined because there are no
    like terms.

1.3 PRACTICE PROBLEMS

1. **(a)** $5 - 2 = 3$ 2. $18 - 11 = ?$
   (b) $6 - 3 = 3$ **(a)** $18 = ? + 11$
   (c) $8 - 7 = 1$ **(b)** $18 = 7 + 11$
   (d) $3 - 1 = 2$ $18 - 11 = 7$

3. **(a)** $8m - 3m = (8 - 3)m$
                 $= 5m$

   (b) $5x - 4x - 2 = (5 - 4)x - 2$
                    $= 1x - 2$ or $x - 2$

   (c) $7y - y - 3b - 2 = 7y - 1y - 3b - 2$
                          $= (7 - 1)y - 3b - 2$
                          $= 6y - 3b - 2$

4. (a) The difference of 9 and n is 1.

$$9 - n = 1$$

(b) x minus 3

$$x - 3$$

(c) x subtracted from 8

$$8 - x$$

5. (a) $(11 - 4) - 1 = 7 - 1$
$$= 6$$

(b) $11 - (4 - 1) = 11 - 3$
$$= 8$$

6.
$600 - 50 = 550$
$600 - 51 = 549$
$600 - 52 = 548$
$600 - 53 = 547$
$600 - 54 = 546$

7.
$$\begin{array}{r} {}^{8\,13}\!\!\not{9}\not{3} \\ -\ 46 \\ \hline 47 \end{array}$$

8.
$$\begin{array}{r} {}^{5\,9\,13}\!\!\not{6}\not{0}\not{3} \\ -\ 278 \\ \hline 325 \end{array}$$

9.
$$\begin{array}{r} {}^{7\,9\,9\,16}\!\!\not{8}\not{0}\not{0}\not{6} \\ -\ 4237 \\ \hline 3769 \end{array} \quad \begin{array}{r} 4237 \\ +\ 3769 \\ \hline 8006 \end{array} \text{ It checks.}$$

10.

Number of yellowtail caught at Dana Point	minus	Number of yellowtail caught at Newport
1810	−	917

$$\begin{array}{r} {}^{10}\\ {}^{17\,0\,10}\\ \not{1}\not{8}\not{1}\not{0} \\ -\ 917 \\ \hline 893 \end{array}$$

The charter from Newport landing caught 893 fewer yellowtail.

11. (a) The profit in 1996 was $7,000,000.

(b)
$$\begin{array}{r} \$4{,}000{,}000 \rightarrow 1992 \\ -\ 3{,}000{,}000 \rightarrow 1994 \\ \hline \$1{,}000{,}000 \end{array}$$
$1,000,000 more profit was made in 1992 than in 1994.

(c) The profits were highest in 1996.

(d)
$$\begin{array}{r} 5 \text{ million} \rightarrow 1993 \\ +\ 3 \text{ million} \rightarrow 1994 \\ \hline 8 \text{ million} \end{array} \qquad \begin{array}{r} 6 \text{ million} \rightarrow 1995 \\ +\ 7 \text{ million} \rightarrow 1996 \\ \hline 13 \text{ million} \end{array}$$
The profits for 1993–1994 were less than the profits for 1995–1996.

12. 1. *Understand the problem.*

Mathematics Blueprint for Problem Solving			
Gather the Facts	What Am I Asked to Do?	How Do I Proceed?	Key Points to Remember
ⓐ	ⓑ	ⓒ	ⓓ

ⓐ Old balance is $569. The deposits are $706 and $234. The bank credited him $22. The withdrawals are $42, $132, $341, and $202.

ⓑ Find his balance this month.

ⓒ 1. Find the total deposits (and credits) and withdrawals made.

2. Add the total of the deposits and the credit to the old balance.

3. From the figure, subtract the total of the withdrawals.

ⓓ Add deposits and credits, then subtract withdrawals.

2. *Solve and state the answer.*

Old balance	+	total deposits	−	total withdrawals	=	new balance

$$\begin{array}{r} \$706 \\ 234 \\ +\ 22 \\ \hline \$962 \end{array} \qquad \begin{array}{r} \$\ 42 \\ 132 \\ 341 \\ +\ 202 \\ \hline \$717 \end{array}$$

$$\$569 \ +\ \$962 \ -\ \$717 \ =\ \$814$$

His balance this month is $814.

1.4 PRACTICE PROBLEMS

1. Since this is a square, we know that if one side measures 15 feet, the other three sides also measure 15 feet. To find the perimeter, we add the lengths of the sides:

$15 \text{ ft} + 15 \text{ ft} + 15 \text{ ft} + 15 \text{ ft} = 60 \text{ ft}$

The perimeter of the square is 60 ft.

2. To find the perimeter of this figure we must add the six sides. Starting at the top and going clockwise, the sides are: 125 ft, 40 ft, 30 ft, 30 ft, 155 ft, and 70 ft.

$125 \text{ ft} + 40 \text{ ft} + 30 \text{ ft} + 30 \text{ ft} + 155 \text{ ft} + 70 \text{ ft}$
$= 450 \text{ ft}$

The perimeter of this shape is 450 ft.

3. (a) The 70–79 bar rises to a height of 14, so there are 14 students who scored a C on the final.

(b) From the histogram, 6 tests were 60–69, 14 tests were 70–79, and 20 tests were 80–89. When we add, $6 + 14 + 20 = 40$, we see that 40 students scored less than 90 on the final.

(c) Since 20 students scored between 80 and 89, and 14 scored between 70 and 79, we subtract $20 - 14 = 6$ to find that 6 more students scored between 80 and 89 than between 70 and 79.

4. **(a)**

Taurus: $\quad$ \$13,988 $\rightarrow$ \$14,000 $\qquad$ \$14,000
Ranger XLT: $\;$ \$12,788 $\rightarrow$ \$13,000 $\quad$ $-$ 13,000
$\qquad\qquad\qquad\qquad\qquad\qquad\qquad$ \$ 1,000

The estimated difference in price is \$1000.

(b) To find the exact difference, we subtract the original prices

$$\begin{array}{r} \$13,988 \\ - \quad 12,788 \\ \hline \$1,200 \end{array}$$

The exact difference, \$1200, is close to the estimated difference, \$1000, so the estimate is reasonable.

5. **(a)**

$$\begin{array}{r} \$1020 \rightarrow \$1000 \\ 106 \rightarrow \quad 100 \\ 494 \rightarrow \quad 500 \\ 186 \rightarrow \quad 200 \\ 105 \rightarrow \quad 100 \\ \hline \$1900 \rightarrow \text{sum of our rounded values} \end{array}$$

The estimated monthly basic expenses are \$1900.

(b)

$$\begin{array}{r} \$2745 \rightarrow \quad \$2700 \\ - \quad 1900 \\ \hline \$800 \end{array}$$

The estimated amount of money left after base expenses is \$800.

1.5 PRACTICE PROBLEMS

1. **(a)** $8 - n$
$\qquad\;\downarrow$
$\quad\; 8 - 3$
$\qquad\; 5$

(b) $8 - n$
$\qquad\;\downarrow$
$\quad\; 8 - 6$
$\qquad\; 2$

2. $P = L + L + W + W$
$\qquad\;\downarrow\;\;\;\downarrow\;\;\;\downarrow\;\;\;\downarrow$
$P = 6\text{ yd} + 6\text{ yd} + 5\text{ yd} + 5\text{ yd}$
$P = 22\text{ yd}$

3. **(a)** What number plus four *is the same as* seven?
$\qquad\qquad\;\downarrow\;\;\;\;\;\downarrow\;\;\;\downarrow\;\;\;\;\;\;\;\;\;\downarrow\;\;\;\;\;\;\;\;\downarrow$
$\qquad\qquad\; x \;\;+\;\; 4 \;\;\;\;\;\;=\;\;\;\;\;\; 7$
$\quad x + 4 = 7$

(b) Three subtracted from six *is equal to* what number?
$\qquad\;\downarrow\;\;\;\;\;\;\;\;\;\;\;\;\;\;\;\;\downarrow\;\;\;\;\;\;\;\;\;\downarrow\;\;\;\;\;\;\downarrow$
$\qquad 6 \;\;\;\;-\;\;\;\;\; 3 \;\;\;\;\;\;=\;\;\;\; x$
$\quad 6 - 3 = x$

(c) The number of cards in a collection plus 20 new cards equals 75 cards.
$\downarrow\;\;\downarrow\;\;\downarrow\;\;\;\;\;\;\;\;\;\;\;\;\;\;\;\downarrow\;\;\;\downarrow$
$x \;\;+\;\; 20 \;\;\;\;\;\;\;\;\;=\;\;\;\; 75$
$x + 20 = 75$

4. Evaluate $x + 8 = 11$ for $x = 5$.
$x + 8 = 11$
$\downarrow$
$5 + 8 \overset{?}{=} 11$
$\quad 13 \overset{?}{=} 11$ false

Since $13 = 11$ is not a true statement, 5 is not a solution to $x + 8 = 11$.

5. Four plus what number is nine? The solution is 5.
$4 + n = 9$
$\qquad\downarrow$
$4 + 5 \overset{?}{=} 9$
$\quad\; 9 = 9$ True

The solution is written $n = 5$.

6. $(3 + x) + 1 = 7$
$(x + 3) + 1 = 7 \qquad$ Commutative property.
$x + (3 + 1) = 7 \qquad$ Associative property.
$x + 4 = 7 \qquad\qquad$ Simplify.

What number plus four is equal to seven? $3 + 4 = 7$
The solution is 3, and is written $x = 3$.

7. $x - 567 = 349$
(a) $x = 349 + 567$
(b) $x = 916$

8. $13 - x = 4$
$\quad 13 = 4 + x$
$\qquad\qquad\downarrow$
$\quad 13 = 4 + 9$
$\qquad x = 9$

9. $x + 46 = 220$
$\quad x = 220 - 46$
$\quad x = 174$

10. **(a)** $x - 34 = 60$
$\qquad x = 60 + 34$
$\qquad x = 94$

(b) $x + 15 = 40$
$\qquad x = 40 - 15$
$\qquad x = 25$

1.6 PRACTICE PROBLEMS

1. **(a)** 60 sec = $\underline{\;1\;}$ min
(b) 1 gal = $\underline{\;4\;}$ qt

2. **(a)** 4 pt = 2 pt + 2 pt
$\qquad\qquad\;\downarrow\;\;\;\;\;\;\;\downarrow$
$\qquad\quad = 1\text{ qt} + 1\text{ qt}$
$\quad 4\text{ pt} = \underline{\;2\;}$ qt

(b) 2 tons = 1 ton + 1 ton
$\qquad\qquad\quad\;\downarrow\;\;\;\;\;\;\;\;\downarrow$
$\qquad\quad = 2000\text{ lb} + 2000\text{ lb}$
$\; 2\text{ tons} = \underline{4000}$ lb

3. **(a)**
$$\begin{array}{r} 12\text{ min } 43\text{ sec} \\ + \quad 4\text{ min } 33\text{ sec} \\ \hline 16\text{ min } 73\text{ sec} \end{array}$$
$\qquad\qquad\;\downarrow$
$16\text{ min } (1\text{ min } 13\text{ sec})$
$= 17\text{ min } 13\text{ sec}$

(b)
$$\begin{array}{r} 4\text{ gal } 1\text{ qt} \\ - \; 2\text{ gal } 3\text{ qt} \\ \hline \end{array} \quad \begin{array}{r} (3\text{ gal} + 4\text{ qt})\, 1\text{ qt} \\ - \quad 2\text{ gal} \\ \hline \end{array} \quad \begin{array}{r} 3\text{ gal } 5\text{ qt} \\ - \; 2\text{ gal } 3\text{ qt} \\ \hline 1\text{ gal } 2\text{ qt} \end{array}$$
$\qquad\qquad\qquad\qquad\qquad\quad 3\text{ qt}$

4. **(a)** 10 cm = 1 $\underline{\;\text{dm}\;}$
(b) 1 dm = 10 $\underline{\;\text{cm}\;}$

5. **(a)** 1 mm < 1 km
(b) 5 m < 5 hm

6. The prefix centi is located to the right of meter on the chart so a centimeter is smaller than a meter.

7. **(a)** 1 mm < 1 in.
(b) 4 m > 4 yd

8. **(a)** 1 L > 1 dL
(b) 1 kg > 1 lb

9. The most reasonable measure for the volume of a bottle of soda is 1 L. The units 1 m and 1 g measure lengths and weight, not volume.

CHAPTER 2

2.1 PRACTICE PROBLEMS

1.

```
       *   *   *
       *   *   *
  5    *   *   *
       *   *   *
       *   *   *
           3
```

2. (a)

	red	blue	green	pink	black
dirt	red dirt bike	blue dirt bike	green dirt bike	pink dirt bike	black dirt bike
racer	red racer bike	blue racer bike	green racer bike	pink racer bike	black racer bike
road	red road bike	blue road bike	green road bike	pink road bike	black road bike

(b) The manufacturer can make 15 different bikes.

3. (a) The product is 63 and the factors are 9 and 7.
(b) The product is z and the factors are x and y.

4. (a) The *product* of a and b.

$$a \cdot b \to ab$$

(b) Double a number.

$$2 \cdot n \to 2n$$

(c) Two *times* what number equals 10?

$$2 \cdot n = 10 \to 2n = 10$$

5. (a) $5y \to$ Five times a number.
(b) $3a = 12 \to$ Three times what number equals 12?

6. (a) $2 \cdot 6 \cdot 0 \cdot 3 = 0$, Multiplication law of zero
(b) $2 \cdot 3 \cdot 1 \cdot 5 = \underline{2 \cdot 3} \cdot \underline{1 \cdot 5}$
$$= \ \ 6 \ \ \cdot \ \ 5$$
$$= 30$$

7. $8(6x) = (8 \cdot 6)x$ Associative property
$\quad\quad = 48x$ Simplify

8. (a) $4(x \cdot 3) = 4(3 \cdot x)$ Commutative property
$\quad\quad\quad = (4 \cdot 3)x$ Associative property
$\quad\quad\quad = 12x$ Simplify
(b) $2(4)(n \cdot 5) = 8(n \cdot 5)$ Simplify
$\quad\quad\quad\quad = 8(5 \cdot n)$ Commutative property
$\quad\quad\quad\quad = (8 \cdot 5)n$ Associative property
$\quad\quad\quad\quad = 40n$ Simplify

2.2 PRACTICE PROBLEMS

1. $8y(40) = 8(40)y$ Commutative property
$\quad\quad = 320y$ Multiply $8(4) = 32$ and attach 1 zero to the product.

2. (a) $2(x - 5) = 2 \cdot x - 2 \cdot 5$
$\quad\quad\quad = 2x - 10$
(b) $4(y + 3) = 4 \cdot y + 4 \cdot 3$
$\quad\quad\quad = 4y + 12$

3. $7(11) = 7(10 + 1)$
$\quad\quad = 7(10) + 7(1)$
$\quad\quad = 70 + 7$
$\quad\quad = 77$

4.
$$\begin{array}{r} \overset{2\,4}{436} \\ \times \ \ 7 \\ \hline 3052 \end{array}$$

5.
$$\begin{array}{r} 87\,2 \\ \times \ \ \ 600 \\ \hline 523{,}200 \end{array}$$

6.
$$\begin{array}{r} 936 \\ \times \ \ 38 \\ \hline 7488 \\ 28080 \\ \hline 35{,}568 \end{array}$$

7.
$$\begin{array}{r} 4651 \\ \times \ \ 203 \\ \hline 13953 \\ 930200 \\ \hline 944{,}153 \end{array}$$

8. 1. *Understand the problem.*

Mathematics Blueprint for Problem Solving			
Gather the Facts	What Am I Asked to Do?	How Do I Proceed?	Key Points to Remember
(a)	(b)	(c)	(d)

(a) Emily averages 7 sales per week. Commission is $55 per sale. The salary would be $1770 per month.

(b) Determine which pay option would earn her more money per year.

(c) 1. Multiply 55×7 to find the commission for 1 week, then multiply by 52 for the yearly commission.

2. Multiply 1770×12 to find the yearly salary.

3. Compare yearly pay for both options.

(d) The phrases per sale and per month indicate multiplication.
I must find yearly pay:
12 months = 1 year
52 weeks = 1 year

2. *Solve and state the answer.*

– Commission for 1 week = $55 \cdot 7 = \$385$
– Commission for 1 year = $385 \cdot 52 = \$20{,}020$
– Salary for 1 year = $1770 \cdot 12 = \$21{,}240$

The salary option would earn Emily more money per year.

3. *Check.* Estimate the answer to see if it is reasonable.

2.3 PRACTICE PROBLEMS

1. (a) $n = n^1$
(b) $6 \cdot 6 \cdot y \cdot y \cdot y \cdot y = 6^2 y^4$
(c) $5 \cdot 5 \cdot 5 \cdot 5 \cdot 5 \cdot 5 \cdot 5 \cdot 5 = 5^8$

2. (a) $x^6 = x \cdot x \cdot x \cdot x \cdot x \cdot x$
   (b) $1^7 = 1 \cdot 1 \cdot 1 \cdot 1 \cdot 1 \cdot 1 \cdot 1$

3. (a) $4^3 = 4 \cdot 4 \cdot 4 = 64$
   (b) $8^1 = 8$
   (c) $10^2 = 10 \cdot 10 = 100$

4. (a) Four to the sixth power $= 4^6$
   (b) x cubed $= x^3$
   (c) Ten squared $= 10^2$

5. (a) *What number cubed* is equal to 8?
$$\downarrow \qquad\qquad \downarrow \quad \downarrow$$
$$x^3 \qquad\quad = \qquad 8$$
   $x^3 = 8$, $x = 2$ because $2^3 = 8$.

   (b) *Four squared* is equal to what number?
$$\downarrow \qquad \downarrow \qquad\qquad \downarrow$$
$$4^2 \qquad = \qquad\quad n$$
   $4^2 = n$, $n = 16$ because $4^2 = 16$.

6. $10^5 = 100,000$
   The exponent is 5; attach 5 trailing zeros

7. $3^2 + 2 - 5 = \underline{3^2} + 2 - 5$
$$= \underline{9 + 2} - 5$$
$$= 11 - 5$$
$$= 6$$

8. $4 \cdot 2^3 = 4 \cdot \underline{2^3}$
$$= \underline{4 \cdot 8}$$
$$= 32$$

9. $2 + 7(10 - 3 \cdot 2) - 4 = 2 + 7(10 - \underline{3 \cdot 2}) - 4$
$$= 2 + 7(\underline{10 - 6}) - 4$$
$$= 2 + \underline{7(4)} - 4$$
$$= \underline{2 + 28} - 4$$
$$= 30 - 4$$
$$= 26$$

2.4   PRACTICE PROBLEMS

1. $4^2 \cdot 4^4 = 4 \cdot 4 \cdot 4 \cdot 4 \cdot 4 \cdot 4 = 4^6$

2. (a) $y^5 \cdot y = y^5 \cdot y^1$ (b) $a^4 \cdot a^5 = a^{4+5}$
$$\quad= y^{5+1} \qquad\qquad\qquad\qquad = a^9$$
$$\quad= y^6$$
   (c) $5^3 \cdot 3^4 = 5^3 \cdot 3^4$ (d) $6^5 \cdot 6^6 = 6^{5+6}$
$$\qquad\qquad\qquad\qquad\qquad = 6^{11}$$

3. $(4y)(5y^2)(y^5) = (4y^1)(5y^2)(1y^5)$
$$= (4 \cdot 5 \cdot 1)(y^1 \cdot y^2 \cdot y^5)$$
$$= 20y^8$$

4. $(4y^3)(3x^2)(5y^2) = (4 \cdot 3 \cdot 5)(y^3 \cdot x^2 \cdot y^2)$
$$= 60x^2 y^5$$

5. $2(x^3 - 6) = 2 \cdot x^3 - 2 \cdot 6$
$$= 2x^3 - 12$$

6. $x^4(x^3 + 6) = x^4 \cdot x^3 + x^4 \cdot 6$
$$= x^{4+3} + 6x^4$$
$$= x^7 + 6x^4$$

2.5   PRACTICE PROBLEMS

1. $150 \div 15$ **2.** $170 \div 5$

3. (a) A number divided by 15
$$\downarrow \qquad\quad \downarrow \qquad\quad \downarrow$$
$$n \qquad \div \qquad 15 \rightarrow n \div 15$$

   (b) 42 items divided equally among n groups
$$\downarrow \qquad\qquad \downarrow \qquad\qquad\qquad \downarrow$$
$$42 \qquad \div \qquad\qquad\qquad n \rightarrow 42 \div n$$

   (c) The quotient of 72 and 8
$$\downarrow \quad\; \downarrow$$
$$72 \;\div\; 8 \rightarrow 72 \div 8$$

   (d) The quotient of 8 and 72
$$\downarrow \quad\; \downarrow$$
$$8 \;\div\; 72 \rightarrow 8 \div 72$$

4. $\dfrac{56}{7} = ?$ $56 = ? \cdot 7$ Thus $\dfrac{56}{7} = 8$
$$56 = \underline{8} \cdot 7$$

5. $\dfrac{4 + 8 \div 2}{7 - 3} = (4 + 8 \div 2) \div (7 - 3)$
$$= (\underline{4 + 4}) \div 4$$
$$= 8 \div 4$$
$$= 2$$

6. $2(x + 7) + 15 \div 3 = 2 \cdot x + 2 \cdot 7 + 15 \div 3$
$$= 2x + 14 + \underline{15 \div 3}$$
$$= 2x + 14 + 5$$
$$= 2x + 19$$

7. $\begin{array}{r} 7\,\text{R1} \\ 6\overline{)43} \\ \underline{42} \\ 1 \end{array}$ Check: $6 \cdot 7 + 1 = 42 + 1 = 43$

8. $\begin{array}{r} 9\,\text{R30} \\ 36\overline{)354} \\ \underline{324} \\ 30 \end{array}$ Check: $36 \cdot 9 + 30 = 354$

9. $\begin{array}{r} 32\,\text{R51} \\ 80\overline{)2611} \\ \underline{240} \\ 211 \\ \underline{160} \\ 51 \end{array}$ Check: $80 \cdot 32 + 51 = 2611$

10. $\begin{array}{r} 403 \\ 37\overline{)14911} \\ \underline{148} \\ 111 \\ \underline{111} \\ 0 \end{array}$ Check: $37 \cdot 403 = 14,911$

11. $100 \div 22 = 4 \,\text{R12}$
   Twelve tickets were donated to the children's home.

2.6   PRACTICE PROBLEMS

1. (a) $9y - 2 = 9(4) - 2$
$$= 36 - 2$$
$$= 34$$

(b) $\dfrac{x}{3} = \dfrac{12}{3} = 4$

2. $6x = 48$

"Six times what number equals 48?"

The answer, or solution, is $x = 8$.

Check: $6x = 48$

$6(8) = 48$

$48 = 48$

3. $\dfrac{x}{4} = 2$

"What number divided by 4 equals 2?"

The answer, or solution, is $x = 8$.

Check: $\dfrac{x}{4} = 2 \;\rightarrow\; \dfrac{8}{4} \overset{?}{=} 2 \;\rightarrow\; 2 = 2 \;\checkmark$

4. $\dfrac{n}{7} = 52$

$n = 52 \cdot 7$

$n = 364$

5. $\dfrac{x}{3^2} = 11 \cdot 4 + 9$

$\dfrac{x}{9} = 44 + 9$ Simplify

$\dfrac{x}{9} = 53$ Simplify

$x = 53 \cdot 9$ Write equivalent multiplication problem

$x = 477$ Simplify

Check:

$\dfrac{477}{3^2} = 11 \cdot 4 + 9$

$53 = 53 \;\checkmark$

6. $9x = 108$ Check: $9(12) = 108$

$x = \dfrac{108}{9}$ $108 = 108 \;\checkmark$

$x = 12$

7. $2(3n) = 30$ Check: $2(3n) = 30$

$(2 \cdot 3)n = 30$ $2(3 \cdot 5) = 30$

$6n = 30$ $2(15) = 30$

$n = \dfrac{30}{6}$ $30 = 30 \;\checkmark$

$n = 5$

8. $8(n \cdot 5) = \dfrac{320}{2}$ Check: $8(4 \cdot 5) = \dfrac{320}{2}$

$8(5 \cdot n) = 160$ $8(20) = 160$

$(8 \cdot 5)n = 160$ $160 = 160 \;\checkmark$

$40n = 160$

$n = \dfrac{160}{40}$

$n = 4$

9. $x \div 35 = 16$ Check: $560 \div 35 = 16$

$x = 16 \cdot 35$ $16 = 16 \;\checkmark$

$x = 560$

2.7 PRACTICE PROBLEMS

1. $P = 4s$

$= 4(11 \text{ yd})$

$= 44 \text{ yd}$

The perimeter of the square is 44 yards.

2. 25 squares

3. The area is $3 \text{ ft} \cdot 4 \text{ ft} = 12$ square feet

4. $A = b \cdot h$

$= (13 \text{ ft})(9 \text{ ft})$

$= (13 \cdot 9)(\text{ft} \cdot \text{ft})$

$= 117 \text{ ft}^2$

5. Total Area $= (7 \text{ yd})^2 + (3 \text{ yd} \cdot 5 \text{ yd}) + (11 \text{ yd} \cdot 5 \text{ yd})$

$= 49 \text{ yd}^2 + 15 \text{ yd}^2 + 55 \text{ yd}^2$

$= 119 \text{ yd}^2$

The area of the shape is 119 yd^2.

6. **(a)** $L = 12 \text{ ft}$

$= 3 \text{ ft} + 3 \text{ ft} + 3 \text{ ft} + 3 \text{ ft}$

$= 4 \times 3 \text{ ft}$

$= 4 \times 1 \text{ yd}$

$= 4 \text{ yd}$

$W = 15 \text{ ft}$

$= 3 \text{ ft} + 3 \text{ ft} + 3 \text{ ft} + 3 \text{ ft} + 3 \text{ ft}$

$= 5 \times 3 \text{ ft}$

$= 5 \times 1 \text{ yd}$

$= 5 \text{ yd}$

$A = L \cdot W$

$= 4 \text{ yd} \cdot 5 \text{ yd}$

$= 20 \text{ yd}^2$

(b) Total cost $= \$11 \text{ per square yard} \cdot 20 \text{ square yards}$

$= \$11 \cdot 20$

$= \$220$

7. $V = L \cdot W \cdot H$

$= 35 \text{ ft} \cdot 25 \text{ ft} \cdot 8 \text{ ft}$

$= 7000 \text{ ft}^3$

They must haul 7000 ft^3 of dirt.

8. Surface Area $=$ side $+$ side $+$ front $+$ back $+$ top $+$ bottom

$= 2$ side area $+ 2$ face area $+ 2$ base area

$= 2 (4 \text{ ft} \cdot 2 \text{ ft}) + 2(3 \text{ ft} \cdot 2 \text{ ft}) + 2(4 \text{ ft} \cdot 3 \text{ ft})$

$= 2(8 \text{ ft}^2) + 2(6 \text{ ft}^2) + 2(12 \text{ ft}^2)$

$= 16 \text{ ft}^2 + 12 \text{ ft}^2 + 24 \text{ ft}^2$

$= 52 \text{ ft}^2$

We need 52 ft^2 of metal material.

2.8 **PRACTICE PROBLEMS**

1.   1.   *Understand the problem.*

Mathematics Blueprint for Problem Solving			
Gather the Facts	What Am I Asked to Do?	How Do I Proceed?	Key Points to Remember
ⓐ	ⓑ	ⓒ	ⓓ

ⓐ   See the invoice for items and prices. There are 2 owners who will split expenses.

ⓑ   Determine each owner's share of expenses.

ⓒ   1. Find the total cost per item by multiplying the number of items × the price per item.

   2. Find the total cost of all items by adding all the numbers in step 1.

   3. Find each owner's share of expenses by dividing the total cost by 2.

ⓓ   Fill in the invoice as each step is completed to help keep the facts organized.

2.   *Solve and state the answer.*
   8 tables at \$230 each = \$1840 cost of tables
   50 chairs at \$25 each = 1250 cost of chairs
   3 ovens at \$910 each = 2730 cost of ovens
   \$5820 total cost

   \$5820 ÷ 2 = \$2910

   Each owner must pay \$2910.

3.   *Check.* Estimate the answer to see if it is reasonable.

2.   1.   *Understand the problem.*

Mathematics Blueprint for Problem Solving			
Gather the Facts	What Am I Asked to Do?	How Do I Proceed?	Key Points to Remember
ⓐ	ⓑ	ⓒ	ⓓ

ⓐ   A customer is awarded 3 frequent flier points for every 2 miles flown. Louie flew 4500 miles.

ⓑ   Determine how many frequent flier points Louie accumulated.

ⓒ   1. Divide 4500 by 2

   2. Multiply the number obtained in step 1 by 3.

ⓓ   Frequent flier points are determined by the number of miles flown.

2.   *Solve and state the answer.*
   $(4500 \div 2) \cdot 3 = 2250 \cdot 3 = 6750$
   Louie earned 6750 points.

3.   *Check.* Estimate the answer to see if it is reasonable.

CHAPTER 3
· · · · · · · · · · · · · · · · · · · ·

3.1 **PRACTICE PROBLEMS**

1.   **(a)** $-5 < 2$
   (b) $-3 > -6$
   (c) $-53 > -218$

2.   **(a)** A property tax increase of \$130: +\$130
   (b) A dive of 7 ft below the surface of the sea: -7 ft

3.   **(a)** $|-67| = 67$
   (b) $|8| = 8$

4. $|-12|$? $|2|$
      12 ? 2
      12 > 2
   $|-12| > |2|$

5.   **(a)** The opposite of -6 is 6.
   (b) The opposite of -1 is 1.
   (c) The opposite of 12 is -12.
   (d) The opposite of 1 is -1.

6.   **(a)** The temperature was $-5°F$ in Fairbanks and $-4°F$ in Bismarck. It was colder in Fairbanks.

   (b) The highest temperature was $11°F$ in Anchorage and the lowest temperature was $-6°F$ in Fargo.

3.2 **PRACTICE PROBLEMS**

1.   **(a)**

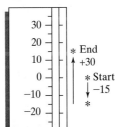

   (b) negative
   (c) $-4 + (-1)$
   (d) -5

2. $-2 + (-4) = -6$
   The answer is negative since the common sign is negative.

3.   **(a)** $-15°F + (+30°F)$
   (b) positive
   (c) $-15°F + 30°F = 15°F$

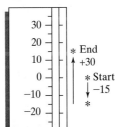

4.   **(a)** $-4 + 7 = +$ 7 is greater than 4 so the answer is positive.

      $-4 + 7 = +3$ or 3   Subtract $7 - 4 = 3$.

   (b) $4 + (-7) = -$ 7 is greater than 4 so the answer is negative.

      $4 + (-7) = -3$ Subtract $7 - 4 = 3$.

5.   **(a)** $-4 + 7 = 3$
   (b) $-4 + (-7) = -11$

6. $-8 + 6 + (-2) + 5 = \underline{-8} + 6 + (-2) + 5$
$= -10 + \underline{6} + \underline{5}$ Add the negative numbers.
$= -10 + 11$ Add the positive numbers.
$= 1$ Add the results.

7. **(a)** $7 + x = -7 + (-5) = -12$

(b) $x + 8 = -2 + 8 = 6$

8.   1st quarter loss + 2nd quarter gain = Net loss
    $-\$40,000$ + $20,000$ $= -\$20,000$

At the end of the second quarter the company had a net loss of $20,000.

3.3 PRACTICE PROBLEMS

1. **(a)** $\$10 - \$20 = -\$10$
(b) $5 - 6 = -1$

2. **(a)** $20 - 10 = 10$
$20 + (-10) = 10$
(b) $5 - 2 = 3$
$5 + (-2) = 3$
(c) $20 - 5 = 15$
$20 + (-5) = 15$

3. **(a)** $-5 - 4$
$-5 + (-4) = -9$
(b) $-9 - (-5)$
$-9 + (5) = -4$

4. **(a)** $7 - 10 = 7 + (-10)$
$= -3$
(b) $(-4) - 15 = (-4) + (-15)$
$= -19$
(c) $8 - (-3) = 8 + 3$
$= 11$
(d) $(-5) - (-1) = (-5) + (1)$
$= -4$

5. $6 - 9 - 2 - 8 = 6 + (-9) + (-2) + (-8)$
$= 6 + (-19)$
$= -13$

6. $-3 - (-5) + (-11) = -3 + (5) + (-11)$
$= -14 + 5$
$= -9$

7. $3800 \text{ ft} - (-895 \text{ ft}) = 3800 \text{ ft} + 895 \text{ ft}$
$= 4695 \text{ ft}$

8. $30,000 - (-20,000) = 30,000 + 20,000$
$= 50,000$
The difference is $50,000.

3.4 PRACTICE PROBLEMS

1. $3(-1) = (-1) + (-1) + (-1) = -3$

2. **(a)** $3(8) = 24$
(b) $3(-8) = -24$
(c) $(-3)(8) = -24$

3. $(-2)(-4) = 8$

4. **(a)** $(-2)(-1)(-4) = \underline{(-2)(-1)}(-4)$
$= 2(-4)$
$= -8$
(b) $(-3)(-2)(-1)(-3) = \underline{(-3)(-2)}(-1)(-3)$
$= \underline{(6)(-1)}(-3)$
$= (-6)(-3)$
$= 18$

5. $(-3)(2)(-1)(6)(-3) = -[(3)(2)(1)(6)(3)]$
$= -108$

6. $(-4)^4 = (-4)(-4)(-4)(-4)$
$= 256$

7. **(a)** $(-2)^3 = -8$
The answer is negative since the exponent, 3, is odd.
(b) $(-2)^6 = 64$
The answer is positive since the exponent, 6, is even.
(c) $(-2)^7 = -128$
The answer is negative since the exponent, 7, is odd.

8. **(a)** $-5^2 = -(5 \cdot 5) = -25$
(b) $(-5)^2 = (-5)(-5) = 25$

9. **(a)** $42 \div 7 = 6$
(b) $42 \div (-7) = -6$
(c) $(-42) \div 7 = -6$
(d) $(-42) \div (-7) = 6$

10. **(a)** $49 \div (-7) = -7$
(b) $4(-9) = -36$
(c) $(-30)(-4) = 120$
(d) $\dfrac{(-54)}{(-9)} = 6$

3.5 PRACTICE PROBLEMS

1. **(a)** $-3b + 5b = (-3 + 5)b$
$= 2b$
(b) $-6y + 8x + 4y = -6y + 4y + 8x$
$= (-6 + 4)y + 8x$
$= -2y + 8x$

2. $7x + 5y - 8x = 7x + 5y + (-8x)$
$= 7x + (-8x) + 5y$
$= [7 + (-8)]x + 5y$
$= -1x + 5y \text{ or } -x + 5y$

3. **(a)** $4 - 6 + 8 = 4 + (-6) + 8$
$= 12 + (-6)$
$= 6$
(b) $4x - 6x + 8x = 4x + (-6x) + 8x$
$= 12x + (-6x)$
$= 6x$

4. **(a)** $(-6a)(-8a) = (-6)(-8)(a \cdot a)$
$= 48a^2$
(b) $(5x^3)(-2y^5) = (5)(-2)(x^3 \cdot y^5)$
$= -10x^3y^5$

3.6 PRACTICE PROBLEMS

1. $-6 + 20 \div 2(-2)^2 - 5 = -6 + 20 \div 2\underline{(-2)^2} - 5$
$$= -6 + 20 \div \underline{2(4)} - 5$$
$$= -6 + \underline{10(4)} - 5$$
$$= -6 + 40 + (-5)$$
$$= 29$$

2. $\dfrac{-10 + 4(-2)}{11 - 20} = \dfrac{-10 + (-8)}{-9}$
$$= \dfrac{-18}{-9}$$
$$= 2$$

3. $s = v + 32t$
$\qquad\quad \downarrow \qquad \downarrow$
$$= -8 + 32(4)$$
$$= -8 + 128$$
$$= 120$$
The speed of the skydiver is 120 ft per sec.

4. 9 oxide + 4 magnesium $= 9(-2) + 4(+2)$
$$= -18 + 8$$
$$= -10$$
The total charge is -10.

CHAPTER 4

4.1 PRACTICE PROBLEMS

1. **(a)** 975 is divisible by "5" (ends in 5) and "3"
$\quad$ ($9 + 7 + 5 = 21$ and 21 is divisible by 3).

$\quad$ **(b)** 122 is divisible by "2" (ends in an even number).

$\quad$ **(c)** 420 is divisible by "2" (ends in an even number),
$\quad$ "3" ($4 + 2 + 0 = 6$ and 6 is divisible by 3), and
$\quad$ "5" (ends in 0).

$\quad$ **(d)** 11,121 is divisible by "3" ($1 + 1 + 1 + 2 + 1$
$\quad = 6$).

2. Prime: 3, 13, 19, 23, 37, 41
$\quad$ Composite: 9, 16, 32, 50
$\quad$ Neither: 0

3. **(a)** $14 = 2 \cdot 7$
$\quad$ **(b)** $27 = 3 \cdot 3 \cdot 3$ or 3^3

4. $50 = 2 \cdot 5 \cdot 5$ or $2 \cdot 5^2$
$$\begin{array}{r} 5 \\ 5\overline{)25} \\ 2\overline{)50} \end{array}$$

5. $96 = 2 \cdot 2 \cdot 2 \cdot 2 \cdot 2 \cdot 3$ or $2^5 \cdot 3$

6. $315 = 3 \cdot 3 \cdot 5 \cdot 7$ or $3^2 \cdot 5 \cdot 7$

7. 36 $\qquad\qquad 36 = 2 \cdot 2 \cdot 3 \cdot 3$ or $2^2 \cdot 3^2$

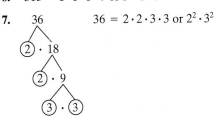

4.2 PRACTICE PROBLEMS

1. **(a)** $\dfrac{7}{16}$ $\qquad$ **(b)** $\dfrac{3}{8}$ $\qquad$ **(c)** $\dfrac{2}{2} = 1$

2. **(a)** $4 \div 0$ $\quad$ Undefined
$\quad$ **(b)** $0 \div 18 = \dfrac{0}{18} = 0$
$\quad$ **(c)** $\dfrac{0}{65} = 0$
$\quad$ **(d)** $\dfrac{65}{0}$ $\quad$ Undefined

3. **(a)** $\dfrac{13}{37}$ $\qquad$ **(b)** $\dfrac{5}{37}$

4. **(a)** Improper fraction
$\quad$ **(b)** Improper fraction, assume $x \neq 0$
$\quad$ **(c)** Mixed number
$\quad$ **(d)** Proper fraction

5. $\dfrac{23}{6} \rightarrow \begin{array}{r} 3 \\ 6\overline{)23} \\ \underline{18} \\ 5 \end{array} \rightarrow 5\dfrac{5}{6}$

6. $8\dfrac{2}{3} = \dfrac{(8 \cdot 3) + 2}{3} = \dfrac{26}{3}$

4.3 PRACTICE PROBLEMS

1. $\dfrac{1}{3} \cdot \dfrac{2}{7} = \dfrac{1 \cdot 2}{3 \cdot 7} = \dfrac{2}{21}$

2. **(a)** $\dfrac{4}{7} = \dfrac{4 \cdot 3}{7 \cdot 3} = \dfrac{12}{21}$
$\quad$ **(b)** $\dfrac{4}{7} = \dfrac{4 \cdot 5}{7 \cdot 5} = \dfrac{20}{35}$

3. $\dfrac{2}{9} = \dfrac{?}{36x}$
$\quad$ Nine times what number equals $36x$? $\quad 4x$.
$\quad \dfrac{2 \cdot 4x}{9 \cdot 4x} = \dfrac{8x}{36x}$

4. **(a)** $\dfrac{4}{22} \overset{?}{=} \dfrac{12}{87}$ $\qquad$ **(b)** $\dfrac{84}{108} \overset{?}{=} \dfrac{7}{9}$
$\quad\quad 87 \cdot 4 \overset{?}{=} 22 \cdot 12$ $\qquad 9 \cdot 84 \overset{?}{=} 108 \cdot 7$
$\quad\quad 348 \neq 264$ $\qquad\qquad 756 = 756$
$\quad\quad \dfrac{4}{22} \neq \dfrac{12}{87}$ $\qquad\qquad \dfrac{84}{108} = \dfrac{7}{9}$

5. **(a)** $\dfrac{6}{13} \, ? \, \dfrac{2}{7}$ $\qquad$ **(b)** $\dfrac{11}{30} \, ? \, \dfrac{9}{20}$
$\quad\quad 7 \cdot 6 \, ? \, 13 \cdot 2$ $\qquad 20 \cdot 11 \, ? \, 30 \cdot 9$
$\quad\quad 42 > 26$ $\qquad\qquad 220 < 270$
$\quad\quad \dfrac{6}{13} > \dfrac{2}{7}$ $\qquad\qquad \dfrac{11}{30} < \dfrac{9}{20}$

6. $\dfrac{7}{8} \ ? \ \dfrac{10}{11}$

$11 \cdot 7 \ ? \ 8 \cdot 10$

$77 < 80$

$\dfrac{7}{8} < \dfrac{10}{11}$

10 out of 11 parcels yields more land.

4.4 PRACTICE PROBLEMS

1. $\dfrac{18}{54} = \dfrac{2 \cdot 3 \cdot 3}{2 \cdot 3 \cdot 3 \cdot 3} = \dfrac{\overset{1}{\cancel{2}} \cdot \overset{1}{\cancel{3}} \cdot \overset{1}{\cancel{3}}}{\underset{1}{\cancel{2}} \cdot \underset{1}{\cancel{3}} \cdot \underset{1}{\cancel{3}} \cdot 3} = \dfrac{1 \cdot 1 \cdot 1}{1 \cdot 1 \cdot 1 \cdot 3} = \dfrac{1}{3}$

2. $\dfrac{28}{60} = \dfrac{2 \cdot 2 \cdot 7}{2 \cdot 2 \cdot 3 \cdot 5} = \dfrac{\overset{1}{\cancel{2}} \cdot \overset{1}{\cancel{2}} \cdot 7}{\underset{1}{\cancel{2}} \cdot \underset{1}{\cancel{2}} \cdot 3 \cdot 5} = \dfrac{1 \cdot 1 \cdot 7}{1 \cdot 1 \cdot 3 \cdot 5} = \dfrac{7}{15}$

3. $\dfrac{90y}{150xy} = \dfrac{2 \cdot 3 \cdot 3 \cdot 5 \cdot y}{2 \cdot 3 \cdot 5 \cdot 5 \cdot x \cdot y} = \dfrac{\overset{1}{\cancel{2}} \cdot \overset{1}{\cancel{3}} \cdot 3 \cdot \overset{1}{\cancel{5}} \cdot \overset{1}{\cancel{y}}}{\underset{1}{\cancel{2}} \cdot \underset{1}{\cancel{3}} \cdot \underset{1}{\cancel{5}} \cdot 5 \cdot x \cdot \underset{1}{\cancel{y}}} = \dfrac{3}{5x}$

4. $\dfrac{80x^2}{140x} = \dfrac{20 \cdot 4 \cdot x \cdot x}{20 \cdot 7 \cdot x} = \dfrac{\overset{1}{\cancel{20}} \cdot 4 \cdot \overset{1}{\cancel{x}} \cdot x}{\underset{1}{\cancel{20}} \cdot 7 \cdot \underset{1}{\cancel{x}}} = \dfrac{4x}{7}$

5. $\dfrac{-84}{14} = -\dfrac{84}{14} = -\dfrac{\overset{1}{\cancel{2}} \cdot 2 \cdot 3 \cdot \overset{1}{\cancel{7}}}{\underset{1}{\cancel{2}} \cdot \underset{1}{\cancel{7}}} = -\dfrac{6}{1} = -6$

4.5 PRACTICE PROBLEMS

1. **(a)** $\dfrac{28 \text{ feet}}{49 \text{ feet}} = \dfrac{4 \cdot 7 \text{ feet}}{7 \cdot 7 \text{ feet}} = \dfrac{4}{7}$

  (b) $\dfrac{27}{81} = \dfrac{3 \cdot 3 \cdot 3}{3 \cdot 3 \cdot 3 \cdot 3} = \dfrac{1}{3}$

2. **(a)** $\dfrac{21}{15} = \dfrac{3 \cdot 7}{3 \cdot 5} = \dfrac{7}{5}$

  (b) $\dfrac{15}{21} = \dfrac{3 \cdot 5}{3 \cdot 7} = \dfrac{5}{7}$

3. $\dfrac{46 \text{ inches}}{29 \text{ inches}} = \dfrac{46}{29}$

4. $\dfrac{\$26}{400 \text{ pounds}} = \dfrac{\$13}{200 \text{ pounds}}$

5. $\dfrac{92 \text{ miles}}{5 \text{ gallons}} = \dfrac{18\frac{2}{5} \text{ miles}}{1 \text{ gallon}}$ or $18\dfrac{2}{5}$ miles per gallon.

6. **(a)** $\dfrac{\text{September}}{\text{August}} = \dfrac{1505}{2185} = \dfrac{301}{437}$

  (b) $\dfrac{6031 \text{ June}}{70 \text{ years}} \rightarrow$ 86 tornados per year in June.

$\begin{array}{r} 6\frac{11}{70} \\ \hline 70\overline{)6031} \end{array}$

7. **(a)** $\dfrac{\text{patients}}{\text{registered nurses}} = \dfrac{40 \text{ patients}}{2 \text{ registered nurses}} = \dfrac{20}{1}$

   = 20 patients per registered nurse

  (b) $\dfrac{\text{patients}}{\text{nurse's aides}} = \dfrac{30 \text{ patients}}{2 \text{ nurse's aides}} = \dfrac{15}{1}$

   = 15 patients per nurse's aide.

  (c) $60 \div 15 = 4$ nurse's aides for 60 patients.

8. **(a)** $\dfrac{\$78}{6} = \$13$ per towel

   $\dfrac{\$108}{9} = \$12$ per towel

  (b) The towels from the Springview Collection are the better buy.

4.6 PRACTICE PROBLEMS

1. $\dfrac{4 \text{ hours}}{144 \text{ miles}} = \dfrac{6 \text{ hours}}{216 \text{ miles}}$

2. **(a)** $\dfrac{14}{45} \overset{?}{=} \dfrac{42}{135}$

   $135 \cdot 14 \overset{?}{=} 45 \cdot 42$

   $1890 = 1890$ This is a proportion.

  (b) $\dfrac{32}{72} \overset{?}{=} \dfrac{128}{144}$

   $144 \cdot 32 \overset{?}{=} 72 \cdot 128$

   $4608 \ne 9216$ This is not a proportion.

3. $\dfrac{240}{3} \overset{?}{=} \dfrac{1240}{18}$

   $18 \cdot 240 \overset{?}{=} 3 \cdot 1240$

   $4320 \ne 3720$ The rates are not equal.

4. $\dfrac{n}{18} = \dfrac{28}{72}$

   $72 \cdot n = 18 \cdot 28$

   $72n = 504$

   $n = \dfrac{504}{72}$

   $n = 7$

5. $\dfrac{\frac{1}{3}}{x} = \dfrac{1}{\frac{1}{5}}$

   $\dfrac{1}{5} \cdot \dfrac{1}{3} = x \cdot 1$

   $\dfrac{1}{15} = x$

4.7 PRACTICE PROBLEMS

1. $\dfrac{18 \text{ desserts}}{15 \text{ people}} = \dfrac{n \text{ desserts}}{180 \text{ people}}$

$$\dfrac{18}{15} = \dfrac{n}{180}$$

$$180 \cdot 18 = 15 \cdot n$$

$$3240 = 15n$$

$$\dfrac{3240}{15} = n$$

$$216 = n$$

The catering company should plan to serve 216 desserts.

2. $\dfrac{12 \text{ feet width of garden}}{16 \text{ feet length of garden}} = \dfrac{18 \text{ feet width of fence}}{x \text{ feet length of fence}}$

$$\dfrac{12}{16} = \dfrac{18}{x}$$

$$x \cdot 12 = 16 \cdot 18$$

$$12x = 288$$

$$x = 24$$

The length of the fence is 24 feet.

3. $\dfrac{3 \text{ Cleo}}{5 \text{ Julie}} = \dfrac{\$2400 \text{ is Cleo's share of profits}}{\$x \text{ is Julie's share of profits}}$

$$\dfrac{3}{5} = \dfrac{2400}{x}$$

$$3x = 12{,}000$$

$$x = 4000$$

Julie's share of the profits is \$4000.

CHAPTER 5
· ·

5.1 PRACTICE PROBLEMS

1. $\dfrac{5}{8} \cdot \dfrac{2}{3} = \dfrac{5 \cdot 2}{8 \cdot 3}$

$$= \dfrac{5 \cdot 2}{2 \cdot 2 \cdot 2 \cdot 3}$$

$$= \dfrac{5 \cdot \cancel{2}}{2 \cdot 2 \cdot \cancel{2} \cdot 3}$$

$$= \dfrac{5}{12}$$

2. $\dfrac{8}{15} \cdot \dfrac{12}{14} = \dfrac{8 \cdot 12}{15 \cdot 14}$

$$= \dfrac{2 \cdot 2 \cdot 2 \cdot 2 \cdot 2 \cdot 3}{3 \cdot 5 \cdot 2 \cdot 7}$$

$$= \dfrac{2 \cdot 2 \cdot 2 \cdot 2 \cdot \cancel{2} \cdot \cancel{3}}{\cancel{3} \cdot 5 \cdot \cancel{2} \cdot 7}$$

$$= \dfrac{16}{35}$$

3. $\dfrac{-12}{24} \cdot \dfrac{-8}{13} = (+)$

$$= \dfrac{12 \cdot 8}{24 \cdot 13}$$

$$= \dfrac{2 \cdot 2 \cdot 3 \cdot 2 \cdot 2 \cdot 2}{2 \cdot 2 \cdot 2 \cdot 3 \cdot 13}$$

$$= \dfrac{\cancel{2} \cdot \cancel{2} \cdot \cancel{3} \cdot \cancel{2} \cdot 2 \cdot 2}{\cancel{2} \cdot \cancel{2} \cdot \cancel{2} \cdot \cancel{3} \cdot 13}$$

$$= \dfrac{4}{13}$$

4. $\dfrac{3x^6}{7} \cdot 14x^2 = \dfrac{3x^6}{7} \cdot \dfrac{14x^2}{1}$

$$= \dfrac{3x^6 \cdot 14x^2}{7 \cdot 1}$$

$$= \dfrac{3 \cdot x^6 \cdot 2 \cdot 7 \cdot x^2}{7 \cdot 1}$$

$$= \dfrac{3 \cdot x^6 \cdot 2 \cdot \cancel{7} \cdot x^2}{\cancel{7} \cdot 1}$$

$$= 6x^8$$

5. (a) $\dfrac{a}{-y} \rightarrow \dfrac{-y}{a}$

(b) $4 \rightarrow \dfrac{1}{4}$

6. $-\dfrac{7}{8} \div \dfrac{5}{13} = -\dfrac{7}{8} \cdot \dfrac{13}{5}$

$$= -\dfrac{7 \cdot 13}{8 \cdot 5}$$

$$= -\dfrac{91}{40}$$

7. $\dfrac{9x^6}{(-21)} \div \dfrac{(-42)}{18x^4} = \dfrac{9x^6}{(-21)} \cdot \dfrac{18x^4}{(-42)}$

The product of a negative and a negative is positive, so our answer is positive.

$$= \dfrac{\cancel{3} \cdot 3 \cdot x^6 \cdot 3 \cdot \cancel{6} \cdot x^4}{\cancel{3} \cdot 7 \cdot \cancel{6} \cdot 7}$$

$$= \dfrac{9x^{10}}{49}$$

8. $28x^5 \div \dfrac{4x}{19} = \dfrac{28x^5}{1} \cdot \dfrac{19}{4x}$

$$= \dfrac{\cancel{4} \cdot 7 \cdot \cancel{x} \cdot x^4 \cdot 19}{1 \cdot \cancel{4} \cdot \cancel{x}}$$

$$= 133x^4$$

9. $\dfrac{2}{13} \text{ of } \$1703 = \dfrac{2}{13} \cdot \dfrac{\$1703}{1} = \$262$

10. $\dfrac{3}{4} \div 2 = \dfrac{3}{4} \cdot \dfrac{1}{2} = \dfrac{3}{8}$

Alice should place $\dfrac{3}{8}$ pound of sugar in each container.

5.2 PRACTICE PROBLEMS

1. (a) multiples of $12x$: $12x, 24x, 36x, 48x, 60x$
      multiples of $20x$: $20x, 40x, 60x, 80x, 100x$

   (b) $60x$

2. multiples of 4: 4, 8, 12, 16, 20, 24, ...
   multiples of 5: 5, 10, 15, 20, 25, 30, ...
   LCM of 4 and 5 is 20.

3. $28 = 2 \cdot 2 \cdot 7$ $\rightarrow$ LCM $= 2 \cdot 2 \cdot 7 \cdot ?$
   $36 = 2 \cdot 2 \cdot 3 \cdot 3$ $\rightarrow$ LCM $= 2 \cdot 2 \cdot 7 \cdot 3 \cdot 3 \cdot ?$
   $70 = 2 \cdot 5 \cdot 7$ $\rightarrow$ LCM $= 2 \cdot 2 \cdot 7 \cdot 3 \cdot 3 \cdot 5$
   The LCM of 28, 36, and 70 $= 2 \cdot 2 \cdot 7 \cdot 3 \cdot 3 \cdot 5 = 1260$

4. $4x = 2 \cdot 2 \cdot x$ $\rightarrow$ LCM $= 2 \cdot 2 \cdot x \cdot ?$
   $x^2 = x \cdot x$ $\rightarrow$ LCM $= 2 \cdot 2 \cdot x \cdot x \cdot ?$
   $10x = 2 \cdot 5 \cdot x$ $\rightarrow$ LCM $= 2 \cdot 2 \cdot x \cdot x \cdot 5 = 20x^2$
   The LCM of $4x$, x^2, and $10x = 2 \cdot 2 \cdot x \cdot x \cdot 5 = 20x^2$

5.3 PRACTICE PROBLEMS

1. $\dfrac{3}{13} + \dfrac{7}{13} = \dfrac{3 + 7}{13}$

   $= \dfrac{10}{13}$

2. $-\dfrac{7}{6} + \left(-\dfrac{21}{6}\right) = \dfrac{-7 + (-21)}{6}$

   $= -\dfrac{28}{6}$

   $= \dfrac{2(-14)}{2(3)}$

   $= -\dfrac{14}{3}$ or $-4\dfrac{2}{3}$

3. (a) $\dfrac{8}{x} - \dfrac{3}{x} = \dfrac{8 - 3}{x}$

   $= \dfrac{5}{x}$

   (b) $\dfrac{y}{9} + \dfrac{5}{9} = \dfrac{y + 5}{9}$

4. (a) the LCD of 4 and 5 is 20.
   (b) the LCD of 6 and 12 is 12.

5. (a) $\dfrac{3}{5} = \dfrac{?}{10}$

   $\dfrac{3 \cdot 2}{5 \cdot 2} = \dfrac{6}{10}$

   (b) $\dfrac{1}{2} = \dfrac{?}{10}$

   $\dfrac{1 \cdot 5}{2 \cdot 5} = \dfrac{5}{10}$

6. (a) $-\dfrac{3}{8} + \dfrac{7}{9}$ LCD $= 72$

   $\dfrac{-3 \cdot 9}{8 \cdot 9} = \dfrac{-27}{72}$ $\dfrac{7 \cdot 8}{9 \cdot 8} = \dfrac{56}{72}$

   $\dfrac{-27}{72} + \dfrac{56}{72} = \dfrac{29}{72}$

   (b) $\dfrac{13}{30} - \dfrac{2}{15}$ LCD $= 30$

   $\dfrac{2 \cdot 2}{15 \cdot 2} = \dfrac{4}{30}$

   $\dfrac{13}{30} - \dfrac{4}{30} = \dfrac{9}{30} = \dfrac{3 \cdot 3}{3 \cdot 10} = \dfrac{3}{10}$

7. (a) $\dfrac{8}{x} + \dfrac{2}{4x}$ LCD $= 4x$

   $\dfrac{8 \cdot 4}{x \cdot 4} = \dfrac{32}{4x}$

   $\dfrac{32}{4x} + \dfrac{2}{4x} = \dfrac{34}{4x} = \dfrac{2 \cdot 17}{2 \cdot 2x} = \dfrac{17}{2x}$

   (b) $\dfrac{7}{y} - \dfrac{4}{x}$ LCD $= xy$

   $\dfrac{7 \cdot x}{y \cdot x} = \dfrac{7x}{xy}$ $\dfrac{4 \cdot y}{x \cdot y} = \dfrac{4y}{xy}$

   $\dfrac{7x}{xy} - \dfrac{4y}{xy} = \dfrac{7x - 4y}{xy}$

8. $\dfrac{8x}{15} + \dfrac{9x}{24}$ LCD $= 120$

   $\dfrac{8x \cdot 8}{15 \cdot 8} = \dfrac{64x}{120}$ $\dfrac{9x \cdot 5}{24 \cdot 5} = \dfrac{45x}{120}$

   $\dfrac{64x}{120} + \dfrac{45x}{120} = \dfrac{109x}{120}$

5.4 PRACTICE PROBLEMS

1. $\begin{array}{r} \dfrac{2}{9} \\ + \dfrac{5}{9} \\ \hline \dfrac{7}{9} \end{array}$

2. $\begin{array}{r} 5\dfrac{1}{3} \cdot \dfrac{5}{5} \\ + 6\dfrac{3}{5} \cdot \dfrac{3}{3} \\ \hline \end{array} = \begin{array}{r} \dfrac{5}{15} \\ + \dfrac{9}{15} \\ \hline 1\dfrac{14}{15} \end{array}$

3. $\begin{array}{r} 7\dfrac{3}{4} \cdot \dfrac{5}{5} \\ + 2\dfrac{4}{5} \cdot \dfrac{4}{4} \\ \hline \end{array} = \begin{array}{r} \dfrac{15}{20} \\ + \dfrac{16}{20} \\ \hline 9\dfrac{31}{20} \end{array} = 9 + 1\dfrac{11}{20} = 10\dfrac{11}{20}$

4.
$$5\frac{3}{12} = 4 + 1\frac{3}{12} = \frac{15}{12}$$
$$-3\frac{5}{12} \qquad\qquad -\frac{5}{12}$$
$$\qquad\qquad\qquad\qquad 1\frac{10}{12} = 1\frac{5}{6}$$

5.
$$6\frac{1}{7}\cdot\frac{4}{4} \qquad \frac{4}{28} \qquad \frac{32}{28}$$
$$-2\frac{3}{4}\cdot\frac{7}{7} \quad = \quad -\frac{21}{28} \quad = \quad -\frac{21}{28}$$
$$\qquad\qquad\qquad\qquad\qquad 3\frac{11}{28}$$

6.
$$9 \qquad\qquad \frac{3}{3}$$
$$-4\frac{1}{3} \quad = \quad -\frac{1}{3}$$
$$\qquad\qquad\qquad \frac{2}{3}$$

7.
$$4\frac{1}{3}\cdot 2\frac{1}{4} = \frac{13}{3}\cdot\frac{9}{4}$$
$$= \frac{13\cdot 3\cdot 3}{3\cdot 4}$$
$$= \frac{39}{4} \text{ or } 9\frac{3}{4}$$

8.
$$2\div\frac{1}{4} = 2\cdot 4$$
$$= 8$$

9.
$$4\frac{1}{2}\div\frac{1}{2} = \frac{9}{2}\div\frac{1}{2}$$
$$= \frac{9}{2}\cdot\frac{2}{1}$$
$$= 9$$

She must fill the $\frac{1}{2}$-tablespoon 9 times.

5.5   Practice Problems

1.
$$\frac{5}{6}+\frac{1}{6}\cdot\frac{3}{4} = \frac{5}{6}+\frac{1\cdot 3}{6\cdot 4}$$
$$= \frac{5}{6}+\frac{3}{24}$$
$$= \frac{5\cdot 4}{6\cdot 4}+\frac{3}{24}$$
$$= \frac{20}{24}+\frac{3}{24}$$
$$= \frac{23}{24}$$

2.
$$\frac{\frac{2}{5}}{\frac{16}{15}} = \frac{2}{5}\div\frac{16}{15}$$
$$= \frac{2}{5}\cdot\frac{15}{16}$$
$$= \frac{2\cdot 3\cdot 5}{5\cdot 2\cdot 8}$$
$$= \frac{3}{8}$$

3.
$$\frac{\frac{x^2}{5}}{\frac{x}{10}} = \frac{x^2}{5}\div\frac{x}{10}$$
$$= \frac{x^2}{5}\cdot\frac{10}{x}$$
$$= \frac{x\cdot\cancel{x}\cdot 2\cdot\cancel{5}}{\cancel{5}\cdot\cancel{x}}$$
$$= 2x$$

4.
$$\frac{\frac{3}{5}+\frac{1}{2}}{\frac{5}{6}-\frac{1}{3}} = \frac{\frac{3\cdot 2}{5\cdot 5}+\frac{1\cdot 5}{2\cdot 5}}{\frac{5}{6}-\frac{1\cdot 2}{3\cdot 2}}$$
$$= \frac{\frac{6}{10}+\frac{5}{10}}{\frac{5}{6}-\frac{2}{6}}$$
$$= \frac{\frac{11}{10}}{\frac{3}{6}}$$
$$= \frac{\frac{11}{10}}{\frac{1}{2}}$$
$$= \frac{11}{10}\div\frac{1}{2}$$
$$= \frac{11}{10}\cdot\frac{2}{1}$$
$$= \frac{11}{5}$$

5.6   Practice Problems

1. **(a)** $\dfrac{4^{11}}{4^7} = 4^{11-7} = 4^4$

   (b) $\dfrac{6^9}{8^{14}}$ We cannot divide.
   The bases are not the same.

   (c) $\dfrac{y^5}{y^9} = \dfrac{1}{y^{9-5}} = \dfrac{1}{y^4}$

2.
$$\frac{25y^5}{45y^8} = \frac{5\cdot 5\cdot y^5}{5\cdot 9\cdot y^8}$$
$$= \frac{5}{9\cdot y^{8-5}}$$
$$= \frac{5}{9y^3}$$

3. $(4^2)^3 = 4^2\times 4^2\times 4^2 = 4^{2+2+2} = 4^6$

4. **(a)** $(3^3)^4 = 3^{(3)(4)} = 3^{12}$

   (b) $(n^0)^7 = n^{(0)(7)} = n^0 = 1$
   or
   $(n^0)^7 = (1)^7 = 1$

5. **(a)** $(xy)^5 = x^5\cdot y^5 = x^5y^5$
   (b) $(4a^3)^4 = 4^4\cdot(a^3)^4 = 4^4a^{12} = 256a^{12}$

6. $\left(\dfrac{x}{3}\right)^3 = \dfrac{x^3}{3^3} = \dfrac{x^3}{27}$

5.7 PRACTICE PROBLEMS

1.   1. *Understand the Problem*

Mathematics Blueprint for Problem Solving			
Gather the Facts	What Am I Asked to Do?	How Do I Proceed?	Key Points to Remember
ⓐ	ⓑ	ⓒ	ⓓ

ⓐ 1. Living room: 22 feet by 15 feet.

2. Rug is being placed in the center of the room.

3. Distance of the rug from walls: $2\frac{1}{2}$ feet.

4. The rug costs $5 per square foot.

ⓑ a) Find the dimensions of the rug.

b) Calculate the cost of the rug.

ⓒ Find the length of the carpet by subtracting each portion of wood floor not covered by the rug from the length of the room.
Find the width of the carpet by subtracting the same amount of wood floor from the width of the room.
Find the area of the rug: $A = L \times W$.
Multiply the area by $5 to find the cost of the rug.

ⓓ We find the sums of the widths of the wood floor on the left and right of the rug, and on the top and bottom of the rug.

2. *Solve and state the answer.*
The length of the carpet:
$$L = 22 - \left(2\frac{1}{2} + 2\frac{1}{2}\right) = 17 \text{ feet}$$

The width of the carpet:
$$W = 15 - \left(2\frac{1}{2} + 2\frac{1}{2}\right) = 10 \text{ feet}$$

The dimensions of the rug are 17 feet × 10 feet.

The area of the rug:
$$A = L \times W$$
$$= 17 \text{ feet} \times 10 \text{ feet} = 170 \text{ ft}^2$$

The cost of the rug $= \dfrac{\$5}{\cancel{ft^2}} \times 170 \ \cancel{ft^2} = \$850$

The oriental rug will cost $850.

3. *Check.* The check is left to the student.

2. First we must find how many $3\frac{1}{8}$ foot shelves we can cut from a 10-foot board.

$$10 \div \ 3\frac{1}{8} = 10 \div \ \frac{25}{8}$$
$$= 10 \cdot \frac{8}{25}$$
$$= \frac{2 \cdot 5 \cdot 8}{5 \cdot 5}$$
$$= \frac{16}{5} \text{ or } 3\frac{1}{5}$$

Three shelves can be cut from each 10-foot board with some wood leftover.

Now we must find how many 10-foot boards are needed for the shelves.

$$8 \div \ 3 = \frac{8}{3} \text{ or } 2\frac{2}{3}$$

Nancy needs to purchase three 10-foot boards.

CHAPTER 6
....

6.1 PRACTICE PROBLEMS

1.
$$x - 19 = -31$$
$$x - 19 \underline{+ 19} = -31 \underline{+ 19}$$
$$x + 0 = -12$$
$$x = -12$$

2.
$$-58 = x + 3$$
$$\underline{+ \ -3 \qquad -3}$$
$$-61 = x$$

3.
$$5 - 8 = x - 2 + 19$$
$$-3 = x + 17$$
$$-3 = x + 17$$
$$\underline{+ \ -17 \qquad -17}$$
$$-20 = x$$

4.
$$\frac{? \cdot x}{-8} = x$$
$$\frac{-8 \cdot x}{-8} = x$$

5.
$$\frac{a}{2} = 17$$
$$\frac{2 \cdot a}{2} = 17 \cdot 2$$
$$a = 34$$

6.
$$5m = -155$$
$$\frac{5m}{5} = \frac{-155}{5}$$
$$m = -31$$

7.
$$65 - 53 = -12y$$
$$65 + (-53) = -12y$$
$$12 = -12y$$
$$\frac{12}{-12} = \frac{-12y}{-12}$$
$$-1 = y$$

6.2 PRACTICE PROBLEMS

1.
$$8x + 9 = 105$$
$$8x + 9 + (-9) = 105 + (-9)$$
$$8x = 96$$
$$\frac{8x}{8} = \frac{96}{8}$$
$$x = 12$$

Check:
$$8x + 9 = 105$$
$$8(12) + 9 \overset{?}{=} 105$$
$$96 + 9 \overset{?}{=} 105$$
$$105 = 105 \ \sqrt{}$$

2.
$$-5m - 10 = 115$$
$$-5m - 10 \underline{+\ 10} = 115 \underline{+\ 10}$$
$$-5m = 125$$
$$\frac{-5m}{-5} = \frac{125}{-5}$$
$$m = -25$$

3.
$$\frac{a}{3} + 7 = 4$$
$$\frac{a}{3} + 7 + (-7) = 4 + (-7)$$
$$\frac{a}{3} = -3$$
$$\frac{3 \cdot a}{3} = -3 \cdot 3$$
$$a = -9$$

4.
$$3x - 1 = 4x - 6$$
$$\underline{+\ -4x \qquad\quad -4x}$$
$$-x - 1 = \qquad -6$$
$$-x - 1 = -6$$
$$\underline{+\qquad 1 \quad\ 1}$$
$$-x \quad = -5$$
$$(-1)(-x) = (-5)(-1)$$
$$x = 5$$

5.
$$5(x + 4) = 40$$
$$5(x) + 5(4) = 40$$
$$5x + 20 = 40$$
$$5x + 20 = 40$$
$$\underline{+\qquad -20 \ -20}$$
$$5x \qquad = 20$$
$$\frac{5x}{5} = \frac{20}{5}$$
$$x = 4$$

6.3   PRACTICE PROBLEMS

1.
$$\frac{x}{5} + \frac{x}{2} = 7$$
$$10\left(\frac{x}{5}\right) + 10\left(\frac{x}{2}\right) = 10(7)$$
$$2x + 5x = 70$$
$$7x = 70$$
$$\frac{7x}{7} = \frac{70}{7}$$
$$x = 10$$
The LCD is 10.

2.
$$-5x + \frac{2}{7} = \frac{3}{2}$$
$$14(-5x) + 14\left(\frac{2}{7}\right) = 14\left(\frac{3}{2}\right)$$
$$-70x + 4 = 21$$
$$-70x + 4 = 21$$
$$\underline{+\qquad\quad -4 \quad -4}$$
$$-70x \qquad = 17$$
$$\frac{-70x}{-70} = \frac{17}{-70}$$
$$x = -\frac{17}{70}$$

3.
$$\frac{x}{4} + x = 5$$
$$4\left(\frac{x}{4}\right) + 4(x) = 4(5)$$
$$x + 4x = 20$$
$$5x = 20$$
$$\frac{5x}{5} = \frac{20}{5}$$
$$x = 4$$

4.
$$\frac{4}{x} + 2 = \frac{1}{7}$$
$$7x\left(\frac{4}{x}\right) + 7x(2) = 7x\left(\frac{1}{7}\right)$$
$$28 + 14x = \qquad x$$
$$28 + 14x = \qquad x$$
$$\underline{+\qquad -14x \qquad -14x}$$
$$28 \qquad = -13x$$
$$\frac{28}{-13} = \frac{-13x}{-13}$$
$$-\frac{28}{13} = x \text{ or } x = -2\frac{2}{13}$$

6.4   PRACTICE PROBLEMS

1. (a) $4x^2$ monomial, one term
    (b) $8x^2 - 9x + 1$ trinomial, three terms
    (c) $5x^3 + 8x$ binomial, two terms

2. $+y^2, -4x^2, +5x, -9y$

3. $(-6x)(3x - 8y - 2) = -18x^2 + 48xy + 12x$

4. $(2y^2 - 5)(-3y^4) = (-3y^4)(2y^2 - 5)$
$$= -6y^6 + 15y^4$$

5. $(x + 2)(x + 5) = x(x + 5) + 2(x + 5)$
$$= x \cdot x + x \cdot 5 + 2 \cdot x + 2 \cdot 5$$
$$= x^2 + 5x + 2x + 10$$
$$= x^2 + 7x + 10$$

6. $(x + 2)(x + 4)$
$$\quad\ \ \overset{F}{} \quad\ \overset{O}{} \quad\quad \overset{I}{} \quad\quad \overset{L}{}$$
$$= x \cdot x + x(+4) + (+2)x + (+2)(+4)$$
$$= x^2 + 4x + 2x + 8$$
$$= x^2 + 6x + 8$$

7.
$$\qquad\qquad\qquad \overset{F}{} \quad \overset{O}{} \quad\ \ \overset{I}{} \quad \overset{L}{}$$
$$(y - 5)(y - 3) = y^2 - 3y - 5y + 15$$
$$= y^2 - 8y + 15$$

8.
$$\qquad\qquad\qquad\quad \overset{F}{} \quad\ \ \overset{O}{} \quad \overset{I}{} \ \ \overset{L}{}$$
$$(3x - 1)(x + 1) = 3x^2 + 3x - x - 1$$
$$= 3x^2 + 2x - 1$$

6.5   PRACTICE PROBLEMS

1. (a) $+(a - 2) = a - 2$
    (b) $-(-4x + 6y - 2n) = 4x - 6y + 2n$

2. $(8z - 9) + (-7z + 4) = 8z - 9 - 7z + 4$
$$= z - 5$$

3. $(4y + 7) - (8y - 6) = 4y + 7 - 8y + 6$
$$= -4y + 13$$

4. $(4x^2 + 6x - 8) - (7x^2 - 6x - 5)$
$\quad = 4x^2 + 6x - 8 - 7x^2 + 6x + 5$
$\quad = -3x^2 + 12x - 3$

5. $4x - 2(3x^2 + 1) - (-3x^2 + x - 6)$
$\quad = 4x - 6x^2 - 2 - (-3x^2 + x - 6)$
$\quad = 4x - 6x^2 - 2 + 3x^2 - x + 6$
$\quad = 3x - 3x^2 + 4$

6.6 PRACTICE PROBLEMS

1. $-5(3x + 2) - 6x = 32$
$\quad -15x - 10 - 6x = 32$
$\quad\quad -21x - 10 = 32$

$$\begin{array}{rcl} -21x - 10 &=& 32 \\ +10 & & +10 \\ \hline -21x &=& 42 \end{array}$$

$$\frac{-21x}{-21} = \frac{42}{-21}$$
$$x = -2$$

Check:
$$-5(3x + 2) - 6x = 32$$
$$-5[3(-2) + 2] - 6(-2) \overset{?}{=} 32$$
$$-5(-6 + 2) + 12 \overset{?}{=} 32$$
$$-5(-4) + 12 \overset{?}{=} 32$$
$$20 + 12 = 32 \;\checkmark$$

2. $2(x + 4) = -7(x - 3) + 2$
$\quad 2x + 8 = -7x + 21 + 2$
$\quad 2x + 8 = -7x + 23$

$$\begin{array}{rcl} 2x + 8 &=& -7x + 23 \\ +7x & & +7x \\ \hline 9x + 8 &=& 23 \end{array}$$

$$\begin{array}{rcl} 9x + 8 &=& 23 \\ -8 & & -8 \\ \hline 9x &=& 15 \end{array}$$

$$\frac{9x}{9} = \frac{15}{9}$$
$$x = \frac{5}{3} \text{ or } 1\frac{2}{3}$$

Check:
$$2(x + 4) = -7(x - 3) + 2$$
$$2\left(\frac{5}{3} + 4\right) \overset{?}{=} -7\left(\frac{5}{3} - 3\right) + 2$$
$$2\left(\frac{17}{3}\right) \overset{?}{=} -7\left(-\frac{4}{3}\right) + 2$$
$$\frac{34}{3} \overset{?}{=} \frac{28}{3} + 2$$
$$\frac{34}{3} = \frac{34}{3} \;\checkmark$$

3. $(4x^2 + 6x + 3) - (4x^2 + 2) = 3x + 1$
$\quad 4x^2 + 6x + 3 - 4x^2 - 2 = 3x + 1$
$\quad\quad\quad 6x + 1 = 3x + 1$

$$\begin{array}{rcl} 6x + 1 &=& 3x + 1 \\ +\; -3x & & -3x \\ \hline 3x + 1 &=& 1 \end{array}$$

$$\begin{array}{rcl} 3x + 1 &=& 1 \\ +\quad -1 & & -1 \\ \hline 3x &=& 0 \end{array}$$

$$\frac{3x}{3} = \frac{0}{3}$$
$$x = 0$$

6.7 PRACTICE PROBLEMS

1. 1. *Understand the Problem*

Mathematics Blueprint for Problem Solving			
Gather the Facts	What Am I Asked to Do?	How Do I Proceed?	Key Points to Remember
ⓐ	ⓑ	ⓒ	ⓓ

ⓐ $L = 9\,\text{m}$
$\quad\quad W = x + 3$
$\quad\quad P = 26\,\text{m}$
$\quad\quad P = 2L + 2W$

ⓑ   Find the width of the rectangle.

ⓒ   1. Substitute $(x + 3)$ and 9 in the formula for perimeter.

$\quad\quad$ 2. Solve for x.

$\quad\quad$ 3. Substitute the value for x in the expression $W = x + 3$, to find the width.

$\quad\quad$ 4. Check the answer.

ⓓ   1. Place the unit 'm' in the answer.

$\quad\quad$ 2. Remember that once I find x, I still need to find W in order to answer the question.

2. *Solve and state the answer.*
$$P = 2L + 2W$$
$$26 = 2(9) + 2(x + 3)$$
$$26 = 18 + 2x + 6$$
$$26 = 24 + 2x$$

$$\begin{array}{rcl} 26 &=& 24 + 2x \\ +\; -24 & & -24 \\ \hline 2 &=& 2x \end{array}$$

$$1 = x$$
$$W = x + 3$$
$$W = 1 + 3$$
$$W = 4$$

The width of the rectangle is 4 m.

2. f = the length of the first side.
$\quad\;\; f + 3$ = the length of the second side.
$\quad\;\; 3f - 12$ = the length of the third side.

3. **(a)** Define the variable expressions:

Let J = Jason's salary

$J - 7400$ = the apprentice's salary

(b) Write the equation:

$J + (J - 7400) = 83,000$

(c) Solve the equation

$$J + J - 7400 = 83,000$$
$$2J - 7400 = 83,000$$
$$2J - 7400 = 83,000$$
$$\underline{+\qquad 7400 \quad + 7\,400}$$
$$2J \qquad = 90,400$$

$J = 45,200$

Jason's salary = $45,200

Apprentice's salary = $45,200 − $7400

$= $37,800$

(d) Check is left to the student.

CHAPTER 7
• •

7.1 PRACTICE PROBLEMS

1. 5.32: Five and thirty-two hundredths

2. Two hundred forty-five and 09/100

3. $8.723 = \dfrac{723}{1000}$

4. $\dfrac{17}{1000} = 0.017$

5. 0.77 ? 0.771
0.770 ? 0.771
0 < 1
0.770 < 0.771

6. 369.26

7. 16,500.0 kilowatt-hours

7.2 PRACTICE PROBLEMS

1.
$$\begin{array}{r} \overset{1}{}50.00 \\ 4.39 \\ + \ 0.70 \\ \hline 55.09 \end{array}$$
Add zeros so each number has the same number of decimal places.

2.
$$\begin{array}{r} \overset{5\ 10}{2\cancel{6}.\cancel{0}1} \\ - \ 5.70 \\ \hline 20.31 \end{array}$$

3. $9.02 - (-5.1) = 9.02 + 5.1$
$$\begin{array}{r} 9.02 \\ + \ 5.10 \\ \hline 14.12 \end{array}$$

4.
$$\begin{array}{r} 4.5y \\ + \ 7.2y \\ \hline 11.7y \end{array}$$

5. We round the cost of each meal to the nearest dollar, then add.

$7 + $9 + $19 + $7 + $9 + $22 = $73

The total cost of meals is approximately $73.

6. **(a)**
$$\begin{array}{r} 1980 \\ 1970 \end{array} \qquad \begin{array}{r} 76.0 \\ - \ 66.4 \\ \hline 9.6 \text{ quadrillion Btu} \end{array}$$

(b)
$$\begin{array}{r} 1980 \\ 1970 \end{array} \qquad \begin{array}{r} 76.0 \\ + \ 66.4 \\ \hline 142.4 \text{ quadrillion Btu} \end{array}$$

7.3 PRACTICE PROBLEMS

1. $0.05 \times 0.07 \rightarrow 5 \times 7 = 35$
0.05 × 0.07 = 0.0035
2 decimal + 2 decimal = 4 decimal
places   places   places

2.
$$\begin{array}{r} 20.1 \\ \times \ 4.32 \\ \hline 402 \\ 6\ 03 \\ 80\ 4 \\ \hline 86.832 \end{array}$$
1 decimal place
2 decimal places

We need 3 decimal places.

3.
$$\begin{array}{r} 6.22 \\ \times \ (-3) \\ \hline -18.66 \end{array}$$
2 decimal places
0 decimal places

We need 2 decimal places.

4. $0.123 \times 100 = 12.3$
↑ ↑
Move the decimal place 2 places to the right.

5. $(0.6944)(10^3) = 694.4$

6. $-11.9 - 6.542 \times 10^2 + 2.7$
$= -11.9 - \mathbf{6.542 \times 10^2} + 2.7$
$= \mathbf{-11.9 - 654.2} + 2.7$
$= -666.1 + 2.7$
$= -663.4$

7. We use the information contained in the Mathematics Blueprint for Example 8, with two changes:

out of state calls: 55 minutes
in-state calls: 185 minutes

Monthly bill = Base fee + 55 minute × $\dfrac{30¢}{\text{minute}}$

$\qquad\qquad + 185 \text{ minute} \times \dfrac{10¢}{\text{minute}}$

$= 4.95 + 55 \times 0.30 + 185 \times 0.10$

$= 4.95 + 16.50 + 18.50$

$= 39.95$

Natasha's average monthly bill of $39.95 is less than her budget of $45.

7.4 PRACTICE PROBLEMS

1. $1.3 \div 2 \rightarrow$
$$\begin{array}{r} .65 \\ 2)\overline{1.30} \\ \underline{1\ 2} \\ 10 \\ \underline{10} \\ 0 \end{array}$$

$1.3 \div 2 = 0.65$

2. $-36.12 \div 14 \rightarrow$

$$14\overline{)\begin{matrix}-2.58\\ -36.12\end{matrix}}$$
$$\begin{array}{r}28\\ \hline 8\,1\\ 7\,0\\ \hline 1\,12\\ 1\,12\\ \hline 0\end{array}$$

2.58 rounded to the nearest tenth is 2.6.

$-36.12 \div 14 \approx -2.6$

3. $14.56 \div 3.5 = \dfrac{14.56}{3.5} \cdot \dfrac{10}{10} = \dfrac{145.6}{35}$

$$35\overline{)145.60}$$
$$\begin{array}{r}4.16\\ \hline 140\\ \hline 5\,6\\ 3\,5\\ \hline 2\,10\\ 2\,10\\ \hline 0\end{array}$$

$14.56 \div 3.5 = 4.16$

4. $1.1 \div 1.8 \rightarrow 1.8\overline{)1.1}$

$$18\overline{)11.000}$$
$$\begin{array}{r}0.611\\ \hline 10\,8\\ \hline 20\\ 18\\ \hline 20\\ 18\\ \hline 2\end{array}$$

$1.1 \div 1.8 = 0.6\overline{1}$

5. $2\dfrac{5}{11} = 2 + \dfrac{5}{11}$

$$11\overline{)5.0000}$$
$$\begin{array}{r}0.4545\\ \hline 4\,4\\ \hline 60\\ 55\\ \hline 50\\ 44\\ \hline 60\\ 55\\ \hline 5\end{array}$$

$\dfrac{5}{11} = 0.\overline{45}$

$2\dfrac{5}{11} = 2.\overline{45}$

6. (a) $35.12 \div 10^3 = 0.03512$

(b) $54.2 \times 10^3 = 54{,}200$

7. $2119 = 2.119 \times 10^3$

8. $0.00345 = 3.45 \times 10^{-3}$

9. (a) $22.5 \div 2.5 = 9$

The janitor needs 9 containers.

(b) $2.5 \times \$5.00 = \$12.50$

One full container of fluid will cost \$12.50.

7.5 PRACTICE PROBLEMS

1. $\begin{aligned} y - 2.23 &= 4.69\\ +2.23\quad &\quad +2.23\\ \hline y &= 6.92\end{aligned}$

2. $5.6x = -25.2$

$\dfrac{5.6x}{5.6} = -\dfrac{25.2}{5.6}$

$x = -4.5$

3. $4(x - 2.5) = 3x + 0.6$

$4 \cdot x - 4 \cdot 2.5 = 3x + 0.6$

$4x - 10 = 3x + 0.6$

$\begin{array}{r}+\,-3x\qquad\quad -3x\\ \hline x - 10 = \quad 0.6\end{array}$

$x - 10 = \quad 0.6$

$\begin{array}{r}+10\quad +10\\ \hline x = \quad 10.6\end{array}$

4. $0.3x + 0.15 = 2.7$

$100(0.3x + 0.15) = 100(2.7)$

$100(0.3x) + 100(0.15) = 100(2.7)$

$30x + 15 = 270$

$30x = 255$

$x = 8.5$

5. We use the information contained in the Mathematics Blueprint for Example 5, with two changes:

— The number of dimes is 4 times the number of quarters.

— The total amount of change is \$3.25.

$(0.25)Q + (0.10)4Q = 3.25$

$0.25Q + 0.40Q = 3.25$

$25Q + 40Q = 325$

$65Q = 325$

$Q = 5 \rightarrow$ There are 5 quarters.

$4Q = 4(5) = 20 \rightarrow$ There are 20 dimes.

7.6 PRACTICE PROBLEMS

1. $\dfrac{42}{100} = 42\%$

42% of the students in the class voted.

2. $\begin{aligned}\text{present cost} &\rightarrow \dfrac{215}{100} = 215\%\\ \text{cost 10 years ago} &\rightarrow \end{aligned}$

3. $\dfrac{0.7}{100} = 0.7\%$

Julio has 0.7% of the solution.

4. (a) $2.66 = 266\%$
(b) $0.001 = 0.1\%$

5.

0.511	51.1%
0.841	84.1%
0.002	0.2%
6.776	677.6%

6. **(a)** $\dfrac{7}{40} \rightarrow 0.175 \rightarrow \underline{17.5\%}$

(b) $\dfrac{7}{40} \leftarrow 0.175 \leftarrow 17.5\%$

7. **(a)** $\dfrac{2}{3} = 0.666\overline{6} \approx 66.67\%$

(b) $\dfrac{15}{33} = 0.45\overline{45} \approx 45.45\%$

7.7 PRACTICE PROBLEMS

1. **(a)** What is 40% of 90?

$n = 40\% \times 90$
$n = 0.40 \times 90$
$n = 36$
$\underline{36}$ is 40% *of* 90.

(b) 36 is 40% of what number?

$36 = 40\% \times n$
$36 = 0.40 \times n$
$90 = n$
36 is 40% of $\underline{90}$.

(c) 36 is what percent of 90?

$36 = \quad n\% \quad \times 90$
$36 = n\% \times 90$
$0.4 = n\%$
$40\% = n$
36 is $\underline{40\%}$ of 90.

2. $80 = n\% \times 20$
$4 = n\%$
$400\% = n$
80 is $\underline{400\%}$ of 20.

3. $x = 22\% \times 60$
$x = 0.22 \times 60$
$x = 13.2$
$\underline{13.2}$ is 22% of 60.

4. 35 is what percent of 50?

$35 = n\% \times 50$
$0.7 = n\%$
$70\% = n$
Alisha earned 70% of the total points on the quiz.

5. \$3.50 is what percent of \$18.55?

$3.50 = n\% \times 18.55$
$0.1887 = n\%$
$18.87\% = n$
\$3.50 is $\underline{18.87\%}$ of \$18.55.

7.8 PRACTICE PROBLEMS

1. Commission = commission rate × value of sale
$= 6\% \times \$145{,}000$
$= 0.06 \times \$145{,}000$
$= \$8700$
The amount of commission you must pay is \$8700.

2. decrease = percent of decrease × original amount
$= 30\% \times \$1640$
$= 0.3 \times \$1640$
$= \$492$
The dining room set was reduced in price by \$492.

3. Amount of raise = percent increase × original salary
$= 3\% \times \$9.45$
$= 0.03 \times \$9.45$
$= \$0.2835$
$= \$0.28$ (rounded to the nearest cent)

New hourly rate = original rate + raise
$= \$9.45 + \$0.28$
$= \$9.73$

Rebecca's new hourly rate, rounded to the nearest cent, is \$9.73.

4. Decrease = percent decrease × original amount
$= 14\% \times 1500$
$= 0.14 \times 1500$
$= 210$

Tickets sold this year = tickets sold last year − decrease
$= 1500 - 210$
$= 1290$

This year 1290 tickets were sold.

5. **(a)** $I = P \times R \times T$
$= \$2000 \times 0.06 \times 1$
$= \$120$
Damon earned \$120 in interest.

(b) Total = principal + interest
$= \$2000 + \$120 = \$2120$
At the end of the year, he will have \$2120.

6. $I = P \times R \times T = \$3100 \times 0.09 \times \dfrac{1}{3} = \$93$

$\left(\dfrac{4}{12} = \dfrac{1}{3} \right)$

7. Interest at the end of the first six months

$= \$7000 \times 0.05 \times \dfrac{1}{2} = \$175$

Balance at the end of the first six months
$= \$7000 + \$175 = \$7175$

Interest at the end of the second six months.

$= \$7175 \times 0.05 \times \dfrac{1}{2} = \$179.375 = \$179.38$
(rounded to the nearest cent)

Balance at the end of the second six months
$= \$7175 + \$179.38 = \$7354.38$

Jerry has \$7354 at the end of one year in his savings account.

CHAPTER 8
....

8.1 PRACTICE PROBLEMS

1. **(a)** Walgreens has approximately
$7 \times 300 = 2100$ stores.

(b) Rite Aid has approximately
$9 \times 300 = 2700$ stores and
CVS has approximately
$5 \times 300 = 1500$ stores. Therefore, Rite Aid has
$2700 - 1500 = 1200$ more stores than CVS.

(c) $600 + 1200 = 1800$

2. **(a)** $9\% + 37\% = 46\%$
(b) 16% of $690,000 = (0.16)(690,000) = 110,400$

3. The bar rises to 250. The number of new cars sold in the fourth quarter of 1996 was 250.

4. $150 - 100 = 50$. Fifty fewer cars were sold.

5. Because the dot corresponding to 1992–93 is opposite 15 and the scale is in hundreds, we have $15 \times 100 = 1500$. Thus, 1500 degrees in visual and performing arts were awarded.

6. The computer science line goes above the visual and performing arts line first in 1990–91. The first academic year with more degrees in computer science was 1990–91.

7.

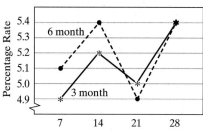

8.

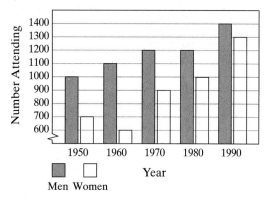

Men Women

8.2 PRACTICE PROBLEMS

1. $\dfrac{88 + 77 + 84 + 97 + 89}{5} = \dfrac{435}{5} = 87$
The mean is 87.

2. $\dfrac{24 + 23 + 25 + 25 + 23 + 26 + 26 + 24}{8} = \dfrac{196}{8}$
$= 24.5$
The mean miles per gallon for Wally, rounded to the nearest mile per gallon, is 25.

3. 3, 3, 4,   5,   6, 10, 13
         ⌣   ↓   ⌣

   three   middle   three
   numbers   number   numbers
The median value is 5 minutes.

4. 88, 90,   100, 105   118, 126
         ⌣   ⌣   ⌣

    two   two middle   two
   numbers   numbers   numbers
$\dfrac{100 + 105}{2} = \dfrac{205}{2} = 102.5$
The median value is 102.5.

8.3 PRACTICE PROBLEMS

1. (year, number of applicants)
           ↓   ↓
        (1986,   1050)
        (1988,   1100)
        (1990,   1050)
        (1992,   1125)

2.

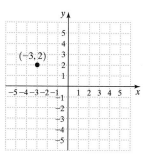

3.

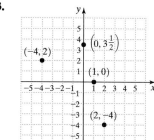

4. **(a)** $S = (-6, 2)$ **(b)** $T = (0, 1)$
        (c) $U = (3, 0)$ **(d)** $V = (7, 1)$
        (e) $W = (-6, -2)$ **(f)** $X = (7, -1)$

5. **(a)** $(5, -2)$
        (b)

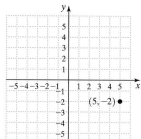

6.

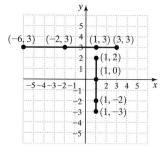

8.4 **PRACTICE PROBLEMS**

1. $(11, 0), (2, 9), (5, 6), (-2, 13)$ (Answers may vary.)

2. (a) number of   number of
    bottles labeled   minutes
$$\downarrow \qquad\qquad \downarrow$$
$$y = 25\,x$$
$$125 = 25x$$
$$\frac{125}{25} = x$$
$$5 = x \text{ or } x = 5 \text{ minutes}$$
Since $x = 5$ when $y = 125$, the ordered pair
(x, y) is $(5, 125)$.

(b) $y = 25x$
$$200 = 25x$$
$$8 = x \text{ or } x = 8 \text{ minutes}$$
Since $x = 8$ when $y = 200$, the ordered pair
(x, y) is $(8, 200)$.

3. $(\underline{\quad}, 0)$ $(\underline{\quad}, 2)$
$$x + 3y = 12 \qquad\qquad x + 3y = 12$$
$$x + 3(0) = 12 \qquad\qquad x + 3(2) = 12$$
$$x + 0 = 12 \qquad\qquad x + 6 = 12$$
$$x = 12 \qquad\qquad\qquad x = 6$$
$$(12, 0) \qquad\qquad\qquad (6, 2)$$

4.

$(x$	$y)$
$(0$	$)$
$($	$8)$
$($	$5)$

$\rightarrow$

$(x$	$y)$
$(0$	$2)$
$(6$	$8)$
$(3$	$5)$

$y = x + 2$ $y = x + 2$ $y = x + 2$
    $\downarrow$ $\downarrow$ $\downarrow$
$y = 0 + 2$ $8 = x + 2$ $5 = x + 2$
$y = 2$ $6 = x$ $3 = x$
$(0, 2)$ $(6, 8)$ $(3, 5)$

5. Answers may vary.

$(x$	$y)$
$(0$	$)$
$(1$	$)$
$(-1$	$)$

$\rightarrow$

$(x$	$y)$
$(0$	$-5)$
$(1$	$-1)$
$(-1$	$-9)$

$y = 4x - 5$ $y = 4x - 5$ $y = 4x - 5$
    $\downarrow$ $\downarrow$ $\downarrow$
$y = 4(0) - 5$ $y = 4(1) - 5$ $y = 4(-1) - 5$
$y = -5$ $y = -1$ $y = -9$
$(0, -5)$ $(1, -1)$ $(-1, -9)$

6. (a) $y = 2x + 1$

$(x$	$y)$
$(0$	$)$
$(1$	$)$
$(-1$	$)$

$\rightarrow$

$(x$	$y)$
$(0$	$1)$
$(1$	$3)$
$(-1$	$-1)$

(b)

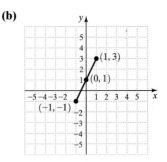

7. (a) $y = -3x - 1$

$(x$	$y)$
$(-1$	$)$
$(0$	$)$
$(1$	$)$

$\rightarrow$

$(x$	$y)$
$(-1$	$2)$
$(0$	$-1)$
$(1$	$-4)$

(b)

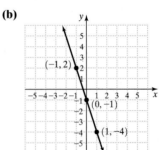

8.

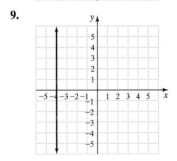

9.

CHAPTER 9

9.1 **PRACTICE PROBLEMS**

1. $420 \text{ min} = 420 \text{ m\cancel{in}} \times \dfrac{1 \text{ hr}}{60 \text{ m\cancel{in}}} = 7 \text{ hr}$

2. $144 \text{ oz} = 144 \text{ o\cancel{z}} \times \dfrac{1 \text{ lb}}{16 \text{ o\cancel{z}}} = 9 \text{ lb}$

3. $7.2 \ \cancel{\text{hr}} \ \times \ \dfrac{60 \ \text{min}}{1 \ \cancel{\text{hr}}} = 432 \ \text{min}$

4. $1.75 \ \cancel{\text{days}} \times \dfrac{24 \ \cancel{\text{hr}}}{1 \ \cancel{\text{day}}} \times \dfrac{1.50 \ \text{dollars}}{1 \ \cancel{\text{hr}}}$

$= 63 \ \text{dollars or} \ \$63$

She paid $63 to park her car.

5. (a) 4 m = 400 cm (move 2 places to the right)
(b) 30 cg = 300 mg (move 1 place to the right)

6. (a) 3 mL = 0.003 L (move 3 places to the left)
(b) 47 cm = 0.00047 km (move 5 places to the left)

7. (a) 389 mm = 0.0389 dkm (move 4 places to the left)
(b) 0.48 hm = 4800 cm (move 4 places to the right)

8. A milliliter is $\dfrac{1}{1000}$ of a liter so the acid would cost

$\dfrac{\$110}{1000} = \$0.11 \ \text{or} \ 11¢ \ \text{per milliliter.}$

9.2   PRACTICE PROBLEMS

1. (a) $17 \ \cancel{\text{m}} \times \dfrac{1.09 \ \text{yd}}{1 \ \cancel{\text{m}}} = 18.53 \ \text{yd}$

(b) $29.6 \ \cancel{\text{km}} \times \dfrac{0.62 \ \text{mi}}{1 \ \cancel{\text{km}}} = 18.352 \ \text{mi}$

(c) $26 \ \cancel{\text{gal}} \times \dfrac{3.79 \ \text{L}}{1 \ \cancel{\text{gal}}} = 98.54 \ \text{L}$

(d) $6.2 \ \cancel{\text{L}} \times \dfrac{1.06 \ \text{qt}}{1 \ \cancel{\text{L}}} = 6.572 \ \text{qt}$

(e) $16 \ \cancel{\text{lb}} \times \dfrac{0.454 \ \text{kg}}{1 \ \cancel{\text{lb}}} = 7.264 \ \text{kg}$

(f) $280 \ \cancel{\text{g}} \times \dfrac{0.0353 \ \text{oz}}{1 \ \cancel{\text{g}}} = 9.884 \ \text{oz}$

2. $180 \ \text{cm} \times \dfrac{0.394 \ \text{in.}}{1 \ \text{cm}} \times \dfrac{1 \ \text{ft}}{12 \ \text{in.}} = 5.91 \ \text{ft}$

3. $\dfrac{88 \ \text{km}}{1 \ \text{hr}} \times \dfrac{0.62 \ \text{mi}}{1 \ \text{km}} = 54.56 \ \text{mi per hr}$

4. 9 mm = 0.9 cm

$0.9 \ \cancel{\text{cm}} \times \dfrac{0.394 \ \text{in.}}{1 \ \cancel{\text{cm}}} = 0.3546 \ \text{in.}$

The bullet is about 0.35 in. wide.

5. $C = \dfrac{5 \times F - 160}{9}$

$C = \dfrac{5 \times 181 - 160}{9}$

$C = \dfrac{745}{9}$

$C \approx 82.77$

To the nearest degree, 181°F is 83°C.

9.3   PRACTICE PROBLEMS

1. (a) 81 is a perfect square because $(9)^2 = 9 \cdot 9 = 81$.
(b) 17 is not a perfect square. There is no whole number or fraction that when squared equals 17.
(c) $\dfrac{1}{9}$ is a perfect square because $\left(\dfrac{1}{3}\right)^2 = \dfrac{1}{3} \cdot \dfrac{1}{3} = \dfrac{1}{9}$.

2. (a) $n^2 = 81$
$n = \sqrt{81}$
(b) What is the positive square root of 81? $n = \sqrt{81}$
(c) $n \cdot n = 81 \ \ n = \sqrt{81}$

3. (a) $\sqrt{25} = 5$ since $5 \cdot 5 = 25$.
(b) $\sqrt{y^2} = y$ since $y \cdot y = y^2$.

4. (a) $\sqrt{\dfrac{81}{100}} = \dfrac{9}{10}$ since $\left(\dfrac{9}{10}\right)\left(\dfrac{9}{10}\right) = \dfrac{81}{100}$
(b) $\sqrt{\dfrac{16}{49}} = \dfrac{4}{7}$ since $\left(\dfrac{4}{7}\right)\left(\dfrac{4}{7}\right) = \dfrac{16}{49}$

5. $A = s^2$
$49 = s^2$
$\sqrt{49} = s$
$7 = s$
The length of the side of the square is 7 feet.

6. (a) $4\sqrt{100} = 4 \cdot 10 = 40$
(b) $7\sqrt{23} \approx 7 \cdot 4.79583$
≈ 33.57081
≈ 33.571

7. (a) $\sqrt{900} = \sqrt{9 \cdot 100}$
$= \sqrt{9} \cdot \sqrt{100}$
$= 3 \cdot 10$
$= 30$
(b) $\sqrt{49y^2} = \sqrt{49 \cdot y^2}$
$= \sqrt{49} \cdot \sqrt{y^2}$
$= 7y$

8. (a) $\sqrt{75} = \sqrt{25 \cdot 3}$
$= \sqrt{25} \cdot \sqrt{3}$
$= 5\sqrt{3}$
(b) $\sqrt{28} = \sqrt{4 \cdot 7}$
$= \sqrt{4} \cdot \sqrt{7}$
$= 2\sqrt{7}$

9. (a) $\sqrt{216y^4} = \sqrt{2 \cdot 2 \cdot 2 \cdot 3 \cdot 3 \cdot 3 \cdot y \cdot y \cdot y \cdot y}$
$= \sqrt{2 \cdot 2} \cdot \sqrt{3 \cdot 3} \cdot \sqrt{y \cdot y} \cdot \sqrt{y \cdot y} \cdot \sqrt{2 \cdot 3}$
$= 2 \cdot 3 \cdot y \cdot y \cdot \sqrt{6}$
$= 6y^2\sqrt{6}$
(b) $\sqrt{160x^3} = \sqrt{16} \cdot \sqrt{10} \cdot \sqrt{x^3}$
$= 4 \cdot \sqrt{10} \cdot \sqrt{x \cdot x \cdot x}$
$= 4 \cdot \sqrt{10} \cdot \sqrt{x \cdot x} \cdot \sqrt{x}$
$= 4 \cdot \sqrt{10} \cdot x \cdot \sqrt{x}$
$= 4x\sqrt{10x}$

10. (a) $7\sqrt{5} - 3\sqrt{5} = 4\sqrt{5}$
(b) $3\sqrt{12} + 2\sqrt{27} = 3\sqrt{4 \cdot 3} + 2\sqrt{9 \cdot 3}$
$= 3 \cdot 2\sqrt{3} + 2 \cdot 3\sqrt{3}$
$= 6\sqrt{3} + 6\sqrt{3}$
$= 12\sqrt{3}$

(c) $\sqrt{75} - \sqrt{32} = \sqrt{25 \cdot 3} - \sqrt{16 \cdot 2}$
$= 5\sqrt{3} - 4\sqrt{2}$
We cannot subtract $5\sqrt{3} - 4\sqrt{2}$ since each term does not contain the same square root.

9.4 PRACTICE PROBLEMS

1. $x + 125 + 15 = 180$
$x + 140 = 180$
$x = 40$
The measure of angle A equals 40°.

2. $P = s + s + s$ or $P = 3s$
$30 = 3s$
$10 = s$
The length of each side is 10 centimeters.

3. $A = \dfrac{bh}{2} = \dfrac{(5 \text{ cm})(16 \text{ cm})}{2} = \dfrac{80 \text{ cm}^2}{2} = 40 \text{ cm}^2$

4. The area of the rectangle is
$A = L \times W = (24 \text{ cm})(11 \text{ cm}) = 264 \text{ cm}^2$
The area of the triangle is
$A = \dfrac{bh}{2} = \dfrac{(11 \text{ cm})(7 \text{ cm})}{2} = \dfrac{77 \text{ cm}^2}{2} = 38.5 \text{ cm}^2$
Total area $= 264 \text{ cm}^2 + 38.5 \text{ cm}^2 = 302.5 \text{ cm}^2$

5. $(\text{Hypotenuse})^2 = (\text{leg})^2 + (\text{other leg})^2$
$c^2 = a^2 + b^2$
$c^2 = 3^2 + 4^2$
$c^2 = 9 + 16$
$c^2 = 25$
$c = \sqrt{25}$
$c = 5$
The length of the hypotenuse is 5 meters.

6. $c^2 = a^2 + b^2$
$c^2 = 12^2 + 15^2$
$c^2 = 144 + 225$
$c^2 = 369$
$c = \sqrt{369}$
$c \approx 19.2$
The ladder is approximately 19.2 feet in length.

9.5 PRACTICE PROBLEMS

1. $C = \pi d$
$\approx (3.14)(5.1 \text{ m})$
$\approx 16.014 \text{ m}$

2. Distance traveled by the larger wheel in 2 complete revolutions is equal to 2 times the circumference.
$2C = (2)(2\pi r)$
$\approx (2)(2)(3.14)(3.75 \text{ ft})$
$\approx 47.1 \text{ ft}$
Distance traveled by the smaller wheel in 1 revolution is equal to the circumference.
$C = 2\pi r$
$\approx (2)(3.14)(1.25 \text{ ft})$
$\approx 7.85 \text{ ft}$
To find the number of revolutions the smaller wheel made to equal the same distance as the larger wheel traveled in two revolutions, we divide
$47.1 \div 7.85 = 6$
The small wheel made 6 complete revolutions.

3. $A = \pi r^2$
$\approx (3.14)(6.1 \text{ cm})^2$
$\approx (3.14)(37.21 \text{ cm}^2)$
$\approx 116.8394 \text{ cm}^2$
The area of the circle is approximately 116.8 cm^2.

4. 1. *Understand the problem.*
   We are given the diameter in inches. We need to find the radius. The cost of the tablecloth is in square feet. We need to change square inches to square feet.

   2. *Solve and state the answer.*
   The diameter is 60 inches so the radius is 30 inches.
   Find the area $A = \pi r^2$
   $\approx (3.14)(30 \text{ in.})^2$
   $\approx (3.14)(900 \text{ in}^2)$
   $\approx 2826 \text{ in}^2$
   Change square inches to square feet.
   $1 \text{ ft} = 12 \text{ in.}$
   $(1 \text{ ft})^2 = (12 \text{ in.})^2$
   $1 \text{ ft}^2 = 144 \text{ in}^2$
   $2826 \text{ in}^2 \times \dfrac{1 \text{ ft}^2}{144 \text{ in}^2} = 19.625 \text{ ft}^2$
   Find the cost $\dfrac{\$8}{1 \text{ ft}^2} \times 19.625 \text{ ft}^2 = \$157$
   The tablecloth will cost $157.

9.6 PRACTICE PROBLEMS

1. $V = \pi r^2 h$
$\approx (3.14)(10.24 \text{ cm})^2(17.78 \text{ cm})$
$\approx 5854.12 \text{ cm}^3$

2. $V = \dfrac{4\pi r^3}{3}$
$\approx \dfrac{4(3.14)(7 \text{ in.})^3}{3}$
$\approx 1436.03 \text{ in}^3$

3. (a) The volume of the inner region
$V_r = \pi r^2 h$
$\approx (3.14)(10 \text{ cm})^2(27 \text{ cm})$
$\approx 8478 \text{ cm}^3$
The thermos can hold 8478 cm^3 of coffee.

   (b) Volume of the entire cylinder
$V_R = \pi R^2 h$
$\approx (3.14)(12 \text{ cm})^2(27 \text{ cm})$
$\approx 12,208.32 \text{ cm}^3$
Volume of the insulated region
$V_R - V_r = 12,208.32 - 8478 = 3730.32$
The volume of the insulation is 3730.32 cm^3.

4. $V = \dfrac{\pi r^2 h}{3} \approx \dfrac{(3.14)(5 \text{ m})^2(12 \text{ m})}{3} \approx 314 \text{ m}^3$

5. **(a)** $B = (6 \text{ m})(6 \text{ m}) = 36 \text{ m}^2$

$$V = \frac{Bh}{3} = \frac{(36 \text{ m}^2)(10 \text{ m})}{3} = 120 \text{ m}^3$$

(b) $B = (7 \text{ m})(8 \text{ m}) = 56 \text{ m}^2$

$$V = \frac{Bh}{3} = \frac{(56 \text{ m}^2)(15 \text{ m})}{3} = 280 \text{ m}^3$$

9.7 PRACTICE PROBLEMS

1. $\dfrac{11}{27} = \dfrac{15}{n}$

$11n = (27)(15)$

$\quad\ = 405$

$\quad n = 36.818$

Side n is approximately 36.8 meters in length.

2. Perimeter of smaller triangle

5 yd + 12 yd + 13 yd = 30 yd

$$\frac{30}{p} = \frac{12}{40}$$

$(40)(30) = 12p$

$\quad 1200 = 12p$

$\quad\ 100 = p$

The perimeter of the larger triangle is 100 yards.

3. $\dfrac{20}{2} = \dfrac{h}{5}$

$(5)(20) = 2h$

$\quad 100 = 2h$

$\quad\ 50 = h$

The building is 50 feet tall.

4. $\dfrac{1.8}{W} = \dfrac{3}{29}$

$(29)(1.8) = 3W$

$\quad 52.2 = 3W$

$\quad 17.4 = W$

The width of the larger rectangle is 17.4 meters.

APPENDIX B

APPENDIX B   PRACTICE PROBLEMS

1. 3792 $\boxed{+}$ 5896 $\boxed{=}$ 9688

2. 7930 $\boxed{-}$ 5096 $\boxed{=}$ 2834

3. 896 $\boxed{\times}$ 273 $\boxed{=}$ 244608

4. 2352 $\boxed{\div}$ 16 $\boxed{=}$ 147

5. 72.8 $\boxed{\times}$ 197 $\boxed{=}$ 14341.6

6. 52.98 $\boxed{+}$ 31.74 $\boxed{+}$ 40.37 $\boxed{+}$ 99.82 $\boxed{=}$ 224.91

7. 0.0618 $\boxed{\times}$ 19.22 $\boxed{-}$ 59.38 $\boxed{\div}$ 166.3
$\boxed{=}$ 0.830730456

8. 3.152 $\boxed{\times}$ $\boxed{(}$ 0.1628 $\boxed{+}$ 3.715 $\boxed{-}$ 4.985 $\boxed{)}$
$\boxed{=}$ −3.4898944

9. 0.5618 $\boxed{\times}$ 98.3 $\boxed{+/-}$ $\boxed{-}$ 76.31 $\boxed{\times}$ 2.98 $\boxed{+/-}$
$\boxed{=}$ 172.17886

10. 3.76 $\boxed{\text{EXP}}$ 15 $\boxed{\div}$ 7.76 $\boxed{\text{EXP}}$ 7 $\boxed{=}$ 48453608.25

11. 6.238 $\boxed{y^x}$ 6 $\boxed{=}$ 58921.28674

12. 132.56 $\boxed{x^2}$ 17572.1536

13. $\boxed{(}$ 0.0782 $\boxed{-}$ 0.0132 $\boxed{+}$ 0.1364 $\boxed{)}$ $\boxed{\sqrt{}}$
0.448776113

APPENDIX C

APPENDIX C   PRACTICE PROBLEMS

1. **(a)** Find 83% of 460.
The value of p is 83.

(b) 18% of what number is 90?
The value of p is 18.

(c) What percent of 64 is 8?
We let p represent the unknown percent number.

2. **(a)** 30% of 52 is 15.6.
$b = 52$ and $a = 15.6$

(b) 170 is 85% of what?
b is unknown and $a = 170$.

3. **(a)** What is 18% of 240?
a is unknown, the value of p is 18, and $b = 240$.

(b) What percent of 64 is 4?
The value of p is unknown, $b = 64$, and $a = 4$.

4. What is 32% of 550?

$\dfrac{a}{550} = \dfrac{32}{100}$

$\dfrac{a}{550} = \dfrac{8}{25}$

$25a = (550)(8)$

$25a = 4400$

$\quad a = 176$

Thus 176 is 32% of 550?

5. Find 340% of 70.

$\dfrac{a}{70} = \dfrac{340}{100}$

$\dfrac{a}{70} = \dfrac{17}{5}$

$5a = (70)(17)$

$5a = 1190$

$\quad a = 238$

Thus 340% of 70 is 238.

6. 68% of what is 476?

$\dfrac{476}{b} = \dfrac{68}{100}$

$\dfrac{476}{b} = \dfrac{17}{25}$

$(25)(476) = 17b$

$\quad 11{,}900 = 17b$

$\quad\ \ 700 = b$

Thus 68% of 700 is 476.

7. 216 is 0.3% of what?

$\dfrac{216}{b} = \dfrac{0.3}{100}$

$(100)(216) = 0.3b$

$\quad 21{,}600 = 0.3b$

$\quad 72{,}000 = b$

Thus 216 is 0.3% of 72,000.

8. What percent of 3500 is 105?

$$\frac{105}{3500} = \frac{p}{100}$$

$$\frac{21}{700} = \frac{p}{100}$$

$$(100)(21) = 700p$$

$$2100 = 700p$$

$$3 = p$$

Thus 3% of 3500 is 105.

9. 42 is what percent of 140?

$$\frac{42}{140} = \frac{p}{100}$$

$$\frac{21}{70} = \frac{p}{100}$$

$$(100)(21) = 70p$$

$$2100 = 70p$$

$$30 = p$$

Thus 42 is 30% of 140.

APPENDIX D

APPENDIX D   PRACTICE PROBLEMS

1. Side LN = side US = 8 in.
    Side NM = side ST = 6 in.
    Side ML = side TU = 10 in.
    All three sides of triangle LNM are equal to corresponding sides of triangle UST. The triangles satisfy SSS and are congruent.

2. Side AB = side OM = 14 cm
    Side AC = side ON = 8 cm
    $\angle A = \angle O = 120°$
    Two sides of triangles ABC and OMN are equal and the angles formed by these sides are also equal. The triangles are congruent by SAS.

3. $\angle T = \angle C = 80°$
    $\angle U = \angle A = 35°$
    Side TU = side CA = 10 cm
    Two angles of triangle TUS equal two angles of triangle CAB, and the sides between these angles are equal. The triangles are congruent by ASA.

4. **(a)** The triangles are congruent by ASA.

    (b) There is not enough information given to determine SSS, ASA, or SAS so the triangles are not necessarily congruent.

CHAPTER 1

CHAPTER 1 PRETEST **1.** $7000 + 30 + 2$ **2. (a)** $10 > 6$ **(b)** $0 < 2$ **3. (a)** 440 **(b)** 600 **4.** $x + 5$
5. (a) $8 + y$ or $y + 8$ **(b)** $11 + x$ or $x + 11$ **6. (a)** 6224 **(b)** 37 **7.** $11y + 6xy$ **8.** $3x - 4$
9. $8 - 2$ **10.** 62 **11.** $375 **12.** 10 ft **13.** 14 students **14.** 4000 mi **15. (a)** 6 **(b)** 8
16. (a) $x = 13$ **(b)** $x = 7$ **(c)** $x = 4$ **17.** $3 + x = 8, x = 5$ **18.** $5 + x = 16, x = 11$ **19.** 6 hr 20 min
20. (a) $2 \text{ cm} < 2 \text{ m}$ **(b)** $3 \text{ mi} > 3 \text{ km}$

EXERCISES 1.1 **1. (a)** Hundreds **(b)** Ones **(c)** Thousands **3. (a)** Thousands **(b)** Hundred thousands
(c) Tens **5. (a)** Millions **(b)** Hundreds **(c)** Ones **7.** $4000 + 900 + 60 + 7$ **9.** $2000 + 400 + 90 + 3$
11. $800{,}000 + 60{,}000 + 7000 + 300 + 1$ **13.** 672; Six hundred seventy-two and 00/100 **15.** 5 hundred-dollar bills,
6 ten-dollar bills, and 2 one-dollar bills **17. (a)** 4 ten-dollar bills and 6 one-dollar bills **(b)** 4 ten-dollar bills,
1 five-dollar bill, and 1 one-dollar bill **19.** $<$ **21.** $<$ **23.** $>$ **25.** $>$ **27.** $>$ **29.** $<$ **31.** $5 > 2$
33. $2 < 5$ **35.** $>$ **37.** 50 **39.** 660 **41.** 16,500 **43.** 823,000 **45.** 38,000 **47.** 13,000 **49.** 5,300,000
51. 9,000,000 **53.** 3 hrs **55.** 560,000,000 **57. (a)** 25,000 **(b)** 25,400 **59. (a)** Eight thousand two
(b) Eight hundred two **(c)** Eighty-two **(d)** One **61.** 5 trillion, 311 billion, 192 million, 809 thousand **63.** 17,000

To Think About 1.2 **1.** $\underline{8 + 5} = \underline{3 + 5} + 5 = 3 + (5 + 5) = 3 + 10 = 13$

EXERCISES 1.2 **1.** $10 + x$ **3.** $m + 2$ **5.** $5 + y$ **7.** $m + 3$ **9.** $m + 7$ **11.** $6 + 0 = 6; 5 + 1 = 6;$
$4 + 2 = 6; 3 + 3 = 6; 2 + 4 = 6; 1 + 5 = 6; 0 + 6 = 6$; We must memorize only three addition facts: $5 + 1, 4 + 2$, and
$3 + 3.$ **13.** $x + 2$ **15.** $n + 8$ **17.** 3758 **19.** 12 **21.** $n + 10$ **23.** $x + 8$ **25.** $x + 2$ **27.** $x + 3$
29. $8 + n$ **31.** $n + 11$ **33.** $x + 10$ **35.** $n + 9$ **37.** $x + 9$ **39.** $n + 9$ **41.** $a + 11$ **43.** $x + 10$
45. 59 **47.** 176 **49.** 50 **51.** 344 **53.** 729 **55.** 893 **57.** 8082 **59.** 6541 **61.** 12,854 **63. (a)** $393
(b) $228 **65.** $2309 **67.** $2x$ **69.** $3a$ **71.** $5x$ and $2x$; $3y$ and $6y$ **73.** $2mn$ and $4mn$ **75.** $11x$ **77.** $11x + 6a$
79. $9xy + 4b$ **81.** $2x + 3xy + 4n + 6$ **83.** $7mn + 6m + 1$ **85.** $10ax + 17x$ **87.** 196,951,000 sq mi
89. Answers may vary. **91.** Answers may vary.

CUMULATIVE REVIEW **1.** Hundreds **2.** Ten thousands **3.** $400 + 8$ **4.** 6 ten-dollar bills and 7 one-dollar bills

To Think About 1.3 **1.** $(0 - 0) - 0 = 0 - (0 - 0)$ and $6 - 6 = 6 - 6$ (subtracting the same numbers)

To Think About 1.3 **1.** We can borrow only from a place value that has a whole number. For example $400. There
are only one-hundred dollar bills to break down (borrow from). **2.** If the sum of numbers in the ones column is 13,
we write 13 as *1 ten and 3 ones* and carry 1 ten to the ten's column.

EXERCISES 1.3 **1.** 4 **3.** 4 **5.** 9 **7.** 9 **9.** 9 **11.** 8 **13.** 18 **15.** 0 **17.** $19 = ? + 7$; 12
19. $18 = ? + 12$; 6 **21.** $15 = ? + 8$; 7 **23.** $16 = 4 + ?$; 12 **25.** $33 = 27 + ?$; 6 **27.** $6y$ **29.** $2a$
31. $7m$ **33.** $3y - 1$ **35.** $12x - 3$ **37.** $6a - 2x - 2$ **39.** $8 - 2 = 6$ **41.** $12 - 9$ **43.** $3 - a$
45. $x - 7$ **47.** $3 - 1$ **49. (a)** 4 **(b)** 14 **51. (a)** 12 **(b)** 6 **53.** 94 **55.** 95 **57.** 73 **59.** 35
61. 19 **63.** 322 **65.** 219 **67.** 424 **69.** 2067 **71.** 4218 **73.** 5335 **75.** 5943 **77.** $4244 **79.** 57
81. $1126; $989; $920; $822; $453 **83.** 5740 mi **85.** $74 **87.** $1716 **89. (a)** 700 **(b)** 100 **(c)** 1993
(d) same **91. (a)** 3000 **(b)** 1000 **93.** $1,808,760,009,000 **95.** 20,157,000
97. When the values of x and y are equal.

CUMULATIVE REVIEW **1.** $>$ **2.** $>$ **3.** 11,829 **4.** 30,428

To Think About 1.4 **1.** Yes, there are many ways to do this. One way is to change 14 to 15. Then the estimate
becomes 170 while the rounded sum stays 170.

EXERCISES 1.4 **1.** 34 in. **3.** 16 ft **5.** 11 in. **7.** 58 ft **9.** 490 in. **11.** 400 ft **13. (a)** 16 **(b)** 32
(c) 4 **15. (a)** 50 **(b)** 100 **(c)** 20 **17.** 27 **19. (a)** $150 **(b)** $153 **(c)** Yes **21.** 200
23. 6000 mi **25. (a)** $920,000, answer may vary depending on accuracy chosen. **(b)** $919,175
(c) Yes, answer may vary depending on accuracy chosen **27.** To estimate, we round first, then do the calculations.
To round, we do the calculations, then round. **29.** 10, 20, 30 or 11, 21, 31; Answers may vary.

CUMULATIVE REVIEW **1.** 28,000 **2.** $x + 11$ **3.** $x + 9$ or $9 + x$ **4. (a)** $>$ **(b)** $<$

PUTTING YOUR SKILLS TO WORK **INDIVIDUAL** $539; yes **GROUP** Yes; two-light fixture (decorative), $25;
oak frame medicine chest, $64; onyx sink (plain), $184; cast iron bathtub (almond), $288

To Think About 1.5 **1.** Answers may vary.

EXERCISES 1.5 **1.** 9 **3.** 1 **5.** 6 **7.** 10 **9.** $5 + n$, 11 **11.** $8 - y$, 5 **13.** 12 ft **15.** 128 mi
17. (a) $442 **(b)** $435 **19.** $24 + x = 50$ **21.** $88 - x = 40$ **23.** $J + 10 = 25$ **25.** $C - 50 = 1480$

27. No   **29.** Yes   **31.** $x = 6$ **33.** $n = 2$ **35.** $n = 1$ **37.** $n = 4$ **39.** $x = 6$ **41.** $x = 12$
43. $x = 3$ **45.** $a = 7$ **47.** $x = 2$ **49.** $y = 5$ **51.** $a = 3$ **53.** $n = 2$ **55.** $x = 4$ **57.** $a = 7$
59. (a) $x = 200 + 133$ **(b)** $x = 333$ **61. (a)** $n = 122 + 988$ **(b)** $n = 1110$ **63. (a)** $21 = 13 + x$
(b) $8 = x$ **65. (a)** $44 = 28 + x$ **(b)** $16 = x$ **67. (a)** $x = 66 - 29$ **(b)** $x = 37$ **69. (a)** $x = 365 - 121$
(b) $x = 244$ **71. (a)** $x = 143$ **(b)** $x = 53$ **73. (a)** $n = 69$ **(b)** $n = 43$ **75. (a)** $a = 361$ **(b)** $a = 79$
77. (a) $4 + x = 8$ **(b)** $x = 4$ **79. (a)** $17 - x = 7$ **(b)** $10 = x$ **81. (a)** $x + 152 = 944$ **(b)** $x = 792$
83. $x = 9456$ **85.** $x = 70$ ft   **87.** $x = 50°$

CUMULATIVE REVIEW   1. $5x$ **2.** $9x$ **3. (a)** $11,000$ **(b)** 1000 **4. (a)** 7000 **(b)** 1000

EXERCISES 1.6   1. 5280 **3.** 1 **5.** 2000 **7.** 1 **9.** 2 **11.** 8 **13.** 3 **15.** $10,560$ **17.** 2 **19.** 2
21. 2 **23.** 4 **25.** 9 ft 7 in.   **27.** 16 ft 3 in.   **29.** 16 yd 2 ft   **31.** 38 ft 3 in.   **33.** 1 c   **35.** Isaac   **37.** 1
39. 10 **41.** 1 **43.** dm   **45.** mg   **47.** cg   **49.** $<$ **51.** $>$ **53.** $<$ **55.** mL   **57.** kg   **59.** $<$
61. $>$ **63.** $>$ **65.** gram   **67.** liter   **69.** (c) liter   **71.** (b) kilograms   **73.** Above   **75.** Warm weather
77. 4 gal

CUMULATIVE REVIEW   1. $15, 17$ **2.** $5x + 1$ **3.** 4859 **4.** 4045

CHAPTER 1 REVIEW   1. (a) Ten thousands   **(b)** Thousands   **(c)** Tens   **2. (a)** Thousands   **(b)** Hundreds
(c) Ten thousands   **3.** $7000 + 600 + 90 + 4$ **4.** $4000 + 300 + 20 + 5$ **5.** $\$341.00$; Three hundred forty-one
and no/100's   **6.** $\$187.00$; One hundred eighty-seven and no/100's   **7.** $<$ **8.** $>$ **9.** $6 > 1$ **10.** $3 < 5$
11. $61,300$ **12.** $382,200$ **13.** $6,400,000$ **14.** $8,100,000$ **15.** 4 hr   **16.** $790,000,000$ **17.** $7 + x$
18. $x + 4$ **19.** $3 + x$ **20.** $6 + x$ **21.** $c + b$ **22.** $z + y$ **23.** $16 + x$ **24.** $n + 5$ **25.** $n + 11$
26. $7 + n$ **27.** $x + 10$ **28.** $x + 8$ **29.** 95 **30.** 95 **31.** 205 **32.** 323 **33.** 1035 **34.** 1649
35. 9025 **36.** 8651 **37.** $\$195$ **38.** 6018 students   **39.** $3xy$ and $2xy$ are like terms. $5y$ and $8y$ are like terms.
40. $4ab$ and $2ab$ are like terms. $3b$ and $8b$ are like terms.   **41.** $13x$ **42.** $16a$ **43.** $9x + 2y$ **44.** $14b + 2a$
45. $4x - 5$ **46.** $6m - 2$ **47.** $11y - x$ **48.** $12a - b$ **49.** $8 - 3$ **50.** $3 - y$ **51.** $x - 10$ **52. (a)** 8
(b) 12 **53. (a)** 9 **(b)** 15 **54.** 62 **55.** 63 **56.** 544 **57.** 525 **58.** 5545 **59.** 3159 **60.** $24,116$
61. $30,528$ **62.** $\$2746$ **63.** $\$7427$ **64.** $30,000$ **65.** $190,000$ **66.** 54 m   **67.** 32 cm   **68.** 28 in.   **69.** 66 m
70. (a) 4 **(b)** 6 **71. (a)** 5 **(b)** 10 **72.** $\$310$ **73.** 13 **74.** 5 **75.** $2 + x = 9$ **76.** $x - 3 = 8$
77. $x = 7$ **78.** $a = 6$ **79.** $n = 6$ **80.** $n = 3$ **81.** $x = 4$ **82.** $n = 1$ **83. (a)** $x = 53 + 34$ **(b)** $x = 87$
84. (a) $y = 93 + 78$ **(b)** $y = 171$ **85. (a)** $15 = 8 + n$ **(b)** $n = 7$ **86. (a)** $16 = 12 + y$ **(b)** $y = 4$
87. (a) $n = 98 - 44$ **(b)** $n = 54$ **88. (a)** $x = 89 - 48$ **(b)** $x = 41$ **89. (a)** $x = 455 - 123$ **(b)** $x = 332$
90. (a) $y = 412 - 321$ **(b)** $y = 91$ **91. (a)** $18 - x = 3$ **(b)** $x = 15$ **92. (a)** $x - 12 = 5$ **(b)** $x = 17$
93. (a) $7 + x = 12$ **(b)** $x = 5$ **94. (a)** $x + 5 = 11$ **(b)** $x = 6$ **95.** 1 **96.** 5280 **97.** 24 **98.** 1
99. 24 **100.** 60 **101.** 2 yd 2 ft   **102.** 1 yd 2 ft   **103.** 4 hr 12 min   **104.** More; 1 c   **105.** 1 **106.** 1
107. 10 **108.** 10 **109.** $>$ **110.** $<$ **111.** $>$ **112.** $<$ **113.** (d) milligrams   **114.** (c) meters
115. (b) kilograms   **116.** (d) 2-liter

CHAPTER 1 TEST   1. $1000 + 500 + 20 + 5$ **2. (a)** $7 > 2$ **(b)** $5 > 0$ **3. (a)** 3000 **(b)** 2900
4. (a) $11 + x$ **(b)** $7 + y$ **(c)** $3 + n$ **5.** 2617 **6. (a)** $7xy + 2y$ **(b)** $3m + 5 + 6mn$ **7.** $2x - 3$
8. $x - 7$ **9.** $\$1290$ **10. (a)** 4 **(b)** 538 **11.** $\$1140$ **12. (a)** 20 in.   **(b)** 30 ft   **13.** 5 ft 9 in.
14. (a) $\$1600$ **(b)** $\$300$ **15. (a)** 3 **(b)** 4 **16. (a)** $x = 15$ **(b)** $y = 3$ **(c)** $x = 7$ **(d)** $b = 11$
17. $B - 155 = 275$ **18.** $7 + x = 9$ **19.** $30 - 3 = x$ **20.** $2 + x = 36; x = 34$ **21.** $x - 6 = 9; x = 15$
22. (a) 5 yd 1 ft   **(b)** 2 ft 10 in.   **23.** Joe's motor oil   **24. (a)** 3 mg < 3 g   **(b)** 6 in. > 6 cm

CHAPTER 2

CHAPTER 2 PRETEST   1. (a) Factors: 8.2; Product: 16 **(b)** Factors: $3, x$; Product: 21 **(c)** Factors: m, n; Product: p
2. (a) 0 **(b)** x **(c)** 0 **3.** $18x$ **4.** $12,000$ **5.** $2x + 6$ **6.** $\$428$ **7.** 3^4 **8.** 8 **9. (a)** x^5 **(b)** 4^2
10. $1,000,000$ **11.** 22 **12. (a)** x^3 **(b)** a^7 **13.** $2x^{10}$ **14.** $z^3 + 2z$ **15.** 3 **16.** 21 R6   **17.** 26 mi/gal
18. 11 **19. (a)** $y = 7$ **(b)** $x = 51$ **(c)** $x = 3$ **20.** 6 **21.** 24 ft   **22.** 16 ft^2 **23.** 6 ft^3 **24.** $66,000$ lb

TO THINK ABOUT 2.1   1. (a) $2(7) + 7 = 21$ **(b)** $5(8) - 8 = 32$ **(c)** $5(8) + 8 = 48$ **(d)** $10(8) - 8 = 72$

EXERCISES 2.1 **1.** ✦✦✦ ✦✦✦

3. (a)

	White	Pink	Blue
Brown	Brown White	Brown Pink	Brown Blue
Black	Black White	Black Pink	Black Blue
Gray	Gray White	Gray Pink	Gray Blue
Dark blue	Dark blue White	Dark blue Pink	Dark blue Blue

(b) 4 times 3, or 12 different outfits **5.** 40
7. 5, 7: factors; 35: product **9.** 7, a: factors;
28: product **11.** $7x = 49$ **13.** $9x$ **15.** ab
17. (a) Seven times y **(b)** Five times what number
equals twenty-five? **19.** 0 **21.** 30 **23.** 20
25. 20 **27.** 0 **29.** 7 **31.** 0 **33.** 120
35. $35b$ **37.** $24x$ **39.** $27c$ **41.** $40x$ **43.** $30z$
45. 0 **47.** $28x$ **49.** $180by$ **51.** $360am$
53. $240ab$ **55.** 9720 **57.** 32,768

CUMULATIVE REVIEW **1.** 7 **2.** 1 **3.** 60

EXERCISES 2.2 **1.** 15,000 **3.** 32,000 **5.** $6300n$ **7.** $3600z$ **9.** $4x + 8$ **11.** $3n - 15$ **13.** $3x - 18$
15. $4x + 16$ **17.** 3; 3; 24; 104 **19.** 4; 4; 8; 208 **21.** 5733 **23.** 4214 **25.** 119,400 **27.** 475,800 **29.** 5168
31. 9306 **33.** 47,432 **35.** 39,130 **37.** 235,458 **39.** 176,688 **41.** 7,674,728 **43.** 1,307,859 **45.** $320
47. 375 trees **49.** 140,800 km **51.** Yes **53.** BLM Accountants **55. (a)** 26°F **(b)** 84°F **57.** 5850 million km
59. The distributive property allows you to rewrite 16 as the sum $6 + 10$, then mentally multiply $6 \cdot 6 + 6 \cdot 10$.

CUMULATIVE REVIEW **1.** $64x$ **2.** 6 **3.** 1538

PUTTING YOUR SKILLS TO WORK **INDIVIDUAL 1.** Plan 1: $4335; Plan 2: $3550; Plan 3: $3695 **2.** Plan 2
3. $3810; Plan 3 **GROUP 1.** Plan 1: $4335; this plan is the same cost, since we count only a 5-day workweek.;
Plan 2: $3790; Plan 3: $4895 **2.** Plan 2

EXERCISES 2.3 **1.** 3^4 **3.** a^5 **5.** 8^1 **7.** 2^2 **9.** 5^2a^3 **11.** 3^2z^5 **13.** 7^3y^2 **15.** n^59^2 **17. (a)** $y \cdot y \cdot y$
(b) $7 \cdot 7 \cdot 7 \cdot 7 \cdot 7$ **19.** 8 **21.** 25 **23.** 1 **25.** 1 **27.** 256 **29.** 10 **31.** 125 **33.** 100,000 **35.** 25
37. 7^3 **39.** 9^2 **41.** $5^2 = x$; $x = 25$ **43.** $2^4 = x$; $x = 16$ **45.** $x^3 = 27$; $x = 3$ **47.** 7 **49.** 51 **51.** 83
53. 45 **55.** 8 **57.** 21 **59.** 13 **61.** 51 **63.** 166 **65.** 79 **67.** 113 **69.** 39 **71.** 100 **73.** 19
75. 19 **77.** 532,822 **79.** 231,788,748 **81.** What number squared is equal to 16? **83.** He should have
multiplied 3 times 2 first, then added 4 to get 10. **85.** The exponent on the 10 determines the number of trailing zeros
we attach to the number.

CUMULATIVE REVIEW **1.** 6841 **2.** 8423 **3.** 75,852 **4.** $2x = 72$

TO THINK ABOUT 2.4 **1. (a)** $8x$ **(b)** $15x^2$ **(c)** $2xy^2$ **(d)** $35x^2y^4$

EXERCISES 2.4 **1. (a)** z^5 **(b)** z^5 **3. (a)** x^6 **(b)** x^6 **5. (a)** z^3 **(b)** z^3 **7.** x^9 **9.** x^9 **11.** x^2
13. x^5 **15.** 3^5 **17.** 4^6 **19.** $8^2 \cdot 7^5$ **21.** x^5y^3 **23.** 7^7 **25.** 3^8 **27.** x^{14} **29.** y^{11} **31.** 3^{10} **33.** $2^5 \cdot 3^2 \cdot 4^7$
35. $24y^{12}$ **37.** $54a^{14}$ **39.** $48x^{13}$ **41.** $2x^7$ **43.** $144a^{10}$ **45.** $28x^{12}$ **47.** $72x^{15}$ **49.** $16x^9$ **51.** $16x^9y^3$
53. $12a^8b^4$ **55.** $20x^2y^7$ **57.** $48x^6y^7$ **59.** $4x + 4$ **61.** $3x^2 + 3$ **63.** $4x^3 + 8$ **65.** $7x^5 + 28$ **67.** $8x^2 - 40$
69. $x^4 + x^2$ **71.** $x^7 + 2x^3$ **73.** $x^{10} + 5x^4$ **75.** $x^3 + x^2$ **77.** $x^3 - 2x$ **79.** $84,065x^{11}$ **81.** $370x$
83. $12x^{10}$; Without the calculator; it is faster to multiply small numbers mentally. **85. (a)** $13cd$ **(b)** $36c^2d^2$
87. (a) $2ab^5$ **(b)** $63a^2b^{10}$

CUMULATIVE REVIEW **1. (a)** $x + 9$ **(b)** $14x$ **2.** 682 **3.** 20

TO THINK ABOUT 2.5 **1.** When the divisor and dividend are the same, $a \div a = 1$

EXERCISES 2.5 **1.** $220 \div 20$ **3.** $66 \div 22$ **5.** $\$n \div 5$ **7.** $27 \div x$ **9.** $42 \div 6$ **11.** $36 \div 3$ **13.** $3 \div 36$
15. (a) $45 = ? \cdot 5$ **(b)** 9 **17. (a)** $27 = ? \cdot 3$ **(b)** 9 **19. (a)** 4 **(b)** 16 **21. (a)** 2 **(b)** 8 **23.** 11
25. 25 **27.** 6 **29.** 3 **31.** 1 **33.** $3x + 22$ **35.** $2x + 17$ **37.** $2x + 8$ **39.** 8 R2 **41.** 371
43. 524 R1 **45.** 422 R2 **47.** 21 R2 **49.** 306 R3 **51.** 51 R2 **53.** 72 R1 **55.** 703 R4 **57.** 340 R11
59. 579 **61.** 515 R101 **63.** 4 tickets **65. (a)** 4 babies **(b)** 400 babies **(c)** 212 people **67.** $17
69. $175 per day **71.** 150,000,000 gal
73. We cannot conclude rules and properties based on a few examples. Answers may vary. **75.** 62 ft

CUMULATIVE REVIEW **1.** $7 + x = 11$ **2.** 946 **3.** 22 ft **4.** 556,000

TO THINK ABOUT 2.6 **1.** Answers may vary.

EXERCISES 2.6 **1. (a)** 40 **(b)** 15 **3. (a)** 5 **(b)** 3 **5. (a)** 19 **(b)** 43 **7.** $x = 2$ **9.** $y = 3$
11. $x = 4$ **13.** $y = 3$ **15.** $x = 2$ **17.** $x = 7$ **19.** $n = 4$ **21. (a)** $x = 6 \cdot 10$ **(b)** $x = 60$
23. (a) $x = 9 \cdot 8$ **(b)** $x = 72$ **25. (a)** $x = 7 \cdot 6$ **(b)** $x = 42$ **27. (a)** $x = 15 \cdot 33$ **(b)** $x = 495$

29. (a) $x = 8 \cdot 9$ **(b)** $x = 72$ **31. (a)** $x = 17 \cdot 16$ **(b)** $x = 272$ **33. (a)** $x = \dfrac{48}{12}$ **(b)** $x = 4$

35. (a) $n = \dfrac{51}{17}$ **(b)** $n = 3$ **37. (a)** $x = \dfrac{38}{19}$ **(b)** $x = 2$ **39. (a)** $x = \dfrac{192}{16}$ **(b)** $x = 12$ **41. (a)** $n = \dfrac{84}{12}$

(b) $n = 7$ **43.** $x = 9$ **45.** $x = 2$ **47.** $x = 3$ **49.** $x = 12$ **51.** $x = 13$ **53.** $x = 1$ **55. (a)** $x = 5$

(b) $x = 55$ **57. (a)** $x = 12$ **(b)** $x = 39$ **59. (a)** $x = 12$ **(b)** $x = 78$ **61.** $9 \cdot x = 81;\ x = 9$

63. $\dfrac{x}{11} = 15;\ x = 165$ **65.** $9x = 72;\ x = 8$ **67.** $2x \cdot 5 = 30;\ x = 3$ **69.** $3x \cdot 2 = 12;\ x = 2$ **71. (a)** 12

(b) $x = 3$ **73. (a)** 25   **(b)** $n = 48$ **75.** 45   **77.** 19,200   **79.** Seven times x, or the product of seven and x.
81. Eight times what number equals 40?   **83. (a)** 30 mph   **(b)** 25 mph   **85.** 30°

CUMULATIVE REVIEW   **1.** 7939   **2.** 1132   **3.** 6 hr 20 min

EXERCISES 2.7   **1.** 18 ft   **3.** 32 ft   **5.** 216 yd   **7.** 36 cm   **9.** 40 yd   **11.** 210 ft   **13.** 160 dm
15. 396 ft^2 **17.** 50   **19.** 40   **21.** 64 in^2 **23.** 108 ft^2 **25.** 4959 m^2 **27.** 259 in^2 **29.** 344 m^2
31. (a) 24 in. by 24 in.   **(b)** 576 in^2 **33. (a)** 12 ft by 9 ft   **(b)** 108 ft^2 **35. (a)** 12 yd^2 **(b)** $96   **37.** $6
39. 29 in^2 **41.** 32 m^2 **43. (a)** 40   **(b)** 40 in^3 **45.** 270 in^3 **47.** 4320 yd^3 **49.** 24 ft^3 **51.** 268 dm^2
53. 152 in^2 **55.** Volume, because we want to find the amount of space inside the pool.   **57.** 2 gal and 1 qt of paint

CUMULATIVE REVIEW   **1.** 2,514,000   **2.** $4x$ **3.** $30x$ **4.** $20x^5$

PUTTING YOUR SKILLS TO WORK   **INDIVIDUAL 1.** $1386   **2.** $252   **GROUP 1. (a)** 196 ft^2 **(b)** 208 ft^2
(c) 652 ft^2 **(d)** $5216

EXERCISES 2.8   **1.** $466   **3.** $3924   **5.** 29 min   **7.** $464   **9.** 245 gal in 1 day, 1715 gal in 1 week
11. (a) $182,202   **(b)** $91,101   **13. (a)** $1674   **(b)** $558   **15.** $53   **17.** $1185   **19.** 45,000   **21.** 1290
23. (a) $250   **(b)** Making the dresses is $85 cheaper.   **25.** $128,760,000,000   **27.** $1300

CUMULATIVE REVIEW   **1.** 120   **2.** 43   **3.** 390

CHAPTER 2 REVIEW   **1.** 3 times 8   **2.** 3 times 2   **3.** 4, x: factors; 32: product   **4.** x, y: factors; z: product   **5.** $6x$
6. What number times 7 equals 63?   **7.** 42   **8.** 60   **9.** $50z$ **10.** $14y$ **11.** $42x$ **12.** $70y$ **13.** 28,000
14. $4800n$ **15.** $2x + 2$ **16.** $4x + 4$ **17.** 40,612   **18.** 12,992   **19.** 1,496,352   **20.** 5,800,872   **21.** 306 mi
22. 504 doors   **23.** 800 sales   **24.** 900 sales   **25.** x^4 **26.** 6^3 **27.** 2^3n^2 **28.** $z^4 5^3$ **29.** $a \cdot a \cdot a \cdot a$
30. $6 \cdot 6 \cdot 6 \cdot 6 \cdot 6$ **31.** 1000   **32.** 1   **33.** 32   **34.** 81   **35.** $4^2 = x;\ x = 16$ **36.** $x^3 = 8;\ x = 2$ **37.** 29
38. 128   **39.** 192   **40.** a^9 **41.** 3^{10} **42.** $2^4 \cdot 4^3$ **43.** x^5 **44.** $15y^8$ **45.** $12x^8$ **46.** $21a^{10}$ **47.** $24z^4 y^{11}$
48. $x^3 + 2x$ **49.** $x^4 - 4x$ **50.** $300 \div 20$ **51.** $\$500 \div n$ **52.** $35 \div y$ **53.** $26 \div 13$ **54. (a)** 3   **(b)** 12
55. 9   **56.** 5   **57.** $4x + 6$ **58.** 451   **59.** 243   **60.** 80 R5   **61.** 50 R6   **62.** 401 R35   **63.** 603 R6
64. $3   **65.** $147   **66. (a)** 4   **(b)** 14   **67.** $x = 3$ **68.** $x = 4$ **69.** $x = 21 \cdot 44;\ x = 924$

70. $y = 19 \cdot 16;\ y = 304$ **71.** $x = 3$ **72.** $y = 2$ **73.** $x = 2$ **74.** $3x \cdot 5 = 30;\ x = 2$ **75.** $\dfrac{x}{12} = 40;\ x = 480$

76. (a) 9   **(b)** $x = 180$ **77.** 26 ft   **78.** 392 ft   **79.** 36 ft^2 **80.** 3744 in^2 **81.** 99 in^2 **82.** 478 m^2
83. 34 in^2 **84.** 180 in.   **85. (a)** 24 yd^2 **(b)** $360   **86.** $8   **87.** 270 in^3 **88.** 750 ft^3 **89.** 125 in^3
90. 198 in^2 **91. (a)** $109,800   **(b)** $27,450   **92.** $678   **93.** $918

CHAPTER 2 TEST   **1.** 12   **2. (a)** Factors: 7, 4; Product: 28   **(b)** Factors: 8, x; Product: 24   **(c)** Factors: r, s;
Product: t **3. (a)** $18x$ **(b)** $21y$ **4. (a)** 0   **(b)** y   **(c)** 0   **5.** 32,000   **6.** $5y + 30$ **7. (a)** $15,340
(b) $7990   **8.** 6^5 **9.** 125   **10. (a)** y^4 **(b)** 7^3 **11.** 100,000   **12.** 32   **13.** 10   **14. (a)** y^5 **(b)** z^4
(c) a^3b^2 **15. (a)** $5x^8$ **(b)** y^4 **16.** $6x + 18$ **17.** 0   **18.** 41   **19.** $16   **20.** 17   **21. (a)** $x = 7$
(b) $x = 48$ **(c)** $x = 9$ **22.** 6   **23.** 22 in.   **24.** 35 ft^2 **25.** 12,000 ft^3 **26.** 7500 points

CUMULATIVE TEST FOR CHAPTERS 1–2   **1.** $>$ **2. (a)** 5000   **(b)** 5300   **3. (a)** $4 + y$ or $y + 4$
(b) $5 + n$ or $n + 5$ **4. (a)** $5xy + 8x + 2$ **(b)** $2x - 4$ **5.** $1035   **6.** $910   **7.** 64 mi   **8.** 30 ft
9. 44 ft^2 **10. (a)** $1300   **(b)** $200   **11.** $x + 385 = 795$ **12. (a)** $x = 7$ **(b)** $y = 4$ **13.** 3 qt   **14.** r
15. 0   **16.** 10,000   **17.** $4x + 20$ **18.** x^6 **19.** 4^3 **20.** 3   **21.** 20   **22.** 22 mi/gal   **23.** 28 in.
24. $7500   **25.** $420

CHAPTER 3

CHAPTER 3 PRETEST   **1. (a)** $<$ **(b)** $>$ **2. (a)** 4   **(b)** 7   **3.** 18   **4.** -7 **5.** 1   **6.** -2 **7.** -10
8. 19   **9.** 1450 ft   **10. (a)** 7   **(b)** -1 **11.** -20 **12.** 18   **13. (a)** 16   **(b)** -16 **14. (a)** -4 **(b)** 4
15. -7 **16.** $2x$ **17.** $18x^3$ **18.** 17   **19.** -12

EXERCISES 3.1   **1.** $<$ **3.** $>$ **5.** $<$ **7.** $<$ **9.** $<$ **11.** $>$ **13.** A   **15.** $+$ **17.** $-$ **19.** $-$
21. $+$ **23.** 8   **25.** 5   **27.** 16   **29.** 44   **31.** $>$ **33.** $<$ **35.** $>$ **37.** $>$ **39.** -33 **41.** 129

43. 5 **45.** −16 **47. (a)** Duluth **(b)** Positive: Topeka and Milwaukee; negative: Omaha, Rapid City, and Duluth
49. Answers may vary.

CUMULATIVE REVIEW **1.** 4751 **2.** 6050 **3.** 23,296 **4.** 152

EXERCISES 3.2 **1.** −10°F + (−5°F) = −15°F **3.** $100 + $50 = $150
5. (a) ·· **(b)** Negative **(c)** −2 + (−2) **(d)** −2 + (−2) = −4
7. (a) −24 **(b)** 24 **9. (a)** −68 **(b)** 68
11. (a) −61 **(b)** 61 **13.** +300 ft + (−400 ft) = −100 ft
15. −$400 + (+500) = $100 **17. (a)** −3 + 10 **(b)** Positive **(c)** −3 + 10 = 7 **19. (a)** 2 + (−4)
(b) Negative **(c)** 2 + (−4) = −2 **21. (a)** −2 **(b)** 2 **23. (a)** 4 **(b)** −4 **25. (a)** 6 **(b)** −6
27. 1 **29.** −3 **31.** 7 **33.** 12 **35.** −38 **37.** −39 **39. (a)** 2 **(b)** −4 **(c)** −2 **41. (a)** −20
(b) −2 **(c)** 2 **43.** 0 **45.** 0 **47.** −5 **49.** −38 **51.** −3 **53. (a)** 1 **(b)** −4 **55. (a)** −4
(b) −13 **57.** Profit: $20,000 **59.** $79 **61.** −225 ft **63.** −3,432,423 **65.** Answers may vary. **67.** 3

CUMULATIVE REVIEW **1.** $10x$ **2.** $24x^2$ **3.** $5x$ **4.** $3x − 12$

EXERCISES 3.3 **1. (a)** −$10 **(b)** −3 **(c)** −2 **3. (a)** −$1 **(b)** −1 **(c)** −1 **5. (a)** 7 + (−4) = 3
(b) 15 + (−7) = 8 **(c)** 10 + (−8) = 2 **7.** −10 **9.** −2 **11.** −9 **13.** −5 **15.** −5 **17.** 6 **19.** −4
21. 14 **23.** −14 **25.** 1 **27.** −22 **29.** 6 **31.** −1 **33.** −6 **35.** −13 **37.** −7 **39.** −6 **41.** −13
43. −15 **45.** −10 **47.** −10 **49.** −1 **51.** −20 **53.** −14 **55.** 19 **57.** 22 **59.** −4 **61.** 5
63. 3706 ft **65.** 344 ft **67.** $40,000 **69. (a)** Brownsville, Texas **(b)** 86°F **71.** −1113 **73.** 1978
75. addition; opposite; add

CUMULATIVE REVIEW **1.** 17 **2.** 3 **3.** 12 **4.** 10

EXERCISES 3.4 **1.** −4 + (−4) + (−4) = −12 **3.** −6 + (−6) + (−6) + (−6) = −24 **5.** −3 + (−3) = −6
7. −4 **9.** −45 **11.** 18 **13.** 18 **15.** −30 **17.** −24 **19. (a)** 8 **(b)** −8 **(c)** −8 **(d)** 8
21. (a) 10 **(b)** 10 **(c)** −10 **(d)** −10 **23.** 72 **25.** 42 **27.** 90 **29.** −40 **31.** −120 **33.** 25
35. −125 **37. (a)** 9 **(b)** −27 **(c)** 81 **(d)** −243 **39. (a)** −1 **(b)** 1 **41. (a)** −16 **(b)** 16
43. −4 **45.** −6 **47.** 2 **49.** 7 **51. (a)** 5 **(b)** −5 **(c)** −5 **(d)** 5 **53. (a)** 5 **(b)** −5 **(c)** −5
(d) 5 **55. (a)** −11 **(b)** −44 **57. (a)** 2 **(b)** 8 **59.** −$700 **61.** −90 or 90 m left of 0. **63.** 403,444
65. −59 **67.** positive **69.** negative **71.** −24 **73.** False

CUMULATIVE REVIEW **1.** 9 **2.** 18 **3.** $15x^6$ **4.** $3x + 18$

EXERCISES 3.5 **1.** −5x **3.** −x **5.** 2x **7.** −9a **9.** 2x **11.** −13x **13.** −6y + 5x **15.** −2x + 2y
17. 4x + 3y **19.** −12x **21.** −14y **23.** 10x **25.** −2a + 2x **27.** −4x + 5y **29.** −2x + 2y
31. 2 + 3a **33.** −4 + 5x **35. (a)** −2 **(b)** −2x **37. (a)** −1 **(b)** −1x or −x **39. (a)** −3 **(b)** −3x
41. $−12y^2$ **43.** $27a^2$ **45.** $−24x^2$ **47.** $−12y^2$ **49.** $−12x^5$ **51.** $56m^6$ **53.** $10n^7$ **55.** $−4x^2$ **57.** $−30x^3$
59. $−96x^3$ **61.** −48,192x **63.** $480,896x^9$ **65. (a)** 1x or x **(b)** 5x; Answers may vary. **67.** −13

CUMULATIVE REVIEW **1.** 18 **2.** 14 **3.** 16 **4.** 20

EXERCISES 3.6 **1.** 11 **3.** −40 **5.** 33 **7.** −23 **9.** 54 **11.** 43 **13.** −14 **15.** 33 **17.** −45
19. 48 **21.** −20 **23.** −32 **25.** −59 **27.** 14 **29.** −88 **31.** −14 **33.** −2 **35.** 2 **37.** 3 sec
39. −33 **41.** +20 **43.** +12 **45.** 1526 **47.** 223 **49.** −4(0) = 0, and 0 is neither positive nor negative.
51. x = 9

CUMULATIVE REVIEW **1.** 2x + 6 **2.** 4a + 8 **3.** 6x + 3y + 8 **4.** 11a + 2b + 4

PUTTING YOUR SKILLS TO WORK **INDIVIDUAL 1.** |0 − (−3)| = 3 hr earlier. 10 A.M.−3 hr is 7 A.M.
2. |0 − 9| = 9 hr later. 2 P.M. + 9 hr is 11 P.M. **3.** |3 − (−7)| = 10 hr earlier. 3 P.M.−10 hr is 5 A.M. **4.** |8 − 5| = 3 hr
earlier. 6 A.M.−3 hr is 3 A.M. **GROUP 1.** 5 P.M. Sunday **2.** Yes, since it takes place at 10 P.M. Monday in Mexico.
3. The call must be made between 8 A.M. and 10 A.M. from the Boston office.
4. The call must be made at noon from Algeria.

CHAPTER 3 REVIEW **1.** < **2.** < **3.** > **4.** > **5.** + **6.** − **7.** 5 **8.** 7 **9.** −23 **10.** 12
11. 16 **12.** −8 **13. (a)** May **(b)** March **14. (a)** January, February, May **(b)** March, April
15. −$500 + (−$200) = −$700 **16.** $100 + $200 = $300 **17. (a)** −59 **(b)** 59 **18. (a)** −66 **(b)** 66
19. Loss **20.** Profit **21. (a)** −32°F + 20°F **(b)** Negative **(c)** −32°F + 20°F = −12°F **22. (a)** −6
(b) 6 **(c)** −10 **23. (a)** 14 **(b)** −14 **(c)** −50 **24.** 4 **25.** −96 **26.** 5 **27.** −5 **28.** −680 ft
29. −290 ft **30.** −12 **31.** −5 **32.** −8 **33.** −4 **34.** −5 **35.** 5 **36.** −2 **37.** −13 **38.** 2
39. $50,000 **40.** $40,000 **41. (a)** 18 **(b)** −18 **(c)** −18 **(d)** 18 **42. (a)** 10 **(b)** −10 **(c)** −10
(d) 10 **43.** 35 **44.** −10 **45.** −12 **46.** 4 **47.** −90 **48.** 64 **49.** −240 **50.** −125 **51.** −4
52. 4 **53. (a)** 7 **(b)** −7 **54. (a)** −6 **(b)** 6 **55. (a)** 11 **(b)** −45 **(c)** 33 **(d)** −5 **56. (a)** −3
(b) −40 **(c)** 24 **(d)** −5 **57.** −2y **58.** 3x + 5y **59.** −9a **60.** x + 9y **61.** 5z − 4 **62.** −1 + 5y

63. $-21x^2$ **64.** $-18y^2$ **65.** $27x^2$ **66.** $-10x^2$ **67.** $-12x^7$ **68.** $-35z^{10}$ **69.** $12a^{14}$ **70.** $-18y^{15}$ **71.** 7
72. 68 **73.** -16 **74.** 0 **75.** $-5°F$

CHAPTER 3 TEST **1.** $<$ **2.** $<$ **3.** -14 points **4. (a)** 12 **(b)** 3 **5.** 8 **6.** -10 **7.** 2 **8.** -10
9. -15 **10. (a)** $-10°F + 15°F$ **(b)** $5°F$ **11.** $15,000 profit **12.** -6 **13.** -12 **14.** 13 **15.** -23
16. (a) 9 **(b)** -14 **17.** -21 **18.** 32 **19.** -30 **20. (a)** 25 **(b)** -125 **(c)** -25 **21. (a)** -4
(b) 4 **22.** -2 **23.** -7 **24.** $-7x + 8y$ **25.** $72x^6$ **26.** 13 **27.** -67 **28.** 2 **29.** IQ $= 115$

CUMULATIVE TEST FOR CHAPTERS 1–3 **1.** 5300 **2.** $5 + 11m$ **3.** 175 years **4.** $x = 60$ **5.** 13
6. $x = 7$ **7.** $x + 9 = 16$ **8.** Centimeters **9.** $42x$ **10.** $24y$ **11.** $2800n$ **12.** $2x + 14 = 28; x = 7$
13. $8^2 = x; x = 64$ **14.** 14,673 **15.** $15y^7$ **16.** $y = 7$ **17.** 144 ft^2 **18. (a)** $81,350 **(b)** $40,675
19. $>$ **20.** $>$ **21.** -1 **22.** -18 **23.** -32 **24.** 2 **25.** -32 **26.** $-4mn + 4m$
27. $-12x^8$ **28.** 22

CHAPTER 4

CHAPTER 4 PRETEST **1.** Divisible by 2 and 3 **2. (a)** Neither **(b)** Prime **(c)** Composite

3. $2 \cdot 2 \cdot 2 \cdot 3 \cdot 5$ or $2^3 \cdot 3 \cdot 5$ **4.** $\frac{7}{10}$ **5. (a)** Mixed number **(b)** Proper fraction **(c)** Improper fraction

6. $2\frac{2}{5}$ **7.** $\frac{11}{4}$ **8.** $\frac{6}{10}$ **9.** $=$ **10.** $\frac{6}{11}$ **11.** $\frac{4}{7}$ **12.** $\frac{\$1}{5 \text{ lb}}$ **13.** 26 mi/gal **14.** Not a proportion

15. $x = 2$ **16.** 7 lb

EXERCISES 4.1 **1.** No; it is not even. **3.** No; the sum of the digits is not divisible by 3. **5.** Yes; the last digit is 5.
7. 2, 3 **9.** 5 **11.** 2, 3, 5 **13.** 3 **15.** 2, 3, 5 **17.** 0, 1: neither; 23: prime; 9, 40, 8, 15, 33: composite
19. (a) Answers may vary. **(b)** $7 \cdot 2 \cdot 2 \cdot 2$ or $7 \cdot 2^3$ **21.** $3 \cdot 3$ **23.** $3 \cdot 2 \cdot 2$ or $2^2 \cdot 3$ **25.** $2 \cdot 2 \cdot 2 \cdot 3$ or $2^3 \cdot 3$
27. $2 \cdot 2 \cdot 3 \cdot 3$ or $2^2 \cdot 3^2$ **29.** $3 \cdot 7$ **31.** $7 \cdot 2 \cdot 5$ **33.** $3 \cdot 5 \cdot 5$ or $3 \cdot 5^2$ **35.** $3 \cdot 3 \cdot 5$ or $3^2 \cdot 5$ **37.** $3 \cdot 3 \cdot 11$ or $3^2 \cdot 11$
39. $5 \cdot 3 \cdot 7$ **41.** $2 \cdot 5 \cdot 11$ **43.** $2 \cdot 2 \cdot 2 \cdot 17$ or $2^3 \cdot 17$ **45.** $2 \cdot 2 \cdot 5 \cdot 11$ or $2^2 \cdot 5 \cdot 11$ **47.** $3 \cdot 3 \cdot 3 \cdot 3 \cdot 2 \cdot 5$ or $2 \cdot 3^4 \cdot 5$
49. $13 \cdot 7$ **51.** $3 \cdot 17 \cdot 11$ **55.** A prime number is only divisible by the number 1 and itself, whereas a composite
number is divisible by other numbers in addition to 1 and itself. **57.** If the last digit is 0, the number is divisible by 10.
59. Answers may vary.

CUMULATIVE REVIEW **1.** $10x^5y$ **2.** $9y^2$ **3.** $8x + 2$ **4.** $30x^2$

EXERCISES 4.2 **1.** $\frac{3}{5}$ **3.** $\frac{3}{4}$ **5.** Undefined **7.** 0 **9.** 1 **11.** 1 **13.** Undefined **15.** 0 **17.** $\frac{7}{15}$

19. $\frac{6}{13}$ **21.** $\frac{9}{26}$ **23.** $\frac{37}{94}$ **25.** $\frac{29}{86}$ **27.** $\frac{58}{135}$ **29.** $\frac{31}{135}$ **31.** Proper fraction **33.** Improper fraction

35. Improper fraction **37.** Mixed number **39.** $1\frac{6}{7}$ **41.** $18\frac{1}{4}$ **43.** $20\frac{1}{2}$ **45.** $6\frac{2}{5}$ **47.** $9\frac{2}{5}$ **49.** 1 **51.** $\frac{59}{7}$

53. $\frac{97}{4}$ **55.** $\frac{47}{3}$ **57.** $\frac{100}{3}$ **59.** $\frac{89}{10}$ **61.** $\frac{261}{30}$ **63.** $\frac{43,037}{32}$ **65.** $\frac{24,401}{612}$

67. denominator, numerator, the same as in the mixed number

CUMULATIVE REVIEW **1.** 40 **2.** -63 **3.** -9 **4.** 9

EXERCISES 4.3 **1.** $\frac{8}{21}$ **3.** $\frac{1}{12}$ **5.** $\frac{4}{35}$ **7.** $\frac{3}{8}$ **9.** $\frac{1}{2x}$ **11.** $\frac{5y}{24}$ **13. (a)** $\frac{28}{36}$ **(b)** $\frac{35}{45}$ **15. (a)** $\frac{16}{44}$

(b) $\frac{20}{55}$ **17. (a)** $\frac{22}{34}$ **(b)** $\frac{33}{51}$ **19. (a)** $\frac{6}{28}$ **(b)** $\frac{9}{42}$ **21.** $\frac{16}{24}$ **23.** $\frac{35}{60}$ **25.** $\frac{21}{49}$ **27.** $\frac{15}{20}$ **29.** $\frac{27}{39}$

31. $\frac{70}{80}$ **33.** $\frac{8y}{9y}$ **35.** $\frac{12y}{28y}$ **37.** $\frac{9a}{18a}$ **39.** $\frac{15x}{21x}$ **41.** $=$ **43.** $>$ **45.** $<$ **47.** $>$ **49.** $=$ **51.** $=$

53. The store $\frac{3}{4}$ mi away **55.** The $\frac{3}{4}$-lb package **57.** The $\frac{3}{4}$ lb of French roast coffee **59.** $>$ **61.** $=$

63. cross, bottom **65.** Because whenever the numerator and denominator of a fraction are equal, the fraction is equal
to 1. Thus, both fractions are equal to 1.

CUMULATIVE REVIEW **1.** $2 \cdot 3 \cdot 11$ **2.** $2 \cdot 2 \cdot 2 \cdot 3 \cdot 3$ or $2^3 \cdot 3^2$ **3.** $2 \cdot 3 \cdot 5 \cdot 7$ **4.** $2 \cdot 2 \cdot 2 \cdot 2 \cdot 7$ or $2^4 \cdot 7$

EXERCISES 4.4 **1.** $\frac{3}{5}$ **3.** $\frac{3}{4}$ **5.** $\frac{5}{6}$ **7.** $\frac{6}{7}$ **9.** $\frac{2}{3}$ **11.** $\frac{6}{17}$ **13.** $\frac{7}{9}$ **15.** $\frac{1}{2}$ **17.** $\frac{7}{5}$ or $1\frac{2}{5}$ **19.** $\frac{5}{4}$ or $1\frac{1}{4}$

21. 3 **23.** $\frac{1}{6}$ **25.** 10 **27.** $\frac{1}{14}$ **29.** $\frac{43}{21}$ or $2\frac{1}{21}$ **31.** $\frac{27}{37}$ **33.** $\frac{2}{3}$ **35.** $\frac{1}{5}$ **37.** $\frac{4x}{9}$ **39.** $\frac{2}{3n}$ **41.** $\frac{4x}{7}$

43. $\dfrac{7}{8y}$ **45.** $\dfrac{5}{11}$ **47.** $\dfrac{2}{3}$ **49.** $\dfrac{8y}{9}$ **51.** $\dfrac{2}{3y}$ **53.** $-\dfrac{7}{8}$ **55.** $-\dfrac{4}{5}$ **57.** $-\dfrac{5}{8}$ **59.** $-\dfrac{7}{9}$ **61. (a)** $-\dfrac{3}{5}$

(b) $-\dfrac{3}{5}$ **(c)** $-\dfrac{3}{5}$ **63.** $\dfrac{13}{23}$ **65.** $-\dfrac{23}{29}$ **67.** $\dfrac{5abc^4}{11}$ **69.** $\dfrac{64xy^2}{75z^5}$

CUMULATIVE REVIEW   **1.** 57   **2.** 4   **3.** $-4a+5$

EXERCISES 4.5   **1.** $\dfrac{3}{13}$ **3.** $\dfrac{19}{47}$ **5.** $\dfrac{7}{2}$ **7.** $\dfrac{5}{14}$ **9.** $\dfrac{17}{6}$ **11.** $\dfrac{2}{5}$ **13.** $\dfrac{17}{41}$ **15.** $\dfrac{121}{423}$ **17. (a)** $\dfrac{3}{7}$ **(b)** $\dfrac{7}{3}$

19. (a) $\dfrac{23}{14}$ **(b)** $\dfrac{14}{23}$ **21.** $\dfrac{800}{837}$ **23.** $\dfrac{296}{337}$ **25.** $\dfrac{425\text{ ft}}{7\text{ sec}}$ **27.** $\dfrac{47\text{ pies}}{340\text{ people}}$ **29.** $\dfrac{253\text{ guests}}{122\text{ spaces}}$ **31.** $\dfrac{205\text{ cal}}{7\text{ g fat}}$

33. $\dfrac{850\text{ kidney-pancreas}}{1099\text{ kidney}}$ **35.** $\dfrac{86\text{ pancreas}}{2029\text{ liver}}$ **37.** $172\dfrac{1}{2}$ gal per hr   **39.** $22\dfrac{1}{2}$ mi/gal   **41.** $13 per hr

43. $3\dfrac{1}{2}$ rotations per sec   **45.** $9 per CD   **47. (a)** 155 clients per agent   **(b)** 62 clients per clerical staff

(c) 15 clerical staff   **49. (a)** 12 sessions are $8 per session; 15 sessions are $9 per session.   **(b)** 12 sessions
51. (a) 4 reams of paper cost $3 per ream. 7 reams cost $3 per ream.   **(b)** They both offer the same unit price.
53. (a) $\dfrac{3161}{2389}$ **(b)** 35 sales per day   **55.** 20 mi/gal   **57.** Answers may vary.

CUMULATIVE REVIEW   **1.** $x=7$ **2.** $n=12$ **3.** $n=16$ **4.** $x=16$

EXERCISES 4.6   **1.** $\dfrac{4}{9}=\dfrac{28}{63}$ **3.** $\dfrac{12}{7}=\dfrac{48}{28}$ **5.** $\dfrac{3}{8}=\dfrac{18}{48}$ **7.** $\dfrac{\frac{1}{3}}{\frac{1}{8}}=\dfrac{\frac{1}{4}}{\frac{3}{32}}$ **9.** $\dfrac{2\text{ printers}}{4\text{ secretaries}}=\dfrac{12\text{ printers}}{24\text{ secretaries}}$

11. $\dfrac{2\text{ cups}}{50\text{ g}}=\dfrac{6\text{ cups}}{150\text{ g}}$ **13.** $\dfrac{6\text{ American dollars}}{4\text{ British pounds}}=\dfrac{264\text{ American dollars}}{176\text{ British pounds}}$ **15.** $\dfrac{3\frac{1}{2}\text{ rotations}}{2\text{ min}}=\dfrac{14\text{ rotations}}{8\text{ min}}$

17. $\dfrac{4\text{ made}}{7\text{ attempts}}=\dfrac{12\text{ made}}{21\text{ attempts}}$ **19.** Yes   **21.** No   **23.** No   **25.** Yes   **27.** No   **29.** $x=20$ **31.** $x=32$

33. $x=20$ **35.** $x=24$ **37.** $x=50$ **39.** $x=35$ **41.** $x=7$ **43.** $x=\dfrac{1}{18}$ **45.** $x=\dfrac{2}{21}$ **47.** $n=6$

49. $n=5$ **51.** $n=80$ **53.** Yes

CUMULATIVE REVIEW   **1.** $2x+6$ **2.** $20\div n=5$ **3.** $x-8=9$ **4.** $4+3x=19$

TO THINK ABOUT 4.7   **1.** A ratio is a fraction. Reducing a fraction does not change the value of the fraction.

EXERCISES 4.7   **1.** 12 hr   **3.** 125 g   **5.** $100   **7.** 150 min or $1\dfrac{1}{2}$ hr   **9.** 20 cups of water   **11.** $384

13. 350 ft   **15.** 1360 shares   **17.** 42 mm   **19.** 20 ft wide   **21.** $240   **23.** $192   **25.** $1290   **27.** $2800
29. 56,000 ft   **31.** 13,981 people   **33.** $W=9$ in. and $H=15$ in.

CUMULATIVE REVIEW   **1.** $\dfrac{1}{12}$ **2.** $\dfrac{2}{35}$ **3.** x^9 **4.** y^{11}

PUTTING YOUR SKILLS TO WORK   **INDIVIDUAL 1.** 576 rolls   **2. (a)** 16 lb   **(b)** 24 lb   **(c)** 24 lb   **3.** 32 baskets
4. 8 cakes   **5.** 10 bags candy   **6.** 108 white daisies   **7.** 20 yd white ribbon   **8.** 48 cups flour   **9.** 24 cups milk
GROUP 1. $36   **2.** $312   **3.** $20   **4.** $112   **5. (a)** 64 cups   **(b)** 128 cups   **(c)** 96 cups   **6.** $96   **7.** $65

CHAPTER 4 REVIEW   **1.** 2, 5   **2.** 2, 3   **3.** 0; neither; 7, 11: prime; 21, 50, 25, 51: composite
4. 1 neither; 7, 13, 41: prime; 32, 12, 50, 6: composite   **5.** $2\cdot2\cdot3\cdot3$ or $2^2\cdot3^2$ **6.** $2\cdot2\cdot2\cdot7$ or $2^3\cdot7$
7. $5\cdot5\cdot17$ or $5^2\cdot17$ **8.** $2\cdot2\cdot2\cdot3\cdot13$ or $2^3\cdot3\cdot13$ **9.** $2\cdot2\cdot3\cdot3\cdot5\cdot5$ or $2^2\cdot3^2\cdot5^2$ **10.** $2\cdot2\cdot2\cdot2\cdot11\cdot5$ or $2^4\cdot11\cdot5$

11. Undefined   **12.** 0   **13.** 1   **14.** 1   **15.** $\dfrac{7}{20}$ **16.** $\dfrac{9}{23}$ **17.** $8\dfrac{2}{5}$ **18.** $9\dfrac{1}{6}$ **19.** 8   **20.** 6   **21.** $\dfrac{7}{3}$

22. $\dfrac{23}{5}$ **23.** $\dfrac{52}{5}$ **24.** $\dfrac{45}{4}$ **25.** $\dfrac{2}{21}$ **26.** $\dfrac{4}{15}$ **27.** $\dfrac{18}{27}$ **28.** $\dfrac{27}{36}$ **29.** $\dfrac{28x}{35x}$ **30.** $\dfrac{18y}{33y}$ **31.** $>$ **32.** $>$

33. $<$ **34.** $=$ **35.** $\dfrac{3}{8}$ mi   **36.** $\dfrac{9}{10}$ mi   **37.** $\dfrac{11}{15}$ **38.** $\dfrac{8}{9}$ **39.** 3   **40.** $\dfrac{7}{3}$ or $2\dfrac{1}{3}$ **41.** $\dfrac{5}{12}$ **42.** $\dfrac{4}{5y}$

43. $-\dfrac{8}{9}$ **44.** $-\dfrac{2}{3}$ **45.** $\dfrac{10}{23}$ **46.** $\dfrac{3}{5}$ **47.** $\dfrac{7}{11}$ **48.** $\dfrac{20}{29}$ **49.** $\dfrac{29}{20}$ **50.** $\dfrac{\$31}{7\text{ washcloths}}$ **51.** $\dfrac{3\text{ in.}}{7\text{ hr}}$ **52.** $\dfrac{105\text{ mi}}{4\text{ hr}}$

53. $\dfrac{1\text{ g protein}}{5\text{ fat calories}}$ **54.** $\dfrac{10\text{ fat calories}}{13\text{ g sugars}}$ **55.** 50 mph   **56.** $20\dfrac{1}{2}$ mi/gal   **57. (a)** 2 legal secretaries per lawyer

(b) 3 paralegals per lawyer   **(c)** 180 paralegals   **58.** $464\dfrac{1}{16}$ lb per ft

59. (a) 6 for $72, $12 per CD; 8 for $96, $12 per CD   **(b)** Both have the same unit price.

60. (a) $\dfrac{24}{31}$ **(b)** $1500 per day   **61.** $\dfrac{2}{7} = \dfrac{14}{49}$ **62.** $\dfrac{\frac{1}{5}}{\frac{1}{9}} = \dfrac{\frac{1}{2}}{\frac{5}{18}}$

63. $\dfrac{2 \text{ in.}}{190 \text{ mi}} = \dfrac{6 \text{ in.}}{570 \text{ mi}}$ **64.** $\dfrac{234 \text{ mi}}{9 \text{ gal}} = \dfrac{468 \text{ mi}}{18 \text{ gal}}$ **65.** Yes   **66.** No   **67.** No   **68.** No   **69.** $x = 12$ **70.** $x = 2$

71. $x = 6$ **72.** $x = 10$ **73.** $x = \dfrac{1}{10}$ **74.** $x = \dfrac{1}{32}$ **75.** $n = 34$ **76.** $n = 5$ **77.** 175 g   **78.** $120

79. 360 mi   **80.** 16 mph   **81.** 8 ft wide   **82. (a)** $51   **(b)** $17

CHAPTER 4 TEST   **1.** 2, 5   **2. (a)** Composite   **(b)** Neither   **(c)** Prime   **3.** $2 \cdot 2 \cdot 3 \cdot 7$ or $2^2 \cdot 3 \cdot 7$

4. $7 \cdot 7$ or 7^2 **5.** 0   **6.** 1   **7.** Undefined   **8.** $\dfrac{17}{36}$ **9.** $\dfrac{16}{28}$ or $\dfrac{4}{7}$ **10.** Improper fraction   **11.** Proper fraction

12. Mixed number   **13.** 4   **14.** $1\dfrac{3}{5}$ **15.** $\dfrac{43}{6}$ **16.** $\dfrac{16}{36}$ **17.** $\dfrac{12y}{27y}$ **18.** $<$ **19.** $\dfrac{3}{4}$-in. head   **20. (a)** $\dfrac{9}{28}$

(b) $\dfrac{1}{2x}$ **21.** $\dfrac{4}{11}$ **22.** $\dfrac{50 \text{ calories}}{1 \text{ g fat}}$ **23.** $7\dfrac{1}{2}$ calories per min   **24.** Yes   **25.** $x = 30$ **26.** 5 lb

CUMULATIVE TEST CHAPTERS 1–4   **1.** $3,825,000   **2.** $26,552,199   **3. (a)** $25 - x = 18$ **(b)** $x = 7$
4. 1000   **5.** 1   **6.** 12   **7.** $15x^3$ **8.** $2x + 6$ **9.** $4y - 8$ **10.** $ab + ac$ **11.** 18   **12.** -18 **13.** 18

14. 4   **15.** -4 **16.** -20 **17.** 20   **18.** -3 **19.** 3   **20.** -1 **21.** -15 **22.** -15 **23.** -1 **24.** $\dfrac{4}{7}$

25. $\dfrac{3}{4}$ cups   **26.** $\dfrac{75 \text{ words}}{1 \text{ min}}$ **27.** 27 g carbohydrates

CHAPTER 5

CHAPTER 5 PRETEST   **1.** $\dfrac{10}{21}$ **2.** $\dfrac{4x}{15}$ **3.** $-\dfrac{1}{9}$ **4.** $-\dfrac{5}{6}$ **5.** 27 bags   **6.** 7, 14, 21, 28   **7.** 60   **8.** $18x^2$

9. 455   **10.** 21   **11.** $\dfrac{2}{3}$ **12.** $\dfrac{38}{7x}$ **13.** $\dfrac{x}{6}$ **14.** $10\dfrac{3}{5}$ **15.** $39\dfrac{3}{8}$ **16.** $4\dfrac{3}{14}$ **17.** $-\dfrac{13}{2}$ or $-6\dfrac{1}{2}$ **18.** 2

19. $\dfrac{5}{18}$ **20.** $\dfrac{3}{2}$ or $1\dfrac{1}{2}$ **21.** $\dfrac{4}{3}$ or $1\dfrac{1}{3}$ **22.** y^8 **23.** $9x^2$ **24.** $\dfrac{x}{2}$ **25. (a)** $30\dfrac{1}{2}$ ft by 19 ft   **(b)** $222\dfrac{3}{4}$

EXERCISES 5.1   **1.** $\dfrac{1}{15}$ **3.** $\dfrac{5}{24}$ **5.** $\dfrac{1}{6}$ **7.** $\dfrac{1}{35}$ **9.** $\dfrac{2}{9}$ **11.** $-\dfrac{1}{7}$ **13.** $\dfrac{6}{7}$ **15.** $-\dfrac{1}{2}$ **17.** $\dfrac{2}{81}$ **19.** 5

21. $\dfrac{x^4}{12}$ **23.** $\dfrac{3}{10x}$ **25.** $-\dfrac{3y}{35}$ **27.** $\dfrac{9x^5}{50}$ **29.** $12x^7$ **31.** $-\dfrac{3}{8y}$ **33.** $\dfrac{7}{3}$ **35.** $\dfrac{8}{1}$ or 8   **37.** $\dfrac{1}{8}$ **39.** $-\dfrac{5}{2}$

41. $-\dfrac{y}{x}$ **43.** $-\dfrac{1}{6}$ **45.** $\dfrac{8}{7}$ **47.** $\dfrac{7}{27}$ **49.** $-\dfrac{1}{9}$ **51.** $\dfrac{7}{27}$ **53.** 35   **55.** $\dfrac{1}{44}$ **57.** $\dfrac{7}{3}$ **59.** $-\dfrac{3}{4}$ **61.** $\dfrac{2x^4}{3}$

63. $\dfrac{3x^6}{4}$ **65.** $-\dfrac{45}{2x}$ **67.** $825   **69.** 16 pipes   **71.** 6 pizzas   **73.** 160 bottles   **75. (a)** 1600 students

(b) 1920 students   **77.** $\dfrac{27}{70}$

CUMULATIVE REVIEW   **1.** $\dfrac{10}{15}$ **2.** $\dfrac{15}{20}$ **3.** $\dfrac{45}{50}$ **4.** $\dfrac{36}{51}$

EXERCISES 5.2   **1. (a)** 6, 12, 18, 24; 8, 16, 24, 32   **(b)** 24   **3. (a)** 2, 4, 6, 8, 10; 5, 10, 15, 20, 25   **(b)** 10
5. (a) $12x, 24x, 36x, 48x; 18x, 36x, 54x, 72x$ **(b)** $36x$ **7.** 18   **9.** 112   **11.** 105   **13.** 120   **15.** 120
17. 70   **19.** $28x$ **21.** $567a$ **23.** $90x^2$ **25.** $44x^3$ **27.** $156x^2$ **29.** 2, 3   **31.** 7, x
33. (a) $660 = 2 \cdot 2 \cdot 3 \cdot 5 \cdot 11; 140 = 2 \cdot 2 \cdot 5 \cdot 7$ **(b)** 4620
35. Because $3 \cdot 4 = 12$ but there is no whole number that we can multiply by 5 to get 12.   **37.** $40x^3y^2$ **39.** $30xyz^2$

CUMULATIVE REVIEW   **1.** $7\dfrac{1}{2}$ **2.** $\dfrac{17}{5}$ **3.** 0

EXERCISES 5.3   **1.** $\dfrac{11}{21}$ **3.** $\dfrac{3}{17}$ **5.** $-\dfrac{6}{7}$ **7.** $-\dfrac{20}{51}$ **9.** $\dfrac{2}{x}$ **11.** $\dfrac{39}{a}$ **13.** $\dfrac{x-5}{7}$ **15.** $\dfrac{y+14}{9}$

17. 28   **19.** 84   **21.** 25   **23.** 63   **25.** $\dfrac{15}{60}$ **27.** $\dfrac{50}{60}$ **29.** $\dfrac{12}{72}$ **31.** $\dfrac{40}{72}$ **33.** $\dfrac{53}{56}$ **35.** $\dfrac{49}{30} = 1\dfrac{19}{30}$

37. $\dfrac{43}{90}$ **39.** $\dfrac{13}{27}$ **41.** $\dfrac{1}{42}$ **43.** $\dfrac{3}{26}$ **45.** $-\dfrac{21}{80}$ **47.** $-\dfrac{15}{28}$ **49.** $\dfrac{57}{100}$ **51.** $\dfrac{26}{175}$ **53.** $\dfrac{37}{60}$ **55.** $\dfrac{1}{32}$

57. $\dfrac{21}{2x}$ **59.** $\dfrac{23}{7x}$ **61.** $\dfrac{7}{3x}$ **63.** $\dfrac{3y + 4x}{xy}$ **65.** $\dfrac{2y - 7x}{xy}$ **67.** $\dfrac{9x + y}{xy}$ **69.** $\dfrac{13x}{15}$ **71.** $\dfrac{9x}{20}$ **73.** $\dfrac{5x}{6}$

75. numerators; denominator **77.** $\dfrac{11}{30}$ **79.** $\dfrac{1}{4}$ **81.** $x = 2$ **83.** $x = 2$

CUMULATIVE REVIEW **1.** (a) $x = 70 + 3$ (b) $x = 73$ **2.** (a) $x = 130 - 8$ (b) $x = 122$
3. (a) $x = 29 + 9$ (b) $x = 38$ **4.** (a) $x = 160 - 5$ (b) $x = 155$

TO THINK ABOUT 5.4 **1.** Answers may vary.

EXERCISES 5.4 **1.** $27\dfrac{5}{7}$ **3.** $16\dfrac{3}{4}$ **5.** $18\dfrac{11}{12}$ **7.** $5\dfrac{7}{10}$ **9.** $3\dfrac{1}{2}$ **11.** $1\dfrac{17}{18}$ **13.** $19\dfrac{11}{21}$ **15.** $21\dfrac{3}{10}$

17. $4\dfrac{19}{25}$ **19.** $6\dfrac{31}{60}$ **21.** $11\dfrac{5}{24}$ **23.** $12\dfrac{1}{12}$ **25.** $6\dfrac{3}{4}$ **27.** $30\dfrac{7}{9}$ **29.** $\dfrac{11}{3}$ or $3\dfrac{2}{3}$ **31.** $\dfrac{39}{4}$ or $9\dfrac{3}{4}$

33. $\dfrac{55}{34}$ or $1\dfrac{21}{34}$ **35.** 2 **37.** $\dfrac{55}{12}$ or $4\dfrac{7}{12}$ **39.** $\dfrac{51}{7}$ or $7\dfrac{2}{7}$ **41.** $\dfrac{26}{3}$ or $8\dfrac{2}{3}$ **43.** $\dfrac{2}{15}$ **45.** 20 **47.** $7\dfrac{1}{2}$

49. 4 pieces **51.** $4\dfrac{1}{2}$ cups flour **53.** 10 cups chocolate chips **55.** $\$1\dfrac{3}{4}$ per share

57. (a) Marcy did not change the mixed numbers to improper fractions before she multiplied. (b) $10\dfrac{2}{15}$

CUMULATIVE REVIEW **1.** 74 **2.** 10 **3.** 2 **4.** 3

EXERCISES 5.5 **1.** $\dfrac{1}{5}$ **3.** $\dfrac{9}{10}$ **5.** $\dfrac{5}{6}$ **7.** $\dfrac{4}{9}$ **9.** $\dfrac{11}{12}$ **11.** $\dfrac{5}{6}$ **13.** $\dfrac{3}{4}$ **15.** $-\dfrac{1}{40}$ **17.** 2 **19.** $-\dfrac{6}{7}$

21. $\dfrac{3}{4}$ **23.** 1 **25.** $\dfrac{4}{3}$ **27.** $2x$ **29.** $\dfrac{3}{x}$ **31.** $\dfrac{2}{x}$ **33.** $\dfrac{25}{18}$ **35.** $\dfrac{14}{33}$ **37.** $\dfrac{2}{11}$ **39.** $\dfrac{8}{7}$ **41.** Multiply

43. $\dfrac{10y}{21x}$ **45.** $\dfrac{y - x}{y + x}$

CUMULATIVE REVIEW **1.** -30 **2.** 15 **3.** -5 **4.** 6

EXERCISES 5.6 **1.** $\dfrac{1}{x^5}$ **3.** 7 **5.** $\dfrac{1}{2^4}$ **7.** x^3 **9.** $\dfrac{9^3}{8^8}$ **11.** $\dfrac{z^8}{y^4}$ **13.** $\dfrac{1}{3^4}$ **15.** z^5 **17.** $\dfrac{1}{y^2 z^3}$ **19.** $\dfrac{m^2}{3}$

21. $\dfrac{a^2}{7^3}$ **23.** $\dfrac{b^2}{9^2}$ **25.** $\dfrac{5y^2}{7}$ **27.** $\dfrac{a}{3}$ **29.** $\dfrac{7x^6}{8}$ **31.** (a) z^6 (b) z^9 **33.** (a) x^4 (b) x^4 **35.** (a) b^{12}

(b) b^{12} **37.** z^{24} **39.** 3^6 **41.** b^6 **43.** 1 **45.** y^6 **47.** 2^{20} **49.** 1 **51.** 7^{27} **53.** $x^2 y^2$ **55.** $5^6 y^6$

57. $3^8 x^{16}$ **59.** $6^9 x^{63}$ **61.** $\dfrac{3^3}{x^3}$ or $\dfrac{27}{x^3}$ **63.** $\dfrac{a^4}{b^4}$ **65.** $\dfrac{x^2}{6^2}$ or $\dfrac{x^2}{36}$ **67.** $\dfrac{3^2}{4^2}$ or $\dfrac{9}{16}$ **69.** (a) $20x^3$ (b) $75x^6$

(c) x^9 (d) $\dfrac{3}{x^2}$ **71.** (a) $12x^3$ (b) $27x^7$ (c) $3^4 x^8$ (d) $\dfrac{1}{3x}$

73. (a) We add coefficients, the variable stays the same. (b) We multiply coefficients, then add exponents of like terms.

(c) We simplify coefficients, then subtract exponents of like bases. **75.** $\dfrac{5y^2 z^4}{27x^5}$ **77.** $\dfrac{13b^2}{12c^9}$

CUMULATIVE REVIEW **1.** $7x$ **2.** $7 + x$ **3.** $x - 7$

EXERCISES 5.7 **1.** 10 lb **3.** (a) $6\dfrac{1}{6}$ gal (b) 148 mi **5.** $58\dfrac{1}{8}$ mph **7.** $7\dfrac{1}{2}$ cups water, $\dfrac{3}{4}$ tsp salt, $1\dfrac{1}{2}$ cups cereal

9. $13\dfrac{3}{4}$ yd **11.** (a) $53{,}333\dfrac{1}{3}$ mi per day (b) $2222\dfrac{2}{9}$ mph **13.** (a) $30\dfrac{1}{2}$ ft by $15\dfrac{1}{2}$ ft (b) $\$207$ **15.** 6 boards

17. 2 in. **19.** $10\dfrac{1}{8}$ in.

CUMULATIVE REVIEW **1.** $x = \dfrac{12}{3}$; $x = 4$ **2.** $x = \dfrac{45}{5}$; $x = 9$ **3.** $x = 6 \cdot 7$; $x = 42$ **4.** $x = 9 \cdot 8$; $x = 72$

PUTTING YOUR SKILLS TO WORK **INDIVIDUAL** **1.** Days 3: $\$3375$; Day 6: $\$3418\dfrac{3}{4}$ **2.** Day 6; $\$3418\dfrac{3}{4}$ **3.** $\$3350$

4. Profit of $\$3\dfrac{1}{8}$ **5.** Yes **6.** $\$5128\dfrac{1}{8}$ **GROUP** **1.** $\$9806\dfrac{1}{4}$ **2.** Profit of $\$175$ **3.** Answers may vary.

4. Answers may vary. **5.** Answers may vary.

CHAPTER 5 REVIEW   **1.** $-\dfrac{9}{2}$ **2.** $\dfrac{1}{7}$ **3.** $-\dfrac{b}{a}$ **4.** $\dfrac{1}{21}$ **5.** $-\dfrac{2}{15}$ **6.** $\dfrac{7}{10x}$ **7.** $-\dfrac{4x}{5}$ **8.** $-10x^7$ **9.** $\dfrac{6}{35}$

10. $-\dfrac{4}{15}$ **11.** $-\dfrac{2}{33}$ **12.** $\dfrac{11x^7}{15}$ **13.** $\dfrac{4x^6}{9}$ **14.** $\dfrac{44}{5}$ **15.** $\dfrac{15}{2}$ **16.** $1000 withheld   **17.** $\dfrac{1}{6}$ lb in each container

18. 14   **19.** 20   **20.** 180   **21.** 84   **22.** $16x$ **23.** $140x$ **24.** $90x^2$ **25.** $100x^2$ **26.** $\dfrac{3}{17}$ **27.** $-\dfrac{34}{27}$

28. $\dfrac{2}{x}$ **29.** $\dfrac{x-5}{7}$ **30.** $\dfrac{23}{18}$ **31.** $\dfrac{7}{32}$ **32.** $\dfrac{5}{42}$ **33.** $\dfrac{31}{6x}$ **34.** $\dfrac{x}{3}$ **35.** $\dfrac{2x}{21}$ **36.** $4\dfrac{3}{10}$ **37.** $9\dfrac{4}{9}$

38. $\dfrac{19}{25}$ **39.** $\dfrac{9}{14}$ **40.** 10   **41.** $\dfrac{77}{40}$ **42.** $\dfrac{66}{125}$ **43.** 18   **44.** $2\dfrac{1}{6}$ **45.** $\dfrac{7}{8}$ ft   **46.** $\dfrac{25}{28}$ **47.** $\dfrac{13}{16}$

48. 6   **49.** $2x$ **50.** $\dfrac{5}{3x}$ **51.** $\dfrac{27}{20}$ **52.** $\dfrac{22}{9}$ **53.** y^2 **54.** $\dfrac{1}{3}$ **55.** $\dfrac{1}{a^3b}$ **56.** $\dfrac{x^3}{y^6}$ **57.** $\dfrac{1}{2^3x^9}$ **58.** $\dfrac{1}{3y^6}$

59. $\dfrac{4}{7x^4}$ **60.** $3y^2$ **61.** y^6 **62.** 2^8 or 256   **63.** x^3y^3 **64.** $9x^4$ **65.** 1   **66.** $\dfrac{3^2}{y^2}$ or $\dfrac{9}{y^2}$ **67.** $\dfrac{x^3}{8}$

68. (a) $29\dfrac{1}{4}$ ft by 16 ft   **(b)** $294\dfrac{1}{8}$ **69.** $1\dfrac{5}{8}$ in.

CHAPTER 5 TEST   **1.** $\dfrac{3}{5}$ **2.** $-\dfrac{1}{10}$ **3.** x^3 **4.** -2 **5.** $\dfrac{3x^4}{11}$ **6.** 5 boards   **7.** 9, 18, 27, 36, 45   **8.** 42

9. $20a^4$ **10.** 1785   **11.** 60   **12.** $\dfrac{15}{x}$ **13.** $\dfrac{19}{20a}$ **14.** $8\dfrac{1}{6}$ **15.** 0   **16.** $\dfrac{7}{24}$ **17.** $\dfrac{7x}{15}$ **18.** $-\dfrac{64}{3}$ or $-21\dfrac{1}{3}$

19. $\dfrac{7}{8}$ **20.** $\dfrac{5}{9}$ **21.** $\dfrac{24}{11}$ or $2\dfrac{2}{11}$ **22.** $-\dfrac{25}{6}$ or $-4\dfrac{1}{6}$ **23.** $\dfrac{19}{24}$ **24.** 4   **25.** 2   **26.** $\dfrac{3}{5}$ **27.** x^{12} **28.** $125x^6$

29. $\dfrac{y^2}{4x}$ **30. (a)** $91   **(b)** Profit   **(c)** $3\dfrac{1}{2}$

CUMULATIVE TEST CHAPTERS 1–5   **1.** $3000 + 400 + 1$ **2.** $10 + x$ **3.** $3x + 5$ **4.** $-13r - 5$
5. $4 + x = 15; x = 11$ **6.** $3x + 3 = 24; x = 7$ **7.** $x - 8 = 18; x = 26$ **8.** $3x^6$ **9.** $5x + 15$ **10.** 58

11. -7 **12.** -5 **13.** 7   **14.** $-50x^3$ **15.** 25   **16. (a)** Composite   **(b)** Prime   **17.** $2^4(3)$ **18.** $\dfrac{16}{21}$

19. $>$ **20.** $<$ **21.** 23 mi/gal   **22.** $x = 15$ **23.** $\dfrac{19}{6x}$ **24.** $\dfrac{12x}{5}$ **25.** x^5 **26.** $\dfrac{1}{4}$ **27.** -20

28. $\dfrac{17}{10}$ or $\dfrac{7}{10}$ **29.** $64x^3$ **30.** $\dfrac{1}{6}$ gal

CHAPTER 6

CHAPTER 6 PRETEST   **1.** $x = 25$ **2.** $x = -11$ **3.** $x = 9$ **4.** $y = 12$ **5.** $y = 4$ **6.** $z = 4$
7. $z = -2$ **8.** $z = 18$ **9.** $z = -6$ **10.** $z = 2$ **11.** $x = 10$ **12.** $x = -1$ **13.** $x = 3$ **14.** $x = -6$
15. Trinomial   **16.** $-12a^2 - 18a$ **17.** $8x^2 + 6x - 10$ **18.** $x^2 - 5x - 14$ **19.** $3x - 2y + 8$
20. $12x^2 + x + 2$ **21.** $11y + 3$ **22.** $x = -4$ **23.** $x = 3$ **24.** $2x - 7 = 33$ **25.** $x = 20$

EXERCISES 6.1   **1.** -4 **3.** 9   **5.** 17   **7.** 28   **9.** $y = 20$ **11.** $x = 9$ **13.** $n = -31$ **15.** $y = -94$
17. $x = 14$ **19.** $y = 18$ **21.** $x = -14$ **23.** $x = 26$ **25.** $a = -18$ **27.** $m = 48$ **29.** $x = 1$
31. $y = -7$ **33. (a)** $x = 30$ **(b)** $x = 14$ **35. (a)** $x = -17$ **(b)** $x = -5$ **37.** 5   **39.** -2 **41.** 6
43. -5 **45.** $y = 180$ **47.** $x = 217$ **49.** $m = -390$ **51.** $x = -84$ **53.** $y = 33$ **55.** $a = 60$

57. $m = -30$ **59.** $a = 80$ **61.** $x = 13$ **63.** $y = -11$ **65.** $x = 4$ **67.** $a = \dfrac{52}{5}$ or $10\dfrac{2}{5}$ **69.** $x = -\dfrac{2}{3}$

71. $x = -4$ **73.** $x = -\dfrac{8}{5}$ or $-1\dfrac{3}{5}$ **75.** $y = -\dfrac{10}{11}$ **77. (a)** $x = 13$ **(b)** $x = 208$ **79. (a)** $x = -13$

(b) $x = -208$ **81. (a)** $x = 28$ **(b)** $x = 21$ **(c)** $x = \dfrac{3}{5}$ **(d)** $x = 55$ **83. (a)** $x = 19$ **(b)** $x = \dfrac{2}{3}$

(c) $x = 54$ **(d)** $x = 23$ **85.** zero   **87.** add   **89.** $x = -27$

CUMULATIVE REVIEW   **1.** 13 R2 or $3\dfrac{2}{11}$ **2.** 8028   **3.** 29,400

EXERCISES 6.2   **1.** $x = 6$ **3.** $x = 7$ **5.** $x = 2$ **7.** $x = \dfrac{17}{5}$ **9.** $y = -14$ **11.** $m = -\dfrac{49}{3}$ or $-16\dfrac{1}{3}$

13. $x = -25$ **15.** $y = -\dfrac{6}{5}$ or $-1\dfrac{1}{5}$ **17.** $x = -1$ **19.** $y = 3$ **21.** $x = 16$ **23.** $m = 36$ **25.** $x = 18$

27. $x = -36$ **29.** $x = -8$ **31.** $y = -56$ **33.** $y = 2$ **35.** $x = \dfrac{8}{11}$ **37.** $x = -3$ **39.** $x = 5$

41. $x = \dfrac{5}{14}$ **43.** $x = -2$ **45.** $x = 18$ **47.** $x = -8$ **49.** $x = 3$ **51.** $x = 4$ **53.** $y = -6$

55. $m = \dfrac{19}{3}$ or $6\dfrac{1}{3}$ **57.** $y = 3$ **59.** $x = 5$ **61.** $x = -6$ **63.** $x = \dfrac{10}{3}$ or $3\dfrac{1}{3}$ **65.** $x = 4$ **67.** $x = -5$

69. Yes   **71.** Answers may vary.   **73.** $x = -\dfrac{1}{5}$

CUMULATIVE REVIEW   **1.** 12   **2.** 20   **3.** $2x$ **4.** $5x$

EXERCISES 6.3   **1.** $x = 16$ **3.** $x = 12$ **5.** $x = 28$ **7.** $x = \dfrac{23}{18}$ or $1\dfrac{5}{18}$ **9.** $x = \dfrac{1}{8}$ **11.** $x = \dfrac{1}{8}$

13. $x = \dfrac{1}{28}$ **15.** $x = \dfrac{1}{8}$ **17.** $x = 5$ **19.** $x = \dfrac{9}{2}$ or $4\dfrac{1}{2}$ **21.** $x = 2$ **23.** $x = -\dfrac{12}{11}$ or $-1\dfrac{1}{11}$ **25.** $x = -6$

27. $x = -\dfrac{15}{4}$ or $-3\dfrac{3}{4}$ **29.** $x = \dfrac{30}{17}$ or $1\dfrac{13}{17}$ **31.** Answers may vary.   **33.** $x = -\dfrac{11}{12}$ **35.** $x = -\dfrac{45}{44}$ or $-1\dfrac{1}{44}$

CUMULATIVE REVIEW   **1. (a)** $11x + 5$ **(b)** $7y + 2x + 8$ **2. (a)** $8x^2$ **(b)** $30y^5$ **3.** $-8x^3$ **4.** $-30y^3$

EXERCISES 6.4   **1. (a)** Binomial   **(b)** Monomial   **(c)** Trinomial   **3.** $-2z^2, +4z, -2y^4, +3$
5. $+6x^6, -3x^3, -3y, -7$ **7.** $16x^2 + 24x - 40$ **9.** $-10y^2 - 15y + 30$ **11.** $-10y^2 - 4y + 16$ **13.** $-6x^2 + 12x$
15. $-8a^2 - 12a$ **17.** $-8x^2 + 4xy - 10x$ **19.** $-10a^2 + 20ab + 35a$ **21.** $8x^2 - 4xy + 16x$ **23.** $-8x^4 + 40x^3$
25. $-4y^8 - 36y^2$ **27.** $-3x^6 - 3x^4$ **29.** $-4y^9 - 24y^5$ **31.** $x^7 + x^6 + x^4$ **33.** $3x^6 + x^5 + 4x^3$ **35.** $a^2 + 3a + 2$
37. $y^2 + 7y + 10$ **39.** $x^2 + 5x + 4$ **41.** $a^2 + 3a - 28$ **43.** $x^2 - 2x - 15$ **45.** $b^2 + b - 6$
47. $m^2 + 2m - 15$ **49.** $z^2 - 4z - 45$ **51.** $3x^2 + 7x + 2$ **53.** $4x^2 + 7x + 3$ **55.** $4y^2 + 2y - 2$
57. $4y^2 - 2y - 2$ **59. (a)** $x^2 + 5x + 4$ **(b)** $x^2 - 3x - 4$ **(c)** $x^2 - 5x + 4$ **(d)** $x^2 + 3x - 4$
61. F First term times first term; O the product of outer terms; I the product of inner terms; L the product of last terms
63. $b = -5$ **65.** $6x^2y - 10xy^2 + 4xy$ **67.** $x^3 + 4x^2 + 6x + 4$

CUMULATIVE REVIEW   **1.** -4 **2.** $2x^2$ **3.** $-4x^2$

EXERCISES 6.5   **1.** $5m - 2$ **3.** $3x - 6z + 5y$ **5.** $-9x - 4$ **7.** $-5x - 2y$ **9.** $8x - 9$ **11.** $-2x - 8y + 3z$
13. $3y + 3$ **15.** $a + 2$ **17.** $-6y - 3$ **19.** $-2x + 9$ **21.** $11x + 9$ **23.** $-12a + 8$ **25.** $-y^2 + 10y + 3$
27. $-4x^2 + 10x + 3$ **29.** $-9z^2 + z + 6$ **31.** $7a^2 + 17a + 3$ **33.** $-9x^2 - 14x - 5$ **35.** $-x^2 + 3x + 4$
37. $-8x^2 - 8x - 12$ **39.** $3x^2 - x + 15$ **41.** $2x^2 + 3x - 16$ **43.** $2x^2 - x - 4$ **45.** $-11x^2 - 4x - 11$
47. $-10x^2 - 9x + 5$ **49.** $-3x^2 - 25x - 25$ **51.** Answers may vary.   **53.** $a = -23$

CUMULATIVE REVIEW   **1.** $x = \dfrac{7}{2}$ **2.** $x = \dfrac{9}{5}$ **3.** $x = -\dfrac{19}{3}$ **4.** $x = -\dfrac{15}{2}$

EXERCISES 6.6   **1.** $x = -3$ **3.** $x = -\dfrac{2}{3}$ **5.** $y = -4$ **7.** $x = -\dfrac{37}{7}$ **9.** $x = \dfrac{1}{4}$ **11.** $y = 1$ **13.** $x = -2$

15. $y = 7$ **17.** $x = 2$ **19.** $x = 3$ **21.** $x = \dfrac{5}{3}$ **23.** $y = -\dfrac{7}{8}$ **25.** $x = \dfrac{14}{11}$ **27.** $x = 1$ **29.** $x = -\dfrac{7}{2}$

31. $x = -3$ **33.** No

CUMULATIVE REVIEW   **1.** $6 + 2x$ **2.** $x - 12$ **3.** $4 + x$ **4.** $2(5 + y)$

EXERCISES 6.7   **1. (a)** $2n + 5 = 15$ **(b)** $n = 5$ **3. (a)** $3n - 4 = 5$ **(b)** $n = 3$ **5. (a)** $2(5 + n) = 12$
(b) $n = 1$ **7.** $L = 18$ m   **9.** First side = 3 cm, second side = 4 cm, third side = 5 cm   **11.** $W = 6$ in.
13. $J =$ Janet's height; $J + 2 =$ Wendy's height; $J - 3 =$ Marci's height   **15.** $F =$ length of the first side;
$F + 4 =$ length of the second side; $2F - 10 =$ length of the third side   **17. (a)** $J =$ Jerome's salary;
$J - 7500 =$ cashier's salary   **(b)** $J + (J - 7500) = 62{,}000$ **(c)** Jerome's salary: \$34,750; Cashier's salary: \$27,250
(d) $34{,}750 + (34{,}750 - 7500) \stackrel{?}{=} 62{,}000; 62{,}000 = 62{,}000$ **19. (a)** $x =$ mi he drove the second day; $x + 85 =$ mi he
drove the first day   **(b)** $x + (x + 85) = 385$ **(c)** Second day: 150 mi; First day: 235 mi   **21. (a)** $s =$ flight time
of the second flight; $\dfrac{1}{2} \times s =$ flight time of the first flight.   **(b)** $s + \left(\dfrac{1}{2} \times s\right) = 15$ or $s + \dfrac{s}{2} = 15$ **(c)** 10 hr, time
of the second flight; 5 hr, time of the first flight   **23. (a)** $L =$ first side of the triangle; $2L =$ second side;
$L + 12 =$ third side   **(b)** $L + 2L + (L + 12) = 120$ **(c)** First side: 27 m; second side: 54 m; third side: 39 m

25. (a) W = width; $3W - 2$ = length   **(b)** $2W + 2(3W - 2) = 68$ **(c)** Width: 9 m; Length: 25 m
27. (a) x = students living off campus; $x + 115$ = students living on campus; $x - 55$ = students living at home
(b) $x + (x + 115) + (x - 55) = 1704$ **(c)** Students living off campus: 548; Students living on campus: 663;
Students living at home: 493   **29.** 3 ft

CUMULATIVE REVIEW   1. 144 R20   **2.** 295,080   **3.** 12 ft^2 **4.** $\dfrac{2}{7}$

PUTTING YOUR SKILLS TO WORK   INDIVIDUAL 1. 350   **2.** $1045   **3.** $295   **4.** 55   **5.** 555
GROUP 1. $C = 2x + 150$ **2.** $R = 6x - 10$ **3.** $P = (6x - 10) - (2x + 150)$ **4. (a)** 40 hinges   **(b)** Less
(c) Answers may vary.   **5.** Model A: $-$40; $-$25; $-$15; $-$10; 0; Model B: $-$100; $-$40; $0; $20; $60
6. (a) There is a loss.   **(b)** Model B, because to produce 45 hinges there is a profit of $20 for Model B and a loss of
$10 for Model A.   **7. (a)** Model A: 55 hinges; Model B: 40 hinges   **(b)** Yes; when the profit is 0, there is a break-even
point, that is, they have covered the cost of production.

CHAPTER 6 REVIEW   1. $y = 45$ **2.** $a = -17$ **3.** $x = -61$ **4.** $y = -20$ **5.** $x = -6$ **6.** $x = -5$

7. $y = 33$ **8.** $y = 10$ **9.** $x = -22$ **10.** $y = -\dfrac{9}{11}$ **11.** $y = -2$ **12.** $x = \dfrac{17}{7}$ **13.** $x = 7$ **14.** $y = -26$

15. $x = -9$ **16.** $y = 77$ **17.** $x = \dfrac{22}{3}$ **18.** $x = -\dfrac{12}{5}$ **19.** $x = 15$ **20.** $x = 6$ **21.** $x = 12$ **22.** $x = \dfrac{5}{8}$

23. $y = -9$ **24.** $x = \dfrac{13}{24}$ **25. (a)** Binomial   **(b)** Monomial   **(c)** Trinomial   **26.** $+2x^2, +5x, -3z^3, +4$

27. $+8a^5, -7b^3, -5b, -4$ **28.** $-24x^2 + 32x - 20$ **29.** $-2y^2 + 12y$ **30.** $27x^2 - 9xy + 6x$ **31.** $20n^2 + 45mn + 35n$
32. $4x^6 - 16x^2$ **33.** $x^9 - 2x^5 - 3x^4$ **34.** $5z^2 - 20z$ **35.** $-6y^2 - 60y$ **36.** $x^2 + 6x + 8$ **37.** $y^2 - 3y - 28$
38. $3x^2 - 2x - 8$ **39.** $5x^2 - 21x + 18$ **40.** $3a - 4$ **41.** $3y - 2z + 3$ **42.** $2x + 7$ **43.** $-4x + 10$

44. $13a^2 + 3a + 6$ **45.** $-8x^2 - 7x - 8$ **46.** $x = -12$ **47.** $x = -3$ **48.** $x = -2$ **49.** $x = \dfrac{1}{8}$ **50.** $y = 5$

51. $x = -\dfrac{6}{5}$ **52. (a)** $2n + 4 = 16$ **(b)** $n = 6$ **53. (a)** $2(16) + 2(x + 3) = 54$ **(b)** 11 ft

54. (a) J = Jamie's salary; $J - 8500$ = cashier's salary   **(b)** $J + (J - 8500) = 59,000$ **(c)** Jamie's salary: $33,750;
Cashier's salary: $25,250   **(d)** $33,750 + (33,750 - 8500) \overset{?}{=} 59,000; 59,000 = 59,000$ **55. (a)** L = fall students;
$L + 85$ = spring students; $L - 65$ = summer students   **(b)** $L + (L + 85) + (L - 65) = 491$
(c) Fall = 157; spring = 242; summer = 92   **(d)** $157 + (157 + 85) + (157 - 65) \overset{?}{=} 491; 491 = 491$

CHAPTER 6 TEST   1. $x = 24$ **2.** $x = -17$ **3.** $x = 0$ **4.** $y = 27$ **5.** $y = -10$ **6.** $y = 4$ **7.** $z = 2$
8. $z = -3$ **9.** $z = 16$ **10.** $z = -5$ **11.** $x = 12$ **12.** $x = -2$ **13.** $y = 6$ **14.** $z = 12$ **15.** Binomial
16. $-14b^2 + 28b$ **17.** $18x^2 - 3x + 3$ **18.** $-8x^5 + 6x^3$ **19.** $x^2 + 14x + 45$ **20.** $x^2 + x - 6$
21. $-4x + 2y + 6$ **22.** $-7x + 7$ **23.** $13x^2 - 2x - 2$ **24.** $2y + 8$ **25.** $-10p - 6$ **26.** $x = -5$
27. $x = 4$ **28.** $x = -2$ **29.** s = length of second side; $s + 2$ = length of first side; $3s$ = length of third side
30. $s + (s + 2) + 3s = 42$ **31.** First side = 10 ft; second side = 8 ft; third side = 24 ft
32. A = Anna's annual salary; $A - 4000$ = sales clerk's annual salary   **33.** $A + (A - 4000) = 61,200$
34. Anna's annual salary: $32,600; salesclerk's annual salary: $28,600.

CUMULATIVE TEST CHAPTERS 1–6   1. 35   **2.** Saturday   **3.** 45   **4. (a)** Factors = 7, x; Product = -21
(b) Factors = x, y; Product: = z **5.** 17   **6.** -12 **7.** -5 **8.** -26 **9.** 0   **10.** >   **11.** <   **12.** >
13. =   **14.** >   **15.** -9 **16.** 12   **17.** 9   **18.** 12   **19.** $\dfrac{1}{2}$ **20.** $\dfrac{27}{4}$ **21.** $\dfrac{5}{8}$ **22.** $\dfrac{3}{16}$ lb   **23.** $\dfrac{1}{2}$ lb

24. $\dfrac{3}{8}$ lb   **25.** s = the length of the first side; $s - 3$ = the length of the second side; $2s$ = the length of the third side

26. $s + (s - 3) + 2s = 29$ **27.** First side = 8 in.; second side = 5 in.; third side = 16 in.
28. J = Juan's annual salary; $J - 5500$ = cashier's annual salary   **29.** $J + (J - 5500) = 52,000$
30. Juan's annual salary: $28,750; cashier's annual salary: $23,250   **31.** $y = -12$ **32.** $x = 4$ **33.** $x = 6$

CHAPTER 7

CHAPTER 7 PRETEST   1. Eighteen and seventy-one thousandths   **2.** $\dfrac{542}{1000}$ **3.** 0.07   **4.** <   **5.** 635.1

6. 3.60   **7.** 11.34   **8.** 16.6   **9.** 32 mi   **10.** $1.43   **11.** 1.236   **12.** 53.6   **13.** 313.1   **14.** $86.90   **15.** 2.125
16. 6.56   **17.** 7.25   **18.** 2.67×10^{-4} **19. (a)** 11 containers   **(b)** $10.65   **20.** $x = 5$ **21.** 86%

22. (a) 37.5%   **(b)** $\dfrac{358}{1000} = \dfrac{179}{500}$ **(c)** 0.0401   **23. (a)** 9.54   **(b)** 26.39   **(c)** 11.22%   **24.** 17.18%   **25.** $384

EXERCISES 7.1 **1.** Five and thirty-two hundredths **3.** Four hundred twenty-eight thousandths **5.** Eight and five tenths **7.** 0.324 **9.** 15.0346 **11.** Twenty-five and 54/100 dollars **13.** One hundred forty-three and 56/100
15. $\frac{7}{10}$ **17.** $3\frac{64}{100}$ **19.** $\frac{1743}{10,000}$ **21.** $100\frac{11}{1000}$ **23.** 0.1 **25.** 12.037 **27.** 0.02 **29.** 0.001 **31.** <
33. > **35.** < **37.** > **39.** 523.72 **41.** 44.00 **43.** 9.1 **45.** 463.0 **47.** 312.952 **49.** 0.0631
51. 10.6 lb, 21.8 lb **53.** $15 **55.** When we change 9 to 10, we write 0 and carry 1 to the 2, the next place value to the
left. Thus the 2 changes to 3. **57. (a)** 0.73, $\frac{7}{10}$, 0.071, 0.007, 0.0069

CUMULATIVE REVIEW **1.** 1427 **2.** 9968 **3.** 86 **4.** 68

PUTTING YOUR SKILLS TO WORK **INDIVIDUAL 1.** Check amounts: $168.96, $43.29, $40, $78.94, $125, $21.56, $231.45;
ATM withdrawals: $100, $100; Deposits: $634.51, $423.62 **2.** $1262.26 **3.** Plan 2
GROUP 1. Checks outstanding: $35.85, $40, $125, $231.45; Deposits outstanding: $423.62; Service Charge: $5.25

EXERCISES 7.2 **1.** 5.57 **3.** 36.2759 **5.** 10.06 **7.** 81.023 **9.** 73.4169 **11.** 28.153 **13.** 33.14
15. 341.82 **17.** 51.443 **19.** 612.88 **21.** 19.84 **23.** 11.87 **25.** −105.343 **27.** −4.57 **29. (a)** −5.5
(b) 15.1 **(c)** −13.3 **31.** 6.2x **33.** 13.5y **35.** 12.6x **37.** 4.9x + 6.2y **39.** 4 in. **41.** $27.69
43. $1492.23 **45.** $385 **47.** Nationwide Mutual, $290.20 **49.** $1479.80 **51.** line up **53.** in line **55.** −4.4
57. 14.43 **59.** 3634.043 **61.** 5625.08

CUMULATIVE REVIEW **1.** 3234 **2.** −108 **3.** −285 **4.** 693

EXERCISES 7.3 **1.** 0.0021 **3.** 0.0032 **5.** 61.669 **7.** 47.12 **9.** 7.3743 **11.** −11.421 **13.** 15.325
15. −34.001 **17.** −170.1891 **19.** 14.98 **21.** 1230 **23.** 855,400 **25.** 308.8 **27.** 240,000 **29.** 930,000
31. 2000 **33.** 5.55 **35.** 15.48 **37.** −2.95 **39.** 46.1 **41.** 3.72 **43.** $143.09 **45.** $83.61 **47. (a)** $4.08
(b) $2.38 **(c)** $1.70 **49.** −29,022.763 **51.** 217,964.9872 **53.** 5 **55.** All Mart: $15.81; A&E Foods: $16.03;
The Market: $16.47 **57.** A&E Foods

CUMULATIVE REVIEW **1.** 95 **2.** 49 **3.** 11 R2 or $11\frac{2}{125}$ **4.** 12 R1 or $12\frac{1}{130}$

PUTTING YOUR SKILLS TO WORK **INDIVIDUAL 1.** 605.05 francs, 179.53 marks, 61.93 pounds
2. 6050.5 francs, 1795.3 marks, 619.3 pounds **3.** 165.3 U.S. dollars **4.** 16.15 U.S. dollars **5.** 55.7 U.S. dollars
GROUP 1. Francs: less; marks: less; pounds: more **2.** Germany **3. (a)** 35.13 pounds **(b)** 284.70 marks
(c) 956.34 francs **(d)** 29.78 marks

EXERCISES 7.4 **1.** 0.62 **3.** 2.16 **5.** 0.0565 **7.** 3.451 **9.** 0.23 **11.** −0.88 **13.** 12.24 **15.** 24.92
17. $2.\overline{7}$ **19.** $0.\overline{63}$ **21.** $-30.\overline{30}$ **23.** 13.5 **25.** 13.5 **27.** 1.83 **29.** 5.33 **31.** 3.4 **33.** 2.36 **35.** 12.133
37. (a) $0.00\overline{3}$ sec **(b)** $0.0\overline{3}$ sec **39. (a)** 0.0456 **(b)** 34,700 **41. (a)** 0.0982 **(b)** 86,000 **43.** 5.46×10^2
45. 3.1235×10^4 **47.** 9.543×10^{-2} **49.** 2.15×10^{-3} **51.** 5×10^3 **53.** 2.38×10^5 **55.** 5.878×10^{12} mi
57. 4.8×10^{-10} electrostatic unit **59. (a)** 26.5 mi/gal **(b)** 56 gal of gas **61. (a)** 18 containers **(b)** $8.25
63. (a) 8.23×10^6 **(b)** 8,230,000 **65.** 2.0271×10^9 **67.** $462\overline{)1270}$ **69.** $33.01 per hr

CUMULATIVE REVIEW **1.** x = 8 **2.** x = 25 **3.** x = 1035 **4.** x = 15

EXERCISES 7.5 **1.** x = 6.1 **3.** y = 9.45 **5.** x = 2.1 **7.** x = −12.1 **9.** x = −10.4 **11.** x = 3
13. x = 5 **15.** x = 5.62 **17.** x = 3.4 **19.** x = −5 **21.** x = 0.9 **23.** x = 17.2 **25.** x = 2.1
27. x = 3.5 **29.** x = 5 **31.** x = 6 **33.** x = 1 **35.** x = 8 **37.** 6 quarters and 30 dimes
39. 4 nickels and 20 dimes **41.** x = 4.5 **43.** x = 18.5648 **45.** x + 1.2 = 2.6 **47.** 4.2x = 8.992 **49.** 52.5 ft/sec

CUMULATIVE REVIEW **1.** 0.8 **2.** 0.625 **3.** $0.3\overline{3}$ **4.** $0.8\overline{3}$

EXERCISES 7.6 **1.** 24% **3.** 31% **5.** 7% **7.** 63% **9.** 113% **11.** 16% **13.** 65% **15.** 0.538
17. 0.24% **19.** 0.0233 **21.** 3.413% **23.** 57.6%; 0.249; 0.003; 154.6% **25.** 374.2%; 0.238; 0.006; 1288.2%
27. 87.5% **29.** $\frac{91}{125}$ **31. (a)** 35% **(b)** $\frac{223}{1000}$ **33.** $\frac{9}{220}$ **35.** 2.5% **37.** 0.8; 80%; $\frac{27}{100}$; 27%; $\frac{7}{1000}$; 0.007;
$4.\overline{3}$; $433.\overline{3}$% **39.** 0.3125; 31.25%; $2\frac{3}{5}$; 260%; $\frac{1}{400}$; 0.0025; 6.5; 650% **41.** 71.43% **43.** 43.86% **45.** 0.468
47. 0.000137 **49.** left; % sign

CUMULATIVE REVIEW **1.** 3x = 48; 16 **2.** 2x = 330; 165 **3.** $\frac{x}{4}$ = 60; 240 **4.** x = $\frac{69}{3}$; 23

EXERCISES 7.7 **1.** x = 0.32 × 84; 26.88 **3.** N = 0.24 × 145; 34.8 **5.** N = 0.46 × 60; 27.6 **7.** $1.94
9. 390 students **11.** 70 = n% × 650; n = 10.77% **13.** 65 = n% × 850; n = 7.65% **15.** 18.75% **17.** 40%
19. 56 = 0.70 × n; n = 80 **21.** 24 = 0.40 × n; n = 60 **23.** 200 **25.** 6.4 mi **27.** n = 0.15 × 25; n = 3.75

29. $n = 0.27 \times 78$; $n = 21.06$ **31.** $1.25 \times 85 = n$; $n = 106.25$ **33.** $44 = 0.5 \times n$; $n = 88$ **35.** $n\% \times 87 = 53.94$; 62%

37. 86 is what percent of 92; 93% **39.** 10% **41.** 234.8097 **43.** 1840 **45.** 35 is much more than $\frac{1}{2}$ of 40.

47. 9 is very close to 10, so it cannot be $\frac{1}{4}$ of 10.   **49.** 7% **51.** $\$162.50$ **53. (a)** $\$9.80$ **(b)** $\$149.80$

CUMULATIVE REVIEW   **1.** $x = 5$ **2.** $x = 5$ **3.** $x = 6$

TO THINK ABOUT 7.8   More

EXERCISES 7.8   **1.** $\$352.50$ **3.** $\$3375$ **5.** $\$379.50$ **7.** $\$38,340$ **9.** 874 **11. (a)** $\$210$ **(b)** $\$3210$
13. $\$20$ **15.** $\$849.06$ **17.** $\$315$ **19.** 140.25 lb   **21.** $\$25,440$ **23. (a)** $\$60,015.05$ **(b)** $\$57,970$
25. (a) $\$999$ **(b)** $\$699.30$ **27. (a)** $15,257.16$ mi   **(b)** $\$4729.72$ may be deducted

CUMULATIVE REVIEW   **1.** 24 cm^2 **2.** 252 cm^3

CHAPTER 7 REVIEW   **1.** Six and twenty-three hundredths   **2.** Six hundred seventy-nine thousandths

3. Seven and eighty-three ten-thousandths   **4.** Forty-six and 85/100   **5.** $\frac{268}{1000}$ **6.** $3\frac{94}{100}$ **7.** 32.761 **8.** 54.026
9. $<$ **10.** $<$ **11.** 842.86 **12.** 359.26 **13.** 521.103 **14.** 406.781 **15. (a)** 8.63 **(b)** -4.964 **(c)** 45.018
16. (a) 8.23 **(b)** -7.3483 **(c)** 89.8669 **17. (a)** 23.65 **(b)** -11.79 **18. (a)** 70.346 **(b)** -11.835
19. (a) -14.6 **(b)** 10.75 **(c)** -0.33 **20.** line up   **21.** 85.0, the decimal points   **22.** $11.8x$ **23.** $12.5x$
24. $\$93$ **25.** $\$24.38$ **26.** 0.00546 **27.** 0.00164 **28.** 40.9528 **29.** 19.2046 **30.** 124.1625 **31.** 50.456
32. five   **33.** seven   **34.** 12.49 **35.** 3240 **36.** $4,100,000$ **37.** 325.5 **38.** 52.11 **39.** $360,000$ **40.** 3.72
41. 23.57 **42.** $\$444.05$ **43.** $\$5498.71$ **44.** 0.72 **45.** 0.89 **46.** -0.64 **47.** 1.18 **48.** $49.15\overline{4}$ **49.** $39.\overline{42}$
50. $721\overline{)2590}$ **51.** $12\overline{)4.349}$ **52.** 10.75 **53.** 23.5 **54.** 1.44 **55.** 5.5 **56.** 9.2 **57.** 0.03166 **58.** 2430
59. 3.48×10^2 **60.** 2.1206×10^4 **61.** 9.2×10^{-2} **62.** 1.359×10^{-1} **63. (a)** 11 **(b)** $\$21.70$ **64.** $x = 10.91$

65. $x = -2.3$ **66.** $x = 3.15$ **67.** $x = 7$ **68.** 40 dimes; 200 nickels   **69.** 85% **70.** $\frac{132}{110} = 120\%$ **71.** 0.057

72. 1.6% **73.** 1.24 **74.** 37.9%; 0.428; 0.005; 347.2% **75.** 124%; 0.035; 0.0025; 56.7% **76.** $33\frac{1}{3}\%$ **77.** $1\frac{3}{10}$

78. 0.375; 37.5%; $\frac{18}{25}$; 72%; 7.4; 740% **79.** 0.75; 75%; $\frac{14}{25}$; 56%; $\frac{1}{500}$; 0.002; 3.4; 340% **80.** 30% **81.** 26.67% **82.** 328
83. 27 **84.** 5% **85.** 300 **86.** 54 **87.** 5% **88.** $\$1190$ **89.** 13.12% **90.** $\$2385$ **91.** 19% **92.** $\$975$ **93.** $\$417.78$

CHAPTER 7 TEST   **1.** Two hundred seven and four hundred two thousandths   **2.** $\frac{13}{1000}$ **3.** 0.51 **4.** $>$
5. 746.14 **6.** 13.14 **7.** 12.64 **8.** -6.1 **9.** 9.888 **10.** 4720 **11.** 4.5 **12.** $6.\overline{3}$ **13.** 10% **14. (a)** 0.8
(b) 80% **15. (a)** $\frac{1}{10}$ **(b)** 10% **16. (a)** $\frac{1}{20}$ **(b)** 0.05 **17.** $\$792$ **18.** 12.5 mL   **19.** 10.7
20. 4.3986×10^4 **21.** 5.61×10^{-2} **22.** $\$5992.06$ **23.** $x = 4$ **24. (a)** 17.4 mi/gal   **(b)** $\$42.04$ **25.** 40
26. 2% **27.** 9.6 **28.** $\$1575$ **29.** 47.9% **30.** 12.5% **31.** $\$45$ **32.** $\$1804$

CUMULATIVE TEST CHAPTERS 1–7   **1.** $3x + 3xy$ **2.** $6x$ **3.** $-3 - 6x - 2xy$ **4.** 32 ft   **5.** 99 sq ft
6. $\$15,000$ profit   **7.** -6 **8.** 38 **9.** 0 **10.** $8\frac{1}{3}$ or 8.33 **11.** $\frac{2}{3y}$ **12.** x^6 **13.** $64x^3y^3$ **14.** $x = 39$
15. $-6x^2 + 15x$ **16.** $3x - 5$ **17.** $x - 12$ **18.** $x = 2$ **19.** $\$1.80$ **20.** 180.27 **21.** 33.76 **22.** 21.71
23. 930 **24.** 3.5 **25.** $\$17.25$ **26. (a)** 0.6 **(b)** 60% **27. (a)** $\frac{30}{100} = \frac{3}{10}$ **(b)** 30% **28. (a)** $\frac{1}{100}$ **(b)** $.01$

CHAPTER 8
....

CHAPTER 8 PRETEST   **1.** 350 people   **2.** 500 people   **3.** 4th quarter of 1996   **4.** 100 people   **5.** 2.7 mi

6. 3 mi   **7–10.**   **11. (a)** Answers may vary.   **(b)**

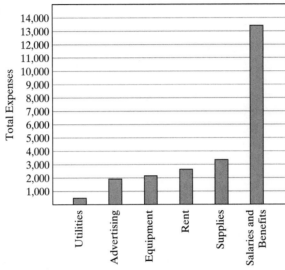

PUTTING YOUR SKILLS TO WORK   **INDIVIDUAL 1.** $24,000

2. Graphs may vary.

3. Salaries and benefits: 56%; equipment: 9%; supplies: 14%; advertising: 8%; rent: 11%; utilities: 2%

4.

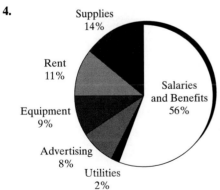

5. $24,960   **GROUP 1.** Salaries and benefits: $14,112; equipment: $864; supplies: $3696; advertising: $1516.80; rent: $2772; utilities: $480

2. (a) $23,440.80   **(b)** $559.20 less   **(c)** 2.33% decrease

3.

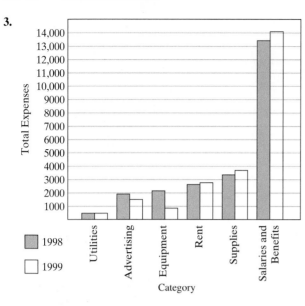

4. (a) Salaries and benefits: 60%; equipment: 4%; supplies: 16%; advertising: 6%; rent: 12%; utilities: 2%

(b)

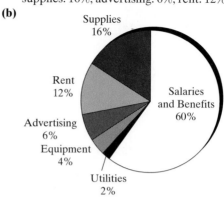

EXERCISES 8.1   1. 2000   **3.** DuPage County   **5.** 2800   **7.** 4,000,000   **9.** 7,200,000   **11.** 15%   **13.** 72%
15. 44,840   **17.** 1,850,000,000   **19.** 37%   **21.** 81%   **23.** 50,000 women   **25.** 1990   **27.** 25,000 more
29. 1930 to 1950   **31.** June   **33. (a)** about 4500 customers   **(b)** increased   **35.** between June and July   **37.** 2.5 in.
39. October, November, and December   **41.** 1.5 in.   **43.**

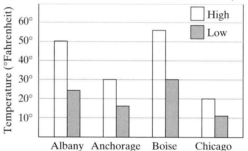

45.

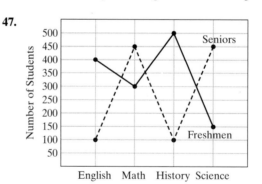

47.

49. 35.7%   **51.** 40%   **53.** The line from March to May goes down the steepest, indicating the greatest decrease.

CUMULATIVE REVIEW   1. 84 in²   **2.** 204 in²   **3.** 16 gal   **4.** 210 in³

EXERCISES 8.2   1. 91.8   **3.** 38   **5.** $87,000   **7.** 5 hr   **9.** 5.6 sales   **11.** 0.375   **13.** 23.7 mi/gal   **15.** 47
17. 1011   **19.** 0.58   **21.** $20,250   **23.** 35.5 min   **25.** $10.99   **27.** 46 actors   **29.** 4850
31. (a) 2,340,400,000 barrels   **(b)** 98,296,800,000 gal   **33. (a)** $2157   **(b)** $1615   **(c)** The median because the
mean is affected by the high amount $6300.   **35.** 60   **37.** two modes, 121 and 150   **39.** $249

CUMULATIVE REVIEW   1. 17   **2.** −4   **3.** 11   **4.** −7

EXERCISES 8.3   1. (1987, 4700), (1988, 4800), (1989, 4950), (1990, 4850)
3. Lena, (60, 4); Janie, (62, 7); Mark, (66, 9)   **5.** Lena, (4, 60); Janie, (7, 62); Mark, (9, 66)

7–13.
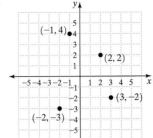

15–21.
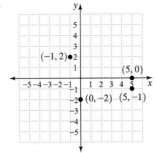

23–29.
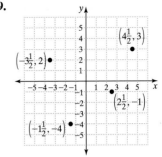

31. $(-3, -3)$ **33.** $(3, 3)$ **35.** $(-1, 4)$ **37.** $\left(-3\frac{1}{2}, 4\right)$ **39.** $\left(1\frac{1}{2}, -3\right)$ **41.** $(5, -1)$

43. (a) $(-4, -2)$
(b)
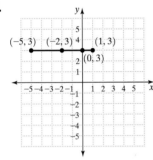

45. (a) $(2, 1)$
(b)
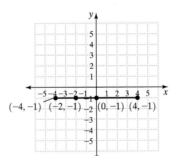

47.
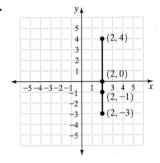

49.
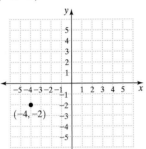

51.
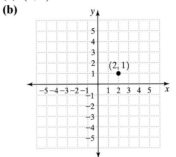

53. origin
55. Starting at the origin, move 2 units right followed by 1 unit down.
57. Since all the x-values are 6, we always move 6 units in the positive x-direction, and this results in a vertical line.

59. (a–d)
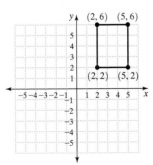

(e) It is a rectangle.

CUMULATIVE REVIEW   **1.** 5   **2.** -13 **3.** -4 **4.** -5

PUTTING YOUR SKILLS TO WORK **INDIVIDUAL 1.** $(1, 3), (2, 6), (4, 12), (6, 18)$

2.

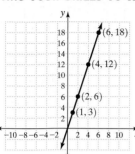

3. Positive correlation **4.** $24 **8.** **9.** $10

5. $P = 3x$ (x is number of items)

6. $60

7. $(1, 3), (2, 4), (3, 5), (4, 6), (6, 8)$

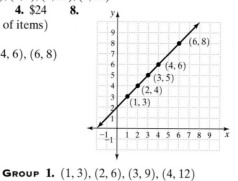

10. $C = x + 2$ (x is number of components) **11.** $52 **GROUP 1.** $(1, 3), (2, 6), (3, 9), (4, 12)$

2, 4.

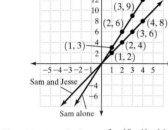

3. $(1, 2), (2, 4), (3, 6), (4, 8)$

5. Sam: $y = 3x$; together: $y = 2x$

6. (a) 36 hr **(b)** 24 hr

7. Sam alone

8. No, because in this situation, negative values don't make sense.

EXERCISES 8.4 **1.** $(0, 4), (4, 0), (1, 3), (-1, 5)$ (Answers may vary.) **3.** $(0, 12), (12, 0), (1, 11), (-1, 13)$
(Answers may vary.) **5. (a)** $(4, 140)$ **(b)** $(8, 280)$ **7. (a)** $(3, 240)$ **(b)** $(5, 400)$ **9.** $(0, 8), (16, 0), (8, 4)$
11. $(3, 2), (0, 5), (1, 4)$ **13.** $(-1, 1), (1, 3), (-2, 0)$ **15.** $(0, 3), (-1, -2), (1, 8)$
17. $(0, -3), \left(\frac{3}{5}, 0\right), (1, 2)$ (Answers may vary.) **19.** $(0, 6), (-6, 0), (1, 7)$ (Answers may vary.)

21.

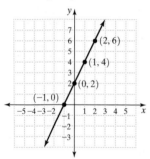

23.

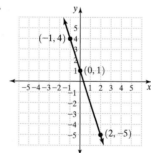

25.

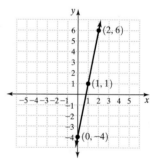

27.

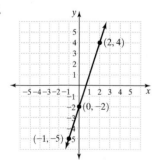

29.

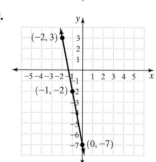

31.

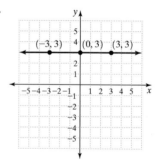

33.

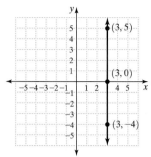

35. Answers may vary.
37. (6, 0); Because this is not on the line formed by the other points.
39. $y = 2x$

CUMULATIVE REVIEW   **1.** 3 ft   **2.** 8 qt   **3.** 2 lb   **4.** 180 min

CHAPTER 8 REVIEW   **1.** 350   **2.** 300   **3.** 150   **4.** 800   **5.** 15%   **6.** 4%   **7.** 57%   **8.** 27%   **9.** $192
10. $360   **11.** $816   **12.** $1008   **13.** 6000   **14.** 4000   **15.** 4th quarter 1994   **16.** 3rd quarter 1993   **17.** 1000
18. 1500   **19.** Yes   **20.** Yes   **21.** 45,000   **22.** 30,000   **23.** 10,000   **24.** 30,000   **25.** 25,000   **26.** 30,000
27. The cooler the temperature, the fewer cones sold.   **28.** The warmer the temperature, the more cones sold.

29.

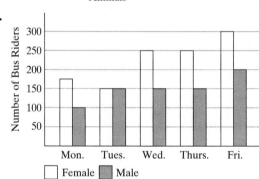

30.

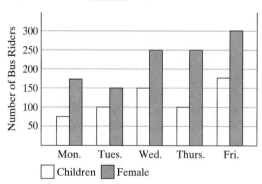

31.

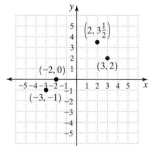

32.

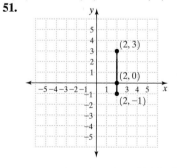

33. 82   **34.** 83   **35.** 19.5   **36.** 18.5   **37.** 90°F   **38.** $114   **39.** 58   **40.** 188
41. (1991, 5000), (1992, 6000), (1993, 5500), (1994, 6250)   **42.** (1995, 7000), (1996, 6500), (1997, 7500), (1998, 7250)
43–46.

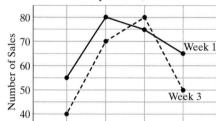

47. (−5, 3)   **48.** (−1, −1)   **51.**

49. $\left(0, 2\frac{1}{2}\right)$ **50.** (4, 1)

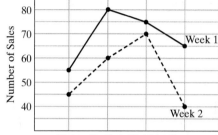

52.

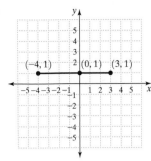

53. $(6, 2), (8, 0), (7, 1)$
(Answers may vary.)
54. $(0, 3), (2, 1), (3, 0)$
(Answers may vary.)
55. **(a)** $(4, 280)$ **(b)** $(5, 350)$
56. $(1, -4), (0, -6), (-1, -8)$
57. $(0, 2), (-1, 8), (1, -4)$

58.

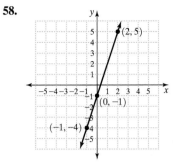

59.

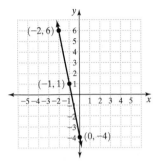

60.

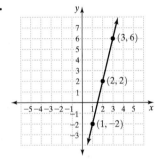

61.

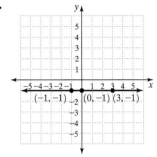

CHAPTER 8 TEST   **1.** Age 18–20   **2.** 77%   **3.** 50%   **4.** 350 students   **5.** 10 years   **6.** 46 years   **7.** 10 years
8. Age 35   **9.** 100 cars   **10.** 50 cars   **11.** 3rd quarter of 1993   **12.** 325 cars   **13.** 350 cars   **14.** 85.7   **15.** 87
16–19.

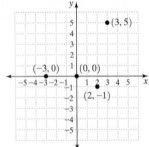

20. $(4, 2)$
21. $(0, -2)$
22. $(3, -1)$
23. $(-1, -3)$

24.

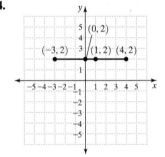

25.

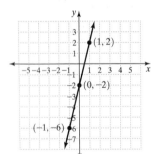

26.

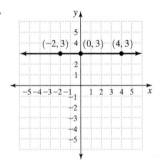

CUMULATIVE TEST CHAPTERS 1–8   **1.** 7 in.   **2.** 5^6 **3.** 64   **4.** 14   **5.** 20   **6.** -8 **7.** $2(7)(7)$ or $2 \cdot 7^2$
8. 23.8 mi/gal   **9.** $\dfrac{y^2}{2}$ **10.** $-\dfrac{1}{11}$ **11.** $-\dfrac{1}{2}$ **12.** $x = 12$ **13.** $x = 36$ **14.** $x = 4$ **15.** $4x^3 - 2x + 12$
16. $4x^2 + 2x - 11$ **17.** 87.5%   **18.** 0.08   **19.** 7.68   **20.** 161.11   **21.** 8.5%   **22.** 31.7 points   **23.** 8 in.
24. 1970, 1980, 1990   **25.** 12 in.   **26.** 12 in.

27.

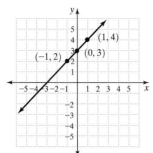

28–31.

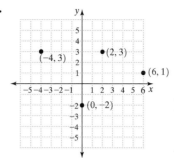

CHAPTER 9

CHAPTER 9 PRETEST   **1.** 6 hr 10 min   **2.** 123 in.   **3.** 5 km   **4.** 114.8 ft   **5.** 0.2758 in.   **6.** $3^2 = n$; $9 = n$
7. $n^2 = 9$; $n = 3$ **8.** $n \cdot n = 4$; $n = 2$ **9.** 6   **10.** $18\sqrt{2}$ **11.** 80°   **12.** 33 in.   **13.** 5 cm   **14.** 21.98 m
15. 28.26 in^2 **16.** 88.2 cm^2 **17.** $20.93   **18.** 904.32 in^3 **19.** 20 m^3 **20.** 15 ft

EXERCISES 9.1   **1.** 12   **3.** 2   **5.** 2000   **7.** 4   **9.** 1   **11.** 60   **13.** 5   **15.** 2   **17.** 4   **19.** 144
21. 7.25   **23.** 26,000   **25.** 28   **27.** 11   **29.** 4200   **31.** 1200   **33.** $9.75   **35.** 14,490 ft   **37.** 440
39. 7600   **41.** 1,760,000   **43.** 0.005261   **45.** 0.294   **47.** 6.328   **49.** 830,000   **51.** 0.018; 0.000018
53. 0.049; 0.000049   **55.** 7.183; 0.007183   **57.** 8.3; 0.083   **59.** 7.812; 0.007812   **61.** 0.0081; 0.0000081   **63.** $573,000
65. (a) 943,800 m   **(b)** 943.8 km   **67. (a)** 0.335 km   **(b)** 33,500 cm   **69.** 26,280 hr   **71.** centi-   **73.** milli-
75. deka-   **77.** 528,000,000   **79.** 24,900,000,000

CUMULATIVE REVIEW   **1.** 20%   **2.** 57.5   **3.** $321.30   **4.** $716.80

EXERCISES 9.2   **1.** 2.14 m   **3.** 22.86 cm   **5.** 15.26 yd   **7.** 28.89 yd   **9.** 9.3 mi   **11.** 21.94 m
13. 132.02 km   **15.** 82 ft   **17.** 6.90 in.   **19.** 218 yd   **21.** 18.95 L   **23.** 5.02 gal   **25.** 4.77 qt   **27.** 14.53 kg
29. 198.45 g   **31.** 35.2 lb   **33.** 4.45 oz   **35.** 5.45 ft or 5.44 ft   **37.** 502.92 cm or 503.25 cm   **39.** 31 mph
41. 96.6 km/hr   **43.** 0.51 in.   **45.** 104°F   **47.** 185°F   **49.** 53.6°F   **51.** 20°C   **53.** 75.56°C   **55.** 30°C
57. 143.87 km   **59.** 18.85 L   **61.** 1397 lb   **63.** 66.2°F; 113°F   **65.** 3420°C   **67.** 21.773 g

CUMULATIVE REVIEW   **1.** 47   **2.** 49   **3.** 91   **4.** 74

PUTTING YOUR SKILLS TO WORK   **INDIVIDUAL 1.** 545.6 mi   **2.** 60.6 L   **3.** meat, 2.27 kg; carrots, 0.91 kg;
apples, 1.36 kg; potatoes, 4.54 kg   **4.** 77°F   **5.** 122.5 marks   **GROUP 1.** 12.74 km/L or 12.75 km/L
2. meat, $3.90/lb or $3.89/lb; apples, $0.91/lb; potatoes, $0.31/lb   **3.** $3.25/gal

EXERCISES 9.3   **1.** No   **3.** Yes   **5.** Yes   **7.** No   **9. (a)** $\sqrt{64}$ **(b)** $\sqrt{64}$ **(c)** $\sqrt{64}$ **11.** 7   **13.** y
15. 8   **17.** n **19.** 12   **21.** b **23.** 15   **25.** $\dfrac{5}{7}$ **27.** $\dfrac{3}{9} = \dfrac{1}{3}$ **29.** $\dfrac{6}{7}$ **31.** 11 ft   **33.** 12 in.   **35.** 5.568
37. 8.307   **39.** 8.944   **41.** 9.487   **43.** 13.856   **45.** 27.839   **47.** 24.739   **49.** $8z$ **51.** $6x$ **53.** $7x$
55. $3\sqrt{5}$ **57.** $7\sqrt{2}$ **59.** $6\sqrt{2}$ **61.** $2x^2\sqrt{15}$ **63.** $2y^2\sqrt{30y}$ **65.** $5z^3\sqrt{5z}$ **67.** $5\sqrt{7}$ **69.** $6\sqrt{5}$ **71.** $2\sqrt{2}$
73. 0   **75.** $30\sqrt{2} + 8\sqrt{3}$ **77.** Answers may vary.

CUMULATIVE REVIEW   **1.** 12 ft   **2.** 22 m   **3.** 32 cm^2 **4.** 36 cm^2

EXERCISES 9.4   **1.** 60°   **3.** 30°   **5.** 90°   **7.** 104 m   **9.** 130 in.   **11.** 10.5 mi   **13.** 22.5 ft^2 **15.** 15.75 in^2
17. 83.125 cm^2 **19.** 14.07 m^2 **21.** 188 yd^2 **23.** 5 in.   **25.** 8.544 yd   **27.** 15.199 ft   **29.** 10.247 km
31. 11.402 m   **33.** 7.071 m   **35.** 3 ft   **37.** 15 in.   **39.** 6.928 yd   **41.** 17 ft   **43.** 5 mi   **45.** 8.7 ft
47. 1740 ft^2 **49.** 52 yd   **51.** right   **53.** Answers may vary.   **55.** You could conclude that all three sides of the
triangle are equal.   **57.** $21,060

CUMULATIVE REVIEW   **1.** $n = 12$ **2.** $n \approx 67.67$ **3.** 716 wait staff   **4.** 96 magazines

EXERCISES 9.5   **1.** 22.5 yd   **3.** 1.9 cm   **5.** 100.5 cm   **7.** 69.1 in.   **9.** 81.6 in.   **11.** 41.9 ft   **13.** 78.5 yd^2
15. 803.8 cm^2 **17.** 201.0 ft^2 **19.** 11,304 mi^2 **21.** $1211.20   **23.** 6.28 ft   **25.** 256.43 ft   **27.** 200
29. 3215.36 in^2 **31.** $226.08   **33.** 0.70057 m   **35.** circumference   **37.** radius   **39. (a)** $0.75/slice, 22.1 in^2/slice
(b) $0.67/slice, 18.8 in^2/slice   **(c)** 15-in. pizza is a better value.

CUMULATIVE REVIEW   **1.** 13.92   **2.** 0.3   **3.** 330 in^3 **4.** 160 in^3

EXERCISES 9.6   **1.** 87.9 m^3 **3.** 267.9 m^3 **5.** 718.0 m^3 **7.** 1017.4 cm^3 **9.** 261.7 ft^3 **11.** 21 m^3
13. 120 m^3 **15.** 1004.8 in^3 **17.** $0.75   **19.** 30.28 ft^3 **21.** $942   **23.** 381,510,000,000,000 mi^3 **25.** 592.1 m^3
27. 95.7 ft^2 **29.** box   **31.** sphere   **33.** pyramid   **35.** 615.44 in^2

CUMULATIVE REVIEW   **1.** $x = 63$ **2.** $x = 15$ **3.** $\frac{135}{16}$ or $8\frac{7}{16}$ **4.** $\frac{25}{14}$ or $1\frac{11}{14}$

EXERCISES 9.7   **1.** 8 m   **3.** 17.5 cm   **5.** 2.2 in.   **7.** 3.4 m   **9.** 2.1 ft   **11.** 36 ft   **13.** 81 ft
15. $n = 1.4$ km   **17.** $n = 16.3$ cm   **19.** 33.3 m   **21.** 11.6 yd^2

CUMULATIVE REVIEW   **1.** 12   **2.** 31   **3.** 42   **4.** 62

CHAPTER 9 REVIEW   **1.** 9   **2.** 36   **3.** 7.5   **4.** 3   **5.** 8000   **6.** 60   **7.** 5.75   **8.** 15.5   **9.** 0.059
10. 560   **11.** 259.8   **12.** 0.778   **13.** 920   **14.** 7000   **15.** 17,000   **16.** 0.473   **17.** 196,000   **18.** 721,000
19. 166.67 mL/jar   **20. a.** 17 ft   **b.** 204 in.   **21.** 92.4   **22.** 9.08   **23.** 4.58   **24.** 54.86 or 54.90   **25.** 368.55
26. 59   **27.** 5.52   **28.** 0   **29.** 32.6 mi   **30.** $2.22   **31.** 8   **32.** y **33.** 6.708   **34.** 8.944   **35.** $10\sqrt{2}$
36. $10\sqrt{3}$ **37.** $3x\sqrt{5x}$ **38.** $4y^2\sqrt{2y}$ **39.** $21\sqrt{2}$ **40.** $5\sqrt{3}$ **41.** $4\sqrt{5} + 3\sqrt{6}$ **42.** 30 ft   **43.** 40°
44. 24.5 m^2 **45.** 5 m   **46.** 5 yd   **47.** 3.61 ft   **48.** 37.7 in.   **49.** 44.0 in.   **50.** 113.04 m^2 **51.** 200.96 ft^2
52. 107.4 ft^2 **53.** 7.2 ft^3 **54.** $74.42   **55.** 129.787 cm^3 **56.** 1728 m^3 **57.** 9074.6 yd^3 **58.** 30 m   **59.** 324 cm

CHAPTER 9 TEST   **1.** 9 lb 1 oz or 9.0625 lb   **2.** 0.162 kg   **3. (a)** 2000 g   **(b)** 4.4 lb   **4.** 392°F   **5.** 20.44 km

6. Yes   **7.** No   **8.** No   **9.** 8 ft   **10.** y **11.** 7   **12.** $\frac{3}{4}$ **13.** 2.45   **14.** 5.29   **15.** $9\sqrt{3}$ **16.** 60°

17. 14 cm   **18.** 60 cm^2 **19.** 700 ft^2 **20.** 10.77 in.   **21.** 6.32 cm   **22.** 17.8 ft   **23.** 16.171 m
24. 5.6 revolutions   **25.** 4.5 cm^2 **26.** $62.80   **27.** 257.72 cm^3 **28.** 2143.57 in^3 **29.** 803.84 cm^3 **30.** 280 m^3
31. 58.5 in.   **32.** 25 ft   **33.** 8 cm

CUMULATIVE TEST FOR CHAPTERS 1–9   **1.** $0 > -5$ **2.** $0.5 > 0.15$ **3.** $-8 > -18$ **4.** $\frac{5}{8} < \frac{2}{3}$ **5.** $\frac{7}{8} < 0.9$

6. $|-6| > |2|$ **7. (a)** 42 ft   **(b)** $105   **(c)** 108 ft^2 **8.** -10 **9.** -16 **10.** 45   **11.** $\frac{19}{30}$ **12.** $-\frac{6}{7}$

13. $3x^2 + 3x - 25$ **14.** $x = 38$ **15.** $x = -15$ **16.** $x = 10$ **17.** $x = 24$ **18.** $x = 13.2$ **19.** $4x + 12$
20. $x^2 + x - 2$ **21.** $x + 10y - 3$ **22.** $207.50   **23.** 12.05%   **24.** 78   **25.** 76
26–30.   **31. (a)** Answers may vary.   **32.** centimeters   **33.** no   **34.** 12

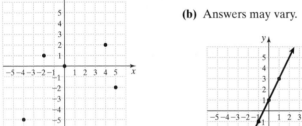

(b) Answers may vary.   **35.** x **36.** $\frac{9}{10}$ **37.** 15.8   **38.** $13\sqrt{7}$

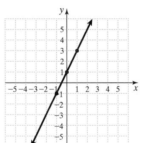

39. 17 ft   **40.** 43.96 in.   **41.** 144.21 cm^3

APPENDIX B

SCIENTIFIC CALCULATOR EXERCISES   **1.** 11,747   **3.** 352,554   **5.** 1.294   **7.** 254   **9.** 8934.7376;
$123.45 + 45.9876 + 8765.3$ **11.** 16.81; $\frac{34}{8} + 12.56$ **13.** 10.03   **15.** 289.387   **17.** 378,224   **19.** 690,284

21. 1068.678   **23.** 1.58688   **25.** 1383.2   **27.** 0.7634   **29.** 16.03   **31.** $12,562.34   **33.** 85,163.7
35. $103,608.65   **37.** 15,782.038   **39.** 2.500344   **41.** 897   **43.** 22.5   **45.** 84.372   **47.** 4.435   **49.** 101.5673
51. 0.24685   **53.** 30.7572   **55.** -13.87849 **57.** -13.0384 **59.** 3.3174×10^{18} **61.** $4.630985915 \times 10^{11}$
63. -0.5517241379 **65.** 15.90974544   **67.** 1.378368601   **69.** 11.981752   **71.** 8.090444982   **73.** 8.325724507
75. 1.08120   **77.** 1.14773

APPENDIX C

EXERCISES   **1.** 75; 660; 495   **3.** 42; 400; a **5.** 49; b; 2450   **7.** p; 50; 30   **9.** p; 25; 10   **11.** 160; b; 400
13. $\frac{a}{300} = \frac{24}{100}$; $a = 72$ **15.** $\frac{a}{30} = \frac{250}{100}$; $a = 75$ **17.** $\frac{a}{8000} = \frac{0.7}{100}$; $a = 56$ **19.** $\frac{45}{b} = \frac{60}{100}$; $b = 75$
21. $\frac{90}{b} = \frac{150}{100}$; $b = 60$ **23.** $\frac{3000}{b} = \frac{0.5}{100}$; $b = 600{,}000$ **25.** $\frac{70}{280} = \frac{p}{100}$; $p = 25$ **27.** $\frac{3.5}{140} = \frac{p}{100}$; $p = 2.5$

29. $\dfrac{90}{5000} = \dfrac{p}{100}$; $p = 1.8$ **31.** $\dfrac{a}{350} = \dfrac{26}{100}$; $a = 91$ **33.** $\dfrac{540}{b} = \dfrac{180}{100}$; $b = 300$ **35.** $\dfrac{82}{500} = \dfrac{p}{100}$; $p = 16.4$; 16.4%

37. $\dfrac{a}{520} = \dfrac{0.7}{100}$; $a = 3.64$ **39.** $\dfrac{16.5}{66} = \dfrac{p}{100}$; $p = 25$; 25% **41.** $\dfrac{68}{b} = \dfrac{40}{100}$; $b = 170$ **43.** $\dfrac{94.6}{220} = \dfrac{p}{100}$; $p = 43$; 43%

45. $\dfrac{a}{380} = \dfrac{12.5}{100}$; $a = 47.5$ **47.** $\dfrac{a}{5600} = \dfrac{0.05}{100}$; $a = 2.8$ **49.** 15.82% **51.** 10.91 **53.** $\dfrac{a}{798} = \dfrac{19.25}{100}$; $a = 153.62$

55. \$118.80

APPENDIX D

EXERCISES   **1.** $8m$ **3.** $\angle F$ **5.** 6 m   **7.** SAS   **9.** SSS   **11.** ASA   **13.** SAS   **15.** SAS
17. Not necessarily congruent   **19.** ASA   **21.** Not congruent   **23.** SSS   **25.** SAS   **27.** ASA
29. Not necessarily congruent

APPLICATIONS INDEX

Agricultural applications:
  dairy cow production, 161
  eggs produced by hens, 161
  farm fence, 347
  farming acres, 477
  lemon trees, 89
  orange grove, 105
  orange trees yield, 162
  rancher's cows in pasture, 133
  Rose Festival, 122
Astronomy applications:
  earth/moon diameters, 38
  Earth's diameter, 133
  earth/sun distances, 39
  Jupiter's diameter, 105
  Jupiter's volume, 609
  Pluto's distance from Sun, 106
  space probe travel, 571
  sun/moon diameters, 38
Automobile/other vehicle applications
  (*See also* Transportation
    applications):
  business mileage, 487
  car maintenance expenses, 22
  car prices, 7
  car sticker prices, 10
  gasoline consumption, 361
  gasoline in metric
    measurements, 578
  kilometers driven, 577
  liters of gas used, 577
  miles driven over four years, 51
  miles per gallon, 167, 175, 268, 269,
    273, 288, 348, 457, 496, 515–16,
    518, 519, 558
  miles per year driving miles, 3
  motor oil purchase, 85
  odometer reading, 438
  sale prices on, 45
  total miles driven, 422, 437
  trip time, 10, 79
  sale prices on, 45
  vehicle total on used car lot, 47

Business applications:
  amusement tickets, 162
  appliance sales, 162
  assembly time, 161
  average number of sales, 518
  baseball glove sale, 209
  bicycle manufacture, 91
  bicycle sales, 12–13
  bicycles sold, 52
  bioengineered flower essence, 571
  bottles of product from vat, 306
  business card colors, 90–91
  car costs, 519
  cars sold, 501, 503, 555
  CDs on sale, 287
  CDs prices, 268

chain pharmacy stores, 501
change in copy machines, 493
circus sales, 551
clients per insurance agent/per staff
  member, 268
color television sales, 547
computer price reduction, 494
computer purchases, 78
computer system sale, 485
cost, revenue, profit on door
  hinges, 410–11
customers at cafe, 65
customers per month, 511
customers per quarter, 550–51
defective parts, 477
dining room set price reduction, 481
employee bonuses, 62
employees in company, 478
flower baskets, 25
high-tech employment, 81
hotel curtains, 105
ice cream cones purchased, 551
in-flight magazines, 597
jar labeling machine, 534, 542
jar sealing machine, 533–34, 542
land subdivisions, 306
legal secretaries/paralegals per
  lawyers, 287
marble in lobby, 571
mean/median salary, 520
mean salary, 517, 519
median value of compact discs, 519
men employees in corporation, 245
monthly production costs, 507–8
net gain, 199
net profits/losses, 222
number of hours/jobs done, 541
office supplies/furniture costs, 171
parking charges, 565, 570
print cartridges sale, 265
printer paper sale, 268
printers to secretaries ratio, 273
production cost, 540
products manufactured, 522
products sold, 553
profit and loss, 189, 225, 227, 497
profits, 33–34, 161, 162, 171, 287,
  505–6
quarterly net losses, 197
quarterly profits, 192
redecorating costs, 158–59
remodeling costs, 162
rental car inquiries, 518
sale price on television, 481–82
sales comparison, 514
sales totals, 168
stereo system sale, 485, 494
suit sales, 269, 490
tanning salon special, 268
television set production, 161
television set reduced price, 627

television sets sold, 521
theater profits, 162
total sales, 39
towel sale, 265
travel allowance, 133
truck sales, 24
typing speed, 542, 554
used car repairs, 276
vending machine change, 461–62,
  463
video store customers, 418
waitstaff to patrons ratio, 597
wine glasses sale, 268

Chemistry applications:
  alcohol in mixture, 495
  antifreeze in containers, 493
  cleaning fluid cost, 569
  containers for fluids, 423, 453–54,
    457, 465–66
  melting point for metallic
    tungsten, 577
  solution in milliliters, 622
  water: chlorine mixture, 266
Construction applications:
  bookcase boards, 347, 350, 358
  bookcase bolts, 348
  building permits, 39
  concrete connector cost, 624
  dirt digging/hauling, 150, 173
  floor tiles, 153
  tile for office building entryway, 149

Education applications:
  ages within student body, 555
  attendees at college, 506
  bachelor's degrees earned, 502, 504
  boys in third grade, 285
  classes for children:adults, 289
  college enrollment figures, 11, 80
  College Service Club
    attendance, 470
  college transfers, 477
  course enrollment, 493, 513
  elementary school enrollment, 11,
    481
  exam scores median values, 552
  final exam scores, 43–44
  freshman to juniors ratio in
    class, 419
  juniors to seniors for field trip, 266
  males to females in class, 230
  male students in class, 289, 290
  mathematics classes enrollment, 550
  mean grade, 518
  mean/median quiz scores, 556
  mean/median test scores, 515, 627
  median value for students taking
    class, 552

Education applications (cont.)
median value of actors for school play, 520
medical school applicants, 522–23
medical school attendance, 470
men to women in class, 262
points in English class, 475–76
PTA profits, 161
quiz points, 476
spelling workbooks cost, 161
student enrollment, 510, 528
student living arrangements, 409
students from California/Texas, 306
students in college, 478
students per instructor/tutor, 268
students planning to take algebra, 423
students taking English class, 409
students taking Spanish course, 415
student to teacher ratio, 278
teachers to children in preschool, 264–65
test questions, 361, 478
test scores, 3, 49
textbooks purchased, 552
tickets for class project contest, 130
women students in engineering, 552
Electrical applications:
kilowatt hours used, 49
speaker wire, 278
summer capacity in kilowatts, 520
Environmental applications:
Great Lakes total area, 502
nature trail length, 478
park visitors, 548
recycling percentages, 478

Financial/economic applications:
American dollars to British pounds, 273, 278
apartment rental costs, 22, 84, 174
auto insurance rates, 438
bill denominations from bank account withdrawals, 5, 9
boat down payment, 37
borrowing money, 31
bronze coins, 245
bus fare, 105, 446
business investment/profits, 277
business loan payments, 486
car costs, 496
car lease/purchase, 495
car loan interest, 486
car payments, 161, 175, 457
car sales tax/total price, 487
certificate account interest, 485
change back, 492
change from dinner bill, 422
change from grocery purchase, 438
checking account balance, 2, 27, 35, 37, 38, 62, 85, 174, 192, 435–36, 438
checking account deposits/debits, 22
coin mix/value, 463
college expense allowance, 38

commission on sales, 423, 480, 485, 490, 494, 496, 572
computer purchase comparison, 50
cost-of-living raises, 485, 486
credit card interest, 163
currency exchanges, 455
custom window shades, 487
decimal notation in checks, 425
Dept. of Agriculture budget, 471
desk lamps, 288
dinner bill, 131, 132, 173
dollar bill conversions, 40
Dow Jones average, 37
employee discount, 486
frequent-flyer program, 159–60, 173
furniture funds, 486
gas meter reading, 163
gasoline expenses, 492
grocery purchases, 552
grocery store comparisons, 446
guitar payments, 161
home office purchases, 82
hourly rate, 481
household budget, 38
house payments, 106, 306
income withholding, 302, 305, 355, 438
inheritance allocations, 456
interest on loan, 482–83, 485, 486, 490, 494
investment choices, 279
job pay comparison, 102–3, 105
land subdivisions, 252, 254
lawn mower cost, 465
loan monthly payments, 169
long-distance calls budget, 443–44
monthly basic expenses, 46
monthly budget allotments, 550
monthly income allocations, 479
monthly living expenses, 85, 86, 174
monthly salary, 63
multimedia notebook price differences, 51
music club account deposit, 169
mutual fund investments, 51
national debt, 11, 40
oil imports, 520
overtime earnings, 445, 492
paint purchase, 123
paycheck deductions, 38, 81, 84, 174, 245–46, 279, 288
percent increase in earnings, 627
personal income/personal spending, 486
phone call costs, 446, 497
price of shares of stock, 268, 269, 278
profits from partnership, 279
property taxes, 280
rental car cost, 446
rents in Orange County, CA, 37
roommate's expenses, 161
rounding prices on menu, 431
salary after raise, 485, 497
salary comparison, 404–6, 408, 415, 417, 419

sales tax on car, 494
savings account balance, 35, 81
savings account contributions, 279, 303, 305
savings account interest, 483–84
school clothes purchases, 445
school supply purchases, 50
shirt sale, 278
spending distribution by two-income family, 513
stock earnings per share, 329, 356
stock gains/losses, 222
stock market investing, 351, 359
stock split, 278
student loan interest, 486
tips for dinners, 423, 476, 477, 494
total pay, 88, 105, 161
truck lease/purchase, 492
truck rental charges, 422
unemployed people, 500
university costs, 51
vacation plan costs, 107
weekly earnings, 171
white-collar/blue-collar jobs salaries, 458
word name for check amounts, 430
yearly salary, 481
Food applications:
beverages ordered in restaurants, 17
calories consumed, 162
candy sharing, 131
candy shipment, 294
carbohydrates in cereal, 273, 278, 288
carbohydrates in yogurt, 292
cereal cost, 622
cereal recipe, 348
cheese package weight, 254
chicken recipe, 419
chili purchase, 73
chocolate chip cookie recipe, 329
chocolate consumption, 431
coffee consumption, 552
coffee purchase, 254
desserts at buffets, 276
desserts at potluck dinner, 285
Easter candy sales, 266
fat/cholesterol in fast foods, 512–13
fat in ice cream, 278
fat in snack bar, 477
flour cost, 270
flour in containers, 355
flour in recipe, 302
french fries calories, 263
grocery items in kilograms, 578
hamburger meat in kilograms, 625
ham package weight, 254
ice cream flavors, 96
juice container measurement, 74
mean soda prices, 518
menu for large events, 281–82
milk for recipe, 83
milk leftover, 73
milk used, 174
mushrooms cost, 570
nutrition facts for cheerios, 286–87

pizza deliveries, 552
pizza for party, 305
pizza slices/pieces, 603, 604
punch concentrate, 278
salad bar items, 245
sandwich types, 172
steak weight, 74
sugar in containers, 303
tomato sauce leftover, 73
vanilla cupcakes, 245

Geometric applications:
angle measures, 623
circle
    area of, 562, 600, 601, 602, 603,
        620, 623, 626
    circumference of, 562, 599, 602,
        603, 623, 626
    radius of, 601, 602
cone
    volume of, 607, 608, 609, 620, 624,
        626
congruent triangles
    length of sides of, A-21
cylinder
    surface area of, 610
    volume of, 605, 606, 608, 609, 620,
        624, 626, 628
equilateral triangle
    length of sides of, 409, 590, 625, 628
    perimeter of, 590, 623
hemisphere
    volume of, 608, 609
irregular shapes
    area, 153, 170
parallelogram
    area, 146, 147, 153, 154, 170, 487,
        514, 587
pyramid
    volume of, 562, 607, 608, 609, 621,
        624, 626
rectangle
    area of, 146, 148, 149, 152, 153,
        154, 407, 497, 514, 587, 625, 627
    dimensions of, 409
    length of, 402–3, 407
    perimeter of, 2, 41–42, 48, 55, 62,
        81, 134, 170, 173, 174, 276–77,
        587
    width of, 407, 415
rectangle/square shape
    perimeter of, 42–43
rectangular box
    volume of, 604, 609
rectangular solid
    surface area of, 155, 171
    volume of, 89, 150, 155, 171, 487,
        514
right triangle
    length of hypotenuse of, 592, 626
    length of legs of, 593, 596
    length of one leg of, 626
    unknown side of, 623, 626
semi-circle
    area of, 602

similar geometric figures
    corresponding sides of, 621
    missing side of, 616
    perimeter of larger of two, 624
    unknown area of, 617
similar rectangles
    missing side of, 616
    width of larger, 614, 626
    width of smaller, 615
similar triangles
    corresponding sides of, 624, 626
    length of sides of, 612, 614, 615
    perimeter of larger
        triangle, 612–13
sphere
    circumference of, 628
    surface area of, 610
    volume of, 605–6, 608, 620, 624,
        626
square
    area of, 89, 152, 153, 170, 227
    length of side of, 409
    perimeter of, 42, 48, 81, 84, 89,
        152, 497
square/rectangle shape
    perimeter of, 84
triangles
    angle measures of, 589, 625
    angles of, 562
    area of, 590, 591, 594, 595, 596,
        597, 619, 623, 625
    lengths of sides of, 403–4, 407,
        408, 409, 417, 419
    missing angles of, 65, 144, 594
    missing sides of, 65
    perimeter of, 48, 81, 587, 594, 622
    unknown side of, 595, 597, 613,
        615, 616
two rectangles
    perimeter of shape with, 81

Health applications. (See also Medical
        applications):
    calories burned, 497
    life expectancy and smoking, 555
    over-the-counter medication, 83
    weight loss, 348
    weight reduction, 486
Home improvement applications:
    baseboard along floor, 303
    bathroom redecoration, 53
    bathroom tiling, 348
    carpet colors, 96
    carpet cost, 175
    carpet purchase, 157
    fencing around flower bed, 349
    fencing around garden, 295, 627
    fencing around property, 49
    fencing around rose garden, 345–46
    interior painting, 50
    molding on kitchen ceiling, 49
    outdoor carpet, 154
    painting family room, 156
    painting supplies purchase, 80
    patio width, 279, 288

pool fence, 279
pool tiles, 349, 357

Medical applications:
    anticancer solution, 571
    lead in child's liver, 577
    medicine based on body weight, 278
    nursing home patient to staff
        ratio, 265
    organ transplants, 267
    vaccine production, 571
    water used in operation, 577
    women physicians in U.S., 509
Miscellaneous applications:
    ages within police force, 547
    animal speeds, 144
    auditorium seating, 485
    banquet tickets, 133
    bedroom measurements, 83
    benches for seating, 273
    blue marbles, 245
    box dimension comparison, 280
    brain weight, 471
    Btu consumption, 434
    bus riders, 552
    carpet for office, 170
    carrots for horses, 131
    chairs in auditorium, 122
    chairs in gym, 131
    children's home donation, 130
    cleaning fluid containers, 454
    concert tickets, 482
    concert ticket sales, 485
    containers for paint, 457
    cross stitching, 134, 162
    custom window blind prices, 163
    dam height, 571
    dance production class, 245
    detailing cars, 73
    door replacements, 167
    dress material amount, 329
    electric toothbrush users, 470
    eye blinking times, 471
    felt for hope chest, 154
    felt-tip pen, 74
    flagpole height, 562
    force applied by torque wrench, 163
    formal dance attendance, 465
    frame dimensions, 280
    fundraising, 37
    garden plants, 105
    German-speaking people, 11
    growth in height, 558
    height comparisons, 403, 407, 408
    homeless shelter donations, 132
    inflating basketball, 562
    IQ determination, 226
    Italian-speaking people, 79
    land plot comparisons, 261
    lawn fertilizer, 230, 290
    lawn treatment, 278
    leisure time activities, 504–5
    local oil production, 547
    marching band, 123
    marinas in California, 286

Miscellaneous applications (*cont.*)
  material for skirt/bodice
    pattern, 349
  mean telephone calls, 518
  mean television viewing time, 518
  miles on map, 273, 278, 288
  minutes on phone calls, 519
  molding clay pieces, 314
  motion detection by fly, 456
  mountain height, 570
  mouse pad dimensions, 446
  nature trail length, 478
  oriental rug placement, 346
  paint for barn, 514
  paintings display, 131
  parking lot spaces, 90
  people playing computer games, 470
  photographic shot set-up, 134
  pipe length, 327
  pipe split into parts, 302, 305
  plant watering, 278
  play tickets, 172
  prize division, 169
  PTA donations, 132
  radio survey, 52
  rope length for tents, 329
  scale drawing of buildings, 278
  scarf material, 302
  seats in auditorium, 478
  shoe size *vs.* height, 528
  skirt yardage, 273
  socket wrenches, 290
  square miles of land/water on
    earth, 23
  suspension bridges, 267
  tank capacity, 348
  telephone calls lengths, 516
  telephone calls received, 548
  ticket prices, 63
  tie colors, 96
  tile for office entryway, 173
  tiles for floor, 105
  trading cards, 245
  typing rates/speed, 274, 288
  water/alcohol mixture, 261–62
  water coverage of states, 470
  water to fill Rose Bowl, 133
  wedding dresses, 163
  wood length, 356
  yellowtail fish catch, 32–33

Physics applications:
  acceleration of object, 464
  pulley rotations, 268, 273
  pulley wheel revolutions, 600
Political applications:
  registered voters, 280

  voting survey, 292
Population applications:
  Las Vegas, 40
  percent increase, 494
  religious faith distribution, 510
  U.S. population net gain, 133
  yearly income in states, 163

Real estate applications:
  agent commissions, 480
  home price increase, 486
  homes/apartment units built, 509
  mean house sale price, 518

Science applications. (See also
    Astronomy applications;
    Chemistry applications):
  ionic charges, 215, 217
  sound travel, 279
Sports applications:
  ages of people swimming, 519
  archery target, 245
  baseball hits, 278
  baseball players' salaries, 291
  base hits, 245
  basketball free throws, 478
  basketball points averaged, 558
  basketball team heights, 50, 477
  basketball wins to losses, 266
  batting average, 278, 518
  bicycle pedal revolutions, 278, 288
  birdie putts, 73
  bowling average, 518
  calories burned playing tennis, 290
  football season injuries, 520
  free throws in basketball, 274
  game tickets, 131
  high-jumps, 486
  javelin throwing, 570
  jogging time, 83
  miles ran, 500
  Olympics: men's high jump, 350
  people playing organized sports, 470
  running on treadmill, 433–34
  rushing yards in football, 161
  soccer goals, 274
  soccer shots, 477
  surfing forecast, 84
  track laps, 305
  track meets, 520
  workout time, 73
  yacht race boats, 287

Temperature applications. (See also
    Weather applications):

  average readings, 552
  Celsius/Fahrenheit
    comparison, 75
  Celsius/Fahrenheit
    interconversions, 464, 577, 625
  differentials between cities, 199
  high/low readings, 106, 200, 512
  positive/negative readings, 182, 183,
    186–87, 191, 223
  total drop in temperature, 209, 216,
    224
Time and distance applications:
  altitude differences, 178, 196, 197,
    199, 223
  distances driven, 408
  distance to travel, 622
  driving speed, 348
  driving time, 278
  flight length, 408
  skydiver speed, 215
  spacecraft travel, 349
  submarine below sea level, 192
  time zone problems, 218–19
  Titanic speed in nautical
    miles, 349
  velocity rate of projectile, 209,
    216
  walking distance, 408
Transportation applications:
  airplane flight time, 8
  airplane travel speed, 105
  boat propeller revolutions, 306
  boat travel, 577
  distance in kilometers, 625
  jet miles flown, 23
  lacquer on plane wings, 597
  miles driven on trip, 51
  railroad line height, 571
  school commute distance, 38
  speed limit in metric
    measurement, 74
  train run, 571
  train trip time, 10
  trip in kilometers/miles, 578
  work commuting time, 8

Weather applications:
  rainfall, 241
  rainfall average, 431
  rainfall comparison, 511–12
  rainfall mean/median, 559
  rainfall total, 330, 437
  snowfall, 262
  tornado occurrences, 264
Weight applications:
  gold nugget, 577
  pounds to kilograms, 577

Subject Index

Absolute value, 220
  defined, 180
  of number, 180–81
Addition:
  application problems involving, 32
  for calculating perimeter, 41–43
  and combining like terms, 12–20
  of decimals, 432–33, 488
  equations solved with, 59–60
  of fractional expressions with
    common denominators, 312–13
  of fractional expressions without
    common denominators, 313–18
  of fractions with common
    denominators, 352
  of fractions without common
    denominators, 352
  key words for expressing, 12
  of measurements, 77
  of mixed numbers, 322–23, 353
  of polynomials, 394–95, 413
  properties of, 12–15
  of signed numbers, 185–89
  of square root expressions, 585, 619
  of two numbers with different
    sign, 220
  of two numbers with same sign, 220
  of whole numbers when carrying is
    needed, 16–18
Addition facts, 16
Addition principle of equality, 375
  equations solved using, 365–67, 411
Addition property of zero, 13
Addition rule:
  for two numbers with different
    signs, 187
  for two numbers with same
    sign, 186
Algebraic expressions, 54
  distributive property for
    multiplying, 119
  evaluating those involving addition
    of signed numbers, 189
  evaluating those involving
    multiplication and division, 135
  evaluating with signed numbers, 220
  multiplying, 117–18, 165, 211, 221
  simplifying, 210–11
Algebraic fractions,
    simplifying, 257–58
Algebraic logic, on calculators, A-2,
    A-3
Angles, 588
  in quadrilaterals or triangles, 588–89
  sum of measure of in triangles, 589
Angle-side-angle (ASA) congruence,
    by triangles, A-16, A-18
Applied problems:
  estimation used for solving, 45–46
  involving addition and
    subtraction, 32–35

involving comparison, 403–6
involving decimals, 434
involving fractions, 345–47
involving geometric figures, 402–3
involving multiplication and
    division, 158–60
involving proportions, 276–77
and metric units of
    measurement, 569
with polynomials, 401–6
with right triangles, 593
solved with more than one
    operation, 215
solved with multiplication, 102–3
solved with proportions, 284
solving those involving addition of
    signed numbers, 189
solving those involving decimals,
    443–44, 453–54, 461–62
solving those involving
    fractions, 301–3
solving those involving ratios and
    rates, 263–65
solving those involving subtraction
    of signed numbers, 196–97
solving with division, 130
solving with mixed numbers, 327
and U.S. units of measurement, 565
Area:
  of circle, 600–601, 620
  of geometric shapes, 145–49
  of parallelogram, 146, 147, 166, 591
  of rectangle, 147, 148, 165
  of rectangular array, 145
  of square, 147, 165
Area problems, containing
    circles/other geometric
    shapes, 601
Arrays, 90
  for area of rectangular region, 145
Associative property:
  to simplify equations, 58
  and subtraction, 28
Associative property of addition, 15, 76
Associative property of
    multiplication, 93, 94
  to simplify algebraic
    expressions, 211
Average (or mean), 515
Axis, 523

Balancing checkbook, 435–36
Base(s), 108, 117
  as denominator of proportion, 475
  and multiplying expressions in
    exponent form, 116
  of triangle, 590
Binomials, 386
  multiplying using FOIL
    method, 388–90, 413

Borrowing:
  with mixed numbers, 323–25
  in subtraction, 30
  units of measurement, 68
Business, profit, cost, and revenue
    in, 410–11

Calculator (See also Graphing
    calculator; Scientific
    calculator):
  adding and subtracting
    decimals, 432
  adding whole numbers, 18
  addition of negative numbers
    on, 188
  checking solutions to equations
    on, 375
  converting temperature on, 575
  division on, 130
  exponents evaluated on, 109
  for multiplying, 102
  negative number operations on, 196
  order of operations on, 112
  percent changed to decimal on, 467
  proportions on, 272
  subtracting whole numbers, 32
  when to use, 203
Casio calculators, A-2
Celsius degrees, converted to
    Fahrenheit degrees, 575, 618
Center, of circle, 598
Centigrams, changed to grams, 568
Centiliters, liters changed to, 568
Centimeters, 69, 71
  meters changed to, 568
Checking account, 421
  balancing, 435–36
Circle:
  area of, 600–601, 620
  area problems containing, 601
  circumference of, 598–600, 620
Circle graphs, 241
  reading those with percentage
    values, 502–3, 547
Circumference, of circle, 598–600,
    620
Class attendance, 43
Clearing the fractions, 381
Coefficient of term, 19
Coefficients, addition or
    multiplication of, 118
Combining like quantities, 18
Combining like terms, 18–20, 76, 221
  in simplifying algebraic
    expressions, 210–11
  and subtraction, 24–30
Commission, 480
Commission problems, solving, 480,
    490
Commission rate, 480

Common denominators:
 adding and subtracting fractional
  expressions without, 313–18
 adding and subtracting fractional
  expressions with, 312–13
Common factors, removing, 297
Common multiple, 307
Commutative property:
 and division, 123
 to simplify equations, 58
Commutative property of
  addition, 13–14, 76
Commutative property of
  multiplication, 93, 94
 to simplify algebraic
  expressions, 211
Comparison, applied problems solved
  involving, 403–6
Comparison line graph, reading and
  interpreting, 503–4, 548
Complex fractions:
 order of operations and, 331
 simplifying, 332–34, 354
Composite numbers, identifying, 232
Compound interest, 483, 484
Cone, volume of, 606–7, 620
Congruent triangles, A-16–A-20
Constant term, 18
Coordinates:
 naming for points, 524–26
 of the point, 523
 plotting point given, 523–24
Corresponding angles, of similar
  triangles, 611
Corresponding sides:
 of similar geometric figures, 614
 of similar triangles, 611
Cost concept, 410
Cross product, 249
Cubed base, 109
Cylinder, volume of, 605–6, 620

Data:
 for making predictions, 540–41
 written as ordered pairs, 522–23
Decimal expressions, word names for
  decimal fractions, 424–25
Decimal notation, 425
 changed to fractional notation, 426
Decimal point, 424, 466
Decimal problems, on calculator,
  A-3–A-4
Decimals:
 adding and subtracting, 432–33, 488
 applied problems solved with, 434,
  443, 453–54, 461–62
 changing between fractions,
  percents and, 468–69
 changing between percents
  and, 466–67, 489
 comparing and ordering, 427–28
 converting between fractions
  and, 425–27
 determination of larger of two, 488
 division of, 447–50, 489

fractions changed to, 450–51, 489
 in metric units, 567–69
 multiplication of, 440–41, 488
 multiplying by power of 10, 441–42,
  488
 order of operations with, 442
 rounding, 428–29, 488
 solving equations involving, 459–61,
  489
 written as fractions, 488
Dekameters, 69
Denominators, 239
 raised to powers, 340
 solving equations containing
  fractions with variables as, 383
Diameter, of circle, 598
Difference:
 estimating, 433–34
 in subtraction, 24, 26
Digits, 4
Discount problems, solving, 480, 490
Distributive property:
 for adding/subtracting radical
  expressions, 585
 and clearing fractions, 381
 and exponent form, 165
 and multiplication, 98–99, 164
 to multiply algebraic
  expressions, 119
 in multiplying binomials using
  FOIL method, 388–90
 and multiplying monomials times
  polynomials, 387
 to remove parentheses, 378, 393, 398
Dividend, 123
Divisibility tests, 231, 282
Division:
 applied problems involving, 130,
  158–60
 and commutative property, 123
 equations solved with, 136–37, 139
 evaluating algebraic expressions
  involving, 135
 of expressions in exponent
  form, 337–38, 354
 of fractional expressions, 299–301
 of fractions, 352
 key words for, 165
 long, 127–30, 165
 meaning of, 122–23
 of mixed/whole numbers, 326, 353
 order of operations with, 125–27
 with signed numbers, 206–7, 221
 of whole number
  expressions, 122–30
 and zero, 240
Division ladder, to find prime
  factors, 233
Division principle of equality,
  equations solved using, 369–70,
  411
Division symbol, 127
Divisor, 123, 127
Double-bar graphs:
 constructing, 504–6, 548
 reading and interpreting, 505, 547

Educated guesses, 58, 338, 536
English phrases, translated into
  symbols, 92
Equality test for fractions, 249–52, 283
Equal signs, in equations, 56
Equations:
 basic, 54–61
 to calculate profit, cost, and
  revenue, 410
 evaluating those with signed
  numbers, 221
 involving division, 136–37
 involving multiplication, 137–39
 for making predictions, 540–41
 percent problems solved
  using, 473–76
 right ordered pair as solution of, 536
 simplifying and solving, 165, 377–78
 solved by inspection, 135–36
 solved with addition, 59–60
 solved with subtraction, 58–59
 solving complicated, 77
 solving those containing
  fractions, 381–83, 412
 solving those involving polynomial
  expressions, 398–99, 413
 solving using addition principle of
  equality, 365–67, 411
 solving using division principle of
  equality, 369–70, 412
 solving using more than one
  principle of equality, 374–76,
  412
 solving using multiplication principle
  of equality, 367–69, 411
 solving with multiplication and
  division, 165
 translating/solving those involving
  multiplication and division, 139
Equations in form $ax = b$,
  solving, 138
Equations in form $ax + b = c$,
  solving, 459
Equilateral triangles, 590
Equivalent fractions, 247–52
 building, 255
 identifying and using, 247–49
Equivalent measures, in U.S. to metric
  interconversions, 573
Equivalent multiplication problems,
  forming, 125
Estimation, 77
 and rounding, 47
 for solving application
  problems, 45–46
 of sums and differences, 433–34
Evaluating expressions, 77, 165
 vs. solving equations, 140
Exams:
 day before, 72
 reviewing for, 68
Expanded notation, whole numbers
  written in, 5
Exponents, 108, 164
 multiplication and division
  properties of, 337–40

multiplication of algebraic
        expressions with, 117
    operations performed with, 116–19
    and order of operations, 108–12
    with signed numbers, 204–6, 221
Expressions:
    evaluating, 77
    multiples of, 307

Factoring, whole numbers, 231–36
Factors, 91, 232
Factor tree, building to find prime
        factors, 235
Fahrenheit degrees, converted to
        Celsius degrees, 575, 618
Feet:
    inches converted to, 563
    yards converted to, 563–64
Floor plan, reading, 157
FOIL method, 389
    multiplying binomials using, 388–90,
        413
Formulas. (See also Symbols):
    for area of circle, 600
    for area of circles and other
        geometric figures, 601
    for area of triangle, 590, 591
    for Celsius/Fahrenheit temperature
        reading interconversions, 575
    for circumference of circle, 599
    for commission problems, 480
    for volume of cone, 607
    for volume of cylinder and
        sphere, 605
    for volume of pyramid, 607
Fractional expressions:
    adding and subtracting those with
        common denominators, 312–13
    adding and subtracting those
        without common
        denominators, 313–18
    dividing in exponent form, 337–38,
        354
    division of, 299–301
    multiplication of, 296–98
    operations on, 293–361
    simplifying, 255–58
Fractional notation, 425
    changed to decimal notation, 426
Fractions. (See also Complex
        fractions; Fractional
        expressions):
    applied problems involving, 301–3,
        345–47
    building, 283
    changed to decimals, 450–51, 489
    changing between decimals,
        percents and, 468–69
    clearing, 381
    converting between fractions
        and, 425–27
    decimals written as, 488
    determining relationship between
        two, 251
    division of, 352

equality test for, 249–52, 283
improper, 241, 242
inverting, 299, 300, 301
meaning of, 239–41
multiplication of, 352
order of operations and, 331, 353
percent changed to, 490
positive attitude toward, 298
proper, 241, 242
raised to powers, 340
reducing, 255–56, 283
solving equations
        containing, 381–83, 412

Gallons, quarts converted to, 564
Geometric figures/shapes, 145–51
    applied problems solved
        involving, 402–3
    area of, 145–49
    perimeter of, 145
Geometric solid, volume and surface
        area of, 149–51
Geometry, 41
Grams, 68, 71, 72, 78
    centigrams changed to, 568
Graphing (See also Graphs):
    horizontal and vertical
        lines, 526–27, 539
    linear equations, 538
    linear equations with two
        variables, 536–38
    straight lines, 549
Graphing calculator:
    adding and subtracting decimals
        on, 432
    addition of negative numbers
        on, 188
    checking solutions to equations
        on, 375
    division on, 130
    exponents evaluated on, 109
    negative number operations on, 196
    percent changed to decimal on, 467
    performing calculations on, A-2
    proportions on, 272
Graphs:
    circle graphs, 502–3, 547
    comparison line, 503–4, 548
    double-bar, 503, 504–6, 547
    interpreting, 43–44
    interpreting and constructing, 501–6
    line, 504–6, 548
    pictographs, 501, 547
    pie, 502
    of production costs, 507–8
Greater than symbol, 6, 76
Greenwich Time (or Universal
        Time), 218
Grouping symbols, simplifying
        polynomials with, 393–94

Height, of triangle, 590
Hewlett-Packard calculators, A-2
Histograms, 44

Horizontal lines, graphing, 526–27, 539
Hours, minutes converted to, 564
Hypotenuse, of right triangle, 592

Identify, calculate, replace
        sequence, 111, 126
    for applications with signed
        numbers, 214
Improper fractions, 241, 242
    changed to mixed numbers, 243, 282
    mixed numbers changed to, 244,
        283, 325, 326
Incentives, solving problems
        involving, 159–60
Inches, converted to feet, 563
Inequality relationships:
    with signed numbers, 179–80
    understanding and using, 6–7
Inequality symbols, 6, 76, 180
    and comparing/ordering
        decimals, 427–28
Inspection, equations solved by, 56–58
Integers, 179
    sign for, in exponent form, 205
Interest, 482
Interest problems, solving, 482–84, 490
Interest rate, 482
Inverting the fractions, 299, 300, 301
Investing in stock market, 351
Isosceles triangle, 589

Key words:
    for division, 123–24, 165
    for exponents, 109–10
    for multiplication, 91–92, 164
Kilograms, 72
Kiloliters, 71
Kilometers, 71, 78
    changed to meters, 567

Large event planning, 281–82
Learning cycle, 43
    and reviewing for exam, 68
    for reviewing tests, 93
Least common denominator
        (LCD), 314, 316–18
    and clearing fractions, 381
    and equations containing
        fractions, 382–83
    finding, 352
Least common multiple (LCM), 308,
        314
    finding, 352
    of expressions, 307–9
Legs, of right triangle, 592
Length:
    metric measurements of, 567
    metric measures of equivalence, 69
    U.S. to metric system
        interconversions, 573
    U.S. unit of measurement for, 66,
        563
Less than symbol, 6, 76

Like terms:
  combining, 18–20
  subtraction and combining, 25–26
Line, 588
Linear equations, graphing, 538
Linear equations with two
      variables, 533–39
  finding solutions to, 533–36
  graphing, 536–38
  graphing horizontal and vertical
      lines, 539
Line graphs:
  constructing, 504–6, 548
  reading, 182, 220
Line segment, 588
List of priorities, and order of
      operations, 111, 126
Liters, 68, 71, 78
  changed to centiliters, 568
Long division, 165
  performing, 127–30
Lowest terms, reducing fraction to, 255

Mathematics, and careers, 376
Mathematics Blueprint for Problem
      Solving:
  checking balance, 35
  computer purchase, 78
  fencing around rose garden, 345–46
  frequent-flyer program, 159–60
  job salary comparison, 102–3
  length of rectangle, 402–3
  money problems, 461–62
  redecorating costs, 158–59
  salary comparison, 404–5
Mean:
  and median, 515–17
  of set of numbers, 515–16, 548
  using, 517
Measurements:
  addition and subtraction of, 77
  converting between metric
      units, 566–69
  converting between U.S.
      units, 563–64
  metric units of, 68–72, 78
  U.S. units of, 66–68, 78
Median value, of set of
      numbers, 516–17, 549
Memorization, of multiplication
      facts, 95
Meters, 68, 71, 78
  changed to centimeters, 568
  kilometers changed to, 567
Metric units of measurement, 68–72,
      78
  applied problems solved with, 569
  changing from larger to smaller, 567
  changing from smaller to larger, 568
  converting between, 566–69
  converting measures between U.S.
      system and, 573–74, 618
Milligram, 72
Milliliter, 71
Millimeter, 71

Minuend, in subtraction, 26, 29
Minus sign, in subtraction, 24
Minutes, converted to hours, 564
Mixed numbers, 241, 242
  addition/addition and subtraction
      of, 322–25, 353
  applied problems solved using, 327
  changed to improper fractions, 244,
      283, 325, 326
  improper fractions changed to, 243,
      282
  multiplication and division of, 326,
      353
  operations with, 322–27
  subtraction of, 353
Mode, of set of data, 517
Money exchange, 455
Monomials, 386
  multiplying by polynomials, 387–88,
      413
Multiplication, 90–95
  of algebraic expressions, 117–18,
      211, 221
  applied problems involving, 158–60
  of decimals, 440–42
  equations solved with, 137–39
  evaluating algebraic expressions
      involving, 135
  in exponent form, 165
  of expressions in exponent
      form, 116–17
  of fractional expressions, 296–98
  of fractions, 352
  key words for, 164
  meaning of, 90–91
  of mixed/whole numbers, 326, 353
  of polynomials, 386–90
  of prime fractions, 247, 283
  properties of, 93–94, 164
  of signed numbers, 220
  to simplify variable
      expressions, 94–95
  symbols and key words for, 91–92
  by unit fraction, 566
  of whole numbers and algebraic
      expressions, 95–103
Multiplication principle of
      equality, 375
  equations solved using, 367–69, 411
Multiplication problems,
      equivalent, 125
Multiplication property for square
      roots, 583
Multiplication property of one, 93, 94,
      248
Multiplication property of zero, 93, 94

Negative correlation, 540
Negative direction, on number
      line, 185
Negative exponents, for scientific
      notation, 453
Negative fractions, multiplying, 297
Negative monomial, polynomial
      multiplied by, 387

Negative numbers, 179
  on calculators, A-5
  fractions reduced with, 257, 258
Negative region, 185
Negative square roots, 581
Note taking, 236
Number line, 6, 179
Number sets:
  mean of, 515–16, 548
  median value of, 516–17, 549
Numerators, 239
  raised to powers, 340
Numerical coefficients, 20, 117, 118
Numerical denominators, solving
      equations containing fractions
      with, 381

Opposites of numbers, 220
  finding, 181
Ordered pairs, writing data as, 522–23
Order of operations, 164. (*See also*
      Addition; Division;
      Multiplication; Subtraction):
  on calculator, A-4
  and complex fractions, 331
  with decimals, 442
  with division, 125–27
  and exponents, 108–12
  following, 111–12
  with fractions, 353
  with signed numbers, 214–15, 221
Origin, 523
  of number line, 179
Ounces, converted to pounds, 564

Parallelograms, 591
  area of, 146, 147, 166, 591
  defined, 146
Parentheses:
  for base of negative number, 205
  for calculations on
      calculators, A-4–A-5, A-6
  in evaluations, 126
  and power rules, 340
  in simplifying/solving
      equations, 377
  solving equations with, 398–99
Partial products, 101
  trailing zeros eliminated in, 102
Percent:
  applied problems involving, 480–84
  changing to fraction, 490
Percent decrease, solving, 480, 490
Percent increase, solving, 480, 490
Percent problems:
  percent proportion used to
      solve, A-11–A-13
  solved with proportion, A-9–A-13
  solving general applied, 475–76
  translating and solving, 473–75, 490
Percent proportion:
  identifying parts of, A-9–A-11
  to solve percent
      problems, A-11–A-13

Percents:
  changing between decimals
    and, 466–67, 489
  changing between fractions,
    decimals and, 468–69
  meaning of, 465–66
Perfect square, defined, 579
Perimeter:
  addition used for calculating, 41–43
  defined, 41
  finding, 77, 589–91
  of geometric shapes, 145
  of similar triangles, 612
  of triangles, 589–90
Perpendicular lines, 588
Pi (π), 598
Pictographs, reading, 501, 547
Pie graphs, 502
Placeholders, 6
Place value, of whole numbers, 4–5
Place-value chart/system, 4
  for decimal fractions, 424
Plus sign, in addition, 12
Points:
  naming coordinates of, 524–26
  plotting given coordinates, 523–24,
    549
Polynomial expressions, solving
    equations involving, 398–99,
    413
Polynomials:
  addition and subtraction of, 394–95,
    413
  applied problems involving, 401–6
  identifying types of, 386
  multiplication of, 386–90
  multiplying monomials by, 387–88,
    413
  simplifying with grouping
    symbols, 393–94
Positive correlation, 540
Positive direction, on number line, 185
Positive fractions, multiplying, 297
Positive numbers, 179
  compared in decimal notation, 427
  in scientific notation, 452
Positive (or principal) square
    roots, 581
Pounds, ounces converted to, 564
Power:
  of exponents, 109
  raising power to, 339–40, 354
Power of 10:
  changing to decimal notation when
    denominator is, 426
  evaluating, 110
  multiplying decimal by, 441
Powers of numbers, on scientific
    calculators, A-5–A-6
Predictions, 499
  data and equations for
    making, 540–41
Prefixes, in metric system, 69, 566
Prime factorizations, and finding least
    common multiples, 308
Prime factors, 282

factor tree for finding, 235
  of whole numbers, 232–34
Prime fractions, multiplication of, 247,
    283
Prime numbers, identifying, 231–32
Principal, 482
Problem solving. (See also Applied
    problems)
  translation for, 60–61
Product, 91, 92
  raised to powers, 340
  reducing to lowest terms, 297
  of signed numbers, 201
Production costs,
    analyzing/constructing graphs
    of, 507–8
Profit concept, 410
Properties of addition, 12–15
  for simplifying, 76
Proportion method, 475
Proportions:
  applied problems involving, 276–77,
    284
  determining if statement is, 270–71,
    284
  finding missing number in, 271–72,
    284
  solving, 566
  writing, 270, 284
Purchases, solving problems
    involving, 158–59
Pyramid, volume of, 607, 620
Pythagoras, 592
Pythagorean Theorem, using, 592, 619

Quadrilaterals, 146
  angles in, 588
Quarts, converted to gallons, 564
Quotient, 123, 127

Radical expressions, simplifying, 583
Radicals, simplifying, 619
Radical sign, 580
Radius, of circle, 598
Rates:
  forming, 283
  solving applied problems
    involving, 263–65
  writing two quantities with different
    units as, 262–63
Ratios:
  solving applied problems
    involving, 263–65
  writing two quantities with same
    units as, 261–62
Reciprocal fractions, 299
Reciprocal numbers, 299
Rectangles, 591
  area of, 147, 148, 165
  defined, 41
  quadrilaterals as, 588
Rectangular coordinate plane,
    solutions to linear equations
    plotted on, 536–38

Rectangular coordinate
    system, 522–27, 523
Rectangular solid, 605
  surface area of, 166
  volume of, 150, 166
Remainder, in long division, 127
Repeating decimals, 450
Repeating factor property for square
    roots, 584
Revenue concept, 410
Reverse Polish notation (RPN), on
    scientific calculator, A-2
Right angles, 41, 588
Right triangles, 590
  applied problems with, 593
Rounding:
  decimals, 428–29
  and estimation, 47
  whole numbers, 7–8, 76
Round-off place, 7
RPN. See Reverse Polish notation

Scientific calculators, A-2–A-6
  adding and subtracting decimals
    on, 432
  addition of negative numbers
    on, 188
  checking solutions to equations
    on, 375
  combined operations on, A-4
  decimal problems on, A-3–A-4
  division on, 130
  exponents evaluated on, 109
  logic used in, A-2
  negative numbers on, 196, A-5
  parentheses used with, A-4–A-5
  percent changed to decimal on, 467
  proportions on, 272
  raising number to power
    on, A-5–A-6
  scientific notation on, A-5
  simple calculations on, A-3
  square roots of numbers found
    on, A-6
Scientific notation, 452
  on calculators, A-5
  writing numbers using, 451–53, 489
Sector, of circle graph, 502
Several-digit number, multiplied by
    several-digit number, 101–2
Sharp calculators, A-2
Side-angle-side (SAS) congruence, by
    triangles, A-16, A-17
Side-by-side (SSS) congruence, by
    triangles, A-16
Sides, of triangle, 588
Signed numbers, 177–227
  absolute value of number, 180–81
  addition of, 185–89, 220
  addition of with different
    sign, 186–88
  addition of with same sign, 185–86
  applied problems solved involving
    more than one operation, 215
  division of, 206–7, 221

Signed numbers (*cont.*)
  evaluating equations with, 221
  evaluating expressions involving
    addition of, 189
  exponents used with, 204–6, 221
  finding product of, 201
  inequality relationship used
    with, 179–80
  multiplication of, 220
  multiplying more than two, 203–4
  multiplying two, 201–2
  opposite of number, 181
  order of operations with, 214–15,
    221
  reading graphs involving, 220
  and reading line graphs, 182
  solving applied problems involving
    addition of, 189
  subtraction of, 193–97, 220
  understanding, 179–82
Sign rule, for integers in exponent
  form, 205
Similar geometric figures,
  corresponding parts of, 611–14,
  620
Similar triangles, corresponding parts
  of, 611–14, 620
Simple interest problems,
  solving, 482–84, 490
Single-digit numbers, multiplying by
  number with several
  digits, 100–101
Slashes, for multiplying factors, 255,
  256
Solutions, to equations, 56, 57
Sphere, volume of, 605–6, 620
Square:
  area of, 147, 165
  defined, 41
Squared base, 109
Square root expressions, adding and
  subtracting, 585, 619
Square roots:
  approximating for number that is
    not perfect square, 582, 619
  meaning of, 579–80
  multiplication property for, 583
  of number that is perfect
    square, 580–82, 619
  repeating factor property for, 584
  on scientific calculator, A-6
  simplifying for expressions, 583–85
  table of, A-1
Squares, 591
Square yards, and square meters, 573
Statistics, mean and median, 515–17
Stock market, investing in, 351
Straight lines, graphing, 549
Study skills:
  class attendance: the learning
    cycle, 43
  day before exam, 72
  evaluating, 429
  final exam preparation, 444
  getting most from time studying, 55
  keep trying, 340–41

mathematics and career, 376
necessity of homework, 160
note taking in class, 236
positive attitude toward
  fractions, 298
previewing new material, 184
problems with accuracy, 252
reading the text, 367
real-life mathematics
  applications, 406
reviewing for exams, 68
reviewing tests, 93
time management, 100
when to use calculator, 203
Subtraction:
  application problems involving, 32
  and combining like terms, 24–30
  of decimals, 432–33, 488
  equations solved with, 58–59
  of fractional expressions with
    common denominators, 312–13
  of fractional expressions without
    common denominators, 313–18
  of fractions with common
    denominators, 352
  of fractions without common
    denominators, 352
  key words for, 26–27
  of measurements, 77
  of mixed numbers, 322, 323–25, 353
  of numbers with two or more
    digits, 29–30
  of polynomials, 394–95, 413
  properties of, 27–29
  of signed numbers, 193–97, 220
  of square root expressions, 585, 619
  understanding, 24–25
  of whole numbers, 76
Subtraction rule, for signed
  numbers, 194
Subtrahend, in subtraction, 26, 29
Sums, estimating, 433–34
Surface area:
  of geometric solid, 150–51
  of rectangular solid, 150–51, 166
Symbols. (*See also* Formulas):
  for degrees of angles, 588
  division, 127
  English phrases for addition
    translated into, 12
  English phrases for subtraction
    translated into, 26
  for greater than/less than, 6
  for multiplication, 91–92
  for positive square root, 580
  for sign of number, 179
  for unequal fractions, 249

Temperature readings, converting
  between Fahrenheit and
  Celsius degrees, 575, 618
Term(s):
  defined, 18
  of polynomial, 386
  sign of, 387, 394

Texas Instruments scientific
  calculators, A-2
Time, 482
  management of, 100
  U.S. measurement of, 66, 563
  zones, 218
Tons, 66
Trailing zeros:
  eliminating in partial
    products, 102
  multiplying by numbers with, 98
Translation, for solving
  problems, 60–61
Traveling, and conversions, 578
Triangles. (*See also* Congruent
    triangles; Right triangles):
  angles in, 588–89
  defined, 41
  perimeter and area of, 589–91
  and Pythagorean Theorem, 588–93,
    619
Trinomials, 386

Unit fractions:
  to convert from one U.S. unit to
    another, 563–64
  multiplying by, 566
Unit rates:
  finding, 263
  forming, 284
Universal Time (or Greenwich
    Time), 218
Unknown angles, measure of, 589
U.S. system of measurement,
    converting measures between
    metric system and, 573–74, 618
U.S. units of measurement:
  applied problems solved with, 565
  converting between, 563–64, 618

Vacation plan cost analysis, 107
Values, of expressions in exponent
    form, 109
Variable expressions:
  defining, 403
  evaluating, 54–55
  multiplication for simplifying, 94–95
Variables, 13, 54, 378
  value of in equations, 140
Variable term, 18
Vertex, of angle, 588
Vertical lines, graphing, 526–27, 539
Volume:
  of cone, 606–7, 620
  of cylinder and sphere, 605–6, 620
  of geometric solid, 149–50
  metric measurements of, 567
  metric measures of equivalence, 71
  of pyramid, 607, 620
  of rectangular solid, 150, 166
  U.S. to metric system
    interconversions, 573
  U.S. units of measurement for, 66,
    563

Weight:
  metric measurements of, 567
  metric measures of equivalence
    for, 71
  U.S. to metric system
    interconversions for, 573
  U.S. units of measurement for, 66, 563
Whole number expressions:
  division of, 122–30
  multiplication of, 90–95
Whole numbers:
  decimals divided by, 447–48
  in expanded notation, 5
  in exponent form, 108–9
  factoring, 231–36

  multiplication of, 95–103, 164
  multiplying fractions by, 298
  place value of, 4–5
  prime factors of, 232–34
  rounding, 7–8, 76
  subtraction of, 76
  understanding, 4–8
Word names, for decimal
    fractions, 424–25
Word problems, 406. (*See also* Applied
    problems)

X-axis, 523, 525
X-coordinate (or value), 523, 527, 536

Yards, 66
  converted to feet, 563–64
Y-axis, 523, 525
Y-coordinate (or value), 523, 527,
    536

Zero:
  division involving, 240
  multiplication by numbers with
    trailing, 98, 164
  as placeholder, 6
  subtraction properties by, 28–29

PHOTO CREDITS

CHAPTER 1   **CO** Ron Chapple/FPG International LLC   **p. 31** Churchill & Klehr Photography   **p. 53** Churchill & Klehr Photography

CHAPTER 2   **CO** Ken Fisher/Tony Stone Images   **p. 107** Chuck Savage/The Stock Market   **p. 122** C Squared Studios/PhotoDisc, Inc.   **p. 134** Jim Cummins/FPG International LLC   **p. 157** Arthur Tilley/FPG International LLC

CHAPTER 3   **CO** Ellis Herwig/Stock Boston

CHAPTER 4   **CO** Kuhn Inc./The Image Bank   **p. 274** Daniel Grogan/Uniphoto Picture Agency   **p. 281** Chuck O'Rear/Westlight/Corbis

CHAPTER 5   **CO** SuperStock, Inc.   **p. 351** David Young Wolff/Tony Stone Images

CHAPTER 6   **CO** Comptroller of Public Accounts   **p. 410** Adam Smith/FPG International LLC

CHAPTER 7   **CO** Jon Riley/Tony Stone Images   **p. 435** Jack Star/PhotoDisc, Inc.   **p. 455** Bill Cardoni/Liaison Agency, Inc.   **p. 461** Churchill & Klehr Photography

CHAPTER 8   **CO** SuperStock, Inc.   **p. 507** Richard Pasley/Stock Boston   **p. 540** John M. Roberts/The Stock Market

CHAPTER 9   **CO** Josef Beck/FPG International LLC   **p. 578** Lois Moulton/Tony Stone Images   **p. 598** David Young Wolff/PhotoEdit   **p. 607** George Holton/Photo Researchers, Inc.